Water Science and Technology Library

Volume 130

The aim of the *Water Science and Technology Library* is to provide a forum for dissemination of the state-of-the-art of topics of current interest in the area of water science and technology. This is accomplished through publication of reference books and monographs, authored or edited. Occasionally also proceedings volumes are accepted for publication in the series. *Water Science and Technology Library* encompasses a wide range of topics dealing with science as well as socio-economic aspects of water, environment, and ecology. Both the water quantity and quality issues are relevant and are embraced by *Water Science and Technology Library*. The emphasis may be on either the scientific content, or techniques of solution, or both. There is increasing emphasis these days on processes and *Water Science and Technology Library* is committed to promoting this emphasis by publishing books emphasizing scientific discussions of physical, chemical, and/or biological aspects of water resources. Likewise, current or emerging solution techniques receive high priority. Interdisciplinary coverage is encouraged. Case studies contributing to our knowledge of water science and technology are also embraced by the series. Innovative ideas and novel techniques are of particular interest.

Comments or suggestions for future volumes are welcomed.

Vijay P. Singh, Department of Biological and Agricultural Engineering & Zachry Department of Civil and Environment Engineering, Texas A&M University, USA
Email: vsingh@tamu.edu

All contributions to an edited volume should undergo standard peer review to ensure high scientific quality, while monographs should also be reviewed by at least two experts in the field.

Manuscripts that have undergone successful review should then be prepared according to the Publisher's guidelines manuscripts: https://www.springer.com/gp/authors-editors/book-authors-editors/book-manuscript-guidelines

Salah Souabi · Abdelkader Anouzla ·
Shalini Yadav · Vijay P. Singh ·
Ram Narayan Yadava
Editors

Wastewater Treatment Plants

Processes, Assessment, Design and Operation

Springer

Editors
Salah Souabi
Faculty of Sciences and Technology
University of Hassan II Casablanca
Mohammedia, Morocco

Abdelkader Anouzla
Faculty of Sciences and Technology
University of Hassan II Casablanca
Mohammedia, Morocco

Shalini Yadav
Department of Civil Engineering
Rabindranath Tagore University
Bhopal, Madhya Pradesh, India

Vijay P. Singh
Texas A&M University
College Station, TX, USA

Ram Narayan Yadava
Research and International Affairs
Mansarovar Global University
Bhopal, Madhya Pradesh, India

ISSN 0921-092X ISSN 1872-4663 (electronic)
Water Science and Technology Library
ISBN 978-3-031-87460-4 ISBN 978-3-031-87461-1 (eBook)
https://doi.org/10.1007/978-3-031-87461-1

This Springer imprint is published by the registered company Springer Nature Switzerland AG
The registered company address is: Gewerbestrasse 11, 6330 Cham, Switzerland

Contents

Part I
Preliminaries

Chapter 1
Operation of Wastewater Treatment Plants: An Overview of Methodology

Salah Souabi, Roukaya Bouyakhsas, Safaa Khattabi Rifi, Abdelkader Anouzla, Shalini Yadav, Ram Narayan Yadava, and Abdelaziz Madinzi

Abstract Water resource management, circular economy, wastewater treatment, and recovery have become necessary in climate, economic, and socio-environmental change and decreasing precipitation. State services, local authorities, and economic actors worldwide are making great efforts to integrate wastewater treatment plants into recycling treated water in irrigation and industry. This will allow concrete and adapted solutions for developing and developed countries to emerge. Indeed, well-adapted treatment channels must be optimized to meet rigorous standards. This makes it possible to improve the quality of treated water, reduce the environmental impacts of treatment activities, and enhance the value of wastewater resources, as well as, subsequently, the application of the circular economy. Indeed, scientists must support public and private actors in the design, sizing, operation, and optimization of wastewater treatment and recovery facilities. To address drought problems, water treatment with recycling is necessary. This requires the integration of a practical methodology to help treatment plant managers properly operate the treatment lines to produce recyclable water. This chapter aims to establish a coherent methodology that constitutes a decision-making aid to improve the operation of treatment plants increasingly. This chapter remains very interesting by specifying the methodology to follow to operate the treatment lines in particular properly: • Collection of essential data at the plant level (assessments already carried out) • Conduct an assessment of the current state • Conduct a report on the current state of the plant based on the database and the assessment carried out at the current state • Proposal of an action plan • Implementation of the action plan • Monitoring of the action plan. In addition, several points should be highlighted in practice to assess a methodology to follow, in particular for: • The

S. Souabi (✉) · R. Bouyakhsas · S. K. Rifi · A. Anouzla · A. Madinzi
Department of GPE, Faculty of Sciences and Technology, Hassan II University, Mohammedis, Casablanca, Morocco
e-mail: salah.souabi@gmail.coml

S. Yadav
Rabindranath Tagore University, Raisen, Madhya Pradesh, India

R. N. Yadava
Mansarovar Global University, Bhopal, Madhya Pradesh, India

S. Souabi et al. (eds.), *Wastewater Treatment Plants*, Water Science and Technology Library 130, https://doi.org/10.1007/978-3-031-87461-1_1

rate of chemical reagents to be integrated during physico-chemical treatment • The aeration rate which makes it possible to minimize the cost of producing well-treated water • The rate of recycling of water from one stage to another • The rate of sludge extraction at the primary and secondary treatment level • Assess the decontamination efficiency by different stages of the plant • Assess the decontamination efficiency of the plant between the inlet and the outlet • Equipment operating time • Waste production rate, compared to the treated flow rate (daily, monthly) • Electricity and reagent consumption rate compared to the treated flow rate (daily, monthly) • Hydraulic operating parameters: • Retention time along the different treatment stages • Operating parameters of the biological process (volume load, mass load, etc.) • Malfunction of the station and proposed solutions. Finally, this allows for the proper optimization of the conditioning of the treatment by using an apparent and practical methodology to produce water that meets industrial standards for irrigation or recycling.

Keywords Wastewater · Treatment station · Methodology · Optimization

1.1 Introduction

Wastewater management is a significant challenge for contemporary societies, exacerbated by the continuing increase in the world's population and rapid urbanization (Hassan Rashid et al. 2018; Arfanuzzaman and Dahiya 2019). With urban growth, the amount of wastewater produced by households, industries, and agriculture has increased considerably. According to United Nations estimates, the global population is projected to reach 9.7 billion by 2050, with the majority living in urban areas (Gu et al. 2021). This rapid growth puts increased pressure on wastewater treatment infrastructures. These facilities are essential in the treatment of wastewater, reducing contaminants before discharge into the environment or reuse for industrial and agricultural applications (Qasim 2017; Salgot and Folch 2018). However, the operation of these plants is not without its challenges, requiring constant optimization of processes to ensure efficiency and sustainability (Lu et al. 2019).

Untreated or inadequately treated wastewater can pose serious risks to public health and the environment, such as the contamination of water sources, the spread of waterborne diseases, and the degradation of aquatic ecosystems. In fact, Guadie et al. (2021) found in their research that vegetables irrigated with untreated wastewater contain significant amounts of heavy metals, some of which can be harmful to the health of both adults and children. On the other hand, according to Edokpayi et al. (2017), the lack or insufficiency of wastewater treatment facilities often leads to the discharge of wastewater and its effluents into surface waters, leading to pollution of the various components of the environment.

Faced with these challenges, optimizing wastewater treatment plants is becoming a priority. Common problems encountered in the operation of these plants include inefficiencies in physico-chemical and biological treatment processes, inadequate sludge management, and high consumption of chemical reagents and energy (Gu et al. 2017; Zhang et al. 2021). These problems can reduce the overall performance of treatment plants, increasing operational costs and lowering the quality of treated effluent. It is, therefore, important to develop and implement solid and consistent ways to improve the operation of these complex systems.

Optimizing wastewater treatment plants involves improving existing processes and identifying and implementing innovative solutions to malfunctions. New technologies and approaches, such as intelligent sensors, advanced automation, and improved biological treatment techniques, offer promising opportunities for increasing the effectiveness of the treatment plants (Zhu et al. 2020; Zhang et al. 2020). In addition, proactive management of malfunctions and proposed solutions ensures that plants continue to operate optimally, minimizing environmental and health impacts.

This article aims to provide an in-depth analysis of the essential elements for optimizing wastewater treatment plants, focusing on the technical and operational criteria determining their performance. It explores the conditions and indicators that signal the need for a plant extension, ensuring it can respond effectively to a growing pollution load. In addition, the article highlights the crucial importance of sludge management, particularly monitoring the sludge index and applying appropriate techniques to maintain the stability of the treatment process.

1.2 Overview of Treatment Plants for Wastewater

Wastewater treatment is essential for preserving the environment and public health (Sikosana et al. 2019). It is generally divided into three main phases: primary treatment, which removes suspended solids; secondary treatment, which uses biological techniques to reduce dissolved organic matter; and tertiary treatment, which aims to remove residual nutrients and contaminants. Various methods are used at each stage, each with its own specific advantages depending on needs and contexts.

Several treatment techniques can be used to treat wastewater.

1.2.1 Activated Sludge Plants

Activated sludge process is an aerobic treatment system using a suspended bacterial culture in an aerated tank to degrade the organic matter in wastewater. The system relies on a continuous oxygen supply to maintain optimal aerobic conditions, thus promoting bacterial activity. Once the organic matter has decomposed, the activated sludge is separated from the purified water by settling. This process is widely used for its efficiency in treating high pollutant loads and flow rates (Demirbas et al. 2017; Jafarinejad 2017; Wilén et al. 2018). Figure 1.1 shows an example of activated sludge wastewater treatment.

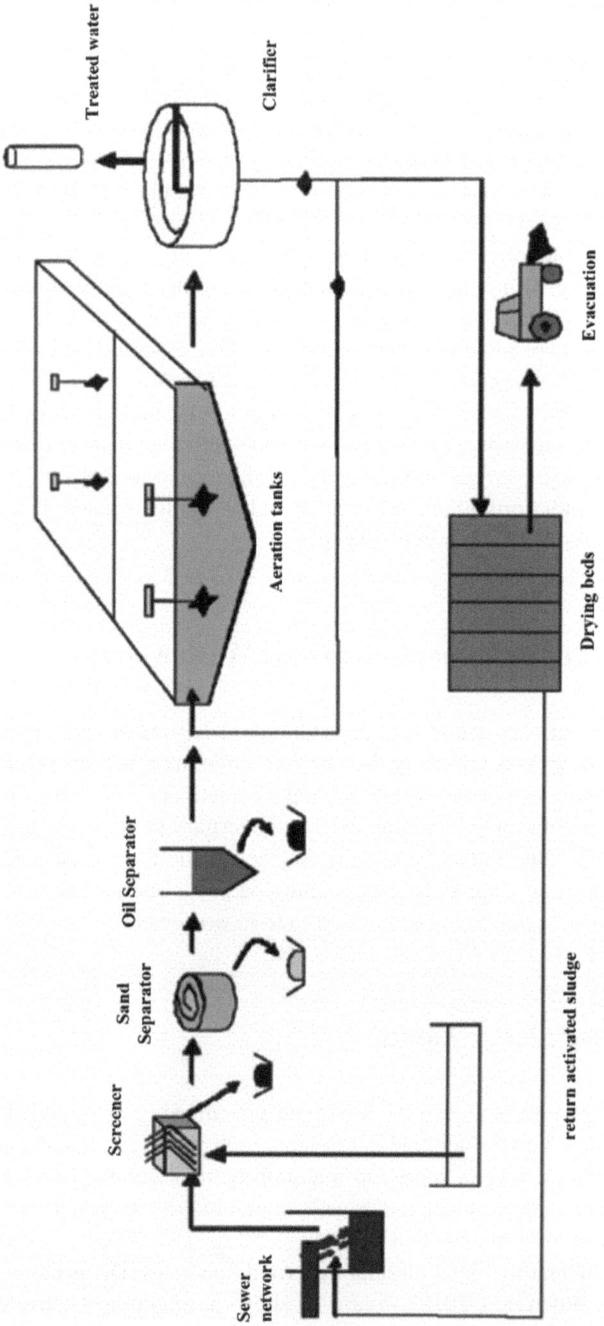

Fig. 1.1 Activated sludge wastewater treatment technique

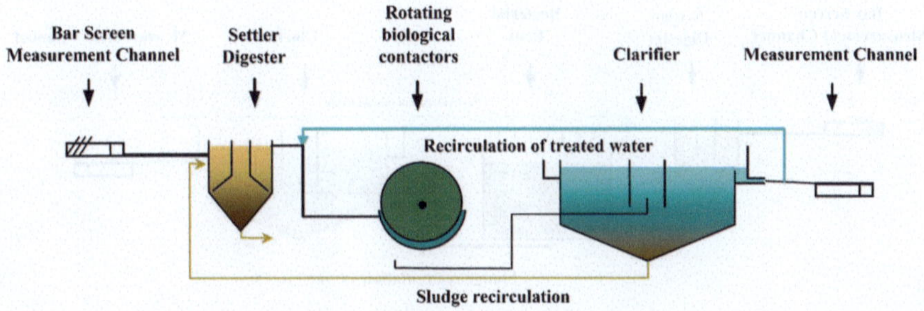

Fig. 1.2 Wastewater treatment by rotating biological contactors

1.2.2 Biological Disc (or Rotating Disc) Stations

Biological disc plants use partially immersed discs that rotate slowly in a basin. A biofilm develops on the disc surfaces, aerobically degrading organic matter. The discs alternate between immersed and exposed phases, allowing natural oxygenation of the biofilm. This system is appreciated for its low energy consumption and low maintenance requirements. It is particularly well suited to small installations and areas where reliability and simplicity are essential (Waqas et al. 2021, 2023). Figure 1.2 illustrates a simple example of this type of treatment.

1.2.3 Bacterial Bed Plants (or Percolator Filters)

In bed filter plants, also known as percolator filters, the effluent is evenly distributed across the surface of a gravel, stone, or synthetic filter bed. A bacterial film develops on these materials and ensures the degradation of organic matter. Water percolates slowly through the bed, where it is purified before being collected at the outlet. Oxygen is supplied by natural diffusion through the bed, maintaining the aerobic conditions necessary for treatment (Guo et al. 2019; Narayanan and Narayan 2019). Figure 1.3 illustrates a simple example of this type of treatment.

1.2.4 Lagoon Plants

Lagoon treatment takes place in low-depth basins where purification is achieved through natural processes involving algae and bacteria. There are several types of lagoons, including aerated lagoons, facultative lagoons, and maturation lagoons, each suited to different levels of treatment and operating conditions. This technique is particularly effective for small communities due to its simplicity and low operating

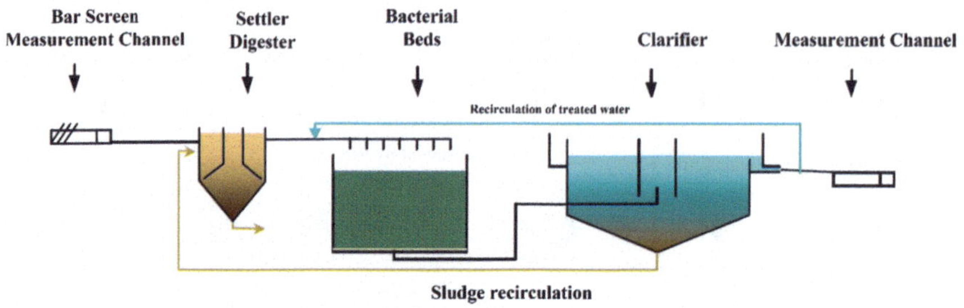

Fig. 1.3 Wastewater treatment by bacterial beds

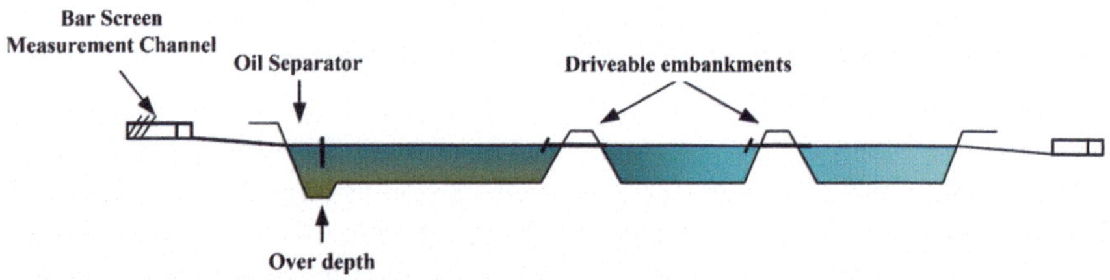

Fig. 1.4 Wastewater treatment by lagoon systems

costs, although the surface area required is generally larger than for other systems (Mohammadi and Fataei 2019; Mancuso et al. 2023).

Figure 1.4 illustrates a simple example of this type of treatment.

1.2.5 Reed Filter Plants

Reed bed filtration systems utilize a sand or gravel bed where wastewater flows either horizontally or vertically. The reeds planted in the bed are crucial in treatment, ensuring physical, chemical, and biological processes. The system effectively treats high organic loads and offers a natural, sustainable solution suitable for small communities. The reed species' diversity can also influence the overall efficiency (Chatterjee et al. 2019; Sasirekha et al. 2020). Figure 1.5 illustrates a simple example of this type of treatment.

The choice of wastewater treatment process depends on many factors, such as the nature of the wastewater, the purification performance required, land restrictions, and the associated costs (Arroyo and Molinos-Senante 2018).

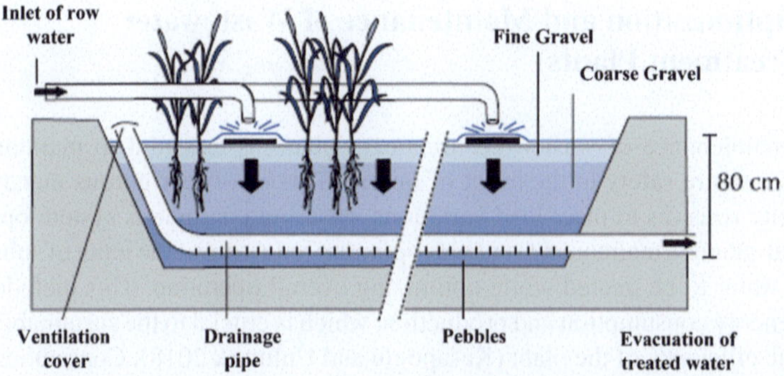

Fig. 1.5 Treatment of less loaded wastewater using the reed bed filtration technique

1.3 Current Challenges and the Importance of Optimization

Many contemporary challenges are facing wastewater treatment plants; they compromise their efficiency and sustainability. Among these challenges are increasing wastewater volumes generated by population growth and rapid urbanization, which put greater pressure on existing infrastructures (Bijekar et al. 2022). Moreover, the variability of wastewater characteristics, particularly concerning pollutant concentration and organic load, complicates treatment and requires constant process adjustments (Baklouti et al. 2018). Increasingly rigorous regulatory requirements for the quality of discharged water also impose additional challenges, particularly concerning the elimination of micropollutants and nutrients.

In addition, wastewater treatment plants face growing economic challenges, including reducing operating costs while maintaining high performance (Qasim 2017). In this context, optimizing treatment processes is crucial. It improves operating efficiency and the quality of treated water and reduces energy and operating costs. Optimization can also enhance system sustainability by reducing environmental impacts, including greenhouse gas emissions and excessive sludge production (Borzooei et al. 2019). By integrating innovative technologies and adaptive management practices, wastewater treatment plant managers can meet today's challenges while guaranteeing reliable service that complies with regulatory requirements. In this way, optimization is not just about technical enhancement but is a strategic necessity to ensure the long-term sustainability of wastewater treatment infrastructure in a constantly evolving world.

1.4 Optimization and Maintenance of Wastewater Treatment Plants

Proper maintenance of wastewater treatment plants is essential to maintain their value and ensure safety in the event of failure. This involves rigorous management of capacity reserves to meet load variations, ensuring continuous system optimization. Well-planned maintenance enables optimum reduction of the input of substances into the water to be treated while optimizing overall operation. This includes optimizing energy consumption and production, which is crucial to the sustainability and economic efficiency of the plant (Kasapoglu and Unluturk 2018). Continuous monitoring of equipment operation enables rapid identification and correction of anomalies, ensuring reliable and efficient operation of treatment plants (Mamandipoor et al. 2020).

The key parameters for monitoring plant performance are listed in Table 1.1.

Table 1.1 gives the main parameters characterizing activated sludge in a wastewater treatment plant. Concentration (C) indicates effluent quality by expressing the mass of matter per unit volume, while flow (or load) (F), the product of flow (Q), and concentration (C), measures the quantity of incoming pollution. The plant's purification efficiency, the ratio of pollution eliminated to pollution received, defines its effectiveness. Mass loading (Cm), the ratio of BOD5 load received to the quantity of sludge present in the aeration tank, and volume loading (Cv), the ratio of BOD5 load to tank volume, are crucial for estimating treatment capacity and biological equilibrium. Low mass and volume loads indicate a high carbon and nitrogen pollution removal level. A high residence time (Ts) in the aeration tank favors this elimination. Sludge age (A), the residence time of sludge in the tank, indicates sludge maturity and its ability to treat pollution. Upwelling velocity (VA), the speed at which water rises in the clarifier, must be kept below the settling velocity of the particles to ensure effective clarification. These interconnected parameters are essential for optimizing and controlling biological treatment processes, ensuring efficient and sustainable wastewater management.

1.4.1 Elements of a WWTP Optimization Program

In order to monitor the operation of a wastewater treatment plant (WWTP), it is crucial to take the following elements into account (Puchongkawarin et al. 2015; Padrón-Páez et al. 2020):

Define the Interests and Objectives of Optimization: It is essential to clarify the specific goals of optimization, whether to improve treatment efficiency, reduce operating costs, or minimize environmental impact.

Table 1.1 Characteristic parameters of activated sludge (MAGE (42) 2006; OFEV 2014)

Parameter	Unit	Meaning	Formula
Flow rate (Q)	m^3/day, m^3/h, l/s	Volume per unit of time	–
Peak flow rate	m^3/h	Maximum hourly flow rate received by the plant	–
Average 24-h flow rate	m^3/h	Average hourly flow rate received by the plant	Qday/24
Concentration (C)	mg/l, g/l, kg/m^3	Mass per unit of volume (quality of the effluent)	–
Flux (or load) (F)	kg/day	Quantity of pollution	$F = C \times Q$
Hydraulic load of the plant	%	Flow rate received compared to the nominal hydraulic capacity	(Flow rate received/ Nominal capacity) × 100
Organic load of the plant	%	Pollution received compared to the nominal capacity	(Pollution received/ nominal BOD5 flux) × 100
Treatment efficiency of the plant and each process stage	%	Pollution removed compared to pollution received	(Pollution received—pollution discharged)/pollution received × 100
Mass load (Cm)	kg BOD5/kg MLVSS/ day	Ratio of BOD5 load received to the amount of sludge present	Cm = kg BOD5 received/kg MLVSS (aeration basin)
Volumetric load (Cv)	kg BOD5/m^3/day	Ratio of BOD5 load received to the volume of the aeration basin	Cv = kg BOD5 received/m^3 (aeration basin)
Retention time (Ts)	Hours	Hydraulic retention time of water in a basin	Ts = V (Basin volume)/Q
Sludge age (A)	Days	Retention time of sludge in the aeration basin	A = (VAB × [SS]AB)/(QE × [SS]E)
Upward velocity (VA)	m/h	Upward velocity of water in the clarifier	VA = Q/S
Sludge production (P)	kg SS/day	Daily sludge production	P = 0.8 × Lr (daily mass of BOD5 removed)

Carry out an Overall Assessment of the WWTP: A complete evaluation of the plant's current state must be carried out, considering each treatment stage's performance and identifying weak points. Several elements must be measured, such as flow rate, pollutant load in COD, BOD5, TSS, mass, and volume loads.

Evaluate the WWTP's Physico-Chemical and Biological Deficiencies: Identify and analyze existing physico-chemical and biological problems to understand the factors limiting the plant's performance.

Implement the Necessary Modifications: Based on the previous assessments, it is essential to make the required improvements, such as extending plant capacity, modernizing treatment stages, or optimizing the addition of chemical reagents. These modifications must be targeted to correct identified weaknesses and improve overall treatment efficiency.

1.4.2 Methodology for Optimizing Treatment Conditions

To evaluate the performance of a wastewater treatment plant, it is essential to carry out an overall assessment of the treatment process, following a clearly defined procedure. Specific calculations must be made, as shown in Fig. 1.6.

First of all, it is important to identify the various stages of the plant's treatment process and pinpoint the sampling points' exact location. Sampling points should

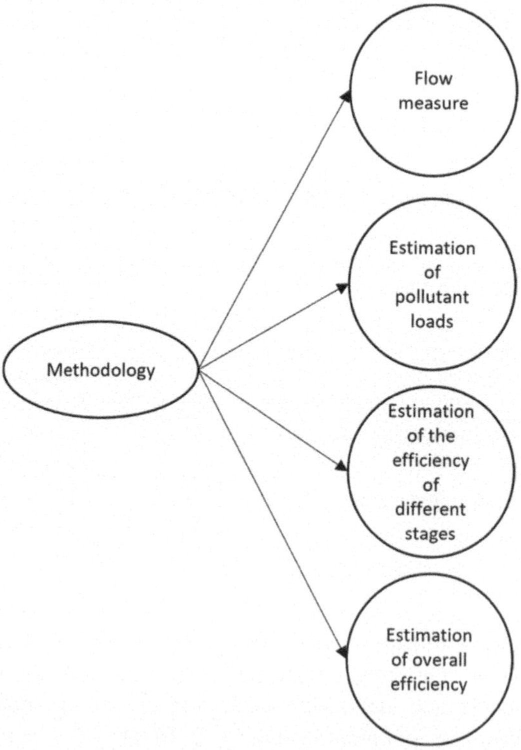

Fig. 1.6 Overview of the methodology for optimization and performance control

include the plant inlet, the inlet and outlet of each treatment stage, and the final plant outlet (Bagheri et al. 2015).

Then, pollution must be quantified at each stage by analyzing several key parameters, such as chemical oxygen demand (COD), 5-day biochemical oxygen demand (BOD5), suspended solids (SS), volatile solids (VS), pH, turbidity, oils, and greases, as well as total Kjeldahl nitrogen (NTK) and nitrate (NO_3-) concentrations (Alsulaili and Refaie 2020). It is also crucial to study the settling of effluents received by the plant and to estimate the pollutant loads at the inlet and outlet of each stage and at the inlet and outlet of the plant itself.

Mass and volume load calculations are used to determine whether the plant needs to be extended and which type of activated sludge plant is most appropriate (Mininni et al. 2015). The sludge index must be evaluated to assess the quality of settling in the classifier and identify any problems associated with the presence of filamentous bacteria. Estimating the age of the sludge in the biological tank is also important, especially in the case of activated sludge, as it enables us to assess its efficiency (Dai et al. 2015).

Sludge recycling must be quantified at each stage of treatment, as must effluent residence time. Estimating the daily quantity of sludge to be extracted and the amounts of waste accumulated by the bar screen and sand collected by the grit chamber is also necessary. Finally, the sludge flow rate needs to be recalculated and optimized to ensure the plant operates optimally (De Vrieze et al. 2016).

The necessary calculations include estimating the pollutant load received based on physico-chemical parameters and flow rate at the plant inlet and calculating the efficiency of each treatment stage. It is also essential to calculate overall plant performance by comparing inlet and outlet data, estimating pollutant loads at each treatment stage in kg/day, and quantifying inlet and outlet sludge and sludge removed by the plant.

1.5 Analysis of Mass and Volume Pollution Loads

Mass and volume loads are essential parameters in the sizing and optimization of wastewater treatment plants. The mass load represents the total quantity of pollutants in the wastewater, generally expressed in kilograms per day (kg/day). It reflects the mass of organic matter or specific substances, such as suspended solids or biological oxygen demand (BOD), the plant must treat that. On the other hand, volume load is defined as the quantity of pollutants per unit volume of the plant, often expressed in kilograms per cubic meter (kg/m^3). It measures the pollution concentration concerning the treatment volume available in the plant's basins or reactors. These two load types play a crucial role in the sizing of treatment plants. The mass load determines the total capacity that the plant must treat daily, while the volume load influences the design of the treatment units and the hydraulic residence time required to guarantee effective purification (Sun et al. 2016; Qasim 2017).

Assessing the current pollution load in a wastewater treatment plant is a fundamental step in ensuring effective wastewater treatment and anticipating capacity and expansion requirements. This assessment relies on analytical methods and continuous measurements to quantify the various pollution parameters. The first method involves regular sampling of wastewater at different points in the plant, such as the inlet and outlet of treatment units. These samples are then analyzed to measure key indicators such BOD5, COD, SS, and nutrients such as nitrogen and phosphorus. These analyses enable us to calculate the mass load, expressed in kg/day, as a function of input flows (Sun et al. 2016).

At the same time, sensors and online measuring instruments can be installed to monitor pollution parameters in real time, providing a continuous assessment of the volume load, which is expressed in kg/m^3. This real-time monitoring is crucial for detecting sudden variations in pollution load, such as concentration peaks due to industrial discharges or rainfall events (Martínez et al. 2020).

Analysis of historical data and flow trends enables us to estimate load trends over time, identifying periods of overload or under-utilization of plant capacity.

Determining critical thresholds beyond which treatment plant expansion becomes necessary is essential to maintain wastewater treatment efficiency. These thresholds are generally based on the evaluation of mass and volume loads, considering the plant's current treatment capacities.

A first exceedance criterion is reached when the mass load exceeds the plant's nominal capacity, i.e., when the total quantity of pollutants (in kg of BOD5 per day, for example) approaches or exceeds the limits for which the plant was designed. This frequently leads to reduced efficiency in the treatment processes, a buildup of pollutants in the basins, and a decline in the quality of the treated water.

At the same time, volumetric load offers another critical indicator. When the pollution concentration per unit volume exceeds the treatment capacity of reactors or settling basins, the hydraulic residence time may become insufficient to ensure adequate degradation of pollutants. This scenario can lead to under-treatment, fluctuations in effluent quality, and overloading of downstream treatment units (Başkurt et al. 2017; Riffat and Husnain 2022).

1.6 WWTPS Sludge

1.6.1 Importance of the Sludge Index

The sludge volume index (SVI) assesses the sludge's suitability for settling. SVI indicates the volume occupied by 1 g of sludge (dry matter) after 30 min of settling in a one-liter test tube. To calculate the SVI, the sludge volume in ml/l is divided by the sludge dry matter concentration (g/l) (Han et al. 2016). The unit of SVI is, therefore, ml/g and can be calculated using the following formulas:

Table 1.2 Interpretation of sludge volume index (SVI)

Sludge volume index	Interpretation	Sludge quality and solutions
Less than 100 ml/g	Compact and well-formed sludge; efficient settling process	Excellent
Approaching 100 ml/g	Good sludge quality	Good
Higher than 100 ml/g	Less compact sludge; poorly formed flocs; increased risk of failures due to filamentous bacteria	Reduced quality; requires intervention Action required: • Extract the maximum amount of sludge from the clarifier • Increase aeration in the biological basin • Increase recycling from the clarifier to the biological basin

$$\mathrm{SVI} = \frac{\text{Volume of settled sludge (ml)}}{\text{Sludge dry matter (g)}} \tag{1.1}$$

$$\mathrm{SVI} = \frac{\text{Sludge volume (ml/l)}}{\text{Concentration of dry matter in sludge (g/l)}} \tag{1.2}$$

Knowledge of SVI is crucial to understanding sludge quality in a treatment system. A high value generally indicates that sludge is less dense and more likely to form larger, less compact flocs, which can lead to problems with clarification and separation of the liquid and solid phases. Conversely, a low value indicates that sludge is more compact, favoring better separation and a more efficient clarification process (Han et al. 2018).

This index also identifies underlying conditions affecting wastewater treatment, such as organic overload or imbalances in sludge digestion. Furthermore, if this index is high, it may be due to problems such as overfeeding conditions, deficiencies in the quality of coagulating agents, or poor health of the sludge microbiome (Table 1.2).

1.6.2 Mohlman Index

Mohlman's index is a crucial parameter for evaluating the settling capacity of sludge in wastewater treatment plants, especially those utilizing activated sludge processes. This index measures the volume occupied by sludge after 30 min of settling, expressed in milliliters per gram of dry suspended solids (SS) (Szeląg et al. 2018).

Mohlman's index is a key indicator of sludge quality. It assesses the compactness of sludge flocs, their ability to separate from water, and, consequently, the efficiency of the effluent clarification process. Regular measurement of Mohlman's index is crucial for monitoring and optimizing treatment plant performance. It enables operators to quickly detect anomalies in the settling process and take corrective action, such as adjusting sludge age, modifying dissolved oxygen concentration, or improving organic load management (Moulkaf et al. 2023).

1.6.3 Sludge Management

The sludge generated during wastewater treatment plays a crucial role in the purification process, requiring careful management and treatment to ensure system efficiency and sustainability. These sludges, resulting from separating solids and liquids, must be treated efficiently to avoid environmental and operational problems (Rorat et al. 2019; Gherghel et al. 2019). The various stages of sludge treatment, from settling and concentration to recirculation and storage, are essential for maintaining treated water quality and optimizing operations.

Optimizing sludge management in a wastewater treatment plant is essential to maintaining efficient, cost-effective performance. Depending on performance indices such as the sludge volume index (SVI), several strategies can be implemented to improve sludge management (Dai et al. 2015).

First of all, adjusting operating conditions plays a crucial role. We can influence their quality by modifying the residence time of sludge in reactors. An appropriate residence time promotes the growth of the micro-organisms responsible for pollutant degradation while avoiding overloading, which can lead to a high SVI. Table 1.3 gives a clarifier's maximum residence times for sludge (Ejhed et al. 2018).

In addition, optimizing the activated sludge recirculation ratio can help balance sludge concentration and improve floc quality. Proper recirculation promotes floc stability and reduces the risk of light sludge formation.

Improving treatment processes is essential for managing sludge effectively. Implementing preliminary treatments such as extended aeration or intensive mixing can enhance flocculation and reduce the sludge volume index (SVI). These techniques help better oxygenate the sludge and promote the formation of more robust flocs. Specific coagulants and flocculants can also play a key role in improving the formation and stability of flocs, thereby reducing the SVI. Properly selecting these chemicals is crucial for optimizing settling performance (Kõrgmaa et al. 2019).

Table 1.3 Maximum solids residence time in a clarifier

Mass load (kg of BOD5 kg of $MLVSS^{-1}$ day^{-1})	Associated carbon treatment time	Maximum retention time (min)
< 0.1	With nitrification and denitrification	100–140
0.1–0.25		90
0.25–0.8	Without nitrogen treatment	30–80
> 0.8		< 30

1.7 Issues Related to Foaming and Filamentous Bacteria

Antifoams in wastewater treatment plants are an important practice for controlling and preventing excessive foam formation, which can compromise the efficiency of wastewater treatment. Antifoams are primarily used when foam becomes a problem due to various factors, such as imbalances in the biological process, organic overload, or the presence of specific substances that promote foam formation (Dlangamandla et al. 2020). Therefore, adding antifoams is crucial when excessive foam disrupts the treatment process, affects effluent quality, or causes operational issues, such as equipment blockages.

The effects of antifoams on the treatment process are numerous. By reducing foam formation, antifoams help maintain a stable and efficient treatment environment, contributing to better separation of solid and liquid phases. This improves the quality of sludge and treated effluents. Additionally, antifoams help protect the plant's equipment by preventing foam accumulations that can cause blockages, damage, or premature wear of components. By controlling foam, the risk of losses of reagents and chemicals is also reduced, and operational safety is improved by minimizing the risks of splashing or overflow (Dlangamandla et al. 2020; Dlangamandla 2022).

Filamentous bacteria represent another major issue in wastewater treatment plants, particularly in activated sludge processes. These bacteria grow in the form of long filaments that can intertwine and accumulate, disrupting the normal formation of sludge flocs. When the filaments become too abundant, they prevent the flocs from compacting properly, leading to inefficient settling and excessive foam formation (Miłobędzka and Muszyński 2014; Li et al. 2020).

To control and limit the growth of filamentous bacteria, it is essential to monitor the operational conditions of the plant carefully. Adjustments in aeration management, changes in organic loads, or the use of specific treatments may be necessary to maintain system balance and minimize the impact of these bacteria on the treatment process. Effective management of filamentous bacteria is crucial to ensure good plant performance and to avoid complications associated with poor settling and excessive foam formation (Miłobędzka et al. 2015).

1.8 General Recommendations

To ensure optimal protection of water resources, it is imperative to establish a dedicated on-call service for managing incidents related to wastewater treatment plants (WWTPs). This service is responsible for responding to malfunctions and failures, even outside regular working hours, to prevent or minimize environmental impacts caused by the discharge of untreated or inadequately treated wastewater. Such incidents often occur at night or during weekends, requiring a rapid and effective response available 24/7 (OFEV 2014). Effective management of wastewater treatment plants (WWTPs) also relies on regular and appropriate maintenance. This includes routine maintenance of facilities to ensure their proper daily operation. It is recommended that a detailed maintenance plan be developed that incorporates all necessary operations to keep equipment in good working condition. This plan should include preventive inspections, repairs, and equipment replacements to minimize service interruptions and extend the lifespan of the facilities (Mamandipoor et al. 2020).

Additionally, each wastewater treatment plant (WWTP) must have capacity reserves to handle variations in pollutant load, which can fluctuate due to factors such as tourism or seasonal discharges of industrial wastewater. These reserves should be defined based on the specific characteristics of the watershed and the vulnerability of the receiving environment. Proper capacity planning allows for effective management of peak loads and ensures optimal treatment even during periods of high demand. The technical and professional management of the WWTP should also aim to optimize overall operations. This involves efficiently using available resources, including personnel, coagulants, energy, and costs (Qasim 2017). It is crucial to ensure that optimization efforts do not compromise water protection by avoiding deterioration in the quality of discharges, increased emissions, or decreased treatment efficiency. Integrating energy aspects into this optimization is particularly important. Proper measurement and recording of energy consumption help identify improvement opportunities in both energy consumption and production, whether electrical or thermal. Monitoring the WWTP's operation is essential to quickly detect anomalies and deviations from normal operating conditions. This monitoring, beyond simple visual inspections, should include collecting and analyzing all relevant operating parameters. The collected data allows for effective plant management, rapid response to extraordinary events, identification of their causes, and implementation of necessary corrective measures. This monitoring also helps meet self-monitoring requirements, report to authorities, and develop national indicators for WWTP management.

Many studies focus on optimizing wastewater treatment plants and improving their efficiency. These research efforts highlight the challenges and potential solutions for better managing the treatment processes.

According to Lu et al. (2019), optimizing wastewater treatment plants requires better alignment between the treatment capacity of the plants and the supporting sewer networks. In particular, in rural areas, where wastewater sources are more dispersed and require more extended sewer networks for wastewater collection compared to urban areas, a sensible approach would be to implement localized and smaller-scale treatment facilities, rather than extending the length of sewers for centralized facilities. For many urban WWTPs, it will be necessary to improve treatment capacity further to address the issue of operational overload. Additionally, strategies to increase the organic strength of wastewater should be promoted to increase the organic content of sludge for more energy-efficient disposal and to enhance the overall energy efficiency of wastewater treatment plants (WWTPs). This can be achieved by constructing more separate sewer networks instead of combined networks (which collect both municipal wastewater and stormwater) and phasing out septic tanks.

On the other hand, Newhart et al. (2019) show in their work that performance analyses based on data offer a significant opportunity for optimizing wastewater treatment plant operations. However, the use of big data in this field is still underutilized, primarily because of a shortage of data science expertise among the staff. Wastewater treatment processes exhibit nonlinear, non-stationary, autocorrelated, and co-correlated behaviors, making them challenging to model with traditional fundamental-based approaches. To address these challenges, various data modeling methodologies, including neural networks, transfer functions, and fuzzy logic, can be used for fault detection, variable prediction, and advanced control. The appropriate implementation of these methods can significantly improve the efficiency of wastewater treatment plants by providing continuous monitoring and real-time process adjustments. Despite progress in techniques for creating environmental decision support systems, knowledge discovery, and management, the research field is still far from providing systems that seamlessly integrate different types of knowledge and various reasoning methods. The article points out several limitations that currently hinder the practical implementation of computational techniques (Corominas et al. 2018).

1.9 Conclusion

With rapid urbanization and the significant increase in wastewater volumes, it is crucial to adopt advanced approaches for wastewater treatment. Effective management of treatment plants becomes essential to meet the growing challenges. To optimize the operation of wastewater treatment plants, it is imperative to address the need to adapt treatment capacities to sewer networks and implement advanced techniques based on data analysis. Optimization strategies should integrate various processes, ranging from preliminary treatments to physico-chemical and biological methods, as well as proper sludge management. Using big data and advanced treatment models offers significant opportunities to improve operational performance. Commitment to these practices enables effective responses to current and future wastewater treatment requirements while protecting receiving environments and supporting sustainable development.

References

Alsulaili A, Refaie A (2020) Artificial neural network modeling approach for the prediction of five-day biological oxygen demand and wastewater treatment plant performance. Water Supply 21:1861–1877. https://doi.org/10.2166/ws.2020.199

Arfanuzzaman Md, Dahiya B (2019) Sustainable urbanization in Southeast Asia and beyond: Challenges of population growth, land use change, and environmental health. Growth and Change 50:725–744. https://doi.org/10.1111/grow.12297

Arroyo P, Molinos-Senante M (2018) Selecting appropriate wastewater treatment technologies using a choosing-by-advantages approach. Sci Total Environ 625:819–827. https://doi.org/10.1016/j.scitotenv.2017.12.331

Bagheri M, Mirbagheri SA, Bagheri Z, Kamarkhani AM (2015) Modeling and optimization of activated sludge bulking for a real wastewater treatment plant using hybrid artificial neural networks-genetic algorithm approach. Process Saf Environ Prot 95:12–25. https://doi.org/10.1016/j.psep.2015.02.008

Baklouti I, Mansouri M, Hamida AB et al (2018) Monitoring of wastewater treatment plants using improved univariate statistical technique. Process Saf Environ Prot 116:287–300. https://doi.org/10.1016/j.psep.2018.02.006

Başkurt M, Kocababuç I, Binici E et al (2017) Life cycle assessment as a decision support tool in wastewater treatment plant design with renewable energy utilization. Desalin Water Treat 93:229–238. https://doi.org/10.5004/dwt.2017.21682

Bijekar S, Padariya HD, Yadav VK et al (2022) The state of the art and emerging trends in the wastewater treatment in developing nations. Water 14:2537. https://doi.org/10.3390/w14162537

Borzooei S, Campo G, Cerutti A et al (2019) Optimization of the wastewater treatment plant: from energy saving to environmental impact mitigation. Sci Total Environ 691:1182–1189. https://doi.org/10.1016/j.scitotenv.2019.07.241

Chatterjee S, Mitra A, Gupta SK, Gupta DK (2019) A review on reed bed system as a potential decentralized wastewater treatment practice. In: Sathishkumar R, Kumar SR, Hema J, Baskar V (eds) Advances in plant transgenics: methods and applications. Springer, Singapore, pp 239–251

Corominas L, Garrido-Baserba M, Villez K et al (2018) Transforming data into knowledge for improved wastewater treatment operation: a critical review of techniques. Environ Model Softw 106:89–103. https://doi.org/10.1016/j.envsoft.2017.11.023

Dai H, Chen W, Lu X (2015) The application of multi-objective optimization method for activated sludge process: a review. Water Sci Technol 73:223–235. https://doi.org/10.2166/wst.2015.489

De Vrieze J, Smet D, Klok J et al (2016) Thermophilic sludge digestion improves energy balance and nutrient recovery potential in full-scale municipal wastewater treatment plants. Biores Technol 218:1237–1245. https://doi.org/10.1016/j.biortech.2016.06.119

Demirbas A, Edris G, Alalayah WM (2017) Sludge production from municipal wastewater treatment in sewage treatment plant. Energy Sources Part A Recovery Utilization Environ Eff 39:999–1006. https://doi.org/10.1080/15567036.2017.1283551

Dlangamandla C (2022) Destabilisation and inactivation of foam and foam-forming microorganisms in poultry slaughterhouse wastewater. Cape Peninsula University of Technology, thesis

Dlangamandla C, Basitere M, Okeleye BI et al (2020) Biofoam formation and defoamation in global wastewater treatment systems. Water Pract Technol 16:1–18. https://doi.org/10.2166/wpt.2020.113

Edokpayi JN, Odiyo JO, Durowoju OS et al (2017) Impact of wastewater on surface water quality in developing countries: a case study of South Africa. In: Water quality. IntechOpen

Ejhed H, Fång J, Hansen K et al (2018) The effect of hydraulic retention time in onsite wastewater treatment and removal of pharmaceuticals, hormones and phenolic utility substances. Sci Total Environ 618:250–261. https://doi.org/10.1016/j.scitotenv.2017.11.011

Gherghel A, Teodosiu C, De Gisi S (2019) A review on wastewater sludge valorisation and its challenges in the context of circular economy. J Clean Prod 228:244–263. https://doi.org/10.1016/j.jclepro.2019.04.240

Gu Y, Li Y, Li X et al (2017) The feasibility and challenges of energy self-sufficient wastewater treatment plants. Appl Energy 204:1463–1475. https://doi.org/10.1016/j.apenergy.2017.02.069

Gu D, Andreev K, Dupre ME (2021) Major trends in population growth around the world. China CDC Wkly 3:604–613. https://doi.org/10.46234/ccdcw2021.160

Guadie A, Yesigat A, Gatew S et al (2021) Evaluating the health risks of heavy metals from vegetables grown on soil irrigated with untreated and treated wastewater in Arba Minch, Ethiopia. Sci Total Environ 761:143302. https://doi.org/10.1016/j.scitotenv.2020.143302

Guo X, Li B, Zhao R et al (2019) Performance and bacterial community of moving bed biofilm reactors with various biocarriers treating primary wastewater effluent with a low organic strength and low C/N ratio. Biores Technol 287:121424. https://doi.org/10.1016/j.biortech.2019.121424

Han H-G, Li Y, Guo Y-N, Qiao J-F (2016) A soft computing method to predict sludge volume index based on a recurrent self-organizing neural network. Appl Soft Comput 38:477–486. https://doi.org/10.1016/j.asoc.2015.09.051

Han H, Wu X, Ge L, Qiao J (2018) A sludge volume index (SVI) model based on the multivariate local quadratic polynomial regression method. Chin J Chem Eng 26:1071–1077. https://doi.org/10.1016/j.cjche.2017.08.007

Hassan Rashid MA, Manzoor MM, Mukhtar S (2018) Urbanization and its effects on water resources: an exploratory analysis. Asian J Water Environ Pollut 15:67–74. https://doi.org/10.3233/AJW-180007

Jafarinejad S (2017) Cost estimation and economical evaluation of three configurations of activated sludge process for a wastewater treatment plant (WWTP) using simulation. Appl Water Sci 7:2513–2521. https://doi.org/10.1007/s13201-016-0446-8

Kasapoglu OA, Unluturk E (2018) Relationship between maintenance and sustainability in waste water treatment plant. Lisbon

Kõrgmaa V, Kriipsalu M, Tenno T et al (2019) Factors affecting SVI in small scale WWTPs. Water Sci Technol 79:1766–1776. https://doi.org/10.2166/wst.2019.177

Li B-B, Peng Z-Y, Zhi L-L et al (2020) Distribution and diversity of filamentous bacteria in wastewater treatment plants exhibiting foaming of Taihu Lake Basin, China. Environ Pollut 267:115644. https://doi.org/10.1016/j.envpol.2020.115644

Lu J-Y, Wang X-M, Liu H-Q et al (2019) Optimizing operation of municipal wastewater treatment plants in China: the remaining barriers and future implications. Environ Int 129:273–278. https://doi.org/10.1016/j.envint.2019.05.057

MAGE (42) (2006) Fiches d'exploitation des stations d'épuration

Mamandipoor B, Majd M, Sheikhalishahi S et al (2020) Monitoring and detecting faults in wastewater treatment plants using deep learning. Environ Monit Assess 192:148. https://doi.org/10.1007/s10661-020-8064-1

Mancuso G, Lavrnić S, Canet-Martí A et al (2023) Performance of lagoon and constructed wetland systems for tertiary wastewater treatment and potential of reclaimed water in agricultural irrigation. J Environ Manage 348:119278. https://doi.org/10.1016/j.jenvman.2023.119278

Martínez R, Vela N, el Aatik A et al (2020) On the use of an IoT integrated system for water quality monitoring and management in wastewater treatment plants. Water 12:1096. https://doi.org/10.3390/w12041096

Miłobędzka A, Muszyński A (2014) Population dynamics of filamentous bacteria identified in Polish full-scale wastewater treatment plants with nutrients removal. Water Sci Technol 71:675–684. https://doi.org/10.2166/wst.2014.512

Miłobędzka A, Witeska A, Muszyński A (2015) Factors affecting population of filamentous bacteria in wastewater treatment plants with nutrients removal. Water Sci Technol 73:790–797. https://doi.org/10.2166/wst.2015.541

Mininni G, Laera G, Bertanza G et al (2015) Mass and energy balances of sludge processing in reference and upgraded wastewater treatment plants. Environ Sci Pollut Res 22:7203–7215. https://doi.org/10.1007/s11356-014-4013-2

Mohammadi M, Fataei E (2019) Comparative life cycle assessment of municipal wastewater treatment systems: lagoon and activated sludge. Caspian J Environ Sci 17:327–336. https://doi.org/10.22124/cjes.2019.3806

Moulkaf S, Elkihel B, Merimi C (2023) Optimization of WWTP of dairy industry. Key Eng Mater 954:3–16. https://doi.org/10.4028/p-Bc7d7o

Narayanan CM, Narayan V (2019) Biological wastewater treatment and bioreactor design: a review. Sustain Environ Res 29:33. https://doi.org/10.1186/s42834-019-0036-1

Newhart KB, Holloway RW, Hering AS, Cath TY (2019) Data-driven performance analyses of wastewater treatment plants: a review. Water Res 157:498–513. https://doi.org/10.1016/j.watres.2019.03.030

OFEV (2014) Exploitation et contrôle des stations d'épuration

Padrón-Páez JI, Almaraz SD-L, Román-Martínez A (2020) Sustainable wastewater treatment plants design through multiobjective optimization. Comput Chem Eng 140:106850. https://doi.org/10.1016/j.compchemeng.2020.106850

Puchongkawarin C, Gomez-Mont C, Stuckey DC, Chachuat B (2015) Optimization-based methodology for the development of wastewater facilities for energy and nutrient recovery. Chemosphere 140:150–158. https://doi.org/10.1016/j.chemosphere.2014.08.061

Qasim SR (2017) Wastewater treatment plants: planning, design, and operation, 2nd edn. Routledge, New York

Riffat R, Husnain T (2022) Fundamentals of wastewater treatment and engineering, 2nd edn. CRC Press, London

Rorat A, Courtois P, Vandenbulcke F, Lemiere S (2019) 8—sanitary and environmental aspects of sewage sludge management. In: Prasad MNV, de Campos Favas PJ, Vithanage M, Mohan SV (eds) Industrial and municipal sludge. Butterworth-Heinemann, pp 155–180

Salgot M, Folch M (2018) Wastewater treatment and water reuse. Curr Opin Environ Sci Health 2:64–74. https://doi.org/10.1016/j.coesh.2018.03.005

Sasirekha P, Ramanikannan R, Rohith M (2020) Waste water treatment by using reed bed system. J Sci Technol (JST) 5:81–86. https://doi.org/10.46243/jst.2020.v5.i4.pp81-86

Sikosana ML, Sikhwivhilu K, Moutloali R, Madyira DM (2019) Municipal wastewater treatment technologies: a review. Procedia Manuf 35:1018–1024. https://doi.org/10.1016/j.promfg.2019.06.051

Sun Q, Li M, Ma C et al (2016) Seasonal and spatial variations of PPCP occurrence, removal and mass loading in three wastewater treatment plants located in different urbanization areas in Xiamen, China. Environ Pollut 208:371–381. https://doi.org/10.1016/j.envpol.2015.10.003

Szeląg B, Gawdzik J, Studziński J (2018) Sludge volume index (SVI) modelling: data mining approach. In: Wilimowska Z, Borzemski L, Świątek J (eds) Information systems architecture and technology: proceedings of 38th international conference on information systems architecture and technology—ISAT 2017. Springer International Publishing, Cham, pp 324–335

Waqas S, Bilad MR, Man ZB et al (2021) An energy-efficient membrane rotating biological contactor for wastewater treatment. J Clean Prod 282:124544. https://doi.org/10.1016/j.jclepro.2020.124544

Waqas S, Harun NY, Sambudi NS et al (2023) A review of rotating biological contactors for wastewater treatment. Water 15:1913. https://doi.org/10.3390/w15101913

Wilén B-M, Liébana R, Persson F et al (2018) The mechanisms of granulation of activated sludge in wastewater treatment, its optimization, and impact on effluent quality. Appl Microbiol Biotechnol 102:5005–5020. https://doi.org/10.1007/s00253-018-8990-9

Zhang W, Tooker NB, Mueller AV (2020) Enabling wastewater treatment process automation: leveraging innovations in real-time sensing, data analysis, and online controls. Environ Sci Water Res Technol 6:2973–2992. https://doi.org/10.1039/D0EW00394H

Zhang J, Shao Y, Wang H et al (2021) Current operation state of wastewater treatment plants in urban China. Environ Res 195:110843. https://doi.org/10.1016/j.envres.2021.110843

Zhu W, Wang Z, Zhang Z (2020) Renovation of automation system based on industrial internet of things: a case study of a sewage treatment plant. Sensors 20:2175. https://doi.org/10.3390/s20082175

Chapter 2
An Overview on Various Wastewater Treatment Plants and Techniques

Prachiprava Pradhan and Ajit P. Rathod

Abstract More risks to the environment and public health arise from the use of wastewater, particularly in the developing nations where large quantities of untreated wastewater are used in agriculture and wastewater treatment is rare. Wastewater treatment is becoming increasingly important. Food, energy, and water security are becoming increasingly important for India and other countries. In India, most river basins have closed or drained completely. Untreated waste water discharged into surface and ground waters is the primary source for creating pollution in India. Due to the effects of urbanization, industrialization, and agricultural growth, the area also experiences moderate to severe water shortages. Approximately, 73,000 MLD (megaliters per day) are generated in the major cities of India. Large amounts of highly polluted wastewater are produced by biorefineries that are used to produce fuel ethanol. Municipal water is another type of waste water collected from homes, offices, shops, and restaurants in urban areas. Loaded waste water collected from biorefineries is first treated by a biological process named anaerobic digestion. Use of algae is another approach to treat waste water. In a similar way, for the treatment of municipal wastewater, constructed wetlands (CWs) provide an affordable alternative. An overview of the research on various waste water treatment plants and techniques is given in this chapter.

Keywords Wastewater treatment plant · Sewage water · Municipal wastewater · Treatment techniques · Membrane filtration

P. Pradhan · A. P. Rathod (✉)
Department of Chemical Engineering, Visvesvaraya National Institute of Technology, Nagpur, Maharashtra, India
e-mail: aprathod@che.vnit.ac.in

S. Souabi et al. (eds.), *Wastewater Treatment Plants*, Water Science and Technology Library 130, https://doi.org/10.1007/978-3-031-87461-1_2

2.1 Introduction

Security of water, electricity, and food is becoming increasingly important for India and the rest of the world. Food, energy, and water security are becoming increasingly important for India and the rest of the world. Due to the effects of urbanization, industrialization, and agricultural production, an increasing number of river basins in India and around the world are closing nowadays. As a result, there is a moderate to severe water shortage in these areas. Even India is having 2.45% of the total world's land area and 4% water resource, still the country is a home to almost 16% of the world's people (Kaur et al. 2012). The primary goals of wastewater treatment are to eliminate pollutants and protect our natural water resources (Introduction To Wastewater Treatment 2024). Microplastic (MP) contamination is pervasive in the world's seas and has sparked worries about the health of people and the environment. These things act as a bridge between the anthropogenic and natural water cycles. Also, the urban water management system which includes water treatment plants (DWTPs) and the wastewater treatment plants (WWTPs) is termed as MP transition points. In WWTPs, the majority of MPs can be moved from wastewater to sewage sludge by primary and secondary treatments. Sewage sludge and aged solid waste had high concentrations of microplastics (MPs); however, digestion and composting significantly reduced and decomposed MPs (Ou et al. 2024). Worldwide manufacturing and consumption of antibiotics are high, and wastewater treatment is still not very effective, which significantly contaminates water. Antibiotic concentrations in the environment are frequently high enough to preferentially pressure clinically significant microorganisms, increasing the prevalence of resistance. In order to prevent the uncontrollably rapid development of antibiotic resistance, efforts must be made to restrict the amount of antibiotics that are released into the environment by better treating contaminated wastewater, as a substantial drop in antibiotic use is not anticipated (Magureanu et al. 2021). These days, more academic and industrial researchers are attracted to sustainable cities' sewage treatment programs. Monitoring and consistent maintenance of sewage treatment systems in cities is essential, particularly because industrial effluent is typically very toxic and can pose a major threat to public health. Implementing an intelligent sewage treatment system that uses Internet of Things sensors and machine learning helps in managing sewage treatment in fine chemical plants (Miao et al. 2021). Both the number of industries and sewage discharges has increased with economic growth. The untreated, uncontrolled discharge of industrial effluent not only harms the environment but also has unforeseen consequences. Many nations currently struggle with industrial water pollution. Industrial sewage is typically very toxic, which can seriously pollute a city and cause health issues for its citizens. This can increase illness and even endanger lives. Apart from that, modern civilizations and smart cities have turned to sustainability as a prominent research topic (Silva et al. 2018; Islam and Jashimuddin 2017). Wastewater treatment plants (WWTPs) uses various range of physical, chemical, and biological techniques, either alone or in combination, to eliminate impurities and bring water quality back to an acceptable level. Wastewater treatment is the process of restoring water to its original state after it has been used and/or contaminated by people or the environment. Chemical, biological, physical, or a combination of these methods are available as treatment options. Any desired level of quality can be achieved for water treatment, but the expense of doing so increases as the level of purity increases. During primary treatment, materials that either float or readily settle by gravity are removed. This process involves physical methods such as sedimentation, grit removal, comminution, and screening.

Water can be used for irrigation, drinking water, or aquatic life, which dictates the necessary quality of the water (Englande et al. 2015). During the initial treatment, materials that are buoyant or readily displaced by gravity are eliminated. The physical water treatment processes that includes sedimentation, grit removal, comminution, and screening. Closely spaced, long and thin metal bars are used to make screens. After big particles and grit are removed from the wastewater, it is sent to primary treatment plants. Less than 0.1% of suspended, dissolved, and settable solids and more than 99.9% of water by volume form wastewater (Introduction To Wastewater Treatment 2024).

Waste water treatment plants have been the focus of quite a bit of investigation and study. This chapter provides an overview and discussion of anaerobic digestion as a method for treating loaded waste water that is collected from biorefineries. This chapter also provides the use of constructed wetlands (CWs) for the treatment of municipal wastewater and also algae for waste water treatment. An overview of the research on various waste water treatment plants and techniques are given in this chapter.

2.2 Research on Different Aspects of Waste Water

2.2.1 Bio-refineries Wastewater Treatment

The biorefinery and wastewater treatment principles are interrelated by the wastewater biorefinery. The goal of a wastewater biorefinery is to enhance resource productivity and restore wastewater to an appropriate stage while simultaneously generating a product or products of enough value to maintain the process's economic viability. Its main goal is to produce clean or "fit for use" water as a by-product while generating value-added products from the organic carbon, nitrogen, phosphorus, and related trace nutrients that exist in the wastewater stream (Pott et al. 2018). Large amounts of extremely contaminated wastewater are produced by biorefineries used for producing fuel ethanol. The first step in waste water treatment process with such high loads includes anaerobic digestion. Currently, the effluent stream that contains more than 90% of Chemical Oxygen Demand (COD) can be removed through anaerobic biological treatment of biorefinery effluents. During this stage, 85–90% of the biological energy is recovered as biogas and 80–90% of the BOD is removed (Satyawali and Balakrishnan 2008; Pant and Adholeya 2007). The anaerobic digestion steps effluent needs additional aerobic treatment to get the BOD down to safe levels. Nevertheless, biological treatment methods by themselves are inadequate to comply with increasingly stringent environmental laws (Pant and Adholeya 2007; Juwar et al. 2024). Using willow plantations to treat primary urban wastewater can be a sustainable way to lessen the financial and environmental costs associated with conventional treatment. It is currently unknown; therefore, how irrigation with wastewater might affect the potential for willow biorefineries. Extractable phytochemicals are another potentially important bioproduct from wastewater-irrigated willow for use in biorefineries. Willow trees produce extremely complicated phytochemical mixtures in their stems for intra- and interspecies signaling, defense against abiotic stress, and defense against pathogens and herbivory. Phytochemicals have vital biological roles that enable them to perform a wide range of potent and diversified bioactive actions, potentially making them valuable as bioproducts that can be extracted (Sas 2021).

Another technique for purifying wastewater is the use of algae. Lowering organic and inorganic loads, increasing dissolved oxygen levels, lowering CO_2 emissions, and producing valuable biomass through the sequential use of heterotrophic and autotrophic algal species are just a few advantages of using algae to treat wastewater. Additionally, this biomass is a great source of "organic" fertilizers. It is well known from eutrophication research that algae can grow in extremely high concentrations of inorganic phosphates and nitrates, which would normally be harmful to other living things. Algae in particular have the ability to clean up very polluted wastewaters.

2.2.2 *Wastewater Treatment Using Constructed Wetlands*

Constructed wetlands, or CWs, are built structures that use the natural processes present in soils, plants, and related microbial communities to help clean wastewater. They are meant to make use of a lot of the same processes seen in naturally occurring wetlands, but in a more controlled environment. Hybrid systems made up of different types of manmade wetlands could be integrated to improve treatment performance, specifically for nitrogen.

2.2.2.1 Surface Constructed Wetlands with Free Water

FWS CW consists of basins with 20–40 cm of water depth and 20–30 cm of rooting soil. Dense emergent vegetation typically covers more than half of the area. Plants that are naturally existing might coexist with cultivated macrophytes. Without the need to harvest plants, denitrification can occur in anaerobic pockets within the litter layer due to the organic carbon in the litter providing the necessary carbon. Area or volume are usually used to estimate FWS CW size. Whereas area-based techniques evaluate pollutant reduction throughout the whole wetland area, volume-based techniques use a hydraulic retention time to quantify the removal of pollutants (Vymazal 2010). Wetland systems with exposed water surfaces to the atmosphere are known as free water surface (FWS) wetlands. The majority of naturally occurring wetlands, such as marshes with emergent macrophytes and primary vegetation grasses, bogs with primary vegetation trees, and swamps with primary vegetation mosses, are FWS systems (United States Environmental Protection Agency 2000).

2.2.2.2 Horizontal Subsurface Flow Constructed Wetlands

Horizontal flow, the materials used for constructing artificial flows include gravel or rock beds that have been covered in wetland plants and an impermeable covering. At the intake, wastewater enters the system and moves through the porous material under the bed's surface in a path that is about horizontal until it reaches the outflow zone, where it is collected and released. Anoxic, aerobic, and anaerobic zones combinedly remove pollution from the filter beds. Also, physical, chemical and microbiological degradation helps in removing pollutant from filter beds. Only the areas around roots where oxygen seeps into the substrate are considered aerobic zones (Vymazal 2010). Different plants can be used in constructed wetland. Macrophytes are one of the most distinguishing characteristics of wetlands. Because of their many characteristics related to the treatment process, the macrophytes that grow in artificial wetlands are an integral part of the design (Vymazal 2011).

2.2.2.3 Vertical Subsurface Flow Constructed Wetlands

Traditional methods require highly qualified staff to be present on the site at all times and are costly for treating sewage discharge on a small scale. Constructed wetlands can keep the degree of pollution control while reducing treatment costs and operational complexity (Abdelhakeem et al. 2016).

Vertical flow constructed wetlands are used for anaerobic septic tank effluents where these tanks are oxygenated by Seidel. However, the vertical flow constructed walls did not spread as quickly as the horizontal flow constructed wetlands, most likely as a result of the increased maintenance and operation costs associated with the need to periodically pump wastewater on the wetland surface. Large batches of water are fed, and the water eventually leaks through the sand layer. Only once all of the water has permeated and the bed is dry is the new batch fed. This allows oxygen in the air to diffuse into the bed. Because of this, vertical flow constructed walls are significantly more aerobic than horizontal flow constructed walls and offer favorable nitrification conditions. Conversely, vertical flow constructed walls don't offer any denitrification. Vertical flow constructed walls are also excellent for eliminating suspended particles and organics. These types of walls are frequently employed to treat wastewater from homes and businesses, particularly in situations where ammonia–nitrogen discharge regulations are in place. On the other hand, a large number of studies on the application of vertical flow constructed walls for various wastewater type, including dairy or cheese manufacturing effluent, composting leachate, airport runoff, and refinery effluent, have been published in the literature (Vymazal 2010).

2.2.2.4 Hybrid Constructed Wetlands

For a better treatment impact, the benefits of different generated wetlands systems could be combined. The majority of hybrid-built wetlands integrate the vertical flow and horizontal flow phases. Even though hybrid systems were rarely used back then, Seidel devised the vertical flow-horizontal flow system since the late 1950s and early 1960s. Wetlands made using vertical flow-horizontal flow hybrid technology were built in France and the UK throughout the 1980s (Vymazal 2010).

2.3 Wastewater Treatment Techniques

Process, by definition, refers to a sequence of occurrences or modifications. Treatment facilities employ a range of techniques that combine to achieve the desired targets for water quality. During these processes, the pollutants present in wastewater are isolated, removed, and disposed of. There are four primary methods or processes used to treat wastewater: chemical, mechanical, biological, and physical, etc. One example of a physical treatment method is the use of tanks and other structures intended to restrict and control wastewater flow in order to aid in the removal of contaminants. Similar to this, mechanical treatment methods make use of machines, both simple and complex in their construction and use. In addition to these methods, there is also the biological method, in which the action of bacteria and other microorganisms permits the removal of contaminants that is not effectively achieved by other means (Introduction To Wastewater Treatment 2024).

Wastewater treatment facilities, or Wastewater Treatment Plants, are essential elements of infrastructure that guarantee our environment is properly protected. Every Wastewater Treatment Plants must include biological treatment, and the most popular biotreatment method for treating sewage and industrial wastewaters is activated sludge. An effective management of Wastewater Treatment Plants operations may result in improved water quality and energy efficiency. The tight rules for the discharged waters have led to a renewed focus on this study field, which is essential to maintaining a clean environment. As the aeration tank consumes more than half of the energy used for waste water treatment plant, the activated sludge process is the most energy-intensive step during the entire process. Therefore, enhancing the aeration process establishes a crucial goal for lowering energy usage and raising energy efficiency (Naşcu and Naşcu 2016).

The layout and selection of various unit functions helps in designing any wastewater treatment plants. Many factors influence the process selection, such as the wastewater characteristics, costs, the availability of land, and the required effluent quality (along with any future limits). The four waste water treatment techniques include: primary or pretreatment, secondary treatment, tertiary treatment, sludge treatment/stabilization, etc. This category also includes technology for residual reuse or final disposal (Englande et al. 2015). The projected water demand by various sectors is shown in Fig. 2.1.

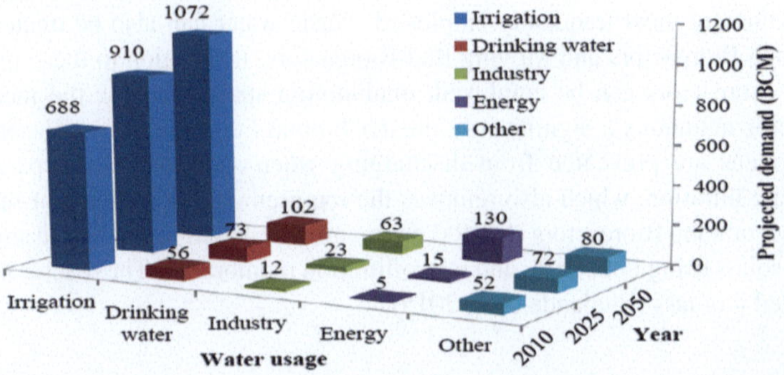

Fig. 2.1 Projected water demand by different sectors (Kaur et al. 2012)

2.3.1 *Primary Treatment*

Initial treatment of municipal wastewater is important before secondary treatment. It involves physical separation, screening or comminution, grit removal, and sedimentation. Large, unwanted solid waste is screened out and frequently eliminated to avoid head loss and flow restriction. A comminutor, which resembles a large "garbage grinder," may be used to homogenize the solids instead of screening them. Grit removal is typically followed by the comminutor to protect the machinery. It is often advised to screen since the homogenized solids add to the organic load at the treatment facility. A comparable mixture of tiny, coarse rock, sand, or other mineral particles is called grit. In order to preserve tank volume capacity and avoid harm to mechanical equipment, it is typically removed. Grit can also be eliminated by adjusting the flow rate through a chamber so that the organic materials stay suspended and the gritty material settles. Usually, the removed grit is cleaned and the land disposed of. Excess oil, grease, etc., can also be removed before primary sedimentation. Removal of excessive oil, grease, and very fine suspended particles is very much essential before going to the sedimentation tank or aeration basin. Processes like flotation and/or gravity separation with skimming can be used to achieve this. Another method for getting rid of settleable solids is primary sedimentation, which involves putting wastewater into it. Before feeding the waste water into a large circular or rectangular tank, pretreatment is done where the solids settle due to gravity (Englande et al. 2015).

2.3.2 *Secondary Treatment*

A biological treatment is used in secondary treatment after pretreatment or initial treatment. Although there are several techniques for developing a biological wastewater treatment process, the trickling filtering and activated sludge processes are the two that are most frequently employed. Waste water can also be treated using Membrane Bioreactors and Moving Bed Bioreactors, in addition to these methods. Smaller bioreactors can be employed, maintaining space, because the membrane bio reactor maintains a significantly greater biomass concentration. The majority of pathogens are prevented from discharging when sediments are separated by membrane filtration, which also removes the requirement for subsequent sedimentation. Membrane Bioreactors (MBRs) are membrane-based biological reactors that remove solids using both ultra- and microfiltration membranes. The reactor includes suspended biomass (Englande et al. 2015).

2.3.2.1 Anaerobic Digestion

Different gases such as methane and carbon dioxide are produced during the biological breakdown of organic waste in anaerobic processes, which lack oxygen. Anaerobic processes are becoming more and more popular as alternatives to aerobic processes for the stabilization of some wastes, especially biological sludge and high-strength industrial wastes (like those from the food industry). This is because anaerobic processes are less expensive than aerobic processes, produce methane, which is beneficial, produce less sludge, and require less energy and nutrients (Englande et al. 2015). Anaerobic digestion is the fermentation process that breaks down organic waste and produces biogas, which is mostly made up of carbon dioxide and methane. Anaerobic digestion processes occur in a range of organic matter-rich and low-redox potential (zero oxygen) environments (Henze 2008). One type of secondary treatment for waste water is anaerobic digestion. The first step in treating wastewater involved separating its constituent elements, resulting in the production of primary effluent and raw or primary sludge. The physical characteristics and organic strength of any product can be determined using biochemical oxygen demand (BOD). Secondary treatments, which are usually biological in nature, are used to further treat these goods. The primary settling provides sludge that goes to digesters, which are large enclosed tanks that are devoid of molecular oxygen that is free. Here, the organic material in the sludge serves as a food source for anaerobic bacteria, which produce carbon dioxide and methane gas. Anaerobic treatment alone is an effective way to remove biodegradable organic molecules, leaving mineralized compounds like S^{2-}, $PO_4{}^{3-}$, and $NH_4{}^+$ in the solution. During anaerobic digestion, the anaerobic bacteria are kept in settings that optimize their biological activity for the duration of the digesting period, which usually lasts between 15 and 28 days. The contents of the digester tank are mixed properly and also heated to a constant temperature of around 35–37 degrees Celsius for promoting the interaction between the organic material and bacteria. Also, this helps in inhibiting the formation of a scum layer. Anaerobic digestion typically occurs in two stages. In order to make the digested biosolids stabilize, the contents of the second stage digestion tanks are periodically left undisturbed. After being extracted, the solids are sent to solids handling facilities for additional processing and the removal of excess water (Introduction To Wastewater Treatment 2024).

2.3.2.2 Activated Sludge

The most popular suspended growth technique for treating municipal wastewater is activated sludge. BOD elimination is the process of treating municipal wastewater. Biological processes, including the suspended growth treatment process, are used to remove BOD. Additional forms of treatment are necessary due to specific characteristics of the organic material present in the primary effluent that overflows the primary settling tanks. The primary effluent's turbid appearance is caused by this organic material, which is made up of suspended or colloidal materials that have been finely divided and dissolved. During the first treatment, the dissolved organic material in the influent stays in solution in the liquid flow. Because of their volume and tiny size, the colloidal particles present in the liquid do not settle during primary treatment. Increasing the wastewater's detention time in the primary tanks in an attempt to eliminate these colloidal solids is not feasible nor feasible. Extended retention periods would encourage the settling tanks' septic conditions to develop, which would lower the effectiveness of the process of eliminating particles (Introduction To Wastewater Treatment 2024).

2.3.2.3 Residual (Sludge) Treatment

At drinking water treatment plants, sludge is formed as a gelatinous particle during the chemical coagulation and softening processes. Similar characteristics are shared by most of the sludge, such as a high water content (sometimes more than 95% by weight). To choose the best management practices and establish facilities that can carry out these operations, it is essential to understand various properties such as physical, chemical, and biological of the sludge produced by water treatment processes. The type and source of the raw water, the rate at which the water is produced, the degree of contamination in the source water treated, the type of coagulant used, the dose used throughout the water treatment process, and the operating parameters of the plant all affect the volume, characteristics, and features of the sludge (Qrenawi and Rabah 2021). Although the amount of sludge produced by wastewater treatment plants (WWTPs) is relatively small in relation to the overall volume of wastewater handled, the expenses associated with managing it can reach 50% of total operating costs. Furthermore, it is now very important to develop a sustainable sludge management strategy (Spinosa et al. 2011). During various separation processes, one of the most common problem with water pollution control is the disposal of the generated solids. The amount of sludge generated through chemical addition will rise, and it can become different from biological sludge. As a result, the type of solids generated may change significantly, requiring several treatment approaches before final disposal (Englande et al. 2015).

2.3.3 *Tertiary Wastewater Treatment*

Many significant tertiary waste water treatment technologies, including membrane filtration, UV irradiation, and chlorination, are available.

2.3.3.1 Membrane Filtration

Several methods are used in membrane filtration processes, with pressure-driven membranes being the most frequently used. In pressure-driven membrane filtration, a semi-permeable membrane is subjected to a pressure differential on both sides, which further defines the sorts of solutes that permeate the membrane. Generally, suspended particles are rejected by microfiltration membranes, which have pore sizes on a scale of 1 m (Zagklis and Bampos 2022). Membrane technology is one of the most widely used techniques for reusing and recycling water from various wastewater sources. Although membrane technology is not new, its complexity and variety of wastewater provide opportunities for advancement in terms of energy, space requirements, efficiency, permeate quality, and technical skill requirements. Larger dissolved molecules can be rejected by ultrafiltration with a narrower pore diameter. Membranes that are categorized as nanofiltration are capable of rejecting smaller dissolved molecules (often up to 200 Da) and divalent ions, with pore diameters as small as 1 nm. In summary, reverse osmosis lacks holes and depends on the distinct speeds of solute diffusion within the membrane's polymer to achieve separation. Monovalent ions can even be rejected by reverse osmosis membranes (Ezugbe and Rathilal 2020). Direct membrane filtration shows better potential in wastewater treatment and resource recovery. It has a vital role while there is no biological treatment in terms of improved treated water quality, nutrient recovery and sustainable operation (Hube, et al. 2020; Pradhan et al. 2023).

2.3.3.2 UV Irradiation

A lot of research has been done on ultraviolet technology as a tertiary treatment method, either by itself or in combination with other degrading processes. In their investigation of the UV process in conjunction with H_2O_2 for secondary effluent treatment, one author found that the cost was assessed at 0.296 USD/m^3, and the ARG reduction was 3.48 log and 2.32 log at pH 3.0 and pH 7.0 respectively (Zhang et al. 2016). When UV light and ozonation were combined to treat surface water, the formation of trihalomethane, haloacetic acid, and TOC was reduced by 50%, 80%, and 70%, respectively (Chin and Bérubé 2005). Combining UV irradiation with H_2O_2 reduced both diclofenac and bezafibrate by 100%, whereas combining three technologies UV/Fe^2+/H_2O_2 produced a 100% reduction in diclofenac. Additionally, the method of UV disinfection of ARB and ARGs found in municipal sewage was studied. It was discovered that there had been a 1.4 log decrease in erythromycin- and tetracycline-resistant bacteria, respectively (Zagklis and Bampos 2022).

2.3.3.3 Chlorination

One common disinfection technique used in the last phases of sewage treatment is chlorination. Chlorine is a potent oxidant that combines with organic compounds but can also produce toxic chlorinated by products that are bad for the environment and human health. Numerous research works have proven the effectiveness of chlorination as a disinfectant. While one author reported a 0.7- to 2.6-log reduction in a number of microbiological markers, including viruses and E. coli (highest log reduction). Another author achieved a 2.5-log reduction in *E. coli*. Since chlorination has no effect on the amount of N and P in secondary effluent, it is not a suitable technology to use when treated wastewater is sent to areas where eutrophication is a risk. The effects of chlorinating pharmaceutical substances, such as sulfamethoxazole, ciprofloxacin, norfloxacin, tetracycline, trimethoprim, and erythromycin were tested and found that the compounds' oxidation was reduced by 43–73% (Zagklis and Bampos 2022).

2.4 Status and Need for the Knowledge and Skills on the Safe Use of Wastewater

Conventional wastewater treatment methods are costly and need sophisticated maintenance and operations. The estimated overall cost of setting up a treatment system for every household wastewater is around Rs. 7560 crores, or roughly ten times the amount that the Indian government has budgeted for. Because of dissolved particles from urban areas that are concentrated by rapid evaporation in dry and tropical regions, wastewater is more saline. Excessive wastewater use in agriculture can lower soil yield and lead to salinity issues. A buildup of harmful chemicals from excessive industrial releases into the environment can promote the growth of weeds, algae, and cyanobacteria, so degrading the quality of groundwater and downstream water sources (Kaur et al. 2012).

The rapid urbanization of the population is the one of the largest issues in the globe today. Balanced urban growth is very complex and it is linked to both the quality of life in different metropolitan areas, the security of food, energy and water supplies. All levels of politicians and urban planners are beginning to emphasize liveability more (Maheshwari et al. 2016).

2.5 Conclusion

Numerous articles that come out every year show how active research is now being done in the area of developing new, affordable, and efficient decontamination techniques. Everyone is now very concerned about preserving the environment, especially with regard to the issue of water pollution. This includes members of the public, entrepreneurs, scientists, researchers, and governmental decision-makers. The problems associated with reusing wastewater in developing countries like India are a result of poor treatment. Therefore, the task is to find low-tech, low-cost, and user-friendly solutions that, although protecting our precious natural resources from degradation, also avoid endangering our heavily dependent lives on wastewater. Constructed wetlands are becoming acknowledged as an effective wastewater treatment technology. Constructed wetlands are easier to maintain for untrained workers, cost less energy and materials than conventional treatment systems, and don't require specific handling for the disposal of sludge. Constructing, operating, and maintaining these systems are cheap because they are powered by the sun, wind, soil, bacteria, plants, and animals. However, for technological and financial reasons, the industrial sector only frequently uses a small number of the many and varied treatment methods. Therefore, it appears that policy decisions and well-thought-out programs incorporating affordable decentralized waste water treatment technology are necessary for the planned, strategic, safe, and sustainable use of wastewater treatment.

References

Abdelhakeem SG, Aboulroos SA, Kamel MM (2016) Performance of a vertical subsurface flow constructed wetland under different operational conditions. J Adv Res 7(5):803–814. https://doi.org/10.1016/j.jare.2015.12.002

Chin A, Bérubé PR (2005) Removal of disinfection by-product precursors with ozone-UV advanced oxidation process. Water Res 39(10):2136–2144. https://doi.org/10.1016/j.watres.2005.03.021

Englande AJ, Krenkel P, Shamas J (2015) Wastewater treatment & water reclamation. In: Reference module in earth systems and environmental sciences. Elsevier. https://doi.org/10.1016/b978-0-12-409548-9.09508-7

Ezugbe EO, Rathilal S (2020) Membrane technologies in wastewater treatment: a review. MDPI AG. https://doi.org/10.3390/membranes10050089

Henze M (2008) Biological wastewater treatment: principles, modelling and design. IWA Pub

Hube S et al (2020) Direct membrane filtration for wastewater treatment and resource recovery: a review. Elsevier B.V. https://doi.org/10.1016/j.scitotenv.2019.136375

Introduction To Wastewater Treatment (2024) Guelph.ca/wp-content/uploads/IntroductionToWastewater.pdf

Islam KMN, Jashimuddin M (2017) Reliability and economic analysis of moving towards wastes to energy recovery based waste less sustainable society in Bangladesh: the case of commercial capital city Chittagong. Sustain Cities Soc 29:118–129. https://doi.org/10.1016/j.scs.2016.11.011

Juwar VA, Rathod AP, Kodape SM, Bhanvase BA (2024) Comparative studies of carbon materials synthesized from agricultural and plastic waste as catalysts for ozonation of real textile wastewater. Clean Technol Environ Policy. https://doi.org/10.1007/s10098-024-02760-y

Kaur R, Wani SP, Singh AK, Lal K (2012) Wastewater production, treatment and use in India
Magureanu M, Bilea F, Bradu C, Hong D (2021) A review on non-thermal plasma treatment of water contaminated with antibiotics. Elsevier B.V. https://doi.org/10.1016/j.jhazmat.2021.125481
Maheshwari B, Singh VP, Thoradeniya B (eds) (2016) Water science and technology library balanced urban development: options and strategies for liveable cities [Online]. http://www.springer.com/series/6689
Miao S, Zhou C, AlQahtani SA, Alrashoud M, Ghoneim A, Lv Z (2021) Applying machine learning in intelligent sewage treatment: a case study of chemical plant in sustainable cities. Sustain Cities Soc 72. https://doi.org/10.1016/j.scs.2021.103009
Naşcu I, Naşcu I (2016) Modelling and optimization of an activated sludge wastewater treatment process. In: Computer aided chemical engineering, vol 38. Elsevier B.V., pp 1159–1164. https://doi.org/10.1016/B978-0-444-63428-3.50198-3
Pradhan P, Rathod AP, Rai SB, Mohapatra SS (2023) An overview of research progress on ceramic-based membranes. Mater Today Proc. https://doi.org/10.1016/j.matpr.2023.03.300
Qrenawi LI, Rabah FKJ (2021) Sludge management in water treatment plants: literature review. Inderscience Publishers. https://doi.org/10.1504/IJEWM.2021.111909
Ou H, Liu R, Liao Z, Zeng EY (2024) Occurrence and fate of microplastics in urban water management systems. Microplastic contamination in aquatic environments: an emerging matter of environmental urgency, pp 181–228. https://doi.org/10.1016/B978-0-443-15332-7.00006-5
Pant D, Adholeya A (2007) Biological approaches for treatment of distillery wastewater: a review. https://doi.org/10.1016/j.biortech.2006.09.027
Pott R et al (2018) Wastewater biorefineries: integrating water treatment and value recovery. In: Green energy and technology, vol 0, no 9783319636115. Springer Verlag, pp 289–302. https://doi.org/10.1007/978-3-319-63612-2_18
Sas E et al (2021) Biorefinery potential of sustainable municipal wastewater treatment using fast-growing willow. Sci Total Environ 792. https://doi.org/10.1016/j.scitotenv.2021.148146
Satyawali Y, Balakrishnan M (2008) Wastewater treatment in molasses-based alcohol distilleries for COD and color removal: a review. J Environ Manage 86(3):481–497. https://doi.org/10.1016/j.jenvman.2006.12.024
Silva BN, Khan M, Han K (2018) Towards sustainable smart cities: a review of trends, architectures, components, and open challenges in smart cities. Elsevier Ltd. https://doi.org/10.1016/j.scs.2018.01.053
Spinosa L et al (2011) Sustainable and innovative solutions for sewage sludge management. Water (Switzerland) 3(2):702–717. https://doi.org/10.3390/w3020702
Vymazal J (2010) Constructed wetlands for wastewater treatment. MDPI AG. https://doi.org/10.3390/w2030530
Vymazal J (2011) Plants used in constructed wetlands with horizontal subsurface flow: a review. https://doi.org/10.1007/s10750-011-0738-9
Wastewater Technology Fact Sheet Free Water Surface Wetlands (2000) United States Environmental Protection Agency
Zagklis DP, Bampos G (2022) Tertiary wastewater treatment technologies: a review of technical, economic, and life cycle aspects. MDPI. https://doi.org/10.3390/pr10112304
Zhang Y, Zhuang Y, Geng J, Ren H, Xu K, Ding L (2016) Reduction of antibiotic resistance genes in municipal wastewater effluent by advanced oxidation processes. Sci Total Environ 550:184–191. https://doi.org/10.1016/j.scitotenv.2016.01.078

Part II
Design and Assessment of Wastewater Treatment Plants

Chapter 3
Study of the Performance of the Wastewater Treatment Plant in the Vegetable Oil Refining Industry

Safaa Khattabi Rifi, Roukaya Bouyakhsass, Mohammed Chatoui, Digua Khalid, Abdelaziz Madinzi, Oussama Hartal, Abdelkader Anouzla, Shalini Yadav, Ram Narayan Yadava, and Salah Souabi

Abstract Vegetable oil refineries generate wastewater loaded with pollutants, oils, fats, and metals, contributing significantly to pollution and aesthetic degradation of aquatic ecosystems. This study has made it possible to diagnose the condition of the Ain Harrouda Crystal-Lesieur sewage treatment plant through activated sludge to combat pollution by oils and fats. The results showed that fluctuating flow and organic load are the most feared enemies for optimal station operation. The percentage of COD, BOD_5, oils, fats, and MES removal shows considerable fluctuations over time, reflecting a variation in the quality of treated waters. High removal efficiencies were obtained, 98% for oil and grease, 82% for COD, and 95% for BOD_5. The removal of pollutant loads in oils and fats in COD and BOD_5 is touched by many factors like the physico-chemical characteristics of wastewater, the station's operation, and the aeration tank's performance. The degradation of organic matter remains satisfactory. However, further optimization is necessary to improve the efficiency of the sewage treatment plant.

Keywords Industrial wastewater · Vegetable oil refinery · Biological treatment · Water treatment plant

---

S. K. Rifi · R. Bouyakhsass · M. Chatoui · D. Khalid · A. Madinzi · O. Hartal · A. Anouzla · S. Souabi (✉)
Laboratory of Process Engineering and Environment, Faculty of Science and Technology, Hassan II University of Casablanca, Mohammedia, Morocco
e-mail: salah.souabi@gmail.coml

S. Yadav
Rabindranath Tagore University, Raisen, Madhya Pradesh, India

R. N. Yadava
Mansarovar Global University, Bhopal, Madhya Pradesh, India

S. Souabi et al. (eds.), *Wastewater Treatment Plants*, Water Science and Technology Library 130, https://doi.org/10.1007/978-3-031-87461-1_3

3.1 Introduction

The production processes in the vegetable oil refining industry demand considerable quantities of resources such as water and chemicals, particularly those involving complexants, degreasers, and surfactants (Hartal et al. 2024). It is regarded as one of the significant contributors to pollution in aquatic ecosystems (Dhanke and Wagh 2020). One of the primary environmental challenges in this sector is the high pollutant load in wastewater generated by industrial activities. These wastewaters contain organic pollutants, including suspended and dissolved substances such as phenols, detergents, oils, and greases (Dhanke and Wagh 2020). Once released into the environment, these pollutants can be assimilated by fauna and flora to varying extents. However, some of these compounds harm humans and ecosystems, potentially accumulating in organisms and leading to biomagnification across food chains (Uddin et al. 2021). Various processes were employed to treat wastewaters with high organic matter, including reverse osmosis (Salehi 2014), biological treatment (Karadag et al. 2015), adsorption (Chaouki et al. 2020), ultrafiltration (Martini et al. 2017), advanced oxidation processes (Joodi 2023), and electrochemical methods (Sharma et al. 2020). The selection of an appropriate treatment method is vital for effective effluent management and must consider both investment and operational costs. Physico-chemical treatments are commonly applied for this type of effluent (Haddaji et al. 2022). However, they present challenges, such as high operational costs and sludge generation that is difficult to manage (Elalami et al. 2019). Combining biological and physico-chemical methods provides a more efficient and cost-effective solution. For instance, biological oily wastewater treatment can be done using conventional activated sludge systems or sequential batch reactors, often in conjunction with physico-chemical treatments (Haddaji et al. 2023).

The design of a treatment system that is as efficient as possible consists of a carefully adapted treatment line and adequate dimensioning of the works of the station (dimensions of the tanks, contact times respected) (Jasim 2020). The rigorous monitoring of treatment conditions requires specific data before and after each stage of purification. Indeed, to consider a thorough and rational cleaning of discharges, it is essential to control the flow of wastewater and pollutant load. The quantification of pollution produced by the industrial unit (COD, BOD_5, TSM) before choosing the most suitable technique to minimize the investment and operating cost remains a necessity.

This study aims to estimate the pollutant loads present in the different discharges received by the station, as well as those eliminated by the various treatment stages, including the overall efficiency of the station and those related to each treatment stage. This allows the station-level challenge to be assessed. This will identify the steps on which the intervention can be carried out to improve further the pollution removal performance in terms of COD, BOD_5, MES, and fats that disrupt the cleaning process.

3.2 Materials and Methods

3.2.1 *Description of Lesieur-Cristal Wastewater Treatment Plant*

To meet the requirements of Moroccan laws concerning industrial wastewater, Lesieur-Cristal has set up activated sludge treatment with a physicochemical process to reduce organic pollution and fat. The plant treats wastewater in two ways: water and sludge sectors. The water sector comprises pretreatment, primary treatment, and secondary treatment. Pretreatment removes solids and regulates the pH of wastewater from pulp and process breakers. The next steps include degreasing and homogenizing water. The primary treatment uses a flotation unit to separate light and heavy particles and a coagulation-flocculation unit to remove suspended and colloidal materials. An SBR basin is used in secondary treatment to eliminate biodegradable pollutants through the action of microorganisms in the presence of oxygen.

Two types of wastewater networks were designed during the construction of the plant. This is the wastewater network of the pulp-breaking activity and the process water network. Figure 3.1 shows the processing line implemented by Lesieur-Cristal.

- **Wastewater from Pulp Breaking**

The pulp extracted in the neutralization step is treated using sulfuric acid (H_2SO_4) to recover the fatty acid. The latter is used for the manufacture of soaps according to the following reaction:

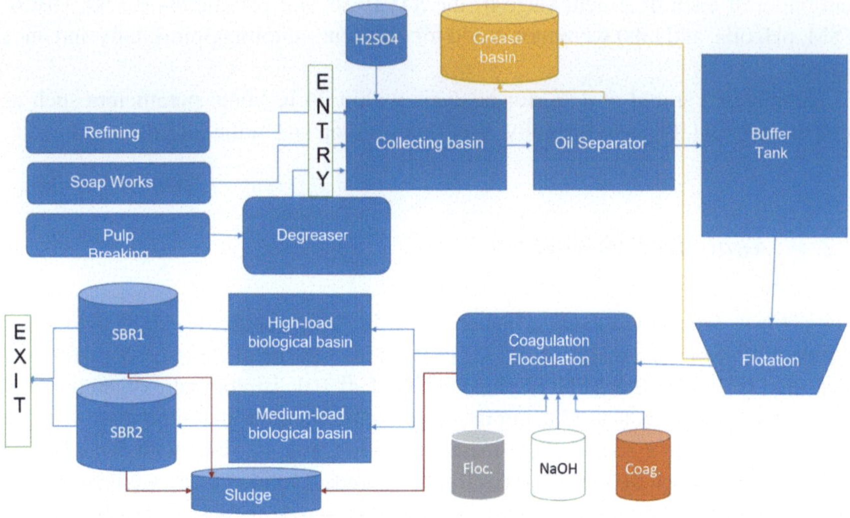

Fig. 3.1 Processing line set up by Lesieur-Cristal

$$\text{pulp} + H_2SO_4 \Rightarrow \text{fatty acid} + \text{waste} + \text{Water (discharge)}$$

The waste from the pulp breakers is sent to the sewage treatment plant, a waste of color that tends to be slightly brown and very acidic.

- **Process Water**

Process waters include soap and refinery discharges:

- **Soap Factory Waste (Laundry)**

During saponification, water is used to wash the soap by removing impurities. This water called the sludge, is then separated and sent to a storage tank before being directed to the treatment plant. These releases are fundamental and contain many suspended matter.

- **Refinery Discharges**

Water from the separation of water and neutralized oil in three refined oil production lines. They are rich in oils and fats, with a yellowish coloration. These waters are collected in a cauldron and sent to the treatment plant.

3.2.2 Sampling

Samples were taken from the raw wastewater produced by the company (Pulp breaking, refining, soap making, process, and STEP inlet) and water at the entrance and outlet of each treatment stage of the STEP. Several parameters (COD, DBO_5, TSM, pH, oils, and fats) were monitored for different sampling points daily and once a week.

Samples are stored at 4 °C for analysis within 24 h. Some parameters such as temperature, pH, O_2, and turbidity were analyzed on site at the STEP.

3.2.3 Analytical Method

The pH was determined using a Fisher Scientific "Accumet basic AB15 pH meter" equipped following the NF T 90-008 standard. Chemical oxygen demand was measured following the AFNOR standard NF T90-101. Biological oxygen demand over five days was measured using the manometric method in compliance with NF EN 1899, utilizing a VELP BOD meter. Turbidity was assessed with NF EN ISO 7027 with a Model 2100N HACH. Total suspended solids (TSS) were measured through centrifugation, following the NF T 90-105 standard. Volatile materials were determined from the TSS sample by placing it in a furnace at 550 °C for two hours.

3.3 Results and Discussions

3.3.1 *Analysis of Wastewater Generated by the Vegetable Oil Refining Industry*

The results of the physicochemical analyses of raw wastewater received by STEP from Lesieur Ain Harrouda are shown in Table 3.1.

The results show that the pH recorded is very acidic, exceeds the limit range set by Moroccan standards, strongly influences raw wastewater quality, and can have several impacts on the station and the receiving environment. pH values range from 0.7 to 4.2, with 1.5 as the mean value. Monitoring the change in turbidity over time showed a significant fluctuation between 300 and 800 NTU. This is related to the quality instability of effluent discharges that are unstable over time. TSS are all the organic and mineral particles contained in wastewater. They modify the turbidity of waters, reduce light penetration endangering photosynthesis, and subsequently self-purification) (Almuktar et al. 2018). The raw wastewater from Lesieur is characterized by an average MES concentration of 4594 mg/L with a maximum concentration of 12,400 mg/L and a minimum concentration of 1600 mg/L. The recorded suspended matter levels are higher than the direct release limit concentration (50 mg/L) and concentration for indirect release (600 mg/L) (Table 3.2) (Ministère de l'environnment du Maroc 2002). Concentrations of TSS are higher than those found by Khouni et al. (2020), with a maximum value of 1116 mg/L.

COD allows the concentration of dissolved or colloidal organic matter to be assessed (Oladeji 2020). COD levels range between 19,900 and 61,200 mg/L, significantly exceeding the 1000 mg/L limit typically allowed for indirect discharge. The COD found in this study is higher than that found by Khalidi-Idrissi et al. (2024), who found a 9000 mg/L value.

BOD_5 varies from 1400 to 5800 mg/L with an average of 2992 mg/L, which remains above 500 mg/L and is considered the indirect release limit. The COD/BOD_5 ratio shows that the effluent received by the STEP is rich in non-biodegradable organic

Table 3.1 Physicochemical characterization of raw wastewater received by the treatment plant

Parameters	Mean	Minimum	Maximum
pH	1.5	0.65	4.22
Temperature (°C)	33	24	39
BOD_5 (mg/L^{-1})	2992	1400	5800
COD (mg O_2 L)	37,723	19,900	61,200
COD/BOD_5	12.60	14.21	10.55
TSS (mg/L)	4594	1600	12,400
Oil and grease (mg/L)	2053	400	6800
Turbidity (UNT)	600	300	800
Sulfate (mg L^{-1})	1080	160	1960

Table 3.2 Values of the treated effluent and Moroccan standard

Parameters	Treated water	Standard
pH	6.1	6.5–8.5
COD (mg/L)	6344	1000
BOD_5 (mg/L)	160	500
TSM (mg/L)	1200	600
Oil and grease (mg/L)	17	20

matter with a value greater than 3. Haddaji et al. (2022) confirm the complex nature and low biodegradability of VORW. Sulfate contents during the different sampling campaigns ranged between 160 and 1960 mg/L with an average value of 1080 mg/L. The raw sewage from the company Lesieur is characterized by average contents in oil and fat oscillating between 400 and 6800 mg/L with an average of 2053 mg/L.

Wastewater from vegetable oil refining industry is highly charged in organic matter and oil and grease, which influences the operation of the processing process. The physicochemical characteristics and daily wastewater flow vary over time (Soap, refining, process, pulp breaking), as Haddaji et al. (2022) reported. A physicochemical process followed by a biological treatment must be used to treat the wastewater from the vegetable oil refining industry.

3.3.2 Performance Study of the Lesieur-Cristal Ain Harrouda STEP Plant

3.3.2.1 Flow Rate Variation

The elimination of pollution by the treatment plant of the company Lesieur-Cristal was monitored by measuring the physico-chemical parameters (TSM, COD, BOD_5). The variation in the flow received by the station over time is shown in Fig. 3.2. The results show that the flow varies over time, with a value of up to 1700 m^3/day. This results in a variation of the pollutant load dependent on the flow rate and the content of pollutants. This requires a neutralization and homogenization pool to ensure good coagulation-flocculation and, subsequently, good biodegradability within the biological basin.

3.3.2.2 Variation of pH and BOD_5

The results (Fig. 3.3) showed pH values at the STEP inlet ranging from 0.65 to 4.22, with a mean value of 1.5. Indeed, as part of the refining process, sulfuric acid is added to cause the separation of free fatty acid (Allen 2018), which can significantly influence the biological treatment type batch sequential reactor (SBR). The pH value

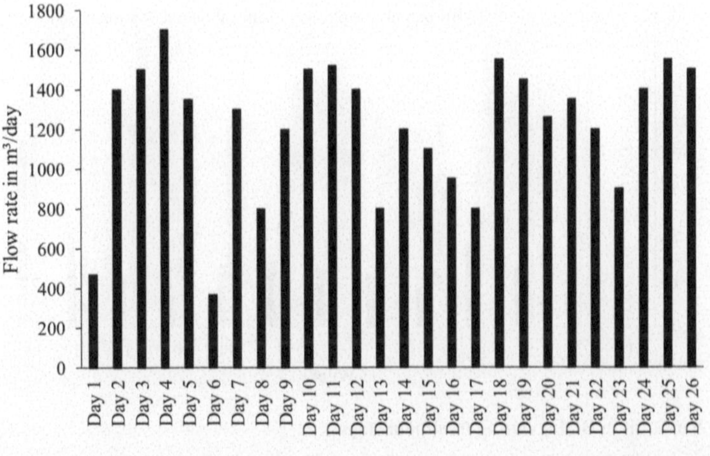

Fig. 3.2 Daily monitoring of the flow of treated wastewater at the Lesieur-Cristal treatment plant

at the SBR reactor ranges from 5.9 to 6.30, with an average value of 6.1, which promotes biomass development and subsequent biological treatment, as shown by Maspolim et al. (2015) and Haddaji et al. (2022).

The results in Fig. 3.4 also show the variation of BOD_5 throughout the treatment steps. BOD_5 at the entrance of the wastewater treatment plant (WWTP) varies between 1400 and 5800 mg/L, with an average of 2992 mg/L and a significant organic charge that fluctuates over time. However, at the FWS, BO_5 ranged from 210 to 320 mg/L, with an average value of 256 mg/L. This significant reduction in FWS illustrates the effectiveness of FWS in treating oil and fat-rich wastewater. Regarding the performance of the physicochemical treatment for BOD_5 elimination, results obtained during several sampling campaigns have shown a notable efficiency of treatment effectiveness. The BOD_5's elimination yield ranged from 84 to 95%,

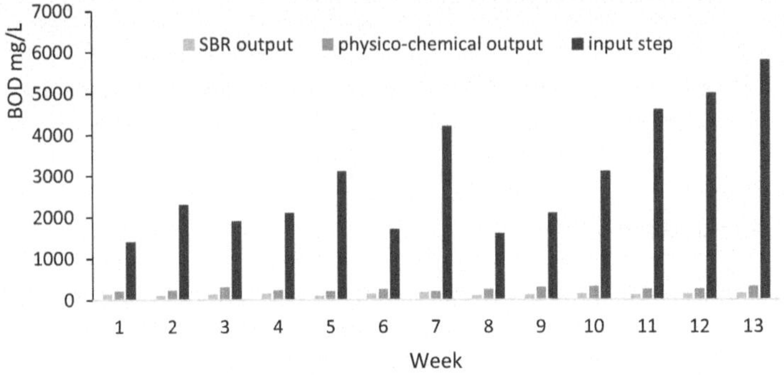

Fig. 3.3 Variation in pH during processing steps

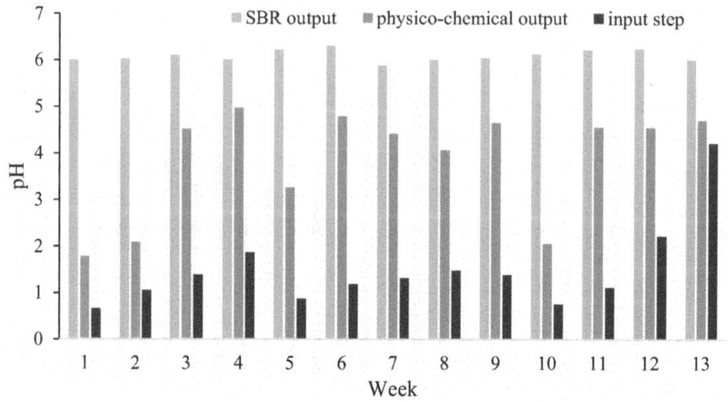

Fig. 3.4 Variation in BOD_5 during treatment stages

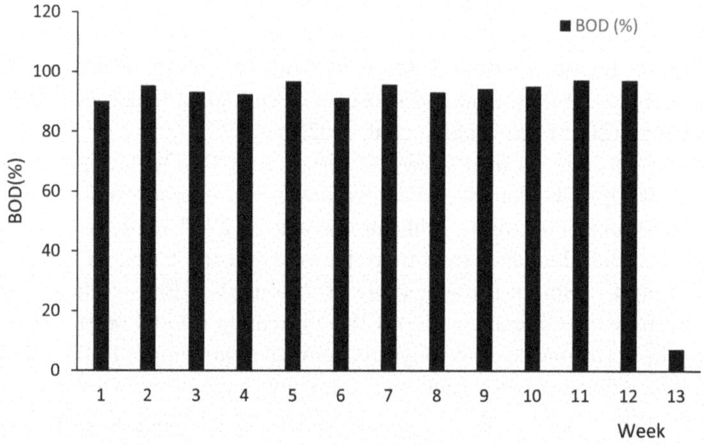

Fig. 3.5 BOD_5 removal efficiency of Lesieur-Cristal WWTP

averaging 89% (Fig. 3.5). This shows that the physicochemical treatment remains effective in reducing the pollutant load present in wastewater rich in oils and fats.

3.3.2.3 Variation in COD and SS

The values in Fig. 3.6 show the variation in COD during the treatment stages. The data revealed a fluctuation in COD at the WWTP inlet, resulting in a fluctuation in the percentage removal. The average COD concentration of the influent and effluent was 37,723 and 6344 mg/L, respectively. Examining the performance of the different treatment stages for COD removal, Table 3.3 and Fig. 3.6 revealed that COD removal by physico-chemical treatment varied from 24 to 83%, with an average value of

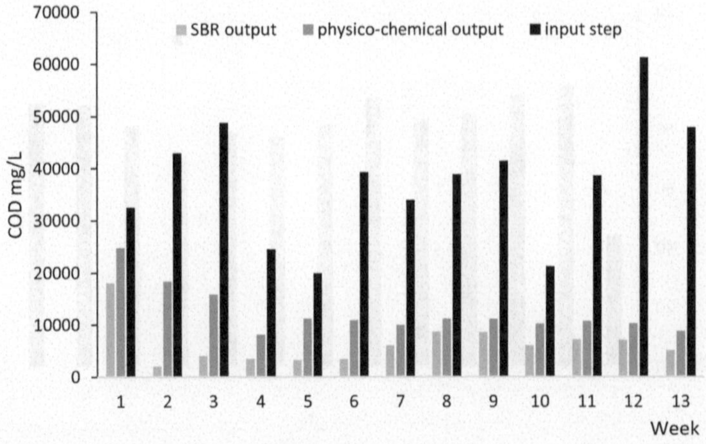

Fig. 3.6 Variation in COD during treatment stages

65%. The percentage of COD removal showed the effectiveness of physico-chemical treatment. The average COD removal at reactor level (SBR) was 47%. The WWTP's COD removal percentage ranged from 22 to 89%, averaging 47% (Fig. 3.7).

Suspended matter concentrations in the various treatment stages are shown in Fig. 3.8. TSS concentration decreased with physico-chemical treatment, but in week 3 and week 10, the physico-chemically treated waters showed an increase in TSS concentration. This may be linked to flocs in the coagulation-flocculation unit, as these flocs are not properly decanted. In addition, the poor performance of this unit is attributed to the low air pressure used. Elimination reached 37% for week 2, 25% for week 3, 33% for week 6, and 27% for week 10. Overall removal from the WWTP averaged 47% (Fig. 3.9). The percentage of elimination varies considerably according to variations in the quantity and physicochemical characteristics of the wastewater. Since wastewater has different characteristics (Kastali et al. 2022; Khalidi-Idrissi et al. 2024). Mkhize Atkinson et al. (2000) observed over 90% removal of SSTs from oil-laden wastewater using an anaerobic/aerobic batch reactor. Fernandes et al. (2013) showed that the SBR reactor can remove 70% of total suspended solids from domestic wastewater. Mahmud and Abbas (2010) found that increasing the operating period of the SBR reactor resulted in better removal of organic matter and suspended solids. The concentrations of water treated by the WWTP compared with

Table 3.3 Oil and grease disposal at the WWTP

Parameters	Primary treatment (%)	Secondary treatment (%)	WWTP (%)
COD	65	47	82
BOD_5	89	45	95
SS	58	10	60
Oils and fats	90	90	98

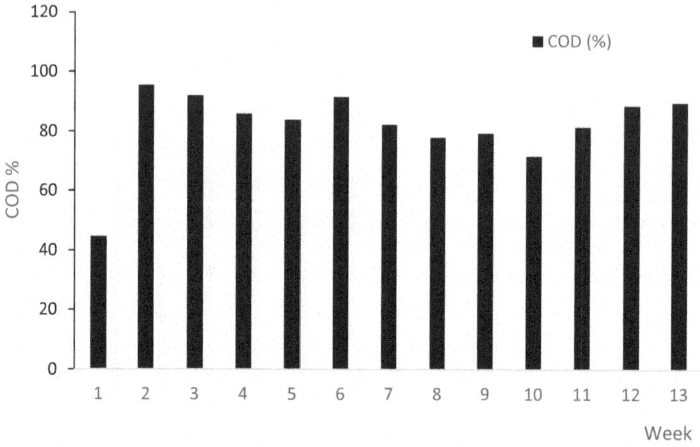

Fig. 3.7 Rendement d'élimination de la DCO par la STEP Lesieur-Cristal

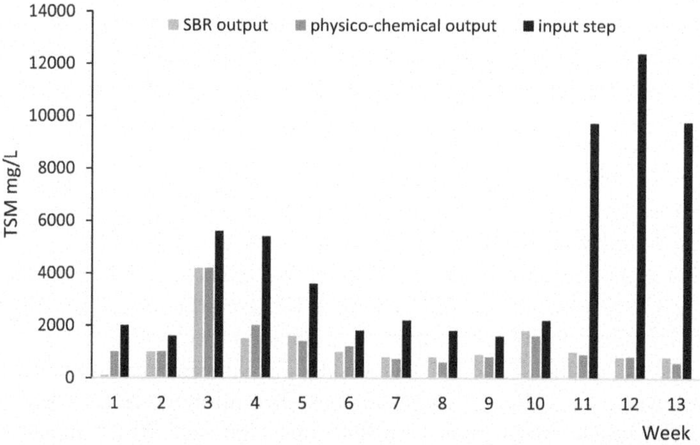

Fig. 3.8 Variation in SST during processing steps

the Moroccan standard are shown in Tables 3.2 and 3.3. The percentage of suspended solids removal is 60%, while the percentage of COD removal is satisfactory (82%), but the treated effluent values exceed the Moroccan standard.

3.3.2.4 Oil and Grease Variations

Figure 3.10 shows the oil and grease concentration in the various treatment stages. The oil and grease concentration in the inlet wastewater was 2053 mg/L. There was considerable fluctuation in the oil and grease concentration in the inlet wastewater.

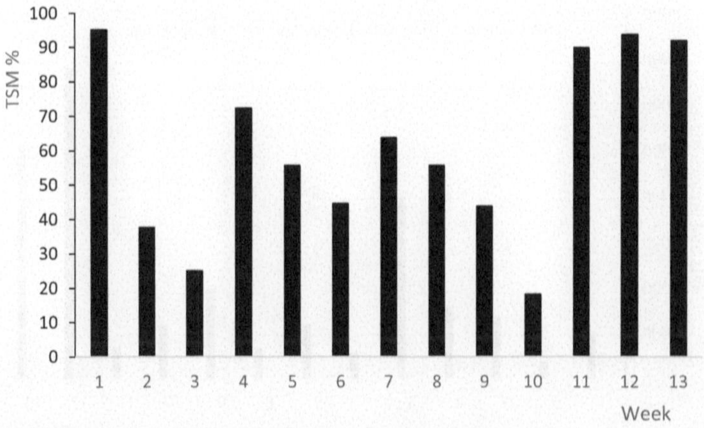

Fig. 3.9 SST elimination efficiency of the Lesieur-Cristal WWTP

High concentrations of oil and grease in wastewater can hinder the removal of solids in biological treatment systems due to their insoluble nature. These substances, being less dense than water, limit the transfer of solids from the solid phase to the liquid phase. Additionally, long-chain fatty acids can inhibit methanogenic activity (Long et al. 2012). The oil concentration of the final effluent shows slight fluctuations during the study period when removal was 99% (Fig. 3.11). Pollution removal from the plant ranged from 98 to 100%. The results showed that a large proportion of the fat was removed by physical and chemical treatment. Mkhize Atkinson et al. (2000) showed that the anaerobic/aerobic SBR reactor removed over 90% of oils and fats. Lee et al (2004) showed that the first SBR stage removed 99.8 $\pm$ 0.1% of oils and fats. The SBR system is a promising option for the VOR industry to adopt as a pretreatment process thanks to its increased efficiency, low cost, and simplicity for removing organic matter and nutrients (Haddaji et al. 2023).

3.3.2.5 Sludge Volume Index in the Biological Basin SBR of the WWTP

In a sequencing batch reactor (SBR) process, bacteria populations are exposed to alternating cycles of aeration, non-aeration, mixing, non-mixing, and feeding to optimize treatment efficiency. However, this operating system can lead to an imbalance in the bacterial ecosystem, favoring certain types of bacteria over others. While this may have a gradual impact on purification performance, it can quickly affect the structure and composition of the bacterial floc. Conventional SBR systems often experience difficulties with liquid–solid separation, indicated by a poor sludge index (Mohan et al. 2005; Yao et al. 2019). This may be linked to the fact that for more than 50% of the operating time, microorganisms grow in an oxygen-depleted environment, generally leading to the development of filamentous building blocks. However, for other authors, on the contrary, the concentration gradient during filling and the

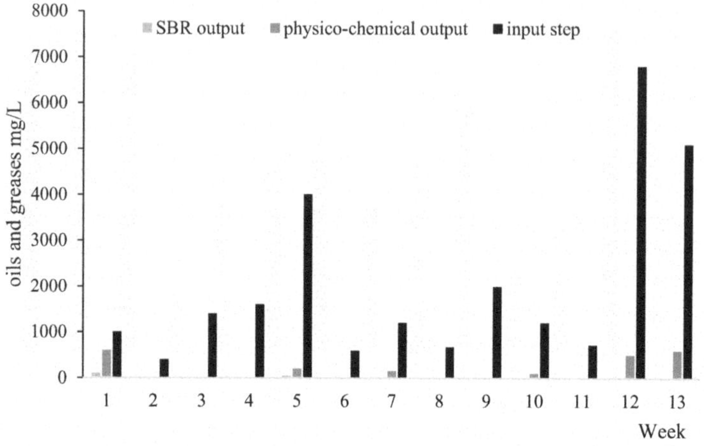

Fig. 3.10 Evolution of oils and fats over time

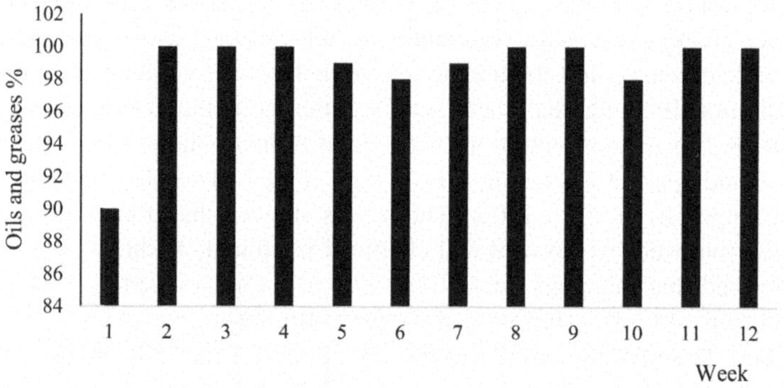

Fig. 3.11 Oil and grease disposal at the WWTP

alternation of feeding and starvation periods is a good way of controlling the development of filamentous bacteria (Dagot et al. 2001; Jagaba et al. 2022). Managing the settling ability of bacterial floc is crucial for optimizing the treatment process. Poor settling can lead to the loss of material in the effluent, which can be detected as total suspended solids (TSS). Additionally, floc disintegration often results in an increase in dissolved substances and a rise in effluent turbidity. Figure 3.12 shows the characteristics of decantation as assessed by measuring the sludge index.

The Sludge Volume Index (SVI) indicates the quality of sludge settling in a wastewater treatment plant. It is generally expressed in milliliters per gram (ml/g) and represents the volume occupied by 1 g of suspended solids after sedimentation (usually 30 min). A high SVI indicates sludge that settles poorly, while a low IB reflects good settling.

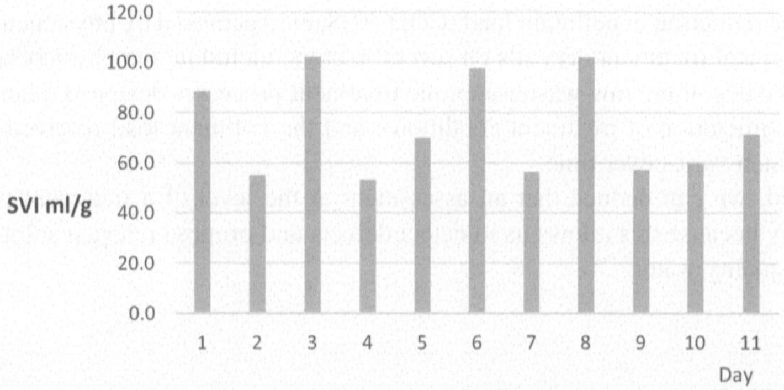

Fig. 3.12 Variation de l'indice de boues en fonction du temps

Figure 3.12 shows variations over the 11 days, with peaks on days 3 and 6, showing episodes of difficult settling. However, after these peaks, SVI tends to return to lower values, suggesting a gradual improvement in sludge quality and settling. These fluctuations could be due to variations in organic load, reactor operating conditions, or other factors influencing sludge production and settling.

The global analysis of Fig. 3.12 shows that the sludge index values generally fall within the range of acceptable values for good settling properties, i.e., between 50 and 150 mL/g (Keller et al. 2001). Despite the good settling results obtained with SBR, some malfunctions may occur. Rodrigues et al. (2001) show that the addition of an anoxic phase upstream of an aerobic anoxic sequence improves settling properties: the sludge index (IB) rises from 200 to 115 mL/g (flocs become larger), and the TSS concentration in the outlet water decreases. For Dagot et al. (2001), the combination of SBR with a continuous process enables the sludge index to be maintained at 100 mL/g. According to many authors, the SBR process offers an ideal situation for settling, and filamentous bulking phenomena are rarely observed (Ai et al. 2021). When mentioned, the sludge index (SI) is most often in the range of values between 50 and 150 mL/g (Zinatizadeh et al. 2011). Although this phase is often considered biochemically inactive, biochemical microbiological phenomena can occur (Keller et al. 2001).

3.4 Conclusion

Wastewater from the vegetable oil refining industry is highly charged with organic matter and oil and grease. This has a significant impact on the operation of the treatment process. The physicochemical characteristics and daily flow rate of wastewater vary from one unit to another (soap-making, refining, processing, pulp breaking, etc.). High removal rates were achieved for oil and grease, COD, and BOD_5, with 98%, 82%, and 95%, respectively. TSS removal efficiency, on the other hand, varied around

60%. The reduction in pollutant load (COD, TSS, etc.) achieved by physicochemical and biological treatment depends on several factors, including the physicochemical characteristics of the raw wastewater, the treatment processes designed, plant operation, optimization of treatment conditions, and the pollutant load received by the plant, which varies over time.

Indeed, we can deduce that an assessment at the level of a treatment plant is necessary because this allows us to detect defects and propose relevant solutions to achieve quality water.

References

Ai S, Du L, Wang Z, Shao L, Kang H, Wang F, Bian D (2021) Effect of controlling filamentous bulking sludge by sequencing batch reactor activated sludge process. In: E3S web of conferences. EDP Sciences, p 4031

Allen DA (2018) Refining. In: Lipid technologies and applications. Routledge, pp 137–167

Almuktar SA, Abed SN, Scholz M (2018) Wetlands for wastewater treatment and subsequent recycling of treated effluent: a review. Environ Sci Pollut Res 25:23595–23623

Chaouki Z, Zaitan H, Nawdali M, Vasarevičius S, Mažeikienė A (2020) Oil removal from refinery wastewater through adsorption on low cost natural biosorbents. Environ Eng Manage J 19

Dagot C, Sommanawan N, Baudu M, Anderson R (2001) Performances et simulation de reacteurs de traitement d'effluents a aeration et decantation sequentielles (SBR) en climat tropical performances and simulation of intermittent decanted extended aeration treatment plants in a tropical climate. Environ Technol 22:1155–1168

Dhanke P, Wagh S (2020) Treatment of vegetable oil refinery wastewater with biodegradability index improvement. Mater Today Proc 27:181–187. https://doi.org/10.1016/j.matpr.2019.10.004

Elalami D, Carrere H, Monlau F, Abdelouahdi K, Oukarroum A, Barakat A (2019) Pretreatment and co-digestion of wastewater sludge for biogas production: recent research advances and trends. Renew Sustain Energy Rev 114:109287

Fernandes H, Jungles MK, Hoffmann H, Antonio RV, Costa RHR (2013) Full-scale sequencing batch reactor (SBR) for domestic wastewater: performance and diversity of microbial communities. Bioresour Technol 132:262–268

Haddaji C, Ennaciri K, Driouich A, Digua K, Souabi S (2022) Optimization of the coagulation-flocculation process for vegetable oil refinery wastewater using a full factorial design. Process Saf Environ Prot

Haddaji C, Chatoui M, Rifi SK, Ettaloui Z, Digua K, Pala A, Anouzla A, Souabi S (2023) Performance of simultaneous carbon, nitrogen, and phosphorus removal from vegetable oil refining wastewater in an aerobic-anoxic sequencing batch reactor (OA-SBR) system by alternating the cycle times. Environ Nanotechnol Monit Manage 20:100827

Hartal O, Khattabi Rifi S, Chatoui M, Haddaji C, Madinzi A, Pala A, Souabi S (2024) Combined natural flotation and chemical precipitation for the treatment of vegetable oil refinery wastewater. Int J Environ Sci Technol 1–12

Jagaba AH, Kutty SRM, Isa MH, Affam AC, Aminu N, Abubakar S, Noor A, Lawal IM, Umaru I, Hassan I (2022) Effect of environmental and operational parameters on sequential batch reactor systems in dye degradation. Dye Biodegrad Mech Tech Rec Adv 193–225

Jasim NA (2020) The design for wastewater treatment plant (WWTP) with GPS X modelling. Cogent Eng 7:1723782

Joodi NS (2023) A review on the treatment of vegetable oils in wastewater and industrial water by advanced oxidation processes. In: AIP conference proceedings. AIP Publishing

Karadag D, Köroӏlu OE, Ozkaya B, Cakmakci M (2015) A review on anaerobic biofilm reactors for the treatment of dairy industry wastewater. Process Biochem 50(2):262–271. https://doi.org/10.1016/j.procbio.2014.11.005

Kastali M, Mouhir L, Chatoui M, Haddaji C, Khattabi S (2022) Study of the pollution generated by wastewater from the refining of vegetable oils. Int J Membr Sci Technol 101–115

Keller C, Kayser A, Keller A, Schulin R (2001) Heavy-metal uptake by agricultural crops from sewage sludge treated soils of the Upper Swiss Rhine Valley and the effect of time. Environ Restor Met Soils USA 273–293

Khalidi-Idrissi A, Hartal O, Madinzi A, El-Abbadi K, Souabi S (2024) Natural flotation and coagulation–flocculation: a dual approach to refinery wastewater treatment. Euro Mediterr J Environ Integr 1–19

Khouni I, Louhichi G, Ghrabi A, Moulin P (2020) Efficiency of a coagulation/flocculation–membrane filtration hybrid process for the treatment of vegetable oil refinery wastewater for safe reuse and recovery. Process Saf Environ Prot 135:323–341. https://doi.org/10.1016/j.psep.2020.01.004

Lee LY, Hu JY, Ong SL, Ng WJ, Ren JH, Wong SH (2004) Two-stage SBR for treatment of oil refinery wastewater. Water Sci Technol 50:243–249

Long JH, Aziz TN, Ducoste JJ (2012) Anaerobic co-digestion of fat, oil, and grease (FOG): a review of gas production and process limitations. Process Saf Environ Prot 90:231–245

Mahmud FY, Abbas WM (2010) Organics and suspended solids removal from hospital wastewater in Mosul City by SBR method. Tikrit J Eng Sci 17:32–37

Martini S, Ang HM, Znad H (2017) Integrated ultrafiltration membrane unit for efficient petroleum refinery effluent treatment. Clean Soil Air Water 45:1600342

Maspolim Y, Zhou Y, Guo C, Xiao K, Ng WJ (2015) The effect of pH on solubilization of organic matter and microbial community structures in sludge fermentation. Bioresour Technol 190:289–298

Ministère de l'environnment du Maroc (2002) Rapport de la quatrième session du CNE: 27 - 28 Février 2002. https://www.environnement.gov.ma/fr/85-cne/378-session-cne-2002

Mkhize Atkinson BW, Bux F (2000) Evaluation of a laboratory-scale biological process for the treatment of edible oil effluent. Water SA 26:555–558

Mohan SV, Rao NC, Prasad KK, Sarma PN (2005) Bioaugmentation of an anaerobic sequencing batch biofilm reactor (AnSBBR) with immobilized sulphate reducing bacteria (SRB) for the treatment of sulphate bearing chemical wastewater. Process Biochem 40:2849–2857

Oladeji SO (2020) Evaluation of physicochemical parameters in wastewater from Muhammad Ayuba dam in Kazaure, Jigawa state, Nigeria. Arch Agric Environ Sci 5:482–488

Rodrigues AC, Brito AG, Melo LF (2001) Posttreatment of a brewery wastewater using a sequencing batch reactor. Water Environ Res 73:45–51

Salehi F (2014) Current and future applications for nanofiltration technology in the food processing. Food Bioprod Process 92:161–177

Sharma S, Aygun A, Simsek H (2020) Electrochemical treatment of sunflower oil refinery wastewater and optimization of the parameters using response surface methodology. Chemosphere 249:126511. https://doi.org/10.1016/j.chemosphere.2020.126511

Uddin MN, Desai F, Asmatulu E (2021) Review of bioaccumulation, biomagnification, and biotransformation of engineered nanomaterials. In: Nanotoxicology nanoecotoxicology, vol 2. Springer, pp 133–164

Yao J, Liu J, Zhang Y, Xu S, Hong Y, Chen Y (2019) Adding an anaerobic step can rapidly inhibit sludge bulking in SBR reactor. Sci Rep 9:10843

Zinatizadeh AAL, Mansouri Y, Akhbari A, Pashaei S (2011) Biological treatment of a synthetic dairy wastewater in a sequencing batch biofilm reactor: statistical modeling using optimization using response surface methodology. Chem Ind Chem Eng Q 17:485–495

Chapter 4
Advancements in Wastewater Treatment Systems: Integrating Simulation and Optimization for Improved Performance

Bayram Arda Kuş and Tolgay Kara

Abstract Advancements in wastewater treatment systems are imperative for mitigating environmental pollution and ensuring the sustainability of water resources. This chapter explores the integration of simulation and optimization techniques to enhance the performance of wastewater treatment systems. By combining advanced simulation models with optimization algorithms, operators can achieve improved efficiency, reduced costs, and enhanced environmental outcomes. This integrated approach enables real-time monitoring, predictive analytics, and optimal decision-making in the management of wastewater treatment processes. Case studies and examples illustrate the practical application and benefits of this approach in various contexts. Overall, this chapter highlights the significance of merging simulation and optimization methodologies to address the evolving challenges of wastewater treatment, paving the way for more sustainable and effective solutions.

Keywords Wastewater simulation and optimization · Predictive analytics · Mathematical models · Historical development of WWT · Computational models · AI integration in wastewater treatment

4.1 Introduction

Due to the environmental laws and increased awareness of people concerning ecological issues during last years, there are numerous analyzes about wastewater treatment (WWT) projects in order for preventing disasters against all living creatures in any receiving environment (Chen et al. 2018). However, regardless of those types

B. A. Kuş (✉) · T. Kara
Department of Electrical and Electronics Engineering, Gaziantep University, Gaziantep, Turkey
e-mail: akus@ilbank.gov.tr

T. Kara
e-mail: kara@gantep.edu.tr

S. Souabi et al. (eds.), *Wastewater Treatment Plants*, Water Science and Technology Library 130, https://doi.org/10.1007/978-3-031-87461-1_4

of investigations, balance between "water treated well enough" and "energy/labor/money consumption," has to be kept (Schölzel 1999).

Wastewater treatment plants (WWTPs) consumes significant amounts of global energy, estimated ratio to be between 1 and 3% (IWA 2021). The energy requirements for the collection and wastewater treatment are anticipated to account for approximately 4% of the total global consumption of electricity. By 2030, the energy consumption specific to wastewater treatment alone is projected to escalate to 680 TWh (Walton 2018).

The electrical energy is needed for operating equipment such as aeration blowers, lift pumps, return pumps, and sludge dewatering in WWTPs. In treatment processes, energy use is distributed across various typical components: sewage lift pumps account for 10–20%, biological treatment for 50–70%, and sludge treatment and disposal for 10–25% of total energy consumption. The set of these three components consumes more than 70% of the total energy (Qiao and Zhou 2018). For instance, in a conventional form of activated sludge system, aeration is the largest energy consumer, using about 60% of the total energy.

With the advancement of emission regulations and standards in wastewater treatment, many plants have adopted advanced treatment processes such as denitrification filters, sand filtration, and ultraviolet (UV) disinfection. WWT is a complex and detailed process. Processes may vary according to climate, geographical structure, flow rate, and wastewater characteristics; for maintenance and efficient operation, proper control systems, and qualified labor are necessary (Sie Chun et al. 2017). Furthermore, optimizing the performance of lift pumps and aeration systems has become essential for achieving energy savings and reducing consumption in sewage treatment plants. Strategies to enhance energy efficiency and decrease consumption include improving existing processes and equipment operations, as well as minimizing operational energy usage (Sandberg et al. 2018). Additionally, optimizing the overall WWT process can further reduce the global energy consumption of treatment plants scale (Yiannopoulos et al. 2008; Gingerich and Mauter 2018).

A variety of mathematical models have been formulated to effectively articulate the process dynamics and energy efficiency in WWT systems (Sie Chun et al. 2017). Within the operational framework of a full-scale WWT process, there exist intricate mathematical interrelationships among the biological, physical, and chemical processes. Ensuring precise consideration of all these factors enhances the understanding and design of any WWTP (Jeppsson et al. 2005; Serdarevic and Dzubur 2016). Interdisciplinary studies are conducted to address wastewater treatment and removal while minimizing environmental impact. The primary goal of designing and implementing WWTPs is to reduce biological phosphorus, organic carbon substances, and biological nitrogen. The design and implementation procedures consist of several subsystems to achieve biological phosphorus, organic carbon substances, and biological nitrogen removal (Gernaey et al. 2004).

4.1.1 Historical Development of WWT and WWTP

The development of sewers and sanitation systems has been a critical aspect of human civilization aimed at maintaining clean and healthy environments. Initially, sewers were constructed to transport wastewater to the nearest discharge points, and the use of cesspools to which home toilets and bathrooms were connected was another sanitation solution. The ancient Greeks advanced this concept further by establishing public toilets and baths, which directed wastewater to collection centers located outside the cities.

The history of sewers dates back to the Mesopotamian Empire, where stormwater drainage systems were developed between 3500 and 2500 BC. Unfortunately, these early collection centers were often overly fertile agricultural areas. Archaeological studies have uncovered four distinct drainage systems in Knossos, at King Minos' Royal Palace with some drains from this system still in use today. In 800 BC, Rome built its centralized sewer system, the Cloaca Maxima (Cooper 2001; Grady et al. 1999).

As urban populations grew, surpassing nature's self-recovery limits, new sanitation challenges emerged. Waste, both solid and liquid, was often discarded from windows or transported to rural areas (Nathanson 2020). Streets were periodically washed, but these rudimentary methods did little to prevent the spread of epidemic diseases, such as the Black Plague (Cooper 2001; Grady et al. 1999). Sir John Harington built two water closets for Queen Elizabeth, a significant advancement in sanitation in 1596 (Jeppsson 1996). However, this innovation was not widely adopted until the 1800s, as connecting toilet pipes to central sewer systems was not yet common practice (Stanwell-Smith 2010).

In 1829, a basic treatment technology was implemented at Craigentinny Meadows in Edinburgh (Mackintosh 1873). The activated sludge model for WWT, based on empirical experience, was later discovered by George W. Fuller in 1914 (Orhon 2015). There is some debate over the exact date of this discovery, with organizations like the IWA and Kalyuzhnyi S. V. citing 1914, while others refer to 1913, based on the studies by Ardern and Lockett. Jeanette Brown, a former president of the Water Environment Association, attributes the discovery to Clark and Gage's work reported in 1913 (Orhon 2015; Ardern and Lockett 1914; Kalyuzhnyi 2003).

The first dynamic model for the activated sludge system evolved from bioreactor models discussed in the 1960s (Sie Chun et al. 2017; Alharbi 2016). By 1974, studies proposing various activated sludge processes were unified. Different reactor types, such as complete mix reactor, plug flow reactor, batch reactor, suspended reactor, membrane bioreactor (MBR), or moving bed biofilm reactor (MBBR), are now used in secondary WWT. Technologies on MBR have become crucial for operational cost assessment (Goodman and Englande 1974), micropollutant removal, and sustainability (Radjenović et al. 2008; Pomiès et al. 2013). Mathematical modeling of MBR technologies highlights their operational advantages and stability (Metcalf and Eddy 2003; Ying Li et al. 2008; Wehner and Wilhelm 1956; Muñoz et al. 2009).

The fill-and-draw aeration process in a WWTP, known as the sequential batch reactor (SBR), dates back to the year 1969 when it was first implemented for large-scale WWTP's for industrial wastewater (Eddy et al. 2003; Beraud et al. 2009; Kowalik et al. 2019). Advances in computer technology and electronically controlled valves have significantly improved SBR systems, making them widely used for removal of organic compounds in the treatment process for both industrial and domestic plants (Cooper 2001; Jeppsson 1996; Kowalik et al. 2019). Properly controlled SBR plants offer numerous benefits, positioning them as a viable alternative to conventional systems (Orhon 2015; Lefebvre and Moletta 2006; Jakubaszek and Stadnik 2019).

Despite the rapid development of process control techniques over the last 40 years, the solutions that were implemented on the WWTP in 1970s are still able to solve the main process control problems (Cooper 2001; Olsson et al. 2014). The industrial computers and electrically controlled equipment such as valves and flowmeters are used in the late 1970s. Traditional methods for WWT rely on intricate mathematical formulas and the calculation of various parameters related to design, process, and operation, must also navigate a range of constraints; including cost, social considerations, regulatory requirements, governmental policies, and technical limitations (Malviya and Jaspal 2021).

In the late 1980s computers (PCs) have gained common use, while the concept of industrial computers evolved to more visualized types. This improvement led researchers to develop more visualized computer models based on mathematical principles of the processes. Being the most popular type of treatment system, activated sludge process is the main content of these models; in these years the anaerobic baffled reactor for WWT has been just developed (Barber and Stuckey 1999). Since the early 1980s, the baffled reactor for WWT has several advantages over well designed and established systems such as the up flow anaerobic filter and the anaerobic sludge blanket.

The activated sludge process, which had previously required fast and powerful computers, benefited significantly from these advancements. Models of the treatments were packaged into simulations to allow the system to operate as a digital twin. The simulation reporting and defining an example of a model library for WWT was performed by Andrews and Graef and in Olsson et al. respectively (Olsson et al. 2014).

In the last two decades, studies are focused to process simulations, which include IAWPRC model (based on COD), WRc STOAT model (based on BOD), ASIM and SSSP and SUMO (Nelson et al. 2021). Research at McMaster University, Ontario, Hamilton, Canada yielded the commercial package GPS-X. The key actors of GPS-X from Hydromantis are Gilles Patry and Imre Takács. A number of other application specific simulator programs have been developed in recent decades, including Aquasim, BioWin, Simba, STOAT, and finally WEST (Olsson et al. 2014; Džubur et al. 2023).

These simulations are essential to observe the theoretical system behavior (Cooper 2001) depending to mathematical expressions in weather conditions and other outside factors, such as dry, stormy and rainy weather conditions, that may affect the treatment process (Gernaey et al. 2004; Caraman et al. 2007; Shen et al. 2008). These variables and disturbances on the system are also take into account in some up-to-date studies using predictive control concept (Vanhooren et al. 2003; Rivas et al. 2008; Shen et al. 2009). Sewer systems, WWTPs, and receiving waters are also mathematically modeled and available on temporary simulators (Gernaey et al. 2004; Neilands et al. 2013). Matlab/Simulink is frequently used for both mathematical modeling and simulators of the WWTPs and other relevant sub-systems because of the elastic and easy-use (Gernaey et al. 2004; Vanhooren et al. 2003; Urena et al. 2011; Sinha et al. 2014; Lee et al. 2002a, b; Olsson and Newell 1999; Stare et al. 2006; Ráduly et al. 2015).

Parameters on both influent and effluent water are critical to the effective operation of a treatment facility. Depending on the residential and industrial components, wastewater might have quite different characteristics. Consequently, throughout the implementation phase, it is crucial to carefully review these components as well as the intended output to the receiving environment (Sinha et al. 2014). The models like the particle swarm optimization, ant colony model, deep learning, genetic algorithms, fuzzy algorithms, and immune algorithms are transforming a number of fields through artificial intelligence (AI) can increase process efficiency and decrease the need for human labor.

In the context of WWT, several parameters including chemical oxygen demand (COD), biological oxygen demand (BOD), emerging pollutants, and heavy metals are critical performance indicators that AI models have been extensively researched to simulate and optimize natural water systems, aiming to predict the removal of targeted or multiple pollutants. Various AI models including neural networks and other hybrid AI models, have been developed to understand the nonlinear patterns of pollutant removal in treatment processes that can achieve more efficient and effective WWT, paving the way for improved environmental sustainability and public health (Malviya and Jaspal 2021).

4.2 Modeling of WWT

The widely held belief that all natural phenomena, whether mechanical, electrical, biological, or otherwise, could be described by mathematical descriptions began to take shape in 1687, when Isaac Newton wrote "Mathematical Principles of Natural Philosophy." Throughout the history of scientific development, classical theory has often fallen short in providing accurate solutions for modern physics-based problems. Compared to a few decades ago, defining and solving such complex models mathematically has become considerably more feasible (Jeppsson 1996).

In WWTPs, hydraulic systems are foundational. The principal governing equation for such systems is the volume balance equation, expressed as the disparity between inflow and outflow, as depicted in Fig. 4.1.

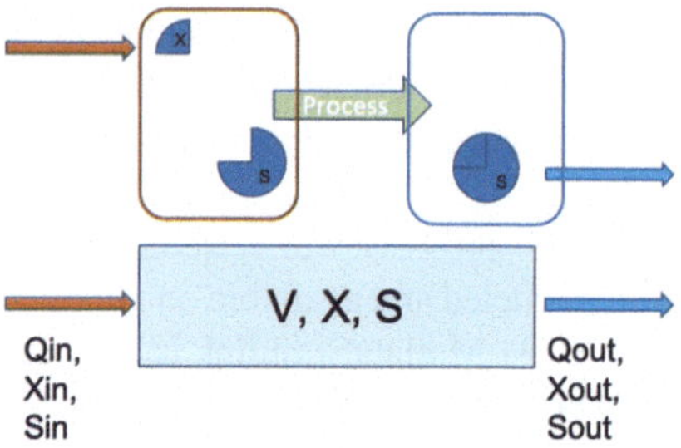

Fig. 4.1 WWT flow diagram

Based on the ratio of food to microorganisms, the activated sludge models may be obtained by (Serdarevic and Dzubur 2016; Mathematics and Perindustrian 2015);

$$\frac{F}{M} = \frac{QS_{in} - QS_{out}}{XV} \tag{4.1}$$

Generally, the food to microorganism ratio quantifies the ratio of incoming food to microorganisms within the system, denoted by Eq. (4.1). The activated-sludge process targets the removal of oxygen-demanding substances from the treatment system. Microorganisms facilitate metabolic reactions such as synthesis-respiration, nitrification (Orhon et al. 2019; Yetilmezsoy 2010). The activated sludge model emphasizes the concentration of biomass and the growth rate of microorganisms (Mathematics and Perindustrian 2015). Schematic flow diagrams in Fig. 4.1 illustrate mass-balance equations. Design principles for all biological treatment reactors rely on achieving equilibrium between biomass and substrate (Serdarevic and Dzubur 2016). Within this framework, a two-state variable system is defined as follows, where S is the Substrate, X is the Biomass.

$$\frac{dX}{dt} = \mu(S)X - bX \tag{4.2}$$

$$\frac{dS}{dt} = -\frac{1}{Y}\mu(S)X \tag{4.3}$$

$$\mu(S) = \frac{\hat{\mu}S}{K_s + S} \tag{4.4}$$

Assuming X and S are measurable, the system characterized by four parameters (K_s, μ, b, Y) exhibits global identifiability. Let $X_0 = X(0)$ and $S_0 = S(0)$.

$$X'(0) = \frac{\hat{\mu}S}{K_s + S}X_0 - bX_0 = X_1 \tag{4.5}$$

$$S'(0) = -\frac{1}{Y}\frac{\hat{\mu}S_0}{K_s + S_0}X_0 = S_1 \tag{4.6}$$

$$\mu = \frac{\hat{\mu} S_0}{K_s + S_0} \tag{4.7}$$

$$X_1 = (\mu - b)X_0 \tag{4.8}$$

$$S_1 = -\frac{\mu}{Y} X_0 \tag{4.9}$$

Perform the first-order Taylor series expansion of Eqs. (4.5) and (4.6) around t = 0.

$$X''(0) = (\mu - b)X_1 + \frac{\mu K_s X_0 S_1}{(K_s + S_0)S_0} = X_2 \tag{4.10}$$

$$S''(0) = -\frac{\mu}{Y}\left(X_1 + \frac{\mu K_s X_0 S_1}{(K_s + S_0)S_0}\right) = S_2 \tag{4.11}$$

Solve (4.8), (4.9), (4.10), and (4.11) for K_s, μ, b, Y (Table 4.1).

$$K_s = \frac{S_2 S_0^2 X_0 - S_1 S_0^2 X_1}{S_1^2 X_0 - S_2 S_0 X_0 + S_1 S_0 X_1} \tag{4.12}$$

$$\mu = \frac{X_2 X_0 S_1 - X_1^2 S_1}{S_2 X_0^2 - S_1 X_1 X_0} \tag{4.13}$$

$$\hat{\mu} = \frac{(X_2 X_0 S_1 - X_1^2 S_1)(K_s + S_0)}{S_2 S_0 X_0^2 - S_1 S_0 X_1 X_0} \tag{4.14}$$

$$b = \frac{S_1 X_2 - S_2 X_1}{S_2 X_0 - S_1 X_1} \tag{4.15}$$

$$Y = \frac{X_1^2 - X_1 X_0}{S_2 X_0 - S_1 X_1} \tag{4.16}$$

Table 4.1 Two-state/One process model in Petersen matrix format (Jeppsson 1996)

Process	Components		Process rate kinetics
	Substrate	Biomass	
Aerobic heterotrophic growth	$-\frac{1}{Y_H}$	1	$\hat{\mu}\left(\frac{S_S}{K_s+S_S}\right)X_H$

4.2.1 Physical Treatment Process

Primary treatment, which targets secondary treatment and the removal of readily settleable particles, which focuses on oxidizing biodegradable waste and nutrient removal, constitute two prevalent phases in WWTPs (Fig. 4.2). The energy consumption of primary treatment is notably lower compared to other stages. While it significantly influences the overall system operation, it is considered less hazardous than the secondary process; biological or chemical stages, for not involving the management of live bacteria. The initial stage of primary wastewater treatment involves the removal of solid waste from the wastewater. In domestic WWTPs, primary treatment is classified as a physical treatment and typically involves physical or chemical processes using primary clarifiers (PCs). These clarifiers operate mainly on the principle of gravity separation to remove solids from the wastewater (Metcalf & Eddy et al. 2014).

To achieve high removal efficiency, it is crucial to use a filter mat alongside sieves. In comprehensive full-scale experiments, the rotating belt filter (RBF) proved to be the most effective solution. Incorporating a small amount of cationic polymer with a static flocculator, prior to using an RBF, yielded favorable performance on wastewater that had previously been considered unsuitable for initial treatment with a fine mesh filter (Metcalf & Eddy et al. 2014). To remove solids from municipal wastewater, various filter systems, including complex moving fine meshfilters, are employed. These systems effectively eliminate a range of contaminants (Sherratt et al. 2019; Ziajahromi et al. 2017). Sedimentation reduces suspended solids (contains organic matter) by at least 20%, as indicated by the BOD over five days (BOD5) (Ziajahromi et al. 2017). BOD5 testing involves measuring dissolved oxygen (DO) levels during a five day period under dark, incubated conditions at 20 °C (Hauser 2018). Furthermore, throughout this phase, there is a minimum 50% reduction in total suspended solids (TSS) (Sincero and Sincero 2003).

In line with the European Union's wastewater treatment regulations, the Norwegian State Pollution Control Agency (SFT) initiated an investigation and testing of various primary treatment systems (Rusten and Ødegaard 2006). Numerous studies have confirmed the effectiveness of rotating belt filters, highlighting their capability to handle significant fluctuations in influent quality while consistently delivering high removal efficiencies. These systems feature a continuously rotating belt with a fine

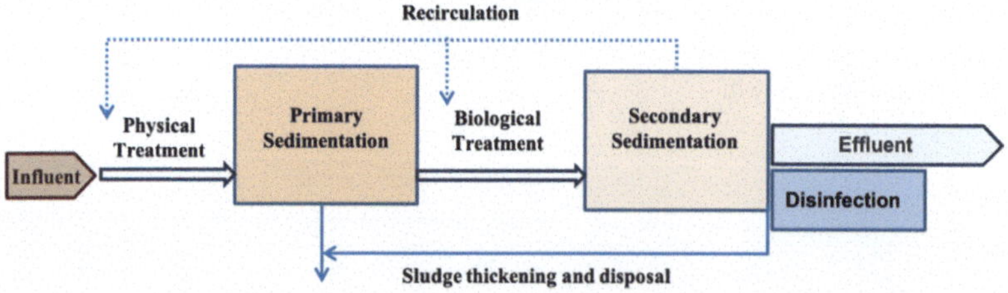

Fig. 4.2 Schematic presentation of the WWTP

mesh that captures particles as the influent flows through. The rotation's mechanical action helps keep the mesh clean, preventing blockages and ensuring reliable performance. Additionally, the use of coagulants and flocculants can further improve the filtration process by enabling the capture of even smaller particles, which boosts the overall efficiency of the treatment system. This combination of mechanical and chemical filtration methods allows rotating belt filters to achieve exceptional contaminant removal, meeting the rigorous standards established by the European Union. Rotating belt filters have emerged as a highly promising alternative in WWT. The advantages of fine mesh filtration include increased sludge energy potential and a reduction in aeration energy requirements, attributed to lower solids loading. These benefits underscore the efficiency of rotating belt filters in optimizing both energy use and sludge management, aligning with advancements in WWT technologies (Ebeling et al. 2006).

In systems employing rotating mesh filters, two primary filtering modes are utilized. The first mode involves the effective screening of larger solids from the influent when the clean filter engages with it. The second mode allows for the removal of solids up to one third in size of the mesh pore size due to the accumulation of cake, which acts as a secondary filter. Furthermore, chemical treatments can be applied to enhance particle removal efficiency in the process of fine mesh filtration. The performance of these systems is significantly influenced by factors such as the TSS in the influent and the speed of filter movement, both of which are closely associated with removal efficiency (Rusten and Ødegaard 2006).

In commercial belt filter systems, the rotational speed is generally regulated to control the thickness of the cake layer and to achieve specific treatment objectives. Reduced rotation speed results in thickening of cake layer, which improves TSS removal but reduces hydraulic capacity. Conversely, increased rotation speed produces a thinner cake layer, leading to lower TSS removal but greater hydraulic capacity (Teoh et al. 2006).

The complex interplay hydrodynamics and removal of solids poses challenges in building accurate models for predicting filtration performance (Ebeling et al. 2006). Existing mathematical models have explored the resistance characteristics of both the mesh and the cake layer, indicating that cake resistance is influenced by factors such as cake porosity, particle diameter, particle density, and cake solidity (Rusten et al. 2017; Franchi and Santoro 2015; Behera et al. 2018). However, these models are generally designed for constant flow rates or pressure differentials, which do not account for the spatial variations in head differences and cake resistances encountered in inclined rotating filters. Moreover, many contemporary models based on Darcy's law assume mesh resistance to be independent from flow velocity. This assumption is valid only at extreme low Reynolds numbers (Rusten et al. 2017; Franchi and Santoro 2015; Behera et al. 2018; Ruiken et al. 2013).

4.2.2 Biological Treatment Process

Biological treatment processes are fundamental in WWT, utilizing microorganisms, primarily bacteria, to degrade and eliminate organic pollutants. These processes are critical for ensuring that treated water is safe for environmental discharge. Among the various biological treatment methods, the activated sludge process is particularly prevalent. This method involves mixing wastewater with a dense population of microorganisms and aerating the mixture to supply oxygen. The oxygen is essential for the microorganisms to thrive and effectively break down the organic matter (Chemicznego and Na 2014; Szaja et al. 2015).

Mathematical models are indispensable in optimizing biological treatment processes. They simulate biological reactions and transport phenomena within the treatment systems, facilitating the design of control strategies, performance prediction, and process efficiency enhancement. Key models in this domain include:

ASM1: Concentrates on carbon oxidation, nitrification, and denitrification.

ASM2 and ASM2d: Expanded models that incorporate biological phosphorus removal.

ASM3: Emphasizes total organic carbon as a significant parameter.

These models assist in devising control strategies, forecasting system performance, and boosting overall process efficiency. The biological treatment process can be segmented into several stages for better understanding and optimization (Chen et al. 2018; Gray 2004).

Aeration System:

Air is bubbled through the wastewater, helping microorganisms consume the organic pollutants and convert them into biomass, carbon dioxide, and water. The highest energy consumption within WWTPs occurs in the aeration systems, constituting between 53 and 68% of total energy consumption, depending on the specific process (Bel et al. 2019).

The modified analytical expression for dissolved salts in the wastewater, encompassing both concentration and composition, can be defined as follows:

$$C_s = C_0.\exp\sum_{i=0}^{n}(K_{iCi})^{-1} \tag{4.17}$$

Considering the nonlinear relationship between water density and temperature (Laizāns and Šnīders 2012):

$$p_h = d.g.h = \left(-0.006\Theta^2 + 0.0365\Theta + 999.91\right).g.h \tag{4.18}$$

$$\begin{aligned} C_S &= C\left(r_{O_2}, p_a, \Theta, K, c_I, h\right) \\ &= \frac{0.668 r_{O_2}.\left(p_a + \left(-0.006\Theta^2 + 0.0365\Theta + 999.91\right).g.h.10^{-3}\right)}{\exp\left(\sum K_i c_i + 0.0219\Theta\right)} \end{aligned} \tag{4.19}$$

The dynamic behavior of DO in a standard aeration tank is represented by the following nonlinear equation (Olsson and Newell 2015):

$$\frac{\Delta So(t)}{\Delta t} = k_{La}(Q_{air}(t))(So_{sat} - So(t)) - \frac{So(t)}{K_o + So(t)} \times R_r(t) \tag{4.20}$$

where the concentration of oxygen saturation is indicated by $So_{sat} = 8.637\ gO_2/m^3$ and the Monod's constant is $K_o = 0.2\ g/m^3$. $k_{La}(Q_{air})$ function determines how oxygen is transferred based on biological circumstances and the aeration system. The function is displayed using several linear models (Santín et al. 2019; SIMBA Home Page):

$$k_{La}(Q_{air}) = \alpha.Q_{air}(t) \tag{4.21}$$

where $\alpha = 0.208\ 1/m^3$. The definition of the cost function J is as follows:

$$J = \sum_{i=1}^{H_p}\left\{So\left(k+\frac{i}{k}\right)S^{ref}\left(k+\frac{i}{k}\right)\right\}^2 + \sum_{i=1}^{H_p}P_c\left(k+\frac{i}{k}\right) \tag{4.22}$$

where the first term of function on the right-hand side (4.22) presents the tracking error, second term denotes the control cost of the system on a prediction horizon H_p.

In the intricate domain of WWT, the optimization and precise control of aeration systems are crucial. Aeration, which ensures the required levels of dissolved oxygen in the system, significantly influences both operational efficacy and energy consumption. This context underscores two fundamental aspects derived from mathematical modeling: the tracking error and the control cost.

Equation (4.22) illustrates these dual aspects, where the tracking error term measures the deviation of the system's actual performance from the intended trajectory. By contrast, the control cost across a prediction horizon represented by the symbol H_p is dealt with in the second term. In this term, equivalent weights are judiciously selected to balance three critical objectives: minimizing the tracking error, limiting the frequency of blower switching, and reducing the energy expenditure for air pumping.

The delicate interplay between these factors necessitates a sophisticated approach to control design. Achieving an optimal compromise involves advanced techniques and a deep understanding of the system's dynamics. This necessitates a precise calibration of each parameter to ensure that the system's overall performance aligns with the desired outcomes (Özturan et al. 2015).

In pursuit of more effective modeling and control strategies, researchers have increasingly adopted electrical analogies. This innovative approach, exemplified by the work of Piotrowski and Krawczyk, endeavors to represent the aeration system through an electrical network model (Sandberg et al. 2018). By translating the physical and operational characteristics of the aeration system into electrical terms, these studies offer a novel perspective that can enhance understanding and facilitate the development of more efficient control methodologies.

The core of this analogy is the premise that the behavior of the aeration system can be mirrored by an equivalent electrical circuit. Such a circuit-based representation not only simplifies the inherent complexity of the physical system but also leverages established electrical principles to analyze and optimize performance. Consequently, the resultant model serves as a powerful tool for simulating various operational scenarios, enabling researchers to examine the impacts of different control strategies without extensive physical experimentation.

Moreover, the electrical network model is well-suited to the application of advanced control techniques (Özturan et al. 2015; Piotrowski et al. 2004, 2008). By incorporating principles from electrical engineering—such as impedance matching and resonance tuning—the model can be finely tuned to achieve superior control performance. This interdisciplinary approach fosters innovation, as insights from one field are applied to address challenges in another.

In conclusion, the integration of tracking error minimization and control cost optimization, alongside the novel application of electrical analogies, marks a significant advancement in the field of wastewater treatment. These methodologies are poised to play a vital role in enhancing the efficiency and efficacy of aeration systems. Through the convergence of mathematical rigor and electrical insights, we pave the way for a future where wastewater treatment processes are not only more efficient but also more sustainable.

Sedimentation:

Biological treatment processes are vital for managing wastewater effectively. The activated sludge process, SBR, and MBR are key methods for treating wastewater biologically. Mathematical modeling and optimization techniques play a vital role in enhancing the performance and efficiency of these processes, ensuring treated water meets environmental standards. Mathematical models serve to depict mechanisms based on the specific requirements of the system. (Olsson and Newell 1999). For instance, achieving higher quality treated water, suitable for purposes like irrigation, necessitates increased removal rates of nitrogen and phosphorus from wastewater (Pokhrel and Viraraghavan 2004). Ammonium ions present in wastewater are converted to nitrite nitrogen during the nitrification stage facilitated by bacteria, which subsequently converts to nitrate. In the denitrification step, water undergoes removal as nitrogen gas under anoxic conditions (Szilvester et al. 2010; Nelson and Sidhu 2009).

In the aerobic growth of heterotrophs, aerobic granules exhibit a more compact structure, enhanced settling capacity, and improved biomass retention compared to traditional sludge flocs. Research indicates the feasibility of utilizing aerobic granules for treating high-strength wastewater while concurrently extracting organic matter, nitrogen, and phosphorus (Ni et al. 2008). Both the substrate (S_S) and soluble oxygen (S_O) can influence the growth process and the alteration in COD (Yu et al. 2010).

During anoxic growth of heterotrophs, anaerobic treatment of wastewater occurs in an oxygen-free environment, facilitated by anaerobic microorganisms that metabolize organic and inorganic substances. In this biological cycle, acidogenic microorganisms convert organic compounds into organic acids. Methanogenic organisms subsequently transform these organic acids into methane and CO_2 gases (Ni et al. 2011).

Secondary treatment processes for wastewater can utilize either the anoxic–oxic (A/O) method or the anaerobic-anoxic–oxic (A/A/O) method. These approaches aim to eliminate carbon-based pollutants while enhancing the removal of nitrogen and phosphorus (Qiao and Zhou 2018). Current surveys indicate that the most prevalent treatment process is the conventional activated sludge method. This method encompasses various techniques, including the anaerobic-anoxic–oxic, sequencing batch reactor (SBR), and oxidation ditch methods (Guo et al. 2019).

After aeration, the mixture is allowed to settle in a secondary clarifier, where the biomass (activated sludge) separates from the treated water. The methods of finite elements are extensively employed to solve the sedimentation equations by discretizing the sedimentation tank into multiple layers. (Jeppsson 1996). The Bürger-Diehl sedimentation model incorporates considerations for dispersion and compression effects (Torfs et al. 2013; Bürger et al. 2012).

With particular reference to the concentrations in the areas between layer j and layer j + 1, the Godunov flow model was utilized (Titarev and Toro 2002).

$$G_j = G_j(C_j, C_{j+1}) := \begin{cases} \min\limits_{C_j \le C \le C_{j+1}} f_{bk}(C) \text{ if } C_j \le C_{j+1} \\ \max\limits_{C_{j+1} \le C \le C_j} f_{bk}(C) \text{ if } C_j > C_{j+1} \end{cases} \tag{4.23}$$

The method determines the unknowns by using the Vesilind velocity function, as f_{bk} exhibits a single local maximum at $\hat{C}$. The layers enable computation of effluent and underflow concentrations (Bürger et al. 2011). The Burger model represents the concentration as follows:

$$\sigma_e(C) = \begin{cases} 0 & \text{for } C < C_c \\ \alpha \ln\left(1 + \frac{C - C_c}{\beta}\right) & \text{for } C \ge C_c \end{cases} \tag{4.24}$$

The stress function paradigm is improved by De Clercq et al. (2005).

4.2.3 Simulation and Optimization in Wastewater Treatment

The behavior of WWT processes can be understood and predicted in large part by using simulation models. These models allow for the detailed representation of physical, chemical, and biological phenomena within treatment systems. By simulating various operational scenarios, operators can gain insights into process dynamics,

Fig. 4.3 Challenge of optimizing processes

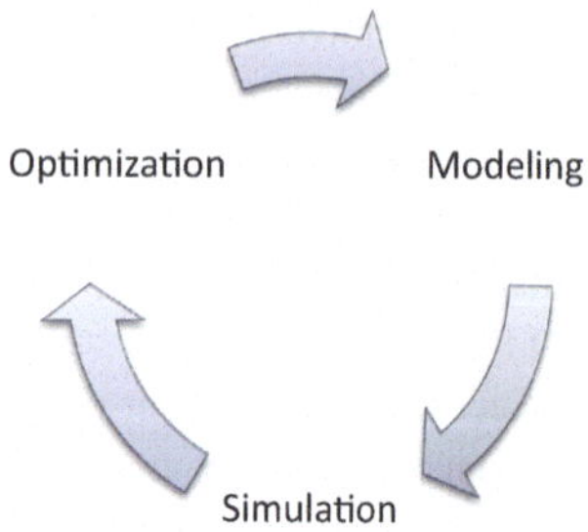

identify potential bottlenecks, and evaluate the impact of different variables on system performance.

By simulating various operational scenarios and optimizing the control strategies, treatment systems can operate more efficiently, reducing energy, and resource consumption. Optimization techniques help in identifying cost-effective operational strategies, minimizing the overall expenditure on treatment processes.

Enhanced Environmental Outcomes: Integrated simulation and optimization facilitate the development of strategies that not only comply with regulatory standards but also minimize environmental impact. The integration of simulation and optimization offers a robust framework for enhancing WWT systems. This combined approach leverages the strengths of both methodologies, enabling operators to achieve superior performance outcomes.

Optimization techniques are essential for improving the efficiency and effectiveness of WWT systems. These techniques involve the use of mathematical algorithms to identify the best possible operational strategies, considering various constraints and objectives (Fig. 4.3).

Optimizing WWT processes presents a complex challenge due to the conflicting objectives of effluent quality, energy consumption, and sludge production. Achieving optimal results becomes particularly difficult when these objectives must be considered simultaneously. Focusing solely on one aspect can lead to suboptimal and unreliable outcomes. For instance, prioritizing effluent quality without considering energy consumption is impractical, given the significant environmental impacts, such as indirect CO_2 emissions, and economic considerations, like pollution taxes. Therefore, a balanced approach is essential to achieve high effluent quality while maintaining manageable energy consumption.

Two main strategies are put out in the literature to optimize bioprocesses in WWTPs. The economic optimality of the process and the dynamic performance around a nominal point are the two primary goals of the first category of optimization attempts. Studies that use disturbance analysis in addition to steady-state calculations to demonstrate this strategy modified the settings of a control law while operating under restrictions to optimize a controller for the nitrification process and lower the noise-to-signal ratio for sensors (Holenda et al. 2007; Fikar et al. 2005; Balku and Berber 2006). Similarly, studies focused on optimizing the control law and design of WWTP bioprocesses to minimize controller errors and investment costs (Vera et al.

2003). However, both studies primarily addressed the rejection of a single perturbation, limiting their applicability to the continuously changing perturbations typical of WWTPs.

Dynamic simulations can be used for optimization since the second category of optimization attempts focuses on a single economic performance metric. This strategy converts operational goals into restrictions, but it frequently leaves room for ambiguity about the trade-offs between operational and economic goals. These restrictions are shown by Rivas et al. (2008) overall plant design.

Various optimization techniques are employed in the field, including dynamic models, linear programming (LP), nonlinear programming (NLP), genetic algorithms (GA), and particle swarm optimization (PSO). Dynamic models simulate the time-dependent aspects of WWT processes capturing transient states and dynamic responses to changes in operational conditions. LP is used for optimizing linear objective functions subject to linear equality and inequality constraints, suitable for problems with linear relationships between variables. NLP addresses optimization problems with nonlinear relationships, which is more complex but necessary for accurately modeling many real-world WWT processes. Genetic algorithms, inspired by the principles of natural selection and genetics, are effective for solving complex optimization problems with large search spaces and multiple objectives.

In summary, balancing the conflicting objectives in WWTPs requires a nuanced approach that considers both operational and economic factors. Optimization techniques must be carefully selected and applied to address the dynamic and multi-faceted nature of WWT processes, ensuring high effluent quality, manageable energy consumption, and minimal sludge production (Rani and Chang 2021).

4.2.4 Primary Treatment Optimization Studies

Water treatment involves a range of methods designed to remove dissolved organic and inorganic substances, as well as solid materials of varying sizes. Physical treatment units, such as grills, screens, and flotation tanks, mechanically separate solid materials from water, while chemical treatments adjust wastewater pH and precipitate substances like suspended solids and heavy metals (Malviya and Jaspal 2021). Biological treatment utilizes microorganisms to decompose organic contaminants, with activated sludge systems aiding in pollutant breakdown and water clarification (Fuentes et al. 2020).

Advanced WWT focuses on efficiently removing nitrogen and phosphorus compounds to produce high-quality treated water suitable for reuse, such as in irrigation. This process involves stages like nitrification, denitrification, and phosphorus removal through chemical or biological means (Yilmaz and Büyüktahtakın 2024). Despite technological advancements, urban sewage treatment plants face challenges such as capacity issues during heavy rains and risks from improper sludge disposal (Metcalf and Eddy 1980).

Innovations in primary treatment, like the use RBFs and fine mesh filters, have greatly improved the efficiency of solids removal and lowered energy consumption. Researches have demonstrated that introducing a small quantity of cationic polymer in combination with a static flocculator prior to an RBF enhances the removal efficiency for wastewater that was previously unsuitable for fine mesh filtration (Metcalf & Eddy et al. 2014). These fine mesh filters can effectively remove a wide range of solids from municipal wastewater, including household waste, toilet paper, ceramics, and mineral solids (Sherratt et al. 2019; Ziajahromi et al. 2017).

These innovative methods also focus on optimizing and controlling filtration systems. By adjusting the speed of filter rotation, the thickness of the cake layer can be managed, which in turn affects total suspended solids (TSS) removal and hydraulic capacity (Ebeling et al. 2006). Although modeling filtration performance is challenging due to the interactions between solids removal and water flow, advances in mathematical modeling and the use of electrical analogies offer promising solutions (Anayochukwu 2013).

Electrical analogies are particularly effective for simulating and optimizing moving fine meshfilter systems. These analogies, which liken hydraulic systems to electrical circuits, help to design efficient control structures and enhance filtration performance (Kuş and Kara 2022). This approach has been successfully applied to simulate and optimize the performance of WWTPs, such as the Tatvan WWTP in Turkey, demonstrating the practical benefits of these innovative methods (Kuş and Kara 2022).

The integration of electrical analogies into WWT modeling provides a robust framework for developing optimized solutions tailored to specific influent characteristics and operational conditions. By employing proportional-integral-derivative (PID) control and other optimization techniques, these models can significantly improve the efficacy of moving fine meshfilter systems, ensuring higher TSS removal and efficient operation under varying conditions.

In conclusion, innovative methods in treatment processes, such as the application of electrical analogies and advanced filtration technologies, play a pivotal role in enhancing the efficiency and sustainability of WWT. These approaches not only address the challenges of energy consumption and treatment efficacy but also pave the way for more resilient and adaptable treatment systems capable of meeting the growing demands of urban populations and stringent environmental standards.

4.3 Optimal On–Off Level Control for Improved Moving Fine Mesh Filter Performance by Electrical Analogy

In modern WWTPs, the efficiency and reliability of filtration systems are critical for ensuring high-quality effluent. One of the innovative approaches to enhancing filtration performance is the implementation of optimal on–off level control designs for moving fine meshfilter systems.

A moving fine meshfilter system operates by continuously rotating a fine mesh screen to capture suspended solids from the influent water. The primary challenge in this system is to maintain an optimal balance between filtration efficiency and operational stability. Traditional continuous control methods often result in suboptimal performance due to the complex dynamics and variable loading conditions of the wastewater. The development of control structures for moving fine meshfilter systems can be effectively conceptualized using electrical analogies. Moreover, the application of such electrical analogies within the water-tree concept can be seamlessly integrated into existing WWTPs as referenced in (Greenslade 2003). This approach ensures the necessary filtration conditions for the removal of total suspended solids (TSS) through the implemented control structure.

The findings highlight a significant correlation between the efficiency of TSS removal and the levels of TSS entering the rotating RBF. This study addresses the complexities the mechanisms of water flow by proposing a tree-water model, inspired by natural processes, to optimize the performance of RBF systems. The electrical analogy applied to the tree-water model aligns well with the behavior of RBF systems.

The aim of this study is to create a simulator capable of identifying the optimal configuration for a moving fine meshfilter system by considering the influent water characteristics and the operational parameters of the filter system. This simulator model dynamically adjusts the on–off cycles or cycle speeds to maintain efficient operation. The tree-water notion is used as an electrical analogy in the frequency domain to find the best solution. Particle size frequencies are categorized, and the dynamic behavior of the actual system is modeled and analyzed using this electrical analogy within the simulator. The results from real case scenarios and simulations of the proposed electrical analogy model are analyzed and discussed.

The study on moving fine meshfilters introduces an electrical model analogous to the moving fine meshfilter system in WWTPs. This model comprises two principal components: tanks and filters, which are represented as a combination of resistors, capacitors, and inductors in an electrical circuit. This representation aims to enhance the efficiency of the entire treatment process and its associated equipment. Utilizing this analogy, the proposed approach seeks to achieve a more uniform and efficient moving fine meshfilter solution through proportional-integral-derivative (PID) control. The conventional moving fine meshfilter system, typically consisting of two separate tanks and a filter, is modeled mathematically. Existing studies serve as the foundation for deriving this mathematical model using electrical analogy.

To attain high removal efficiency, it is critical to use a filter mat in conjunction with sieves. Full-scale experiments have demonstrated that the rotating belt filter performs optimally. For wastewater that was previously thought to be inappropriate for primary treatment using fine mesh filters, the results are particularly beneficial when a modest quantity of cationic polymer is paired with a static flocculator prior to a rotating belt filter (Rusten et al. 2017).

Figure 4.4 elucidates the simulation model that dynamically defines the on–off cycle or cycle speed, aiming to achieve optimal performance for the two tank interaction system (TTIS) concept. Using an electrical analogy, the tree-water notion is carefully used in the frequency domain to get the desired result. This approach

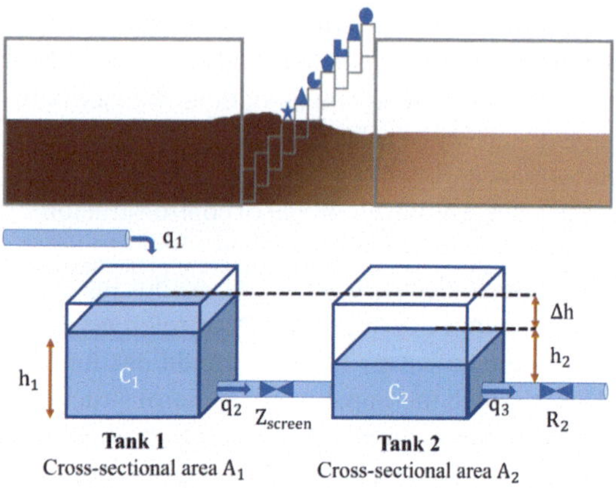

Fig. 4.4 Two tank interacting system

involves categorizing particle sizes by frequency and subsequently observing the system's dynamic behavior through the electrical analogy model implemented in the simulator.

The model is essentially comprised of two main elements: tanks and filters, which are symbolized as a circuit integrating resistors, capacitors, and inductors. This complex setup aims to replicate the functionality of the moving fine mesh-filter system. The primary goal of using this analogy is to enhance the efficiency of subsequent treatment processes and related equipment. Through the application of proportional-integral-derivative (PID) control, this method seeks to provide a more uniform and effective solution for the moving fine meshfilter system.

Two tanks and a filter are included in the mathematical model of the traditional moving fine meshfilter construction that is being presented. By leveraging the electrical analogy, this model is derived based on extant studies and further adapted to align with dynamic behavior through numerical simulations. The model's accuracy, objectives, boundaries, and time scales are meticulously defined to ensure comprehensive understanding and precise implementation. Computer simulations of the optimized level-controlled moving fine meshfilter system have been performed, with the Tatvan WWTP in Bitlis Province, Turkey, serving as a comprehensive case study for comparison and validation.

Through the use of advanced simulation techniques, the study reveals significant insights into the operation and optimization of the mesh (fine) filter system. The dynamic simulation model offers a profound understanding of how the system responds to various operational parameters, thereby facilitating the identification of optimal control strategies. The PID control, in particular, plays a pivotal role in enhancing system performance by maintaining the desired operational states with high precision.

Moreover, the findings from the Tatvan WWTP case study underscore the practical applicability and effectiveness of the proposed model. By validating the model against real-world data, the study demonstrates the feasibility and advantages of using electrical analogies and advanced control techniques in wastewater treatment.

This innovative approach not only improves the efficiency and reliability of the treatment process but also provides a scalable solution that can be adapted to different contexts and requirements. By employing a comprehensive electrical analogy model and leveraging sophisticated control techniques, the study not only enhances our understanding of the dynamic behavior of moving fine meshfilters but also provides practical solutions for optimizing their performance in real-world applications.

In summary, the integration of electrical analogies and PID control in the simulation model represents a significant advancement in the optimization of wastewater treatment systems. The detailed depiction of the TTIS concept in Fig. 4.4, along with the comprehensive mathematical modeling and simulation results, highlights the potential of this approach to revolutionize the field of wastewater treatment. Future research should continue to explore and refine these methodologies, ensuring that they remain at the forefront of technological innovation and environmental sustainability (Kuş and Kara 2022).

Additionally, the research addresses an optimization problem inherent to the moving fine meshfilter system, which is characterized by its highly nonlinear and multimodal nature. Identifying an optimal or even sub-optimal solution amidst numerous complex constraints and conflicting objectives is a significant challenge. The issue, particularly regarding the durations of the moving fine meshfilter's on–off cycles, has been redefined to simplify the problem. This refined approach offers a valuable alternative for effectively monitoring and managing the operations of the moving fine meshfilter system.

4.4 Secondary Treatment Optimization Studies

Advanced wastewater treatment focuses on efficiently removing nitrogen and phosphorus compounds to produce high-quality treated water suitable for reuse, such as in irrigation. This process involves stages like nitrification, denitrification, and phosphorus removal through chemical or biological means (Balku and Berber 2006). Despite technological advancements, urban sewage treatment plants face challenges such as capacity issues during heavy rains and risks from improper sludge disposal (Wang et al. 2023; Carrette et al. 2001).

Following this overview of treatment processes, the next step involves mathematical modeling of WWTPs. Mathematical models are essential tools for simulating and optimizing plant operations, predicting pollutant removal efficiencies, and assessing environmental impacts. These models integrate biological, chemical, and physical processes, providing insights into plant performance under varying conditions and aiding in decision-making for plant design and operation improvements.

Several studies aimed to minimize energy consumption by manipulating aeration sequences while maintaining effluent quality (Balku and Berber 2006). The optimal aeration profile derived from simulations may not be suitable for other disturbances, as the open-loop control law used does not allow for corrective actions. A study sought to reduce the total nitrogen concentration in the effluent by adjusting the aeration sequence (Fikar et al. 2005). Researchers employed a nonlinear programming solver for dynamic optimization, incorporating state values in the optimization variables to ensure initial and final state equality, thus enabling closed-loop operation to reject disturbances. However, the study's limitation lies in its focus on only one day of dry weather, excluding rainy conditions.

Holenda and colleagues used a genetic algorithm to optimize both effluent quality and aeration energy simultaneously. Although this method showed excellent performance for the specific perturbations considered, its effectiveness may diminish rapidly with changes in perturbations. Moreover, the assumption regarding initial states related to the number of cycles per day lacks accuracy (Holenda et al. 2007).

In recent years, the integration of machine learning algorithms with traditional optimization techniques has emerged as a promising approach to enhancing the performance of WWTPs. These advanced algorithms possess the capability to process vast amounts of data and identify complex patterns, thus enabling more accurate predictions and robust control strategies. By leveraging real-time data and adaptive learning capabilities, machine learning can continuously optimize plant operations, adapting to changing conditions and thereby improving overall efficiency.

Moreover, the advent of advanced sensors and monitoring technologies has revolutionized the capacity to track and control various parameters within WWTPs. These sensors provide high-resolution data on water quality, flow rates, and other critical variables, facilitating more precise and timely adjustments to treatment processes (Vanrolleghem and Lee 2003). The amalgamation of advanced monitoring technologies with sophisticated modeling and optimization techniques can lead to significant improvements in treatment efficiency, energy consumption, and effluent quality (Longo et al. 2016). This integration paves the way for more sustainable and resilient wastewater management systems.

4.5 Optimal Unified Tank Model (UTM) for Aeration-Sedimentation Processes

Aeration and sedimentation are fundamental processes in wastewater treatment, essential for the removal of organic matter, nutrients, and suspended solids. The design and operation of these processes can be significantly enhanced through the development of an UTM. The UTM integrates the aeration and sedimentation processes into a single, cohesive model, optimizing the performance of both processes simultaneously. This unified approach allows for better coordination and control, leading to improved treatment efficiency and reduced operational costs.

The activated sludge, primarily comprising a mixture of wastewater and bacterial populations, is a critical component in WWTPs. The aeration process, which promotes microbial activity, accounts for approximately 60% of total energy consumption in a WWTP. Oxygenation is essential for microbial growth and the breakdown of organic pollutants within the wastewater. Control and optimization of

DO levels are crucial for effective nitrogen reduction processes. Furthermore, the secondary clarifier facilitates the separation of purified water from activated sludge through sedimentation, functioning as an integrated process with the aeration tank from a control perspective.

In order to precisely forecast the behavior of the activated sludge process, research on mathematical modeling is started, with the ASM1 model coming first. Subsequent models such as ASM2 and ASM3 incorporate biological and chemical phosphorus removal mechanisms, with parameters such as COD used to quantify organic pollution. Computer-aided design and simulations play a significant role in optimizing WWT processes, with mathematical models serving as the backbone for process understanding and improvement. This chapter proposes a mathematical model of a unified tank encompassing both aeration and sedimentation processes, facilitating performance evaluation and optimization.

Various optimization techniques, including metaheuristic approaches inspired by biological systems, are employed to enhance WWTP efficiency. The Whale Optimization Algorithm (WOA), inspired by the feeding behavior of humpback whales, is utilized in this study to minimize production costs or maximize efficiency in aeration systems within WWTPs.

4.5.1 UTM Concept

To streamline the intricate and detailed process of WWT, Fig. 4.5 presents the diagram of a typical wastewater treatment plant (WWTP). The primary focus of this research is the secondary treatment process, particularly the aeration process depicted in the figure.

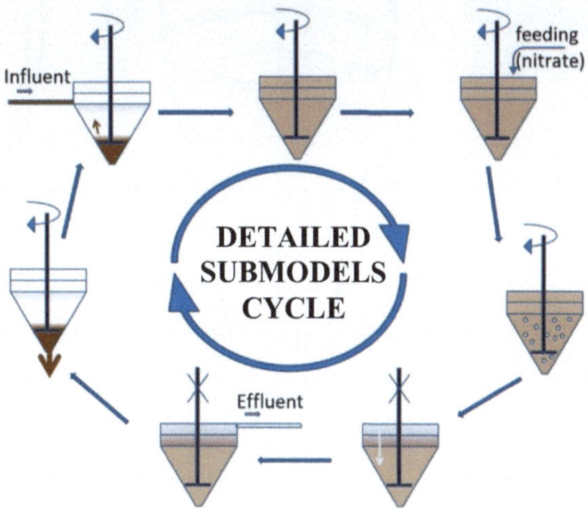

Fig. 4.5 Batch reactor operating system with sequencing tailored for UTM (Kuş and Kara 2020)

The objective is to integrate these two distinct processes into a single tank, resembling the structure of a traditional sedimentation tank. The algorithm for the conventional sequencing batch reactor (SBR) involves several steps carried out sequentially within a tank, as illustrated in Fig. 4.5. This sequence is systematically maintained in the UTM.

The proposed unified treatment model (UTM) aims to enhance the efficiency and effectiveness of the aeration and sedimentation processes by consolidating them into a single system. This approach not only simplifies the overall treatment process but also potentially reduces the space and resources needed for separate tanks. By continuously preserving the SBR sequence within the UTM, the established benefits of the conventional system are retained while introducing the advantages of an integrated approach.

The system's inputs are the qualities of the influent water, and its outputs are the waste sludge produced after the procedure and the treated water that overflows the conical tank. The treated water is discharged through multiple level taps, with the discharge cycle initiating only after sufficient sedimentation has been achieved. Figure 4.6 provides a detailed depiction of the UTM's structural configuration.

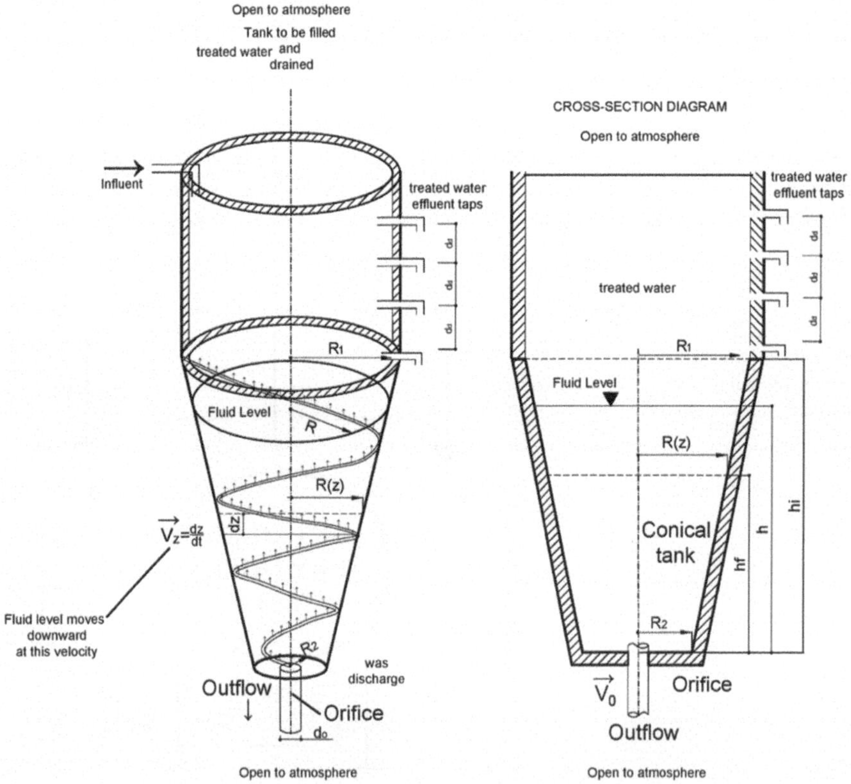

Fig. 4.6 Unified tank model (Kuş and Kara 2020)

By maintaining a constant level of RAS, the UTM simplifies the traditional setup by reducing the operational complexity associated with the sludge return pumps. This streamlined approach prioritizes the quality of the influent water and the efficient separation of treated water and waste sludge. The design ensures that treated water is effectively discharged only after optimal sedimentation has occurred, thereby enhancing the overall efficiency and reliability of the treatment process. The system structure, as outlined in Fig. 4.6, exemplifies the streamlined and integrated nature of the UTM, underscoring its potential advantages over conventional systems.

The simplification achieved by keeping the RAS constant minimizes the need for frequent adjustments and monitoring of the sludge return process, which is typically a labor-intensive aspect of conventional systems. Furthermore, by focusing on the influent water quality and its subsequent treatment, the UTM ensures a higher degree of consistency and effectiveness in the separation and discharge phases. This holistic approach not only improves operational efficiency but also contributes to the sustainability and cost-effectiveness of wastewater treatment.

The unified tank model signifies a groundbreaking shift from traditional systems by employing a single tank configuration for both processes of aeration and sedimentation. This design not only manages the physical charging and discharging but also alternates between aerobic, anaerobic, and sedimentation phases. This integrated approach is intended to model the activated sludge process within a single tank, enabling both biological reactions and separation of solids from the liquid in a periodic manner (Nasr et al. 2011).

Building on the inherent advantages of sequencing batch reactors (SBRs) such as operational flexibility, cost-effectiveness, and a small footprint, the UTM offers a sequence of consecutive stages that can be seamlessly applied to WWTPs with minimal modifications, which ensures the necessary conditions for the growth of various microorganisms, effectively removing contaminants from wastewater.

In this research, an optimized aeration system model utilizing UTM is introduced, designed to enhance aeration efficiency and homogeneity through the optimization of DO levels, facilitated by the WOA. The conventional structure, comprising two distinct tanks, serves as a reference for the mathematical model presented. Leveraging insights from existing research on conventional WWTPs, we derive and dynamically adapt the UTM mathematical model, followed by a series of numerical simulations to evaluate its performance.

The goals, bounds, accuracy, purpose, and time frame are all clearly stated in the functional constraints and objectives. To give a more realistic picture, the aeration system has been represented using an electrical analogy. Additionally, the study includes experimental research and models related to the biological processes involved.

In conclusion, the UTM represents a significant advancement over traditional methods, offering a more cohesive and efficient framework for wastewater treatment. By maintaining constant RAS levels and optimizing sedimentation and discharge processes, the UTM enhances both the operational simplicity and the overall performance of the treatment system. The structural configuration presented in Fig. 4.6 highlights the innovative and integrated design of the UTM, paving the way for future developments and applications in wastewater treatment technology. This comprehensive approach not only enhances the efficiency of the treatment process but also provides a robust framework for future applications and developments in wastewater treatment technology (Kuş and Kara 2020).

4.5.2 Key Features of the UTM

Integrated Process Modeling: The UTM combines the dynamics of aeration and sedimentation into a single mathematical framework. This allows for the simultaneous optimization of oxygen transfer, biomass growth, and solids settling.

Unified Tank Model Design: The study introduces a novel unified tank model that amalgamates the aeration and sedimentation phases within a singular tank structure. This model is distinguished by its utilization of an Archimedean spiral geometry for the aeration tank, which is strategically embedded within the main tank. This design contrasts sharply with the conventional approach that typically employs separate tanks for aeration and sedimentation processes.

Optimization via Whale Optimization Algorithm (WOA): A rigorous optimization framework is established to determine the optimal placement of diffusers within the UTM. The WOA is employed to enhance the efficiency of both aeration and sedimentation processes. This method allows for the precise calibration of diffuser locations to maximize the operational effectiveness of the UTM.

Simulation and Validation: Extensive computer simulations are conducted to assess the operational performance of the proposed UTM design. These simulations reveal substantial improvements in the performance metrics of the wastewater treatment system. The efficacy of the design is demonstrated through these computational experiments, showcasing its potential to outperform traditional systems.

4.5.3 Case Study Application

The practical applicability of the optimized UTM design is further substantiated through its implementation at the Oğuzeli WWTP. A comprehensive case study is performed, wherein the simulation results are meticulously compared with empirical data collected over a one-year period. This comparative analysis confirms the advantages of the proposed design relative to standard models, underscoring its practical value.

Implications and Future Research:

According to the study, the ideal UTM design that has been suggested can be a good substitute for the design and execution of aeration-sedimentation processes in WWT systems. The study indicates that practical applications in various WWT systems as well as laboratory-scale trials may be used to validate and improve the approach further. These future research directions aim to solidify the method's practical utility and to extend its applicability across different operational contexts.

These key features delineate the innovative aspects of the proposed UTM design, the methodological rigor of the optimization process, the empirical validation through simulations and case studies, and the potential future directions for research, thereby highlighting the study's contributions to the field of wastewater treatment.

The unified treatment model introduces a conical tank configuration that distinguishes it from conventional systems by combining aeration and sedimentation processes within a single tank. This design allows for the alternating modeling of aerobic, anaerobic, and sedimentation phases, effectively handling the activated sludge process through periodic biological reactions and solid–liquid separation.

In addition to the well-known advantages of sequencing batch reactors (SBRs) such as operational flexibility, cost-effectiveness, and compact footprint, the UTM offers a series of successive phases that can be easily integrated into existing plants with minimal modifications. This adaptability ensures the optimal conditions for the growth of various microorganisms needed to remove contaminants from wastewater.

The purpose of this study is to propose an optimum aeration system model, or UTM, that uses the WOA to improve DO in order to produce a more uniform and effective aeration solution. As a guide, the two distinct tanks in the mathematical model for the traditional system are given. The UTM model is derived from existing research on conventional WWTPs, dynamically adapted, and validated through numerical simulations.

An electrical analogy has been employed to model the aeration system, providing a more realistic representation. Additionally, experimental research and models related to the biological processes are discussed. For validation and comparison, computer simulations of the UTM have been conducted using data from the Oğuzeli WWTP in Gaziantep Province, Turkey.

Drawing inspiration from the whale feeding behavior, which involves hunting in progressively shrinking circles and adjusting positions accordingly, the optimization of the UTM emphasizes two primary aspects: aeration duration and achieving the target DO value of 2 mg O_2/L. To assess the UTM's performance in terms of aeration time duration and DO tracking error, both in optimized and standard designs, the root mean square error (RMSE) index is utilized.

By integrating these innovative approaches, the UTM not only optimizes oxygen transfer, biomass growth, and solids settling but also provides a robust and adaptable solution for modern wastewater treatment challenges.

$$\text{RMSE} = \sqrt{\Sigma(\text{observed value} - \text{desired value})^2/(\#\text{ of samples})} \tag{4.25}$$

The performance of the designed UTM and the impact of optimization are evaluated using root mean square error (RMSE) index values, as summarized in Table 4.2.

Table 4.2 Root mean square error (RMSE) values for traditional and optimized UTM designs

Parameters	RMSE values		Improvement in RMSE (%)
	Conventional	Optimal	
t	0.138	0.068	51
DO	0.236	0.049	79

As evidenced by the root mean square error (RMSE) index values presented in Table 4.2, the duration of aeration and the DO in the proposed UTM design, optimized through the WOA, have shown substantial improvements. The results pertaining to the optimal WOA solution demonstrate an exemplary tracking performance, characterized by a precise and well-timed aeration process that closely follows the desired DO values with notable accuracy.

The enhanced tracking performance, as revealed by the optimal WOA solution, underscores the efficacy of the proposed UTM design. The system's ability to maintain the desired DO levels within the specified range over the aeration period is indicative of its improved control mechanisms. The fine-tuning of aeration time, facilitated by WOA optimization, ensures that the DO variation remains within acceptable limits, thereby contributing to the overall efficiency and effectiveness of the wastewater treatment process.

The RMSE index values serve as a quantitative measure of the model's performance, reflecting the reduced error margins and increased precision in maintaining DO levels. These improvements are a testament to the robustness of the UTM design, particularly when enhanced by advanced optimization techniques such as WOA. The alignment of the aeration duration with the optimal DO values not only enhances the treatment process but also ensures a more stable and predictable operational framework.

Furthermore, the application of WOA optimization in the UTM design provides a valuable methodology for achieving optimal control over aeration processes. By effectively minimizing deviations in DO levels, the system can operate with higher accuracy and reliability. This optimization approach is instrumental in addressing the inherent complexities and dynamic behaviors of wastewater treatment systems, paving the way for more advanced and efficient treatment solutions.

In conclusion, the significant improvements in aeration time duration and DO variation accuracy, as evidenced by the RMSE index values, highlight the superior performance of the UTM design optimized through WOA. The findings demonstrate the potential of advanced optimization algorithms in refining and enhancing wastewater treatment processes, thereby offering a more precise and efficient solution for maintaining optimal DO levels. This study underscores the importance of integrating robust optimization techniques in the design and operation of modern wastewater treatment systems to achieve superior performance and reliability.

4.6 Future Trends and Research Directions

These models have been employed to predict and optimize various aspects of the WWT process (Tsalas et al. 2024). Previous research has shown that AI models are highly effective in both prediction and optimization tasks. They have been successfully applied to various aspects of WWT, including process design, water quality monitoring, optimization of process parameters, and performance prediction. These studies underscore the potential of AI as a powerful and efficient tool in wastewater management (Vanrolleghem and Lee 2003; Longo et al. 2016; Martínez et al. 2020).

With the rapid advancement of computer technology, machine learning (ML), a fundamental subset of artificial intelligence (AI), has emerged as a prominent field. ML utilizes data, algorithms, statistics, and mathematical optimization to replicate human learning processes, progressively improving accuracy and contributing to the broader capabilities of AI. The rise of big data and increasingly powerful supercomputing resources have further accelerated the adoption of ML, leading to its successful application across diverse sectors such as industry, agriculture, medicine, environmental protection, and scientific research. Despite these advancements, most review articles on AI applications in WWT tend to concentrate on specific techniques or aspects of process design—such as adsorption processes, membrane bioreactors, membrane technologies, and WWTPs. There remains a notable gap in comprehensive reviews that encompass AI applications throughout the entire WWT process, including water quality monitoring, laboratory research, and real-world process design. Additionally, few reviews provide a thorough introduction to the commonly used AI models in WWT, along with a detailed discussion of their advantages, disadvantages, and potential improvements.

This lack of comprehensive reviews underscores the need for a more holistic examination of AI applications in WWT, bridging the gap between individual techniques and overarching process design. By addressing this need, we can better understand the full potential and limitations of AI in enhancing the efficiency and effectiveness of WWT (Wang et al. 2023).

4.7 AI Integration in Wastewater Treatment

In the realm of WWTPs, the integration of AI represents a transformative approach aimed at enhancing operational efficiency and environmental sustainability. Mathematical models play a pivotal role by simulating the intricate biological, chemical, and physical reactions within these systems, enabling engineers to predict how different operational adjustments impact treatment outcomes.

Recent advancements in AI, such as machine learning algorithms and predictive analytics, have revolutionized WWTP operations. These technologies analyze extensive datasets from sensors and historical records to identify patterns and optimize resource allocation. By doing so, AI facilitates real-time adjustments to treatment processes, improving efficiency, and reducing energy consumption and chemical usage.

AI's application extends to optimizing critical components like aeration systems, which consume a substantial portion of a WWTP's energy. Through AI-driven algorithms, aeration can be fine-tuned to balance energy efficiency with treatment effectiveness, responding dynamically to incoming wastewater characteristics. Furthermore, AI aids in predictive maintenance by analyzing sensor data to foresee equipment failures, thereby minimizing downtime and operational costs.

The utilization of AI in WWTPs not only enhances current processes but also fosters the development of innovative treatment technologies. By modeling pollutant removal dynamics and predicting the behavior of complex water systems, AI contributes significantly to environmental sustainability and public health. As AI technology evolves, its integration in WWT holds promise for further optimizing operations and advancing treatment methodologies.

In conclusion, AI's integration in WWTPs marks a paradigm shift toward more efficient and sustainable WWT practices, driven by advanced modeling and predictive capabilities. Future research aims to leverage AI's potential fully, addressing ongoing challenges and expanding its application across diverse facets of wastewater treatment engineering.

The treatment of wastewater involves intricate process design and operational conditions. Artificial intelligence (AI) models offer significant advantages in simplifying and managing these complexities. They excel in establishing complex relationships between various input and output variables. Common input variables include time, temperature, pH, initial pollutant concentrations, and influent water quality parameters. Output variables typically focus on the efficiency of contaminant removal or adsorption, as well as the quality parameters of treated effluent.

AI models have proven effective across multiple facets of wastewater treatment. They are widely used for predicting effluent water quality and overall treatment performance, as well as optimizing energy consumption and operational parameters (Rahmani et al. 2015). In terms of process design for wastewater treatment, AI applications are primarily concentrated on optimizing process parameters and forecasting process performance (Ataei 2010). These models play a crucial role in streamlining the complexities involved in wastewater treatment processes, making them more efficient and adaptable to varying conditions.

Artificial intelligence (AI) has emerged as a transformative force in the field of wastewater treatment, offering significant advancements in managing complexities and optimizing operational efficiency. Over recent decades, the application of AI in various facets of wastewater management has expanded substantially, reflecting its growing importance in the industry. The integration of AI into wastewater treatment processes encompasses several critical areas:

1. **Performance Prediction**: AI models facilitate the forecasting of wastewater treatment performance under varying operational conditions. By leveraging

historical and real-time data, these models enable the anticipation of performance outcomes, thereby allowing for preemptive adjustments to optimize system efficiency.

2. **Effluent Quality Estimation**: Advanced AI algorithms provide accurate predictions of the quality of treated effluent. These models analyze data inputs to estimate key parameters such as chemical oxygen demand (COD), biological oxygen demand (BOD), and total suspended solids (TSS), ensuring compliance with environmental regulations.
3. **Optimization of Operating Parameters**: AI techniques are employed to fine-tune operational parameters, including chemical dosing, aeration rates, and sludge management. This optimization not only enhances treatment efficacy but also contributes to cost reduction and resource conservation.
4. **Design of Treatment Units**: AI supports the design and development of treatment units by simulating various configurations and identifying optimal designs based on specific treatment needs and constraints. This capability improves the efficiency and effectiveness of treatment facilities.
5. **Design of Sensors**: The advancement of sensor technology, aided by AI, enhances the detection and measurement of critical components in wastewater. AI-driven sensors provide more accurate and reliable data, which is essential for effective monitoring and control.
6. **Micro-pollutants and Emerging Contaminants**: AI models address the challenge of micro-pollutants and emerging contaminants by optimizing treatment processes. These models improve the removal efficiency of pollutants that are difficult to treat with conventional methods.
7. **Maintenance Strategies and Fault Diagnosis**: AI facilitates predictive maintenance by identifying patterns and anomalies in system performance. This proactive approach helps in scheduling maintenance activities, preventing unforeseen breakdowns, and extending equipment lifespan.
8. **Operating Instructions Based on Performance**: Real-time data analysis through AI generates operational instructions tailored to current performance metrics. This capability enables operators to make informed decisions and adjustments swiftly.
9. **Energy Optimization**: AI contributes to energy efficiency by optimizing processes such as aeration, pumping, and heating. This optimization reduces energy consumption, which is a major operational cost in wastewater treatment facilities.
10. **Automation**: The automation of treatment processes, driven by AI, enhances system efficiency and reduces the need for manual intervention. This automation leads to more consistent and reliable operations.

Each AI-driven method offers distinct advantages and limitations, necessitating careful selection based on the specific operational mechanisms of the wastewater treatment system and its intended objectives. Employing the most suitable AI techniques ensures optimal performance, cost-effectiveness, and environmental compliance, thereby advancing the overall effectiveness of wastewater treatment processes.

4.8 Conclusion

Mathematical modeling plays a central role in the simulation and optimization of WWTP operations. Models integrating biological, chemical, and physical processes provide comprehensive and detailed insights into plant performance under various conditions, aiding in the development of effective operational strategies. Advanced modeling approaches, such as tree-water analogies and electrical analogies, have proven particularly effective in optimizing the performance of RBF systems.

The detailed challenges and progress in optimizing aeration and filtration processes that the works underscore the importance of adaptive control strategies and predictive maintenance in managing energy consumption and achieving high effluent quality. However, limitations remain, such as the inability of open-loop control systems to handle variable disturbances and the potential degradation of solution effectiveness under changing conditions.

Overall, the integration of innovative filtration technologies and sophisticated optimization techniques provides a robust foundation for enhancing WWT performance. These approaches can reduce energy consumption while making treatment processes more efficient and resilient. Future research should continue to refine these methods, improving their accuracy, adaptability, and practical applicability across diverse operational settings. This will enable the development of treatment systems better suited to meet the growing demands of urban populations and stringent environmental regulations.

The integration of simulation and optimization techniques marks a significant advancement in the field of WWT. By harnessing the power of these methodologies, operators can achieve enhanced system performance, reduced operational costs, and improved environmental outcomes. The synergy between simulation and optimization not only facilitates a deeper understanding of complex treatment processes but also enables the fine-tuning of operational parameters to achieve optimal performance. The implementation of advanced simulation models allows for the accurate prediction of system behavior under various conditions, while optimization algorithms identify the best strategies for achieving desired outcomes. This dual approach ensures that treatment systems can be operated more efficiently, with reduced energy consumption and minimized environmental impact.

Future research should prioritize the development of more sophisticated models and algorithms, as well as their application across diverse treatment contexts. The continual evolution of these technologies will be crucial in meeting the increasing demands for sustainable water resource management and environmental protection. As the challenges associated with WWT become more complex, the role of integrated simulation and optimization techniques will become even more pivotal.

Moreover, advancing these technologies involves not only enhancing their computational efficiency but also improving their robustness and adaptability to various treatment scenarios. This includes developing algorithms capable of handling the nonlinearities and uncertainties inherent in real-world WWT systems. Additionally, expanding the application of these techniques to include emerging contaminants and innovative treatment processes will further bolster the effectiveness and sustainability of WWT operations.

In conclusion, the integration of simulation and optimization techniques represents a transformative step forward in the realm of wastewater treatment. By enabling more precise and efficient system management, these methodologies contribute significantly to achieving sustainable water management and environmental justice goals. Continued research and innovation in this field are essential to keep pace with the growing environmental and regulatory demands, ensuring that WWT systems remain capable of delivering high-quality treatment outcomes in an increasingly challenging landscape.

References

Alharbi AOM (2016) The biological treatment of wastewater: mathematical models. Bull Aust Math Soc 94(2):1–2. https://doi.org/10.1017/S0004972716000411

Anayochukwu AV (2013) Analogous electrical model of water processing plant as a tool to study 'The Real Problem' of process control. Int J Control Theory Comput Model 3(1):1–18. https://doi.org/10.5121/ijctcm.2013.3101

Ardern E, Lockett WT (1914) Experiments on the oxidation of sewage without the aid of filters. J Soc Chem Ind 33(10):523–539. https://doi.org/10.1002/jctb.5000331005

Ataei A (2010) Wastewater treatment: energy-conservation opportunities. search.proquest.com [Online]. https://search.proquest.com/openview/a611bd65b9a49aea06054f42c6701d5f/1.pdf?pq-origsite=gscholar&cbl=41910. Accessed 30 July 2024

Balku S, Berber R (2006) Dynamics of an activated sludge process with nitrification and denitrification: start-up simulation and optimization using evolutionary algorithm. Comput Chem Eng 30(3):490–499. https://doi.org/10.1016/j.compchemeng.2005.10.014

Barber WP, Stuckey DC (1999) The use of the anaerobic baffled reactor (ABR) for wastewater treatment: a review. Water Res 33(7):1559–1578. https://doi.org/10.1016/S0043-1354(98)00371-6

Behera CR, Santoro D, Gernaey KV, Sin G (2018) Organic carbon recovery modeling for a rotating belt filter and its impact assessment on a plant-wide scale. Chem Eng J 334:1965–1976

Bel A, Avil L, Vel C, Llorens M (2019) Methodology for energy optimization in wastewater treatment plants. Phase I : control of the best operating conditions

Beraud B, Lemoine C, Steyer J (2009) Multiobjective genetic algorithms for the optimisation of wastewater treatment processes multiobjective genetic algorithms for the optimisation of wastewater treatment processes, no March 2015. https://doi.org/10.1007/978-3-642-01888-6

Bürger R, Diehl S, Nopens I (2011) A consistent modelling methodology for secondary settling tanks in wastewater treatment. Water Res. https://doi.org/10.1016/j.watres.2011.01.020

Bürger R, Diehl S, Farås S, Nopens I (2012) On reliable and unreliable numerical methods for the simulation of secondary settling tanks in wastewater treatment. Comput Chem Eng 41:93–105. https://doi.org/10.1016/j.compchemeng.2012.02.016

Caraman M, Sbarciog S, Barbu M (2007) Predictive control of a wastewater treatment process. Int J Comput Commun Control 2(2):132–142. https://doi.org/10.1016/S0043-1354(97)00215-7

Carrette R, Bixio D, Thoeye C, Ockier P (2001) Full-scale application of the IAWQ ASM no. 2d model. Water Sci Technol 44(2–3):17–24

Chemicznego F, Na Z (2014) Chemical oxygen demand fractionation of reject water from municipal wastewater treatment plant. 8(2):23–26. https://doi.org/10.2429/proc.2014.8(2)051

Chen X, Al R, Behera CR, Sin G (2018) Process synthesis, design, and control of wastewater treatment plants. In: Reference module in chemistry, molecular sciences and chemical engineering, no January. Elsevier, pp 1–14. https://doi.org/10.1016/B978-0-12-409547-2.14345-8

Cooper PF (2001) Historical aspects of wastewater treatment. Decent Sanit Reuse Concepts Syst Implementations 11–38 [Online]. http://www.bvsde.paho.org/bvsacd/leeds/cooper.pdf

De Clercq J et al (2005) Detailed spatio-temporal solids concentration profiling during batch settling of activated sludge using a radiotracer. Water Res. https://doi.org/10.1016/j.watres.2005.03.023

Džubur A, Serdarević A, Šuvalija S (2023) Modelling steps for dynamic simulation of wastewater treatment processes. Lect Notes Networks Syst (LNNS) 539:122–137. https://doi.org/10.1007/978-3-031-17697-5_10

Ebeling JM, Welsh CF, Rishel KL (2006) Performance evaluation of an inclined belt filter using coagulation/flocculation aids for the removal of suspended solids and phosphorus from microscreen backwash effluent. Aquac Eng 35(1):61–77

Fikar M, Chachuat B, Latifi MA (2005) Optimal operation of alternating activated sludge processes. Control Eng Pract 13(7):853–861. https://doi.org/10.1016/j.conengprac.2004.10.003

Franchi A, Santoro D (2015) Current status of the rotating belt filtration (RBF) technology for municipal wastewater treatment. Water Pract Technol 10(2):319–327. https://doi.org/10.2166/wpt.2015.038

Fuentes R, Molinos-Senante M, Hernández-Sancho F, Sala-Garrido R (2020) Analysing the efficiency of wastewater treatment plants: the problem of the definition of desirable outputs and its solution. J Clean Prod 267. https://doi.org/10.1016/j.jclepro.2020.121989

Gernaey KV, van Loosdrecht MC, Henze M, Lind M, Jørgensen SB (2004) Activated sludge wastewater treatment plant modelling and simulation: state of the art. Environ Model Softw 19(9):763–783. https://doi.org/10.1016/j.envsoft.2003.03.005

Gingerich DB, Mauter MS (2018) Air emission reduction benefits of biogas electricity generation at municipal wastewater treatment plants. Environ Sci Technol 52(3):1633–1643. https://doi.org/10.1021/acs.est.7b04649

Goodman BL, Englande AJ (1974) A unified model of the activated sludge process. J Water Pollut Control Fed 46(2):312–332 [Online]. http://www.jstor.org/stable/25038126

Grady CPL, Daigger GT, Lim HC (1999) Biological wastewater treatment. Hazard Waste October(19):1076 [Online]. http://www.loc.gov/catdir/enhancements/fy0647/98037263-d.html

Gray NF (2004) Biology of wastewater treatment. 4(January):2004. https://doi.org/10.1142/p266

Greenslade TB Jr (2003) The hydraulic analogy for electric current. Phys Teach 41(8):464–466

Guo Z, Sun Y, Pan SY, Chiang PC (2019) Integration of green energy and advanced energy-efficient technologies for municipal wastewater treatment plants. Int J Environ Res Public Health 16(7):1–29. https://doi.org/10.3390/ijerph16071282

Hauser BA (2018) Biochemical oxygen demand (BOD5). In: Practical manual of wastewater chemistry. CRC Press, pp 63–68. https://doi.org/10.1201/9780203744123-7

Holenda B, Domokos E, Rédey Á, Fazakas J (2007) Aeration optimization of a wastewater treatment plant using genetic algorithm. Optim Control Appl Methods 28(3):191–208. https://doi.org/10.1002/oca.796

IWA (2021) Circular economy: tapping the power of wastewater. International Waster Association. https://iwa-network.org/learn/circular-economy-tapping-the-power-of-wastewater/. Accessed 27 Jun 2024

Jakubaszek A, Stadnik A (2019) Efficiency of sewage treatment plants in the sequential batch reactor. Civ Environ Eng Rep 28(3):121–131. https://doi.org/10.2478/ceer-2018-0040

Jeppsson U (1996) Modeling aspects of wastewater treatment processes, no 1

Jeppsson U, Baky A, Hellström D, Kärrman E (2005) The URWARE wastewater treatment plant models the mistra programme urban water. Urban Water

Kalyuzhnyi SV (2003) Wastewater sludge management in the Russian Federation, pp 73–80. https://doi.org/10.2166/wpt.2007.085

Kowalik R, Gawdzik J, Gawdzik A, Gawdzik B (2019) Innovative sequential batch reactor solutions for wastewater treatment. Struct Environ 11(2):141–152. https://doi.org/10.30540/sae-2019-011

Kuş BA, Kara T (2020) Modeling and simulation of an optimal unified tank model for aeration-sedimentation processes in wastewater treatment plants. J Water Process Eng 37:101487. https://doi.org/10.1016/j.jwpe.2020.101487

Kuş BA, Kara T (2022) Optimal on-off level control design by electrical analogy for improved moving fine mesh filter system performance in wastewater treatment plants. J Clean Prod 350:131497. https://doi.org/10.1016/j.jclepro.2022.131497

Laizāns A, Šnīders A (2012) Analytical model and simulation of oxygen solubility in wastewater. Renew Energy Energy Effic 109–114

Lee DS, Jeon CO, Park JM, Chang KS (2002a) Hybrid neural network modeling of a full-scale industrial wastewater treatment process. Biotechnol Bioeng 78(6):670–82. https://doi.org/10.1002/bit.10247

Lee Y, Cho J, Seo Y, Lee JW, Ahn KH (2002b) Modeling of submerged membrane bioreactor process for wastewater treatment. Desalination 146(1–3):451–457. https://doi.org/10.1016/S0011-9164(02)00543-X

Lefebvre O, Moletta R (2006) Treatment of organic pollution in industrial saline wastewater: a literature review. Water Res 40(20):3671–3682. https://doi.org/10.1016/j.watres.2006.08.027

Longo S et al (2016) Monitoring and diagnosis of energy consumption in wastewater treatment plants. A state of the art and proposals for improvement. Appl Energy 179:1251–1268. https://doi.org/10.1016/J.APENERGY.2016.07.043

Mackintosh A (1873) Craigentinny meadows near Edinburgh. BMJ 2(665):389–389. https://doi.org/10.1136/bmj.2.665.389

Malviya A, Jaspal D (2021) Artificial intelligence as an upcoming technology in wastewater treatment: a comprehensive review. Environ Technol Rev 10(1):177–187. https://doi.org/10.1080/21622515.2021.1913242

Martínez R, Vela N, El Aatik A, Murray E, Roche P, Navarro JM (2020) On the use of an IoT integrated system for water quality monitoring and management in wastewater treatment plants. mdpi.com. https://doi.org/10.3390/w12041096

Mathematics A, Perindustrian T, Cemerlang N (2015) A mathematical model for wastewater treatment, vol 2015, pp 1–13. https://doi.org/10.1016/j.ica.2005.12.075

Metcalf E, Eddy H (1980) Wastewater engineering: treatment, disposal and reuse. Adv Water Resour 3(3):146. https://doi.org/10.1016/0309-1708(80)90067-6

Metcalf and Eddy (2003) Metcalf & Eddy, Inc. wastewater engineering teatment and reuse. J Wastewater Eng

Metcalf & Eddy, Burton FL, Stensel HD, Tchobanoglous G (2003) Wastewater engineering treatment and reuse. McGraw Hill

Metcalf & Eddy et al (2014) Wastewater engineering: treatment and resource recovery. McGraw Hill Education

Muñoz R, Köllner C, Guieysse B (2009) Biofilm photobioreactors for the treatment of industrial wastewaters. J Hazard Mater. https://doi.org/10.1016/j.jhazmat.2008.03.018

Nasr MS, Moustafa MAE, Seif HAE, El Kobrosy G (2011) Modelling and simulation of German BIOGEST/EL-AGAMY wastewater treatment plants—Egypt using GPS-X simulator. Alexandria Eng J 50(4):351–357. https://doi.org/10.1016/j.aej.2011.05.003

Nathanson JA (2020) Solid-waste management | Definition, Methods, and Facts | Britannica. https://www.britannica.com/technology/solid-waste-management. Accessed 17 July 2020

Neilands R et al (2013) Building and renovation high strength wastewater treatment process simulation, pp 68–76

Nelson MI, Sidhu HS (2009) Analysis of the activated sludge model (number 1). Appl Math Lett 22(5):629–635. https://doi.org/10.1016/j.aml.2008.05.003

Nelson MJ, Nakhla G, Zhu J (2021) The circulating fluidized bed bioreactor as a biological nutrient removal process for municipal wastewater treatment: process modelling and costing analysis. J Environ Manage 299:113604. https://doi.org/10.1016/j.jenvman.2021.113604

Ni BJ, Yu HQ, Sun YJ (2008) Modeling simultaneous autotrophic and heterotrophic growth in aerobic granules. Water Res 42(6–7):1583–1594. https://doi.org/10.1016/j.watres.2007.11.010

Ni BJ et al (2011) Heterotrophs grown on the soluble microbial products (SMP) released by autotrophs are responsible for the nitrogen loss in nitrifying granular sludge. Biotechnol Bioeng 108(12):2844–2852. https://doi.org/10.1002/bit.23247

Olsson G, Newell B (1999) Modelling, diagnosis and control. Model Diagn [Online]. http://test.iwaponline.com/wio/2005/04/pdf/wio200504RF1900222159.pdf

Olsson G, Newell B (2015) Wastewater treatment systems: modelling, diagnosis and control. Water Intell Online 4:9781780402864–9781780402864. https://doi.org/10.2166/9781780402864

Olsson G et al (2014) Instrumentation, control and automation in wastewater—from London 1973 to Narbonne 2013. Water Sci Technol 69(7):1373–1385. https://doi.org/10.2166/wst.2014.057

Orhon D (2015) Evolution of the activated sludge process: the first 50 years. J Chem Technol Biotechnol 90(4):608–640. https://doi.org/10.1002/jctb.4565

Orhon D, Allı B, Sözen S (2019) Which activated sludge configurations qualify for maximizing energy conservation—Why? J Chem Technol Biotechnol 94(2):556–568. https://doi.org/10.1002/jctb.5802

Özturan PM, Bozanta A, Basarir-Ozel B, Akar E, Coşkun M (2015) A roadmap for an integrated university information system based on connectivity issues: case of Turkey. Int J Manage Sci Inf Technol 17(17):1–23. https://doi.org/10.14313/JAMRIS

Piotrowski R, Duzinkiewicz K, Brdys MA (2004) Dissolved oxygen tracking and control of blowers at fast time scale. IFAC Proc 37(11):281–287. https://doi.org/10.1016/S1474-6670(17)31625-7

Piotrowski R, Brdys MA, Miotke D (2008) Centralized dissolved oxygen tracking at wastewater treatment plant: Nowy Dwor Gdanski case study. IFAC Proc 43(8):292–297. https://doi.org/10.3182/20100712-3-FR-2020.00049

Pokhrel D, Viraraghavan T (2004) Treatment of pulp and paper mill wastewater—a review. Sci Total Environ 333(1–3):37–58. https://doi.org/10.1016/j.scitotenv.2004.05.017

Pomiès M, Choubert JM, Wisniewski C, Coquery M (2013) Modelling of micropollutant removal in biological wastewater treatments: a review. Sci Total Environ. https://doi.org/10.1016/j.scitotenv.2012.11.037

Qiao J, Zhou H (2018) Modeling of energy consumption and effluent quality using density peaks-based adaptive fuzzy neural network. IEEE/CAA J Autom Sin 5(5):968–976. https://doi.org/10.1109/JAS.2018.7511168

Radjenović J, Matošić M, Mijatović I, Petrović M, Barceló D (2008) Membrane bioreactor (MBR) as an advanced wastewater treatment technology. In: Emerging contaminants from industrial and municipal waste. Springer Berlin Heidelberg, Berlin, Heidelberg, pp 37–101. https://doi.org/10.1007/698_5_093

Rahmani AR, Nematollahi D, Azarian G, Godini K, Berizi Z (2015) Activated sludge treatment by electro-Fenton process: parameter optimization and degradation mechanism. Korean J Chem Eng 32(8):1570–1577. https://doi.org/10.1007/s11814-014-0362-2

Rani A, Chang CT (2021) Modeling and optimization of wastewater treatment processes. In: Current developments in biotechnology and bioengineering: strategic perspectives in solid waste and wastewater management, pp 373–396. https://doi.org/10.1016/B978-0-12-821009-3.00006-3

Ráduly B, Capodaglio AG, Vaccari DA (2015) Empirical modelling of wastewater treatment processes—an approach to model reduction and integration, pp 1–7

Rivas A, Irizar I, Ayesa E (2008) Model-based optimisation of wastewater treatment plants design. Environ Model Softw 23(4):435–450. https://doi.org/10.1016/j.envsoft.2007.06.009

Ruiken CJ, Breuer G, Klaversma E, Santiago T, van Loosdrecht MCM (2013) Sieving wastewater—cellulose recovery, economic and energy evaluation. Water Res 47(1):43–48. https://doi.org/10.1016/j.watres.2012.08.023

Rusten B, Ødegaard H (2006) Evaluation and testing of fine mesh sieve technologies for primary treatment of municipal wastewater. Water Sci Technol 54(10):31–38. https://doi.org/10.2166/wst.2006.710
Rusten B, Rathnaweera SS, Rismyhr E, Sahu AK, Ntiako J (2017) Rotating belt sieves for primary treatment, chemically enhanced primary treatment and secondary solids separation. Water Sci Technol 75(11):2598–2606. https://doi.org/10.2166/wst.2017.145
Sandberg M, Venkatesh G, Granström K (2018) Experimental study and analysis of the functional and life-cycle global warming effect of low-dose chemical pre-treatment of effluent from pulp and paper mills. J Clean Prod 174:701–709. https://doi.org/10.1016/j.jclepro.2017.10.214
Santín I, Barbu M, Pedret C, Vilanova R (2019) Dissolved oxygen control in biological wastewater treatments with non-ideal sensors and actuators. Ind Eng Chem Res 58(45):20639–20654. https://doi.org/10.1021/acs.iecr.9b02572
Schölzel, Bower (1999) Small scale wastewater treatment plant. SOPAC Tech Rep 288(July):47 [Online]. http://www.pacificwater.org/userfiles/file/tr0288.pdf
Serdarevic A, Dzubur A (2016) Wastewater process modeling. Coupled Syst Mech 5(1):21–39. https://doi.org/10.12989/csm.2016.5.1.021
Shen W, Chen X, Corriou JP (2008) Application of model predictive control to the BSM1 benchmark of wastewater treatment process. Comput Chem Eng 32(12):2849–2856. https://doi.org/10.1016/j.compchemeng.2008.01.009
Shen W, Chen X, Pons MN, Corriou JP (2009) Model predictive control for wastewater treatment process with feedforward compensation. Chem Eng J 155(1–2):161–174. https://doi.org/10.1016/j.cej.2009.07.039
Sherratt A, DeGroot CT, Straatman AG, Santoro D (2019) Numerical modeling and control of solids separation using continuously moving fine mesh filters. Chem Eng Sci 195:881–893. https://doi.org/10.1016/j.ces.2018.10.033
Sie Chun T, Malek MA, Ismail AR (2017) A review of wastewater treatment plant modelling: revolution on modelling technology. Rev Wastewater Treat Plant Model Revolut Model Technol Am J Environ Resour Econ 2(1):22–26. https://doi.org/10.11648/j.ajere.20170201.13
SIMBA Home Page. http://www.inctrl.ca/software/simba
Sincero AP, Sincero GA (2003) Physical-chemical treatment of water and wastewater. Choice Rev Online 40(06):40–3411–40–3411. https://doi.org/10.5860/CHOICE.40-3411
Sinha SK, Sinha VK, Pandey SK, Tiwari A (2014) A study on the waste water treatment technology for steel industry : recycle and reuse. Am J Eng Res 3(4):309–315
Stanwell-Smith R (2010) Public toilets down the drain? Why privies are a public health concern. The early history of public conveniences. Public Health. https://doi.org/10.1016/j.puhe.2010.07.002
Stare A, Hvala N, Vrecko D (2006) Modeling, identification, and validation of models for predictive ammonia control in a wastewater treatment plant–a case study. ISA Trans 45(2):159–74. https://doi.org/10.1016/S0019-0578(07)60187-6
Szaja A, Łagód G, Aguilar JA (2015) Estimation of chemical oxygen demand fractions of municipal wastewater by respirometric method—case study. Rocz Ochr Sr 17(1):289–299
Szilvester S, Raduly B, Abraham B, Lanyi S, Niculae DR (2010) Mathematical models for domestic biological wastewater treatment process. Environ Eng Manage J 9(5):629–636
Teoh S, Tan RBH, Tien C (2006) Analysis of cake filtration data—a critical assessment of conventional filtration theory. AIChE J 52(10):3427–3442
Titarev VA, Toro EF (2002) ADER: arbitrary high order godunov approach. J Sci Comput. https://doi.org/10.1023/A:1015126814947
Torfs E et al (2013) Towards improved 1-D settler modelling: calibration of the Bürger model and case study. https://doi.org/10.2175/193864713813685430
Tsalas N, Golfinopoulos SK, Samios S, Katsouras G, Peroulis K (2024) Optimization of energy consumption in a wastewater treatment plant: an overview. Energies 17(12):2808. https://doi.org/10.3390/en17122808

Urena F, Alfaro VM, Arrieta O, Vilanova R (2011) MATLAB based graphical interface simulator for wastewater treatment plant control by using the benchmark BSM1. In: 2011 IEEE Global Engineering Education Conference, pp 585–591. https://doi.org/10.1109/EDUCON.2011.5773196
Vanhooren H, Meirlaen J, Amerlinck Y, Claeys F, Vangheluwe H, Vanrolleghem PA (2003) WEST: modelling biological wastewater treatment. J Hydroinformatics 27–50
Vanrolleghem PA, Lee DS (2003) On-line monitoring equipment for wastewater treatment processes: state of the art. Water Sci Technol 47(2):1–34. https://doi.org/10.2166/WST.2003.0074
Vera J, Torres NV, Moles CG, Banga J (2003) Integrated nonlinear optimization of bioprocesses via linear programming. AIChE J 49(12):3173–3187. https://doi.org/10.1002/aic.690491217
Walton M (2018) Commentary: the energy sector should care about wastewater. www.iea.org [Online]. https://www.iea.org/newsroom/news/2018/november/the-energy-sector-should-care-about-wastewater.html
Wang Y et al (2023) A review on applications of artificial intelligence in wastewater treatment. Sustainability 15(18):1–28. https://doi.org/10.3390/su151813557
Wehner JF, Wilhelm RH (1956) Boundary conditions of flow reactor. Chem Eng Sci. https://doi.org/10.1016/0009-2509(56)80014-6
Yetilmezsoy K (2010) Modeling studies for the determination of completely mixed activated sludge reactor volume: steady-state, empirical and ann applications. Neural Netw World 20(5):559–589
Yiannopoulos AC, Manariotis ID, Chrysikopoulos CV (2008) Design and analysis of a solar reactor for anaerobic wastewater treatment. Bioresour Technol 99(16):7742–7749. https://doi.org/10.1016/j.biortech.2008.01.067
Yilmaz D, Büyüktahtakın E (2024) An expandable machine learning-optimization framework to sequential decision-making. Eur J Oper Res 314(1):280–296. https://doi.org/10.1016/j.ejor.2023.10.045
Ying Li R, Zhang T, Fang HHP (2008) Characteristics of a phototrophic sludge producing hydrogen from acetate and butyrate. Int J Hydrogen Energy. https://doi.org/10.1016/j.ijhydene.2008.02.055
Yu JJ, Gu GW, Esposito G, Fabbricino M, Wang SP, Sun LP (2010) Determination of textile dyeing wastewater COD components by comparison with respirometry and full-scale data. Environ Technol 31(11):1191–1201. https://doi.org/10.1080/09593330903527872
Ziajahromi S, Neale PA, Rintoul L, Leusch FDL (2017) Wastewater treatment plants as a pathway for microplastics: development of a new approach to sample wastewater-based microplastics. Water Res 112:93–99. https://doi.org/10.1016/j.watres.2017.01.042

Chapter 5
Maintenance of Wastewater Treatment Plants

Zain Muhammad and Abid Mahmood

Abstract Keeping wastewater treatment plants running smoothly requires proper maintenance. This chapter explains why maintenance is important, the different types (like checking regularly and fixing problems), and how to plan for it effectively. We discuss the various parts of a treatment plant that need maintenance, from big machines to buildings and pipes. The chapter also explains how to create a maintenance schedule, keep track of spare parts, and document everything that's done. Different techniques are covered, like regular cleaning and checks, using tools to predict problems before they happen, and fixing things when they break. We also talk about the skills and training needed by the people who maintain the plants, and how to manage the costs involved. Challenges that can arise during maintenance, like equipment failures or unexpected events, are addressed along with solutions and best practices for continuous improvement. Ultimately, the chapter emphasizes the importance of a proactive and sustainable approach to keep wastewater treatment plants functioning efficiently and reliably.

Keywords Maintenance · Techniques · Plants · Wastewater · Treatment

5.1 Introduction

5.1.1 Importance of Effective Maintenance in Wastewater Treatment

A wastewater treatment plant is an essential utility that contains biological and chemical processes for the removal of pollutants from the wastewater before discharging the treated water into the environment (Hernández-Chover et al. 2021). These facilities are of great importance, especially in the aspect of safeguarding the health of the public and our common natural water sources by eradicating pollutants in wastewater

Z. Muhammad · A. Mahmood (✉)
Department of Environmental Sciences, Government College University, Faisalabad, Pakistan
e-mail: drabid@gcuf.edu.pk

S. Souabi et al. (eds.), *Wastewater Treatment Plants*, Water Science and Technology Library 130, https://doi.org/10.1007/978-3-031-87461-1_5

before it is returned to the natural water systems. If treatment plants do not undergo maintenance, they are prone to equipment malfunction and inefficiency and may also emit hazardous substances and violate laws; these factors all lead to risks of causing harm to the public's health (Khan 2013). Maintenance can reduce the time machines are out of service, prolong the useful lifespan of tools, and observe the laws protecting the environment.

5.1.2 Types of Maintenance (Preventive, Predictive, Corrective)

There are three primary types of maintenance strategies employed in wastewater treatment plants:

Preventive Maintenance: It entails carrying out scheduled maintenance work as a way of avoiding equipment breakdowns. Tasks like inspections, cleaning, oiling tasks, and replacement of parts among others that are supposed to be done periodically fall under this category (Wireman 2005).

Predictive Maintenance: This strategy entails the use of data and technology to ascertain the right time when a particular equipment is most likely to develop a fault, exactly when it is about to fail and needs the attention of a technician. Methodologies are vibration analysis, thermal imaging, and oil analysis which aid in predicting the state of equipment and prevent faults (Jardine et al. 2006).

Corrective Maintenance: This is also known as reactive maintenance, and is the technique of mending or replacing machinery only after it has begun to malfunction. However, it is noted that in most cases using corrective maintenance can be expensive and disruptive (Mobley 2002).

5.1.3 Benefits of a Proactive Maintenance Program

Implementing a proactive maintenance program that combines preventive and predictive maintenance strategies offers several benefits:

Increased Reliability and Efficiency: Maintenance guarantees efficiency by reducing equipment failures, which lowers the plant's productivity; all the equipment must be in good condition.

Cost Savings: Since maintenance helps to avoid major equipment breakdowns and prolongs the service life of a company's assets, it can be concluded that it has a high economic effect in the long term. This entails cost-saving concerning exigent rectifications, time loss, and replacement expenses (Pintelon and Parodi-Herz 2008).

Enhanced Safety: Specifically, safe work equipment is less likely to fail in a manner that can put danger to workers and members of the public, making safety results better.

Regulatory Compliance: Regular maintenance helps to ensure that the plant remains in compliance with environmental regulations, avoiding costly fines and legal issues.

Sustainability: The management of the plant should be able to consider regular maintenance as a critical planning factor for sustainability to minimize expenses incurred in cases of violation of environmental laws and legal complications (Smith 2004).

5.2 Key Equipment and Systems Requiring Maintenance

5.2.1 Unit Processes in Wastewater Treatment

Wastewater treatment plants are made up of several unit processes, all of which perform a unique part in the process of treating wastewater (Azizi and Davarnejad 2023). These processes must be maintained regularly to ensure they function efficiently:

Screening: This is the first stage in wastewater treatment where solid particles including sticks, rags, and other enormous matters in the wastewater are filtered out. This means that screens require constant cleaning and maintenance so that there is ease in the passage of water (Metcalf & Eddy, Inc., 2014).

Grit Removal: The silt remediation process is one way of sieving out sand, gravel, and other worn parts that take a toll on machinery. Grit chambers and classifiers should be properly checked and cleaned to work optimally (Tchobanoglous et al. 2003).

Primary Treatment: This involves the sheer process of getting rid of suspendable substances through sedimentation. Primary clarifiers shall be desludging and inspected often since there is an accumulation of solids (Spellman 2013).

Secondary Treatment: Organic matter in water gets adsorbed by microorganisms and this is one of the biological treatment methods (Behrouzeh et al. 2022). Some of these commonly used methods include aeration tanks, secondary clarifiers, and biological reactors and these require maintenance often to facilitate microbial action and prevent failures in the system (Droste 1997).

Disinfection: Disinfection is the last process in wastewater treatment and it aims at the elimination of pathogens. However, it was observed that all these systems—chlorine, UV light, or ozone need to be closely checked and frequently maintained to remove pathogens effectively (Tchobanoglous et al. 2003).

5.2.2 Specific Equipment Examples

Various pieces of equipment within these unit processes require regular maintenance:

- **Pumps**: Pumps are utilized for transporting the wastewater through the treatment channel; hence, they require regular checkups for wear and tear, greasing of moving parts and replacement of seal and bearings if necessary (Coelho and Langevin 2020).
- **Motors**: Engine systems of different equipment should be checked for signs of overheating, alignment, and electrical problems. Maintenance of the motor components is important and simple procedures such as lubrication and cleaning should be done frequently (Mobley 2002).
- **Valves**: The valves are used to enable or to hinder the flow of water; these should be checked for signs of leakage, or corrosion or to confirm if they are operating correctly. It also states that maintenance by way of often cleaning and changing worn-out parts is required (Tullis et al. 2017).
- **Mixers**: As a piece of equipment utilized in several treatment processes, specific to aid in the suspension of solids, mixers must have their blades and shafts checked frequently for wear, bearings greased or lubricated, and the entire machine should be checked for any signs of vibration (Morris 2015).
- **Instruments**: Maintenance programs of monitoring and control instruments that include flow meters, pH meters, and dissolved oxygen sensors require periodic calibration, cleaning and change of faulty subcomponents (Metcalf & Eddy, Inc. 2014).

5.2.3 Infrastructure and Buildings

In addition to the treatment processes and equipment, the infrastructure and buildings that house these systems also require maintenance:

- **Pipelines**: The wastewater collection system which includes pipes conveying wastewater and treated effluent also requires assessment in terms of leakages, blockages, or corrosion. There is a need to frequently clean and repair these pipelines for the integrity of the pipeline system to be maintained (Hammer and Hammer 2012).
- **Electrical Systems**: The power system of the treatment plant involves several wiring systems, circuit breakers, and control panels to be inspected to check out for any electrical failures that could have occurred (Spellman 2013).
- **Control Rooms**: These rooms contain the hardware and instrumentation for controlling and supervising the processes. Thus, preventive maintenance of the HVAC systems, proper cleaning of the equipment, and proper working of the control systems are very important for the proper functioning of the plant (Morris 2015).

5.3 The Maintenance Planning Process

5.3.1 *Developing a Maintenance Schedule and Frequency*

The development of efficient strategies for emissions reduction is necessary to create an optimized system for regular maintenance of the wastewater treatment plants (Jasim 2020). The schedule, therefore, should depend on plant equipment and processes, equipment manufacturer prescribes, working conditions, and performance history data (Mobley 2002). Maintenance frequency can be different greatly, some components can need daily checks, but the other one can need only annual checks. The management of the maintenance works or programs can be easily planned, scheduled, and monitored through the help of a computerized maintenance management system (CMMS) (Pintelon and Parodi-Herz 2008).

5.3.2 *Spare Parts Inventory Management*

Inventory management of spare is also crucial so that there will be less time that the equipment will be out of service while waiting for its spare parts to arrive. This requires stocking and tracking of operational spare parts and other consumables they use within the firm, and ordering for new ones when they run low. Prominent media integrate inventory control and help to anticipate which parts will be required depending on maintenance plans and previous usage patterns to avoid part shortages since this usually leads to delays (Wireman 2005).

5.3.3 *Documentation and Record-Keeping Practices*

Documentation is central in maintenance management since it must be accurate and encompass as much detail as possible. This must encompass keeping records on all the maintenance done, histories of equipment, or any problem faced while maintenance. Documentation proves useful in evaluating equipment usage, review the common barriers, and even determine repair or replacement decisions (Smith 2004). It also helps in the compliance of regulatory requisites and hence serves as a good source of information during audits and inspections.

5.4 Maintenance Procedures and Tools

5.4.1 Standard Operating Procedures (SOPs) for Equipment Maintenance

It is recommended that SOPs are implemented to help maintain consistency in the method in which a variety of maintenance tasks are conducted. SOPs contain anti-static measures to be taken, the operational procedures of equipment and machines to be followed, the parts to be cleaned or replaced and the tool and equipment to be used etc. These procedures assist in the development of some form of standard operation procedures throughout the plant, which would make it difficult to make mistakes and at the same time adhere to the set standards as prescribed by various authorities (Hale et al. 1998).

5.4.2 Preventive Maintenance Techniques

Maintenance planning entails various routine activities of equipment whose purpose is to avert equipment breakdowns in the future. Common preventive maintenance techniques include:

- **Cleaning**: Cleaning dirt and accumulations on equipment contact surfaces to avoid corrosion and other related problems.
- **Lubrication**: Using greases on some parts of the machine that undergo friction in an endeavor to lessen the wear and tear.
- **Adjustments**: As with all operating systems, there is always a level of checking and adjusting of equipment settings periodically.
- **Inspections**: Carry out first visual assessments and evaluate the operational condition to define existing problems (Mobley 2002).

5.4.3 Predictive Maintenance Techniques

Predictive maintenance is the application of complex methods to determine the condition of the equipment and its failures. This approach relies on data collected through various monitoring methods, such as:

- **Condition Monitoring**: Detecting the state of functioning of machines and equipment to analyze them in real-time with the help of sensors and diagnostic equipment.
- **Vibration Analysis**: Dynamics and resonance as a way of finding problems with balancing, alignment, and other mechanical problems.

- **Thermography**: Infrared cameras to see patterns of temperature that is different from normal levels that will serve as warning signs of problems (Mobley 2002).

5.4.4 *Corrective Maintenance Procedures*

Corrective maintenance is done on the equipment that has already failed or is not performing as it is required to. This involves:

- **Troubleshooting**: Analysis of the cause of failure through a process of scrutiny and/or experimentation.
- **Repairs**: The process of repairing defective or worn-out parts to maintain the working status of the equipment.
- **Post-repair Testing**: Making sure that when the repair is done, the equipment is okay or does not have other problems (Smith 2004).

5.4.5 *Maintenance Tools and Equipment*

Effective maintenance requires a variety of tools and equipment, including:

- **Hand Tools**: Small pieces of equipment that can be used day to day and include but are not limited to the following; Wrenches, screwdrivers, pliers, etc.
- **Power Tools**: Impact tools which are generally used in activities that require more force such as drills, grinders, and the like.
- **Diagnostic Tools**: Ammeters, cyclic meters, and AC/DC multimeters for monitoring the condition of the equipment and detecting faults.
- **Lifting Equipment**: Lifting devices that are used in the moving and orienting of heavy and large equipment in the course of maintenance (Bloch and Geitner 1997).

5.5 Maintenance Personnel and Training

5.5.1 *Skills and Qualifications Required for Maintenance Staff*

Employees of wastewater treatment plants perform basic responsibilities and require certain qualifications, first of all, among the maintenance staff. These include:

- **Technical Knowledge**: Getting familiar with the generalized concepts of wastewater treatment processes and functional aspects of the equipment.
- **Mechanical Skills**: Mechanical repair skills, and use of tools in handling mechanical and automobile systems.

- **Electrical Skills**: Comprehension of electrical systems and the skills in solving issues with the electrical parts.
- **Analytical Skills**: Proficiency in analyzing the causes of certain issues and in finding the right solutions for the matter.
- **Safety Awareness**: The learners' knowledge of safety measures and the practice of the use of personal protective equipment (PPE) (Higgins 2010).

5.5.2 Training Programs and Continuing Education

Continuing education is important to providing a competent staff in the maintenance division. Training programs should cover:

- **Basic Training**: Some of the basic concepts concerning wastewater management and utility practices.
- **Advanced Training**: Sub-courses that focus on certain pieces of equipment, new and sophisticated diagnostic methods, and new technologies.
- **Safety Training**: Safety lectures and training along with practice fires, ways of protective equipment, and emergency measures.
- **Continuing Education**: In the same way, maintenance personnel can attend workshops, seminars, and certifications that enhance their update on the current standards and technology (Kolb 1984).

5.5.3 Importance of Safety and Personal Protective Equipment (PPE)

Maintenance activities involve risks and thus have to be approached with much care. Ensuring the well-being of maintenance staff involves:

- **Personal Protective Equipment (PPE)**: Supervising the employees to ensure that they use the right personal protective gear; gloves, safety glasses, helmets, protective clothing, etc.
- **Safety Protocols**: Safe working practices throughout the organization that will reduce cases of accidents and/or injuries.
- **Safety Audits**: Safety checks should be done often so that one can be able to find out different areas that may pose a danger to the firms' safety norms.
- **Emergency Preparedness**: Creating awareness about emergency procedures among the employees and making sure that the emergency products are within easy access (Goetsch 2011).

5.6 Managing Maintenance Costs and Budgets

5.6.1 Cost Estimation and Budgeting for Maintenance Activities

The primary stage in the effective management of maintenance costs entails the aspects of cost estimation and cost control. This involves:

- **Identifying Maintenance Activities**: Enumerate all the maintenance requirements and schematization.
- **Estimating Costs**: Deducing the expenses that would be incurred on labor, parts and tools as well as additional services that must be availed.
- **Budget Allocation**: This can also be a function of the level of priority and the degree of criticality of types of equipment and processes for which funds are to be provided. As critical products may be allocated more attention and guard in order not to break and cost high, they are regarded as having a high criticality.
- **Contingency Funds**: Performing adequate maintenance work in a timely and proper manner to decrease financial risks or other future costs such as the amount of money to be set aside for repairing and other cases that may arise (Pintelon and Parodi-Herz 2008).

5.6.2 Strategies for Cost Optimization and Resource Management

To optimize maintenance costs and manage resources efficiently, the following strategies can be implemented:

- **Preventive Maintenance**: It is therefore recommended that organizations spend some amount in ordering check-ups for their machinery, this way, they end up using less on repairs and also most probably, their machinery would have a longer life.
- **Predictive Maintenance**: Employing prescriptive maintenance methods that help estimate the probable time to the probability of a failure and take measures to prevent such failures and their resultant downtime.
- **Efficient Spare Parts Management**: It also involves the management of spare parts for not to have a large stock which is costly while at the same time it is also important to avoid having a small stock which will be a hindrance to the normal running of the organization.
- **Energy Management**: Applying traditional changes aimed at lowering energy-related expenses on item usage as well as maintenance and repair work (Tsang 2002).

5.6.3 *Utilizing Technology for Maintenance Management*

Advanced technologies play a crucial role in modern maintenance management:

- **Computerized Maintenance Management Systems (CMMS)**: Maintenance management systems assist in developing scheduling of the maintenance work and tracking schedules to guide the usage of resources and time in performing the tasks.
- **Condition Monitoring Tools**: Some portable monitoring equipment includes vibration analyzers, thermographic cameras, and ultra-sonic detectors which can give real-time of the condition of equipment.
- **Mobile Technology**: Introducing the use of gadgets and apps to allow every maintenance team staff to have physical access to the SOPs, equipment manuals and even records while on the field.
- **Data Analytics**: Using data analysis to suggest possible breakdowns of equipment and using data analysis to adjust the maintenance strategy depending on its history and tendencies (Ben-Daya et al. 2009).

5.7 Common Maintenance Challenges and Solutions

5.7.1 *Identifying and Addressing Equipment Failures*

All the equipment in wastewater treatment plants is another challenge which has been described as equipment failures. It is highly important to rapidly detect these failures and respond to them because they affect the smooth operation of the facility. The steps include:

- **Routine Inspections**: These involve pressing the equipment to determine various uses or the possible point of weakness or failure.
- **Failure Analysis**: Investigating equipment failures and their causes to detect the true possibility of a failure.
- **Timely Repairs**: Taking timely corrective measures wherein appropriate processes require immediate corrections concerning the issues that have been identified.
- **Training Staff**: Maintenance personnel should be educated to look at specific signs that could signal equipment breakdown and act appropriately (Mobley 2002).

5.7.2 *Spare Parts Availability and Management Issues*

To ensure that service and downtime are low spare parts inventory needs to be managed well. Challenges and solutions include:

- **Inventory Optimization**: Adequate stocking of spare parts so as not to be congested with the parts or have shortages of the parts.
- **Supplier Relationships**: Developing a good relationship with trustworthy suppliers to facilitate availableness of parts.
- **Inventory Tracking Systems**: Implementing integrated computerized stock control systems to inventory the use/consumption of spare parts and the right time to order.
- **Critical Parts Identification**: Subsequently, the classification and categorization of practically important spare parts that are vital in the operation of the plant is done (Kennedy 2011).

5.7.3 Managing Emergencies and Unforeseen Events

This means that there are varieties of complications that may hinder the scheduled manner of performing the maintenance. Strategies to manage these situations include:

- **Emergency Response Plans**: Prescribing guidelines for different kinds of emergencies and revising the emergency plans now and then.
- **Training and Drills**: Training and practicing the staff on different emergencies to be able to handle them effectively in the future.
- **Resource Allocation**: Maintaining stocks of spare parts, equipment and employees for on-call in case of emergencies.
- **Communication Systems**: Implementation and strengthening of efficient communication channels to facilitate efficient communication in emergencies (Hollnagel 2014).

5.7.4 Ensuring Environmental Compliance During Maintenance Activities

Adherence to environmental standards is important when carrying out such exercises. Key considerations include:

- **Regulatory Awareness**: To be aware of all the current rules and regulations about the environmental impact and make sure that all the maintenance works are done keeping these in mind.
- **Waste Management**: Disposing of the waste that is produced during the maintenance exercises in a manner that is friendly to the natural environment.
- **Documentation**: Documenting the activities about maintenance and disposal of wastes which will serve as evidence.
- **Environmental Training**: Educating the members of the maintenance department on proper sustainable behaviors and introducing them to all the legislation concerning this area (Cairns 2013).

5.8 Continuous Improvement in Maintenance Practices

5.8.1 Performance Monitoring and Evaluation

Sustained practice improvement in the area of maintenance is important for improving the sustainability of wastewater treatment plants. Performance monitoring and evaluation play a crucial role in this process:

- **Key Performance Indicators (KPIs)**: Some of the parameters include the percentage of the use of equipment, the costs of maintenance and the time taken to respond to maintenance needs.
- **Regular Audits**: Carry out the performance review for the maintenance programs to see their effectiveness in enhancing the performance of maintenance activities and plan for any adjustments that may be needed.
- **Feedback Loops**: Establishing communication systems to obtain feedback from the maintenance employees and incorporating these into the improvement system.
- **Data Analysis**: Examining maintenance data to establish trends and opportunities for improvement and causes of breakdowns and problems (Parida and Kumar 2006).

5.8.2 Benchmarking and Sharing Best Practices

Benchmarking and sharing best practices help in identifying successful maintenance strategies and implementing them across the organization:

- **Internal Benchmarking**: Comparing maintenance practices and performance metrics within different parts of the organization to identify best practices.
- **External Benchmarking**: Engaging with other wastewater treatment facilities and industry organizations to compare performance and learn from their experiences.
- **Knowledge Sharing**: Creating platforms for maintenance staff to share their experiences, solutions to common problems, and innovative ideas.
- **Continuous Learning**: Encouraging a culture of continuous learning and improvement through regular training sessions and workshops (Camp 1989).

5.8.3 Utilizing Technology and Automation for Efficient Maintenance

Leveraging technology and automation can significantly enhance maintenance efficiency and effectiveness:

- **Automation Tools**: Implementing automation tools such as robotic process automation (RPA) for routine maintenance tasks.
- **IoT and Sensors**: Utilizing Internet of Things (IoT) devices and sensors for real-time monitoring of equipment conditions and predictive maintenance.
- **Data Analytics**: Employing advanced data analytics to predict equipment failures and optimize maintenance schedules.
- **Digital Twin Technology**: Using digital twin technology to create virtual models of equipment and processes for better planning and troubleshooting (Lee et al. 2015; Wang et al. 2024).

5.9 Conclusion

5.9.1 Importance of a Proactive and Sustainable Approach to Maintenance

Effective maintenance of wastewater treatment plants is vital for ensuring their long-term functionality and environmental compliance. Adopting a proactive and sustainable maintenance approach involves:

- **Preventive and Predictive Maintenance**: Emphasizing preventive and predictive maintenance to identify and address issues before they escalate into significant problems.
- **Sustainable Practices**: Incorporating sustainable maintenance practices that minimize environmental impact, such as efficient energy use and proper waste management.
- **Continuous Improvement**: Engaging in continuous improvement initiatives to refine maintenance processes, increase efficiency, and reduce costs.
- **Stakeholder Involvement**: Ensuring that all stakeholders, including maintenance personnel, management, and the community, are involved in the maintenance planning and execution process (Smith 2011).

5.9.2 Conclusion and Recommendations

In conclusion, effective maintenance of wastewater treatment plants is a complex undertaking that requires comprehensive planning. The main recommendations include:

Developing a Comprehensive Maintenance Plan: Creating a detailed maintenance plan that includes preventive, predictive, and corrective maintenance activities.

- **Investing in Training**: Providing ongoing training and development opportunities for maintenance staff to ensure they have the necessary skills and knowledge.

- **Utilizing Advanced Technologies**: Leveraging advanced technologies, such as CMMS, IoT, and data analytics, to enhance maintenance efficiency and effectiveness.
- **Fostering a Culture of Continuous Improvement**: Encouraging a culture of continuous improvement to regularly evaluate and enhance maintenance practices.
- **Ensuring Compliance**: Maintaining strict adherence to environmental regulations and standards to ensure sustainable operation.

References

Azizi J, Davarnejad R (2023) Electro-Fenton technique associated with core-shell Fe@ Fe_2O_3 nanoparticles-bentonite catalyst for pre-treating a super-pollutant industrial wastewater. J Water Process Eng 51. https://doi.org/10.1016/j.jwpe.2022.103385

Behrouzeh M, Parivazh MM, Danesh E, Dianat MJ, Abbasi M, Osfouri S, Rostami A, Sillanpää M, Dibaj M, Akrami M (2022) Application of photo-fenton, electro-fenton, and photo-electro-fenton processes for the treatment of DMSO and DMAC wastewatersArab. J Chem 15(11). https://doi.org/10.1016/j.arabjc.2022.104229

Ben-Daya M, Duffuaa SO, Raouf A, Knezevic J, Ait-Kadi D (2009) Handbook of maintenance management and engineering. Springer

Bloch HP, Geitner FK (1997) Practical machinery management for process plants: Volume 3: Machinery component maintenance and repair. Gulf Professional Publishing

Cairns J (2013) Regulatory frameworks for pollution control: a comparison of practices in the UK and US. Edward Elgar Publishing

Camp RC (1989) Benchmarking: The search for industry best practices that lead to superior performance. Quality Press

Coelho ST, Langevin J (2020) Wastewater pump station maintenance. IWA Publishing

Droste RL (1997) Theory and practice of water and wastewater treatment. Wiley

Goetsch DL (2011) Occupational safety and health for technologists, engineers, and managers. Prentice Hall

Hale AR, Heming BHJ, Carthey J, Kirwan B (1998) Modeling of safety management systems. Safety Sci 29(2):131–150

Hammer MJ, Hammer MJ, Jr (2012) Water and wastewater technology. Prentice Hall

Hernández-Chover V, Bellver-Domingo Á, Hernández-Sancho F (2021) The influence of oversizing on maintenance cost in wastewater treatment plants. Process Saf Environ Prot 147:734–741. https://doi.org/10.1016/j.psep.2020.12.033

Higgins LR (2010) Maintenance engineering handbook. McGraw-Hill

Hollnagel E (2014) Safety-I and Safety-II: the past and future of safety management. CRC Press

Jardine AKS, Lin D, Banjevic D (2006) A review on machinery diagnostics and prognostics implementing condition-based maintenance. Mech Syst Signal Process 20(7):1483–1510

Jasim NA (2020) The design for wastewater treatment plant (WWTP) with GPS X modelling. Cogent Eng 7(1)

Kennedy R (2011) A guide to the implementation of the spare parts management system. Springer

Khan A (2013) Maintenance management in water and wastewater treatment plants. Procedia Eng 70:1243–1251

Kolb DA (1984) Experiential learning: experience as the source of learning and development. Prentice-Hall

Lee J, Bagheri B, Kao HA (2015) A cyber-physical systems architecture for industry 4.0-based manufacturing systems. Manuf Lett 3:18–23

Metcalf & Eddy, Inc. (2014) Wastewater engineering: treatment and resource recovery. McGraw-Hill Education
Mobley RK (2002) An introduction to predictive maintenance. Butterworth-Heinemann
Morris J (2015) Industrial wastewater treatment. Butterworth-Heinemann
Parida A, Kumar U (2006) Maintenance performance measurement (MPM): issues and challenges. J Qual Maint Eng 12(3):239–251
Pintelon L, Parodi-Herz A (2008) Maintenance: an evolutionary perspective. In: Complex system maintenance handbook. Springer, pp 21–48
Smith R (2004) Plant maintenance management set: planning and scheduling. Elsevier
Smith R (2011) Maintenance planning and scheduling handbook. McGraw Hill Professional
Spellman FR (2013) Handbook of water and wastewater treatment plant operations. CRC Press
Tchobanoglous G, Burton FL, Stensel HD (2003) Wastewater engineering: treatment and reuse. McGraw-Hill
Tsang AHC (2002) Strategic dimensions of maintenance management. J Qual Maint Eng 8(1):7–39
Tullis JP, Hobbs D, Finnemore EJ (2017) Hydraulic valves. CRC Press
Wang A-I, Li H, He Z, Tao Y, Wang H, Yang M, Savic D, Daigger GT, Ren N (2024) Digital twins for wastewater treatment: a technical review. Engineering 36:21–35
Wireman T (2005) Developing performance indicators for managing maintenance. Industrial Press Inc.

Chapter 6
Problems of Extension of Wastewater Treatment Plants

Tahira Maqsood and Abid Mahmood

Abstract The environment and human activity interact through wastewater treatment plants which provide an ideal environment for the selection and distribution of antibiotic resistance genes and antibiotic resistant bacteria. Large amounts of ARGs and ARBs are ultimately released into the water environment by wastewater treatment plants effluents since existing treatment techniques are unable to completely eliminate ARGs. This ultimately causes them to propagate and may transfer to people through water and the food chain. We studied the consequences of discharge by wastewater treatment plants on the bacterial community ecology in the sediment of a small, low-gradient stream in South Australia. Nitrous oxide (N_2O), carbon dioxide (CO_2), and methane (CH_4) are examples of greenhouse gases (GHG) that are directly released from biological processes when wastewater treatment plants are in operation. In usual wastewater treatment systems, 15 to 40% of the operational costs contribute to the extensive consumption of resources, mainly energy, in the wastewater treatment process. When wastewater is treated biologically, some harmful microbes develop that release bad odor and gases. The discharge of treated effluents such as water into bodies of water is not allowed because it has the ability to alter the chemical and physical characteristics of water which may result in a decrease in the population of aquatic animals.

Keywords Wastewater · Environmental · Treatment · Plants · Antibiotic

6.1 Introduction

Basic engineering is to remove contaminants from wastewater treatment plants so that an environmentally friendly effluent is produced. This integrates concerns about wastewater, human health, and environmental hazards. However, a significant amount of energy and resources, including land and infrastructure, are required for the treatment of wastewater. During this procedure, a significant amount of sludge is

T. Maqsood · A. Mahmood (✉)
Department of Environmental Sciences, Government College University, Faisalabad, Pakistan
e-mail: drabid@gcuf.edu.pk

S. Souabi et al. (eds.), *Wastewater Treatment Plants*, Water Science and Technology Library 130, https://doi.org/10.1007/978-3-031-87461-1_6

produced, which might be hazardous to the environment. Fortunately, techniques to remediation have developed, and efforts to make wastewater treatment plants more environmentally friendly are receiving increased attention. This might be explained by the rising market value of the elements found in wastewater, which are predicted to be recovered and include fresh water, metals, nutrients like ammonium, and energy (Ye et al. 2018).

Across the world, wastewater treatment plants are used to enhance the quality of wastewater before it is discharged into groundwater or surface and re-enters sources of water (Muhaba et al. 2024). Numerous countries have made efforts over the past 50 years to decrease the amount of untreated wastewater that is released into streams and rivers by closely checking and continuously enhancing municipal and industrial wastewater treatment plants (Riffat and Husnain 2022). In contrast to unregulated discharges, wastewater treatment plants are preferred because they impose physical modifications on the receiving system and don't discharge water of the same quality as the body of water that receives it. Aquatic systems may be significantly influenced by effluent discharges in a number of forms, such as number of bacterial species (Petersen et al. 2005), The quantity of organic carbon (McConnell 1980), the efficiency of nutrient absorption (Marti et al. 2004), hydrologic features (Dennehy et al. 1998).

From the point of plant construction to the point of equipment demobilization, the energy consumption of the water industry is high. Treatment plants utilize the majority of this energy, that normally represents 5–30% of the plants' worldwide operational costs (Lorenzo-Toja et al. 2016). Electrical energy is the primary source of power for valves, pumps, compressors, and other components (Wang et al. 2016). Energy consumption is linked to environmental issues such as greenhouse gas (GHG) productions in addition to cost. For example, a conventional treatment system can emit up to 1400 tons of CO_2 while it is operating (Muga and Mihelcic 2008). Life cycle assessment (LCA) and other methods have been used to determine such effects (Balkema et al. 2002), a tool standardized by ISO 14044. Wastewater systems and traditional water consume an excessive amount of energy, are expensive, and are often inefficient (Mahgoub et al. 2010). One of the main human sources of greenhouse gases is energy use, which also has some of the most significant effects on global warming (Vilanova and Balestieri 2015). According to Soares et al. (2017), 57% of global warming is caused by energy usage. The main objective of the city plan is lowering carbon emissions and ensuring the security of electricity and water (Singh and Kansal 2018). Enhancing a wastewater treatment plant's (WWTP) energy self-sufficiency can lead to increased profitability (Metcalf and Eddy 2015) and reduce GHG emissions.

The three stages of traditional wastewater treatment are primary, secondary, and sometimes tertiary. At each stage of the treatment procedure, multiple biological and physicochemical techniques are used. The objective of primary treatment is to lower the wastewater's solid content such as grease, sand, grit, oils and fats, and settle able particles. That process, which is standard at all WTPs, is performed out fully mechanically using filtration and sedimentation. Modern municipal wastewater treatment plants use various kinds of biological treatments, but among those most frequently used is conventional activated sludge (CAS), and moving bed biofilm reactors (MBBR), membrane bioreactors (MBR) and fixed bed bioreactors (FBR) are less generally employed. The secondary technique, on the other hand, can vary greatly. Anaerobic or aerobic systems are usually applied in a biological process to eliminate nutrients and/or organic waste. Under specific conditions, Activated sludge plants significantly decrease nitrogen and organic matter by promoting the growth of floating organisms using dissolved oxygen.

To eliminate phosphorus via rainfall and elements on a filter, tertiary wastewater treatment procedures might be used as the final option (Batt et al. 2007). Before being discharged into the environment, the effluent from specific WTPs is further disinfected, usually using UV or chlorination radiation.

6.2 Wastewater

Water is necessary for all human activities. On a daily basis, the domestic, industrial, and agricultural sectors generate huge quantities of wastewater due to over population. The supplies of freshwater are being depleted to keep up with the increasing water demands of people worldwide. As a result, the different sectors are now illegally competing with one another for the little freshwater resources available. That's why a lot of people don't have access to drinkable water, particularly in countries that are underdeveloped.

Additionally, the lack of sufficient water resources for year-round irrigation and raising livestock has a major effect on the production of agriculture. Similar circumstances have been observed around the world, but especially in Asia, Africa, the Middle East, and the United States.

More than four billion people experience severe water shortages for at least one month of the year, and 2.1 billion people lack access to clean water to consume at home. The facts are alarming (Water 2019).

A significant quantity of organic and inorganic waste are found in wastewater, which depletes freshwater's dissolved oxygen. Consequently, the direct dumping of raw sewage into water bodies causes a significant reduction in oxygen levels and a high rate of aquatic species death. Rich in nutrients like phosphorus and nitrogen, untreated wastewater promotes the growth of aquatic plants that are out of control and eutrophicates bodies of water (Chahal et al. 2016). Pathogens that cause waterborne illnesses that harm the health of humans and animals can also be found in raw sewage.

6.3 Wastewater Treatment Plant

All populated areas need WWTPs, and where they are built and situated be determined by number of intricate causes, for example available funding, the quantity and source of incoming wastewater, and particular management goals for effluent discharge concentrations (Tchobanoglous 2003). Since there are numerous biological, physical, and chemical processes that can be used to maximize efficiency, wastewater treatment plants employ a variety of techniques to either meet or exceed regulatory standards regarding effluent discharges either to fulfill or surpass legal requirements for wastewater emissions. The three major categories of primary, secondary, and advanced tertiary treatment can be utilized to classify all treatments, regardless of the techniques employed at any given wastewater treatment plants. The degree of technology employed in each treatment group affects expenses linked to the initial capital investment required to construct or renovate wastewater treatment facilities that are associated with management and operation. (Tchobanoglous 2003) present a comprehensive overview of wastewater treatment, with a focus on the principal procedures. Flotation, settling, and screening processes are used in first treatment to filter out big particles from entering wastewater. Primary clarifiers eliminate both inorganic and organic suspended materials. Smaller solids that are inorganic like sand, are eliminated using sedimentation basins and grit chamber. Anaerobic digesters receive the sludge, or collected and removed solids, from sedimentation basins and use them for further processing. Once the initial treatment step is completed, secondary treatment addresses any remaining organic matter in wastewater and may involve removing dissolved nutrients.

Trickling filters are one type of secondary treatment technique. These filters consist of stones covered with microorganisms that break down organic contaminants found in wastewater. At the secondary treatment step, organic materials are additionally removed using activated sludge, which has a high bacterial content. Bacteria in the sludge are better able to consume organic contaminants when the treatment effluent is exposed to the atmosphere. Solids can settle down it can be extracted from the effluent by the use of secondary clarifiers. The wastewater may still include dissolved and suspended particles even after secondary treatment, and extensive tertiary treatment could be necessary to comply with regulations safeguarding receiving waterways.

A number of biological and chemical nutrient removal (BNR) methods are existing to lessen effluent quantity before release, nutrients and metals may be targeted during advanced treatment (Pagilla et al. 2006). Around thirty years ago, BNR first appeared, and due to its inherent benefits over conventional chemical treatments, it has mostly been included into standard wastewater treatment techniques (Tchobanoglous, 2003). Compared to chemical treatment procedures, BNR technologies actually use less chemical input, generate a reduction in waste, and use a smaller amount of energy. Their usage of microbes in either anaerobic or aerobic environments to absorb nutrients through attached and/or suspended growth processes describes this.

Wastewater treatment plants acquire wastewater from a variety of sources, and the ultimate release of ARGs is influenced by the composition of the influent from these sources. ARG concentrations are greater in some industrial wastewater treatment systems, such as those used in pharmaceutical manufacturing, dyeing, and pesticide manufacturing, than in residential wastewater treatment systems. These systems include those in hospitals, animal farms, and other healthcare facilities (Semedo and Song 2023). Clinical settings and the environment are connected by the environment, that contains more high-risk ARGs that cause resistance to standard clinic antibiotics, including last-resort drugs, and wastewater from hospitals in particular (Zhu et al. 2023). The livestock organizations use large amounts of antibiotics for livestock to support growth in livestock as well as avoid diseases. As a result, farm wastewater becomes a prominent AMR hotspot that poses a concern to public health (Zhao et al. 2021). Containing organic contaminants such as polycyclic aromatic hydrocarbons, quaternary ammonium compounds, and the benzene series, industrial wastewater can elevate the risk of effluent discharge by enhancing the conjugation transfer of plasmids that are resistant to antibiotics (Su et al. 2023).

Wastewater that has not been treated is received by wastewater treatment plants from sanitary sewage networks, and multiple unit processes are used to remove pollutants until the water fulfills Clean Water Act regulations for discharge or reuse (Copeland 2001). Municipal wastewater treatment starts with physical procedures like grit trapping and screening to remove big particles and debris from the raw wastewater. Biological treatment comes next. Among biological waste water treatment methods, the most widely used is the traditional sludge activation procedure.

Diverse communities of microorganisms in sludge activation system are maintained by aeration and cycling of biologically-active solids, also known as "bio solids" or "solids," to break down a variety of organic chemicals and nutrients. After treated water and bio solids are separated by clarification (gravity settling), the environment is disinfected and the treated water is released. Advanced treatment (e.g., diffusive membrane technology or advanced oxidation techniques) may also be necessary, depending on the original quality of the water, to remove salts or pollutants of rising concern (e.g., pharmaceuticals, items for personal use, synthetic organic compounds) (Newhart et al. 2019). Wastewater treatment infrastructure collect sewage and wastewater from many different aspects of human culture, acting as a connection between human activity and the environment. When wastewater treatment plants release their effluent into aquatic environments, they create an environment that contains abundant nutrients, high microorganism's density, and sub-inhibitory quantities of antimicrobial agents that contribute to the selection and spread of antibiotic resistance genes. This is hazardous because environmental antibiotic resistance genes and antibiotic resistant bacteria create a risk to public health environments because they may be transmitted from the environment to humans through direct contact, the food chain, or drinking water (Berglund et al. 2023; Hernando-Amado et al. 2019).

6.4 Problems of Extension of Wastewater Treatment Plants

6.4.1 Sludge Production

Sewage sludge is a primary unused material obtained from wastewater anaerobic bio digestion; if improperly treated, it can pose a management challenge. Water treatment plants can provide a diversity of sewage sludge types, including primary and secondary, and the characteristics of sewage sludge can differ depending on a number of variables, including the time of day, the plant's location, the type of treatment procedures used, etc.

Because of this, a wide range of compounds, including compounds of nitrogenous and phosphorous, diseases causing agents, microorganisms, and a significant quantity of water, are observed. Sewage sludge is fascinating because of its high moisture concentration, which includes bound, free, interstitial, and vicinal water. This can result in an increase in costs if further procedures, such as drying, are necessary for the sludge's valorization.

Some of the more intriguing applications of biogas slurries that have been 20 of 35 considered recently include adding deer waste to them during anaerobic fermentation to increase the quantity of methane produced by biogas. Landfilling is one of the multiple traditional uses for sewage sludge, which could potentially damage the environment (Wang et al. 2018).

6.4.2 Energy Consumption

According to estimates, residential wastewater, which comes in chemical, thermal, and hydraulic forms, contains ten times more energy than is needed for treatment (Shen et al. 2015). Despite the large energy content, only a portion of it is recovered owing to thermodynamic and technical limitations (Gikas 2017) and there are still obstacles that need to be solved before all of the energy can be collected in a fashion that is economically feasible. Between 39 and 76% of the energy used in the waste water treatment plant may be obtained from the biogas created by the sludge's anaerobic breakdown (Silvestre et al. 2015).

6.4.3 Putrid Odor

The unpleasant odor of the air and the smell of treated water can adversely impact the quality of life in the community since the municipal wastewater treatment procedure, which reduces odor emissions (Agus et al. 2012). The wide surface area of wastewater treatment plants, which may extend from several to more than twelve hectares, causes them to be often assigned for odor emissions. Although the majority of the odors are not released by the plants, huge plants are frequently blamed for the release of undesirable smells (Bruszkiewski and Skorupski 1999).

Numerous industrial companies may discharge odorous materials directly into the sewerage system, or they may form in the sewers under anaerobic conditions owing to high temperatures, high oxygen demand, long sewers, and slight gradients, depositions in the sewers, inadequate maintenance, or any combination of these factors. In the wastewater treatment plant itself, odorous materials form primarily under regular or irregular anaerobic conditions, and once present, they may be released into the atmosphere.

Pressure, temperature, and the concentrations of the compounds in the fluid and gaseous phases all affect the inclination to switch to the air, which is reliant on the partial pressure variation of the substances in the water/sludge and air. The transition disposition and the physical circumstances determine how quickly a fluid phase changes to a gaseous phase and vice versa (Frechen 1988).

6.4.4 High Installation Cost

Insufficient infrastructure refers to a waste water treatment plants where the actual operation scale is significantly smaller than the policy and manufacture scale. Based on our research, we discovered that this issue affects several China's wastewater treatment systems. The main causes of this issue are as follows: In some wastewater treatment plants, the actual treatment capacity exceeds the designed treatment capacity because of factors like population growth, urban economic development, and the structure of comprehensive drainage systems.

The majority of urban China's wastewater treatment systems do not have primary sedimentation tanks and have inadequate pretreatment facilities. For prolonged periods of time, wastewater treatment facilities run at high or ultra-high operational load rates, which not only puts more stress on the plants but also affects the water body's health. Due to its continuous, long-term operation, activated sludge often has low activity. As China builds an ecological civilization, the country's pollution discharge standards are getting stricter. Some wastewater treatment plants are unable to operate within the current standard, necessitating the addition of more pretreatment facilities or the adjustment of process operation parameters. Sediment accumulation at the bottom is severe, and the ratio of the combined liquor volatile suspended solid to combined liquor suspended solid is normally between 0.25 to 0.5, much lower than the average level of 0.75 (Zhang et al. 2022).

6.4.5 Environmental Footprint

One of the most significant obstacles to activated sludge treatment is the footprint it requires. Activated sludge factories require a significant amount of land and are expensive to build. Significant and expensive settling tanks and aeration basins are required for primary and secondary processes, which need significant expanses of land. The capacity of municipal wastewater treatment facilities must also be expanded due to the ongoing population growth. GHG Emissions of greenhouse gases from wastewater treatment plants can be categorized as direct or indirect. When wastewater and sludge are treated, direct, non-biogenic greenhouse gas emissions—also known as scope 1 emissions—occur (Maktabifard et al. 2022). Because the direct CO_2 emissions are biogenic, meaning they come from the breakdown of organic molecules in wastewater, they are regarded as carbon neutral. This strategy differs from others, such transportation and energy, where CO_2 derived from fossil fuels is the main cause of greenhouse gas emissions. Conversely, the high global warming potential (GWP) of non-CO_2 direct emissions, such as nitrous oxide (N_2O) and methane (CH_4), is a major cause for worry. The wastewater treatment industry is known to have a substantial impact on emissions of CH_4 and N_2O, with estimates ranging from 7 to 10% for each gas (Zhang et al. 2021). However, these percentages might vary based on site-specific factors, such as treatment methods, plant size, and regional variances in wastewater management techniques. With reference to the most recent assessment from the Intergovernmental Panel on Climate Change (IPCC) (Masson-Delmotte et al. 2021), In a century, the global warming potential (GWP) for non-fossil origin CH4, fossil origin CH4, and N2O is 27.2, 29.8, and 273, respectively.

The energy and electricity consumption by wastewater treatment plants is the source of the indirect GHG emissions, or scope 2 emissions. The transportation and chemical manufacture outside the plant is the source of the remaining indirect emissions, or scope 3 emissions (Ren et al. 2022). The carbon footprint (CF) is the total amount of greenhouse gas emissions from all perspectives, and it is stated using the carbon dioxide equivalent (CO2e) unit (Chen et al. 2020).

There are fewer research on net-zero carbon pathways in the wastewater sector than in the energy and transportation sectors (Pagilla 2022). Nonetheless, a lot of nations are starting to operate at a lower carbon level and are taking steps to decarbonize municipal water management (Marinelli et al. 2021). Near-term strategies and plans have net-zero aims, and the conversation around the condition has recently changed to reflect the goals of the Paris Agreement (Jeudy-Hugo et al. 2021). Tchobanoglous (2003) defined standards for a net-zero transition in wastewater treatment, such as the overall energy balance, greenhouse gas emissions from processes, chemical use, and sludge disposal. As a result, the net-zero paths in the wastewater sector are more complicated and varied than those in other sectors.

6.4.6 Habitat Disruption

Environmental issues of long-term sustainability such as rising soil salinity and sodium content, are of particular concern. Because of their osmotic and toxic effects, high soil sodium and salinity adsorption ratios degrade soil structure, and decrease crop yields, reduce soil permeability.

Increased runoff and soil erosion may result from the salt content, sealing, and sodium buildup caused by irrigation with treated wastewater (Halliwell et al. 2001). Anaerobic films in the treatment plant tank containing treated wastewater of poor quality or treated wastewater being drawn from them may give off a strong odor that is bothersome.

Inadequate management of treated wastewater irrigation may result in a drop in crop quality and output.

The existence of microorganisms and hazardous chemical substances in wastewater is the cause of adverse consequences (Jiménez 2006). The use of treated wastewater for irrigation presents several health concerns to humans due to the possible ingestion or contact with toxic organic compounds, heavy metals, and pathogenic microbes. Consequently, the prevailing consensus is that diseases causing microorganisms represent the biggest risk to human life. Sewage from houses contains a large amount of organic materials along with hazardous microorganisms as viruses, bacteria, and parasites. A wide range of illnesses, including diarrhea, typhoid, and malabsorption and vomiting, are linked to these infections. Any interaction between humans and the treated wastewater can be dangerous (Fattal et al. 1981). The harmful microorganism content is decreased, but not entirely eliminated, by the procedure at the treatment plants. Desalinating the treated wastewater is one way to fix this issue, but it's a costly procedure that's often not needed for agricultural use.

There are several risks to human health associated with different crops that are watered by treated wastewater (Al-Mahdawi and Al-Kenany 2009).

6.4.7 Greenhouse Gases Emissions

During the biological wastewater treatment processes of wastewater treatment, nitrous oxide, methane and carbon dioxide are released. Additionally, CO_2 is emitted during the energy generation procedure that powers the plant. Carbon dioxide emissions from energy consumption may be directly reduced by increasing the energy efficiency of treatment plants for wastewater. By increasing the energy savings, it is possible to achieve both the simultaneous reduction of environmental effects and the decrease in treatment expenses. Nitrification and denitrification are processes used to remove nitrogenous chemicals from wastewater, and they produce the N_2O released by each GHG source. Sludge and grit storage tanks provide the remaining 10% of its production, with activated sludge units accounting for 90% of it (Czepiel et al. 1995). An intermediary in biological processes like heterotrophic denitrification and nitrification is N_2O gas. It forms when hazardous chemicals or low amounts of dissolved oxygen (DO) are present in the medium during denitrification, which is carried out at low pH levels (Desloover et al. 2012). Campos et al. (2009) found out that nitrifying bacteria may create N_2O in both aerobic and anoxic environments. Methane accounted for around 1% of the chemical oxygen demand (COD) that the wastewater treatment plants received, according to research done in 2012 (Daelman et al. 2012).

By using the biogas produced during anaerobic digestion, more carbon dioxide emissions were prevented than in this case. These authors observed the main causes of methane: the centrifuge, the primary sludge thickener, the buffer tank containing the digested sludge, the storage tank containing the dewatered sludge, and the exhaust gas from the cogeneration plant.

The remaining emissions from the wastewater treatment plants biological reactors are mostly attributable to the methane dissolved in wastewater that cannot be completely wiped out by the biological system. The energy use and biological treatment plants are the two primary variables that contribute to the emission of carbon dioxide. Wastewater's organic carbon is converted to carbon dioxide or combined with biomass in the wastewater treatment plants main stream. During the burning of biogas in the sludge line, methane is eventually oxidized to carbon dioxide, where it performs anaerobic digestion and mostly changes into carbon dioxide and methane (Campos et al. 2009).

6.4.8 Eutrophication

Excessive inputs of nitrogen and phosphorus can lead to eutrophication, a situation where the water becomes too enriched in these nutrients. These nutrients are vital sources of nourishment for the marine food chain. The natural food chain is unable to keep up with the increase of phytoplankton and algae once additional nitrogen and phosphorus from industrial, municipal, and agricultural runoff enter the environment. The creatures below are unable to get light because of the algal layer that forms above the water as a result of these organisms consuming the nutrients and multiplying in number. Eventually, these creatures die and are broken down by bacteria that take up oxygen; this process frequently results in oxygen While excessive algal growth in the ocean due to eutrophication may disturb ecological balances, in a regulated setting, algae can be an effective wastewater treatment method and even a source of raw materials for biofuel (Mee 2006).

Algal are a resource that may be used in wastewater treatment to lower the amounts of nutrients like phosphorus and nitrogen in runoff because they promote algal development. The phosphorus precipitate that is produced as a result of a common wastewater treatment process is usually disposed of in landfills or used to make sludge fertilizer. This process removes a significant amount of nitrogen and phosphorus from the water, preventing eutrophication downstream (Pittman et al. 2011). Removing phosphorus, nitrogen, and heavy metals from wastewater without creating a sludge byproduct is a feasible option offered by microalgae, which carry out a comparable function without the need for chemical treatment (Cai et al. 2013). Algae production in this type of wastewater efficiently eliminates the extra nutrients since agricultural runoff from fertilizers and manure has high amounts of nitrogen and phosphorus.

6.4.9 Bacterial Imbalance

Wastewater treatment plants that use activated sludge worldwide encounter the most prevalent and intricate sludge separation issues, known as filamentous bulking and foaming, which stem from the overabundance of filamentous bacteria in the wastewater treatment plants (Liu et al. 2016) there should be a perfect equilibrium between the development of floc-forming and filamentous bacteria (Guo and Zhang 2012). Because they provide a structural foundation for strong floc formation, filamentous bacteria, when present in moderation, improve AS settle-ability in clarifiers (Burger et al. 2017). However, depending on the kind of filamentous microorganisms present, filamentous overgrowth may lead to the existence of either particles are highly packed or not highly packed. In the former case, water is trapped inside the flocs due to the growth of many filaments inside poorly consolidated flocs. In order to avoid individual flocs from compaction, the later happens when filaments emerge from the flocs and develop bridges between them (Liao et al. 2004). A number of physico-chemical parameters or modifications to process conditions that circuitously affect the AS's biochemical state can encourage the development of filamentous bacteria excessively (Schuler and Jassby 2007). According to studies employing laboratory-scale sequencing batch reactors with synthetic wastewater under low dissolved oxygen conditions, the type and amount of extra polymeric substances present during excessive filamentous bacterial growth significantly affect settling behavior (Shen et al. 2020).

When a significant amount of solids and foams build up on the surface of settling tanks and airing sinks, it is known as foaming sludge. The abiotic foaming in wastewater and sludge that results from large concentrations of surface-active substances like oil and grease must be clearly differentiated from the biotic foaming that is produced by certain species of bacteria. The presence of microbes with hydrophobic cell wall, usually caused by the existence of mycolic acid, is known to cause biological foaming, a well-known AS operating issue. The cells are able to float on the surface of the varied fluid only because of their hydrophobic qualities. While filamentous bacteria are often thought to be the cause of biological foaming It might also be caused by non-filamentous bacteria with hydrophobic walls of cells (Sam et al. 2022).

6.4.10 Disturbance of Aquatic Life

Numerous kinds of personal care products and pharmaceuticals are found in most aquatic settings, according to a worldwide assessment. Regretfully, these facts are mostly based on research done in Scandinavia, North America, Europe, and a few other locations. The scarcity of study on personal care products and pharmaceuticals and several other endocrine disturbing pollutants or contaminants in African countries, particularly in the aquatic system of South Africa, explains why such data are available in those regions of the world (Petrie et al. 2015). Although data are few, reports have shown that some personal care products and pharmaceuticals have been found in ambient waters besides wastewater treatment plant effluents at concentrations that occasionally exceed μg/L (Matongo et al. 2015). Pollutant removal from toxic and polluted aqueous matrices, in potable treatment facilities, or during wastewater treatment before discharge to surface water is yet largely unexplored and perhaps insufficiently studied.

Unfortunately, a lot of infrastructure that treat wastewater are unable to entirely remove the majority of personal care products and pharmaceuticals, hormones, and their metabolites, which leaves them as residues in surface waters, sludge, sediments, and wastewaters. This inefficiency may be caused by the widespread integrated treatment techniques applied by wastewater treatment plants for biological treatment and microbial decomposition, which breaks down complex organic matter, sludge that is activated to remove pollutants from water and wastewater. Regretfully, while most wastewater treatment plants have removal efficiencies ranging from 50 to 90%, pharmaceuticals and personal care products residues, hormones, and their metabolites are not completely eliminated (Zhang et al. 2013). Wastewater treatment plants utilize ultraviolet (UV) light or chlorine as a last step before discharge to decrease harmful bacteria that might still be residing in the effluent because of public health concerns about wastewater. Concerns across the country for the preservation of water resources for purposes other than aquatic and public health (Ongerth and Ongerth 1982).

Nutrient concentrations and ratios in streams and rivers subjected to emission can be drastically changed, and the chemical composition of wastewater treatment plants effluents frequently deviates from the surroundings of the water source. Average NO3-N and SRP absorptions were found in water samples taken from the Bush River in South Carolina's tributary and headwater areas to be both 1.6 and 0.3 mg per liter, respectively. However, NO_3-N concentrations rose significantly downstream of the watershed's two WWTPs to be 29 mg per liter and 72 miligram per liter, and SRP concentrations to 2.9 mg per liter and 4.4 miligram per liter (Andersen et al. 2004).

As a result, nutrients from wastewater treatment plants effluents are held in receiving waters at a distance greater than local discharge regions. It is widely known that through equilibrium interactions, Phosphorus that is biologically available can be bound and released into the water column by floating and benthic sediments (Ekka et al. 2006). Therefore, by releasing stored phosphorus, sediments can replenish decreasing nutrient contents (Gibson and Meyer 2007). Any management plan to lower nutrient concentrations should take this feature of nutrients in natural aquatic systems into account.

6.4.11 Low Water Quality

Irrigation with treated wastewater may bring some toxins, such chlorides, into the groundwater.

There is a risk of accumulation since the contamination enters the water supply systems, travels to the treatment plants, and then returns to the groundwater. These kinds of dangers are hard to assess and have lasting effects. Irrigation with wastewater that is treated, is not allowed near drinking water wells, under the health regulation (save for non-hazardous effluent) (Shakir et al. 2017).

6.5 Conclusion

Nitrous oxide from denitrification and nitrification processes (three percent of Nitrous oxide emissions from all sources), carbon dioxide from aerobic (oxidation) activities, and methane from anaerobic activities (three to nineteen percent of global anthropogenic methane emissions) can all be released into the atmosphere during wastewater treatment.

The primary causes of plant issues include low capacity construction, inadequate operating and maintenance budgets, and inadequate influent loading supplied to treatment facilities. Additionally, industrial wastewater treatment facilities do not have a control system. The intended values and the operational circumstances differ. The study's advice is to enforce rules pertaining to wastewater treatment user fees and to raise the level of facility utilization. Plans for managing operational issues with wastewater treatment plants should be prepared by each local government and industrial facility.

References

Agus E, Zhang L, Sedlak DL (2012) A framework for identifying characteristic odor compounds in municipal wastewater effluent. Water Res 46(18):5970–5980

Al-Mahdawi GT, Al-Kenany K (2009) Efficiency of using treated wastewater in agriculture analytical study for Al-Rustumia station for sewage clarifying–third expansion. Fifth scientific conference "strategies for spatial development and investment in Iraq", The Institute of Urban and Regional Planning for Graduate Studies, University of Baghdad

Andersen CB, Lewis GP, Sargent KA (2004) Influence of wastewater-treatment effluent on concentrations and fluxes of solutes in the Bush River, South Carolina, during extreme drought conditions. Environ Geosci 11(1):28–41

Balkema AJ, Preisig HA, Otterpohl R, Lambert FJD (2002) Indicators for the sustainability assessment of wastewater treatment systems. Urban Water 4(2):153–161

Batt AL, Kim S, Aga DS (2007) Comparison of the occurrence of antibiotics in four full-scale wastewater treatment plants with varying designs and operations. Chemosphere 68(3):428–435

Berglund F, Ebmeyer S, Kristiansson E, Larsson DGJ (2023) Evidence for wastewaters as environments where mobile antibiotic resistance genes emerge. Commun Biol 6(1):321

Bruszkiewski H, Skorupski W (1999) Air pollution caused by a concentrated liquid waste treatment plant. Chem Ekol Eng 10:979–987

Burger W, Krysiak-Baltyn K, Scales PJ, Martin GJO, Stickland AD, Gras SL (2017) The influence of protruding filamentous bacteria on floc stability and solid-liquid separation in the activated sludge process. Water Res 123:578–585

Cai T, Park SY, Li Y (2013) Nutrient recovery from wastewater streams by microalgae: status and prospects. Renew Sustain Energy Rev 19:360–369

Campos JL, Arrojo B, Vázquez-Padín JR, Mosquera-Corral A, Méndez R (2009) N 2 O production by nitrifying biomass under anoxic and aerobic conditions. Appl Biochem Biotechnol 152:189–198

Chahal C, Van Den Akker B, Young F, Franco C, Blackbeard J, Monis P (2016) Pathogen and particle associations in wastewater: significance and implications for treatment and disinfection processes. Adv Appl Microbiol 97:63–119

Chen K-H, Wang H-C, Han J-L, Liu W-Z, Cheng H-Y, Liang B, Wang A-J (2020) The application of footprints for assessing the sustainability of wastewater treatment plants: a review. J Clean Prod 277:124053

Copeland C (2001) Clean water act. Nova Science Press, Huntington NY

Czepiel P, Crill P, Harriss R (1995) Nitrous oxide emissions from municipal wastewater treatment. Environ Sci Technol 29(9):2352–2356

Daelman MRJ, van Voorthuizen EM, van Dongen UGJM, Volcke EIP, van Loosdrecht MCM (2012) Methane emission during municipal wastewater treatment. Water Res 46(11):3657–3670

Dennehy KF, Litke DW, Tate CM, Qi S, McMahon PB, Bruce BW, Kimbrough RA, Heiny JS (1998) Water quality in the South Platte River basin, Colorado, Nebraska, and Wyoming, 1992–95. US Geological Survey; US Geological Survey, Information Services [distributor]

Desloover J, Vlaeminck SE, Clauwaert P, Verstraete W, Boon N (2012) Strategies to mitigate N2O emissions from biological nitrogen removal systems. Curr Opin Biotechnol 23(3):474–482

Ekka SA, Haggard BE, Matlock MD, Chaubey I (2006) Dissolved phosphorus concentrations and sediment interactions in effluent–dominated Ozark streams. Ecol Eng 26(4):375–391

Fattal B, Shuval HI, Wax Y, Davies AM (1981) Study of enteric disease transmission associated with wastewater utilization in agricultural communities in Israel [Water source for irrigation]

Frechen FB (1988) Odour emissions and odour control at wastewater treatment plants in West Germany. Water Sci Technol 20(4–5):261–266

Gibson CA, Meyer JL (2007) Nutrient uptake in a large urban River 1. JAWRA J Am Water Resour Assoc 43(3):576–587

Gikas P (2017) Towards energy positive wastewater treatment plants. J Environ Manage 203:621–629

Guo F, Zhang T (2012) Profiling bulking and foaming bacteria in activated sludge by high throughput sequencing. Water Res 46(8):2772–2782

Halliwell DJ, Barlow KM, Nash DM (2001) A review of the effects of wastewater sodium on soil physical properties and their implications for irrigation systems. Soil Res 39(6):1259–1267

Hernando-Amado S, Coque TM, Baquero F, Martínez JL (2019) Defining and combating antibiotic resistance from one health and global health perspectives. Nat Microbiol 4(9):1432–1442

Jeudy-Hugo S, Re L. Lo, & Falduto C (2021) Understanding countries' net-zero emissions targets

Jiménez B (2006) Irrigation in developing countries using wastewater. Int Rev Environ Strateg 6(2):229–250

Liao J, Lou I, de los Reyes III, F. L. (2004) Relationship of species-specific filament levels to filamentous bulking in activated sludge. Appl Environ Microbiol 70(4):2420–2428

Liu J, Yang Q, Wang D, Li X, Zhong Y, Li X, Deng Y, Wang L, Yi K, Zeng G (2016) Enhanced dewaterability of waste activated sludge by Fe (II)-activated peroxymonosulfate oxidation. Biores Technol 206:134–140

Lorenzo-Toja Y, Vázquez-Rowe I, Amores MJ, Termes-Rifé M, Marín-Navarro D, Moreira MT, Feijoo G (2016) Benchmarking wastewater treatment plants under an eco-efficiency perspective. Sci Total Environ 566:468–479

Mahgoub ME-SM, van der Steen NP, Abu-Zeid K, Vairavamoorthy K (2010) Towards sustainability in urban water: a life cycle analysis of the urban water system of Alexandria City, Egypt. J Cleaner Prod 18(10–11):1100–1106

Maktabifard M, Awaitey A, Merta E, Haimi H, Zaborowska E, Mikola A, Mąkinia J (2022) Comprehensive evaluation of the carbon footprint components of wastewater treatment plants located in the Baltic Sea region. Sci Total Environ 806:150436

Marinelli E, Radini S, Foglia A, Lancioni N, Piasentin A, Eusebi AL, Fatone F (2021) Validation of an evidence-based methodology to support regional carbon footprint assessment and decarbonisation of wastewater treatment service in Italy. Water Res 207:117831

Marti E, Aumatell J, Godé L, Poch M, Sabater F (2004) Nutrient retention efficiency in streams receiving inputs from wastewater treatment plants. J Environ Qual 33(1):285–293

Masson-Delmotte V, Zhai P, Pirani A, Connors SL, Péan C, Berger S, Caud N, Chen Y, Goldfarb L, Gomis MI (2021) Climate change 2021: the physical science basis. Contribution of working group I to the sixth assessment report of the intergovernmental panel on climate change 2(1):2391

Matongo S, Birungi G, Moodley B, Ndungu P (2015) Pharmaceutical residues in water and sediment of Msunduzi River, kwazulu-natal, South Africa. Chemosphere 134:133–140

McConnell JB (1980) Impact of urban storm runoff on stream quality near Atlanta, Georgia (vol 1). Municipal Environmental Research Laboratory, Office of Research

Mee L (2006) Reviving dead zones. Sci Am 295(5):78–85

Metcalf L, Eddy HP (2015) Tratamento de efluentes e recuperação de recursos. McGraw Hill Brasil

Muga HE, Mihelcic JR (2008) Sustainability of wastewater treatment technologies. J Environ Manage 88(3):437–447

Muhaba S, Darun MR, Oumer AN (2024) Life cycle assessment of an industrial-scale petrochemical wastewater treatment plant. J Cleaner Prod 475. https://doi.org/10.1016/j.jclepro.2024.143673

Newhart KB, Holloway RW, Hering AS, Cath TY (2019) Data-driven performance analyses of wastewater treatment plants: a review. Water Res 157:498–513

Ongerth HJ, Ongerth JE (1982) Health consequences of wastewater reuse. Annu Rev Public Health 3(1):419–444

Pagilla KR (2022) Pathways to water sector decarbonization, carbon capture and utilization. Pathways to water sect decarbonization. Carbon Capture Util 51–66

Pagilla KR, Urgun-Demirtas M, Ramani R (2006) Low effluent nutrient technologies for wastewater treatment. Water Sci Technol 53(3):165–172

Petersen TM, Rifai HS, Suarez MP, Stein AR (2005) Bacteria loads from point and nonpoint sources in an urban watershed. J Environ Eng 131(10):1414–1425

Petrie B, Barden R, Kasprzyk-Hordern B (2015) A review on emerging contaminants in wastewaters and the environment: current knowledge, understudied areas and recommendations for future monitoring. Water Res 72:3–27

Pittman JK, Dean AP, Osundeko O (2011) The potential of sustainable algal biofuel production using wastewater resources. Biores Technol 102(1):17–25

Ren ZJ, Schnoor JL, Pagilla KR (2022) Toward a net zero circular water economy. Pathways to water sector decarbonization, carbon capture and utilization, 1

Riffat R, Husnain T (2022) Fundamentals of wastewater treatment and engineering. Crc Press

Sam T, Le Roes-Hill M, Hoosain N, Welz PJ (2022) Strategies for controlling filamentous bulking in activated sludge wastewater treatment plants: the old and the new. Water 14(20):3223

Schuler AJ, Jassby D (2007) Filament content threshold for activated sludge bulking: artifact or reality? Water Res 41(19):4349–4356

Semedo M, Song B (2023) Sediment metagenomics reveals the impacts of poultry industry wastewater on antibiotic resistance and nitrogen cycling genes in tidal creek ecosystems. Sci Total Environ 857:159496

Shakir E, Zahraw Z, Al-Obaidy AHMJ (2017) Environmental and health risks associated with reuse of wastewater for irrigation. Egypt J Pet 26(1):95–102

Shen Y, Linville JL, Urgun-Demirtas M, Mintz MM, Snyder SW (2015) An overview of biogas production and utilization at full-scale wastewater treatment plants (WWTPs) in the United States: challenges and opportunities towards energy-neutral WWTPs. Renew Sustain Energy Rev 50:346–362

Shen Y, Huang D-M, Chen Y-P, Yan P, Gao X (2020) New insight into filamentous sludge bulking during wastewater treatment: Surface characteristics and thermodynamics. Sci Total Environ 712:135795

Silvestre G, Fernández B, Bonmatí A (2015) Significance of anaerobic digestion as a source of clean energy in wastewater treatment plants. Energy Convers Manage 101:255–262

Singh P, Kansal A (2018) Energy and GHG accounting for wastewater infrastructure. Resour Conserv Recycl 128:499–507

Soares RB, Memelli MS, Roque RP, Gonçalves RF (2017) Comparative analysis of the energy consumption of different wastewater treatment plants. Int J Archit Arts Appl 3(6):79–86

Su Z, Wen D, Gu AZ, Zheng Y, Tang Y, Chen L (2023) Industrial effluents boosted antibiotic resistome risk in coastal environments. Environ Int 171:107714

Tchobanoglous G (2003) Tchobanoglous G, Burton FL, Stensel HD (2003) Wastewater engineering, treatment and reuse. McGraw-Hill Companies, New York 14:1505–1532

Vilanova MRN, Balestieri JAP (2015) Modeling of hydraulic and energy efficiency indicators for water supply systems. Renew Sustain Energy Rev 48:540–557

Wang H, Yang Y, Keller AA, Li X, Feng S, Dong Y, Li F (2016) Comparative analysis of energy intensity and carbon emissions in wastewater treatment in USA, Germany, China and South Africa. Appl Energy 184:873–881

Wang H, Xu J, Sheng L, Liu X (2018) Effect of addition of biogas slurry for anaerobic fermentation of deer manure on biogas production. Energy 165:411–418

Water UN (2019) World water Day-Factsheet. United Nations, New York, NY, USA

Ye Y, Ngo HH, Guo W, Liu Y, Chang SW, Nguyen DD, Liang H, Wang J (2018) A critical review on ammonium recovery from wastewater for sustainable wastewater management. Biores Technol 268:749–758

Zhang DQ, Gersberg RM, Hua T, Zhu J, Goyal MK, Ng WJ, Tan SK (2013) Fate of pharmaceutical compounds in hydroponic mesocosms planted with Scirpus validus. Environ Pollut 181:98–106

Zhang J, Wang H, Shao Y, Liu G, Qi L, Dang W, Yuan J, Li Y, Xia Z (2022) Analysis on common problems of the wastewater treatment industry in urban China. Chemosphere 291:132875

Zhang Q, Smith K, Zhao X, Jin X, Wang S, Shen J, Ren ZJ (2021) Greenhouse gas emissions associated with urban water infrastructure: what we have learnt from China's practice. Wiley Interdiscip Rev Water 8(4):e1529

Zhao Y, Yang QE, Zhou X, Wang F-H, Muurinen J, Virta MP, Brandt KK, Zhu Y-G (2021) Antibiotic resistome in the livestock and aquaculture industries: status and solutions. Crit Rev Environ Sci Technol 51(19):2159–2196

Zhu L, Yuan L, Shuai X-Y, Lin Z-J, Sun Y-J, Zhou Z-C, Meng L-X, Ju F, Chen H (2023) Deciphering basic and key traits of antibiotic resistome in influent and effluent of hospital wastewater treatment systems. Water Res 231:119614

Chapter 7
The Effect of the Capacity Increase in the Organized Industrial District Wastewater Treatment Plant on the Treatment Performance

İzel Kenan and Gökhan Ekrem Üstün

Abstract With the industrial revolution, the increase in the use of fossil fuels and natural resources brought environmental pollution. Organized industrial district (OID) began to be established rapidly in these processes to ensure sustainable production, shorten supply chains, support regional development, and gather industrialists in a common organization to minimize the environmental impact of production. In OIDs, common wastewater treatment plants (WWTPs) are used to treat industrial wastewater generated because of production processes and domestic wastewater generated because of employee activities. In this study, an advanced biological WWTP that treated the wastewater of a mixed OID located in Bursa, the fourth largest city of Turkey, was examined. The effect of the capacity increase in the plant (approximately 17% increase in the amount of wastewater) on the removal efficiency of pollutant parameters, the amount of energy consumed, and chemical consumption was investigated. In this context, the changes in total nitrogen (TN), total phosphorus (TP), chemical oxygen demand (COD), and total chromium 6 (Cr^{+6}), iron (Fe), and copper (Cu) parameters of wastewater and the changes in electricity consumption changes and chemical consumption were examined over a two-year period. Additionally, it was evaluated statistically whether there was a difference in removal efficiency with capacity increase.

Keywords Organized industrial district · Wastewater treatment · Removal efficiency · Capacity increase

---

İ. Kenan · G. E. Üstün (✉)
Department of Environmental Engineering, Faculty of Engineering, Bursa Uludağ University, Nilüfer, Bursa, Turkey
e-mail: gokhaneu@uludag.edu.tr

S. Souabi et al. (eds.), *Wastewater Treatment Plants*, Water Science and Technology Library 130, https://doi.org/10.1007/978-3-031-87461-1_7

7.1 Introduction

Since the beginning of the eighteenth century, humanity has gradually moved from being an agricultural society to industrialization with the increased use of machines, which has led to an increase in natural resource consumption and environmental pollution. Since the industrial revolution in world history, epidemics, wars, natural disasters, development and differentiation of production techniques, globalization and problems caused by natural resource consumption have caused economic crises. The economic downturn has caused unemployment and disruption of socio-economic order and slowed down the development of national economies in countries. Some local economies in America and Europe have resisted all these economic negativities and have quickly turned the situation to their advantage and gained economic growth momentum. Since many of these local economies operate within a similar organizational framework, they have attracted the attention of scientists. The concept of Organized Industrial District (OID) emerged because of academic Alfred Marshall's studies and observations on highly localized industries in England and Germany during the nineteenth-century industrial revolution (Belussi and Caldari 2008). An important development tool, OIDs, later played a role in the development of Northern Italy. This concept is still relevant today and is accepted by academics and politicians as an important economic development tool for developing and underdeveloped countries (Demirköprülü 2023). The concept of OID started to attract attention in Turkey in the second half of the twentieth century. Taking the industrial zone models of other developed and developing countries as an example, Turkey tried to copy the successful examples of the Marshallian/Northern Italian industrial zone model as an economic development tool. It was first supported by state authorities in the 1960s to regulate uncontrolled industrial activities and provide space for capital, and the relevant public authorities were authorized to establish industrial zones to meet the needs of small and medium-sized enterprises (SMEs) (Demirköprülü 2023; Dursun et al. 2019). The Turkish Government has determined the management system, organization, and management of OIDs in detail by making important legal regulations such as the OID Law No. 4562 in 2000 and the OID Implementation Regulation in 2002. With this law, OID management, planning, functions, and types have been determined. Governorships, municipalities, and chambers of industry play an active role in the establishment of OIDs. Today, 282 OIDs are active in Turkey and 121 OIDs are in the planning and construction phase (Anonymous 2024). Since OIDs gather industrial establishments under one organization, they are organizations that support sustainable production by preventing uncontrolled growth and resource consumption and by providing wastewater infrastructure services. OIDs produce different wastewater and pollutant combinations depending on the industries in their structures. The discharge of wastewater originating from different industrial establishments in OIDs into the environment without treatment causes ecosystem deterioration and many diseases such as cancer, mutagenic changes, etc. with pollutants reaching humans through the food chain (Balaji et al. 2005; Hesselmann et al. 2000; Rahimi et al. 2020). WWTP performance assessment is necessary to evaluate wastewater quality in treatment plants and/or to meet higher treatment requirements and to find out whether it is possible to handle higher hydraulic and organic loadings (Metcalf & Eddy 1974; Silva et al. 2014; Bashar 2018). The main purpose of this study is to evaluate the differences in the removal efficiency of conventional pollutant parameters, chemical, and energy consumption of the plant, as the flow rate of the OID WWTP changes over the years.

7.2 Materials and Methods

7.2.1 Study Area

The wastewater originating from 156 enterprises serving in different sectors within the borders of Hasanaga Organized Industrial District (HOID) is treated with coarse screen and inlet pumping station, automatic fine screen, sand and grease retainer unit, equalization tank, anaerobic tanks, aeration tanks, sedimentation tanks, return pump station, sludge dewatering unit, Parshall weir units in the WWTP with an installed capacity of 3500 m^3/day. WWTP is basically designed to remove carbon, nitrogen, and phosphorus simultaneously, like the five-stage Bardenpho process. The wastewater collected from users with pumping stations completes the physical treatment processes by passing through the coarse screens and the compactly designed fully automatic fine screen and aerated sand and grease retainer. Figure 7.1 shows the process flow diagram of HOID WWTP. The wastewater, whose flow rate and pollution concentration have stabilized during the 2 h retention period in the balancing tank, is promoted to the anaerobic tank, which is the first biological treatment step. In this process, where biological phosphorus removal, which is a complex process, takes place, microorganisms are exposed to an airless environment (Grady et al. 1999), and phosphate-accumulating microorganisms store phosphorus above their phosphorus storage capacity and take more phosphorus into their bodies in an airy environment (Özdemir 2004).

$FeCl_3$ is also dosed to the anaerobic tank to support phosphorus removal. The point where sludge is recycled in the system is the anaerobic tank. In the system that continues to flow with its own gravity, the aeration tank designed as a Carousel ditch has anaerobic, anoxic, and aerobic zones, respectively. By creating an oxygenated and partially oxygenated environment in the aeration tank, it is aimed to basically perform carbon–nitrogen removal and very little phosphorus removal. When nitrogen removal is targeted, the only method to be applied is nitrification and denitrification processes. Nitrification, which is the first step of this two-stage process, requires wastewater to be kept in aeration tanks for long periods (long aeration activated sludge system). In the sedimentation tank, the suspended activated sludge coming from the aeration tank with water is separated from the treated water by settling under the effect of gravity during the appropriate waiting period. The treated water collected with a single-sided weir system around the sedimentation tank is discharged under

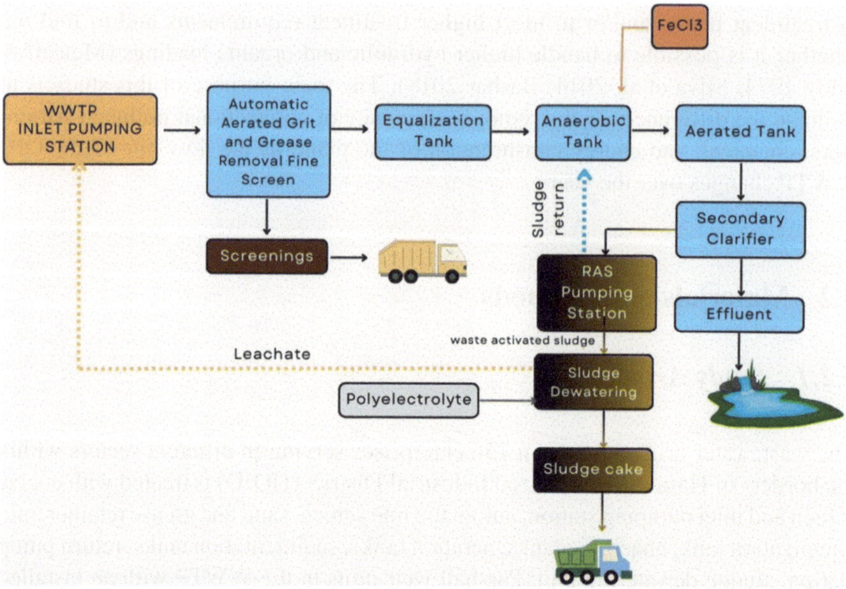

Fig. 7.1 HOID WWTP process flow diagram

Fig. 7.2 HOID WWTP location

the effect of gravity. A decanter is used to remove excess sludge from the system. The treated water flowing through the final sedimentation basin is conveyed to the Parshall Weir by its own gravity and leaves the facility from there (Fig. 7.2).

7.2.2 Chemical Consumption

Phosphorus can be found in various chemical forms in wastewater (Ren et al. 2020). In advanced biological WWTPs where domestic industrial wastewater is treated, $FeCl_3$ is dosed in the plant to treat phosphorus at peak concentration and to support phosphorus removal (Ren et al. 2020; Oehmen et al. 2007). In addition to removing phosphorus by chemical precipitation, iron salts can also be used as coagulants to improve sludge dewatering (Wei et al. 2018). In anaerobic treatment, iron salts can prevent hydrogen sulfide emissions, and iron salts are used to reduce phosphorus levels in wastewater pumping stations and WWTPs, and to control odor and corrosion (Bai et al. 2010). Various chemical forms of phosphorus generally require different treatment methods. Many methods such as chemical precipitation, sludge-based processes and ion exchange have been developed to remove inorganic phosphorus from aqueous solutions (Chu et al. 2018; de-Bashan and Bashan 2004; Galarneau and Gehr 1997; Melia et al. 2017; Morse et al. 1998; Özacar and Şengil 2002; Paul et al. 2001). The most used technique is chemical precipitation using iron or aluminum salts because this method is both easy to operate and low in cost.

7.2.3 Energy Consumption

Electrical energy is used to operate pumps, mixers, blowers, decanters, and other mechanical equipment in the WWTP and to meet office needs in the business building. Blowers and decanters constitute the largest energy consumers in the plant.

7.2.4 Monitored Wastewater Parameters

HOID WWTP is obliged to provide the discharge standards specified in Table 19 of the Water Pollution Control Regulation (WPCR) (Anonymous 2004). In this context, conventional pollution parameters and heavy metals are monitored at different periods in the laboratory located in the WWTP to monitor both the quality of the effluent and to determine the operating conditions. All parameters are carried out using analysis kits on a HACH brand DR4900 model spectrophotometer. The parameters monitored in the wastewater, the monitoring frequency, and the methods used are given in Table 7.1.

7.2.5 Statistic

Statistical evaluation of the data of the parameters examined in the study was evaluated using the IBM SPSS v16 statistical package program. Kolmogorov–Smirnov normality test and Pearson correlation test were applied to all data.

Table 7.1 Monitoring period and analysis methods of parameters measured in wastewater

Wastewater parameter	Influent	Effluent	Method
pH	Daily	Daily	Direct measurement
Conductivity	Daily	Daily	Direct measurement
Total nitrogen	Daily	Daily	LCK 350-318 (Kit)
Nitrate	Weekly	Daily	LCK 339 (Kit)
Nitrite	Weekly	Weekly	LCK 341 (Kit)
Total Kjeldahl nitrogen	Weekly	Daily	–
Total phosphorus	Daily	Daily	LCK 348-350 (Kit)
Fluoride	Weekly	Weekly	LCK 323 (Kit)
Iron	Daily/weekly	Weekly	LCK 321 (Kit)
Copper	Weekly	Weekly	LCK 329 (Kit)
Zinc	Weekly	Weekly	LCK 360 (Kit)
Total Cr^{+6}	Daily/weekly	Weekly	LCK 313 (Kit)
Sulfate	Weekly	Weekly	LCK 353 (Kit)
COD	Weekly	Weekly	LCK 314-614 (Kit)
Chloride	Weekly	Weekly	LCK 310 (Kit)

7.3 Results and Discussion

7.3.1 Change in Flow Rate According to the years of Measurement

As the number of companies and employees in the OID increases, it is expected that the wastewater flow rate will also increase. As shown in Table 7.2, the flow rate increases between 2022 and 2023 was 16.9% within the scope of the study period, while the flow rate increases in the first 6 months between 2023 and 2024 was 73.2% with the acceptance of wastewater from a different OID to the WWTP. Another operating parameter of the facility, the increase in the amount of sludge transferred back to the system, was determined to be 25.3% between 2022 and 2023 and 64.5% when the first 6 months between 2023 and 2024 were compared. The amount of sludge dewatered and removed from the system with the help of a decanter in the WWTP has changed over the years depending on the pollution load entering the facility. It was determined that there was a 26.2% increase between 2022 and 2023, and a 46% increase when the first 6 months of 2023–2024 were compared.

Table 7.2 Monthly flow rates and waste sludge in WWTP between 2022 and 2024

Month	Influent (m^3/month)			Return sludge (m^3/month)			Waste sludge (ton/month)		
	2022	2023	2024	2022	2023	2024	2022	2023	2024
January	37,451	32,111	64,310	47,846	40,102	86,508	29.3	47.9	75.4
February	39,400	38,326	63,055	43,510	43,025	82,400	44.9	52.8	84.1
March	40,409	37,447	67,532	45,751	49,929	85,177	55.4	70.9	107.4
April	37,936	32,638	52,361	48,180	46,037	79,487	48.9	55.0	66.7
May	32,289	30,162	62,105	48,416	49,884	73,967	57.3	55.8	83.0
June	32,663	32,384	42,381	52,101	54,192	58,467	55.2	50.8	69.9
July	29,922	40,157	–	46,636	73,016	–	47.2	53.8	
August	36,452	42,622	–	53,134	78,262	–	52.3	45.1	
September	33,020	49,072	–	53,933	75,051	–	42.5	59.4	
October	32,515	52,182	–	57,099	71,804	–	45.1	76.3	
November	41,276	63,058	–	52,838	75,130	–	54.3	88.7	
December	38,485	54,341	–	44,008	87,241	–	53.6	83.3	
Total	431.818	504.500	351.744	593,452	743,673	466,006	586.1	739.8	486.6
	16.9% (year)			25.3% (year)			26.2%		
	73.2% (6 month)			64.5% (6 month)			46% (6 month)		

7.3.2 *Energy-Flow-Pollution Load Change*

Energy consumption in kWh for the measured years 2022, 2023, and 2024 are given in Table 7.3. While the increase in energy consumption between 2022 and 2023 was 5.79%, when the first six months of 2023 and 2024 were compared, it was seen that there was an increase of 9.82% in energy consumption.

In order to examine the relationship between the differences in inlet flow rate and energy amount used in the WWTP according to the years in the monitored period, as a first step, it was determined that the inlet flow rate difference and energy difference data showed normal distribution according to the Kolmogorov–Smirnov test, and according to the Wilcoxon test, a significant relationship was found in the context of flow-energy differences (Asymp. Sig. $0.016 < 0.05$). To examine the relationship between the inlet flow rates and the energy consumed for a total of 30 months, Pearson correlation was used, and it was determined that $r^2 = 0.33$. To determine the effect of pollution parameters on energy consumption in the WWTP, the average pollution loads of COD, TP, and TN parameters were calculated monthly and were associated with monthly energy consumption using Pearson correlation. The r^2 values obtained between energy consumption and pollution load are COD $r^2 = 0.25$, TP $r^2 = 0.31$ and TN $r^2 = 0.30$, respectively. Sludge removal from the system is a widely used and highly effective method for biological phosphorus removal (Morse et al. 1998). When the relationship between the amount of sludge removed from the system in

Table 7.3 Monthly energy consumption in WWTP between 2022 and 2024

Month	2022 (kwh)	2023 (kwh)	2024 (kwh)
January	30.606	38.2002	36.6940
February	29.503	31.8193	35.7082
March	33.3052	35.2918	39.5450
April	33.1588	30.1936	33.2326
May	31.3452	29.2550	35.7662
June	29.6374	30.3730	33.3398
July	31.572	34.953	
August	32.6122	39.5852	
September	34.2302	36.9558	
October	36.1258	37.9796	
November	38.9708	38.73520	
December	38.8828	39.80260	
Total	399.9494	423.14425	214.2858
	5.79% (year)		9.82% (6 month)

WWTP according to years and the pollution load in terms of TP was investigated, since both data groups did not show normal distribution (WAS p value 0.001 < 0.005 and TP pollution load p value 0.002 < 0.005), the non-parametric Wilcoxon test was applied and a significant relationship was found between WAS and TP pollution load (Asymp. Sig. 0.000 < 0.005).

7.3.3 Pollution Load

The pollution loads of conventional pollution parameters COD, TP, and TN in kg/day were 68.36%; 51.44%, and 8.41% in kg/day between 2022 and 2023, respectively (Fig. 7.3). When the first six months of 2023 and 2024 were compared, it was found that the increase in pollution load was 134%; 96.22%, and 28.55% TN, respectively. The Pearson correlation values of COD, TP, and TN parameters against the total flow rate were associated as 0.84; 0.75, and 0.40, respectively. Fe, Cr^{+6}, and Cu originating from industrial wastewater were monitored for 30 months during the study period and pollution loads were calculated. The increase in heavy metal pollution loads between 2022 and 2023 was 126.1%; 70.9%, and 25.2%. When the first six months of 2023 and 2024 were compared, the increase in pollution load was determined to be 191.8%; 122.9%, and 92.9%, respectively (Fig. 7.3). The Pearson correlation values of the Fe, Cr^{+6}, and Cu parameters against the total flow rate were associated as 0.30; 0.63, and 0.35 respectively.

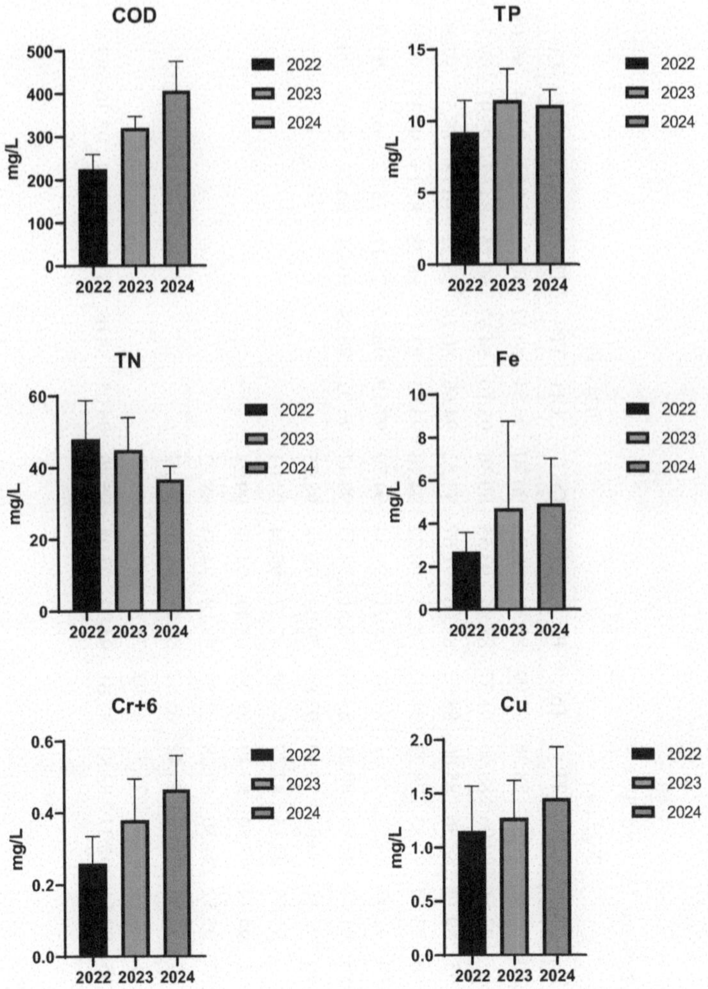

Fig. 7.3 Monthly average COD, TP, TN, and heavy metal concentrations in influent between 2022 and 2024

7.3.4 Treatment Efficiency

To follow the process in the WWTP and to monitor the effluent quality, the treatment efficiency of the facility is calculated on a parameter basis. Table 7.4 shows the monthly changes in the treatment efficiencies of COD, TP, TN, Fe, Cr^{+6}, and Cu for the years 2022, 2023, and 2024.

The changes in the annual average removal efficiencies of pollutant parameters are given in Fig. 7.4.

Table 7.4 Monthly average removal efficiencies between 2022 and 2024

Month	2022 (%)						2023 (%)						2024 (%)					
	COD	TP	TN	Fe	Cr^{+6}	Cu	COD	TP	TN	Fe	Cr^{+6}	Cu	COD	TP	TN	Fe	Cr^{+6}	Cu
January	91.83	87.02	94.26	87.02	91,30	56.64	91.53	88.38	92.09	88.38	86.36	90.70	96.54	84.12	40.53	84.12	94.44	99.16
February	89.57	87.65	83.20	87.65	9.63	93.51	88.06	88.96	91.37	88.96	93.55	89.36	95.10	84.67	60.25	84.67	95.92	99.29
March	89.27	90.30	82.51	90.30	78.95	89.74	88.36	85.06	88.93	85.06	91.67	93.52	94.66	80.18	88.22	80.18	88.57	97.56
April	90.16	90.90	91.62	90.90	80.00	93.59	92.55	78.23	91.62	78.23	90.32	94.16	96.82	87.15	88.78	87.15	93.33	96.18
May	82.03	91.24	83.66	91.24	90.00	92.00	89.51	78.21	87.37	78.21	94.34	94.70	97.19	96.12	75.11	86.12	91.07	95.60
June	84.08	90.74	84.34	90.74	92.86	91.67	94.86	80.51	83.23	80.51	94.55	99.47	96.02	96.24	91.74	96.24	96.49	99.08
July	88.55	89.89	88.64	89.89	92.86	94.44	90.91	79.60	93.85	79.60	91.18	98.75						
August	84.88	87.42	83.41	87.42	90	95.65	92.13	87.37	91.62	87.37	95.24	94.03						
September	90.43	90.97	90.29	90.97	85,71	94.44	95.50	85.49	88.90	85.49	94.59	99.24						
October	90.29	90.56	90.87	90.56	93.55	95.52	94.50	89.28	91.53	89.28	94.34	98.72						
November	88.20	91.44	87.29	91.44	91.67	94.85	94.60	86.41	65.52	86.41	91.10	98.57						
December	88.84	86.15	91.28	86.15	91,11	94.44	95.04	78.57	47.97	78.57	97.73	99.07						
Average	88.18	89.52	87.61	92.30	89.05	90.54	92.27	83.83	87.61	94.99	93.08	95.86	96.05	88.08	69.35	98.70	93.30	97.81

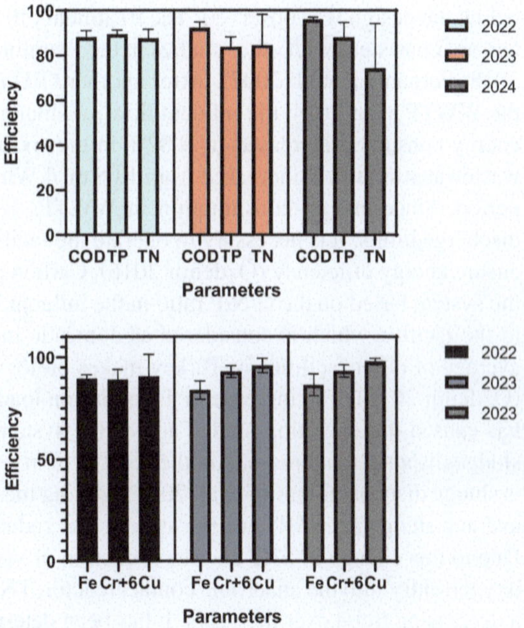

Fig. 7.4 Annual average removal efficiencies between 2022 and 2024

7.3.5 Chemical Consumption

Phosphorus removal in WWTP is carried out biologically, but $FeCl_3$ chemical is dosed at the anaerobic tank entrance to increase removal efficiency. The consumption amount of the chemical was 8.3 m^3 in 2022, 11.9 m^3 in 2023, and 4.2 m^3 in the first six months of 2024 during the study period. The increase rate in $FeCl_3$ consumption between 2022 and 2023 was 43.4%. In WWTP, excess sludge is removed from the system by pumping the sludge in the return pumping station directly to the decanter. Cationic polyelectrolyte solution is used to dewater the sludge more easily during the sludge dewatering process. This chemical, supplied in powder form, is turned into solution in the automatic polyelectrolyte preparation unit and pumped into the decanter. The consumption amount of cationic polyelectrolyte during the study period was 1125 kg in 2022, 1875 kg in 2023, and 1250 kg in the first six months of 2024.

7.4 Conclusions

In this study, the effects of the increase in influent flow rate observed in a WWTP over a 30-month period on the pollutant load, energy, and chemical amounts used were investigated. The plant is operated at approximately 60% occupancy and since it operates within the installed capacity, no decrease in pollutant removal efficiencies below the design was observed. The treatment efficiencies in all monitored parameters are consistent with other studies in the literature (Baeza et al. 2004; Bashar et al. 2018; Fernández et al. 2007). After another OID made a wastewater connection to the WWTP as of 2024, the influent flow rate increased by 73.2% and the amount of energy consumed increased by 9.82% in the six-month period. The energy cost in wastewater treatment increased from 1.08 m^3/kWh to 1.64 m^3/kWh during the study period. Since energy consumption in WWTPs is directly dependent on ensuring discharge limits, it is necessary to operate the facilities at the most optimum level to ensure energy efficiency (Özdemir 2016). Carbon plays a role as a limiting factor in the system based on the C/N/P ratio in the influent. The fact that the TP load coming to the facility, which is considered as domestic industrial wastewater, is above the average of other facilities in Turkey makes biological phosphorus removal difficult (Özdemir 2016). The increase in TP pollution load in HOID WWTP over the years has caused the operating sludge age of the system to be reduced and the need for sludge disposal to increase. Another cost item in conventional wastewater treatment is sludge disposal (Domini et al. 2022). During the 30-month monitoring period, the average sludge age (SRT) of the system was reduced from 9.66 days to 9.40 days. Due to this change in SRT and the reduction of ventilation in the system to prevent oxygen entry into the anaerobic contact reactor, TN removal efficiencies have shown a decreasing trend over the years. It has been determined that when the WWTPs are operated at flow rates within the projected flow limits, no significant difference is observed in the pollutant removal efficiencies in the WWTP even if the flow rates increase, but there is a significant increase in energy and chemical consumption amounts due to the increase in flow rates.

Acknowledgements The authors would like to thank the Hasanaga Organized District for the support and funding.

References

Anonymous (2004) National water pollution control regulation (revised), Official Gazette, Reference no: 25687, dated: 31 December 2004

Anonymous (2024) Central organization of organized industrial zones. www.osbuk.org: https://osbuk.org/view/osb/osbliste.php

Baeza JA, Gabriel D, Lafuente J (2004) Effect of internal recycle on the nitrogen removal efficiency of an anaerobic/anoxic/oxic (A2/O) wastewater treatment plant (WWTP). Process Biochem 39(11):1615–1624

Balaji V, Datta S, Bhattacharjee C (2005) Evaluation on biological treatment for industrial wastewater. Biomed Environ Sci 85(18):320–326

Bai S, Zhu Y, Zhang X, Zhang H, Gong Y (2010) Enhanced phosphorus removal from municipal wastewater by coagulation with alum and iron salts. In: 2010 4th international conference on bioinformatics and biomedical engineering. IEEE, pp 1–4

Bashar MA (2018) Biogas quality improvement using water wash and phosphorus recovery as struvite in Jones Island WWTP. Master's thesis, The University of Wisconsin-Milwaukee

Bashar R, Gungor K, Karthikeyan KG, Barak P (2018) Cost effectiveness of phosphorus removal processes in municipal wastewater treatment. Chemosphere 197:280–290

Belussi F, Caldari K (2008) At the origin of the industrial district: Alfred Marshall and the Cambridge School. Cambridge J Econ 33:335–355 (Advance Access publication 18 November 2008)

Chu YB, Li M, Liu JW, Xu W, Cheng SH, Zhao HZ (2018) Molecular insights into the mechanism and the efficiency-structure relationship of phosphorus removal by coagulation. Water Res 147:195e203

de-Bashan LE, Bashan Y (2004) Recent advances in removing phosphorus from wastewater and its future use as fertilizer (1997-2003). Water Res 38(19):4222e4246

Demirköprülü HC (2023) Buıldıng organızed ındustrıal zones ın türkiye: performance and socıal capıtal. Master's thesis, Middle East Technical University

Domini M, Abbà A, Bertanza G (2022) Analysis of the variation of costs for sewage sludge transport, recovery and disposal in Northern Italy: a recent survey (2015–2021). Water Sci Technol 85(4):1167–1175

Dursun M, Goker N, Tulek BD (2019) Efficiency analysis of organized industrial zones in Eastern Black Sea Region of Turkey. Socioecon Plann Sci 68:100659

Fernández FJ, Villaseñor J, Rodríguez L (2007) Effect of the internal recycles on the phosphorus removal efficiency of a WWTP. Ind Eng Chem Res 46(22):7300–7307

Galarneau E, Gehr R (1997) Phosphorus removal from wastewaters: experimental and theoretical support for alternative mechanisms. Water Res 31(2):328e338

Grady CPL, Daigger GT, Lim HC (1999) Biological wastewater treatment, 2nd edn, Marcel Dekker, Inc., New York

Hesselmann RPX, Von Rummell R, Resnick SM, Hany R, Zehnder AJB (2000) Anaerobic metabolism of bacteria performing enhanced biological phosphate removal. Water Res 34(14):3487–3494

Melia PM, Cundy AB, Sohi SP, Hooda PS, Busquets R (2017) Trends in the recovery of phosphorus in bioavailable forms from wastewater. Chemosphere 186:381e395

Metcalf & Eddy, United States, Environmental Protection Agency, Office of Technology Transfer (1974) Process design manual for upgrading existing wastewater treatment plants, vol 625. US Environmental Protection Agency, Technology Transfer

Morse GK, Brett SW, Guy JA, Lester JN (1998) Review: phosphorus removal and recovery technologies. Sci Total Environ 212(1):69e81

Oehmen A, Lemos PC, Carvalho G, Yuan Z, Keller J, Blackall LL, Reis MA (2007) Advances in enhanced biological phosphorus removal: from micro to macro scale. Water Res 41(11):2271–2300

Özacar M, Sengil IA (2002) Enhancing phosphate removal from wastewater by using polyelectrolytes and clay injection. J Hazard Mater 100(1e3):131e146

Özdemir B (2004) Biological phosphorus removal in wastewater treatment plants. Master's thesis, Marmara Universitesi, Turkey

Özdemir O (2016) Revısıon and energy effıcıency of malatya advanced wastewater treatment plant for sustaınable operatıon. J Eng Sci Adıyaman Univ 3(5):9–20

Paul E, Laval ML, Sperandio M (2001) Excess sludge production and costs due to phosphorus removal. Environ Technol 22(11):1363e1371

Rahimi S, Modin O, Mijakovic I (2020) Technologies for biological removal and recovery of nitrogen from wastewater. Biotechnol Adv 43:107570

Ren J, Li N, Wei H, Li A, Yang H (2020) Efficient removal of phosphorus from turbid water using chemical sedimentation by FeCl3 in conjunction with a starch-based flocculant. Water Res 170:115361

Silva C, Quadros S, Ramalho P, Alegre H, Rosa MJ (2014) Translating removal efficiencies into operational performance indices of wastewater treatment plants. Water Res 57:202–214

Wei H, Gao BQ, Ren J, Li AM, Yang H (2018) Coagulation/flocculation in dewatering of sludge: a review. Water Res 143:608e631

Chapter 8
Utilizing Performance Parameters in Wastewater Treatment Plants for the Implementation of Wastewater-Based Epidemiology

Eric Morales-Mora, Erick Centeno-Mora, Mary Luz Barrios-Hernández, and Luz Chacon

Abstract This chapter aims to describe the application framework of wastewater-based epidemiology (WWBE) as a pivotal source for public health surveillance. It seeks to bridge environmental engineering with public health, providing a comprehensive overview of WWBE's potential in monitoring and responding to epidemiological surveillance and healthcare needs. A multidisciplinary approach has been applied to develop the WWBE framework. The chapter introduces WWBE, incorporating fundamental microbiology, chemistry, surveillance, engineering, and epidemiology concepts. This approach includes monitoring techniques, agent descriptions, laboratory analyses, data analysis methods, and epidemiologic considerations. Finally, the chapter illustrates the theory through study cases and future perspectives. It is focused on the application of laboratory, surveillance, statistical, and epidemiological approaches. Mainly, it considers the critical points for a successful WWBE surveillance system. Furthermore, it illustrates how these advanced analytical techniques can significantly improve public health surveillance systems' accuracy and predictive capacity, allowing for timely and informed decision-making based on wastewater data. The chapter concludes by showing a practical application and implementation of WWBE in real-world cases, explicitly focusing on public health decision-making uses through engineering and epidemiology methods. In addition,

E. Morales-Mora (✉)
Health Technologies School, University of Costa Rica, San José, Costa Rica
e-mail: eric.morales@ucr.ac.cr

E. Centeno-Mora
Civil Engineering School, University of Costa Rica, San José, Costa Rica

M. L. Barrios-Hernández
Environmental Protection Research Center, Costa Rica Institute of Technology, Cartago, Costa Rica

L. Chacon
Health Research Institute, University of Costa Rica, San José, Costa Rica

S. Souabi et al. (eds.), *Wastewater Treatment Plants*, Water Science and Technology Library 130, https://doi.org/10.1007/978-3-031-87461-1_8

this section includes future perspectives based on the presented information to integrate WWBE into routine public health surveillance. It also highlights the need for ongoing technological advancements and interdisciplinary collaboration to maximize its efficacy and sustainability.

Keywords Wastewater-based epidemiology · Wastewater surveillance · Health surveillance · Epidemiology · Wastewater management

8.1 Introduction

Wastewater-based epidemiology (WBE) is an emerging tool for public health surveillance. It leverages the analysis of wastewater to monitor and detect various health-related indicators within a population. Contrasting traditional epidemiological methods, WBE avoids limitations such as participant recruitment and study selection biases, providing a comprehensive overview of community health. WBE is significant in public and environmental health as it enables the early detection of infectious diseases, substance use, and pollutant exposure. By analyzing the contents of wastewater, health officials can monitor the presence of pathogens, pharmaceutical residues, and biomarkers, thereby gaining insights into community health behaviors and conditions.

This chapter outlines the principles and applications of WBE, highlighting its role in modern public health surveillance. It discusses the methodology of WBE, its implementation in different sanitation models, and its potential to monitor various health-related parameters. The structure of the chapter is as follows: The fundamentals of WBE explore the theoretical basis and practical applications of WBE in health surveillance. Sewerage infrastructure and WBE application describe different sewage management models and their relevance to WBE. Techniques for pathogen surveillance in wastewater detail the methods for detecting pathogens in wastewater, including sampling and laboratory analysis. Monitoring health-related chemical compounds in wastewater discusses the surveillance of pharmaceuticals, illicit drugs, and biomarkers through WBE. Statistical analysis and machine learning applied to WBE cover using statistical and machine learning techniques to interpret WBE data. Finally, implementing the WBE program provides a step-by-step guide to establishing and operating a WBE surveillance program (Fig. 8.2).

8.2 Fundaments of WBE

The WBE has become a novel and relevant tool in public health surveillance, offering an approach without the limitations of traditional epidemiological methods such as participant recruitment, study selection bias, or effect detection (Vitale et al. 2021). By systematically analyzing wastewater contents, the WBE provides an overview

Fig. 8.1 Wastewater-based epidemiology approach illustration

of the population regarding the target agent, enabling the early detection of infectious diseases, substance use, and pollutant exposure (Maréchal et al. 2023). This method depends on collecting and examining agents present in wastewater, reflecting a community's health-related behaviors and conditions (Gatica-Bahamonde et al. 2023; Jiang et al. 2023).

The WBE focuses on the understanding that human waste carries information about community status that reflects health conditions and behaviors (Medina et al. 2022). This information includes pathogens' presence or concentration, residues from pharmaceuticals and drugs, and biomarkers for diseases and metabolic processes (Devianto and Sano 2023; Fontenele et al. 2023; Laimou-Geraniou et al. 2023; Mtetwa et al. 2023). The technique involves sampling wastewater from treatment facilities or sewer networks and analyzing these samples for specific indicators (Gonçalves et al. 2022). Then, the collected data is analyzed in terms of population health. In the case of illness detection, this approach offers a unique advantage: It captures data from a broad population, including asymptomatic individuals who do not seek medical care, thus providing a more comprehensive overview of community health status (Gonçalves et al. 2022; Vitale et al. 2021). Furthermore, WBE has been mainly applied to determine illicit drug consumption or drug potential abuse (Santana-Viera et al. 2023).

WBE applies some principles of epidemiology, such as study design, outbreak detection, health trends, and forecasting analysis, to the wastewater matrix (Amin et al. 2023; Kannan et al. 2023). It focuses on wastewater agents' data collection for community-wide assessment, enabling public health surveys to track the spread of diseases and the emergence of pathogens and assess the efficacy of interventions across entire populations (Sim et al. 2023). This method facilitates a proactive approach to public health management, allowing timely responses to emerging health threats before they escalate into more significant public health concerns (Maréchal et al. 2023) (Fig. 8.2).

8.2.1 Sewerage Infrastructure and WBE Application

Different models for managing sewage are possible, depending on the population density, infrastructure development, and sustainability considerations in a city or region (Crites and Tchobanoglous 1998; Roefs et al. 2017; Tchobanoglous et al. 2014) (Fig. 8.3).

As shown in Fig. 8.3, three main models can be depicted:

(1) **Individual or in-situ sanitation model**: suitable for low-density, such as rural or suburban areas, where each household manages its wastewater through septic systems (or other technologies) and infiltration trenches. Thus, sewage is not connected to a sewerage system for transportation to wastewater treatment plants (WWTP). It requires sufficient space for each system to operate effectively and safely.
(2) **Decentralized sanitation model**: apt for medium-density regions. Each user is connected to a simple sewer network system that transports sewage to a communal WWTP, where it is usually treated by extensive technologies such as lagoons or constructed wetlands. This model bridges the gap between the individual management of sewage and the highly centralized systems, providing a community-level solution that can be more cost-effective and easier to maintain than extensive sewer networks.
(3) **Centralized sanitation model**: the most complex system designed for high-density urban areas. An extensive and complex system of collection networks connected to trunk sewers transports the sewage to a large-scale WWTP. This system requires substantial infrastructure and investment but is necessary for the sanitation needs of densely populated cities.

WBE has been effectively implemented across various sanitation models, including on-site systems (Jakariya et al. 2022) and on a reduced or small scale

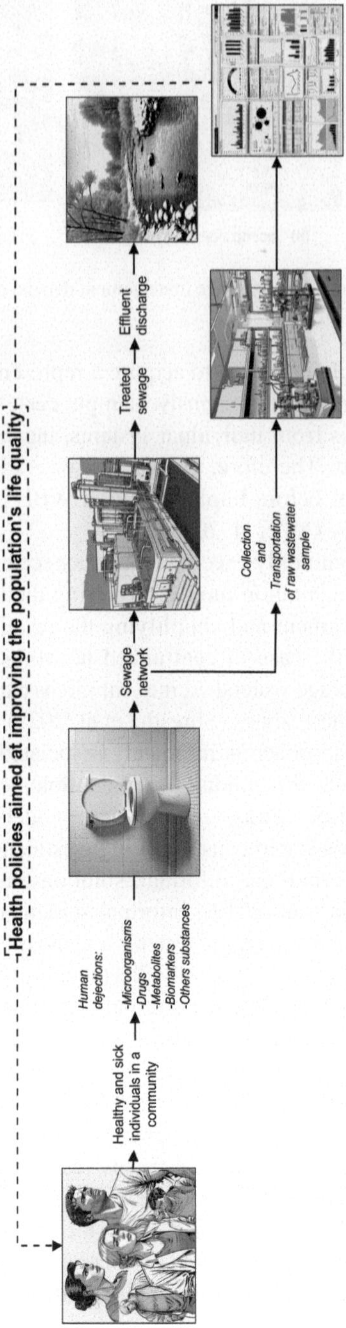

Fig. 8.2 Wastewater-based epidemiology concept

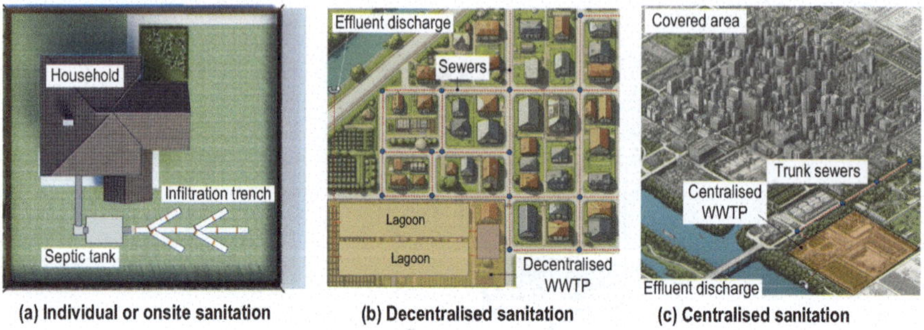

Fig. 8.3 The main management models for sewage management depend on the population density

(Schmiege et al. 2024). In such instances, to acquire a representation of the population's health status in a region, a comprehensive sample design might necessitate a substantial number of samples from individual systems, increasing the complexity and costs of this study phase. Therefore, evaluating sewerage coverage and user connectivity is very important before implementing a WBE program to affirm its representativeness and efficacy (Yu et al. 2024).

However, WBE is more commonly executed on larger scales (e.g., decentralized and, particularly, centralized sanitation models), offering the advantage of considering a broader population segment and simplifying the acquisition of a representative population sample. In the case of centralized networks, sampling at varied points and subdividing the sewage watershed into sub-areas can yield more detailed epidemiological insights for these zones (Shrestha et al. 2021).

The efficacy of the WBE approach is intimately associated with the infrastructure of sewer systems, which act as a conduit for epidemiological data. In numerous developing countries, centralized sewage management models may be infrequent, even in high-density regions, owing to constraints in investment, institutions, governance, and alternative (both formal and informal) solutions are employed (Cetrulo et al. 2019; Kelkar and Seetha Ram 2019). Informal settlements in urban areas or areas not connected to the main sewerage system, a standard in many developing countries, can impact the representativeness of WBE models in centralized sanitation systems (Shrestha et al. 2021). In such cases, alternative methods for data collection in these areas, such as using portable wastewater collection and testing systems or other types of environmental surveillance, should be considered.

Hence, the complexity and coverage of the sewer network determine the representability of the collected data (Rauch et al. 2022). Understanding the network's population density, flow rates, and industrial versus residential wastewater distribution is critical for accurate sample collection and data modeling (Lizot et al. 2023; Steenbeek et al. 2022). Effective WBE requires a collaborative effort between epidemiologists, environmental engineers, and urban planners to optimize sampling strategies and ensure that the data accurately reflects the health dynamics of the population served.

Once the wastewater system is identified, the direct relationship between human health and wastewater composition is suitable for applying the WBE approach. This relationship enables the identification and early detection of public health threats by analyzing excreted substances (Ceolotto et al. 2024). For example, the presence of SARS-CoV-2 RNA in wastewater provided early warnings of COVID-19 outbreaks, illustrating how the WBE approach can provide an early alert system for infectious diseases; this topic will be expanded later (Fontenele et al. 2023; Mtetwa et al. 2023). Also, detecting pharmaceuticals and illicit drugs offers insights into substance use patterns, while industrial chemicals can show environmental exposures and their potential health impacts (Gatica-Bahamonde et al. 2023). However, the WBE approach allows surveilling biomarkers to measure the prevalence of illnesses such as cardiovascular disease or Alzheimer's (Amin et al. 2023; Armenta-Castro et al. 2024).

The WBE approach presents numerous advantages over traditional health surveillance methods. Its non-invasive and observational method ensures that data collection does not infringe on individual privacy and captures information across all population segments (Maréchal et al. 2023). The advantage of WBE over classical epidemiological techniques, retrospective data analysis, surveys, and direct monitoring of many individuals is that it accomplishes rapid, even real-time, data collection, making WBE an early warning tool (Vitale et al. 2021). The approach's suitability in providing early warnings of disease outbreaks supports public health authorities in implementing interventions quickly, potentially reducing the spread and impact of infectious diseases and chemical substance dispersion (Lizot et al. 2023). Additionally, the WBE approach is cost-effective, allowing for surveilling large populations without the logistical and financial burdens associated with individual testing (Li et al. 2024a, b, c). This broad-spectrum monitoring capability makes the WBE approach a powerful tool for public health decision-makers, allowing the simultaneous tracking of multiple health indicators.

Integrating the WBE approach into public health policy represents a paradigm shift in epidemiological surveillance and environmental health pollution tracking. Policymakers can apply the WBE approach to inform evidence-based decision-making, guide resource allocation, and implement targeted interventions to address public health challenges (Gatica-Bahamonde et al. 2023). By promoting collaborations between governmental agencies, research institutions, and the private sector, policies can support the development of WBE infrastructure, allowing data sharing and encouraging the adoption of best survey practices (Dehghan Banadaki et al. 2023; Zhang et al. 2023). Furthermore, the approach could be a tool to ensure environmental justice, focusing on health actions in the most vulnerable populations due to the WBE results (Medina et al. 2022). This collaborative approach ensures that WBE findings are effectively translated into actionable public health strategies, enhancing the resilience of communities to emerging health threats. Even the WBE approach could support other health approaches, such as One Health, combining actions with the potential to provide more integrative health interventions to decrease the environmental and health burden from hazardous chemicals and pathogens (Kasprzyk-Hordern et al. 2023a, b). The One Health concept emphasizes that health is not exclusively a human concern. It posits that our surroundings, including animals and the environment, influence our health. As they affect our health, we also impact their health. One Health promotes the idea that to maintain our health, we must consider the health of our entire environment (Harrison et al. 2019).

Finally, the WBE approach has emerged as a critical component of modern public and environmental health surveillance, offering a window into the health of populations that is both broad and deep in terms of reactive actions to improve the population's health status. As this field evolves with the integration of new technologies such as machine learning or remote sensors, the potential for the WBE approach to inform and transform public health practice and policy grows (Jiang et al. 2023; Mao et al. 2021). The strategic integration of the WBE approach into the public health system enhances the ability to detect, respond to, and mitigate the impacts of infectious diseases, substance use, and environmental exposures, promoting a more proactive and preventive approach to public and environmental health.

8.3 Techniques for Pathogen Surveillance in Wastewater

The surveillance of pathogens through WBE offers a highly beneficial approach, providing near real-time data from a representative population sample. This method has the potential to mitigate biases often associated with data obtained from surveys, retrospective analyses, and healthcare services (Mao et al. 2020; Sims and Kasprzyk-Hordern 2020; Vitale et al. 2021). Additionally, the methodology is highly sensitive; a previous report demonstrated that individuals excreting 10^{11} viral particles per day could be detected during a wastewater sampling monitoring program, highlighting its potential as an early warning system (Hellmér et al. 2014).

The pioneering studies focused on the surveillance of poliovirus. In 1939, during an epidemic outbreak, John Pilot and his research group were the first to successfully detect poliovirus in sewage samples from Charleston, USA (Paul et al. 1940). Böttiger et al. in 1978, conducted a novel study to isolate pathogenic viruses from sewage. They successfully retrieved various enterovirus types through culture and identified them using serological techniques. Furthermore, Böttiger advocated for this methodology to detect circulating poliovirus strains in countries like Sweden, Finland, and the Netherlands, where effective vaccination programs employing inactivated polio vaccine were implemented, and minimal poliovirus circulation was expected (Böttiger 1978).

Later, the pathogen surveillance continued with other enteric viruses such as rotavirus (Bates et al. 1984), hepatitis A virus (Pintó et al. 2007), norovirus (Lodder and de Roda Husman 2005), and recently respiratory viruses such as SARS-CoV-2 (Ahmed et al. 2020; Medema et al. 2020; Schenk et al. 2023), and Influenza A (Ahmed et al. 2023). Additionally, efforts have been made regarding arboviruses such as Mengovirus detection (Cervantes-Avilés et al. 2021). All these approaches are aimed at viral early detection of pathogen circulation in the population, correlating the viral charge with the number of cases, tracking the emergence of a new variant of interest and its spread in communities, among others (Ahmed et al. 2023; Fontenele et al. 2023; Gonçalves et al. 2022).

8.3.1 Selection of Target Pathogens/Resistance Genes

Wastewater surveillance is a highly adaptable public health tool due to the prevalent shedding of pathogens in feces or urine, which subsequently find their way into sewage systems. Public health practitioners strategically prioritize pathogens for surveillance programs based on prevailing threats. Amid the COVID-19 pandemic, SARS-CoV-2 emerged as a paramount pathogen for monitoring. Previously, poliovirus garnered attention because of the potential environmental emergence of wild-type strains, particularly in regions with incomplete vaccination coverage or compromised sanitation infrastructure (Centers of Disease Control and Prevention 2023). Additionally, surveillance extends to pathogens such as hepatitis A, norovirus (Hellmér et al. 2014), and influenza A, given their propensity to spur outbreaks (Wolfe et al. 2022).

Furthermore, there is an increasing need to monitor additional biohazards, such as antibiotic resistance genes within WWTP. Recent studies highlight the significance of these genetic elements as emerging biological pollutants (Sanderson et al. 2016). Importantly, these genes can horizontally transfer among bacteria, particularly in environments abundant in nutrients like WWTP. Among the extensively studied mobile genetic elements are integrons, notably *intI-1*. Bacteria exposed to higher levels of antibiotics tend to exhibit heightened levels of these genes compared to less exposed counterparts (Gaze et al. 2005). Integrons are crucial in capturing exogenous DNA. Within the realm of antibiotic resistance, these elements facilitate the acquisition of genes capable of counteracting the efficacy of antibiotics against bacteria (Boucher et al. 2007). Although the list of genes associated with antibiotic resistance is extensive, the prevailing approach is to include in surveillance programs genes linked to resistance against specific classes of antimicrobial agents: aminoglycosides, beta-lactams, macrolides, quinolones, sulfonamides, tetracyclines, and trimethoprim (Nguyen et al. 2021).

8.3.2 Sampling Recommendations

The sampling design for pathogen surveillance will vary depending on the monitoring goals and the region's WW system. If the community possesses a WWTP, the most common approach involves sampling once weekly using a composite sample collected over 24 h, programmed with an autosampler. While considering the flow rate to obtain a flow-weighted sample is advantageous, it is not always necessary, especially when the flow is stable. Without a WWTP, the recommendation is to sample specific "hot points", such as effluent discharges from facilities where people converge or polluted surface water. Microbiological samples must be transported to laboratory facilities as soon as possible and under refrigeration conditions. For pathogenic viral analysis, immediate analysis is recommended to prevent viral degradation in the sample, as well for bacteria and parasites. However, samples for antibiotic resistance genes can be frozen at – 70 ºC and analyzed later (Gonçalves et al. 2022; Maréchal et al. 2023; Nguyen et al. 2021).

8.3.3 Laboratory Analysis

The method chosen for analyzing pathogens will vary depending on the specific microorganism. Consequently, aligning the laboratory analysis strategy with the selection of pathogens is the initial step.

8.3.3.1 Pretreatment

Wastewater is a highly contaminated sample, and achieving precise pathogen recovery necessitates the inclusion of a pretreatment step. Since different pathogens display varying degrees of sensitivity to pretreatment procedures, optimizing these steps for the specific target organisms is required to ensure maximum recovery and accurate results that align with the goals of the monitoring program.

For viral pathogens, these steps may include pasteurization, solids removal (via centrifugation, filtration, or both), or chemical addition (such as salt acids to decrease pH and PEG) (Pecson et al. 2021). Each of these methods has its advantages and disadvantages. For example, pasteurization is often included due to biosafety protocols and does not alter viral detection (Wu et al. 2020). Adding chemicals like NaCl and polyethylene glycol (PEG) can aid in precipitating organic matter and facilitating subsequent concentration steps. However, viruses may be adsorbed to the precipitated organic matter, leading to diminished recovery in the supernatant liquid. Alternatively, another approach incorporates a parallel step to recover viruses from precipitated organic matter (Masachessi et al. 2018; Sapula et al. 2021).

Finally, acidification of the sample has been effective for non-enveloped viruses. Yet, given that current monitoring may include enveloped viruses such as Influenza A and SARS-CoV-2, this approach may demonstrate low performance. Extreme pH changes can damage viral particles, leading to loss of genetic material (Barrantes et al. 2023; Sapula et al. 2021).

Other targets, such as bacteria, parasites, and antibiotic-resistance genes, typically do not require pretreatment.

8.3.3.2 Concentration

The following analysis stage involves sample concentration due to the large original sample volume (usually exceeding 100 mL) and the minimal volume needed for subsequent steps. For viral pathogens, common concentration techniques include filtration followed by precipitation (Baird et al. 2017), ultracentrifugation, or precipitation followed by centrifugation (Calgua et al. 2013).

Parasites such as *Cryptosporidium* sp. and *Giardia* sp. require a concentration step, typically achieved through flocculation or filtration followed by immunomagnetic separation for further analysis (US-EPA 2005).

Bacteria usually do not need a concentration step. However, the increasing relevance of monitoring antimicrobial resistance has led to the need to detect antimicrobial-resistant bacteria or their antibiotic-resistance genes. In this regard, sample filtration to concentrate bacterial population on the filter can be used as a basis for subsequent DNA extraction (Philo et al. 2024).

8.3.3.3 Detection and Quantification

In the case of bacteria, they are commonly detected as contamination indicators, such as *Escherichia coli* or *Salmonella* sp. Detection protocols for bacteria are well-established by international institutions such as the International Organization for Standardization (ISO), the US Environmental Protection Agency (EPA), and the American Public Health Association (APHA). These quantification methods include bacterial growing using the multiple tube technique or membrane filtration (Baird et al. 2017; World Health Organization 2006).

Since some pathogenic viruses cannot be isolated and cultured using traditional methods, which often require significant time, specialized facilities, and costly resources, molecular methods are the chosen option for their detection (Pecson et al. 2021). For this reason, viral detection commonly involves molecular analysis. The most widely used technique is PCR (polymerase chain reaction), based on the specific amplification of a target DNA region. This region can be explicitly related to a gene or microorganism. The DNA amplification requires short synthetic DNA sequences called primers, which flank the DNA region of interest and guide the DNA polymerase enzyme to amplify the target exclusively. One variation of PCR is qPCR (quantitative PCR), which allows the quantification of the target's presence using fluorescent molecules that can be detected by specialized equipment (Bustin et al. 2009).

For more comprehensive information, qPCR is the most recommended methodology. Certain government agencies or global institutions, such as the World Health Organization (WHO), may advocate using specific virus detection primers to enhance comparability across various monitoring initiatives. This advocacy is exemplified in the context of poliovirus detection (World Health Organization 2023).

However, certain pathogenic viruses that cause diarrhea, such as Rotavirus, Norovirus, Hepatitis A, or respiratory viruses like SARS-CoV-2 and Influenza A, possess RNA as their genetic material. Since qPCR analysis requires DNA as a template, an additional step to convert RNA to DNA must be undertaken for RNA microorganisms. Other considerations in this step include inhibitor removers, as the wastewater matrix can contain substances that interfere with PCR enzymatic reactions (Barrantes et al. 2023; World Health Organization 2003).

On the other hand, bacteria can also be identified by molecular methods, utilizing specific target genes or through *rRNA16S* sequencing. Similarly, this approach is applicable for detecting antibiotic resistance genes, which can also be identified by qPCR or dPCR (Sims et al. 2023) or metagenomic analysis (Miłobedzka et al. 2022).

If the monitoring program includes parasites, they can be detected by qPCR or microscopic observation using immunofluorescent staining (US EPA Office of Water 2004). Both methodologies help quantify the parasite load; however, microscopic observation can identify viable forms using adequate staining techniques.

8.3.4 Quality Control

Generating reliable results depends on selecting a robust methodology and implementing quality control measures. Each step of the process must be carefully controlled and monitored. However, there are no standard criteria regarding which quality controls should be employed. Therefore, the general guidelines are outlined in this section.

For bacterial analysis by culture, the usual quality controls are related to ensuring the growth of the correct bacteria and their accurate identification. This involves using certified reference bacterial strains to verify the suitability of the culture media, temperature, and processes used for biochemical characterization. If the certified strain shows the expected behavior, the materials and methods are suitable for recovering these microorganisms (Baird et al. 2017).

For any pathogen, when a molecular procedure is involved and includes a quantification process such as qPCR, it is recommended to use quantified DNA fragments to construct a calibration curve and interpolate the gene concentration. In 2004, the US EPA published guidelines for quality assurance in molecular methods for environmental samples (US EPA Office of Water 2004). Additionally, specific guidelines for qPCR experiments, known as the MIQE Guidelines, were developed (Bustin et al. 2009). These guidelines require qPCR validation, including parameters such as specificity, the values of slope and y-intercept, efficiency, coefficient of determination (r^2) for the calibration curve, linear dynamic range, quantification cycle (Cq) variation at the detection limit (LOD), and in multiplex reactions, the efficiency and LOD of each assay, Cq determination, number and stage of technical replicates, repeatability, reproducibility, among others (Bustin et al. 2009).

As mentioned, since wastewater samples include many chemical compounds, some of which can affect enzyme performance, internal controls for genetic material extraction and amplification are advised. One approach is to add artificial DNA material as an internal control (Hata et al. 2011), using a microorganism with well-known behavior as a proxy (Hsu et al. 2022), or spiking the sample with a microorganism regularly absent (Pecson et al. 2021).

Moreover, it is imperative to validate the recovery of pathogens whenever methodologies involving a concentration step are employed. This validation is facilitated by the inclusion of spiked internal controls, which enable the estimation of the recovery rate after the procedure. This calculation involves comparing the measured value of the spiked internal control to the expected value and applying the necessary corrections to the target value accordingly (Barrantes et al. 2023; Cervantes-Avilés et al. 2021). For parasites such as *Cryptosporidium* sp. and *Giardia* sp., spiked internal control is required (Solano Barquero et al. 2019; US-EPA 2005).

8.3.5 Results Interpretation

Once data is obtained, the next step is analyzing this information. A good practice is maintaining the original aim established during the surveillance program design. However, the most used approach is to detect changes in circulation patterns of the target and identify if these changes are related to increases or decreases in the clinical cases reported (Cervantes-Avilés et al. 2021; Chacón et al. 2021; Rauch et al. 2022; Schenk et al. 2023; Wang et al. 2023).

Other approaches can include monitoring the presence of a pathogen of particular interest such as the poliovirus (Deshpande et al. 2003; Hovi et al. 2012; World Health Organization 2023), or the presence of antibiotic resistance genes (Gupta et al. 2018; Huijbers et al. 2019; Johar et al. 2023; Nguyen et al. 2021). Another valuable approach is to use the values of the surveillance target and compare the quantity before and after the treatment, analyzing the efficiency of the plant to remove it and the associated risk with the discharge (Adegoke et al. 2018; Chacón et al. 2020; Gantzer et al. 1998; Victoria et al. 2014). Finally, surveillance results can be used for risk assessment related to work or environmental exposure to pathogens present in wastewater and to implement better safety measures (Morales-Mora et al. 2023; Zaneti et al. 2021).

8.3.6 Challenges, Limitations, and Conclusion in Pathogens Surveillance

Epidemiological surveillance of pathogens using wastewater represents a powerful and adaptable tool with immense potential for sanitation service providers and public health professionals. However, its widespread adoption faces significant challenges, particularly in low- and middle-income countries where inadequate infrastructure affects its wider implementation. The lack of comprehensive sewage systems limits the representativeness of samples collected for surveillance, thereby restricting the population coverage in the monitoring efforts (WHO and UN-Habitat 2021).

Moreover, the successful implementation of WBE hinges upon establishing standardized methodologies to ensure the reliability and comparability of data across different regions and settings. Without uniform protocols and procedures, the interpretation and comparison of surveillance findings becomes challenging, hindering the ability to draw meaningful conclusions and formulate evidence-based public health interventions (Jiang et al. 2023; Mao et al. 2020; Maréchal et al. 2023; Miłobedzka et al. 2022; Philo et al. 2024).

Despite these challenges, the utility of WBE in disease tracking and pandemic control cannot be overstated. The approach has demonstrated remarkable versatility and effectiveness, particularly during the unprecedented challenges of the SARS-CoV-2 pandemic. By providing near real-time data on pathogen circulation within communities, WW analysis has enabled public health authorities to monitor the dynamics of infectious diseases, identify emerging outbreaks, and implement targeted interventions to mitigate transmission (Iwu-Jaja et al. 2023; Randazzo et al. 2020; Schenk et al. 2023; Wang et al. 2023; Xu et al. 2023).

Addressing the challenges and limitations of WBE will require concerted efforts and investments in infrastructure development, capacity building, and research. Collaborative initiatives involving governments, international organizations, academia, and industry stakeholders are essential to overcome barriers to implementation, standardize methodologies, and enhance the scalability and sustainability of WW monitoring programs worldwide.

In conclusion, while the challenges and limitations of epidemiological surveillance using WW are significant, the potential benefits are immense. By leveraging the power of WBE, it is possible to strengthen the current capability to detect, monitor, and control infectious diseases, ultimately safeguarding public health and well-being globally.

8.4 Techniques for the monitoring of health-related chemical compounds in wastewater

Chemical surveillance is vital in monitoring pharmaceuticals, illicit drugs, and biomarkers, providing timely insights into usage trends and their environmental impacts. The surveillance aids in detecting emerging health threats, evaluating intervention effectiveness, and directing resources for targeted health initiatives (Gatica-Bahamonde et al. 2023). Additionally, it informs evidence-driven public health strategies by investigating connections between environmental exposures and health outcomes (Maal-Bared et al. 2023).

8.4.1 Selection of Target Chemical Compounds

WBE estimates drug consumption by analyzing human excretion products from municipal wastewater. Drug concentrations change during transportation from human bodies to WWTPs due to hydrolysis, absorption, repartition, and microbial decay (Gao et al. 2023). Per capita, mass drug loads can be back-calculated using substance balance equations based on their concentrations in WWTP-influent and additional correction factors (Gao et al. 2023). Considerations in selecting target compounds include their prevalence in the population and their detectability and stability in the wastewater matrices (Gatica-Bahamonde et al. 2023).

Legally manufactured, processed, and distributed substances as pharmaceutical drugs are meant for medical purposes. They include medications healthcare professionals prescribe to treat various conditions (Wen et al. 2024). Targeted pharmaceutical classes include analgesics, antibiotics, antidepressants, anticonvulsants, and other medicines for chronic conditions as hypertension (Li et al. 2024a, b, c; Wen et al. 2024). A list of common legal substances is shown in Table 8.1. Its monitoring mainly helps to detect disease outbreaks or emerging health issues early as well as misuse, overuse, or underuse of those medications (Li et al. 2024a, b, c).

Illicit drugs encompass substances unauthorized for recreational use. They are prohibited by law due to their potential for abuse and harm (Huizer et al. 2021). It also comprises legally manufactured pharmaceutical drugs abused illicitly. A range of substances, both natural and synthetic, are targeted due to their societal impact and prevalence, including cocaine, heroin, methamphetamine, and marijuana in regions where it is illegal (Shao et al. 2023), among others.

Most illicit drugs are stable, except for cocaine, heroin, and tetrahydrocannabinol, which may require substitution by their metabolites as biomarkers. Generally, the stability of biomarkers during transportation in sewers and storage significantly impacts the back-calculation of population consumption, making it challenging to identify. They can be adsorbed by solids present in water or show instability due to hydrophobicity (Wen et al. 2024). Moreover, as is the case for some antidepressants, they can experience losses during frozen storage (van Nuijs et al. 2015).

On the other hand, cardiovascular and antidiabetic drugs can exhibit high stability among common biomarkers and can be reliably used as indicators (Kasprzyk-Hordern et al. 2023a, b). Tables 8.1 and 8.2 list the most common drugs monitored in wastewater and their strongest metabolite based on their stability in sewage.

8.4.2 Sampling Techniques

As previously discussed in the techniques for the pathogen surveillance section, researchers discern spatial patterns, seasonal variations, and emerging health trends within communities by analyzing wastewater from different locations and over time. Active sampling includes grab, composite, and automatic sampler devices. Grab

Table 8.1 Some detected pharmaceuticals in wastewater influent matrix, use, and stability in sewage

Compound	Use	Stability[a]	Reference
Alprazolam	Sedative-hypnotic medication		Massano et al. (2023)
Atenolol	Cardiovascular drugs		
Bezafibrate	Cardiovascular drugs	Medium	Li et al. (2024a, b, c), Massano et al. (2023)
Carbamazepine	Anticonvulsant		
Citalopram	Antidepressant	High	Gao et al. (2023)
Diclofenac	Nonsteroidal anti-inflammatory drug		
Diazepam	Sedative and hypnotic drugs	Medium	Gao et al. (2023), Li et al. (2024a, b, c)
Fluoxetine	Antidepressant	Medium	
Ibuprofen	Pain reliever		Gao et al. (2023), Meyer et al. (2024)
Metoprolol	Cardiovascular drugs		
Naproxen	Pain reliever		Gao et al. (2023), Li et al. (2024a, b, c)
Nitrazepam	Sedative-hypnotic medication	High	
Oxazepam	Sedative-hypnotic medication	High	Gao et al. (2023)
Paracetamol (Acetaminophen)	Pain reliever		
Propranolol	Cardiovascular drugs		Gao et al. (2023), Meyer et al. (2024)
Venlafaxine	Antidepressant		

[a]Based on Gao et al. (2023)

sampling offers precise snapshots at designated moments, which help understand wastewater composition at that specific instance. Even though this method proves economic advantages for biological targets (Kmush et al. 2022), it may not be thoroughly recommended for chemical surveillance since it does not record temporal variations in highly dynamic aquatic environments (Nitti et al. 2022). In that case, composite sampling is preferable. This technique collects numerous grab samples over a predefined duration, typically 24 h, presenting an average combination of chemical concentrations (Kmush et al. 2022).

For a more extensive picture, passive sampling devices are desired. According to Nitti et al. (2022), the most applied passive sampling devices for monitoring drugs in an aquatic environment are the polar organic chemical integrative sampler (POCIS), a diffusive gradient in thin film (DGT), and Chemcatcher®. POCIS and

Table 8.2 Detected illicit drugs, their metabolite, and the stability level found in influent wastewater

Compound	Metabolite	Stability[a]	Reference
Amphetamines	–	High	
Cocaine	Cocaethylene		González-Mariño et al. (2019)
	Benzoylecgonine (BEG)	High	
Ecstasy	3,4-methylenedioxymethamphetamine (MDMA)	High	
Methylone	–	High	
Methadone	–	Medium	Baz-Lomba et al. (2016)
	EDDP	High	
Methamphetamine	Amphetamine	High	
Opioids:	Monoacetylmorphine (6-MAM)	Low	
	Heroin	Low	
	Morphine	High	
Cannabis	THC	Low	NIDA (2024)
	THC-COOH	High	
Ketamine	–	Medium	
	Nor-ketamine	High	
Synthetic cannabinoids	JWH-018, AM-2201	–	
New psychoactive substances (NPS)	Methylenedioxypyrovalerone (MDPV)	High	Baz-Lomba et al. (2016), Celma et al. (2019), Fogarty et al. (2023), Pasin et al. (2017)
	Methoxetamine (MTO)	High	
	N, *N*-Dimethylpentylone	Unknown	
	4-Methoxy-methamphetamine (PMMA)	Medium	
	4-Methoxy-amphetamine (PMA)	Unknown	
	Methiopropamine (MPA)	Medium	
	Mephedrone (MEPH)	Low	
	Methylone (METL)	Low	

[a]Based on Gao et al. (2023)

Chemcatcher® are valuable tools for assessing water-soluble (polar or hydrophilic) organic chemicals. They differ from each other in terms of design, sorbent material, and deployment characteristics (Gravell et al. 2020). For example, Chemcatcher® uses a Horizon Atlantic™ HLB-L sorbent disk as a receiving phase instead of loose powder; the sorbent disk makes the collection technique more reliable since it does not need a recovery factor to estimate sorbent losses. Regarding deployment duration, both devices are reliable over a week to a month, whereas Chemcatcher®'s kinetic sampling is usually possible in one week (Gravell et al. 2020). The optimal deployment duration may depend on the specific study objectives and the target chemicals.

Alternatively, DGT relies on the diffusion of solutes through a thin gel layer and their selective binding to a resin or binding agent, where a labile (readily available) fraction of analytes solutes accumulates in the binding agent (Challis et al. 2016). The analytes commonly used are trace metals and nutrients such as phosphate and nitrate in water (Shakallis et al. 2022). The customary membrane is removed to sample organic compounds (o-DGT), and a resin is used. Higher recovery has been reported when using HLB as a binding gel over others, such as XAD 18 and XDA-1 resins, specifically for illicit drugs (Liu et al. 2021). However, the sorbent material will depend on the target analyte and its polarity.

8.4.3 Sample Treatment

Analytical methods employed to identify prescribed and unprescribed drugs and their metabolites in WBE typically include a solid-phase extraction (SPE) pre-concentration method, which is based on the adsorption principles, partitioning, and selective retention to separate and purify analytes from complex matrices (Daniels et al. 2020). Standard procedure steps like cleaning, drying, elution, and extract preparation of the collection device are standard, with some differences in specific details such as device disassembly, drying time, extraction method, and handling of residual moisture, reflecting variations in equipment design and operational needs (Gravell et al. 2020). For example, POCIS dries the exposed HLB sorbent overnight. In contrast, Chemcatcher® dries the HLB-L disk for about 1 h on a vacuum manifold.

After that, downstream procedures are similar. Methanol is the most used solvent to elute the sequestered compounds from the sorbent material. The eluate is collected into glass vials for further processing. The eluate is concentrated (reduced in volume) using methods like vacuum evaporation to reduce the volume of methanol. The extracts are then further processed (diluted and re-solvated) before analysis to minimize matrix effects during instrumental analysis. A mobile phase (e.g., an aqueous solution containing 10% methanol, 5 mM ammonium formate, and 0.01% formic acid) can be used (Gravell et al. 2020).

8.4.4 Detection Methods

Ultrahigh-performance liquid chromatography-tandem mass spectrometry (UHPLC-MS/MS), liquid chromatography coupled to tandem mass spectrometry (LC-MS/MS), ultrahigh-performance liquid chromatography coupled to quadrupole time-of-flight mass spectrometry (UHPLC-QTOF) are standard analytical techniques for drugs identification and quantification (Celma et al. 2019; González-Mariño et al. 2019). See Table 8.3 for a detailed description of the instruments.

Table 8.3 Advanced analytical instrumentation: UHPLC-MS/MS and UHPLC-QTOF systems

Equipment	Description	Key components	Detection range limits ng/L	Reference
UHPLC-MS/MS or LC-MS/MS	The combination enables rapid separation of complex mixtures and precise quantification of analytes	UHPLC system: provides high-pressure liquid chromatography for efficient separation Tandem mass spectrometer (MS/MS): detects and quantifies analytes based on their mass-to-charge ratios	5–15	Bade et al. (2017), Celma et al. (2019), Massano et al. (2023)
UHPLC-QTOF	UHPLC-QTOF is employed for comprehensive profiling and identification of unknown compounds	UHPLC system: provides efficient chromatographic separation QTOF-HRMS: high-resolution mass spectrometer for accurate mass determination	< 5	

The main difference between identifying pharmaceuticals using UHPLC-MS/MS versus UHPLC-QTOF lies in the capabilities and performance characteristics of each setup's mass spectrometry component. UHPLC-MS/MS is, on average, used for targeted analysis of specific pharmaceuticals with high sensitivity and specificity, relying on predefined mass transitions for identification and quantification (Bade et al. 2017; Celma et al. 2019). However, matrix effects pose significant challenges to the accuracy and reliability of analytical techniques, mainly when dealing with complex samples.

In UHPLC-MS/MS, these effects stem from co-eluting compounds in the sample matrix, which can interfere with ionization or impact mass spectrometric response. Several strategies are employed to mitigate these effects, including thorough sample preparation techniques such as the SPE to remove matrix components selectively or new methods such as a direct sample injection (Li et al. 2024a, b, c). Additionally, using stable isotope-labeled internal standards aids in correcting for matrix effects during quantification and employing multiple reaction monitoring to minimize interference (Bang et al. 2023). Calibrating instruments with matrix-matched standards improves accuracy, contributing to high sensitivity and specificity in targeted analysis. Similarly, LC-MS/MS faces matrix effects due to co-eluting compounds, necessitating similar mitigation strategies. Employing multiple reaction monitoring transitions helps minimize matrix interference, offering adaptability for various applications, including pharmaceutical research and clinical analysis (Cortese et al. 2020).

UHPLC-QTOF offers versatility in targeted and non-targeted screening, leveraging high-resolution accurate mass data for comprehensive compound identification and characterization, including unknowns (Baz-Lomba et al. 2016). UHPLC-QTOF benefits from high-resolution, accurate mass data, inherently reducing matrix interference. Mitigation strategies include leveraging the QTOF mass spectrometer's high resolution for better differentiation between analytes and matrix components and data-dependent acquisition (DDA) to minimize interference by selecting precursor ions for fragmentation based on intensity. Exact mass data aids in identifying compounds even in complex matrices, rendering UHPLC-QTOF suitable for comprehensive compound identification, including unknowns (Yang et al. 2023).

8.4.5 Quality Control

Rigorous quality control measures ensure the reliability and accuracy of analytical results. Matrix spiking and internal standards help validate analytical methods. At the same time, adherence to Quality Assurance/Quality Control (QA/QC) protocols maintains data integrity. These protocols cover everything from sample collection and storage to instrument calibration and data reporting.

8.4.6 Results Interpretation

Quantitative analysis interpretations involve comparing concentrations against reference values, assessing temporal and spatial variations, and identifying significant trends over time. It considers the presence or absence of specific compounds, indicating societal behaviors, consumption patterns, and environmental pollution. Correlating chemical data with epidemiological or socio-demographic information enhances understanding of health-related trends and behavioral patterns within communities.

Back-calculating results in analytical methods like UHPLC coupled with MS/MS or QTOF involves leveraging the calibration curve established during method validation (Gao et al. 2023). First, a calibration curve must be established by analyzing standard solutions with known concentrations of target analytes. This creates a plot of instrument response versus concentration. Next, the samples are run through the UHPLC-MS/MS or UHPLC-QTOF system to record the instrument response, such as peak area or height, for the analyte in each sample.

To determine the analyte concentration in the sample, the calibration curve is used by applying the instrument response from the sample analysis. Matrix-matched or internal standards correct potential interference if matrix effects are suspected. When internal standards are used, the analyte concentration in the sample is calculated by comparing the response ratio of the analyte to that of the internal standard.

The formula for back-calculation using a calibration curve is given in the following equation (Dolan 2009; Verheyden et al. 2009):

$$\text{Concentration} = \frac{\text{Sample response} - \text{Intercept}}{\text{Slope}}$$

When an internal standard is utilized, the formula is adjusted to:

$$\text{Concentration} = \frac{\text{Sample response}}{\text{Internal Standard Response}} * \text{Internal standard concentration}$$

The sample response must fall within the calibration curve's range to maintain accuracy. Regular checks and validations of the calibration curve are necessary to ensure the ongoing accuracy and precision of the back-calculated results.

8.4.7 Challenges, Limitations, and Conclusion in Chemical Surveillance

There is a need to gather necessary information about sewage sampling, substance stability in wastewater, pharmacokinetics, and analytical method performance during WBE studies (van Nuijs et al. 2015). Regarding sampling, methods may not accurately reflect population size or consumption behaviors, introducing uncertainties in interpreting drug concentration data. Furthermore, the limited number of sampling locations further restricts the scope and representativeness of data collection efforts, undermining the reliability of consumption estimates derived from wastewater analysis. Careful design of sampling protocols is essential to collect aggregate data without targeting specific demographics or households, safeguarding privacy and anonymity while maintaining the integrity of public health interventions based on WBE findings.

Selecting stable parent compounds or biomarkers is crucial for consistent results over time and across various environments. With its myriad chemical compounds, sewage's inherent complexity introduces uncertainty that can compromise the reliability of temporal and spatial analyses. Consequently, estimations of population exposure based on biomarker concentrations in wastewater must be approached cautiously, acknowledging the potential for error. By addressing these challenges, WBE can significantly contribute to proactively managing public health concerns.

8.5 Statistical Analysis and Machine Learning Applied to WBE

WBE has become a crucial tool for monitoring public health and detecting viral pathogens like SARS-CoV-2, chemical pollutants, and biomarkers indicative of community health, as previously explained in this chapter. A central piece in the WBE approach is the statistic and data analysis strategy that allows pattern recognition, forecasting, and early alert of the possible outcomes of the agent's presence (Brookmeyer and Stroup 2004; Rauch et al. 2022). Furthermore, applying statistical analysis and machine learning to WBE data enhances the interpretability of complex datasets, predicts health trends, and effectively informs public health decisions (Lai et al. 2023).

Statistical analysis in WBE involves multiple techniques to extract meaningful information from the data collected through wastewater sampling. Essential tasks include associating wastewater data with health indicators and forecasting potential public health outcomes (Devianto and Sano 2023). This activity requires a moderate understanding of the data's statistical properties and the application of appropriate models to estimate relationships and predict trends. Techniques range from simple linear regression to more complex multivariate analyses or machine learning algorithms, adapting to the particulars of wastewater data, which is often non-linear and influenced by various external factors.

8.5.1 Data Preprocessing and Feature Selection

The initial phase in the WBE data analysis process is data integration, which combines multiple data sources to create a comprehensive dataset that supports the modeling of causality, temporality, and spatial factors (Ahrens and Pigeot 2009; Brookmeyer and Stroup 2004; Devianto and Sano 2023). For instance, in the early stages of the WBE framework for infectious diseases, it is essential to integrate data on infectious disease prevalence and wastewater infectious agents to train models that allow carried-out predictions and forecasting in later stages (Kitajima et al. 2020). This integration requires supporting the temporal and spatial scopes of the data collected from wastewater sampling and health data, making it possible to merge all the data into a single dataframe. The successful merging of these datasets focuses on using standard spatial and temporal identifiers, allowing continuous integration into a unified dataframe.

After data integration, the next step involves data depuration and exploratory data analysis. This stage is critical for ensuring the quality and integrity of the data. It includes scrutinizing the data to identify any missing, incorrect, or anomalous values (outliers) that may need to be addressed through imputation or removal, depending on the specific circumstances. Additionally, this phase involves analyzing trends and distributions within the data.

Data must be preprocessed before implementing analytical models, usually forecasts or predictive models, to ensure quality and relevance. This process includes several vital data and mathematical management steps:

Normalization: This step adjusts for population variations within the catchment area to ensure that the data accurately reflects the actual conditions of the study area. Normalization is crucial when comparing data across different scales, distributions, regions, or times, as it accounts for external factors that could skew the analysis or cause inequitable contribution of some parameter (James et al. 2013).

Outlier detection: Identifying and addressing outliers is essential for maintaining data integrity. Outliers can distort the results of the analysis and lead to incorrect conclusions. Outliers may be removed or adjusted depending on their nature and the study context (Bebchuk and Wittes 2012).

Resampling and interpolation: Managing time series data often requires resampling techniques or interpolating to ensure that the data points are evenly spaced over time. This is particularly important in WBE, where data collection intervals might vary due to logistical or technical restrictions, so statistical resampling could be helpful (Lansley and Cheshire 2016; Peng and Dominici 2008).

Feature selection: Selecting the most relevant features is critical in WBE. This involves identifying the variables most indicative of the studied outcomes, such as specific pathogens, chemical compound concentrations, or other biomarkers (Ding et al. 2022; Schenk et al. 2023). Another relevant feature could be state parameters such as temporal lags, physicochemical, weather conditions, or demographic characteristics (Brookmeyer and Stroup 2004; Li et al. 2024a, b, c; Zhang et al. 2023). The behavior of this second group of variables could vary between wastewater systems and influence the model performance in different ways. Given this variability, not all the variables are suitable in all WBE models, so selecting the proper model features is a fundamental step for a successful implementation. Effective feature selection enhances the accuracy and efficiency of the machine learning models by focusing on the most informative attributes (Ding et al. 2022). Some typical methods to perform this process are K-means clustering, principal component analysis, hierarchical clustering, or decision trees (Hoffman et al. 2008; James et al. 2013).

8.5.2 Modeling and Machine Learning in the WBE Context

Machine learning (ML) offers powerful tools for handling the complex and often non-linear relationships in WBE data. It enables prediction, classification, and pattern recognition, which is invaluable for identifying emerging health threats from wastewater datasets (Chaudhary and Naaz 2018; Schenk et al. 2023). Techniques such as neural networks, support vector machines, and ensemble methods can uncover patterns in the data that traditional statistical methods might miss, offering insights into public health dynamics (Rauch et al. 2022).

Machine learning has emerged as a transformative technology, offering sophisticated tools to analyze complex datasets derived from wastewater samples (Hino et al. 2018). This technology facilitates the identification of patterns and trends that are not readily observable through traditional analytical methods (Hino et al. 2018; James et al. 2013). The application of ML in WBE allows the development of predictive models that can forecast health outcomes, enabling proactive responses to potential public health threats (Lai et al. 2023).

Predictive modeling is crucial in forecasting health outcomes based on wastewater data analysis. Several modeling techniques could be applied in the WBE. The model selection requires a posterior performance analysis to choose the best fit, so no unique technique exists. Some popular ML models that could be used in WBE are detailed in Table 8.4 (Lai et al. 2023; Rauch et al. 2022; Xu et al. 2019).

To validate the effectiveness of the predictive models, the dataset will be divided into training and testing subsets using a data-splitting strategy. This division allows the assessment of how well the model generalizes to new data. Additionally, a technique known as boosting sampling will be utilized to enhance the predictive performance of the models (Haas et al. 2014; Kuhn 2022). Boosting sampling is an ensemble

Table 8.4 Machine learning models that could be used in the WBE context

Model	Purpose	Mechanism	Effect	References
Lasso regression	Enhances model simplicity and interpretability	Incorporates a penalty equal to the absolute value of the coefficients to the loss function	Coefficients of less important variables are reduced to zero, reducing complexity and preventing overfitting	Schenk et al. (2023)
Random forest	Increases predictive accuracy and controls overfitting	Builds multiple decision trees during training, outputs the mode or mean of predictions from these trees	Averages multiple deep decision trees to mitigate overfitting and improve generalization across varied subsets of data	Aberi et al. (2021), Dejus et al. (2023), Li et al. (2023), Vaughan et al. (2023)
Support vector machines (SVM)	Effectively separates data into classes in high-dimensional spaces	Identifies the optimal separating hyperplane that maximizes the margin between the closest members of different classes	Can model linear separations with linear kernels or complex non-linear boundaries with non-linear kernels as RBF	Aberi et al. (2021), Rauch et al. (2022), Schenk et al. (2023)

technique that focuses on building a series of models so that each subsequent model attempts to correct the errors from the previous ones (Lai et al. 2023).

Cross-validation will be implemented to ensure the robustness and applicability of the models (Xu et al. 2019). This technique involves partitioning the data into subsets, training the model on some subsets while validating the remaining ones, and then rotating which subsets are used for training and validation (Lai et al. 2023). This process helps minimize bias and gives a better insight into how the model will perform in practice.

Performance metrics such as accuracy, precision, recall, F1-score, the area under the ROC curve, Akaike Information Criterion (AIC), the root-mean-squared error of residuals (RMSE), and Bayesian Information Criterion (BIC) can be used to evaluate and compare the efficacy of the different models (Bebchuk and Wittes 2012; Kuhn 2022). These metrics provide comprehensive details on various aspects of model performance, including its error characteristics and complexity (Kuhn 2022; Sau and Bhakta 2019). The performance analysis allows the choice of the model that best fits the specific WBE context and the model's reliability.

After the model selection, improving the model performance is a good practice, mainly when focusing on boosting techniques. Boosting involves adjusting the weights of an ensemble's constituents to focus more on those instances in which previous models were misclassified (Kuhn 2022; Lai et al. 2023). This technique not only improves the accuracy but also enhances the generalization abilities of the model.

The iterative refinement of models using boosting and rigorous validation techniques ensures that the final predictive model is accurate and reliable, capable of effectively predicting health outcomes from WBE data (Rauch et al. 2022). This process exemplifies a structured approach to machine learning in WBE, ensuring the derived insights are scientifically robust and practically applicable in public health surveillance. Figure 8.4 shows the complete data processing flow.

8.5.3 Challenges, Limitations, and Conclusion

ML significantly boosts WBE but faces challenges such as data inconsistencies, complex model interpretability, and high computational demands. Integrating varied data types also presents technical challenges that can impede model performance.

Despite these barriers, ML can improve the WBE, enabling precise interventions through advanced data analysis and predictive modeling. The advantages of applying an ML approach include enhancing efficacy in public health surveillance and early alerts related to population health outcomes.

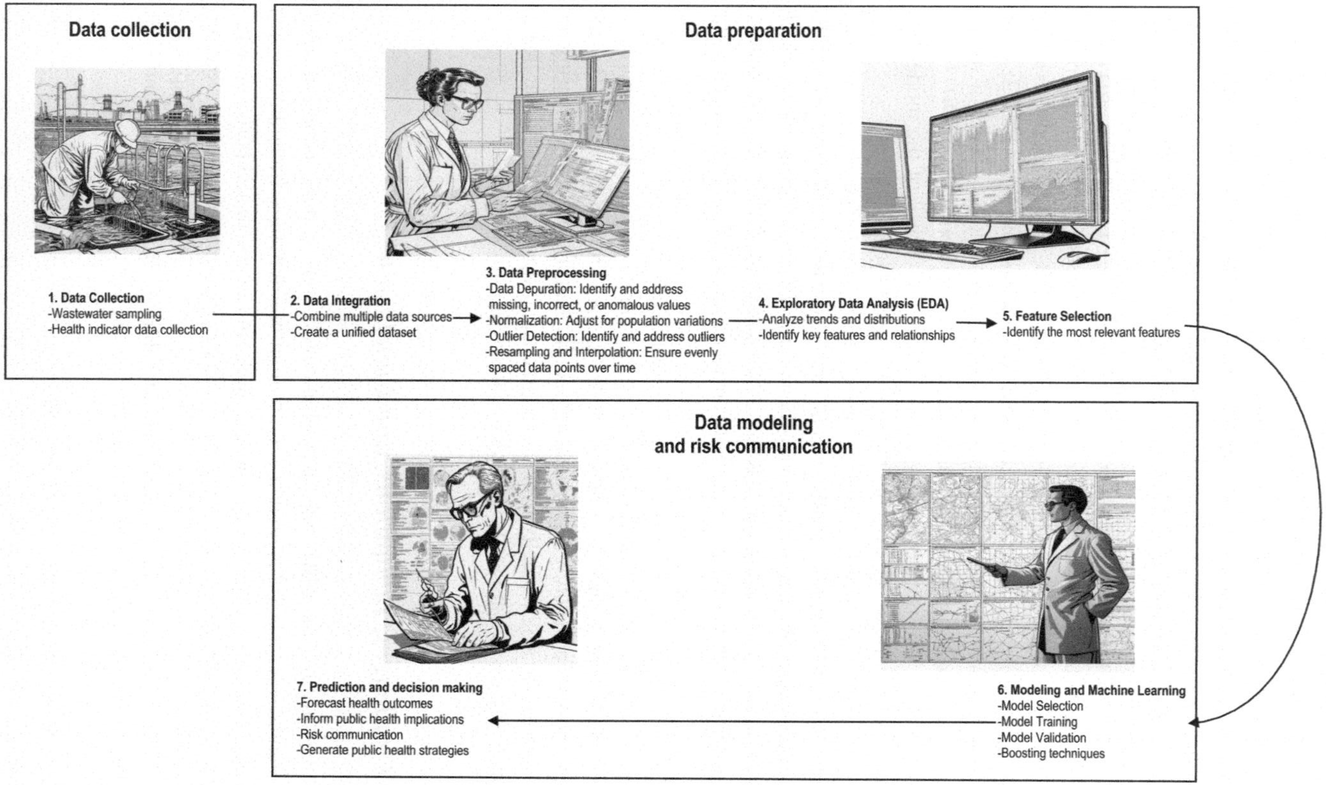

Fig. 8.4 Wastewater-based epidemiology data analysis process

8.6 Implementation of the WWBE Program

Figure 8.5 shows a flow diagram for the implementation of a WWBE program. According to Fig. 8.5, the first phase of implementing a WBE program involves evaluating the existing infrastructure and available resources. During this stage, it is essential to define the targeted biomarkers, microorganisms, or chemical compounds the program aims to monitor. Ensuring laboratories are equipped with the necessary equipment and qualified personnel are available to handle the anticipated sample load and perform the required analyses is also crucial. Additionally, assessing financial resources is vital to guarantee the program's sustainability and long-term success (National Academies of Sciences 2024).

For instance, the utility of WBE for timely outbreak detection and public health assessment was firmly established during the COVID-19 pandemic. The potential for global adoption to monitor virus transmission was highlighted. Key findings demonstrated the viability of virus recovery, RNA extraction, RT-qPCR, and deep learning (DL) model analysis. Despite these promising results, challenges remain, such as technical difficulties, infrastructure feasibility, network sampling points, population size estimation, and the integration of household WBE with advanced sensor technologies. These factors complicate the consideration and implementation of WBE projects (Chea et al. 2024).

Another crucial consideration is the target of future WBE efforts. A 2022 study in China found that illicit use of sildenafil (SIL) exceeded prescription use, particularly in North China. SIL consumption correlated with medications for gout, hypertension, nicotine, and alcohol. Wastewater samples from 33 cities were analyzed to estimate SIL consumption and its correlation with comorbidities and lifestyle factors. The study noted uncertainties in evaluating unregistered SIL and reliance on previous excretion rate studies as limitations (Shao et al. 2023). If a wide range of outcomes and behaviors correlate with the target, the potential applicability of the approach could be challenging.

Nevertheless, linking wastewater measures with specific real-world outcomes, such as infections, drug use, or illnesses, is highly recommended. For example, Alzheimer's urine biomarkers can potentially be used to diagnose the disease and integrate into WBE platforms. Alzheimer's disease biomarkers in urine and peripheral fluids could be used to calculate the biomarker abundance and integrate WBE data with clinical reports for public health assessment. However, there is still limited experimental evidence regarding endogenous biomarkers' presence, stability, and detectability in sewage systems. The low concentrations of biomarkers in wastewater samples and the complexity of wastewater as a matrix pose significant challenges and require more research to ensure a generic application (Armenta-Castro et al. 2024).

The second phase focuses on mapping and plan design. This involves collecting detailed maps of the urban sewage system, including sewer networks and the locations of WWTPs. Identifying key areas, such as hospitals, high-density residential zones, and industrial regions, ensures strategic planning and adequate monitoring (Yu et al. 2024). This phase is critical for establishing representative sampling points that

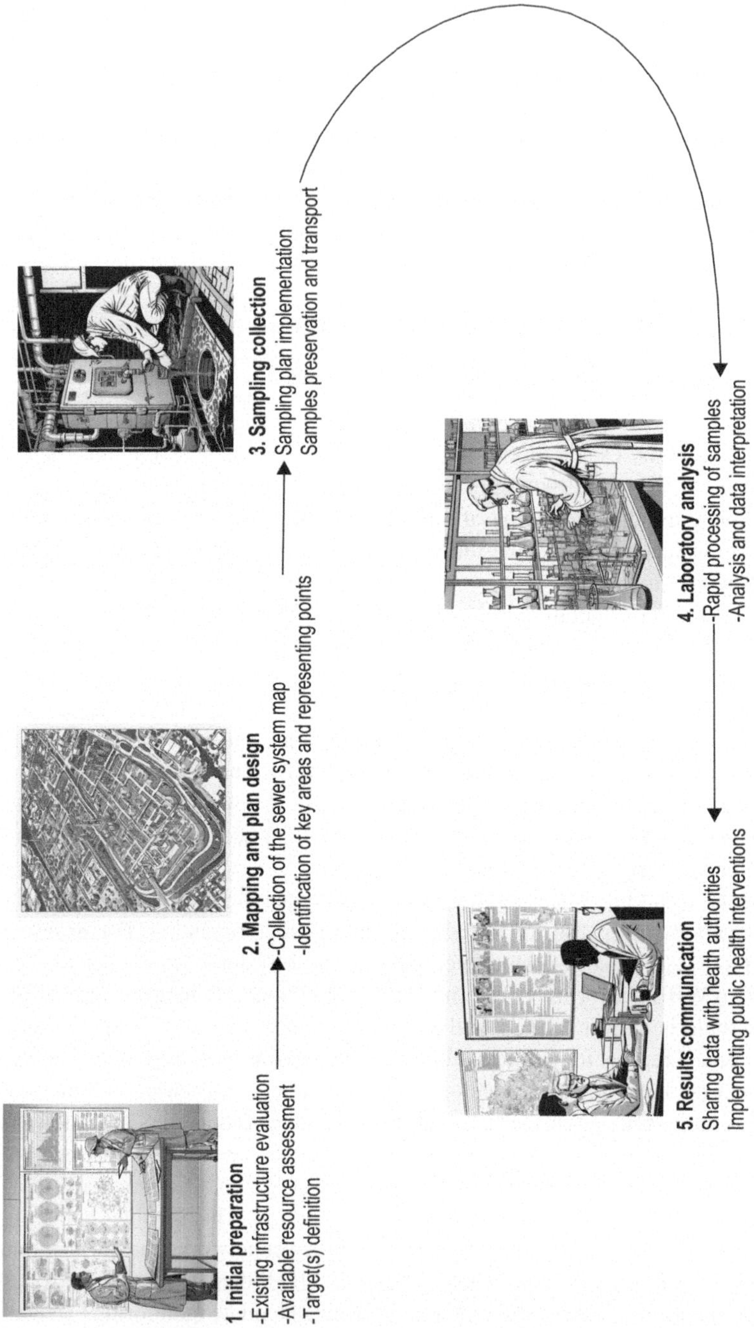

Fig. 8.5 Flow diagram for a WWBE program implementation

provide comprehensive geographic and demographic data. In this phase, identifying non-sewered areas, especially in more vulnerable areas, is important in considering alternatives for collecting samples from this population (Shrestha et al. 2021).

The spatial delimitation of the wastewater system plays a significant role in the WBE. Firstly, the spatial and temporal resolution of the sampling allows for the description of social and health events with the potential to be modified to improve the public health status (Boogaerts et al. 2021). For example, spatial-temporal cocaine use in 13 European cities identified trends of use, availability, price, and use per week. This specific analysis promotes drug use and sales control, providing spatial and temporal information about the use hotspots (Steenbeek et al. 2022). Another possible approach that considers spatial and temporal information is the pharmaceutical use analysis by suburbs and regions that considers the spatial granularity of the wastewater system (Boogaerts et al. 2021). For example, the degree of segmentation of the sample collection could be a relevant factor to identify the target hotspots that allow accurate intervention actions.

The third phase, sampling collection, entails implementing the sampling plan by selecting representative sampling points across different social strata and areas with varying vulnerabilities and demographic characteristics (Schmiege et al. 2024). Proper planning for the logistics of sample collection, preservation, and transport to the laboratory is essential. Using automatic samplers can enhance efficiency and safety, ensuring consistent and well-preserved samples for analysis (for more details, review Sects. 3.2 and 4.2).

In the fourth phase, laboratory analysis, the collected samples should be rapidly processed to ensure timely and accurate results. Standardized methods for the concentration of analytes, RNA extraction, and detection using techniques like RT-PCR are employed for posterior data analysis. For instance, in 2024, SARS-CoV-2, IFV-A, and RSV RNA were successfully detected in wastewater samples from Kathmandu, demonstrating the potential of WBE for disease surveillance in low- and middle-income countries. The methodology included viral concentration using PEG precipitation, RNA extraction, and RT-qPCR detection, with statistical analysis comparing daily clinical COVID-19 cases and SARS-CoV-2 RNA concentrations in wastewater. Challenges included recovering viral RNA of enveloped viruses, lack of a gold standard for virus concentration methods, and cost implications of new technologies. Future research should optimize and validate virus concentration methods, compare different approaches, and investigate the feasibility of WBE for monitoring other respiratory viruses. Long-term surveillance should also be conducted to monitor seasonal trends and prevent outbreaks (Raya et al. 2024).

Data analysis is the core of the WBE approach. It establishes the relationship between the outcome event (disease, infection, or drug use) and the wastewater explanatory variable (biomarker, pathogen, or drug). For example, in 2023, a time series machine learning (TSML) method was developed for varied nowcast and forecast lengths of SARS-CoV-2, demonstrating superior performance compared to ARIMAX features, typically used to forecasts, particularly in feature selection for specific prediction tasks (Lai et al. 2023).

In other cases, higher norovirus (NoV) infections in Japan were estimated in Winter–Spring compared to Summer–Autumn using WBE in 2023. One-step reverse transcription-quantitative PCR was used for NoV quantification, with estimates calculated using a mass balance equation and Monte Carlo Simulation. In this study, applying the Monte Carlo technique allowed uncertain propagation to obtain applied results for the context, preventing a potential underestimation of the cases (Shrestha et al. 2024).

Finally, the results communication and public health action phase involves sharing the analyzed data with public health authorities and stakeholders. Effective communication of the findings is crucial for informing policy decisions and public health interventions, such as targeted vaccination campaigns or public health advisories. By integrating WBE data with other surveillance sources, health authorities can make informed decisions to protect public health and respond promptly to emerging threats.

Effective risk communication prioritizes the public's right to know about health and environmental concerns, ensuring that results are conveyed to the population. The communication strategy should address the following: What information should be presented? How will the information be included? Who is the audience, and what is their prior knowledge of the topic? How can knowledge gaps in the audience be addressed? What actions can be taken to prevent or improve the current situation? (Robson et al. 2023).

WBE is a valuable tool for public health surveillance, providing comprehensive insights into community health through wastewater analysis. This chapter covered the fundamentals of WBE, its application in different sewage systems, pathogen and chemical surveillance techniques, and statistical and machine learning methods for data analysis. Implementing WBE programs requires careful planning, infrastructure evaluation, strategic mapping, efficient sample collection, and robust laboratory analysis. The COVID-19 pandemic showed WBE's potential for early outbreak detection, highlighting its importance in modern public health. Integrating WBE into public health strategies can lead to proactive health management, better environmental stewardship, and greater resilience to emerging health threats.

References

Aberi P, Arabzadeh R, Insam H, Markt R, Mayr M, Kreuzinger N, Rauch W (2021) Quest for optimal regression models in SARS-CoV-2 wastewater based epidemiology. Int J Environ Res Public Health 18(20). https://doi.org/10.3390/ijerph182010778

Adegoke AA, Amoah ID, Stenström TA, Verbyla ME, Mihelcic JR (2018) Epidemiological evidence and health risks associated with agricultural reuse of partially treated and untreated wastewater: a review. Front Public Health 6. https://doi.org/10.3389/fpubh.2018.00337

Ahmed W, Angel N, Edson J, Bibby K, Bivins A, O'Brien JW, Choi PM, Kitajima M, Simpson SL, Li J, Tscharke B, Verhagen R, Smith WJM, Zaugg J, Dierens L, Hugenholtz P, Thomas KV, Mueller JF (2020) First confirmed detection of SARS-CoV-2 in untreated wastewater in Australia: a proof of concept for the wastewater surveillance of COVID-19 in the community. Sci Total Environ 728:138764. https://doi.org/10.1016/j.scitotenv.2020.138764

Ahmed W, Bivins A, Stephens M, Metcalfe S, Smith WJM, Sirikanchana K, Kitajima M, Simpson SL (2023) Occurrence of multiple respiratory viruses in wastewater in Queensland, Australia: potential for community disease surveillance. Sci Total Environ 864:161023. https://doi.org/10.1016/j.scitotenv.2022.161023

Ahrens W, Pigeot I (2009) Handbook of epidemiology (Second). Springer Science + Business Media. https://doi.org/10.1198/jasa.2006.s89

Amin V, Bowes DA, Halden RU (2023) Systematic scoping review evaluating the potential of wastewater-based epidemiology for monitoring cardiovascular disease and cancer. Sci Total Environm 858(August 2022):160103. https://doi.org/10.1016/j.scitotenv.2022.160103

Armenta-Castro A, Núñez-Soto MT, Rodriguez-Aguillón KO, Aguayo-Acosta A, Oyervides-Muñoz MA, Snyder SA, Barceló D, Saththasivam J, Lawler J, Sosa-Hernández JE, Parra-Saldívar R (2024) Urine biomarkers for Alzheimer's disease: a new opportunity for wastewater-based epidemiology? Environ Int 184(January):108462. https://doi.org/10.1016/j.envint.2024.108462

Bade R, Bijlsma L, Sancho JV, Baz-Lomba JA, Castiglioni S, Castrignanò E, Causanilles A, Gracia-Lor E, Kasprzyk-Hordern B, Kinyua J, McCall AK, van Nuijs ALN, Ort C, Plósz BG, Ramin P, Rousis NI, Ryu Y, Thomas KV, de Voogt P, Hernández F et al (2017) Liquid chromatography-tandem mass spectrometry determination of synthetic cathinones and phenethylamines in influent wastewater of eight European cities. Chemosphere 168(November 2016):1032–1041. https://doi.org/10.1016/j.chemosphere.2016.10.107

Baird R, Eaton A, Rice E, Bridgewater L (2017) Standard methods for the examination of water and wastewater In: Bair R, Eaton A, Rice E, Bridgewater L (eds), 23rd edn. American Public Health Association (APHA)

Bang HY, Kim YK, Kim H, Baek EJ, Na T, Sim KS, Kim HJ (2023) Ultra-high-performance liquid chromatography quadrupole time-of-flight mass spectrometry for simultaneous pesticide analysis and method validation in sweet pepper. Molecules 28(14). https://doi.org/10.3390/molecules28145589

Barrantes K, Chacón-Jiménez L, Rivera-Montero L, Segura-Villalta A, Badilla-Aguilar A, Alfaro-Arrieta E, Rivera-Navarro P, Méndez-Chacón E, Santamaría-Ulloa C (2023) Challenges detecting SARS-CoV-2 in Costa Rican domestic wastewater and river water. Sci Total Environ 897:165393. https://doi.org/10.1016/j.scitotenv.2023.165393

Bates J, Goddard MR, Butler M (1984) The detection of rotaviruses in products of wastewater treatment. J Hyg 93(3):639–643. https://doi.org/10.1017/S0022172400065219

Baz-Lomba JA, Reid MJ, Thomas KV (2016) Target and suspect screening of psychoactive substances in sewage-based samples by UHPLC-QTOF. Anal Chim Acta 914:81–90

Bebchuk J, Wittes J (2012) Fundamentals of biostatistics. In: Clinical trials in neurology: design, conduct, analysis. https://doi.org/10.1017/CBO9781139032445.005

Boogaerts T, Ahmed F, Choi PM, Tscharke B, O'Brien J, De Loof H, Gao J, Thai P, Thomas K, Mueller JF, Hall W, Covaci A, van Nuijs ALN (2021) Current and future perspectives for wastewater-based epidemiology as a monitoring tool for pharmaceutical use. Sci Total Environ 789:148047. https://doi.org/10.1016/j.scitotenv.2021.148047

Böttiger M (1978) Isolation of viruses from sewage, with special regard to poliovirus: an evaluation of the sensitivity of three tissue culture methods. Bull World Health Organ 56(6):937–943. http://www.ncbi.nlm.nih.gov/pubmed/216501

Boucher Y, Labbate M, Koenig JE, Stokes HW (2007) Integrons: mobilizable platforms that promote genetic diversity in bacteria. Trends Microbiol 15(7):301–309. https://doi.org/10.1016/j.tim.2007.05.004

Brookmeyer R, Stroup D (2004) Monitoring the health of populations. Oxford University

Bustin SA, Benes V, Garson JA, Hellemans J, Huggett J, Kubista M, Mueller R, Nolan T, Pfaffl MW, Shipley GL, Vandesompele J, Wittwer CT (2009) The MIQE guidelines: minimum information for publication of quantitative real-time PCR experiments. Clin Chem 55(4):611–622. https://doi.org/10.1373/clinchem.2008.112797

Calgua B, Rodriguez-Manzano J, Hundesa A, Suñen E, Calvo M, Bofill-Mas S, Girones R (2013) New methods for the concentration of viruses from urban sewage using quantitative PCR. J Virol Methods 187(2):215–221. https://doi.org/10.1016/j.jviromet.2012.10.012

Celma A, Sancho JV, Salgueiro-González N, Castiglioni S, Zuccato E, Hernández F, Bijlsma L (2019) Simultaneous determination of new psychoactive substances and illicit drugs in sewage: potential of micro-liquid chromatography tandem mass spectrometry in wastewater-based epidemiology. J Chromatogr A 1602:300–309. https://doi.org/10.1016/j.chroma.2019.05.051

Centers of Disease Control and Prevention (2023) Considerationns for wastewater testing for Poliovirus in the United States. https://www.cdc.gov/polio/us/lab-testing/wastewater-testing/considerations.html#print

Ceolotto N, Dollamore P, Hold A, Balne B, Jagadeesan KK, Standerwick R, Robertson M, Barden R, Kasprzyk-Hordern B (2024) A new wastewater-based epidemiology workflow to estimate community wide non-communicable disease prevalence using pharmaceutical proxy data. J Hazard Mater 461(August 2023):132645. https://doi.org/10.1016/j.jhazmat.2023.132645

Cervantes-Avilés P, Moreno-Andrade I, Carrillo-Reyes J (2021) Approaches applied to detect SARS-CoV-2 in wastewater and perspectives post-COVID-19. J Water Process Eng 40:101947. https://doi.org/10.1016/j.jwpe.2021.101947

Cetrulo TB, Marques RC, Malheiros TF (2019) An analytical review of the efficiency of water and sanitation utilities in developing countries. Water Res 161:372–380. https://doi.org/10.1016/j.watres.2019.05.044

Chacón L, Barrantes K, Santamaría-Ulloa C, Solano M, Reyes L, Taylor L, Valiente C, Symonds EM, Achí R (2020) A somatic coliphage threshold approach to improve the management of activated sludge wastewater treatment plant effluents in resource-limited regions. Appl Environ Microbiol 86(17):1–16. https://doi.org/10.1128/AEM.00616-20

Chacón L, Morales E, Valiente C, Reyes L, Barrantes K (2021) Wastewater-based epidemiology of enteric viruses and surveillance of acute gastrointestinal illness outbreaks in a resource-limited region. Am J Trop Med Hyg 105(4):1007–1015. https://doi.org/10.4269/ajtmh.21-0050

Challis JK, Hanson ML, Wong CS (2016) Development and calibration of an organic-diffusive gradients in thin films aquatic passive sampler for a diverse suite of polar organic contaminants. Anal Chem 88(21):10583–10591. https://doi.org/10.1021/acs.analchem.6b02749

Chaudhary S, Naaz S (2018) Use of big data in computational epidemiology for public health surveillance. In: 2017 international conference on computing and communication technologies for smart nation, IC3TSN 2017, 2017-Octob, pp 150–155. https://doi.org/10.1109/IC3TSN.2017.8284467

Chea TG, Quoie GDS, Yang Y (2024) Application and challenge of wastewater-based epidemiology for the COVID-19 epidemic control in countries at different developing levels. J Water Process Eng 58(February):104911. https://doi.org/10.1016/j.jwpe.2024.104911

Cortese M, Gigliobianco MR, Magnoni F, Censi R, Martino PD (2020) Molecules compensate for or minimize matrix effects ? strategies for overcoming matrix effects in liquid. Molecules 25:3047

Crites RW, Tchobanoglous G (1998) Small and decentralized wastewater management systems (Series in). McGrawHill

Daniels KD, Park M, Huang Z, Jia A, Flores GS, Lee HK, Snyder SA (2020) A review of extraction methods for the analysis of pharmaceuticals in environmental waters. Crit Rev Environ Sci Technol 50(21):2271–2299

de Lizot LF, Bastiani MF, Hahn RZ, Meireles YF, Freitas M, Bondan AP, do Nascimento CA, Quevedo DM, Linden R (2023) Risk assessment of a Brazilian urban population due to the exposure to pyrethroid insecticides during the COVID-19 pandemic using wastewater-based epidemiology. Chemosphere 345(October):140526. https://doi.org/10.1016/j.chemosphere.2023.140526

Dehghan Banadaki M, Torabi S, Strike WD, Noble A, Keck JW, Berry SM (2023) Improving wastewater-based epidemiology performance through streamlined automation. J Environ Chem Eng 11(2):109595. https://doi.org/10.1016/j.jece.2023.109595

Dejus B, Gudra D, Dejus S, Ustinova M, Roga A, Strods M, Kibilds J, Boikmanis G, Ortlova K, Krivko L, Birzniece L, Skinderskis E, Berzins A, Fridmanis D, Juhna T (2023) Science of the total environment wastewater-based prediction of COVID-19 cases using a random forest algorithm with strain prevalence data: a case study of fi ve municipalities in Latvia. Sci Total Environ 891(May):164519. https://doi.org/10.1016/j.scitotenv.2023.164519

Deshpande JM, Shetty SJ, Siddiqui ZA (2003) Environmental surveillance system to track wild poliovirus transmission. Appl Environ Microbiol 69(5):2919–2927. https://doi.org/10.1128/AEM.69.5.2919-2927.2003

Devianto LA, Sano D (2023) Systematic review and meta-analysis of human health-related protein markers for realizing real-time wastewater-based epidemiology. Sci Total Environ 897(June):165304. https://doi.org/10.1016/j.scitotenv.2023.165304

Ding X, Yang F, Ma F (2022) An efficient model selection for linear discriminant function-based recursive feature elimination. J Biomed Inform 129(April):104070. https://doi.org/10.1016/j.jbi.2022.104070

Dolan J (2009) Calibration curves, Part 4: choosing the appropriate model

Fogarty M, Krotulski A, Papsun D, Walton S, Lamb M, Truver M, Chronister C, Goldberger B, Logan B (2023) N,N-Dimethylpentylone (dipentylone)—a new synthetic cathinone identified in a postmortem forensic toxicology case series. J Anal Toxicol 47(8):753–761. https://doi.org/10.1093/jat/bkad037

Fontenele RS, Yang Y, Driver EM, Magge A, Kraberger S, Custer JM, Dufault-Thompson K, Cox E, Newell ME, Varsani A, Halden RU, Scotch M, Jiang X (2023) Wastewater surveillance uncovers regional diversity and dynamics of SARS-CoV-2 variants across nine states in the USA. Sci Total Environ 877(January):162862. https://doi.org/10.1016/j.scitotenv.2023.162862

Gantzer C, Maul A, Audic JM, Schwartzbrod L (1998) Detection of infectious enteroviruses, enterovirus genomes, somatic coliphages, and bacteroides fragilis phages in treated wastewater. Appl Environ Microbiol 64(11):4307–4312. https://doi.org/10.1128/AEM.64.11.4307-4312.1998

Gao Z, Li P, Lin H, Lin W, Ren Y (2023) Biomarker selection strategies based on compound stability in wastewater-based epidemiology. Environ Sci Pollut Res 30(3):5516–5529. https://doi.org/10.1007/s11356-022-24268-y

Gatica-Bahamonde G, Godynyuk EA, Neicun J, Roberts E, Tangerli MM, van Kessel R, Czabanowska K, Humphreys K, Roman-Urrestarazu A (2023) Analysing the use trends of new psychoactive substances using wastewater-based epidemiology in Europe: a systematic review. Emerg Trends Drugs Addictions Health 3(August):100053. https://doi.org/10.1016/j.etdah.2023.100053

Gaze WH, Abdouslam N, Hawkey PM, Wellington EMH (2005) Incidence of class 1 integrons in a quaternary ammonium compound-polluted environment. Antimicrob Agents Chemother 49(5):1802–1807. https://doi.org/10.1128/AAC.49.5.1802-1807.2005

Gonçalves J, Torres-Franco A, Rodriguéz E, Diaz I, Koritnik T, da Silva PG, Mesquita JR, Trkov M, Paragi M, Muñoz R, García-Encina PA (2022) Centralized and decentralized wastewater-based epidemiology to infer COVID-19 transmission—a brief review. One Health 15(March). https://doi.org/10.1016/j.onehlt.2022.100405

González-Mariño I, Estévez-Danta A, Rodil R, Da Silva KM, Sodré FF, Cela R, Quintana JB (2019) Profiling cocaine residues and pyrolytic products in wastewater by mixed-mode liquid chromatography–tandem mass spectrometry. Drug Test Anal 11(7):1018–1027. https://doi.org/10.1002/dta.2590

Gravell A, Fones GR, Greenwood R, Mills GA (2020) Detection of pharmaceuticals in wastewater effluents—a comparison of the performance of Chemcatcher® and polar organic compound integrative sampler. Environ Sci Pollut Res 27(22):27995–28005. https://doi.org/10.1007/s11356-020-09077-5

Gupta SK, Shin H, Han D, Hur H-G, Unno T (2018) Metagenomic analysis reveals the prevalence and persistence of antibiotic- and heavy metal-resistance genes in wastewater treatment plant. J Microbiol 56(6):408–415. https://doi.org/10.1007/s12275-018-8195-z

Haas CN, Rose JB, Gerba CP (2014) Quantitative microbial risk assessment, 2nd edn. Wiley

Harrison S, Kivuti-Bitok L, Macmillan A, Priest P (2019) EcoHealth and one health: a theory-focused review in response to calls for convergence. Environ Int 132(July):105058. https://doi.org/10.1016/j.envint.2019.105058

Hata A, Katayama H, Kitajima M, Visvanathan C, Nol C, Furumai H (2011) Validation of internal controls for extraction and amplification of nucleic acids from enteric viruses in water samples. Appl Environ Microbiol 77(13):4336–4343. https://doi.org/10.1128/AEM.00077-11

Hellmér M, Paxéus N, Magnius L, Enache L, Arnholm B, Johansson A, Bergström T, Norder H (2014) Detection of pathogenic viruses in sewage provided early warnings of hepatitis A virus and norovirus outbreaks. Appl Environ Microbiol 80(21):6771–6781. https://doi.org/10.1128/AEM.01981-14

Hino M, Benami E, Brooks N (2018) Machine learning for environmental monitoring. Nat Sustain 1(10):583–588. https://doi.org/10.1038/s41893-018-0142-9

Hoffman FM, Hargrove WW, Mills RT, Mahajan S, Erickson DJ, Oglesby RJ (2008) Multivariate spatio-temporal clustering (MSTC) as a data mining tool for environmental applications. In: Proceedings of the IEMSs 4th biennial meeting—international congress on environmental modelling and software: integrating sciences and information technology for environmental assessment and decision making, IEMSs 2008, vol 3, June 2014, pp 1774–1781

Hovi T, Shulman LM, Van der Avoort H, Deshpande J, Roivainen M, De Gourville EM (2012) Role of environmental poliovirus surveillance in global polio eradication and beyond. Epidemiol Infect 140(1):1–13. https://doi.org/10.1017/S095026881000316X

Hsu SY, Bayati M, Li C, Hsieh HY, Belenchia A, Klutts J, Zemmer SA, Reynolds M, Semkiw E, Johnson HY, Foley T, Wieberg CG, Wenzel J, Johnson MC, Lin CH (2022) Biomarkers selection for population normalization in SARS-CoV-2 wastewater-based epidemiology. Water Res 223(August):118985. https://doi.org/10.1016/j.watres.2022.118985

Huijbers PMC, Flach C-F, Larsson DGJ (2019) A conceptual framework for the environmental surveillance of antibiotics and antibiotic resistance. Environ Int 130:104880. https://doi.org/10.1016/j.envint.2019.05.074

Huizer M, ter Laak TL, de Voogt P, van Wezel AP (2021) Wastewater-based epidemiology for illicit drugs: a critical review on global data. Water Res 207(September):117789. https://doi.org/10.1016/j.watres.2021.117789

Iwu-Jaja C, Ndlovu NL, Rachida S, Yousif M, Taukobong S, Macheke M, Mhlanga L, van Schalkwyk C, Pulliam JRC, Moultrie T, le Roux W, Schaefer L, Pocock G, Coetzee LZ, Mans J, Bux F, Pillay L, de Villiers D, du Toit AP, McCarthy K et al (2023) The role of wastewater-based epidemiology for SARS-CoV-2 in developing countries: cumulative evidence from South Africa supports sentinel site surveillance to guide public health decision-making. Sci Total Environ 903(July). https://doi.org/10.1016/j.scitotenv.2023.165817

Jakariya M, Ahmed F, Islam MA, Al Marzan A, Hasan MN, Hossain M, Ahmed T, Hossain A, Reza HM, Hossen F, Nahla T, Rahman MM, Bahadur NM, Islam MT, Didar-ul-Alam M, Mow N, Jahan H, Barceló D, Bibby K, Bhattacharya P (2022) Wastewater-based epidemiological surveillance to monitor the prevalence of SARS-CoV-2 in developing countries with onsite sanitation facilities. Environ Pollut 311(February):1–10. https://doi.org/10.1016/j.envpol.2022.119679

James G, Witten D, Hastie T, Tibshirani R (2013) An introduction to statistical learning with applications in R. In: Tibshirani R (ed) Robust SRAM designs and analysis, 8th edn. Springer Science+Business Media B.V. https://doi.org/10.1007/978-1-4614-7138-7

Jiang G, Liu Y, Tang S, Kitajima M, Haramoto E, Arora S, Choi PM, Jackson G, D'Aoust PM, Delatolla R, Zhang S, Guo Y, Wu J, Chen Y, Sharma E, Prosun TA, Zhao J, Kumar M, Honda R,

Meiman J et al (2023) Moving forward with COVID-19: future research prospects of wastewater-based epidemiology methodologies and applications. Curr Opin Environ Sci Health 33:100458. https://doi.org/10.1016/j.coesh.2023.100458

Johar AA, Salih MA, Abdelrahman HA, Al Mana H, Hadi HA, Eltai NO (2023) Wastewater-based epidemiology for tracking bacterial diversity and antibiotic resistance in COVID-19 isolation hospitals in Qatar. J Hosp Infect 141:209–220. https://doi.org/10.1016/j.jhin.2023.08.011

Kannan A, Sims N, Hold AJ, Jagadeesan K, Standerwick R, Barden R, Kasprzyk-Hordern B (2023) The burden of city's pain treatment—a longitudinal one year study of two cities via wastewater-based epidemiology. Water Res 229(July 2022):119391. https://doi.org/10.1016/j.watres.2022.119391

Kasprzyk-Hordern B, Béen F, Bijlsma L, Brack W, Castigioni S, Covaci A, Martincigh B, Mueller J, van Nuijs ALN, Oluseyi T, Thomas KV (2023) Wastewater-based epidemiology for the assessment of population exposure to chemicals: the need for integration with human biomonitoring for global one health actions. J Hazard Mater 450(131009). https://doi.org/10.1016/j.jhazmat.2023.131009

Kasprzyk-Hordern B, Sims N, Farkas K, Jagadeesan K, Proctor K, Wade MJ, Jones DL (2023) Wastewater-based epidemiology for comprehensive community health diagnostics in a national surveillance study: mining biochemical markers in wastewater. J Hazard Mater 450(January):130989. https://doi.org/10.1016/j.jhazmat.2023.130989

Kelkar V, Seetha Ram K (2019) A literature review evaluating new approaches to resolving the sanitation challenge in developing Asia, vol 976. ADBI Working Paper Series. https://www.adb.org/publications/evaluating-

Kitajima M, Ahmed W, Bibby K, Carducci A, Gerba CP, Hamilton KA, Haramoto E, Rose JB (2020) SARS-CoV-2 in wastewater: state of the knowledge and research needs. Sci Total Environ 739:139076. https://doi.org/10.1016/j.scitotenv.2020.139076

Kmush BL, Monk D, Green H, Sachs† DA, Zeng T, Larsen DA (2022) Comparability of 24-hour composite and grab samples for detection of SARS-2-CoV RNA in wastewater. FEMS Microbes 3(December 2021):1–5. https://doi.org/10.1093/femsmc/xtac017

Kuhn M (2022) Caret: classification and regression training. https://cran.r-project.org/pack%0Dage=caret

Lai M, Wulff SS, Cao Y, Robinson TJ, Rajapaksha R (2023) An interpretable time series machine learning method for varying forecast and nowcast lengths in wastewater-based epidemiology. MethodsX 11(September):102382. https://doi.org/10.1016/j.mex.2023.102382

Laimou-Geraniou M, Heath D, Heath E (2023) Analytical methods for the determination of antidepressants, antipsychotics, benzodiazepines and their metabolites through wastewater-based epidemiology. Trends Environ Anal Chem 37(September 2022):e00192. https://doi.org/10.1016/j.teac.2022.e00192

Lansley G, Cheshire J (2016) An introduction to spatial data analysis and visualisation in R. Geogr Anal 38:5–22. https://doi.org/10.1111/j.0016-7363.2005.00671.x

Li X, Liu H, Gao L, Sherchan SP, Zhou T, Khan SJ, van Loosdrecht MCM, Wang Q (2023) Wastewater-based epidemiology predicts COVID-19-induced weekly new hospital admissions in over 150 USA counties. Nat Commun 14(1):1–14. https://doi.org/10.1038/s41467-023-40305-x

Li J, Choi PM, Gao J, Ren J, O'Brien JW, Thomas KV, Mueller JF, Thai PK, Jiang G (2024a) In-sewer stability of 31 human health biomarkers and suitability for wastewater-based epidemiology. Water Res 249(August 2023):120978. https://doi.org/10.1016/j.watres.2023.120978

Li K, Hu Y, Jiang Y, Han X, Liu X, Du M (2024b) Direct-injection UHPLC-MS/MS method for simultaneous determination of 78 illegal drugs and psychoactive substances in domestic wastewater. Water (Switzerland) 16(9). https://doi.org/10.3390/w16091315

Li Z, Li J, Hu Y, Yan Y, Tang S, Ma R, Li L (2024c) Evaluation of pharmaceutical consumption between urban and suburban catchments in China by wastewater-based epidemiology. Environ Res 250(February):118544. https://doi.org/10.1016/j.envres.2024.118544

Liu X, Zhang R, Cheng H, Khorram MS, Zhao S, Tham TT, Tran TM, Minh TB, Jiang B, Jin B, Zhang G (2021) Field evaluation of diffusive gradients in thin-film passive samplers for wastewater-based epidemiology. Sci Total Environ 773:145480. https://doi.org/10.1016/j.scitotenv.2021.145480

Lodder WJ, de Roda Husman AM (2005) Presence of noroviruses and other enteric viruses in sewage and surface waters in The Netherlands. Appl Environ Microbiol 71(3):1453–1461. https://doi.org/10.1128/AEM.71.3.1453-1461.2005

Maal-Bared R, Brisolara K, Knight M, Mansfeldt C (2023) To sample or not to sample: a governance-focused decision tree for wastewater service providers considering participation in wastewater-based epidemiology (WBE) in support of public health programs. Sci Total Environ 905(July):167128. https://doi.org/10.1016/j.scitotenv.2023.167128

Mao K, Zhang H, Pan Y, Yang Z (2021) Biosensors for wastewater-based epidemiology for monitoring public health. Water Res 191. https://doi.org/10.1016/j.watres.2020.116787

Mao K, Zhang K, Du W, Ali W, Feng X, Zhang H (2020) The potential of wastewater-based epidemiology as surveillance and early warning of infectious disease outbreaks. Curr Opin Environ Sci Health 17:1–7. https://doi.org/10.1016/j.coesh.2020.04.006

Maréchal V, Maday Y, Wallet C, Cluzel N, Borde C (2023) Wastewater-based epidemiology: retrospective, current status, and future prospects. Anaesth Crit Care Pain Med 42(5). https://doi.org/10.1016/j.accpm.2023.101251

Masachessi G, Pisano MB, Prez VE, Martínez LC, Michelena JF, Martínez-Wassaf M, Giordano MO, Isa MB, Pavan JV, Welter A, Nates SV, Ré V (2018) Enteric viruses in surface waters from argentina: molecular and viable-virus detection. Appl Environ Microbiol 84(5). https://doi.org/10.1128/AEM.02327-17

Massano M, Salomone A, Gerace E, Alladio E, Vincenti M, Minella M (2023) Wastewater surveillance of 105 pharmaceutical drugs and metabolites by means of ultra-high-performance liquid-chromatography-tandem high resolution mass spectrometry. J Chromatogr A 1693:463896. https://doi.org/10.1016/j.chroma.2023.463896

Medema G, Heijnen L, Elsinga G, Italiaander R, Medema G (2020) Presence of SARS-Coronavirus-2 in sewage. Methods sewage samples. MedRxiv. https://doi.org/10.1101/2020.03.29.20045880

Medina CY, Kadonsky KF, Roman FA, Tariqi AQ, Sinclair RG, D'Aoust PM, Delatolla R, Bischel HN, Naughton CC (2022) The need of an environmental justice approach for wastewater based epidemiology for rural and disadvantaged communities: a review in California. Curr Opin Environ Sci Health 27:100348. https://doi.org/10.1016/j.coesh.2022.100348

Meyer C, Stravs MA, Hollender J (2024) How wastewater reflects human metabolism–suspect screening of pharmaceutical metabolites in wastewater influent. Environ Sci Technol 58(22):9828–9839. https://doi.org/10.1021/acs.est.4c00968

Miłobedzka A, Ferreira C, Vaz-Moreira I, Calderón-Franco D, Gorecki A, Purkrtova S, Bartacek J, Dziewit L, Singleton CM, Nielsen PH, Weissbrodt DG, Manaia CM (2022) Monitoring antibiotic resistance genes in wastewater environments: the challenges of filling a gap in the one-health cycle. J Hazard Mater 424:127407. https://doi.org/10.1016/j.jhazmat.2021.127407

Morales-Mora E, Badilla-Aguilar A, Rivera-Navarro P, Alfaro-Arrieta E, Symonds EM, Barrantes-Jiménez K, Ruepert C, Crowe J, Chacon L (2023) A novel approach to biological risk surveillance in occupational hygiene: a case study on the use of quantitative microbial risk assessment to determine occupational risks for wastewater treatment plants workers in Costa Rica. Microbial Risk Anal 24(100262). https://doi.org/10.1016/j.mran.2023.100262

Mtetwa HN, Amoah ID, Kumari S, Bux F, Reddy P (2023) Exploring the role of wastewater-based epidemiology in understanding tuberculosis burdens in Africa. Environ Res 231(P1):115911. https://doi.org/10.1016/j.envres.2023.115911

National Academies of Sciences (2024) Wastewater-based disease surveillance for public health action. https://nap.nationalacademies.org/resource/26767/interactive/

Nguyen AQ, Vu HP, Nguyen LN, Wang Q, Djordjevic SP, Donner E, Yin H, Nghiem LD (2021) Monitoring antibiotic resistance genes in wastewater treatment: current strategies and future challenges. Sci Total Environ 783:146964. https://doi.org/10.1016/j.scitotenv.2021.146964

NIDA (2024) Prescription opioids DrugFacts. https://nida.nih.gov/publications/drugfacts/prescription-opioids

Nitti F, Kapitan OB, Ola PD, Siswanta D (2022) Passive sampling techniques for monitoring of pharmaceuticals and personal care products in water matrix: trends from 2016 to 2020. In: Kumar M, Mohapatra S (eds) Impact of COVID-19 on emerging contaminants: one health framework for risk assessment and remediation. Springer Nature Singapore, pp 17–44. https://doi.org/10.1007/978-981-19-1847-6_2

Paul JR, Trask JD, Gard S (1940) II. Poliomyelitic virus in urban sewage. J Exp Med 71(6):765–777. https://doi.org/10.1084/jem.71.6.765

Pasin D, Cawley A, Bidny S, Fu S (2017) Current applications of high-resolution mass spectrometry for the analysis of new psychoactive substances: a critical review. Anal Bioanal Chem 409(25):5821–5836. https://doi.org/10.1007/s00216-017-0441-4

Pecson BM, Darby E, Haas CN, Amha YM, Bartolo M, Danielson R, Dearborn Y, Di Giovanni G, Ferguson C, Fevig S, Gaddis E, Gray D, Lukasik G, Mull B, Olivas L, Olivieri A, Qu Y, SARS-CoV-2 Interlaboratory Consortium (2021) Reproducibility and sensitivity of 36 methods to quantify the SARS-CoV-2 genetic signal in raw wastewater: findings from an interlaboratory methods evaluation in the U.S. environmental science: Water Res Technol 7(3):504–520. https://doi.org/10.1039/D0EW00946F

Peng RD, Dominici F (2008) Statistical methods for environmental epidemiology with R. In: Gentleman R, Hornik K (eds) A case study in air pollution and health. Springer

Philo SE, De León KB, Noble RT, Zhou NA, Alghafri R, Bar-Or I, Darling A, D'Souza N, Hachimi O, Kaya D, Kim S, Gaardbo Kuhn K, Layton BA, Mansfeldt C, Oceguera B, Radniecki TS, Ram JL, Saunders LP, Shrestha A, Delgado Vela J et al (2024) Wastewater surveillance for bacterial targets: current challenges and future goals. Appl Environ Microbiol 90(1). https://doi.org/10.1128/aem.01428-23

Pintó RM, Alegre D, Domínguez A, El-Senousy WM, Sánchez G, Villena C, Costafreda MI, Aragonés L, Bosch A (2007) Hepatitis A virus in urban sewage from two Mediterranean countries. Epidemiol Infect 135(2):270–273. https://doi.org/10.1017/S0950268806006753

Randazzo W, Truchado P, Cuevas-Ferrando E, Simón P, Allende A, Sánchez G (2020) SARS-CoV-2 RNA in wastewater anticipated COVID-19 occurrence in a low prevalence area. Water Res 181:115942. https://doi.org/10.1016/j.watres.2020.115942

Rauch W, Schenk H, Insam H, Markt R, Kreuzinger N (2022) Data modelling recipes for SARS-CoV-2 wastewater-based epidemiology. Environ Res 214(P1):113809. https://doi.org/10.1016/j.envres.2022.113809

Raya S, Malla B, Shrestha S, Sthapit N, Kattel H, Sharma ST, Tuladhar R, Maharjan R, Takeda T, Kitajima M, Tandukar S, Haramoto E (2024) Quantification of multiple respiratory viruses in wastewater in the Kathmandu Valley, Nepal: potential implications of wastewater-based epidemiology for community disease surveillance in developing countries. Sci Total Environ 920(February):170845. https://doi.org/10.1016/j.scitotenv.2024.170845

Robson MG, Toscano WA, Meng Q, Kaden DA (2023) Risk assessment for environmental health. In: Risk assessment for environmental health (Second). Taylor & Francis Group, LLC. https://doi.org/10.1201/9780429291722

Roefs I, Meulman B, Vreeburg JHG, Spiller M (2017) Centralised, decentralised or hybrid sanitation systems? Economic evaluation under urban development uncertainty and phased expansion. Water Res 109(2017):274–286. https://doi.org/10.1016/j.watres.2016.11.051

Sanderson H, Fricker C, Brown RS, Majury A, Liss SN (2016) Antibiotic resistance genes as an emerging environmental contaminant. Environ Rev 24(2):205–218. https://doi.org/10.1139/er-2015-0069

Santana-Viera S, Lara-Martín PA, González-Mazo E (2023) High resolution mass spectrometry (HRMS) determination of drugs in wastewater and wastewater based epidemiology in Cadiz Bay (Spain). J Environ Manage 341(January). https://doi.org/10.1016/j.jenvman.2023.118000

Sapula SA, Whittall JJ, Pandopulos AJ, Gerber C, Venter H (2021) An optimized and robust PEG precipitation method for detection of SARS-CoV-2 in wastewater. Sci Total Environ 785:147270. https://doi.org/10.1016/j.scitotenv.2021.147270

Sau A, Bhakta I (2019) Screening of anxiety and depression among the seafarers using machine learning technology. Inform Medi Unlocked 16(December 2018). https://doi.org/10.1016/j.imu.2018.12.004

Schenk H, Heidinger P, Insam H, Kreuzinger N, Markt R, Nägele F, Oberacher H, Scheffknecht C, Steinlechner M, Vogl G, Wagner AO, Rauch W (2023) Prediction of hospitalisations based on wastewater-based SARS-CoV-2 epidemiology. Sci Total Environ 873(February). https://doi.org/10.1016/j.scitotenv.2023.162149

Schmiege D, Haselhoff T, Thomas A, Kraiselburd I, Meyer F, Moebus S (2024) Small-scale wastewater-based epidemiology (WBE) for infectious diseases and antibiotic resistance: a scoping review. Int J Hyg Environ Health 259(January):114379. https://doi.org/10.1016/j.ijheh.2024.114379

Shakallis AG, Fallowfield H, Ross KE, Whiley H (2022) The application of passive sampling devices in wastewater surveillance. Water (Switzerland) 14(21). https://doi.org/10.3390/w14213478

Shao XT, Liu SY, Zhao YT, Jiang B, Lin JG, Wang DG (2023) Evaluation of eight psychoactive drugs used in Chinese cities by wastewater-based epidemiology. Sci Total Environ 855(September 2022):158982. https://doi.org/10.1016/j.scitotenv.2022.158982

Shrestha S, Malla B, Haramoto E (2024) Estimation of norovirus infections in Japan: an application of wastewater-based epidemiology for enteric disease assessment. Sci Total Environ 912(December 2023):169334. https://doi.org/10.1016/j.scitotenv.2023.169334

Shrestha S, Yoshinaga E, Chapagain SK, Mohan G, Gasparatos A, Fukushi K (2021) Surveillance of COVID-19 in low- and middle-income countries: challenges and opportunities. Water (Switzerland) 13(20):2897. https://www.mdpi.com/2073-4441/13/20/2897

Sim W, Park S, Ha J, Kim D, Oh JE (2023) Evaluation of population estimation methods for wastewater-based epidemiology in a metropolitan city. Sci Total Environ 857(October 2022):159154. https://doi.org/10.1016/j.scitotenv.2022.159154

Sims N, Kannan A, Holton E, Jagadeesan K, Mageiros L, Standerwick R, Craft T, Barden R, Feil EJ, Kasprzyk-Hordern B (2023) Antimicrobials and antimicrobial resistance genes in a one-year city metabolism longitudinal study using wastewater-based epidemiology. Environ Pollut 333(June):122020. https://doi.org/10.1016/j.envpol.2023.122020

Sims N, Kasprzyk-Hordern B (2020) Future perspectives of wastewater-based epidemiology: monitoring infectious disease spread and resistance to the community level. Environ Int 139:105689. https://doi.org/10.1016/j.envint.2020.105689

Solano Barquero M, Morales Mora E, Chacón Jiménez L, Cordero Jara E, Reyes Lizano L, Barrantes Jiménez K, Achí R (2019) Low-cost internal controls for detection of Giardia cysts in water samples. Parasitol Int 71:177–179. https://doi.org/10.1016/j.parint.2019.04.012

Steenbeek R, Emke E, Vughs D, Matias J, Boogaerts T, Castiglioni S, Campos-Mañas M, Covaci A, de Voogt P, ter Laak T, Hernández F, Salgueiro-González N, Meijer WG, Dias MJ, Simões S, van Nuijs ALN, Bijlsma L, Béen F (2022) Spatial and temporal assessment of crack cocaine use in 13 European cities through wastewater-based epidemiology. Sci Total Environ 847(June). https://doi.org/10.1016/j.scitotenv.2022.157222

Tchobanoglous G, Burton F, Stensel D (2014) Wastewater engineering: treatment and resource recovery. In: Metcalf & Eddy, Inc. (Fifth). McGrawHill

US-EPA (2005) Method 1623: cryptosporidium and Giardia in water by Filtration/IMS/FA. In: US-EPA

US EPA Office of Water (2004) Quality assurance/quality control guidance for laboratories performing PCR analyses on environmental samples. https://nepis.epa.gov/Exe/ZyNET.exe/P100993V.TXT?ZyActionD=ZyDocument&Client=EPA&Index=2000+Thru+2005&Docs=&Query=&Time=&EndTime=&SearchMethod=1&TocRestrict=n&Toc=&TocEntry=&QField=&QFieldYear=&QFieldMonth=&QFieldDay=&IntQFieldOp=0&ExtQFieldOp=0&XmlQuery=

van Nuijs ALN, Covaci A, Beyers H, Bervoets L, Blust R, Verpooten G, Neels H, Jorens PG (2015) Do concentrations of pharmaceuticals in sewage reflect prescription figures? Environ Sci Pollut Res 22(12):9110–9118. https://doi.org/10.1007/s11356-014-4066-2

Vaughan L, Zhang M, Gu H, Rose JB, Naughton CC, Medema G, Allan V, Roiko A, Blackall L, Zamyadi A (2023) An exploration of challenges associated with machine learning for time series forecasting of COVID-19 community spread using wastewater-based epidemiological data. Sci Total Environ 858(October 2022):159748. https://doi.org/10.1016/j.scitotenv.2022.159748

Verheyden K, Vanhaeke L, Vanden Bussche J, Brabander H, Scoutson F (2009) UHPLC coupled with fourier transform orbitrap for residue analysis

Victoria M, Tort LFL, García M, Lizasoain A, Maya L, Leite JPG, Miagostovich MP, Cristina J, Colina R (2014) Assessment of gastroenteric viruses from wastewater directly discharged into Uruguay River, Uruguay. Food Environ Virol 6(2):116–124. https://doi.org/10.1007/s12560-014-9143-7

Vitale D, Morales Suárez-Varela M, Picó Y (2021) Wastewater-based epidemiology, a tool to bridge biomarkers of exposure, contaminants, and human health. Curr Opin Environ Sci Health 20:100229. https://doi.org/10.1016/j.coesh.2021.100229

Wang T, Wang C, Myshkevych Y, Mantilla-Calderon D, Talley E, Hong P-Y (2023) SARS-CoV-2 wastewater-based epidemiology in an enclosed compound: a 2.5-year survey to identify factors contributing to local community dissemination. Sci Total Environ 875:162466. https://doi.org/10.1016/j.scitotenv.2023.162466

Wen J, Duan L, Wang B, Dong Q, Liu Y, Chen C, Huang J, Yu G (2024) In-sewer stability assessment of 140 pharmaceuticals, personal care products, pesticides and their metabolites: implications for wastewater-based epidemiology biomarker screening. Environ Int 184(January):108465. https://doi.org/10.1016/j.envint.2024.108465

WHO, UN-Habitat (2021) Progreso en el tratamiento de las aguas residuales. Estado mundial y Necesidades de aceleración Del indicador 6.3.1. De los ODS. https://www.unwater.org/publications/progress-on-wastewater-treatment-631-2021-update/

Wolfe MK, Duong D, Bakker KM, Ammerman M, Mortenson L, Hughes B, Arts P, Lauring AS, Fitzsimmons WJ, Bendall E, Hwang CE, Martin ET, White BJ, Boehm AB, Wigginton KR (2022) Wastewater-based detection of two influenza outbreaks. Environ Sci Technol Lett 9(8):687–692. https://doi.org/10.1021/acs.estlett.2c00350

World Health Organization (2003) Assessing microbial safety of drinking water In: Dufour A, Snozzi M, Koster W, Bartram J, Ronchi E, Fewtrell L (eds). IWA Publishing

World Health Organization (2006) WHO Guidelines for the safe use of wastewater, excreta and greywater in Agriculture. https://www.who.int/water_sanitation_health/publications/gsuweg4/en/

World Health Organization (2023) Field guidance for the implementation of environmental surveillance for poliovirus. WHO

Wu F, Zhang J, Xiao A, Gu X, Lee WL, Armas F, Kauffman K, Hanage W, Matus M, Ghaeli N, Endo N, Duvallet C, Poyet M, Moniz K, Washburne AD, Erickson TB, Chai PR, Thompson J, Alm EJ (2020) SARS-CoV-2 titers in wastewater are higher than expected from clinically confirmed cases. MSystems 5(4). https://doi.org/10.1128/mSystems.00614-20

Xu X, Deng Y, Ding J, Zheng X, Wang C, Wang D, Liu L, Gu H, Peiris M, Poon LLM, Zhang T (2023) Wastewater genomic sequencing for SARS-CoV-2 variants surveillance in wastewater-based epidemiology applications. Water Res 244(August):120444. https://doi.org/10.1016/j.watres.2023.120444

Xu X, Liu Y, Liu S, Li J, Guo G, Smith K (2019) Real-time detection of potable-reclaimed water pipe cross-connection events by conventional water quality sensors using machine learning methods. J Environ Manage 238(March):201–209. https://doi.org/10.1016/j.jenvman.2019.02.110

Yang Y, Yang L, Zheng M, Cao D, Liu G (2023) Data acquisition methods for non-targeted screening in environmental analysis. TrAC Trends Anal Chem 160:116966. https://doi.org/10.1016/j.trac.2023.116966

Yu QQ, Olesen SW, Duvallet C, Grad YH (2024) Assessment of sewer connectivity in the United States and its implications for equity in wastewater-based epidemiology. PLOS Glob Public Health 4(4):1–24. https://doi.org/10.1371/journal.pgph.0003039

Zaneti RN, Girardi V, Spilki FR, Mena K, Westphalen APC, da Costa Colares ER, Pozzebon AG, Etchepare RG (2021) Quantitative microbial risk assessment of SARS-CoV-2 for workers in wastewater treatment plants. Sci Total Environ 754:142163. https://doi.org/10.1016/j.scitotenv.2020.142163

Zhang S, Shi J, Li X, Tiwari A, Gao S, Zhou X, Sun X, O'Brien JW, Coin L, Hai F, Jiang G (2023) Wastewater-based epidemiology of Campylobacter spp.: a systematic review and meta-analysis of influent, effluent, and removal of wastewater treatment plants. Sci Total Environ 903(August):166410. https://doi.org/10.1016/j.scitotenv.2023.166410

Chapter 9
Effective Maintenance Practices for Wastewater Treatment Plants

Wafa Hassen, Bilel Hassen, and Abdennaceur Hassen

Abstract The treatment of wastewater is now an important issue for the quality of the environment and the quality of the water resource. The transport of wastewater through the pipes of a sanitation network, also known as "collectors," is generally done by gravity, but can sometimes be carried out by pumping, under pressure or under vacuum through pumping processes and lifting stations. To protect the network against clogging and corrosion, a pretreatment including screening, sieving, desanding, and degreasing is carried out before the discharge of water into the separate or combined sanitation network. The regulation of the flow of wastewater and rainwater, in the case of a combined network, is ensured by equipment designed to temporarily keep significant and sudden water inflows, in order to not disrupt the proper functioning of the treatment plants and to limit the risk of flooding. It is therefore recommended and possible to significantly increase the capacity of wastewater treatment plants by optimizing them to primarily improve the quality of the treated effluent and reduce the operating costs of these facilities. This current state-of-the-art rule summarizes how to approach optimizing the operation of an existing wastewater treatment plant by designing and scheduling equipment maintenance related to the various wastewater treatment processes, which must be well established and scheduled to maintain the performance of the recommended treatment system by releasing treated wastewater of satisfactory quality and in compliance with current standards, thus ensuring the sustainability of the different sectors and notably reducing the treatment cost. When envisioning the future wastewater treatment plant as a key component of the circular economy, it becomes apparent that the continuously advancing treatment facility must be properly equipped and staffed with skilled personnel. This,

W. Hassen · A. Hassen (✉)
Laboratory of Treatment and Wastewater Valorization, Center of Research and Water Technologies (CERTE), Soliman, Tunisia
e-mail: abdohas@gmail.com

W. Hassen
e-mail: hassen.wafa@gmail.com

B. Hassen
Laboratories of Microbiology and Pathology of Aquatic Organisms, National Institute of Navy Sciences and Technologies (INSTM), Salammbô, Tunisia
e-mail: bilel.hassen@fst.utm.tn

S. Souabi et al. (eds.), *Wastewater Treatment Plants*, Water Science and Technology Library 130, https://doi.org/10.1007/978-3-031-87461-1_9

combined with coordinated efforts of various kinds, will facilitate the production and valorization of treated water, helping to generate energy, raw materials, and high-quality recycled water. These resources can then support specific industrial sectors, water green areas and golf courses, or irrigate crops, marking the transition into the circular economy era.

Keywords Wastewater collection and processing · Treatment plant · System maintenance · Full recycling center · Circular economy concept

9.1 Introduction

Among the resources that contribute to the development of human activities, water has several features that distinguish it from all others: it is essential for life; it is omnipresent (making up 65% of the human body and covering 70% of the Earth's surface); it is available in strictly fixed quantities, dictated by the laws of conservation and the water cycle. That it is essential for life on Earth makes it a coveted resource more than any other does: its scarcity keeps entire populations trapped in poverty and fuels political conflicts that can eventually escalate to armed conflict. It is omnipresent, but the quantity useful for human activities is distributed extremely unevenly. Out of the approximately 1300 million cubic kilometers of water on the planet, 97.2% comprises unusable saltwater for human activities; of the remaining 2.8%, 2.15% is "trapped" in polar ice caps; leaving 0.65%, of which 0.62% is in underground form.

The water cycle ensures a terrestrial precipitation level of about 113,000 km^3 per year, from which 72,000 km^3 of evaporation must be subtracted to obtain the available net flow; out of the remaining total, 32,900 km^3 are considered geographically accessible, but the timing of these precipitations must also be considered. Most are concentrated in short periods of time and lead to floods; leaving 9000 km^3 actually accessible; when adding the precipitation waters kept in dams, there are ultimately 12,500 km^3 of freshwater available each year for human uses, representing over 5000 L per person per day globally (Baechler 2012).

When all the parameters determining the conditions of articulation between water supply and demand are combined, we ultimately conclude that access to water is essentially a development problem. The world population is still increasing, especially in developing countries, drawing from a water supply at best constant, most of the time in regions already poorly endowed with water due to climatic and geological reasons (Fig. 9.1). The quantity of accessible water is not the only issue, as poor water quality poses problems that are at least as serious. These factors mean that the most serious scarcity situations are found in arid, densely populated, and poor areas, that is, unable to provide the necessary investment efforts to ensure the availability of clean water (Noin 1998).

Fig. 9.1 Example of water supply difficulties in developing countries

Development or conflict issues arising from water access difficulties are more about managing scarce resources than the availability of the resource. These problems take various forms depending on the situations at hand: it may involve lack of investment in infrastructure to improve water access, lack of cooperation in transboundary water resource management, wastage in various uses of the resource. All these situations require clear identification of the causes of water access difficulties to design appropriate solutions for each case.

These development issues or conflicts arising from difficulties in accessing water have compelled policymakers and human populations to adopt a water management policy and consider a circular water management policy, which involves encouraging individuals to recycle and reuse water in various communal, agricultural, and industrial activities, etc. This includes wastewater, its availability, collection in networks, volume, treatment, quality, transportation to storage areas, mode of use, etc. (Al Hamedi et al. 2023).

Therefore, man must think toward consuming and reusing this increasingly expensive and scarce neo-resource in a continuous quest for a sustainable resource. The quality criterion of this neo-resource takes precedence and is at the top of the list of other less important criteria. Thus, when we talk about quality, we must discuss the treatment of this neo-resource, its collection, and its transport to a specific location: the purification plant. So, to be reused by man in the various activities mentioned earlier, this neo-resource must be well treated or purified to make it non-harmful to humans and without deleterious risks to their environment and living conditions. The strengthening of regulatory requirements and compliance with discharge standards defined within the framework of territorial, national, and international regulations have been put in place to compel companies to maintain and control their treatment plants and ensure the production and release of effluents of satisfactory quality that are not harmful to the living conditions of humans and their environment. The purification plant, where techniques and procedures for treating this liquid residual resource, wastewater, are implemented, must have reliable and reassuring tools and operating systems to release treated wastewater or effluents of quality under current standards and without risks to humans and their general living conditions (Bdour et al. 2009; Monchy 2000).

Therefore, wastewater treatment plants are complex structures for which concrete essentially provides suitable, recognized, and sustainable technical solutions. They must meet functional, environmental, and structural requirements related to the specificities of treatment processes and increasingly stringent regulatory developments, in a logic of landscape and architectural integration.

9.2 Wastewater Treatment Plants

Guaranteeing the quality control of effluents and the dependable functioning of the wastewater treatment facility and its equipment for production are essential aspects of the maintenance service's responsibilities. These tasks are crucial for safeguarding public health and environmental protection, encompassing activities from managing the sewage collection network to the final discharge of treated wastewater, including its appropriate treatment, storage, utilization, or disposal figure (Rajaei and Nazif 2022).

A wastewater treatment plant is supplied by a collective sanitation network, either a combined system that evacuates wastewater and rainwater in the same pipeline, or a separate system that collects the two types of water in separate pipelines. This network includes pipelines, usually made of concrete, that collect wastewater and rainwater and transport them by gravity to the plant where they are treated. The objective of wastewater treatment is to purify them so they do not degrade the quality of the natural environment into which they will be discharged later.

It is a fundamental phase in the water cycle. It addresses two essential concerns:

- preserving the natural heritage and quality of life by playing a role in the sanitary protection of populations and contributing to the maintenance of environmental quality and water-related activities;
- preserving water resources.

That means wastewater treatment is a public health and environmental protection issue regulated by an increasingly demanding international regulation, reinforced by continuous daily amendments in treatment technologies and the quality of wastewater produced.

The wastewater treatment plant mainly consists of two treatment processes: the water treatment process which purifies water through three main phases (primary, secondary, and tertiary treatments) and the sludge treatment process which treats the solid residues of purification known as residual sludge resulting from various treatment procedures.

The establishment of a preventive maintenance and servicing plan for wastewater treatment is mainly based on three important steps: a functional analysis: the raw wastewater collection facilities known as sewage networks, proper and correct procedures for treating these waters and their tools for proper functioning; a dysfunctional analysis of these systems and finally a definition of preventive measures and procedures to be implemented in the event of incidents or malfunctions.

Preventive maintenance typically includes a range of tasks such as regular oil changes, applying lubricants and protective coatings to equipment, checking for chips, cracks or surface defects in concrete reinforcement or steel structures, performing calibrations and adjustments, and maintaining a regular cleaning schedule. In addition to the traditional preventive maintenance and maintenance measures of the sewage system and wastewater treatment plant, we are also interested in carrying out some predictive measures, such as inspection with infrared cameras, airborne/structure-borne noise, engine circuit, and oil analysis (United States Steel Corporation 2018).

9.3 Upkeep and Maintenance of Wastewater Collection Networks

As previously mentioned, a sanitation network includes all systems and equipment dedicated to wastewater and rainwater in a combined network. It ensures the transport of these waters to treatment centers, the purification station, and allows their evacuation from consumption points. These points can be individual or collective housing, company buildings, commercial spaces, or production sites. To ensure their proper functioning, the maintenance of sanitation networks remains essential.

The maintenance of sanitation networks begins and is mainly done with the help of vacuum trucks or hydro-cleaning trucks. These vehicles can remove sludge and deposits present in the networks. They are also used for cleaning pipelines and areas at risk of clogging. Manual intervention is sometimes required in areas that are not easily accessible to vehicles and in very narrow areas. The collection of wastewater has a set of collectors and pipelines that ensure the evacuation and sanitation of wastewater and rainwater in the case of separate or combined networks, to make them "healthier, purer". Wastewater treatment essentially involves two main steps: the proper collection of raw wastewater, the disposal of harmful water, and purification to reduce health risks. Therefore, the maintenance of sanitation networks can be preventive or curative. The maintenance of municipal sanitation networks is generally the responsibility of municipalities or specialized public or private companies (Alm et al. 2021).

This operation must also be conducted on all community networks that follow a well-respected regulation and periodicity. The maintenance of sanitation networks is carried out using pump trucks or hydro-cleaning trucks and sometimes manually following handling interventions. These vehicles can remove sludge and deposits present in the networks. They also clean the pipes and areas at risk of blockages. Recently, preventive maintenance of sanitation networks has introduced new technologies and implementation techniques such as cameras inserted into the pipes to check their proper functioning and control any anomalies not directly discernible by humans. This is why the inspection of drainage pipes and sanitation networks is important, both at the time of their design, to make sure they are laid in a compliant way, and throughout their existence, to expect operational problems and any maintenance work on sanitation networks and repairs that may be required (Phyu Hnin and Khaodhiar 2021).

Regardless of the sanitation system, whether mixed or separate, maintaining the network is essential in order to:

- Ensure the evacuation of wastewater: as the collection and transmission, pipes become less suitable for the transport and carrying of solid elements. Over time, these residues will collect, forming lime scale deposits on the walls of the pipes. The water cannot flow, and the pipes can become blocked;
- Prolong the lifespan of your sanitation system and ensure its efficiency;
- Ensure a good ecological state of the water and prevent the poor flow of effluents;
- Preserve the natural environment, specifically the soil and the environment, following various accidental leaks and unintentional accumulation of muddy solid residues;
- And finally, protect and defend against the harmful effects on health by the spread of pathogenic bacteria and viruses, diarrheal and contagious diseases, etc.

Therefore, the maintenance and upkeep of the sanitation network is considered a sanitary and ecological necessity to ensure the proper collection and arrival of wastewater at the treatment plant. Pollutants released into the environment can degrade ecosystems and human health. Thus, treatment is necessary to preserve the quality of life and the environment. Furthermore, this process requires great knowledge and skill. Therefore, it is necessary to seek the help of a professional to install, maintain, and renovate collection networks. It is also worth noting that legislation in most countries requires the use of a professional to monitor the sanitation network. Why should one turn to a specialized and expert company for the maintenance of sanitation networks? (WHO and UNICEF 2021).

For the control of sanitation networks and for maintenance, the intervention of an experienced and expert professional company is required. They must have the knowledge and equipment to carry out this service. Among the services offered by these expert companies in this specific field, maintenance and sanitation network maintenance mainly involve high-pressure cleaning or cleaning through televised inspection of the installation and drainage pipes, as well as air and water tightness tests, etc. The high-pressure cleaning method involves sending a flexible hose inside the collector to maintain the pipes and various sanitation systems. Cleaning or high-pressure cleaning helps to remove waste build-ups that could block the networks and solve potential obstructions causing blockages in the drainage and transport of wastewater to the treatment plant.

It is determinant to emphasize that before carrying out hydro jetting, it is necessary to conduct a camera inspection to identify the source of the problem and locate the malfunction. Common issues include pipe breakage due to various factors such as road remodeling, installation of other electrical, telephone, drinking water networks, root accumulation, and residue buildup. In recent years, the frequency of cleaning and maintenance operations for sanitation networks has significantly increased due to the growing use of wipes, which are often flushed down toilets. These elements combine with fats to cause significant damage and overflows.

It is essential to maintain the sewage collection network and ensure the regularity of inspections and maintenance of purification systems and sanitation networks (Salihu et al. 2022). This regular maintenance plays an important role in preserving the environment by ensuring proper wastewater disposal and helps optimize the management of health risks, particularly regarding soil pollution and water quality. It ensures the long-term integrity of the system. The regulations about the inspection of pipelines and drainage networks have not been overlooked, as well as the civil engineering of water conveyance structures, which have been subject to strict regulations.

9.4 The Different Treatment Processes in a Wastewater Treatment Plant and Their Maintenance and Upkeep

Since water is a precious commodity and resource, and to protect our natural waters and support the production of drinking water, all wastewater must be cleaned or purified of its residues and pollutants before being reintegrated into natural clean water cycles. To achieve this, various processes are often used to treat wastewater and achieve the highest possible water quality level.

As often, these waters carry a random composition of toxic or other compounds and must undergo two main types of treatment. First, toxic substances are removed from wastewater through bio-physico-chemical cleaning processes, such as the removal of metals known as more or less toxic, such as chromium, copper, cobalt, cadmium, and mercury, the elimination of iron and manganese, which are less toxic but their removal is sometimes desired, phosphates and nitrates, which are often desired in the case of agricultural reuse of treated wastewater, disinfection or elimination of opportunistic and pathogenic microorganisms, which is always required, and desalination or softening, which are highly desirable in the case of reuse in an industrial cooling process, such as heavy industries and nuclear power plants. In certain special conditions of wastewater treatment, substances may be added to improve quality and influence parameters such as pH or conductivity (Marzougui et al. 2022).

The diverse wastewater treatment methods for human, agricultural, and industrial sources are essential for two main reasons: firstly, to recycle and repurpose wastewater to meet the future demand for water and alleviate water scarcity, and secondly, to guarantee adherence to wastewater discharge regulations for the sustainability of the environment, while also minimizing contamination of groundwater and soil. These processes encompass (Nishat et al. 2023):

- Physical processes for mechanical preparation such as aeration, sedimentation, or thermal influence. This also includes the use of screens, filters, and sieves;
- Biological processes such as anaerobic wastewater treatment, biochemical oxidation, or sludge digestion;
- Chemical processes such as neutralization, disinfection, flocculation, and precipitation;
- Membrane processes such as filtration, osmosis, and nanofiltration.

It is in municipal wastewater treatment plants that the largest volumes of water to be treated are seen, so more diversified solutions and procedures that are more efficient must be implemented. The preferred procedures depend on the wastewater treatment plant. The conventional treatment processes in common wastewater treatment plants are divided into different categories.

9.4.1 Mechanical Water Treatment Processes

The initial category of wastewater treatment processes focuses on the mechanical treatment of water, in which 20–30% of the suspended solids that may be present in raw wastewater will be removed primarily through mechanical means such as screening installations. Here, a screen, a grid, or a drum sieve will filter out coarse impurities like leaves, plastic, paper, and textiles. The water will pass through various screens, ranging from coarse screens several centimeters wide with about 15 cm to fine mesh screens only a few centimeters wide, flowing at different speeds and gradually filtering out coarse materials. The mechanically collected debris is squeezed, transported, and then disposed of in an incineration plant. At this primary and coarse level of treatment, the pre-purified water must then pass through a sand trap, a device designed to remove sand often carried by wastewater during transport in the sewer system.

This system is a wastewater treatment device including a sedimentation tank to remove coarse particles, such as stones, glass shards, or sand and organic matter that has not been pre-filtered by screens. The flow moves at a relatively high speed of about 0.3 m/s. In wastewater treatment plants, there are different types of grit chambers based on their design and location, including non-aerated long grit chambers, aerated long grit chambers, also known as cylindrical grit chambers, and round grit chambers. Additionally, the aerated grit chamber helps remove fats and oils from raw wastewater, a process known as degreasing. This method is supported by the injection of compressed air, creating a rolling motion in the water that brings lighter substances, such as oils and fats, to the surface. Once on the surface, they can be easily removed through mechanical scraping. In a round grit chamber, substances in the wastewater are separated using centrifugal force before being evacuated by suction. After being cleaned in the grit chamber, the debris will undergo a second cleaning process to remove all solid waste traces before being packaged and disposed of in landfills or destroyed in incineration plants.

9.4.2 Primary Wastewater Treatment

The next step in wastewater treatment is represented by the primary treatment basin, which is the immediate phase of wastewater treatment. The dimensions and number of this basin differ according to the sizing of the station and the flow rate to be treated.

The conveyance, flow, and treatment of mechanically pre-treated wastewater are considerably slower in the basin than in the grit chamber. Simply increasing the diameter of the basin is enough to decrease the flow rate. This process requires a low flow rate to let the finest particles settle, whether at the bottom or on the water surface, and to promote the development of primary biological treatment processes. The present microorganisms begin to decompose the biodegradable organic matter present in the wastewater, which can contribute to the removal of suspended solids and the biotransformation, known as bio flocculation and the formation of active sludge in aerobic treatment zones, or pellets in anaerobic treatment zones.

The primary sludges are formed by the sedimentation of solid particles at the bottom of the tank, under the effect of physical sedimentation and biological action. They generally consist of organic matter. The primary sludges are collected by a scraper toward a fresh sludge hopper. The floating substances are transferred to a floating sludge pipe. A pump transports the fresh sludges to a digester.

This anaerobic digestion system will promote the production of methane through four successive steps: hydrolysis, acidogenesis, acetogenesis, and finally methanogenesis; the methane obtained is then converted into electricity in a thermal power plant, thus providing energy to an on-site facility. The digestion or anaerobic transformation process of the raw sludge material in the digester typically takes about four weeks. The residues include an odorless sludge known as stabilized sludge, which is often used for agricultural purposes after dehydration by a centrifuge or filter (Fig. 9.2).

After leaving the primary clarifier, the wastewater is called primary wastewater, and at this stage of treatment, about 30–40% of pollutants will be removed from the wastewater during these two previous steps of mechanical pretreatment and primary treatment. At this point in the wastewater treatment plant, the wastewater then reaches the next important and central treatment phase.

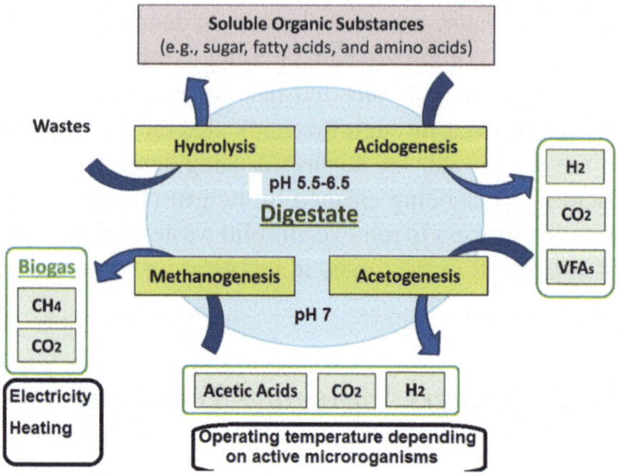

Fig. 9.2 Process of anaerobic fresh sludge digestion system leading to the production of methane biogas

9.4.3 Biological Activation Basins

In most wastewater treatment plants equipped with conventional treatment processes, the pre-purified water from the mechanical treatment phase is directed to basins equipped with aeration systems, which often are circulation basins. This is where the biological cleaning process takes place. These basins are called secondary treatment basins where the water is subjected to agitation, mixing, stirring, and circulation through the supply of oxygen and the use of agitators. These are the biological activation basins.

The goal in these basins is to create and provide more or less ventilated zones to establish varied conditions for the development and growth of aerobic heterotrophic bacteria and microorganisms, ensuring an intense removal of the various pollutants present in the pre-purified water. Thus, the microorganisms, as the main biological agents responsible for treating the different pollutants, along with some other bio-physico-chemical conditioning factors such as temperature, pH, and the availability of dissolved oxygen primarily, can feed on the residual organic contaminants and convert them into inorganic mineral substances in various gaseous, solid, and soluble liquid states.

During this significant development and growth, bacteria essentially form activated sludge flocs that float freely in the water, also known by water treatment experts as flocs. The flocs are formed after a complex bio-physico-chemical coagulation-flocculation process. The introduction of oxygen into the treatment basins stimulates the growth and proliferation of heterotrophic bacteria, thus promoting the formation of a complex matrix known as activated sludge. Therefore, this biological wastewater treatment process is also called the activated sludge process. This process is the most common method of wastewater treatment. The mixed wastewater containing activated sludge is then discharged into the secondary wastewater treatment basin. Once again, the system reduces the flow rate of wastewater. Sedimentation can then occur: The activated sludge settles at the bottom of the purified water, where it can be extracted by mechanical cleaning devices positioned at the end of the basin. A portion is then transferred to the digester, which is supplied with more biomass. The other portion of the sludge, also called "return sludge," is sent back to the aeration basin to ensure a sufficient level of microorganisms for the decomposition of impurities. Once the biological treatment is complete, nearly 90% of the wastewater has been cleaned of biodegradable substances.

The biological cleaning step, which involves supplying a large amount of oxygen, is therefore the most energy-intensive part of the entire process. Once the water has reached regulatory quality, it can be reintegrated into clear water systems, such as a river (Fig. 9.3).

9.4.4 Alternatives to Biological Purification: Chemical Treatment

In many other cases, biological purification proves to be insufficient. In such instances, other wastewater treatment processes are necessary, such as a chemical treatment taking over. Chemical additives are used to enhance water sanitation, known as chemical wastewater treatment. At this stage of wastewater treatment, chemical processes are employed. These chemical compounds are used to meet the values prescribed by the law. Chemical treatments in treatment plants include neutralization, disinfection, phosphate precipitation or dephosphatation, elimination of nitrogen forms like NO_3, NH_3, and demagnetization through cold. pH change and neutralization are sometimes used to achieve the prescribed pH value, where an acid, such as hydrochloric acid, or a base, such as lime milk, is added (Samer 2015).

During the disinfection process, pathogens are eliminated by the addition of a halogen, often chlorine or chlorine dioxide, bromine, etc. Irradiating wastewater with UV rays is a good alternative to adding chemicals, although it is less often implemented. The removal of phosphates or dephosphatation of wastewater is carried out in certain situations since wastewater is often known to be contaminated with phosphates from detergents, fertilizers, food additives, and other fecal matter. If they are still present in wastewater, they lead to over-fertilization of water bodies, resulting in the harmful phenomenon of eutrophication for the aquatic ecosystem and a nutrient enrichment that can lead to good fertilization of soils, plant growth, and proliferation of plants and algae (Hidri et al. 2014).

Among the prominent negative effects that can result from the release of wastewater into the natural environment, especially water bodies, is eutrophication. This is a process of water quality degradation that occurs when nutrients, particularly nitrates and phosphates, accumulate in large quantities in these water bodies such as freshwater lakes and rivers because of good sunlight and temperature conditions. This nutrient enrichment triggers a surge in the growth and proliferation of algae and macrophyte, leading to malodorous effects through excessive absorption of dissolved

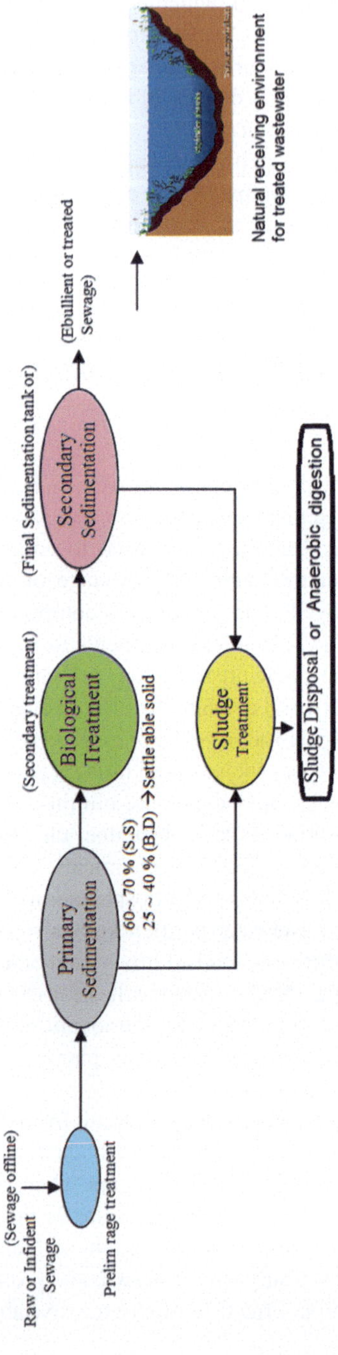

Fig. 9.3 Common schematic diagram of a municipal wastewater treatment plant

oxygen and resulting in the depletion that leads to the asphyxiation and death of aquatic ecosystems (Gilbert 2023).

Phosphates carried by wastewater are removed through a process of chemical precipitation or flocculation. The precipitation of phosphates will be partially triggered by the addition of aluminum or iron salts in the coagulant sand trap, or in the secondary basin of wastewater treatment. During this secondary clarification, the metal-phosphate flakes formed will then be removed from the wastewater with activated sludge. Depending on the process configuration available at the site of the wastewater treatment plant, it will be possible to "capture" phosphate-using microorganisms drawn from the wastewater. This scenario corresponds to a biological removal of phosphorus, a technique still rarely used. The chemical purification of water also includes the removal of nitrogen: through this process, harmful nitrogen compounds in water, including ammonia and ammonium, are removed from wastewater.

When rejected into the water bodies of natural or engineered receiving environments, nitrogen compounds absorb the available and important dissolved oxygen for living organisms such as algae, plants, and water, leading to the death and suffocation of fish, for example. Nitrogen is removed by nitrification and denitrification processes: In the nitrification phase, ammonia is oxidized to nitrite by a reaction with anaerobic bacteria and oxygen; then, in a second step, nitrite is oxidized to nitrate. The denitrification that follows is also triggered by the addition of anaerobic microorganisms (Hassen et al. 2023). They decompose the nitrate into gaseous nitrogen (N_2) through enzymatic activities; this is then released into the atmosphere. Deferrization involves reducing the iron content of wastewater to meet standard threshold values for discharge by adding oxygen, which triggers the oxidation of reduced forms of ferrous iron cations (Fe^{2+}) to oxidized forms of ferric iron (Fe^{3+}). To trigger the oxidation process, it is necessary to add caustic soda to the wastewater (Mazuelos et al. 2000).

Demanganization involves oxidizing the soluble divalent manganese (Mn^{2+}) with oxygen or other oxidizing agents to form insoluble manganese oxide (MnO_2). The manganese oxide can then be filtered out by sand filters or other methods. The process of demanganization involves the removal of manganese through this specific method; this element is present in wastewater essentially as manganese hydrogen carbonate. The addition of oxygen forms weakly soluble and settleable manganese IV compounds, which are easy to remove from water through the settling effect (El-Faramawy et al. 2004). Lastly, another type of specific water purification processes is suitable for certain types of wastewater, namely membrane filtration processes and nanofiltration processes. Water treatment by membrane, ultrafiltration, or nanofiltration is a wastewater purification process that uses microporous membranes that only allow particles smaller than the pore size of the membrane to pass through. This mechanical technique requires no chemical reagents. Membrane water filtration is mainly applied specifically for certain industrial wastewater or drinking water sources.

These waters may have highly toxic elements and micro pollutants such as pesticides and other organic micro pollutants like hydrocarbons. The most well-known are: (1) polycyclic aromatic hydrocarbons (PAHs) formed during the incomplete combustion of organic matter unintentionally present in many materials such as tars and fuels; (2) polychlorinated biphenyls (PCBs), synthetic molecules with special properties of thermal stability, resistance to oxidation, widely used as insulators in capacitors and electrical transformers; polybrominated biphenyls (PBBs) used as flame retardants; (3) polychlorinated dibenzodioxins (PCDDs) or dioxins, by-products generated during the synthesis of other products and during the combustion of certain waste and petroleum products; (4) polychlorinated dibenzofurans (PCDFs) or furans, also by-products generated unintentionally during the synthesis of other compounds; (5) and finally, detergents or surfactants with surface-active properties; phthalate esters, substances used to improve the flexibility of plastic, which are easily released, explaining their frequent presence in the environment; various pharmaceutical substances such as hormones like estrogen, testosterone, antibiotics and their degradation products, anesthetics, etc.

Some organic micro pollutants such as PCBs, dioxins, furans, and certain pesticides are part of persistent organic pollutants (POPs) such as pentachlorophenol (PCP). These extremely toxic and poorly biodegradable compounds can be transported over long distances by currents and collect in tissues along the food chain. Additionally, mineral micro pollutants are essentially represented by trace metal elements (TMEs). The main TMEs found in polluted waters and environments are cadmium (Cd), lead (Pb), chromium (Cr), mercury (Hg), arsenic (As), copper (Cu), zinc (Zn), and nickel (Ni). These elements are naturally present in rocks and soils, but current concentration levels result mostly from various human activities such as steelmaking, tanneries, road transport, and agricultural effluents. In some regions, pedo-geochemical background noise can also be responsible for high contamination levels, such as lead in the Plumbers region, for example, and in the environments of certain heavily trafficked roads. Organometallic micro pollutants are mixed molecules in which a metal ion is bound to an organic group, such as ethyl mercury and cadmium methyl (Khessairi et al. 2014; Werheni Ammeri et al. 2022).

9.4.5 Ultrafiltration or Nanofiltration Using Membranes

Ultrafiltration and nanofiltration using membranes with pores close to 0.001 μm can keep elements such as bacteria, organic molecules, algae, pollen, or protozoa. This water disinfection process can be integrated into a broader pollution control system or used on its own.

The final phase of wastewater treatment will involve processes using membranes and filters coupled with chemical precipitation and flocculation processes. Precipitants and flocculants are added to the wastewater, causing the substances to flocculate and separate. The wastewater with the flocculated material is then channeled to a membrane or sand filter. Nanofiltration works in a similar way. Unlike normal filtration, water is pressurized through a membrane that keeps the tiniest dissolved particles, such as molecules or heavy metal ions. This is a phenomenon similar to

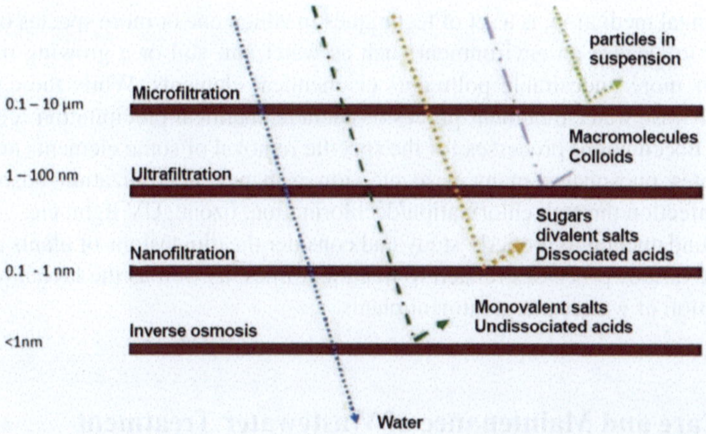

Fig. 9.4 Membrane technologies at different size: microfiltration (MF), ultrafiltration (UF), nanofiltration (NF), and reverse osmosis (RO). According to Tas-Koehler and Lerch (2022)

reverse osmosis, where high pressures and thin membranes are used. Figure 9.4 shows the filtration levels of each technology and the different levels of purity that can be achieved.

To effectively separate solids from complex effluents based on their composition, temperature, acidity, etc., such as waste liquors from metallurgy, dyes, pigments, high-density oils, and drilling oils, using tangential filtration membranes made from durable materials like graphite or ceramics is necessary. In these scenarios, small volumes of highly concentrated and aggressive effluents undergo treatment and filtration through high-speed loop circulations, along with a sophisticated countercurrent washing system that eliminates materials causing cyclic membrane clogging. Lastly, the pollutants collected during the filtration, nanofiltration, and reverse osmosis phases are filtered during sludge treatment, as filter sludge, through the primary wastewater treatment tank. The water finally reaches the last section of the treatment plant, the treated water storage tank.

In conclusion, there are numerous technologies commonly utilized in biological and chemical wastewater treatment processes. The most significant ones have been mentioned, discussed, and evaluated above, such as: (1) Bioremediation, a biochemical process that utilizes living organisms to eliminate various pollutants, compounds, and toxins from soil, water, and other environments. These organisms may consist of microorganisms, primarily bacteria, microphytes or algae, and plant macrophytes. Wastewater bioremediation involves aerobic treatment methods (2) like oxidation ponds, aeration lagoons, aerobic bioreactors, activated sludge, percolating or trickling filters, biological filters, rotating biological contactors, biological nutrient removal, and anaerobic treatment methods such as anaerobic bioreactors and anaerobic lagoons; (3) Phytoremediation, which encompasses constructed wetlands, rhizofiltration, rhizodegradation, phytodegradation, phytoaccumulation, phytotransformation, and hyper accumulators; and finally (4) Mycoremediation, sometimes also called fungal mediation, is a set of techniques in which one or more species of fungi are used to cleanse an environment such as water, air, soil or a growing medium of one or more undesirable pollutants or chemical elements. While the common chemical wastewater treatment processes include chemical precipitation (coagulation and flocculation) processes for the specific removal of some elements and salts like nitrates, phosphates, manganese, etc., ion exchange, neutralization, adsorption, and disinfection through chlorination/dechlorination, ozone, UV light, etc.

It is fundamental to properly study and consider the dimensions of plants and the sizing of various processes related to treatment lines, as well as the layout, design, and location of wastewater treatment plants.

9.5 Care and Maintenance of Wastewater Treatment Systems

As previously mentioned, both combined and separate sewer networks in public sanitation and wastewater treatment systems at a treatment plant require regular maintenance for preventive and/or corrective purposes and occasional maintenance in case of blockages or any malfunctions. This maintenance is vital not only for sanitation networks but also for the different treatment processes carried out at a wastewater treatment plant. It guarantees the smooth operation of the treatment processes, prevents flow problems, and ensures efficient water transfer between the processes. This, in turn, allows for the necessary time for effective treatment by maintaining the residence time in each treatment system.

The urban wastewater collectors transport highly heterogeneous and often bulky materials, especially in combined sewer systems, and a pretreatment of raw water at the wastewater treatment plant is strongly necessary to protect the raw water pumping, pipes against blockages, and other treatment devices against abrasion, and more generally to eliminate anything that could hinder subsequent treatments.

The proper upkeep and functioning of the wastewater pretreatment system requires a sequential execution of screening, sieving, shredding, desanding, and sand removal, as well as degreasing and removal of fats and oils, often in conjunction with desanding, and regular treatment of by-products such as sands, fats, and cleaning products from the network and station sands. Fats and cleaning products are routinely disposed of in landfills for inert products or recycling stations for reusable products like fats. The latter must be recycled and ecologically disposed of in appropriate heating and biodiesel transformation plants.

The primary, secondary, and tertiary processes are continuously and strictly monitored to avoid disruptions to the various treatment systems at the wastewater treatment plant site. It is critical to ensure the consistent and proper supply of wastewater to these systems, as well as the effective distribution and distribution of compressed air. This ensures an oxidative environment in the primary and secondary reactors and pools that facilitates chemical and biological oxidation and promotes microbial development essential for degradation and biological conversion.

Improper execution of control, maintenance, and monitoring operations in the primary and secondary reactors and basins can result in the significant growth of filamentous organisms, causing issues with settling, sludge bulking, and foaming. This can indirectly affect the overall purification efficiency and lead to a decline in discharge quality. Various factors contribute to these phenomena, including high sludge age, low temperatures, low oxygen levels, the presence of anoxic or anaerobic zones, and malfunctioning grease traps. These issues are closely linked to inadequate maintenance and control, particularly in biological processes as activated sludge, which is the most widely, used method for wastewater treatment globally.

To achieve rapid and efficient results, adjusting the control and maintenance parameters of the station's filters (such as water flow control, sludge extraction, and aeration) is commonly paired with an oxidizing treatment that can negatively affect the overall microorganisms. Specialized products have been introduced in the global market like combined aluminum and iron salts utilized for shock curative treatment or continuous preventive treatment. These products aid in swiftly rebalancing the purifying biomass in scenarios of filamentous bacteria bloom, along with the use of rapid examination kits to identify the filamentous species inhabiting reactor waters and select the most appropriate reagent and method of application. These filamentous bacteria can impede the purification process and even lead to the shutdown of a purification plant. Hence, conducting almost daily assessments to detect these bacteria in the treated waters is decisive.

Microthrix parvicella, a filamentous bacterium, is frequently found in activated sludge wastewater treatment plants under various conditions and aeration basin setups. Numerous studies worldwide have highlighted its role in causing issues related to solid–liquid separation, such as bulking and foaming. Extensive efforts have tackled the challenges it poses in WWTPs. *Microthrix parvicella* is a long, thin, non-branched, and unsheathed bacterium with a diameter of 0.6–0.8 μm. The presence of filamentous bacteria, particularly *Microthrix parvicella*, can lead to operational difficulties in activated sludge treatment plants. Chlorination is the most commonly used method to address this issue, although it is not specific and affects the overall activated sludge biology. An alternative approach involves the use of aluminum salts, which target *M. parvicella* specifically while preserving the microfauna and nitrifying bacteria (Fig. 9.5).

Since most wastewater treatment plants worldwide operate on the activated sludge principle, around a quarter of them suffer from biological operating problems known as sludge bulking. This malfunction is characterized by a sludge index (IB) greater than 200 ml g and by excessive development of filamentous bacteria, giving rise to various operating problems, particularly (Durban et al. 2012).

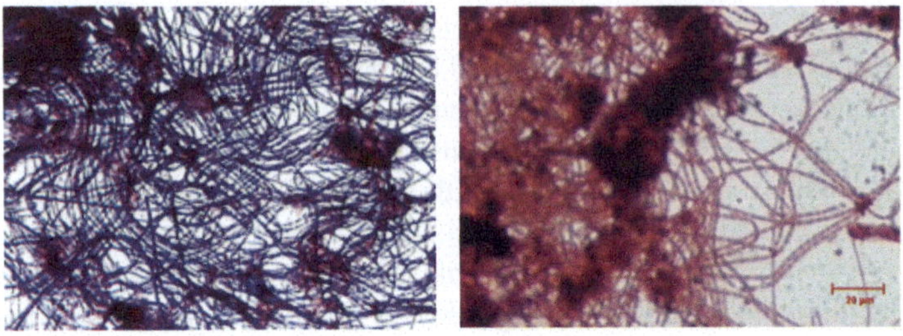

Fig. 9.5 Illustration of *Microthrix parvicella* filamentous bacteria infesting wastewater during treatment

9.6 Qualification of Wastewater Treatment Plant Personnel

It is important to have a specialized administrative and technical team in water and sanitation, as well as a laboratory team specialized in various analyses to ensure the proper functioning and maintenance of a wastewater treatment plant. Specific technical activities include monitoring and maintaining the plant and its facilities, carrying out and monitoring self-control measures, maintaining equipment such as pumps and adjusting control points, managing disinfection and extraction processes, treating sludge produced, logistics and supply chain management, technical control, maintenance, and repair of equipment and materials, as well as compliance with hygiene, health, and safety rules at the station site.

The staff working at a wastewater treatment plant must have more or less these skills or knowledge:

- Knowledge of water treatment principles (processes);
- Basic computer skills;
- Use and maintenance of measuring equipment;
- Water treatment techniques and equipment;
- Equipment maintenance techniques: cleaning, minor repairs, valve operation, unloading, water treatment facility maintenance, reagent injection devices and circuits;
- Knowledge of water chemistry and microbiology;
- Principles of automation;
- Role, use, dosage, and precautions for water treatment products;
- Sludge manufacturing processes;
- Sludge composition, volume reduction techniques;
- Sludge stabilization techniques, sludge, and by-product disposal methods;
- Regulatory knowledge related to sludge production, transportation, and disposal;
- Risks associated with equipment, chemicals used, and treated water;
- Parameters to analyze, role of measurements, and their interpretation.

The working conditions of the staff at the wastewater treatment plant must consider both indoor and outdoor work, with the possibility of frequent travel between different sites if multiple installations need to be managed. It is also important to consider on-call duties during nights and weekends, as well as challenging hygiene conditions during various sanitation operations. Additionally, physical work involving the transportation of bags or barrels of products, as well as handling dangerous chemical reagents, should be expected. The required specializations in major wastewater treatment plants include laboratory analysis technicians, electro mechanics, automation specialists, mechanics, and electricians. It is essential to maintain regular contact with laboratory technicians to track the evolution of self-control parameters.

9.7 Evolution of the Wastewater Treatment Profession and the Importance of a Functional Wastewater Treatment Plant

The main factors influencing the evolution of the wastewater treatment profession at a treatment plant include technical and technological advancements such as automation of plants, the complexity of treatment processes, using smart networks like remote monitoring, geolocation, and connected tools, deepening technical knowledge on different pollutants, life cycle analysis of sludge from treatment plants, changes in public policies, institutional and regulatory framework with stricter standards and enhanced self-monitoring, increasing requirements in terms of treated water quality, consideration of hygiene and safety standards at work, as well as the adoption of circular economy policies, energy production, alternative solutions, and biodiversity preservation.

However, all solid and liquid human waste and discharge must go somewhere, and while this generally refers to garbage and food waste, it also includes wastewater from kitchens and bathrooms. Usually, humans dispose of their water waste by dumping it into rivers or the ocean. However, if they do not treat the waste, they reintroduce contaminated water into our water sources, which can harm wildlife, the ecosystem, and even humans.

In addition to harming wildlife and flora, the solid materials in the water undergo a natural decomposition process. When solid materials decompose, they consume oxygen in the environment. This can be dangerous, as a lack of oxygen in the water can have a significant impact on underwater plants and wildlife, which play an important role in maintaining our ecosystems.

Therefore, a wastewater treatment plant has an important role to play. These plants are responsible for treating all wastewater from an area and making sure it is not reintroduced into the oceans and rivers until it has been well cleaned and treated to remove all solid particles and contaminants.

The operational procedures on a municipal wastewater treatment plant mainly generate and released two types of residues, liquid residues as treated wastewater and solid residues, as sewage sludge (Fig. 9.6).

There are two main destinations for sewage sludge:

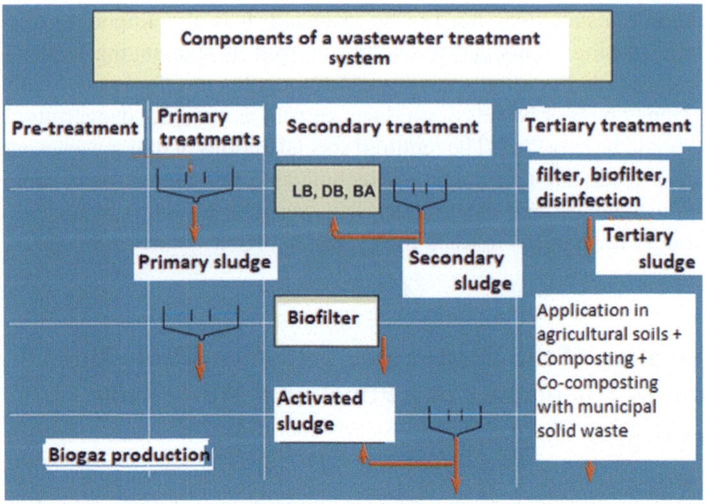

Fig. 9.6 Different procedures operational on the site of a municipal wastewater treatment plant

1. Recycling: mainly in agriculture, with or without supplementation, but also in the rehabilitation of degraded soil (quarries, road embankments, and re-vegetation of landfills), in forestry, in urban landscaping. Recycling allows the main components of the sludge (carbon, nitrogen, and phosphorus) to be reintegrated into the major geochemical cycles through the soil, using their fertilizing power (Marzougui et al. 2022).
2. The elimination or destruction of organic matter with energy recovery through thermal processes (dedicated or co-incineration, pyrolysis/gasification, wet oxidation, or treatment of dehydrated sludge): the goal is to completely and economically oxidize the organic matter in the sludge to recover only an ultimate mineral residue. This residue can, according to current regulations, be valorized in materials engineering or disposed of in a landfill with or without the obligation of inerting depending on the results of leaching tests.

In Europe as reference countries, waste is categorized into three classes for disposal: class 1 for hazardous waste, class 2 for household and similar waste, and class 3 for inert waste. These wastes are deposited in technical landfill sites (TFS) or waste storage centers (WSC). Various constraints related to residual sludge treatment have arisen in the past ten years in many developed countries in Europe and America. These constraints can be procedural, legislative, environmental, economic, etc., and vary depending on the destination of the sludge.

9.8 Extended Conclusion

Considered as mere infrastructure projects for a long time, the construction or renovation of wastewater treatment plants has recently evolved to incorporate various new elements. Public awareness of environmental issues, especially the preservation of nature and the value of water, along with the political part of these projects, has made them more visible and even appealing, potentially attracting visitors for industrial tourism purposes. Consequently, they must now meet new requirements for representation, both in terms of technology and ecology. Downstream from our modern societies, this challenging yet noble task of treating polluted waters has led to the emergence of new types of public facilities that need to be planned with a coherent approach to territorial development, in line with a genuine sustainable development strategy.

For a long time, treated as mere infrastructure projects, the construction or upgrading projects of wastewater treatment plants are undergoing a recent evolution that incorporates several new factors: the public's awareness of environmental issues, particularly the preservation of nature and the value of water; these projects are taking on a political dimension, no longer hidden, but instead highlighted, even visited as part of industrial tourism, thus subject to a new requirement for representation, both technological and ecological.

Downstream of our modern societies, this thankless yet noble function of purifying polluted waters is giving rise to new types of public facilities that must be placed in context and space, within a coherent vision of territorial planning, following a true logic of sustainable development.

These projects have the potential to acquire significant qualitative and cultural value like other public facilities in the city. When choosing the treatment process, it is increasingly common to integrate criteria of urban and landscape analysis in the impact study:

- Long-term urban integration of the project.
- Precise definition of the land area assigned to the project to ensure its integration.
- Consideration of landscape and architectural parts in the design and implementation of structures, their access points, and their environmental impacts, as well as in the cost estimation of the structures.
- Energy optimization and material valorization (cycles).

The study of a wastewater treatment plant project therefore requires close collaboration between engineers, architects, and landscapers and the idea of circular economy, considering wastewater as water resources, must be integrated by decision-makers, legislators, and users. Finally, by considering the wastewater treatment plant as a factory producing a valuable liquid, namely treated wastewater, it is important to work toward strengthening the reuse of this resource in areas like agriculture, water resources, and industry. It is also important to improve water governance by promoting water use planning and coordinating actions aimed at using this resource. It is appropriate to promote a general culture of recycling and circular economy to encourage the emergence of new resource-friendly development models. Furthermore, it is essential to encourage the sharing and pooling of facilities and equipment, especially when consulting with various stakeholders, to support the implementation and monitoring of other water-saving measures, such as centralized irrigation control and water meter monitoring.

References

Al Hamedi FH, Kandhan K, Liu Y, Ren M, Jaleel A, Alyafei MAM (2023) Wastewater irrigation: a promising way for future sustainable agriculture and food security in the United Arab Emirates. Water 15:2284. https://doi.org/10.3390/w15122284

Alm J, Paulsson A, Jonsson R (2021) Capacity in municipalities: infrastructures, maintenance debts and ways of overcoming a run-to-failure mentality. Local Econ 36(2):81–97. https://doi.org/10.1177/02690942211030475

Baechler L (2012) «La bonne gestion de l'eau: un enjeu majeur du développement durable», L'Europe en Formation 3(365):3–21. https://www.cairn.info/revue-l-europe-en-formation-2012-3-page-3.htm. https://doi.org/10.3917/eufor.365.0003

Bdour Ahmed N, Hamdi Moshrik R, Tarawneh Z (2009) Perspectives on sustainable wastewater treatment technologies and reuse options in the urban areas of the Mediterranean region. Desalination 237:162–174, ISSN 0011-9164. https://doi.org/10.1016/j.desal.2007.12.030

Durban N, Juzan L, Krier J, Héduit A (2012) Sylvie Gillot. Treatment of filamentous overgrowth caused by Microthrix parvicella using metal salts. Water Information Days, Sep 2012. Poitiers, France, p 16. ffhal-00801882ff

El-Faramawy H, Mattar T, Eissa M, El-Fawakhry K (2004) Demanganization of high manganese pig iron to produce high manganese slag. Iron Making Steelmaking 31(1):23–30. https://doi.org/10.1179/030192304225012123

Gilbert P (2023) Eutrophisation de l'eau: définition et solutions. Publié le mercredi 21 juin. Aquagir, la Banque des Territoires « Eutrophisation des milieux aquatiques » Écotoxicologie, janvier 2012

Hassen, W., Ben Rejab, A., Hassen, B., Jedidi, N., & Hassen, A. (2023). Investigation of a basic nitrification–denitrification biofiltration system for primary wastewater treatment. Water Quality Research Journal, 58(2), 153–168.

Hidri Y, Fourti O, Eturki S, Charef A, Hassen A (2014) Effects of 15-year application of municipal wastewater on microbial biomass, fecal pollution indicators, and heavy metals in a Tunisian calcareous soil. J Soils Sediments 14(1). https://doi.org/10.1007/s11368-013-0801-4

Khessairi, A., Fhoula I., Jaouani A., Turki Y., Cherif A., Boudabous A, Ouzari H. & Hassen A (2014). Pentachlorophenol Degradation by Janibacter sp., a New Actinobacterium Isolated from Saline Sediment of Arid Land. BioMed Research International. 09/2014; 2014:296472

Marzougui N, Ounalli N, Sabbahi S, Fezzani T, Abidi F, Jebari S, Melki S, Berndtsson R, Oueslati W (2022) How can the use of sewage sludge in sustainable Tunisian agriculture be increased? Sustainability 14:13722. https://doi.org/10.3390/su142113722

Mazuelos R, Alfonso F, Romero CR (2000) High efficiency reactor for the bio-oxidation of ferrous iron. Hydrometallurgy 58(3):269–275. https://doi.org/10.1016/S0304-386X(00)00141-9

Monchy F (2000) Maintenance, méthode et organization. Dunod, Paris, p 512

Nishat A, Yusuf M, Qadir A, Ezaier Y, Vambol V, Khan MI, Ben Moussa S, Kamyab H, Sehgal Satbir S, Prakash C, Yang H, Ibrahim H, Sayed ME (2023) Wastewater treatment: a short assessment on available techniques. Alexandria Eng J 76:505–516, ISSN 1110-0168. https://doi.org/10.1016/j.aej.2023.06.054

Noin D (1998) Population et environnement dans le monde aride. In: Espace, populations, sociétés, 1998–1. Populations et environnement dans le monde aride—populations and the environment in arid countries, pp 13–22. www.persee.fr/doc/espos_0755-7809_1998_num_16_1_1818. https://doi.org/10.3406/espos.1998.1818

Phyu Hnin E, Khaodhiar S (2021) Inspection and understanding of sewer network condition in Dindaeng District, Bangkok, Thailand. Appl Environ Res 43(2):1–23 (LicenseCC BY-NC 4.0). https://doi.org/10.35762/AER.2021.43.2.4

Rajaei M, Nazif S (2022) Improving wastewater treatment plant performance based on effluent quality, operational costs, and reliability using control strategies for water and sludge lines. Process Saf Environ Prot 167:398–411, ISSN 0957-5820. https://doi.org/10.1016/j.psep.2022.09.012

Salihu C, Hussein M, Mohandes SR, Zayed T (2022) Towards a comprehensive review of the deterioration factors and modeling for sewer pipelines: a hybrid of bibliometric, scientometric, and meta-analysis approach. J Clean Prod 351:131460

Samer M (2015) Biological and chemical wastewater treatment processes. Wastewater treatment engineering. InTech. https://doi.org/10.5772/61250

Tas-Koehler S, Lerch A (2022) Chapter 2—application of computational fluid dynamics technique in microfiltration/ultrafiltration processes. In: Basile A, Ghasemzadeh K (eds) Current trends and future developments on (Bio-) membranes. Elsevier, pp 27–61, ISBN 9780128222942. https://doi.org/10.1016/B978-0-12-822294-2.00004-7

United States Steel Corporation (2018) Wastewater treatment, manual and preventative maintenance program plan, manual for the midwest plant's wastewater treatment facilities, pp 1–184

Werheni AR, Di Rauso Simeone G, Hidri Y, Abassi MS, Mehri I, Costa S, Hassen A, Rao MA (2022) Combined bioaugmentation and biostimulation techniques in bioremediation of pentachlorophenol contaminated forest soil. Chemosphere 290(2022):133359. https://doi.org/10.1016/j.chemosphere.2021.133359

WHO, UNICEF (2021) Progress on household drinking water, sanitation and hygiene 2000–2020: five years into the SDGs. World Health Organization (WHO) and the United Nations Children's Fund (UNICEF), Geneva, Switzerland

Chapter 10
Environmental Assessment Methods at the Wastewater Treatment Plants

Zain Muhammad and Abid Mahmood

Abstract Protecting the environment while treating wastewater is crucial. This chapter explains how environmental assessments (EAs) help ensure wastewater treatment plants operate responsibly. We start by highlighting why EAs are important, including following regulations and minimizing environmental impact. The chapter then discusses key concerns like water quality, air emissions, and land use changes from these plants. Next, we explore the different steps involved in an EA, like setting the scope and evaluating potential impacts. Specific tools used, like computer models and field studies, are also explained. Recognizing the importance of public input, the chapter emphasizes involving stakeholders in the EA process. Strategies for effective communication and collaboration are also discussed. Finally, the chapter underlines the significance of clear reports summarizing the EA findings. These reports play a role in permitting decisions, ultimately contributing to more sustainable and responsible wastewater treatment plants. By following these steps and engaging stakeholders, this chapter aims to equip readers with a basic understanding of conducting environmental assessments for wastewater treatment plants.

Keywords Environmental assessment · Impacts · Water · Air · Soil · Wastewater treatment plant

10.1 Introduction

10.1.1 Importance of Environmental Assessment in Wastewater Treatment

Environmental assessment (EA) is critical in wastewater treatment because it offers an understanding of the likely environmental effects due to the change in the status quo before they take place. Perhaps, it is better to regard EA as an attempt to prevent

Z. Muhammad · A. Mahmood (✉)
Department of Environmental Sciences, Government College University, Faisalabad, Pakistan
e-mail: drabid@gcuf.edu.pk

S. Souabi et al. (eds.), *Wastewater Treatment Plants*, Water Science and Technology Library 130, https://doi.org/10.1007/978-3-031-87461-1_10

the emergence of problems, much like regular monitoring of one's health with the goal of early detection of diseases. It is very important in society since it helps in the elimination of wastewater which I so deadly to both human beings and the environment. However, if not managed, it can harm the quality of water, air, and soil and affect the local ecosystem. By performing an EA, it is thus possible to ensure that the negative impacts of developments are reduced and appropriately controlled (Glasson et al. 2012).

For instance, a small town seeking to improve its wastewater treatment plant is a project that requires a large capital investment. An EA may provide information on challenges like; in the future, the rivers in the locality may receive concentrations of nutrients that can cause algal blooms and fish mortality. This way the town shall be in a position to implement approaches aimed at minimizing nutrient release into the river, thus conserving the ecosystem and water quality (Adhikari et al. 2023).

10.1.2 Regulatory Framework and Legal Requirements

As with any other water treatment project, understanding and applying the relevant and existing regulations is always critical. Various hosts have their legal requirements that define how they want the environmental effects of such projects controlled. In the United States, to control the pollutants in Sewage Discharge, the Clean Water Act (CWA) controls the amount by setting effluent standards for wastewater treatment plants also need a permit. Likewise, the EU Water Framework Directive calls for holistic protection of water and efficient use of water resources by all member nations (Khan and Rosenthal 2013).

This acts as a measure to make sure that wastes in wastewater treatment projects do not affect the environment or human health in this way. Any deviation from these standards may attract legal consequences as well as time and money losses as observed in various projects. Thus, realizing and incorporating such legal considerations into the project formulation and implementation phases plays a crucial role in project delivery.

10.1.3 Scope and Objectives of Environmental Assessment

More to, the purpose and the goals of an EA for a wastewater treatment project are the detailed elaboration of the environmental consequences that a given project can have, the determination of their importance, and, finally, the elaboration of a plan for excluding or minimizing the potentially adverse effects. This comprises an assessment of the impact the project might have on the environmental vectors of water, air, soil, and biodiversity. The main goals are to meet environmental standards, safeguard the health of the population, and support development (Noble 2015).

For example, the EA procedure concerning a brand-new wastewater treatment plant may involve an investigation of the possibility of the effluent harming downline water quality; an appraisal of the stench released by the plant and the influence on the neighbouring communities; and the examination of an influence of the plant's construction and functioning on regional wildlife. Planning the assessment helps also in setting efficient objectives and delineating the sphere of concern, which will allow the project planners to concentrate on the most crucial ecological problems, thus saving resources.

10.2 Key Environmental Impacts of Wastewater Treatment Plants

10.2.1 Water Quality Impacts

They help in the improvement or otherwise of water quality depending on whether the effluent is treated or the raw sewage is directly discharged into rivers, controlled lakes, or the ocean. The discharge can introduce such things as nutrients (nitrogen and phosphorus), chemicals, and pathogens into water sources. They cause eutrophication, a situation that makes plants, including excessive algae growth, and use up much oxygen in the water leading to the formation of dead zones that have diverse impacts on water-bound lives (Smith 2003).

For instance, there has been excessive eutrophication in the Baltic Sea caused by nutrients from agricultural drainage and wastewater, resulting in the formation of dead zones that are fatal to marine life and affect fishing businesses (Conley et al. 2009).

10.2.2 Air Quality Impacts

Consequently, the quality of air is directly affected by wastewater treatment plants mainly in terms of odour and GHG emissions including methane (CH_4) and carbon dioxide (CO_2) (Burgués et al. 2022; Pasquarelli et al. 2024). Odour emissions can be felt to be very annoying to the neighbours in the neighbourhood, and hence, their quality of life is affected. Further, the treatment process can emit GHGs into the atmosphere, hence affecting climate change exercise. These problems require efficient odour control measures as well as good GHG measures (Tseng and Wu 2014).

For example, a wastewater treatment plant in California had to integrate a high level of odour control measures after several citizens' complaints. Such measures included putting in place biofilters and chemical scrubbers to keep off foul smells since they are signal negative odour-producing compounds (Wang et al. 2008).

10.2.3 Land Use and Soil Quality Impacts

Wastewater treatment plants also bring changes in the usage of land and soil characteristics since they are built and operated to deal with wastewater. These facilities may use a huge extent of land which affects the land use in the area in one way or the other. Conversely, sludge which is a product that results from the treatment process influences soil quality in case of improper disposal. Principal elements that may be found in sludge include heavy metals and other organic or inorganic pollutants which if applied to land and unprocessed then they cause pollution of the soil and health problems (Fytili and Zabaniotou 2008).

Some measures like anaerobic digestion and composting help in the reduction of the volume of sludge and also in the stabilization of the sludge and hence can be used as conditioners—fertilizers.

10.2.4 Noise and Vibration Impacts

Sounds and oscillations resulting from wastewater treatment facilities are inconvenient for people and animals. These impacts are mostly realized during construction but they might also be evident when the plant is in operation. Similarly, high noise produces stress and affects sleep among the human population leading to health complications. Preventing and reducing the impacts of noise pollution include the use of barriers, insulation, and low/noise emitting equipment (Canter 1996).

For example, a treatment plant near a residential area in the UK installed acoustic enclosures around noisy equipment and scheduled loud activities during daytime hours to minimize disturbances to nearby residents (Walker et al. 2015).

10.3 Environmental Assessment Methods

10.3.1 Screening and Scoping: Identifying Potential Impacts and Defining the Assessment Scope

Evaluating the environment is the first process of EA while mapping or scanning can be considered as the second process. Scoping entails an assessment of whether a certain venture necessitates a comprehensive EA given the influential environmental effects. This step makes sure that the available resources are properly used by directing more detailed assessments to projects that may have large effects. Scoping sets the limits of the EA by indicating which environmental problems and consequences will be studied and analysed. It entails the process of engaging stakeholders, assessing the compliance guidelines, and considering the peculiarity of the given project (Morgan 2012).

For instance, a proposed project such as a wastewater treatment plant goes through a screening process to recommend which of them needs a full EA. If the plant intends to discharge effluent into a critical or protected water body, then, it would go to scoping. In scoping, potential issues like the quality of water smells and effects on the fauna of the surrounding areas would be discovered and further examined.

10.3.2 Baseline Assessment: Establishing Existing Environmental Conditions

The identification of the state of the environment before the start of a particular project is important for any assessment. This includes gathering of information concerning the condition of environmental factors, including water and air, soil type, and flora and fauna within the said project. This is helpful as it will ensure that one has a starting point whereby any other changes which may happen in the future can be ascribed to the specific effect of the intervention (Treweek 2009).

For example, basic data regarding water quality may involve concentration levels of nutrients, solutions' PH level, dissolved oxygen, and the existence of contaminants in a river which is to receive the treated wastewater. It assists in the determination of how this project may potentially adjust these characteristics and in the formulation of measures to counter this.

10.3.3 Impact Assessment: Predicting and Evaluating Potential Environmental Impacts

The technique of impact assessment includes anticipating and estimating the likely environmental effects of the contemplated project. This step employs the application of modelling and risk analysis to gauge the degree and dimension of effects on the different environmental components. The goal is to understand how the project will interact with the environment and to identify any adverse effects (Canter 1996).

For example, using hydrological models, the impact of effluent discharge on downstream water quality can be predicted. These models can simulate different scenarios, such as varying levels of effluent treatment, to evaluate their potential impacts on aquatic ecosystems. The assessment helps in making informed decisions about project design and operation to minimize negative impacts.

10.3.4 *Mitigation and Monitoring: Developing Measures to Reduce Impacts and Monitoring Effectiveness*

Management consists of the identification of measures that will be necessary to prevent, reduce, or compensate the described environmental effects during the impact assessment. Strategies for minimizing the environmental impacts of wastewater treatment plants are an important factor that warrants this paper's focus. Evaluation is as crucial as it checks whether the carried-out measures are being implemented properly and if they are attaining the intended impacts (Morrison-Saunders and Arts 2004).

For example, a treatment plant to reduce nutrient loading from effluent discharge may need to upgrade its technology with nutrient removal processes. Programs would then continually analyse water quality downstream to verify that nutrient levels are at appropriate ranges and use the data for adjusting them.

10.4 Specific Assessment Tools and Techniques

10.4.1 *Modelling Tools: Predicting Impacts on Water, Air, and Soil Quality*

In general, the pro forma models are critical in environmental assessment and forecasting the effects of WWT plants on water, air, and soil qualities. These tools help to predict how each of the environmental factors will behave in response to the various activities of the project. For example, in water quality models, one can forecast the level of pollutants in rivers after discharging effluents into the water body, or the high-risk areas that might be affected and measures to put in place (Chapra 2008).

Models of air dispersion, on the other hand, can predict the concentration of odour and greenhouse gases from treatment plants to develop efficient odour control systems and measures to address greenhouse-effective gases. Soil models also can determine how sludge disposal alters soil characteristics and possible hazards to agriculture and groundwater issues (Lai et al. 2011).

10.4.2 *Field Studies: Collecting Data on Baseline Conditions and Monitoring Impacts*

Field surveys are important as they facilitate the collection of evidence-based data on the initial environmental state and assessment of WWT plant effects. These studies entail the assessment and quantification of the environmental characteristics of the project area. Baseline studies typically cover aspects such as water and air quality, soil properties, and biodiversity, providing a comprehensive snapshot of existing conditions before project implementation (Hill et al. 2005).

For example, field studies might involve sampling water from rivers and lakes to measure baseline nutrient levels and assess the presence of contaminants. Similarly, air quality monitoring might include measuring baseline concentrations of odour-causing compounds and greenhouse gases. These baseline data are essential for detecting changes and assessing the effectiveness of mitigation measures.

10.4.3 Risk Assessment: Evaluating the Likelihood and Severity of Potential Impacts

In this case, it involves the assessment of the likelihood that the impact will occur and the severity if the impact is to occur. Risk assessment is defined as a method of identifying the probability and consequences of the occurrence of adverse environmental effects originating from WWTW (Gosset et al. 2021). This procedure entails recognizing foreseeable threats, defining the degree of susceptiveness of environmental receptors, and evaluating the likelihood of the rate of undesirable effects. The concept aims to identify such risks and then work on creating plans to address each of them (Kaplan and Garrick 1981).

For example, risk assessment of the wastewater treatment plant may include an assessment of the risk of polluting the sources of drinking water. The assessment can then determine the degree of risk to people's health using parameters like concentration of contaminant, distance from water intakes, and existing treatment measures to suggest adequate measures to be taken.

10.4.4 Life Cycle Assessment: Assessing Environmental Impacts Throughout the Entire Lifecycle

It is a comprehensive methodology called the life cycle assessment (LCA) that enables evaluation of WWTP's environmental impacts all over its life cycle beginning with the construction stage up to the reconstruction and demolition stage. LCA focuses on the four Rs of a business encompassing resource extraction, energy, emission, water, and waste usage in a project. This long-term focus aids in the determination of ways to enhance sustainability and decrease the amount of detrimental impact on the environment (Hospido et al. 2004).

For example, an LCA of a wastewater treatment plant might reveal that most of the environmental impact occurs during the operational phase due to energy consumption and chemical use. By identifying these hotspots, project planners can focus on optimizing energy efficiency and exploring alternative, less harmful treatment chemicals.

10.5 Public Participation and Stakeholder Engagement

10.5.1 Importance of Public Involvement in Environmental Assessment

Stakeholders play a very central and crucial role in the environmental assessment of projects thus the EA process with the construction of wastewater treatment plants. Public involvement increases the chances of initiating more acceptable and sustainable projects since the social views of the population possibly affected by the project are taken into consideration. Public participation also improves the levels of transparency as well as trust and helps in decreasing conflict since people have a way of being involved in the decision-making processes (Petts 2003).

For example, in the development of a new wastewater treatment plant in a small town, early and meaningful public engagement can help address community concerns about potential odours, water quality impacts, and land use changes. By involving residents in discussions and decision-making, the project proponents can better tailor their plans to mitigate adverse effects and foster community support.

10.5.2 Strategies for Effective Stakeholder Engagement

Strategic management of stakeholder relationships is an important process based on an adequate mapping of the key stakeholder groups and taking into account their opinion and feedback. Key strategies for successful engagement include:

Early Engagement: Consulting stakeholders on the issue early enough, when planning the project to be able to observe issues that may be a cause for concern according to the project plan.

Inclusive Participation: Making sure that people inside the neighbourhood, environmental organizations, and relevant government and non-government organizations and industries have the chance to express themselves.

Transparent Communication: Informing the public about the project details, its possible effects, and what EA is all about. This can range from public assembly, awareness creating distributing of fliers, production of photocopy newspapers, production of newspapers and magazines, production of newsletters, Websites, and Facebook among others.

Facilitation and Mediation: They include having independent persons conducting public affairs meetings and moderating people's interactions, to solve differences and establish unity.

Feedback Mechanisms: Creating avenues through which the stakeholders can give their input and making sure that their inputs are recorded and acted upon.

For instance, a wastewater treatment project in Ontario, Canada, was able to involve the stakeholders through stakeholder workshops, stakeholder open houses, and stakeholder online questionnaires. These methods assisted in the process of collecting rather useful information from the community that was incorporated into the project to respond to the community's concerns as well as to enhance the impact on the surroundings (Sinclair and Diduck 2001).

10.6 Environmental Assessment Reporting and Decision-Making

10.6.1 Structure and Content of Environmental Assessment Reports

The EA report is the summary, conclusion, and recommendations document of the assessment process that has been carried out. This way it functions as a valuable instrument for decision-makers, interested parties, and the public as it contains information on the likely effects on the environment and the measures, aimed at preventing or minimizing such effects, planned for a proposed project. An effective EA report typically includes the following sections (Glasson et al. 2012):

Executive Summary: A concise overview of the report's key findings and recommendations.

Introduction: Background information on the project, its purpose, and the scope of the EA.

Project Description: Detailed description of the proposed project, including location, design, construction, and operational details.

Methodology: Explanation of the methods and tools used to conduct the EA, including screening, scoping, baseline studies, impact assessment, and mitigation planning.

Baseline Conditions: Comprehensive description of the existing environmental conditions in the project area, covering water, air, soil, biodiversity, and socio-economic aspects.

Impact Assessment: Detailed analysis of the potential environmental impacts of the project, including both direct and indirect effects.

Mitigation Measures: Description of the proposed measures to mitigate adverse impacts, including design changes, operational controls, and monitoring programs.

Public and Stakeholder Engagement: Summary of the public and stakeholder engagement process, including key concerns raised and how they were addressed.

Conclusion and Recommendations: Summary of the overall findings, conclusions, and recommendations for project approval or modification.

10.6.2 Role of Environmental Assessment in Permitting and Approval Processes

It is worth mentioning that the EA report becomes important when obtaining the permit for wastewater treatment plants and other significant projects. The authorities need the report to assess compliance of the intended undertaking with environmental legislation and guidelines to approve it, amend it, or reject it. Every decision made regarding an EA is supported by data that came from previous assessments, therefore minimizing the risks of making wrong decisions due to a lack of understanding of the environmental effects of a project and the actions that will be taken to prevent or counter them (Wood 2003).

For instance, in the United States, the National Environmental Policy Act (NEPA) mandates federal agencies in the evaluation of environmental implications of their planned actions by producing Environmental Impact Statements (EIS). The EIS process entails the participation of the public and interagency consultations to pave the way for environmental impacts into federal actions and decisions. There are other similar processes in other countries; for instance, the Environmental Impact Assessment Directive in the EU and its member states require the conduct of EAs on some specific public and private projects (Sheate 2010).

By providing a transparent and systematic approach to evaluating environmental impacts, EA reports help to ensure that projects are developed in a manner that protects environmental quality and public health. They also facilitate informed decision-making, allowing stakeholders to understand the potential benefits and trade-offs associated with proposed projects.

10.7 Conclusion

10.7.1 Importance of Continuous Improvement and Future Trends in Environmental Assessment

Sustainability is indispensable in EA because it helps wastewater treatment plants and other projects update on emerging environmental problems and changing standards. EA is not stagnant and has continued to experience development in aspects such as technology, methodology, and stakeholders. Adopting such innovations could incredibly help measure positive effects and responsible solutions to reduce these impacts for the general improvement of environmental and social performances (Bond et al. 2013; Heredia et al. 2024).

Further enhancement can be operated by deploying modern technologies including remote sensing, GIS, and AI. The following technologies can be used to increase the accuracy and expedite the processes of collecting baseline data and impact modelling and monitoring. For instance, remote sensing also gives real-time information on the dynamics of change in land use and water quality, and AI could further assist in assessing big data and come up with meaning patterns and predictive analytics for the impact assessment (Bastin et al. 2019).

Another trend proposes a consideration of comprehensive sustainability and the principles of the circular economy in EA. This requires not only the application of procedures limiting the negative potential effects of treated wastewater and biosolids on the environment but also searching for possibilities of obtaining positive results that can include resource recovery and reuse. An integration of some sort of sustainability indicators and life cycle assessment into EA can assist the project proponents to come up with more sustainable wastewater treatment systems which advance environmental and social values (Bengtsson et al. 2010).

10.7.2 Conclusion and Recommendations

Consequently, environmental assessment is a valuable means of regulating the possibility of polluting wastewater treatment plants to protect the environment and the population's health. For this reason, EA provides a methodical approach to assess, additionally analyse and manage those consequences on water, air, soil, and other components of the environment considering potential risks and stakeholders' involvement in decision-making. According to the techniques used in EA, effective tools like modelling, field studies, risk assessment, and life cycle assessment add more strength to EA in regards to the predictive measures of the consequences that are likely to impact the environment.

Subsequently, practitioners have to pave their way towards embracing change and follow the progressive enhancement best practices and stay updated in the light of advancements in the field of technology. This also implies beginning to adopt the various ideas regarding data collection and analysis, embracing the principles relating to sustainability, and improving on the strategies used in the management of stakeholders. Thus, the EA process can be improved for the serious and dynamic environmental issues of wastewater treatment and global infrastructure.

Recommendations for practitioners and policymakers include:

- Investing in training and capacity building to ensure that EA practitioners are equipped with the latest tools and techniques.
- Promoting collaboration and knowledge sharing among stakeholders to foster innovation and best practices in EA.
- Encouraging the integration of sustainability and circular economy principles into EA processes to maximize positive environmental and social outcomes.

- Enhancing public and stakeholder engagement to ensure that diverse perspectives are considered and that projects reflect community values and needs.

By implementing these recommendations, the field of environmental assessment can continue to evolve and play a vital role in promoting sustainable development and environmental stewardship.

References

Adhikari S, Kumar R, Driver EM, Bowes DA, Ng KT, Sosa-Hernandez JE, Oyervides-Muñoz MA, Melchor-Martínez EM, Martínez-Ruiz M, Coronado-Apodaca KG, Smith T, Bhatnagar A, Piper BJ, McCall KL, Parra-Saldivar R, Barron LP, Halden RU (2023) Occurrence of Z-drugs, benzodiazepines, and ketamine in wastewater in the United States and Mexico during the Covid-19 pandemic. Sci Total Environ 857. https://doi.org/10.1016/j.scitotenv.2022.159351

Bastin L, Buchanan G, Beresford AE, Pekel JF, Dubois G (2019) Remote sensing and GIS for environmental assessment. Environ Model Softw 121:104509

Bengtsson M, Lundin M, Molander S (2010) Life cycle assessment of wastewater systems: influence of system boundaries and scale on environmental performance. J Environ Manage 91(1):61–71

Bond A, Pope J, Morrison-Saunders A, Retief F, Gunn JAE (2013) Impact assessment: eroding benefits through streamlining? Environ Impact Assess Rev 45:46–53

Burgués J, Doñate S, Esclapez MD, Saúco L, Marco S (2022) Characterization of odour emissions in a wastewater treatment plant using a drone-based chemical sensor system. Sci Total Environ 846. https://doi.org/10.1016/j.scitotenv.2022.157290

Canter LW (1996) Environmental impact assessment. McGraw-Hill

Chapra SC (2008) Surface water-quality modeling. Waveland Press

Conley DJ, Paerl HW, Howarth RW et al (2009) Controlling eutrophication: nitrogen and phosphorus. Science 323(5917):1014–5

Fytili D, Zabaniotou A (2008) Utilization of sewage sludge in EU application of old and new methods—a review. Renew Sustain Energy Rev 12(1):116–40

Glasson J, Therivel R, Chadwick A (2012) Introduction to environmental impact assessment. Routledge

Gosset A, Wiest L, Fildier A, Libert C, Giroud B, Hammada M, Hervé M, Sibeud E, Vulliet E, Polomé P, Perrodin Y (2021) Ecotoxicological risk assessment of contaminants of emerging concern identified by "suspect screening" from urban wastewater treatment plant effluents at a territorial scale. Sci Total Environ 778. https://doi.org/10.1016/j.scitotenv.2021.146275

Heredia IB, Belén González-Gaya B, Zuloaga O, Garrido I, Acosta T, Etxebarria N, Ruiz-Romera E (2024) Occurrence of emerging contaminants in three river basins impacted by wastewater treatment plant effluents: spatio-seasonal patterns and environmental risk assessment. Sci Total Environ 946. https://doi.org/10.1016/j.scitotenv.2024.174062

Hill MJ, Furlong C, Young R (2005) Environmental field techniques and laboratory methods. Wiley-Blackwell

Hospido A, Carballa M, Moreira MT, Omil F (2004) Environmental assessment of anaerobically digested sludge reuse in agriculture: potential impacts of emerging micropollutants. Water Res 44(10):3225–33

Kaplan S, Garrick BJ (1981) On the quantitative definition of risk. Risk Anal 1(1):11–27

Khan FI, Rosenthal A (2013) Sustainable environmental assessment: concepts, methods, and applications. Springer

Lai SH, Masi F, Manios T (2011) Soil modeling for sludge disposal impact assessment. Desalin Water Treat 30(1–3):252–7

Morgan RK (2012) Environmental impact assessment: the state of the art. Impact Assess Proj Apprais. 30(1):5–14

Morrison-Saunders A, Arts J (2004) Assessing impact: handbook of EIA and SEA follow-up. Earthscan

Noble BF (2015) Introduction to environmental impact assessment: a guide to principles and practice. Oxford University Press

Pasquarelli F, Oliva G, Mariniello A, Buonerba A, Li C-W, Belgiorno V, Naddeo V, Zarra T (2024) Carbon neutrality in wastewater treatment plants: an integrated biotechnological-based solution for nutrients recovery, odour abatement and CO_2 conversion in alternative energy drivers Chemosphere 354. https://doi.org/10.1016/j.chemosphere.2024.141700

Petts J (2003) Barriers to participation and deliberation in risk decisions: evidence from waste management. J Risk Res 6(2):115–33

Sheate WR (2010) Tools, techniques and approaches for sustainability: collected writings in environmental assessment policy and management. World Scientific

Sinclair AJ, Diduck AP (2001) Public involvement in EA in Canada: a transformative learning perspective. Environ Impact Assess Rev 21(2):113–36

Smith VH (2003) Eutrophication of freshwater and coastal marine ecosystems: a global problem. Environ Sci Pollut Res Int 10(2):126–39

Treweek J (2009) Ecological impact assessment. Wiley-Blackwell

Tseng LY, Wu CH (2014) Greenhouse gas emissions from wastewater treatment plants: a case study. Environ Eng Sci 31(7):353–8

Walker S, Simmons P, Evans D (2015) Noise pollution mitigation in a residential area adjacent to a wastewater treatment plant. J Environ Manage 150:215–23

Wang LK, Shammas NK, Hung YT (2008) Advanced physicochemical treatment technologies. Humana Press

Wood C (2003) Environmental impact assessment: a comparative review. Prentice Hall

Chapter 11
Assessment of the Condition of a Wastewater Treatment Plant

Ananya Khasnabis, Chibibharath Parthasarathy, Ankita Rani, Vaishnavi Rameshwar Kokurde, and Senthilmurugan Subbiah

Abstract This chapter focuses on the assessment of wastewater treatment plants (WWTPs) and the various techniques and methodologies used to evaluate their condition and performance. Several techniques are employed, including physical inspections, maintenance records review, process monitoring and control, performance evaluation and benchmarking, structural and mechanical integrity assessments, advanced data analysis and modeling, and environmental and health impact assessments. Physical inspections and maintenance records review are crucial for identifying signs of wear, damage, or systemic problems that may affect the plant's operation. Process monitoring and control, including online monitoring and laboratory analysis, help ensure optimal operating conditions and assess treatment efficiency. Performance evaluation and benchmarking involve assessing treatment efficiency and capacity utilization, and comparing performance indicators against industry standards or similar plants. Structural and mechanical integrity assessments utilize non-destructive testing techniques and comprehensive evaluation programs to prioritize maintenance and rehabilitation efforts. Advanced data analysis and modeling, such as data analytics and hydraulic and process modeling, are employed to analyze historical operational data, predict future performance, and optimize process design and operation. Environmental and health impact assessments, including life cycle assessment and public health impact studies, evaluate the environmental impacts of WWTPs throughout their lifecycle and assess the effectiveness of pathogen and toxic compound removal to minimize health risks. Compliance audits and regulatory reviews ensure adherence to environmental regulations, permits, and standards. Overall, this chapter emphasizes the importance of conducting comprehensive assessments that consider environmental and operational aspects to enhance sustainability and flexibility in wastewater treatment.

Keywords WWTP assessment · Performance evaluation · Process monitoring · Environmental impact assessment · Sustainability in wastewater treatment

---

A. Khasnabis · C. Parthasarathy · A. Rani · V. R. Kokurde · S. Subbiah (✉)
Department of Chemical Engineering, Indian Institute of Technology Guwahati, Guwahati, Assam, India
e-mail: senthilmurugan@iitg.ac.in

S. Souabi et al. (eds.), *Wastewater Treatment Plants*, Water Science and Technology Library 130, https://doi.org/10.1007/978-3-031-87461-1_11

11.1 Introduction

The assessment of the condition of wastewater treatment plants (WWTPs) is crucial for ensuring their effective operation and compliance with environmental regulations. Wastewater, as defined by the World Health Organization (WHO), includes any water that has been adversely affected in quality by anthropogenic influence. This encompasses any liquid waste discharged from commercial properties, domestic residences, industry, and agriculture, and can include an extensive array of potential contaminants and concentrations (Onu et al. 2023).

Modern WWTPs must address a variety of contaminants to protect public health and the environment. These contaminants are categorized into physical, chemical, and biological parameters. Physical contaminants include suspended solids that affect water clarity and can harbor harmful microorganisms. Chemical contaminants cover a broad spectrum, including heavy metals, nutrients like nitrogen and phosphorus, and a variety of organic compounds. Biological contaminants mainly consist of pathogens such as bacteria, viruses, and protozoa, which can pose significant health risks if not adequately treated.

In addition to traditional contaminants, emerging contaminants include over 3000 types of compounds and their derivatives, such as pesticides, fertilizers, heavy metals, microplastics, pharmaceutically active compounds (PhACs), personal care products (PPCPs), natural and synthetic hormones usually ending up in the environment. Pharmaceuticals and antibiotics, widely used in human and veterinary medicine, often pass through conventional treatment processes and enter water bodies, potentially leading to antibiotic resistance and ecological harm. Microplastics, originating from various sources such as cosmetic products, synthetic textiles, and plastic waste degradation, persist in the environment and pose significant risks to aquatic life and human health. The assessment and removal of these emerging contaminants require advanced treatment technologies and continuous monitoring to ensure they do not compromise water quality and safety (Puri et al. 2023).

Energy efficiency has become a critical component in the operation of WWTPs. As energy-intensive facilities, WWTPs consume substantial amounts of electricity, primarily for aeration and pumping processes. Therefore, energy efficiency measures such as the integration of variable frequency drives (VFDs), advanced aeration systems, and real-time monitoring of energy consumption are essential to reduce operational costs and environmental impact. Lowering energy consumption and prices can reduce WWTP running costs. At the same time, determining the best use of energy during the wastewater treatment process may result in reduced levels of GHG emissions into the environment. This results in assessments of the energy efficiency of the wastewater treatment process as well as quantification of the energy saved.

The growing urbanization and industrialization worldwide have led to increased wastewater generation, necessitating the development of advanced treatment technologies and robust assessment methodologies. Effective WWTPs are integral to sustainable urban development, protecting water bodies from pollution and ensuring the safe reuse of treated effluents. Water reuse and recycling are vital aspects of modern wastewater management, aimed at conserving water resources and reducing the environmental footprint of water usage. Advanced treatment processes, including membrane filtration and advanced oxidation, enable the production of high-quality effluent suitable for various reuse applications, such as agricultural irrigation and industrial processes.

Assessment techniques for WWTPs involve physical inspections, maintenance records review, and process monitoring and control. Performance evaluation and benchmarking are also essential, involving the comparison of treatment efficiency and capacity utilization against industry standards or similar facilities. Structural and mechanical integrity assessments use non-destructive testing (NDT) techniques to prioritize maintenance and rehabilitation efforts.

Advanced data analysis and modeling, including hydraulic and process modeling, are employed to analyze historical operational data, predict future performance, and optimize process design and operation. Environmental and health impact assessments evaluate the impacts of WWTPs throughout their lifecycle, ensuring the removal of pathogens and toxic compounds to minimize health risks.

Condition assessment techniques encompass a variety of methods aimed at ensuring optimal performance, longevity, and sustainability of treatment facilities. These methods not only ensure the efficient and sustainable operation of treatment plants but also contribute to environmental protection and resource conservation. The use of real-time sensors and instrumentation for continuous monitoring of key parameters, such as chemical concentrations, dissolved oxygen, pH, and turbidity facilitates immediate detection of deviations from normal operating conditions, enabling prompt corrective actions.

This chapter emphasizes the importance of comprehensive assessments that consider environmental and operational aspects to enhance the sustainability and flexibility of wastewater treatment processes. Ensuring compliance with regulatory standards through audits and reviews is also highlighted as a critical component of effective WWTP management.

Overall, the integration of energy-efficient practices, water reuse, advanced monitoring technologies, and the assessment of emerging contaminants in the condition assessment of WWTPs is essential for achieving sustainable wastewater management and protecting public health and the environment.

11.2 Contaminants in Wastewater: Sources, Levels, and Detection

This section delves into the sources, concentration ranges, and measurement techniques for various contaminants in wastewater. It highlights the differences in feasibility and implementation of these techniques in developed and developing countries.

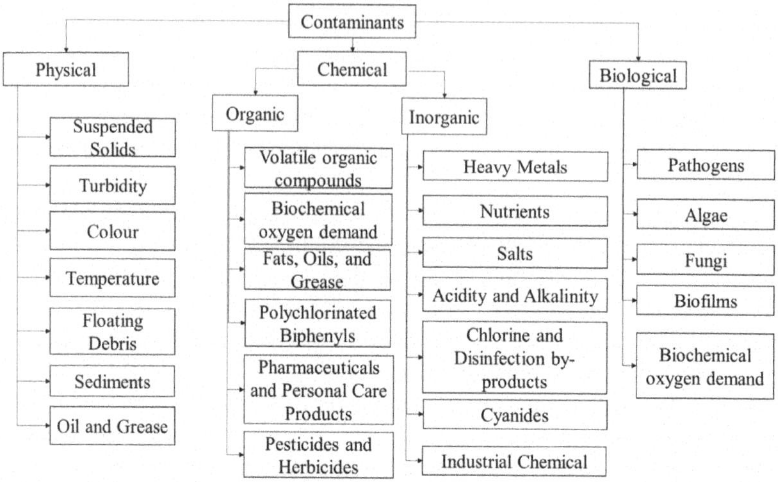

Fig. 11.1 Classification of contaminants in wastewater

In wastewater, the contaminants can be broadly categorized as physical, chemical, and biological contaminants (Fig. 11.1) (Palansooriya et al. 2020; Binnie et al. 2009).

11.2.1 Physical Contaminants

In wastewater, physical contaminants primarily include suspended solids, turbidity, temperature variations, microplastics, and color. These contaminants are typically the most visible and may immediately affect the physical properties of water (Shah et al. 2023).

Suspended solids from urban runoff, industrial discharge, agricultural runoff, and domestic wastewater are typically measured using gravimetric analysis, turbidimetry, and real-time sensors. Developed countries utilize high-cost, high-precision instruments, whereas developing countries rely on more cost-effective manual methods.

11.2.2 Chemical Contaminants

Chemical contaminants in wastewater encompass a wide range of substances, including nutrients, heavy metals, pesticides, pharmaceuticals, oil, grease, chlorides, sulphates, phenols, and cyanides (Garg et al. 2022). These contaminants can have significant environmental and health impacts if not adequately treated.

Heavy metals such as cadmium, chromium, and lead are prevalent in industrial discharges and agricultural runoff (Mitra et al. 2022). Measurement techniques like AAS, ICP-MS, and ASV vary in cost and precision, with advanced techniques being more common in developed countries. Emerging organic contaminants (EOCs) from personal care and pharmaceutical products are detected using LC-MS, ELISA, and electrochemical sensors. Developed countries favor LC-MS for its sensitivity, while ELISA and electrochemical sensors are more practical in developing regions. Microplastics, originating from synthetic textiles and degraded plastic waste, are identified using filtration, microscopy, FTIR, and Raman spectroscopy. Chemical contaminants such as pesticides, phenolic compounds, and radioactive materials can originate from sources like industrial discharge and agricultural runoff. Pesticides, typically absent, are detected using methods like LSLE, GC/MS/ECD, SPE, and HPLC-UV. Phenolic compounds and radioactive materials are analyzed using spectrophotometry, Alpha counter and Geiger Muller counting systems. Detailed analysis techniques are more feasible in developed countries, while simpler methods are preferred in developing areas.

11.2.3 Biological Contaminants

Biological contaminants in wastewater primarily include pathogens (bacteria, viruses, and protozoa), chemical oxygen demand (COD), biochemical oxygen demand (BOD), and algae. These contaminants pose risks to the environment, and human health if not properly managed.

Pathogens from domestic sewage and agricultural runoff are measured using CFU counting, PCR, and qPCR. While PCR and qPCR offer high accuracy and speed and are standard in developed countries, CFU counting is widely used in developing countries due to its low cost and simplicity.

By understanding the sources and measurement techniques for these contaminants, WWTPs can better design their processes to protect public health and comply with environmental regulations (Table 11.1). The feasibility of implementing these techniques varies significantly between developed and developing countries, highlighting the need for cost-effective and practical solutions in resource-limited settings.

11.3 Condition Assessment Techniques for Wastewater Treatment Plants

The assessment of the condition of WWTPs is essential for ensuring their optimal performance, longevity, and sustainability. Over the period, various methodologies are employed to evaluate the structural, mechanical, and operational aspects of these facilities (Fig. 11.2). They commonly follow a similar progression of steps, which

Table 11.1 Summary of common water contaminants and their measurement techniques

Contaminant	Standards for discharge*	Concentration range	Measurement techniques**
Physical contaminants			
• Suspended solids	< 100 mg/L	100–350 mg/L	Gravimetric analysis, turbidimetry, real-time sensors
• Oil and grease	< 10 mg/L		Partition gravimetric method, partition infrared method, Soxhlet extraction method
Chemical contaminants			
• Total residual chlorine	< 1.0 mg/L		Titrimetric methods, spectrophotometry
• Ammoniacal nitrogen	< 50 mg/L		Spectrophotometry
• Chloride	< 1000 mg/L		Titrimetric methods
• Fluoride	< 2 mg/L		Spectrophotometry
• Sulphates	< 1000 mg/L		Spectrophotometry
• Dissolved phosphates	< 5.0 mg/L		Spectrophotometry
• Heavy metals	Arsenic (< 0.2 mg/L) Mercury (< 0.01 mg/L) Lead (< 0.1 mg/L) Cadmium (< 2.0 mg/L) Total Chromium (< 2.0 mg/L) Hexa. Chromium (< 0.1 mg/L) Copper (< 3.0 mg/L) Zinc (< 1.0 mg/L) Selenium (< 0.05 mg/L) Nickel (< 3.0 mg/L) Boron (< 2.0 mg/L)	Lead (0.005–0.2 mg/L), Cadmium (0.001–0.05 mg/L), Chromium (0.005–0.5 mg/L)	AAS, ICP-OES/MS, ASV
• Cyanide	< 0.2 mg/L		Cyanide ion selective electrode
• Surfactants	1–100 mg/L		Surface tension measurement, HPLC, colorimetry
• Emerging organic contaminants (EOCs)	–	ng/L to μg/L	LC-MS, ELISA, Electrochemical Sensors

(continued)

Table 11.1 (continued)

Contaminant	Standards for discharge*	Concentration range	Measurement techniques**
• Microplastics	–	0.1–100 particles/L	Filtration and microscopy, FTIR, Raman spectroscopy
• BOD and COD	BOD: < 30 mg/L COD: < 250 mg/L	BOD: 100–400 mg/L, COD: 250–1000 mg/L	Titrimetric methods, spectrophotometry, real-time sensors
• pH	5.5–9.0	pH: 6.0–9.0	pH meters, pH strips, real-time sensors
• TDS	< 2100 mg/L	500–1500 mg/L	Gravimetric analysis, conductivity meters, real-time sensors
• Pesticides	Absent		Liquid–solid/liquid extraction, capillary column gas chromatography/mass spectrometry/electron capture detector, solid phase extraction and high-performance liquid chromatography with UV detection
• Phenolic Compounds	< 1 mg/L		Spectrophotometry
• Radioactive materials • Alpha emitters • Beta emitters	Alpha: < 10^{-7} (Microcurie/ml) Beta: < 10^{-6} (Microcurie/ml)		Alpha counter and Geiger Muller counting system
• Volatile organic compounds (VOCs)	Varies by compound; e.g., < 0.005 mg/L for benzene (EPA)	µg/L to mg/L	Gas chromatography-mass spectrometry (GC-MS)
• Chlorine (residual)	< 0.1 mg/L (EPA for aquatic life)	0.01–5.0 mg/L	Colorimetric method (DPD)
Biological contaminants			
Pathogens		*E. coli*: 10^2 to 10^6 CFU/100 mL	CFU counting, PCR, real-time PCR (qPCR)
Pathogens (bacteria, viruses, protozoa)	Zero tolerance for drinking water	Varies by pathogen	Membrane filtration, polymerase chain reaction (PCR), immunofluorescence assay (IFA)
Algae	Varies by local standards	< 10 µg/L	Fluorometry

(continued)

Table 11.1 (continued)

Contaminant	Standards for discharge*	Concentration range	Measurement techniques**
E. coli and coliforms	Zero tolerance for drinking water	0–10,000 CFU/ 100 ml	Most probable number (MPN) method
Cryptosporidium and *Giardia*	Zero tolerance for drinking water	Presence/absence	Immunofluorescence assay (IFA), polymerase chain reaction (PCR)

*Central pollution control board (CPCB) standards for discharge: inland surface water—general standards for discharge of environmental pollutants Part-A: effluent
**Standard methods for the examination of water and wastewater APHA, AWWA and WEF and methods of sampling and test (physical and chemical) for water and wastewater

includes setting objectives for condition assessment, identification of assets and available data, asset inspection, data analytics, and decision-making. This section explores the primary techniques used in condition assessment for WWTPs.

11.3.1 Visual Inspection and Scoring Systems

Visual inspections are a crucial element in assessing the condition of WWTPs. These inspections involve examining the physical state of plant components and assigning scores based on observed defects. For example, some assessments use the Visual Degradation Index (DI), which ranges from 1 (no noticeable defects) to 5 (extensive deterioration or defects), providing a quick overview of the plant's condition. This method is effective for identifying areas that need immediate maintenance or upgrades.

Visual Inspection and Scoring Systems are essential techniques in WWTP condition assessment. These methods include detailed evaluations of the physical state of various plant components, offering critical insights into their current condition and potential issues. This section covers the benefits, standards, global practices, challenges, reliability, and necessary instruments for visual inspections.

11.3.1.1 Benefits of Visual Inspection and Scoring Systems

- **Early Detection of Defects**: Visual inspections allow for early identification of physical defects such as cracks, corrosion, wear, and other signs of deterioration. Early detection can prevent more severe damage and costly repairs.
- **Operational Efficiency**: Regularly assessing the condition of components allows proactive maintenance, reducing unplanned downtime and enhancing overall operational efficiency.

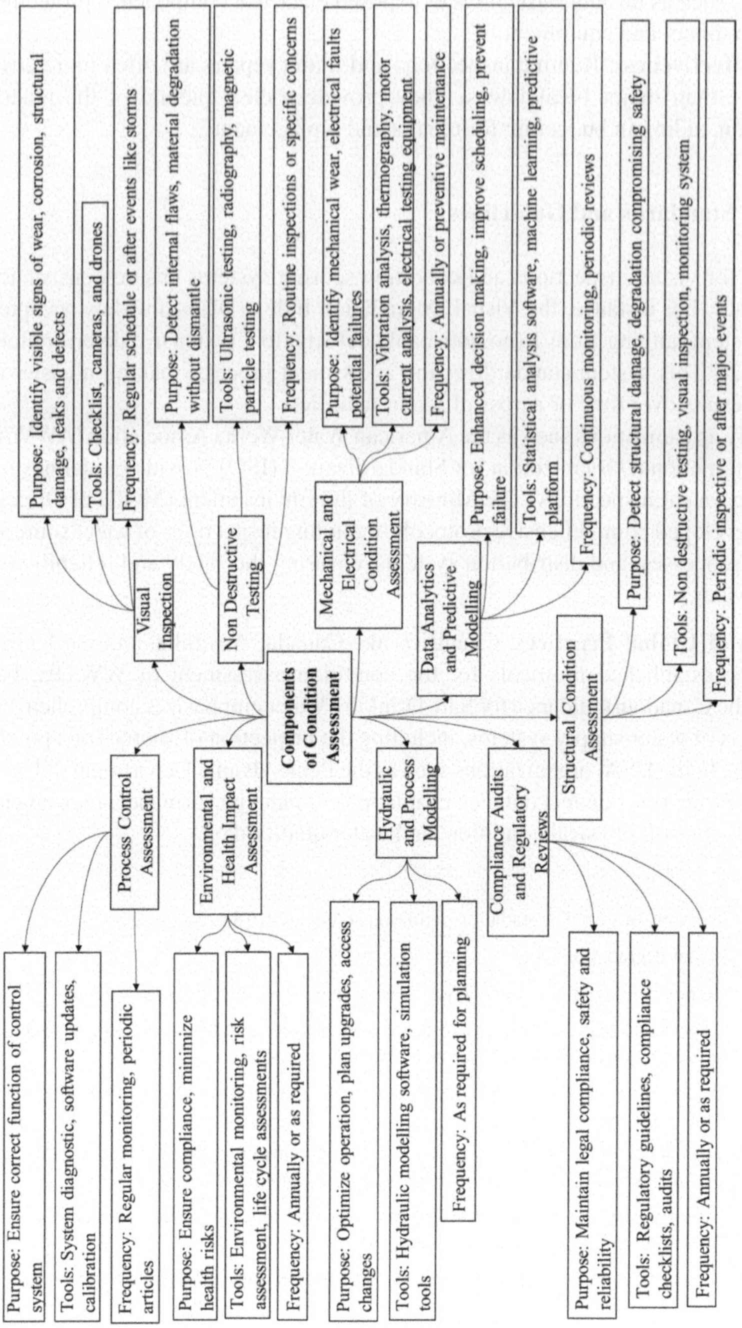

Fig. 11.2 Components of condition assessments

- **Safety Assurance**: Routine visual inspections help identify and address safety hazards, such as unstable structures or exposed electrical components, protecting both personnel and equipment.
- **Cost-Effectiveness**: Routine inspections and minor repairs are often more cost-effective than major breakdowns. They provide a clear picture of the plant's condition, aiding in budgeting for repairs and replacements.

11.3.1.2 Standards and Guidelines

Standards for visual inspections and condition scoring systems ensure consistency and accuracy. For instance, the Visual Degradation Index (DI) is a widely accepted scoring system ranging from 1 (no noticeable defects) to 5 (extensive deterioration) (Table 11.2). This system standardizes the assessment process, making it easier to compare results over time or across different facilities.

Globally, organizations such as the American Water Works Association (AWWA) and the International Organization for Standardization (ISO) provide guidelines for conducting visual inspections. The Ministry of the Environment (MOE) of Ontario has also developed comprehensive protocols, including inspections of water sources, treatment processes, and distribution systems, ensuring the safety and reliability of drinking water.

Examples of Global Practices: Countries like Canada, Australia, and the United States have established protocols for the condition assessment of WWTPs. For example, the Canadian Guidance for Safe Drinking Water emphasizes comprehensive assessments of water supply systems, including the inspection of sources and potential hazards. In the USA, organizations such as the Pennsylvania Department of Environmental Protection conduct detailed evaluations of plant performance, focusing on operational control, physical condition, and water quality data.

Table 11.2 Overview of visual degradation definitions (Asquer 2018)

Score	Visual degradation index
1	No noticeable defects; some aging or wear may be visible; fully functional
2	Only minor deterioration or defects are evident; noticeable wear or aging is visible; fully functional; minor maintenance may be required
3	Deterioration or defects are evident; function is not significantly affected. Minor repairs are required
4	Serious deterioration or defects in at least some portion of the asset; function is significantly affected; extensive repairs or replacement are required
5	Extensive deterioration or defects in the asset. Not functional, barely functional, beyond repair, or a dangerous safety condition exists
NA	This asset was not evaluated and/or was not present during the assessment

11.3.1.3 Challenges and Reliability

While visual inspections are invaluable, they do face challenges. Subjectivity can be a significant issue, as different inspectors may interpret conditions differently. Additionally, visual inspections are limited to detecting surface-level issues and may not identify underlying structural problems.

To enhance reliability, it is essential to complement visual inspections with other assessment techniques, such as NDT methods. These combined approaches offer an added comprehensive understanding of the plant's condition.

11.3.1.4 Instruments Required for Visual Inspections

- High-resolution cameras for documenting defects and conditions.
- Inspection mirrors for viewing hard-to-reach areas.
- Ultrasonic thickness gauges for measuring material thickness and detecting corrosion.
- Flashlights and portable lights for illuminating dark areas.
- Binoculars for inspecting high or distant components.

These tools enable inspectors to conduct thorough evaluations and accurately record findings.

11.3.1.5 Methodology for Visual Inspection and Scoring Systems

The methodology for visual inspection and scoring systems in WWTPs involves a systematic process to evaluate the physical and operational condition of plant components. This process includes planning, conducting the inspection, data analysis, reporting, and follow-up (refer to Fig. 11.3).

Planning and Preparation

Proper planning and preparation are essential to conducting an effective visual inspection of WWTP components. This phase ensures that the inspection is well-organized, comprehensive, and aligned with the overall goals of the condition assessment. The following steps outline the key activities involved in the planning and preparation process:

- Define Objectives: The first step in the planning process is to clearly define the objectives of the inspection. This involves outlining the specific purpose of the inspection, such as identifying possible defects, evaluating the overall condition of the plant's components, or prioritizing maintenance and repair activities. By

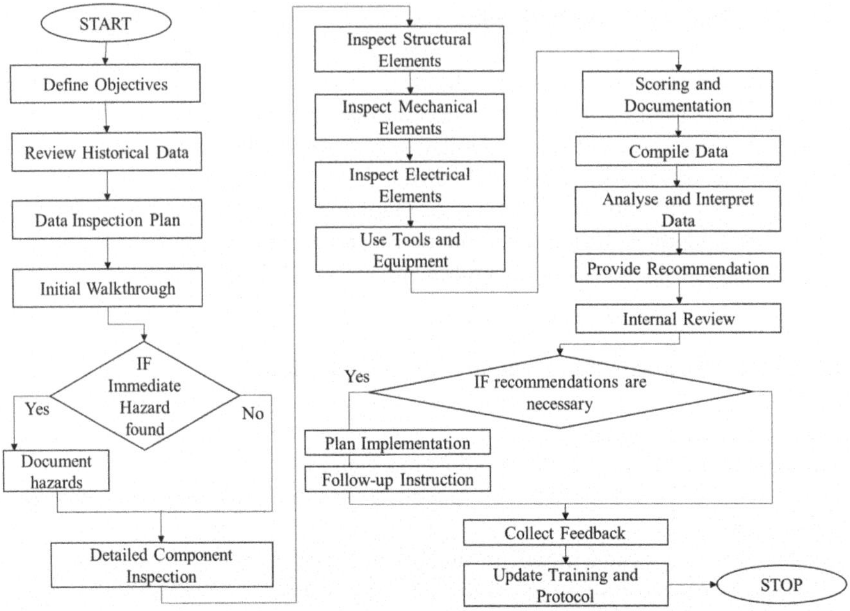

Fig. 11.3 Methodology for visual inspection and scoring systems

establishing clear objectives, the inspection team can focus their efforts on gathering the most relevant data and avoid unnecessary evaluations. A well-defined objective serves as a guiding framework for the entire inspection process, ensuring that all activities are aligned with the desired outcomes.

- Review Historical Data: Before initiating the physical inspection, it is crucial to gather and review all relevant historical data. This includes previous inspection reports, maintenance records, design drawings, construction documents, and operational logs. Reviewing this data provides a valuable understanding of the plant's past performance, any recurring issues, and areas that may require special attention during the inspection. Historical data can also help the inspection team identify trends, such as gradual deterioration or frequent maintenance needs, which can inform the current assessment and future planning.
- Develop Inspection Plan: With the objectives defined and historical data reviewed, the subsequent step is to develop a comprehensive inspection plan. This plan should detail the specific components of the WWTP that will be inspected, such as tanks, basins, pumps, and control systems. It should also outline the sequence of inspection activities to ensure a systematic and thorough evaluation of the entire plant.

The inspection plan should include a list of the required tools and equipment, such as cameras, inspection checklists, measuring devices, and personal protective equipment (PPE). Ensuring that all necessary tools are available and in good working order is critical for conducting an effective inspection. In addition to technical considerations, the inspection plan must also address safety measures. Inspectors should be aware of potential hazards, such as confined spaces, chemical exposures, or structural instabilities, and take appropriate precautions. The plan should include specific safety protocols, such as the use of PPE, emergency procedures, and guidelines for safely accessing difficult-to-reach areas. Hence, by carefully planning and preparing for the inspection, the team can ensure that the process is efficient, thorough, and aligned with the overall goals of the condition assessment. This preparation sets the stage for a successful visual inspection and accurate scoring of the WWTP components.

Conducting Inspection

The inspection process is a critical step in assessing the condition of WWTP components. This phase involves a systematic evaluation of the plant's infrastructure, using both observation and specialized tools to identify potential issues. The following steps outline the key activities involved in conducting a thorough inspection:

- Initial Walkthrough: Gain a general understanding of the layout and identify immediate hazards.
- Detailed Component Inspection: Assess structural, mechanical, and electrical systems.
- Use of Tools and Equipment: Utilize cameras, mirrors, binoculars, and measuring devices.
- Scoring and Documentation: Assign scores using a predefined system and document defects with photographs.

Data Analysis and Reporting

- Compile Data: Organize inspection notes, scores, and photographs.
- Analysis and Interpretation: Identify trends and areas needing attention.
- Recommendations: Provide maintenance and repair recommendations, prioritizing actions based on severity.

Review and Follow-Up

- Internal Review: Ensure the accuracy and completeness of the inspection report.
- Implementation of Recommendations: Schedule repairs and allocate resources.
- Follow-Up Inspections: Verify the effectiveness of recommended actions.

Continuous Improvement

- Feedback Loop: Improve future inspection processes based on feedback.
- Training and Updates: Ensure personnel are trained and updated on the latest techniques and safety protocols.

The methodology for visual inspection and scoring systems in WWTPs is a comprehensive process that requires meticulous planning, execution, and follow-up. Adhering to a structured methodology ensures thorough, accurate, and useful assessments, supporting proactive maintenance and long-term planning. This process contributes to the overall efficiency and safety of the WWTP, maintaining plant integrity and performance. Table 11.3 summarizes a conventional visual and scoring system for WWTP condition assessment for tanks and basins.

11.3.2 Non-destructive Testing (NDT) Techniques

Non-Destructive Testing (NDT) techniques are crucial for evaluating the condition of WWTPs without inflicting any damage to the structures or components under examination (Shiotani et al. 2024; Inês Silva et al. 2023; Gdoutos and Konsta-Gdoutos 2024; Razali et al. 2018). Based on principles, NDT methods can be broadly categorized into five categories (1) visual inspection, (2) acoustic wave–based techniques, (3) optical techniques, (4) imaging-based techniques, and (5) electromagnetic field–based techniques (Fig. 11.4). These techniques offer a dependable way to detect internal defects, corrosion, material degradation, and other structural issues that visual inspections alone cannot reveal. This section details the methodology, benefits, examples, challenges, reliability, and necessary instruments for NDT techniques.

11.3.2.1 Methodology for NDT Techniques

The flowchart provides a comprehensive overview of the NDT methodology, designed to assess materials and components' integrity in WWTPs without causing damage. The chart outlines the process from planning and preparation through execution, data analysis, and reporting (refer to Fig. 11.5).

Planning and Preparation

Objective Definition: Clearly outline the purpose of the NDT, such as identifying internal defects, measuring material thickness, or assessing structural integrity.

Table 11.3 Visual and scoring system for WWTP condition assessment (tank/basins) (Kiliç et al. 2023)

Category	Parameter	Scoring scale (1–9)	Description
Physical (design and construction stage)	Design considerations	1 (sub-standard)–9 (special)	Considerations of durability aspects, joint details, and uniformities affecting service life
	Material type	1 (wood)–9 (reinforced concrete)	Type and quality of materials used will impact the long-term condition
	Size and capacity	1 (< 5000 m^3/day)–9 (> 10,000 m^3/day)	Larger sizes might experience more stress, influencing their condition
Physical (service stage)	Age	1 (old)–9 (new)	Condition typically deteriorates over time
	Corrosion	1 (severe)–9 (none)	Presence and severity of corrosion as a condition indicator
	Coating/paint	1 (poor)–9 (excellent)	Quality of coating or paint affects longevity
	Cracks and flaws	1 (significant)–9 (none)	Presence of cracks or flaws, their size, and location are significant indicators
Environmental	Type of soil	1 (poor)–9 (optimal)	The type of soil influences the foundation type and potential for structural issues
	Quality of source water	1 (poor)–9 (good)	Water quality impacts treatment processes and infrastructure condition
	Vibration (including earthquake)	1 (high)–9 (none)	Vibration effects, including those from seismic activity, on structural integrity
	Weather condition	1 (harsh)–9 (mild)	Effects of seasonal variations, freeze–thaw cycles, and thermal expansion
Operational	Chemicals doses	1 (high)–9 (low)	The type and quantity of chemicals used affect the condition of components

(continued)

Table 11.3 (continued)

Category	Parameter	Scoring scale (1–9)	Description
	Operation and maintenance practices	1 (poor)–9 (good)	Deviation from standard practices accelerates deterioration
	Manager and operators experience	1 (inexperienced)–9 (experienced)	Experience levels of staff affect maintenance and operational outcomes
	Control system	1 (manual, defective)–9 (automated)	Quality and reliability of control systems impact operational efficiency and risk of damage

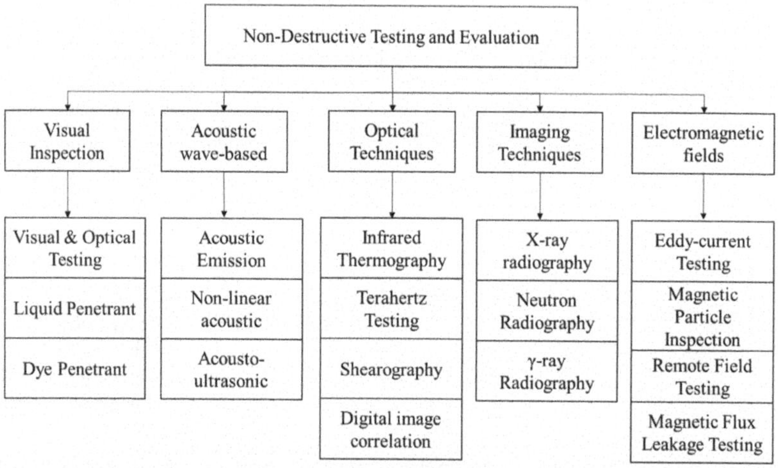

Fig. 11.4 Non-destructive testing and evaluation techniques

Selection of NDT Methods: Select NDT methods according to the material type, structure, and anticipated defect types. Common techniques include radiographic testing, ultrasonic testing, eddy current testing (ECT), and magnetic particle testing (MPT).

- Material Type: The appropriateness of certain NDT methods depends on the material. While the MPT is suitable for ferromagnetic materials, the ECT is more appropriate for non-ferromagnetic materials.
- Expected Defect Location: Surface defects are generally assessed using ECT or MPT, whereas Ultrasonic or Radiographic Testing is more effective for detecting subsurface or internal defects.

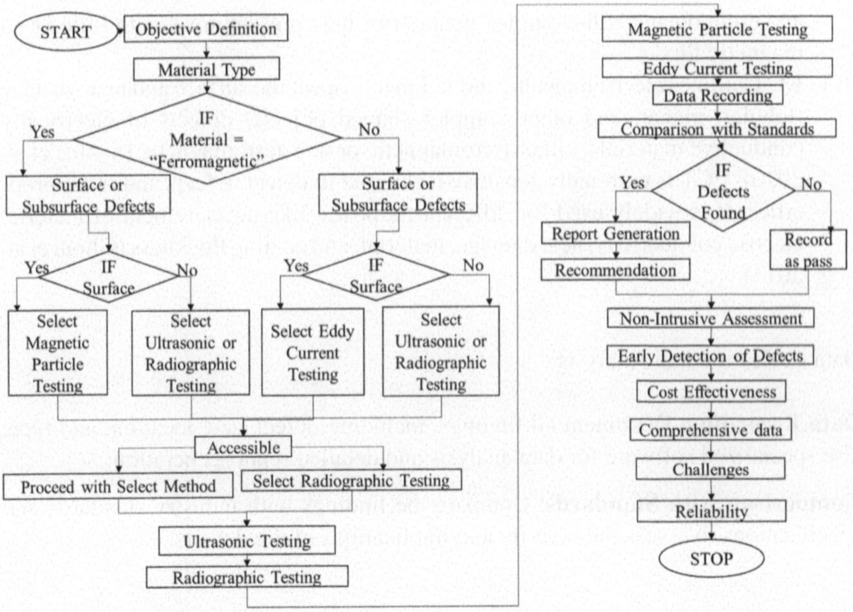

Fig. 11.5 Flow chart for NDT methodology

- Accessibility: The accessibility of the area can affect the choice of method. For hard-to-reach areas, Radiographic Testing may be the most suitable option.
- Preparation of the Area: Ensure the area is clean and accessible. Remove any surface coatings that might interfere with the NDT process.

Execution of NDT

(i) **Ultrasonic Testing** uses ultrasonic waves to detect internal flaws and measure the thickness of material by sending high-frequency sound waves into the material and analyzing the reflected signals to identify defects.

(ii) **Radiographic Testing** employs high-energy electromagnetic radiation, with very short wavelength rays (such as X-rays or gamma rays) to generate internal image of the structure, useful for identifying internal flaws such as cracks, voids, and inclusions.

(iii) **MPT** is an NDT technique that is used to inspect the surface and subsurface discontinuities in components made of ferromagnetic material. The MPT is based on the phenomenon that when an object with high magnetic permeability is subjected to a magnetic field, a magnetic flux leakage occurs in presence of defects in the components. Depending upon the intensity of the magnetic flux leakage, and mobility of strongly ferromagnetic particles attracted to the region with magnetic flux leakage, the detectability and resolution of MPT may vary (Inês Silva et al. 2023). MPT does not provide information on the defect depth and components with complex geometries may present areas with little or no magnetic flux.

(iv) **ECT** utilizes electromagnetic induction to inspect the surface and near-surface (tubular, planar, and other complex-shaped objects) defects of electrically conductive materials, either ferromagnetic or non-ferromagnetic (Santos et al. 2020). ECT is extremely sensitive technique to detect defects and other properties. It is widely used for different purposes, like measurement of material electric conductivity, heat damage, material, and coating thickness (Ghoni et al. 2014).

Data Collection and Analysis

Data Recording: Document all findings, including defect size, location, and type. Use specialized software for data analysis and detailed report generation.

Comparison with Standards: Compare the findings with industry standards and specifications to assess the severity and implications of the defects.

Reporting and Recommendations

Report Generation: Prepare a comprehensive report detailing the findings, analysis, and recommendations for repairs or further investigation.

Recommendations: Provide actionable recommendations based on the severity of the detected defects, which may include repair, replacement, or further monitoring.

11.3.2.2 Benefits of NDT Techniques

(i) **Non-intrusive Assessment**: NDT methods do not require physical damage to the components being tested, preserving the integrity of the structures.
(ii) **Early Detection of Defects**: NDT allows for the early detection of internal defects, preventing possible failures and reducing the risk of catastrophic events.
(iii) **Cost-Effectiveness**: By detecting issues early, NDT can prevent costly repairs and unplanned downtime.
(iv) **Comprehensive Data**: NDT offers detailed information about the internal condition of components, aiding in informed decision-making for maintenance and repairs.

11.3.2.3 Examples and Standards

(i) **ASME Boiler and Pressure Vessel Code (BPVC)**: Provides guidelines for NDT methods in pressure vessel and boiler inspections.

(ii) **International Organization for Standardization (ISO) standards** such as:

a. ISO 55000 series covers the principles and terminologies for asset management;
b. ISO 24510 focuses on the performance assessment of drinking water and wastewater services;
c. **ISO 9712**: Standard for the qualification and certification of NDT personnel.

(iii) **ASTM Standards**: Include various standards for specific NDT methods, such as:

a. ASTM E1444 for MPT;
b. ASTM E165 for liquid penetrant testing, and
c. ASTM E213 for ultrasonic testing of metal pipe and tubing.

(iv) American Water Works Association (AWWA) Standards, such as:

a. AWWA M77 provides methodologies for assessing the condition of infrastructure, adaptable to WWTPs.
b. AWWA M70 includes asset management practices including condition assessment for water and wastewater utilities.

(v) National Association of Clean Water Agencies (NACWA) provides framework for assessing the condition of wastewater treatment infrastructure.

(vi) Water Environment Federation (WEF):

a. WEF Manual of Practice No. 32 detailed guidance on asset management, including condition assessment.
b. WEF Manual of Practice No. 8 includes information on maintenance and condition assessment.

(vii) U. S. Environment Protection Agency (USEPA) provides an overview of tools and guidance for managing assets, including condition assessment in water and wastewater utilities.

(viii) Australian Standards such as AS/NZS 4360 provide a risk management framework, applicable to condition assessment processes.

(ix) Canadian Standards Association (CSA) such as CSA S478 provides guidelines on durability assessment, applicable to structure within WWTPs.

(x) Local and regional organizations such as municipalities, regional authorities, and industry organization may have specific guidelines or framework for the condition assessment of WWTPs based on local requirements and conditions.

Table 11.4 Challenges and reliability of non-destructive testing (NDT) techniques

NDT technique	Challenges	Reliability factors
Magnetic particle testing	Limited to ferromagnetic materials; not effective for deeper subsurface defects	Reliable for surface and near-surface defects in ferromagnetic materials; requires a proper magnetization
Radiographic testing	Safety concerns due to radiation exposure; expensive; requires skilled operators	High reliability for detecting internal defects; affected by material density and thickness; image quality critical
Ultrasonic testing	Requires coupling medium; limited by material grain size and surface condition	Reliable for detecting internal and subsurface defects; highly dependent on operator skill and equipment calibration
Eddy current testing	Limited to conductive materials; depth of penetration is shallow	Reliable for detecting surface and near-surface defects in conductive materials; accuracy affected by probe design

11.3.2.4 Challenges and Reliability

- **Challenges**: NDT techniques may encounter challenges like access limitations, surface conditions, and result interpretation. The effectiveness of NDT depends on the operator's ability and the quality of the equipment used.
- **Reliability**: NDT methods are highly reliable when performed by certified personnel using calibrated equipment. However, false positives or negatives can occur, necessitating complementary methods or repeat testing.

Table 11.4 highlights the challenges faced by each NDT technique along with the factors that influence their reliability in detecting defects in materials.

11.3.2.5 Instruments Required

Ultrasonic Flaw Detectors

Ultrasonic Flaw Detectors are specialized devices used in ultrasonic testing (UT) to identify internal defects within materials, such as cracks, voids, and inclusions. These detectors work by emitting high-frequency sound waves into the material. As these waves travel through the material, they reflect off any flaws or discontinuities. The detector captures the reflected waves, which analyzes the data to provide insights about the internal structure of the material. Ultrasonic flaw detectors are commonly used in industries where safety and integrity are critical, such as in pipelines, pressure vessels, and structural components in wastewater treatment plants. These detectors are crucial for preventive maintenance, ensuring that potential issues are identified and addressed before they lead to failures.

X-Ray and Gamma-Ray Equipment

X-ray and Gamma-ray equipment are primarily used for Radiographic Testing (RT) in WWTPs. This method is crucial for inspecting welds, pipes, and concrete structures for internal defects. The high-energy radiation penetrates the material, allowing for the creation of an image that reveals internal flaws, such as cracks, voids, inclusions, and corrosion that are not visible on the surface.

- X-rays are generated electronically using an X-ray tube, this method is preferred in environments where electronic control and lower radiation exposure are essential. They are typically used in situations where high-resolution images are required, such as in the inspection of welds in metal pipes, tanks, and pressure vessels within the WWTP. The prime advantage of X-ray is that it provides detailed imaging of internal structures, making it ideal for detecting small and fine defects. It is also safer compared to gamma rays due to the ability to control the exposure electronically.
- Gamma rays are emitted from radioactive isotopes such as Cobalt-60 or Iridium-192. This method is highly effective in penetrating thick materials. Gamma-ray equipment is used when deep penetration is necessary, such as in inspecting thick-walled pipes and concrete structures. These equipment are portable and does not require electricity, making it suitable for field inspections in remote areas of the WWTP.

Magnetic Particle Testing Kits

MPT is used to detect surface and near-surface defects in ferromagnetic materials, such as steel components commonly found in WWTPs. These components may include storage tanks, pipelines, pumps, and valves. The process involves magnetizing the component and then applying magnetic particles (either dry or in a liquid suspension) to the surface. The particles gather at the locations of defects, such as cracks, seams, or inclusions, making them visible under proper lighting conditions.

Eddy Current Testing Equipment

ECT is used primarily to examine non-ferrous metals in WWTPs, though it can also be used on ferrous materials. This method is particularly useful for detecting surface and near-surface flaws, such as corrosion, cracks, and thickness variations, in components like heat exchanger tubes, pipelines, and other critical structures.

Table 11.5 Summary of non-destructive techniques and their application in WWTPs

Non-destructive techniques	Application	Purpose
Ultrasonic testing	Pipe and tank thickness measurement	To measure wall thickness of corroded sludge pipelines to prevent leaks
Magnetic particle testing	Weld inspection on steel structure	To inspect welds on a steel tank to detect cracks or imperfections
Radiographic testing	Inspection of welds and pipe integrity	To check weld integrity in large-diameter pipes carrying treated water
Liquid penetrant testing	Detection of surface crack in metal component	To detect surface cracks on a pump impeller handling abrasive sludge
Eddy current testing	Detection of corrosion and surface cracks	To inspect heat exchanger tubes for surface cracks and corrosion
Visual testing	Inspection of equipment and infrastructure	To visually inspect clarifiers and aeration tanks for signs of wear or corrosion
Thermography	Monitoring of electrical systems and equipment condition	To detect hot spots in electrical panels indicating potential overload
Acoustic emission testing	Monitoring of structural integrity of tanks and vessels	To detect crack propagation in a sludge storage tank during operation
Leak testing	Detection of leaks in pipelines and sealed systems	To conduct pressure decay tests to ensure pipelines are leak-free
Ultrasonic flow measurement	Monitoring flow rates in pipelines	To measure the flow rate of treated effluent being discharged

An ECT system typically includes a signal generator, a probe (coil), and a display unit. The probe induces an alternating magnetic field into the material, and variations in the resulting eddy currents are detected and analyzed.

ECT is highly sensitive to small surface defects and can be used on coated or painted surfaces without the need for extensive surface preparation. It provides immediate results, which is beneficial for maintenance and repair decisions in a WWTP setting. Additionally, ECT does not require contact with the material, making it a non-intrusive and versatile inspection method.

In summary, NDT techniques are invaluable for maintaining the safety and integrity of WWTPs (Table 11.5). They offer a non-invasive means of detecting internal defects, ensuring the continued safe operation of the facility. By adhering to established standards and employing skilled personnel, the reliability and effectiveness of NDT can be maximized, providing essential data for proactive maintenance and repair planning.

11.3.3 Non-invasive Methods for WWTP Condition Assessment

To effectively evaluate the condition WWTPs, it is essential to consider three distinct scenarios based on the level of sensor integration and data acquisition capabilities. Each scenario necessitates a customized approach to gather and analyze data, resulting in an accurate evaluation of the plant's condition. The following outlines the methodology for each scenario, providing a comprehensive approach to condition assessment.

11.3.3.1 Scenario 1: Existing Sensors with Historical Data for Wastewater Treatment Plant Optimization

In this scenario, considering a WWTP equipped with comprehensive array of sensors and data acquisition systems. These sensors continuously monitor various critical parameters such as dissolved oxygen, chemical concentrations, pH, turbidity, and other indicators of plant performance. Over the years, these sensors have accumulated a vast repository of historical data which can be leveraged for in-depth analysis, predictive modeling, and overall optimization of the plant's operations.

Step 1—Data Collection and Review: The first step in leveraging this existing sensor network involves a systematic approach to data collection and a rigorous review process to ensure the quality and usability of the data.

(a) **Compile Historical Data**: The process begins by aggregating data from the plant's extensive network of sensors. These sensors have been continuously collecting data points across various parameters over a significant period. The collected data might include, but is not limited to:

- pH Levels: Monitoring the acidity or alkalinity of the water, which is crucial for the effective functioning of biological treatment processes.
- Turbidity: Measuring the cloudiness or haziness of the water, which can indicate the presence of suspended solids and is an essential factor in assessing the clarity and quality of treated water.
- Dissolved Oxygen (DO): Crucial for the survival of aerobic microorganisms that are involved in the breakdown of organic matter during the treatment process.
- Chemical Concentrations: Tracking the levels of various chemicals, such as chlorine, ammonia, and nitrates, which are used or produced during the treatment process.

This aggregation process involves compiling data from different sources, which might include real-time sensor data, logged historical data stored in plant databases, and records maintained in data acquisition systems. The goal is to create a comprehensive dataset that reflects the plant's operational history.

(b) **Data Quality Check**: After compiling data, the data undergoes a stringent quality check to ensure its completeness and accuracy. This step is critical because the reliability of subsequent analysis hinges on the quality of the input data. The quality check process typically involves:

- Completeness Check: Ensuring that the data set is complete and that no crucial data points are missing. Gaps in data can occur due to sensor malfunctions, communication errors, or other technical issues. These gaps need to be identified and addressed, possibly through data interpolation or other methods.
- Accuracy Verification: Cross-referencing sensor data with manual readings or other reliable sources to verify its accuracy. This step ensures that the data accurately reflects the real-world conditions of the plant at the time of collection.
- Data Cleaning: Removing anomalies and outliers that could skew the analysis. Anomalies might arise from temporary sensor malfunctions, calibration errors, or other unforeseen factors. Data cleaning ensures that these outliers do not distort the analysis, leading to erroneous conclusions.

By the end of this stage, the plant will have a high-quality dataset that accurately represents its historical operational conditions. This dataset serves as the foundation for the next steps in the assessment process.

(c) **Data Analysis and Predictive Modeling**: With a robust, high-quality dataset in place, the next phase involves performing advanced data analysis and developing predictive models. These models can provide invaluable insights into the plant's operations, helping to identify trends, forecast future conditions, and pre-emptively address potential issues.

- Trend Analysis: Trend analysis provides a deeper understanding of how various parameters have evolved over time and how they interact with each other. For example:
  - Long-Term pH Stability: Analyzing historical pH data can reveal how stable the pH levels have been over the years and whether any fluctuations are linked to specific events or operational changes. Consistent deviations from expected pH levels might indicate underlying issues, such as changes in influent composition or problems with chemical dosing systems.
  - Turbidity Variations: Monitoring turbidity trends can help identify periods when the plant struggled to maintain water clarity, potentially indicating issues with the filtration system or the efficiency of sedimentation processes.
  - Dissolved Oxygen Levels: DO data can reveal how effectively the biological treatment processes have been operating. Fluctuations in DO levels might suggest problems with aeration equipment or shifts in the biological load that the plant needs to process.

The trend analysis also helps in understanding seasonal variations, the impact of operational changes, and the effect of external factors such as weather conditions. By identifying these trends, the plant can better understand its performance history and prepare for future challenges.

- Predictive Modeling: After trend analysis, the next step is to use this data to develop predictive models. Machine learning algorithms are particularly well-suited for this task because they can analyze large datasets and recognize complex patterns that might not be immediately apparent through traditional analysis method.
  - Algorithm Selection: The choice of machine learning algorithm depends on the specific objectives of the predictive modeling. Common algorithms include decision trees, regression models, support vector machines, and neural networks. Each algorithm has its strengths and is suited for different types of predictive tasks, such as forecasting time-series data, classifying operational states, or predicting equipment failures.
  - Model Training: The selected algorithm is trained using historical data. During this training process, the model learns to recognize patterns and relationships between different parameters and their outcomes. For example, the model might learn that certain combinations of pH, turbidity, and DO levels are indicative of a high risk of biological process failure.
  - Validation and Testing: After training, the model is validated and tested using a separate subset of the historical data. This step ensures that the model is accurate and reliable when applied to real-world scenarios. The model's performance is evaluated using metrics such as precision, accuracy, recall, and F1-score, which help determine its effectiveness in predicting future conditions.

Predictive models can be used to predict potential issues before they arise, enabling the plant to implement predictive maintenance strategies. Predictive maintenance involves performing maintenance activities based on the predicted condition of equipment or processes, rather than on a fixed schedule. This approach has several benefits:

- Minimizing Downtime: By addressing potential issues before they escalate into serious problems, the plant can avoid unplanned shutdowns and minimize operational downtime.
- Optimizing Maintenance Costs: Predictive maintenance allows the plant to focus its resources on areas that need attention, rather than performing routine maintenance on all equipment regardless of its condition. This targeted approach can reduce maintenance costs and improve overall efficiency.
- Extending Equipment Life: By addressing wear and tear before it leads to failure, the plant can extend the lifespan of its equipment, delaying the need for costly replacements.

For example, if the predictive model indicates that a particular aeration system is likely to fail based on declining DO levels and increased energy consumption, the plant can schedule a maintenance check to address the issue before it leads to a complete breakdown.

(d) **Performance Evaluation**: The final stage of the assessment involves a comprehensive evaluation of the plant's performance. This evaluation compares the plant's current and historical performance against industry standards and best practices, providing insights into areas where improvements can be made.

- Benchmarking: Benchmarking is a critical process that involves comparing the plant's performance metrics against both its historical data and industry standards. Key performance indicators (KPIs) such as effluent quality, energy consumption, chemical usage, and overall treatment efficiency are evaluated to assess how well the plant is performing. By benchmarking these KPIs, the plant can identify areas where it is performing well and areas that require improvement. This comparison also helps the plant set realistic performance targets and develop strategies to achieve them.
- Reporting: The final step in the performance evaluation process is to compile the findings into detailed reports. These reports serve as a comprehensive record of the plant's performance, providing actionable insights and recommendations for future operations.

Summary of Scenario 1: In this scenario, the wastewater treatment plant is equipped with a sophisticated network of sensors that have accumulated a wealth of historical data. By systematically collecting, analyzing, and leveraging this data, the plant can significantly enhance its operational efficiency, reduce maintenance costs, and ensure compliance with environmental standards. The process involves not only examining past performance but also using advanced predictive models to anticipate future challenges and proactively address them. Ultimately, this approach allows the plant to continuously improve its operations, ensuring long-term sustainability and resilience.

11.3.3.2 Scenario 2: Semi-automated System with Partial Sensor Coverage

In this scenario, the WWTP operated with a semi-automated system that includes some existing sensors, but coverage is not comprehensive. Further, to optimize the plant's operations, additional sensors may be installed to fill in the gaps in monitoring, enabling more effective data collection and analysis.

- **Initial Assessment and Planning**: The first step in enhancing the semi-automated system is a thorough assessment of the existing sensor network and planning for improvements.

  - **Review Existing Sensor Data**: The process begins with a detailed review of the data currently being collected by the existing sensors. This review aims to evaluate the effectiveness of the current monitoring setup, identifying which parameters are being adequately tracked and where there are significant gaps. Common parameters that might already be monitored include pH levels, dissolved oxygen, and temperature, while gaps might exist in areas like chemical concentrations or flow rates. During the review, a gap analysis is conducted to pinpoint areas of the plant where critical parameters are not being monitored effectively. For example, if dissolved oxygen is monitored in only one section of the biological treatment process, additional sensors may be needed in other sections to ensure comprehensive coverage.
  - **Determine Additional Sensor Needs**: Once gaps in the monitoring coverage have been identified, the next step is to determine what additional sensors are required to achieve a more complete and reliable data collection system. This involves selecting sensors that can monitor parameters critical to the plant's performance and environmental compliance. The decision on where to place new sensors is driven by the need to cover all critical points in the treatment process. For instance, if ammonia levels are a concern, sensors might be placed at various stages of the nitrification process. The goal is to ensure that every key process and potential point of failure is being monitored.

- **Sensor Installation and Integration**: With a fully integrated sensor network, the plant can now engage in real-time monitoring and advanced data analysis to optimize its operations

  - **Install Additional Sensors**: The installation of new sensors is a crucial step in closing the gaps identified during the initial assessment. These sensors need to be strategically placed at locations where they can provide the most value, such as at the inlet and outlet of various treatment units, or in areas where previous data collection was insufficient.

    After installation, each sensor must be carefully calibrated to ensure it provides accurate readings. Calibration is essential for integrating the sensors into the existing system, as it aligns the new data streams with the parameters and standards already in place.
  - **Data Integration**: Once the new sensors are installed and calibrated, their data streams need to be merged with the existing data acquisition system. This integration process involves configuring the system to handle inputs from multiple sources, ensuring that data from both old and new sensors is collected in a unified manner. By integrating data from all sensors into a single platform, the plant can perform more comprehensive analysis and gain a holistic view of its operations. This platform acts as the central hub for monitoring, where all sensor data is aggregated, stored, and made available for further analysis.

- **Real-Time Monitoring and Analysis**: With a fully integrated sensor network, the plant can now engage in real-time monitoring and advanced data analysis to optimize its operations.

- **Continuous Data Collection**: The enhanced system enables continuous, real-time data collection across all critical parameters. This real-time monitoring is crucial for detecting deviations from normal operating conditions as soon as they occur. To ensure prompt response to any issues, automated alerts can be set up. These alerts are triggered when sensor readings exceed predefined thresholds, such as a sudden drop in dissolved oxygen or a spike in turbidity, enabling operators to take immediate corrective action.
- **Advanced Analytics**: Beyond basic monitoring, advanced data analytics tools can be employed to analyze the collected data in real time. These tools use algorithms to detect patterns, identify anomalies, and predict potential failures before they occur. The insights gained from advanced analytics can be used to implement predictive maintenance strategies. For example, if data analytics indicate that a certain pump is likely to fail based on its operating conditions, maintenance can be scheduled proactively to prevent unplanned downtime.

- **Condition Assessment**: The final stage involves a comprehensive assessment of the plant's overall condition, utilizing both real-time and historical data.

  - **Comprehensive Analysis**: With the integrated and enhanced data collection system, the plant can now conduct a thorough analysis that combines real-time data with historical trends. This analysis provides a detailed understanding of the plant's performance, highlighting areas of strength and potential weaknesses. By comparing current data with historical performance, operators can identify trends and deviations that may indicate underlying issues. For instance, a gradual increase in energy consumption could signal inefficiencies in the aeration system that need to be addressed.
  - **Actionable Recommendations**: Based on the findings from the comprehensive analysis, the plant can develop actionable recommendations for improving operations. These recommendations might include specific maintenance tasks, process optimizations, or even upgrades to equipment or systems.
  - **Maintenance and Optimization**: Recommendations might focus on fine-tuning the plant's operations to improve efficiency, such as adjusting chemical dosing strategies based on real-time feedback or optimizing aeration settings to reduce energy consumption.
  - **Potential Upgrades**: If the analysis reveals that certain parts of the plant's infrastructure are outdated or underperforming, recommendations might include upgrading to more modern, efficient equipment.

Summary of Scenario 2: In a semi-automated wastewater treatment plant with partial sensor coverage, enhancing the system through the strategic installation of additional sensors and the integration of their data can significantly improve monitoring and operational efficiency. This approach allows for comprehensive real-time monitoring, advanced data analysis, and proactive maintenance strategies, leading to better overall plant performance and more reliable compliance with environmental regulations. By continuously assessing and optimizing the system, the plant can ensure its long-term sustainability and operational excellence.

11.3.3.3 Scenario 3: Manual Operation with No Existing Sensors

In this scenario, the wastewater treatment plant currently operates entirely through manual processes, with no existing sensors to monitor key operational parameters. The transition to a sensor-based operation involves the design, installation, and integration of a complete sensor network and data acquisition system. This upgrade enables real-time monitoring, enhanced operational efficiency, and data-driven decision-making.

(i) **Baseline Assessment**: The first step in transitioning from manual to sensor-based operations is to conduct a comprehensive baseline assessment of the plant and design an appropriate monitoring system.

- **Site Survey**: A thorough site survey is essential to understand the current operational setup and to identify the critical areas within the plant that require monitoring. The survey should focus on identifying the key processes and equipment where sensor data would provide the most value in terms of improving efficiency, ensuring compliance, and preventing failures.
  - Identifying Critical Areas: The survey should pinpoint areas where the lack of real-time data has the most significant impact on plant performance. For example, areas where chemical dosing is critical, or where flow rates vary significantly, should be prioritized for sensor installation.
  - Determining Sensor Requirements: Based on the critical areas identified, the next step is to determine the types and numbers of sensors needed. This could include sensors for measuring flow rates, chemical concentrations, temperature, pressure, and mechanical vibrations. Each type of sensor plays a specific role in monitoring the conditions that directly affect the plant's operational efficiency and compliance with regulatory standards.
- **Designing the Monitoring System**: Once the site survey is complete, the next step is to design a comprehensive monitoring system that meets the plant's specific needs. This system should be tailored to the plant's processes and should be scalable to accommodate future expansions or upgrades.
  - System Architecture: The design should outline the overall architecture of the monitoring system, including the placement of sensors, the type of data acquisition units required, and the integration with data management and visualization tools. The architecture should ensure that all critical parameters are monitored and that data is collected efficiently and accurately.
  - Data Management Infrastructure: The system should include a robust data management infrastructure capable of storing, processing, and analyzing large volumes of data. This infrastructure will serve as the backbone of the monitoring system, enabling real-time analysis and long-term data storage for historical trend analysis.

(ii) **Sensor Installation and System Setup**: With the design finalized, the next phase involves the physical installation of the sensors and the setup of the data acquisition system.

- Installation of Sensors: The installation process begins by deploying sensors throughout the plant covering all identified critical areas. The sensors are selected based on the specific parameters it needs to monitor.
  - Strategic Deployment: The sensors should be installed at strategic locations to ensure comprehensive coverage. For example, flow rate sensors might be placed at key points along the influent and effluent streams, while temperature and pH sensors are installed in biological treatment tanks. Vibration sensors might be deployed on critical mechanical equipment such as pumps and blowers to detect early signs of wear or malfunction.
  - Calibration and Configuration: After installation, each sensor must be calibrated to ensure accurate readings. Calibration involves adjusting the sensors to match known standards, ensuring that the data collected is reliable. Additionally, sensors should be configured to communicate with the data acquisition system, ensuring seamless data transmission.
- Data Acquisition System: Once the sensors are in place, the data acquisition system (DAS) must be installed and configured to collect and manage the incoming data.
  - Data Collection and Storage: The DAS should be capable of collecting data from multiple sensors simultaneously, and storing it in a centralized database. This system should be designed to handle both real-time data for immediate analysis and long-term data storage for historical review.
  - Integration with Analytics Tools: To maximize the utility of the data, the DAS should be integrated with data visualization and analytics tools. These tools allow operators to view data in real-time, detect trends, and analyze historical data to make informed decisions. Dashboards can be created to display key performance indicators (KPIs) and alert operators to any anomalies or deviations from normal operating conditions.

(iii) **Data collection and monitoring**: With the system fully installed, the plant can now begin data collection and monitoring, establishing a new baseline for normal operations and integrating offline data sources.

- Initial Data Collection: With the system fully installed, the plant can now begin data collection and monitoring, establishing a new baseline for normal operations and integrating offline data sources.
  - Establishing Normal Operating Conditions: By continuously collecting data across all monitored parameters, the plant can define what constitutes normal operation. This involves tracking metrics such as flow rates,

chemical concentrations, temperature, and equipment performance under typical load conditions.
  - Immediate Insights: Even in the early stages of data collection, real-time monitoring can provide valuable insights. For example, if sensors detect that chemical dosing is inconsistent or that a piece of equipment is operating outside its normal parameters, operators can take immediate corrective action.

- Offline Data Integration: In addition to real-time sensor data, the offline data (laboratory analyses, and manual readings) source should also be integrated into the overall monitoring system during routine inspections, and historical records.
  - Offline data Collection: Operators should regularly collect offline data, such as lab results for effluent quality or manual measurements of sludge levels in tanks. This data should be recorded in a standardized format and integrated into the central database alongside real-time sensor data.
  - Comprehensive Data Set: By combining real-time and offline data, the plant can create a more comprehensive dataset that reflects both the continuous operation of the plant and specific measurements that are not captured by sensors. This integration enhances the accuracy and reliability of condition assessments and operational decisions.

(iv) **Comprehensive Condition Assessment**: With data flowing from both real-time sensors and offline sources, the plant can now conduct a comprehensive condition assessment to evaluate its performance and identify areas for improvement.

- Analyzing Data for Condition Assessment: Using advanced data analytics and modeling techniques, the collected data is analyzed to assess the overall condition of the plant. This analysis focuses on identifying abnormalities, inefficiencies, and potential issues that could impact the plant's operations.
  - Data Analysis: Using data analytics tools, operators can detect patterns and correlations in the data that might not be immediately apparent. For example, a gradual increase in energy consumption might be correlated with a decline in equipment efficiency, indicating that maintenance is needed.
  - Modeling Technique: Predictive modeling can be employed to forecast future conditions based on current data trends. For instance, if vibration data from a pump suggests that it is nearing the end of its operational life, the model might predict when the pump is likely to fail, thereby allowing for proactive maintenance scheduling.

- Maintenance and Optimization Plan: Based on the findings from the condition assessment, the plant can develop a targeted maintenance and optimization plan. This plan should prioritize actions based on the criticality of the issues identified and their potential impact on the plant's performance.

  - Prioritizing Maintenance: Based on the severity of detected issues and the importance of the affected equipment or processes, the maintenance tasks should be prioritized. For example, a pump showing signs of imminent failure should be prioritized for maintenance over equipment that is performing within normal parameters.
  - Optimization Operations: The assessment may also reveal opportunities to optimize plant operations, such as adjusting chemical dosing strategies, improving energy efficiency, or enhancing process control. These optimizations can lead to significant cost savings and improved treatment performance.

(v) **Reporting and Follow-Up**: The final stage involves documenting the assessment findings, implementing recommended actions, and setting up continuous monitoring for ongoing improvement.

- Detailed Reporting: The entire assessment process, along with its findings and recommendations, should be thoroughly documented in a detailed report. This report serves as a record of the plant's current condition and provides a roadmap for future improvements.
  - Comprehensive Report: The report should include an overview of the assessment process, a summary of the data collected, an analysis of the findings, and detailed recommendations for maintenance and optimization. It should also include visual aids such as charts and graphs to help stakeholders understand the data and its implications.
  - Management Communication: The report should be presented to plant management, providing them with the information needed to make informed decisions about investments in maintenance, upgrades, and process improvements.
- Implementation and Monitoring: Following the report, the recommended actions should be implemented, with a focus on continuous monitoring to ensure their effectiveness.
  - Implementation Recommendation: Plant operators should follow the action plan outlined in the report, addressing the most critical issues first and gradually implementing optimization strategies.
  - Continuous Monitoring: Once the recommendations are in place, continuous monitoring should be maintained to evaluate the effectiveness of these actions. Regularly reviewing sensor data will help the plant to adapt and refine its strategies, ensuring long-term operational excellence and compliance with environmental regulations.

Summary of Scenario 3: Transitioning from a manual operation to a sensor-based system in a wastewater treatment plant is a significant upgrade that enhances monitoring capabilities, operational efficiency, and overall plant performance. By carefully planning, installing, and integrating a comprehensive sensor network, the plant can shift to a data-driven approach, enabling real-time monitoring, predictive maintenance, and continuous improvement. This proactive approach not only helps in maintaining compliance with regulatory standards but also optimizes resources, reduces operational costs, and ensures the long-term sustainability of the plant's operations.

These three scenarios provide a comprehensive framework for assessing the condition of WWTPs using sensor-based methodologies. Each scenario requires a unique approach, depending on the existing infrastructure and data availability. By leveraging historical data, enhancing sensor coverage, or installing a new sensor network, plant operators can achieve accurate and actionable insights into plant performance and condition. This holistic approach ensures that maintenance, optimization, and future planning are based on reliable data, leading to improved operational efficiency and sustainability.

11.3.4 Advanced Data Analysis and Modeling for WWTP Condition Assessment

Advanced data analysis and modeling techniques are pivotal for understanding the intricate dynamics of WWTP operations. These methods facilitate the analysis of historical operational data, prediction of future performance, and optimization of process design and operation. By identifying trends and anomalies, data analytics significantly enhance informed decision-making. Two primary modeling approaches: First Principle Models and Machine Learning Models each offer unique advantages and challenges, and their integration can lead to superior predictive accuracy and interpretability.

11.3.4.1 First Principle Models

First Principle Models, also known as white-box or mechanistic models, are grounded in the fundamental physical, chemical, and biological laws governing treatment processes. These models require an in-depth understanding of the system's internal processes and are constructed using governing equations.

Principles and Applications

Mass and Energy Balances: These models use mass and energy balance equations to describe the behavior of various processes within the WWTP. For instance, the Activated Sludge Model (ASM) series, developed by the International Water Association (IWA), details the biochemical processes in activated sludge systems.

Hydrodynamic Modeling: Computational Fluid Dynamics (CFD) models flow and mixing patterns within reactors, aeration tanks, and clarifiers, predicting how changes in operating conditions impact hydrodynamics and overall plant performance.

Reaction Kinetics: These models simulate pollutant transformations, such as organic matter degradation, nitrification, and denitrification, incorporating reaction kinetics based on experimental data and calibrated with plant-specific parameters.

Advantages and Challenges

Advantages: High accuracy and reliability when system dynamics are well understood. They predict system responses to operational changes and design modifications.

Challenges: Require detailed system knowledge and significant computational resources. Sensitive to parameter inaccuracies and complex to develop and calibrate.

Use Cases in Condition Assessment

(i) **Process Optimization**: One of the primary applications of First Principle Models in WWTPs is the simulation of treatment processes. These models simulate how different components of the plant, such as reactors, aeration tanks, and clarifiers, behave under various operational conditions.

- Design and Retrofitting: Engineers use First Principle Models during the design phase of new treatment plants or when retrofitting existing facilities. By simulating different design configurations, they can optimize the size and operation of reactors, settling tanks, and other components to ensure efficient treatment while minimizing costs.
- Operational Scenarios: The models allow for the simulation of various operational scenarios, such as changes in influent load, temperature variations, or adjustments in chemical dosing. By understanding how these factors affect the treatment process, operators can make informed decisions to maintain optimal conditions and prevent issues like overloading or underperformance.
- Predicting Process Outcomes: For example, in an activated sludge process, First Principle Models based on the Activated Sludge Model (ASM) series can predict how changes in influent characteristics or aeration rates will impact the growth of microbial populations and the removal of nutrients, and organic matter. This predictive capability is crucial for adjusting operations to maintain effluent quality.

(ii) **Diagnosing Process Inefficiencies**: First Principle Models can be used to diagnose inefficiencies in the treatment process by comparing actual plant performance with model predictions. When discrepancies arise, they can indicate potential issues such as equipment malfunctions, improper operational settings, or unexpected changes in influent quality.

- Troubleshooting: For instance, if a First Principle Model predicts a certain level of dissolved oxygen in an aeration tank but the actual measured level is significantly lower, this could indicate a problem with the aeration system, such as blocked diffusers or insufficient blower capacity. Operators can then investigate and address the specific cause of the discrepancy.
- Energy Efficiency: These models can also help identify areas where energy consumption could be reduced without compromising treatment performance. By simulating the impact of reducing aeration intensity or optimizing the operation of pumps and blowers, the model can guide adjustments that improve energy efficiency while maintaining or even enhancing treatment outcomes.

(iii) **Condition-Based Maintenance**: First Principle Models contribute to condition-based maintenance strategies by providing insights into the operational health of various plant components. These models predict how equipment and infrastructure should perform under ideal conditions, making it easier to detect deviations that might indicate wear, damage, or the need for maintenance.

- Predicting Equipment Degradation: For example, in a sludge digestion process, a First Principle Model might predict gas production rates based on substrate characteristics and operating conditions. If actual gas production falls below the predicted levels, this could signal issues such as digester fouling or suboptimal mixing, prompting maintenance before a complete failure occurs.
- Maintenance Scheduling: By simulating the impact of different maintenance schedules on process performance, operators can optimize maintenance intervals, balancing the need to prevent equipment failure with the goal of minimizing downtime and operational costs.

(iv) **Regulatory Compliance and Reporting**: Regulatory agencies often require detailed reports on the performance of WWTPs, including proof that the plant consistently meets effluent quality standards. First Principle Models play a key role in ensuring and demonstrating compliance.

- Compliance Verification: These models are used to simulate worst-case scenarios, such as peak loading conditions or equipment failures, to ensure that the plant remains compliant under all circumstances. If the model predicts that the plant might exceed effluent limits under certain conditions, operators can take pre-emptive actions to mitigate the risk.
- Documentation and Reporting: The results from First Principle Models provide a scientific basis for compliance reports, helping to justify operational decisions and demonstrate that the plant's processes are designed and operated according to best practices. This is particularly important in audits and when applying for permits or renewals.

(v) **Risk Assessment and Contingency Planning**: First Principle Models are valuable tools for risk assessment and contingency planning, allowing WWTPs to prepare for potential emergencies or failures.

- Simulating Extreme Conditions: By simulating extreme conditions, such as toxic spills or equipment malfunctions, the models help operators understand the plant's resilience and identify critical vulnerabilities. This knowledge is essential for developing contingency plans and ensuring that the plant can respond effectively to emergencies.
- Risk Mitigation: The models can also help quantify the risks associated with different operational strategies, guiding decisions on how to mitigate those risks. For instance, a model might show that operating a reactor at maximum capacity increases the risk of process upset, prompting the decision to build additional capacity or implement more conservative operating procedures.

First Principle Models are powerful tools in the condition assessment of WWTP. This is done by simulating the behavior of treatment processes under various conditions, these models provide valuable insights into the plant's performance, help diagnose inefficiencies, support real-time process control, and guide maintenance strategies. Additionally, they play a crucial role in ensuring regulatory compliance, conducting risk assessments, and planning for contingencies. When integrated with modern data analysis techniques, such as machine learning, First Principle Models offer a comprehensive approach to optimizing WWTP operations and ensuring long-term sustainability and efficiency.

11.3.4.2 Machine Learning Models

Machine Learning models, also known as black-box or data-driven models, learn from historical data to identify relationships and patterns within the system. These models don't require detailed knowledge of underlying processes.

Principles and Applications

Supervised Learning: Training models on labeled data with known outcomes. Algorithms include Support Vector Machines, Linear Regression, Neural Networks, Decision Trees, and Random Forests, predicting continuous outputs (regression) or categorical outcomes (classification).

Unsupervised Learning: Clustering and anomaly detection without labeled data. Algorithms such as principal component analysis, K-means clustering, and autoencoders, identify patterns, and detect anomalies.

Reinforcement Learning: Learning optimal actions based on rewards and penalties, useful for process control and optimization, where the model maximizes a predefined reward function.

Advantages and Challenges

Advantages: Handle large datasets, identify complex nonlinear relationships, and adapt to changing conditions. Easier to implement without needing detailed process knowledge.

Challenges: Require large amounts of high-quality data for training. Less interpretable and prone to overfitting and underfitting, which need addressing through proper validation and regularization techniques.

Use Cases in Condition Assessment

Predictive Maintenance: Predicting equipment failures by analyzing historical data and identifying breakdown-associated patterns, enabling timely maintenance and reducing downtime.

Anomaly Detection: Unsupervised learning techniques detect unusual sensor data patterns, indicating potential faults or inefficiencies.

Optimization: Reinforcement learning develops adaptive control strategies optimizing operational parameters in real-time, enhancing process efficiency and stability.

Integration and Hybrid Modeling Approaches

Combining First Principle Models and Machine Learning Models leverages the strengths of both approaches, providing accurate predictions with physical interpretability. For example, a First Principle Model can simulate a baseline process, while an ML model fine-tunes predictions using real-time data.

Model Calibration and Validation

Calibration: Adjusting model parameters to match observed data using optimization techniques like gradient descent or genetic algorithms.

Validation: Evaluating model accuracy by comparing predictions with a separate validation dataset. Metrics such as Mean Squared Error (MSE), R-squared, and confusion matrix (for classification tasks) are used for evaluation.

Data Requirements and Preprocessing

Data Collection: Continuous collection of high-quality sensor data, including parameters like flow rates, chemical concentrations, temperature, and pressure.

Preprocessing: Cleaning, normalizing, and transforming data to ensure consistency. Feature engineering may involve creating new variables capturing important process aspects.

Implementation Considerations

Scalability: Models must be scalable to accommodate expanding data streams and computational demands.

Interoperability: Integrating with existing SCADA systems and data management platforms ensures seamless data flow and real-time monitoring.

User Interface: A user-friendly interface for operators to interact with models, visualize data, and receive actionable insights.

Advanced data analysis and modeling techniques are vital for WWTP condition assessment. By integrating First Principle Models and Machine Learning Models, operators can comprehensively understand the system, optimize processes, and implement predictive maintenance strategies. This holistic approach ensures efficient and sustainable wastewater treatment facility operations.

11.3.4.3 Assessment of Operational and Maintenance

This section outlines the methodologies and best practices on operational and maintenance (O&M) practices, regulatory compliances, and overall management effectiveness. These assessments not only highlight current deficiencies but also provides insights into future predictive operational challenges, ensuring the plant remains resilient against potential issues.

Assessments of O&M Practices

Overview of Inspection Visits

Inspection visits are fundamental to the condition assessment process, providing an in-depth review of each key treatment process and equipment unit within the wastewater treatment plant (WWTP). These visits involve a thorough examination of both existing documentation and real-time operations to uncover any inconsistencies or areas that may require attention. The findings from these visits enable a structured classification of deficiencies into either minor or major issues. Minor deficiencies, which typically can be corrected with minimal effort, are recorded along with suggestions for enhancements. Major deficiencies, however, might necessitate more substantial repairs or equipment replacements and could have a considerable impact on the plant's performance if not addressed in a timely manner.

Review of Existing Reports and Historical Data

A thorough review of existing reports and historical operational data is a critical first step in the assessment process. This review includes analyzing the original design basis, historical influent and effluent flow rates, and pollutant loads. By examining this data, assessors can identify patterns or deviations from expected performance, which might indicate underlying issues that need to be addressed. This historical perspective is essential for contextualizing the current state of the plant and informing future maintenance and upgrade strategies.

Stage-Wise Operations Monitoring

Monitoring the operational performance of each treatment stage is vital for ensuring that all processes function as intended. This involves detailed inspections of equipment, observation of process flows, and verification of operational parameters against the original design specifications. This approach outlines the need for a systematic approach to categorize any deficiencies found during this monitoring phase. These deficiencies are then classified as either minor, requiring routine maintenance, or major, necessitating more significant intervention. This step is crucial for maintaining the integrity of the treatment process and ensuring that the plant can continue to meet environmental discharge standards.

Assessment of Plant Performance Against Design Capacity

Assessing the plant's performance relative to its design capacity is a fundamental aspect of the condition assessment. This involves comparing current operational data with the original design specifications to determine whether the plant is functioning at its intended capacity. By implementing a long-term sampling program and conducting detailed tests on influent, effluent, and process stages, assessors can evaluate whether the plant is operating within acceptable parameters. Ensuring that the plant meets these standards is critical for regulatory compliance and for safeguarding public and environmental health.

Assessment of O&M Management

Organizational Structure

A well-structured organization is pivotal for the efficient management of WWTP operations. The assessment of the organizational structure involves evaluating how well the roles and responsibilities of staff align with the operational needs of the plant. This emphasizes that this evaluation should consider the adequacy of staff levels, the clarity of job descriptions, and the effectiveness of communication channels within the organization. A robust organizational structure not only enhances operational efficiency but also ensures that the plant can respond swiftly to any issues that arise, minimizing downtime and maintaining treatment quality.

Staffing and Duties

Staffing and the allocation of duties are critical factors in the successful operation of a WWTP. The assessment process involves reviewing the qualifications, experience, and responsibilities of the plant's staff to ensure they are well-equipped to handle the operational demands, ensuring they have a clear understanding of how staff duties are assigned and whether these assignments align with the plant's operational goals. This assessment helps identify any gaps in staffing or skill levels that could potentially affect the plant's performance, providing a basis for targeted training or restructuring.

Unit-Wise Activity Assessment

Daily and weekly activities within the plant need to be carefully assessed to ensure that all units operate effectively and efficiently. This assessment should include a review of maintenance schedules, operational logs, and the effectiveness of current practices in sustaining plant performance. By examining each unit individually, assessors can identify specific areas where improvements are needed, whether in maintenance routines, operational procedures, or equipment functionality. This detailed approach ensures that every aspect of the plant's operation is optimized, contributing to overall system reliability.

Training Programs

Continuous education and training are essential for maintaining high standards of operation and safety in WWTPs. These programs should cover the latest technologies, regulatory updates, and best practices in plant management. By keeping staff well-informed and skilled, the plant can operate more smoothly, with fewer disruptions due to human error or outdated practices. Effective training programs are also critical for ensuring that staff can respond appropriately to emergencies, thereby safeguarding both the plant and the surrounding environment.

11.3.5 Performance Evaluation, Benchmarking, and Impact Assessment

One of the main objectives of condition assessment of WWTPs is to evaluate their performance, benchmark their efficiency, and assess their environmental and health impacts. These aspects are interrelated and require a holistic approach that considers the entire life cycle of the treatment processes, from raw material acquisition to disposal. The following techniques are commonly used for this purpose:

- Performance evaluation involves analyzing the efficiency and capacity utilization of the treatment processes. This includes comparing performance indicators such as effluent quality, energy consumption, and CO_2 emissions against industry standards or similar facilities. Performance evaluation helps identify areas for improvement and supports the optimization of operational strategies.
- Benchmarking is the process of comparing the performance of a WWTP with that of other WWTPs or with best practices in the industry. Benchmarking helps identify gaps, set targets, and implement best practices. Benchmarking can also enable comparative analysis and peer learning among WWTPs.
- Environmental and health impact assessment, including life cycle assessment (LCA), is essential for evaluating the broader effects of WWTPs on the environment and human health. LCA considers the environmental impacts associated with all stages of a product's life cycle (i.e., from raw material acquisition to disposal). LCA helps identify pollutants, optimize resource management, and promote sustainable practices. LCA can also support decision-making, regulatory compliance, and continuous improvement.

By integrating these techniques, the condition assessment of WWTPs can provide a comprehensive and holistic picture of their performance, efficiency, and sustainability. This can help operators understand the system, optimize processes, and implement predictive maintenance strategies.

11.4 Life Cycle Assessment

LCA is crucial for assessing the environmental impacts of WWTPs by providing a comprehensive analysis of environmental impacts, detailed impact profiles, monthly impact assessment, components analysis, and uncertainty analysis. LCA helps in identifying pollutants, evaluating environmental impacts, optimizing resource management, supporting decision-making, enabling comparative analysis, ensuring regulatory compliance, and promoting continuous improvement (Corominas et al. 2020; Toniolo et al. 2021). By incorporating LCA findings, the review paper on condition assessment of WWTPs can highlight the holistic environmental impact, guide sustainable practices, and enhance operational efficiency and sustainability. The LCA methodology for WWTPs typically follows the framework outlined in the ISO 14040 standards, which includes four main phases (Fig. 11.6) (Finkbeiner et al. 2006):

(i) Goal and Scope Definition: This phase involves defining the purpose of LCA study, including the specific environmental issues to be addressed, the boundaries of the system under study, and the functional unit, which is often a cubic meter of treated wastewater. The scope also includes the selection of impact categories, such as GHG emissions, energy consumption, water use, and the production of solid waste and sludge.

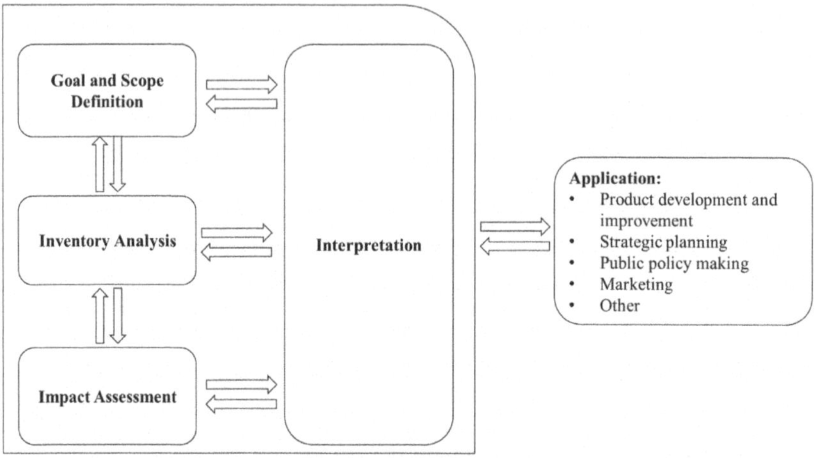

Fig. 11.6 Stages of life cycle assessment (ISO 14040, 2006) (Tayyebi et al. 2023)

(ii) Life Cycle Inventory (LCI): The phase involves collection of data on all inputs and outputs associated with the WWTP's operation. This includes inputs like energy, chemicals, and materials used in the construction and maintenance of the plant, as well as outputs such as treated effluent, emissions to solid waste, and air. The accuracy and comprehensiveness of this data are crucial for the subsequent assessment.

(iii) Life Cycle Impact Assessment (LCIA): During the LCIA phase, the data collected in the LCI is analyzed to evaluate the potential environmental impacts. This involves categorizing and quantifying the effects of emissions and resource use across different environmental impact categories, such as global warming potential, resource depletion eutrophication, and acidification. The LCIA provides insights into which stages of the WWTP's life cycle and which processes are the most environmentally burdensome (Tabesh et al. 2019).

(iv) Interpretation: The final phase involves interpretation of the LCIA results to draw conclusions and provide recommendations (Sala et al. 2020). This interpretation should address the goal and scope defined earlier and identify areas where environmental performance can be improved. The results may suggest changes in operational practices, the adoption of more efficient technologies, or the need for policy adjustments to minimize environmental impacts.

LCA methodologies can be classified into different types based on their approach, scope, and the kind of impacts they assess (Toniolo et al. 2021).

- Attributional LCA (ALCA) focuses on the environmental impacts directly attributing to the life cycle of a process, product, or system. It typically uses average data and reflects the static environmental burdens associated with the production and use of a specific service or product. This is commonly used for product labeling, environmental product declarations, and regulatory compliance.

- Consequential LCA assesses the environmental impacts of changes in a system by considering both direct and indirect effects. It aims to understand how changes in production or consumption affect overall environmental outcomes. This is commonly used in policy-making, strategic planning, and evaluating the broader impacts of decisions.
- Process-based LCA focuses on specific processes within the life cycle, often at a very detailed level. It examines the inputs and outputs of individual processes, such as specific unit operations within a manufacturing facility. They are conventionally used in detailed environmental assessments of specific technologies or processes.
- Input–Output LCA uses economic input–output tables to estimate the environmental impacts of economic activities associated with the production of goods and services. It is broader in scope and includes indirect effects across the economy. They are useful for assessing the impacts of complex supply chains and understanding the broader economic effects of environmental policies.
- Hybrid LCA combines process-based LCA with input–output LCA to leverage the strengths of both methods. It provides a more comprehensive assessment by integrating detailed process data with broader economic data. They are widely used for comprehensive environmental assessments that require both detailed process-level data and broader economic context.
- Streamlined LCA is a simplified version of the traditional LCA, focusing on the most significant processes and environmental impacts. It reduces the time and resources required for a full LCA. They are used in early-stage product development, rapid assessments, and decision-making where time and resources are limited.
- Dynamic LCA considers changes in environmental impacts over time, incorporating temporal variations in both the life cycle processes and the environmental impacts. They are applied in studies where the timing of impacts is critical, such as climate change assessments.
- Monte Carlo LCA uses probabilistic modeling to account for uncertainty in LCA results. It generates a range of possible outcomes based on the variability in input data and assumptions. They are used in LCA studies where uncertainty is a significant concern, such as risk assessments and policy evaluations.
- Life cycle costing LCA integrates economic costs with environmental impacts, providing a combined assessment of both the financial and ecological aspects of a product or process. They are used in projects where both cost-effectiveness and environmental performance are critical, such as in sustainable procurement.
- Environmental Input–Output LCA is a variant of IOLCA that focuses specifically on the environmental impacts associated with the economic inputs and outputs of a system. They are used for high-level assessments of industries and sectors, particularly in policy analysis and large-scale economic planning.

In context to WWTPs, Table 11.6 summarizes the strengths and limitations of various types of LCA.

LCA is widely applied in the wastewater sector to compare the environmental performance of different treatment technologies, such as activated sludge systems,

Table 11.6 Overview of the various LCA methodologies applied in the context of wastewater treatment

LCA methods	Description	Strength	Limitations
Attributional LCA	Focuses on the environmental impacts directly attributable to the WWTP processes	Provides a clear, static picture of the environmental impact	Does not account for changes over time or external factors
Consequential LCA	Considers the broader consequences of changes in the system, including indirect effects	Accounts for changes in the system, providing a dynamic perspective	Complex modeling required; results can be sensitive to assumptions
Hybrid LCA	Combines process-based LCA with economic input–output (EIO) analysis to cover the whole life cycle	Captures both direct and indirect environmental impacts	Data-intensive and computationally demanding
Input–output LCA	Uses economic input–output data to estimate the environmental impacts of economic activities related to the WWTP	Can handle complex systems with numerous interactions	May lacks specificity for WWTP operations, leading to less accurate results
Process-based LCA	Focuses on specific processes within the WWTP, typically on a unit operation basis	Highly detailed and specific to the WWTP processes	May miss broader system-level impacts; can be time-consuming and data-intensive
Streamlined LCA	Simplified LCA focusing on key processes and significant environmental impacts	Less resource-intensive and quicker to conduct; good for initial assessments	May overlook some environmental impacts due to simplifications
Monte Carlo LCA	Uses probabilistic modeling to assess uncertainty in LCA results	Provides a range of possible outcomes, reflecting uncertainty in data and assumptions	Computationally intensive; requires expertise in statistical modeling
Dynamic LCA	Accounts for changes over time in environmental impacts, reflecting the temporal dimension	Captures temporal variations in environmental impacts; more realistic for long-term assessments	More complex to model; requires time-dependent data, which may not always be available
Life cycle costing LCA	Integrates economic costs with environmental impacts in the LCA framework	Provides a comprehensive view that includes both economic and environmental aspects	Balancing environmental and economic outcomes can be challenging; requires detailed financial data

membrane bioreactors, and natural treatment systems like constructed wetlands. For example, studies have shown that while activated sludge systems are effective at removing organic pollutants, they also tend to be energy-intensive and generate high levels of sludge, which can be environmentally challenging to manage. On the other hand, natural treatment systems, though less energy-intensive, may require more land and may not be as effective in removing certain contaminants. Additionally, LCA can be used to assess the impact of various operational strategies, such as energy recovery from sludge through anaerobic digestion or the use of alternative disinfection methods like ozone or ultraviolet (UV) radiation. Each of these strategies has different environmental trade-offs, and LCA helps in quantifying these impacts to make informed decisions.

One of the main challenges in applying LCA to WWTPs is the variability in data quality and availability, particularly for newer technologies where long-term operational data may be scarce. Moreover, LCAs can be resource-intensive to conduct, requiring significant expertise in both environmental science and process engineering. Despite these challenges, LCA remains a powerful tool for improving the sustainability of wastewater treatment. As environmental regulations become more stringent and the demand for sustainable water management practices grows, the role of LCA in the design and operation of WWTPs is likely to expand. Future developments may include the integration of real-time data and more sophisticated modeling techniques, allowing for dynamic and more accurate assessments of environmental impacts.

Overall, LCA is an indispensable tool in the assessment of wastewater treatment plants, providing a comprehensive understanding of their environmental impacts. By guiding the selection of more sustainable practices and technologies, LCA contributes to the development of WWTPs that are not only effective in treating wastewater but also minimize their environmental footprint.

11.5 Preventing and Predictive Maintenance

WWTPs are critical infrastructure for maintaining public health and protecting the environment by treating sewage and industrial effluents before they are released into water bodies. To ensure the continuous and efficient operation of these facilities, maintenance is essential. In this section, two key maintenance strategies preventive and predictive maintenance strategies are discussed.

11.5.1 Preventive Maintenance

Preventive (or preventative) maintenance is a proactive strategy that involves regularly scheduled maintenance activities aimed at preventing equipment failures before they occur (Jia et al. 2021). In WWTPs, the preventive maintenance is crucial for ensuring the continuous and efficient operation of various mechanical, electrical, and structural components. By adhering to a preventive maintenance schedule, plant operators can reduce the likelihood of unexpected breakdowns, extend the lifespan of critical equipment, and maintain consistent compliance with environmental regulations (Zhang et al. 2024). WWTPs operate under conditions that expose equipment to harsh environments, including corrosive substances, fluctuating temperatures, and continuous mechanical stress. The components such as pumps, blowers, aerators, and filtration systems are essential to the treatment process and are frequently subjected to wear and tear over time. Without regular maintenance, these components are more likely to fail, leading to costly repairs, environmental non-compliance, and potential service disruptions. Preventive maintenance helps mitigate these risks by ensuring that equipment is inspected, cleaned, lubricated, and adjusted at regular intervals. By addressing potential issues before they result in equipment failure, WWTPs can maintain operational reliability and avoid the high costs associated with emergency repairs and unplanned downtime. A well-structured preventive maintenance program in a WWTP typically includes the following components (Viveros et al. 2021):

a. Maintenance Schedules: A preventive maintenance program is built around a detailed schedule that specifies when maintenance activities should be performed. Schedules are based on the manufacturer's recommendations, historical performance data, and the criticality of the equipment. Common intervals include daily, weekly, monthly, quarterly, and annual maintenance tasks.
b. Standard Operating Procedures (SOPs): SOPs provide detailed instructions for each maintenance task, ensuring consistency and thoroughness. These procedures may include steps for inspecting equipment, cleaning components, replacing worn parts, lubricating moving parts, and testing system performance.
c. Inspection and Monitoring: Regular inspections are a key part of preventive maintenance. These inspections may include visual checks, performance tests, and monitoring of operational parameters such as temperature, pressure, and flow rates. Inspections help identify early signs of wear, corrosion, or other issues that could lead to equipment failure.
d. Record Keeping: Comprehensive records of all maintenance activities are essential for tracking the condition of equipment over time. These records help identify recurring issues, assess the effectiveness of the maintenance program, and inform future maintenance planning.
e. Training and Staff Development: Effective preventive maintenance relies on well-trained personnel who are familiar with the equipment and the specific maintenance procedures. This ensures that staff are up-to-date with the latest maintenance techniques and technologies.

The implementation of a preventive maintenance program in a wastewater treatment plant offers several key benefits. The regular maintenance minimizes unplanned downtime by ensuring that the potential problems are identified and resolved prior they lead to equipment failure. Similarly, the regular maintenance helps keep equipment in good condition, extending its operational lifespan and delaying the need for costly replacements. Although preventive maintenance involves ongoing costs, it is generally less expensive than the costs associated with emergency repairs, equipment replacements, and unplanned downtime. The preventive maintenance ensures that equipment operates within design specifications, helping WWTPs maintain compliance with environmental regulations and avoid penalties. Also, regular maintenance reduces the risk of accidents and safety incidents by ensuring that equipment is functioning correctly and safely. While preventive maintenance offers many advantages, there are also challenges to its implementation, such as:

a. Resource Allocation: Preventive maintenance requires time, labor, and materials. For WWTPs with limited resources, it can be challenging to allocate the necessary resources for regular maintenance activities.
b. Over-Maintenance: In some cases, equipment may be serviced more frequently than necessary, leading to over-maintenance. This can result in unnecessary costs and may even cause wear and tear from excessive handling.
c. Balancing Maintenance and Operations: Scheduling maintenance activities without disrupting the continuous operation of the plant can be difficult. Careful planning is required to balance maintenance needs with operational demands.
d. Data Management: Keeping accurate and comprehensive records of all maintenance activities is essential but can be challenging, especially in large plants with numerous pieces of equipment.

To implement an effective preventive maintenance program in a WWTP, the following strategies are recommended:

- **Step 1—Develop a Comprehensive Maintenance Plan**: The plan should outline all maintenance tasks, schedules, procedures, and responsibilities. It should be tailored to the specific needs and conditions of the WWTP.
- **Step 2—Prioritize Critical Equipment**: Focus preventive maintenance efforts on the most critical equipment—those components whose failure would have the greatest impact on plant operations and compliance.
- **Step 3—Utilize Maintenance Management Software**: Computerized Maintenance Management Systems (CMMS) can help track maintenance schedules, manage work orders, and store maintenance records, improving the efficiency and effectiveness of the maintenance program.
- **Step 4—Regularly Review and Update the Program**: Maintenance needs can change over time due to factors such as equipment aging, changes in plant operations, or the introduction of new technologies. Regularly review and update the preventive maintenance program to ensure it remains effective.
- **Step 5—Engage and Train Staff**: Ensure that all maintenance personnel are properly trained and engaged in the maintenance process. Encourage a culture of proactive maintenance where staff are motivated to identify and address issues before they escalate.

The preventive maintenance is an essential strategy for ensuring the reliable and efficient operation of wastewater treatment plants. By proactively addressing potential equipment issues before they lead to failure, WWTPs can reduce downtime, extend equipment life, and maintain compliance with environmental regulations (Hernández-Chover et al. 2020). While implementing a preventive maintenance program requires careful planning and resource allocation, the long-term benefits far outweigh the costs, making it a critical component of effective wastewater treatment plant management.

11.5.2 Predictive Maintenance

In the context of wastewater treatment plants (WWTPs), predictive maintenance plays a critical role in ensuring the reliability and efficiency of operations, minimizing unplanned downtime, and optimizing maintenance activities. By predicting when maintenance should be performed based on the actual condition of equipment, PdM offers a more targeted and cost-effective alternative to traditional maintenance strategies. Wastewater treatment plants operate continuously, processing large volumes of sewage and industrial effluents daily. The reliability of key equipment—such as pumps, blowers, aerators, and filtration systems are crucial to maintaining the efficiency of the treatment process and ensuring compliance with environmental regulations. The unexpected equipment failures can lead to costly repairs, environmental violations, and disruptions in service. The predictive maintenance addresses these challenges by continuously monitoring the condition of critical assets and predicting their failure points. Reportedly, this proactive approach allows maintenance teams to perform necessary interventions just in time, preventing failures and extending the life of equipment.

In WWTPs, the predictive maintenance involves the following steps:

a. Condition Monitoring: Sensors and other monitoring devices are installed on critical equipment to continuously measure parameters such as vibration, temperature, pressure, flow rates, and electrical consumption. These sensors collect data in real time, providing a constant stream of information about the equipment's operating condition.
b. Data Analysis and Modeling: The data collected from sensors is analyzed using advanced algorithms, often incorporating machine learning models. These models can detect patterns and trends that indicate impending failures. For example, an increase in vibration levels in a pump might signal bearing wear, which could lead to failure if not addressed.
c. Predictive Alerts: When the analysis detects signs of potential failure, the system generates alerts that notify maintenance personnel. These alerts provide detailed information about the likely cause of the issue and the recommended course of action, allowing for timely maintenance interventions.
d. Maintenance Planning: Based on the predictive alerts, maintenance tasks can be scheduled and prioritized according to the severity of the potential failure and the criticality of the equipment. This ensures that resources are used efficiently

and that maintenance activities are conducted with minimal disruption to plant operations.

e. Feedback Loop: After maintenance is performed, the system updates its models and algorithms based on the new data, continually improving its predictive accuracy.

The appropriate implementation of predictive maintenance in WWTPs offers the following advantages, such as:

a. Reduced Unplanned Downtime: By predicting equipment failures before they occur, PdM minimizes unexpected downtime, ensuring continuous operation of the treatment plant.
b. Optimized Maintenance Scheduling: PdM allows maintenance to be performed only when necessary, reducing the frequency of maintenance activities and focusing efforts where they are most needed.
c. Cost Savings: Predictive maintenance can reduce maintenance costs by preventing catastrophic failures that require expensive repairs and by extending the life of equipment through timely interventions.
d. Improved Equipment Reliability: Continuous monitoring and data-driven maintenance decisions enhance the reliability of critical equipment, reducing the risk of operational disruptions.
e. Enhanced Environmental Compliance: By maintaining equipment in optimal condition, PdM helps ensure that the treatment plant consistently meets environmental regulations and avoids fines or penalties for non-compliance.

While predictive maintenance offers numerous advantages, there are certain challenges linked with its implementation in WWTPs, which includes the following:

a. Initial Investment: Setting up a predictive maintenance system requires an initial investment in sensors, data analytics tools, and possibly new software platforms. Additionally, training staff to use these new tools effectively is essential.
b. Data Management: The continuous monitoring of equipment generates large volumes of data that need to be stored, processed, and analyzed. Managing this data efficiently requires robust IT infrastructure and expertise.
c. Integration with Existing Systems: PdM systems must be integrated with existing plant management systems, such as SCADA (Supervisory Control and Data Acquisition), to provide a seamless flow of information and enhance decision-making.
d. Technical Expertise: Implementing and managing a predictive maintenance program requires specialized knowledge in data analysis, machine learning, and the specific operational characteristics of WWTP equipment.

Thus, predictive maintenance represents a significant advancement in the management of wastewater treatment plants. By leveraging real-time data and advanced analytics, WWTPs can move from a reactive or preventive maintenance approach to a more proactive and efficient strategy. Although the initial setup of a PdM system may involve substantial investment and technical challenges, the long-term benefits—such as reduced downtime, cost savings, and improved operational reliability—make it a valuable addition to the maintenance practices of modern wastewater treatment facilities. As technology continues to evolve, predictive maintenance is likely to become an integral part of the maintenance landscape in the wastewater industry.

11.6 Summary

The assessment of the condition of WWTPs is essential for maintaining their operational efficiency and ensuring compliance with environmental standards (Table 11.7). This chapter has explored various methodologies for evaluating the condition of WWTPs, including physical inspections, maintenance record reviews, process monitoring and control, performance evaluation, structural and mechanical integrity assessments, advanced data analysis, and environmental impact assessments.

Physical inspections and maintenance record reviews are crucial for identifying early signs of wear, damage, or systemic issues that could compromise the plant's performance. Process monitoring, through both online systems and laboratory analysis, plays a vital role in maintaining optimal operating conditions and ensuring treatment efficacy. Performance evaluation and benchmarking against industry standards enable plants to measure their efficiency and capacity utilization, guiding improvements and investments in technology. Structural and mechanical integrity assessments, often using non-destructive testing techniques, help prioritize maintenance and rehabilitation efforts, thereby extending the lifespan of critical infrastructure. Advanced data analysis and modeling, including hydraulic and process simulations, provide deeper insights into operational trends and future performance, supporting data-driven decision-making for process optimization. Environmental and health impact assessments, such as life cycle assessments and pathogen removal studies, highlight the broader implications of WWTP operations on public health and ecosystems. These assessments underscore the importance of effective treatment processes in reducing environmental footprints and protecting community health.

Reportedly, the regular condition assessments can result in significant operational cost reductions. For example, systematic maintenance practices and early detection of issues can reduce energy costs by 10–15%, while optimizing biogas production and equipment efficiency (Mishra et al. 2017). Similarly, proactive maintenance strategies have contributed to operational cost reductions of up to 20% by minimizing equipment downtime and preventing major breakdowns (Crippen et al. 2018).

Overall, a comprehensive approach to condition assessment, integrating both operational and environmental perspectives, is essential for enhancing the sustainability, resilience, and adaptability of WWTPs. By prioritizing continuous evaluation and adopting best practices, wastewater treatment facilities can better navigate future challenges, including regulatory changes, population growth, and climate impacts, ensuring they continue to serve their vital role in urban and environmental management. The potential for cost savings through regular condition assessments further

Table 11.7 Overall summary of wastewater treatment plant condition assessment techniques

Assessment technique	Description	Parameters/indicators
Physical inspection	Regular visual inspections to identify wear, damage, or signs of failure in plant infrastructure	Structural integrity, corrosion, leaks, visible wear
Maintenance records review	Analysis of maintenance logs to assess the frequency, scope, and effectiveness of maintenance activities	Maintenance frequency, types of repairs, downtime
Process monitoring and control	Continuous or periodic monitoring of key operational parameters to ensure the plant operates within design specifications	BOD, COD, TSS, pH, flow rates, ammonia, temperature, DO levels
Performance evaluation and benchmarking	Comparison of the plant's performance against industry standards or similar facilities to identify inefficiencies or areas for improvement	Treatment efficiency, energy consumption, sludge production, compliance with discharge standards
Structural and mechanical integrity assessments	Use of non-destructive testing techniques and engineering analysis to evaluate the condition of critical components	Integrity of tanks, pipes, pumps, mixers, structural components
Advanced data analysis and modeling	Use of data analytics and modeling tools to predict future performance and optimize operations	Predictive maintenance schedules, process optimization scenarios, hydraulic and process modeling results
Environmental and health impact assessments	Evaluation of the plant's impact on the surrounding environment and public health, including compliance with environmental regulations	Emission levels, pathogen removal efficiency, heavy metals and toxin concentrations in effluent, GHG emissions
Compliance audits and regulatory reviews	Periodic audits to ensure compliance with all relevant environmental regulations, permits, and operational standards	Audit findings, non-compliance issues, regulatory updates
Life cycle assessment (LCA)	Assessment of environmental impacts associated with the entire life cycle of the WWTP, from construction to decommissioning	GHG emissions, energy use, resource depletion, waste generation, environmental impact scores

emphasizes the value of ongoing maintenance and monitoring in securing long-term operational efficiency.

References

Asquer A (2018) Regulation of infrastructure and utilities. Springer International Publishing, Cham. https://doi.org/10.1007/978-3-319-67735-4

Binnie C, Kimber M, Clements P (2009) Basic water treatment. Royal Society of Chemistry

Corominas L, Byrne DM, Guest JS, Hospido A, Roux P, Shaw A, Short MD (2020) The application of life cycle assessment (LCA) to wastewater treatment: a best practice guide and critical review. Water Res 184:116058. https://doi.org/10.1016/j.watres.2020.116058

Crippen J, Sawsan Bataineh E, Yousef EA, Minkwitz R (2018) Ramtha WWTP expansion condition assessment report (Task 5) Final

Finkbeiner M, Inaba A, Tan R, Christiansen K, Klüppel H-J (2006) The new international standards for life cycle assessment: ISO 14040 and ISO 14044. Int J Life Cycle Assess 11:80–85. https://doi.org/10.1065/lca2006.02.002

Garg S, Chowdhury ZZ, Faisal ANM, Rumjit NP, Thomas P (2022) Impact of industrial wastewater on environment and human health, pp 197–209. https://doi.org/10.1007/978-3-030-83811-9_10

Gdoutos E, Konsta-Gdoutos M (2024) Nondestructive testing (NDT), pp 201–225. https://doi.org/10.1007/978-3-031-45990-0_8

Ghoni R, Dollah M, Sulaiman A, Ibrahim FM (2014) Defect characterization based on eddy current technique: technical review. Adv Mech Eng 6:182496. https://doi.org/10.1155/2014/182496

Hernández-Chover V, Castellet-Viciano L, Hernández-Sancho F (2020) Preventive maintenance versus cost of repairs in asset management: an efficiency analysis in wastewater treatment plants. Process Saf Environ Prot 141:215–221. https://doi.org/10.1016/j.psep.2020.04.035

Inês Silva M, Malitckii E, Santos TG, Vilaça P (2023) Review of conventional and advanced non-destructive testing techniques for detection and characterization of small-scale defects. Prog Mater Sci 138:101155. https://doi.org/10.1016/j.pmatsci.2023.101155

Jia Y, Wang S, Huang A, Gao Y, Wang J, Zhou W (2021) A comparative long-term effectiveness assessment of preventive maintenance treatments under various environmental conditions. Constr Build Mater 273:121717. https://doi.org/10.1016/j.conbuildmat.2020.121717

Kiliç S, Firat M, Özdemir Ö (2023) A novel current condition assessment system for sustainable management and operation of wastewater treatment plants. J Environ Eng 149. https://doi.org/10.1061/JOEEDU.EEENG-7252

Mishra N, Reddy R, Kuila A, Rani A, Nawaz A, Pichiah S (2017) A review on advanced oxidation processes for effective water treatment. Curr World Environ 12:469–489. https://doi.org/10.12944/cwe.12.3.02

Mitra S, Chakraborty AJ, Tareq AM, Bin Emran T, Nainu F, Khusro A, Idris AM, Khandaker MU, Osman H, Alhumaydhi FA, Simal-Gandara J (2022) Impact of heavy metals on the environment and human health: novel therapeutic insights to counter the toxicity. J King Saud Univ Sci 34:101865. https://doi.org/10.1016/j.jksus.2022.101865

Onu MA, Ayeleru OO, Oboirien B, Olubambi PA (2023) Challenges of wastewater generation and management in sub-Saharan Africa: a review. Environ Challenges 11:100686. https://doi.org/10.1016/j.envc.2023.100686

Palansooriya KN, Yang Y, Tsang YF, Sarkar B, Hou D, Cao X, Meers E, Rinklebe J, Kim K-H, Ok YS (2020) Occurrence of contaminants in drinking water sources and the potential of biochar for water quality improvement: a review. Crit Rev Environ Sci Technol 50:549–611. https://doi.org/10.1080/10643389.2019.1629803

Puri M, Gandhi K, Kumar MS (2023) Emerging environmental contaminants: a global perspective on policies and regulations. J Environ Manage 332:117344. https://doi.org/10.1016/j.jenvman.2023.117344

Razali N, Sultan MTH, Jawaid M (2018) Impact damage analysis of hybrid composite materials, durability and life prediction in biocomposites. Fibre Reinf Compos Hybrid Compos 121–132. https://doi.org/10.1016/B978-0-08-102290-0.00006-4

Sala S, Laurent A, Vieira M, Van Hoof G (2020) Implications of LCA and LCIA choices on interpretation of results and on decision support. Int J Life Cycle Assess 25:2311–2314. https://doi.org/10.1007/s11367-020-01845-2

Santos TG, Oliveira JP, Machado MA, Inácio PL, Duarte VR, Rodrigues TA, Santos RA, Simão C, Carvalho M, Martins A, Nascimento M, Novais S, Ferreira MS, Pinto JL, Fernandes FB, Camacho E, Viana J, Miranda RM (2020) Reliability and NDT methods, pp 265–295. https://doi.org/10.1007/978-3-030-44522-5_8

Shah A, Arjunan A, Baroutaji A, Zakharova J (2023) A review of physicochemical and biological contaminants in drinking water and their impacts on human health. Water Sci Eng 16:333–344. https://doi.org/10.1016/j.wse.2023.04.003

Shiotani T, Ogura N, Okude N, Watabe K, Van Steen C, Tsangouri E, Lacidogna G, Czarnecki S, Chai HK, Yang Y, Verstrynge E, Aggelis DG (2024) Non-destructive inspection technologies for repair assessment in materials and structures. Dev Built Environ 18:100443. https://doi.org/10.1016/j.dibe.2024.100443

Tabesh M, Feizee Masooleh M, Roghani B, Motevallian SS (2019) Life-cycle assessment (LCA) of wastewater treatment plants: a case study of Tehran, Iran. Int J Civ Eng 17:1155–1169. https://doi.org/10.1007/s40999-018-0375-z

Tayyebi F, Golabi M, Jaafarzadeh N (2023) Life cycle assessment, a decision-making tool in wastewater treatment systems: a case study wastewater treatment plant of Ahvaz, Iran. Appl Water Sci 13:145. https://doi.org/10.1007/s13201-023-01958-7

Toniolo S, Borsoi L, Camana D (2021) Life cycle assessment: methods, limitations, and illustrations. In: Methods in sustainability science. Elsevier, pp 105–118. https://doi.org/10.1016/B978-0-12-823987-2.00007-6

Viveros P, Miqueles L, Mena R, Kristjanpoller F (2021) Opportunistic strategy for maintenance interventions planning: a case study in a wastewater treatment plant. Appl Sci 11:10853. https://doi.org/10.3390/app112210853

Zhang W, Gan J, He S, Li T, He Z (2024) An integrated framework of preventive maintenance and task scheduling for repairable multi-unit systems. Reliab Eng Syst Saf 247:110129. https://doi.org/10.1016/j.ress.2024.110129

Chapter 12
Life Cycle Assessment of Constructed Wetland for Wastewater Treatment

Shweta D. Jaiswar and Pratham Arora

Abstract This book chapter provides a comprehensive overview of constructed wetlands (CWs) and the application of life cycle assessment (LCA) to evaluate their environmental footprint. Constructed wetlands are increasingly acknowledged as an economical and sustainable technology for wastewater treatment. The chapter reviews literature favoring constructed wetlands and identifies research gaps and future directions to enhance environmental performance and sustainability through LCA-driven optimization strategies. A methodological approach is applied to examine two CWs as case studies: a 150-kilo liter per day (KLD) sewage treatment plant and a 1.2 million liter per day (MLD) effluent treatment plant during their construction and operation phases, with the functional unit considered as 1 m^3 of treated wastewater. The results reveal the total potential impact for the climate change category as 0.523 kg CO_2 eq/m^3 of treated wastewater for the ETP and 0.372 kg CO_2 eq/m^3 of treated wastewater for the STP, with higher emissions observed during the operation phases. The chapter underscores the importance of integrating LCA into the design, operation, and management of constructed wetlands, fostering a deeper understanding of their environmental benefits and supporting their widespread adoption as an eco-friendly solution for wastewater treatment in the era of climate change.

Keywords Constructed wetland · Life cycle assessment · Wastewater treatment · Environment

Abbreviations

ADP Abiotic depletion potential

S. D. Jaiswar · P. Arora (✉)
Department of Hydro and Renewable Energy, Indian Institute of Technology Roorkee, Roorkee, India
e-mail: pratham.arora@hre.iitr.ac.in

S. D. Jaiswar
e-mail: shweta_djaiswar@hre.iitr.ac.in

S. Souabi et al. (eds.), *Wastewater Treatment Plants*, Water Science and Technology Library 130, https://doi.org/10.1007/978-3-031-87461-1_12

AP	Acidification potential
CC/GWP	Climate change/global warming potential
CPCB	Central Pollution Control Board
CW	Constructed wetland
EP	Eutrophication potential
FAETP	Freshwater aquatic ecotoxicity
FCP	Freshwater consumption
FDP	Fossil depletion potential
GHG	Greenhouse gases
HFCW	Horizontal flow constructed wetland
HHCP	Human health carcinogenic
HTC	Human toxicity, cancer
HTNC	Human toxicity, non-cancer
KLD	Kilo liter per day
LU	Land use
MAETP	Marine aquatic ecotoxicity
MLD	Million liter per day
ODP	Ozone depletion potential
PE	Population equivalent
PED	Primary energy demand
PO	Photochemical oxidation
POFP	Photochemical ozone formation potential
RCC	Reinforced cement concrete
RD	Resource depletion
RI	Respiratory inorganics
TETP	Terrestrial ecotoxicity potential
VFCW	Vertical flow constructed wetland
WC	Water consumption
WWTP	Wastewater treatment plant

12.1 Introduction

Water is essential for life, yet its scarcity with the expanding population is a growing concern. Reusing water conserves this precious resource, ensuring its availability for present and future generations and mitigating water stress. Since the mid-twentieth century, the global population has tripled, increasing from over 2.5 billion in 1950 to nearly 7.9 billion in 2021 (United Nations). With the rise in the population, there has been tremendous growth in industrialization and urbanization, which have led to increasing usage of resources (https://ourworldindata.org/population-growth, https://nap.nationalacademies.org/read/9148/chapter/5). The existing freshwater reserve is significantly lower compared to the vast mass of saltwater covering nearly 70% of the Earth. Furthermore, the availability of freshwater distribution is uneven and scarce (http://www.agriculture-biodiversite-oi.org/en/Nature-agriculture/News-from-the-field/Environment-and-Biodiversity/Preserving-natural-resources/Water-a-scarce-and-valuable-resource). Disruptions in weather patterns caused by climate change result in harsh weather occurrences, irregular water supply, exacerbated water shortages, and pollution of water systems (https://www.unicef.org/stories/water-and-climate-change-10-things-you-should-know). Conserving and sustaining water resources (surface or groundwater) is one of humanity's most pressing concerns today. This challenge has become increasingly acute with the rise of urban, industrial, and agricultural activities that generate significant amounts of wastewater (Breida et al. 2019).

India, the largest global consumer of groundwater, relies on around 30 million access points to meet 85% of its rural drinking water needs and 48% of its urban water demand (https://www.unicef.org/india/what-we-do/clean-drinking-water). Despite having only 4% of the world's water resources while accommodating 18% of the global population, the country faces severe water stress. According to a recent evaluation by NITI Aayog, the Indian government's policy think tank, a large portion of the population is experiencing critical to extreme levels of water scarcity (https://www.worldbank.org/en/country/india/brief/world-water-day-2022-how-india-is-addressing-its-water-needs). Safeguarding high-quality wastewater treatment and sanitation services in urban areas has long been a priority, given their profound influence on public health, environmental conservation, and the promotion of sustainable development. Proper sanitation measures help to safeguard the environment, prevent water pollution, and maintain the health of ecosystems (Tsangas et al. 2023). In the fiscal year 2020–2021, urban areas in India generated 72,368 million liters per day (MLD) of sewage, as per the Central Pollution Control Board (CPCB) (Bassi et al. 2023). In India, the total installed capacity for treating urban wastewater (domestic sewage) is 44%, amounting to 31,841 MLD, whereas the daily sewage generation is approximately 72,368 MLD (CPCB 2021). The actual treatment capacity is only 28 percent, amounting to 20,236 MLD (Bassi et al. 2023).

In recent years, the emphasis on improving the performance of wastewater treatment plants (WWTPs) has changed from producing high-quality effluent to guaranteeing long-term survival. WWTPs are widely recognized as the main municipal energy users (https://www.worldbank.org/en/country/india/brief/world-water-day-2022-how-india-is-addressing-its-water-needs). Wastewater treatment involves screening, sedimentation, biological treatment, filtration, nutrient removal, disinfection, and sludge treatment. Treated water can be reclaimed for irrigation, industry, or potable supply reuse. On one hand, sewage treatment reduces direct environmental impacts. On the other, it increases energy and chemical usage for operating treatment plants and wastewater infrastructure, potentially leading to increased environmental burdens (Negi and Chandel 2021). The principal environmental problems related to WWTPs are greenhouse gas (GHG) emissions from biological treatment processes, such as methane (CH_4), carbon dioxide (CO_2), and nitrous oxide (N_2O). CO_2 is released during the energy production required for plant operation (Tsangas et al. 2023). India is the third-largest greenhouse gas emitter globally. However, when considering emissions per capita, it has the lowest level among G20 nations (Chateau et al. 2023). Traditional wastewater treatment systems demand a lot of energy for mechanical components and have significant operational and investment expenses (Almuktar et al. 2018). One of the wastewater treatment methods focused in the current study is constructed wetlands (CWs), among other available options. CWs are increasingly recognized as a sustainable alternative for wastewater treatment, and they present notable benefits compared to traditional technologies, particularly in terms of low energy consumption (Andreo-Martínez et al. 2021).

12.2 Constructed Wetlands

Wastewater treatment comprises multiple stages: preliminary treatment (screening and grit removal), primary treatment (sedimentation), secondary treatment (biological processes and clarification), optional tertiary treatment (filtration and disinfection), and advanced treatment (nutrient removal and chemical treatment). CWs are engineered wastewater treatment techniques designed to mimic the processes seen in a natural wetland ecosystem. It is a secondary treatment process, comprising a basin that is stuffed with sand, and gravel (substrates), and planted with vegetation that is capable of surviving saturated conditions. The substrate (gravel) and root zone serve as a home for microbes that help in the breakdown and treatment of wastewater contaminants. In addition, the plants absorb and transpire water, which helps to eliminate excess moisture from the system and increase oxygen levels in the root zone. CWs are a natural, low-cost, environmentally friendly biological wastewater treatment technique. Using solar energy along with the strength of nature, the entire process is done by interactions between wastewater, microbes, plants, soil/gravel media, and the atmosphere. The wastewater is directed from one end, infiltrates through gravel beds, and gets discharged from the other end at a lower height. Microbial nitrification and denitrification are the most common removal methods in CWs (Ombretta 2007). To prevent soil contamination, the basins are waterproofed with a plastic lining.

CWs effectively can treat a variety of wastewater types, including urban, agricultural, animal, industrial, and mining discharges, along with petroleum and municipal waste. The efficiency of a wetland system for WWT is significantly impacted by variables such as its nature, design, type of plant, microbiological activity, and local weather conditions (Almuktar et al. 2018). The low operation, construction, and maintenance costs of constructed wetlands (CWs) make them particularly preferable. The effluent received after the treatment can be disposed of easily or put to use if met local regulations, or else it can be further treated in tertiary treatment for disinfection such as chlorination or UV radiation, and ozonation. These treatment procedures are both visually appealing and effective in removing odors by keeping the water level below the top of the gravel indicating subsurface movement as shown in Fig. 12.1.

CWs are made of several key components designed to facilitate the treatment of wastewater as discussed below:

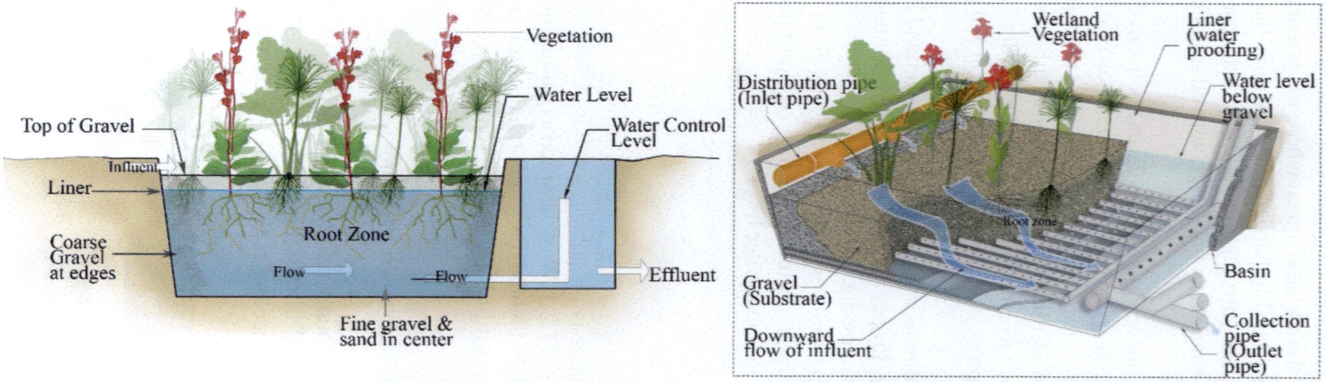

(a) Cross section of Constructed Wetland

(b) Perspective view of Constructed Wetland

Fig. 12.1 **a** Cross section and **b** perspective view of wetland unit

- **Basin**: The basin is the main structure of the wetland and is a shallow ditch filled with gravel and support plant growth, typically excavated or constructed to hold the wastewater. It provides the physical space where treatment processes occur.
- **Liner**: The liner is an impermeable barrier placed within the basin to prevent leakage of wastewater into the surrounding soil or groundwater. Common liner materials include geomembranes, clay liners, specialized sealants, and bituminous membranes.
- **Substrate**: The substrate supports plant growth and helps in pollutant removal. Typical substrates include gravel and sand. They offer a surface area for microbial attachment, enhancing treatment efficiency via biological processes. The substrate nourishes the root zone and facilitates denitrification and bioremediation.
- **Vegetation**: Vegetation, such as emergent aquatic plants, plays an important role in pollutant removal using processes like phytoremediation and provides a habitat for microorganisms.
- **Inlet Pipe**: The inlet pipe directs wastewater into the wetland basin, distributing it evenly across the surface. Proper distribution helps maximize contact between the wastewater and treatment media.
- **Outlet Pipe**: The outlet pipe removes treated wastewater from the wetland basin once treatment processes have occurred. It may be equipped with a control structure to regulate flow rates and water levels.

12.2.1 Types of CW

The focus of the current chapter is on subsurface flow CWs, and their types are described in detail below in Fig. 12.2. CW is classified as surface flow and subsurface flow based on the water flow. UN-Habitat has categorized subsurface CWs into three categories based on water flow within the substrate (Ombretta 2007).

12.2.1.1 Horizontal Flow Constructed Wetland

The untreated water flows horizontally through a gravel media (substrate), typically gravel or sand, facilitating natural filtration and biological degradation of contaminants. The course gravels are placed at the sides to protect the inlet (distribution)

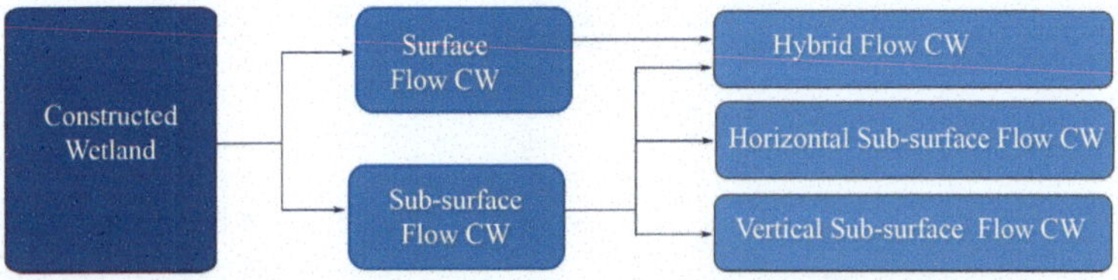

Fig. 12.2 Classification of CWs

and outlet pipes. The central portion of the treatment area is filled with sand and gravel that supports the wetland plants and forms a medium for the microbes to grow in it, contributing to pollutant removal. The substrate depth in this system varies from 30 to 80 cm based on the vegetation species and root depth, with a slope of 1 to 3% to support the gravity flow of the applied influent (Almuktar et al. 2018). The distribution or inlet pipe is placed at the top, whereas the outlet pipe is placed at the bottom to collect the treated water. Throughout this process, the wastewater will encounter a network of aerobic, anoxic, and anaerobic zones (Ombretta 2007). This type of system requires more space and thus has a larger land footprint when compared to vertical flow CW, but it is more economical as it requires no pumping system for influent transfer. The components of horizontal flow are almost the same as vertical flow CWs.

12.2.1.2 Vertical Flow Constructed Wetland

In a vertical flow system, the wastewater is flowing vertically downward into the system (filtering media), and thus, it is called VFCW. In this type, the wastewater inlet (distribution) pipe is positioned above the substrate level, and a pumping system periodically distributes pre-treated water over its whole top surface ensuring no direct contact with the environment. Microbes grow in the space between plant roots and substrate and help to break down the pollutants present in wastewater. Oxygen, needed for aerobic bacteria, is provided by water-distributing splash plates and plant roots. Treated water is then collected through a drainage pipe.

12.2.1.3 Hybrid Flow Constructed Wetland

To increase treatment efficiency, different types of CWs can be integrated while utilizing the benefits of each unique system. A combination of a VFCW and a HFCW is known as a hybrid CW. Many nations across the world currently have hybrid CW in use. The three types of subsurface flow CW are presented in Figure 12.3.

12.2.2 The Design of Constructed Wetland

A constructed wetland is typically composed of a bar screen chamber, wet well, holding tank, wetland unit, and treated water tank. The entire structure is constructed of reinforced cement concrete (RCC). Refer to Fig. 12.4, which illustrates a typical cross section of an aerated CW system.

- **A bar screen chamber** located before the wet well, uses a mechanical filter to remove large particles, preventing clogging and system breakdowns.

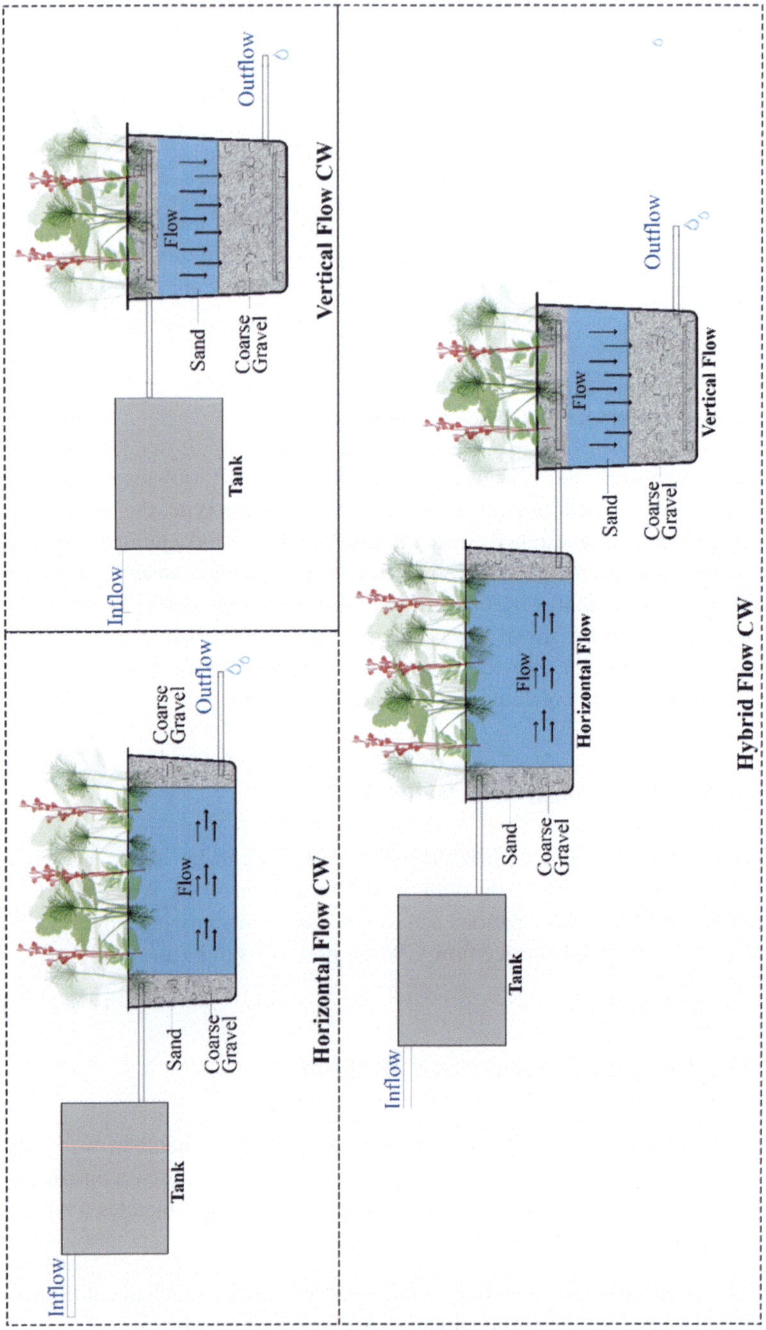

Fig. 12.3 Types of subsurface flow CW

Fig. 12.4 Cross section of a typical aerated CW system

- **A wet well** ensures continuous flow to the holding tank by transferring sewage with a raw sewage pump. It maintains a water retention time of 3–4 h.
- **The holding tank** provides 24 to 36 h of retention for microbial treatment of wastewater, with its size matching the plant capacity. It is divided into chambers and dosed with 3–5 g of microbes per 1 KLD to enhance organic matter consumption.
- **The wetland unit** varies in size depending on its type and technology. It uses gravel and ornamental plants for wastewater treatment, with flow driven by gravity and enhanced by forced bed aeration.
- **The treated water tank** has a 6-h retention time and stores water before disinfection through methods like chlorination, ozonation, or UV rays.

12.2.3 Vegetation

Plants are integral to CWs for wastewater treatment, serving multiple functions. Various types of plants are used in CWs based on their ability to thrive in waterlogged conditions and their capacity for pollutant uptake. Through phytoremediation, they absorb and metabolize contaminants, improving water quality by removing nutrients and pollutants. Additionally, plants oxygenate the root zone via photosynthesis, fostering aerobic conditions beneficial for microbial activity and organic matter decomposition. Their roots provide a habitat for beneficial microbes, stabilize the wetland substrate, and create a diverse ecosystem. Plants also contribute to the wetland's aesthetic appeal, provide habitat for wildlife, and aid in evapotranspiration, reducing water levels. Overall, their presence enhances the effectiveness, biodiversity, and ecological sustainability of CWs. Some of the plant species used in CW are *Canna indica, Papyrus, Iris Pseudacorus, Alocasia, Spider Lily, Acorus calamus, Phragmites Australis, Schoenoplectus* spp., *Phalaris Arundinacea, Typha Latifolia, Vetiver, Heliconia, Arundo Donax, Cyperus Alternifolius, and Thalia Dealbata.* Some of the ornamental plants are shown in Table 12.1.

12.3 Life Cycle Assessment

Life cycle assessment (LCA) is an extensive approach for evaluating the environmental impacts of a product, process, or activity, encompassing every stage from raw material extraction to final disposal. It examines all lifecycle stages to comprehensively understand environmental impact, assisting in informed sustainability and resource management decisions. LCA considers resource extraction, production, use, and end-of-life treatment to quantify environmental burdens and benefits. The following Fig. 12.5 illustrates a linear model of manufacturing, often referred to as a cradle-to-grave life cycle.

Table 12.1 Plants used in CWs

 Scientific Name: Alocasia spp. Common Name: Elephant Ear	 Scientific Name: Hymenocallis Carolina Common Name: White Spider Lily	 Scientific Name: Canna indica Common Name: Indian Shot
 Scientific Name: Cyperus Alternifolius Common Name: Umbrella Palm, Umbrella Papyrus	 Scientific Name: Cyperus Papyrus Common Name: Papyrus	 Scientific Name: Thalia dealbata Common Name: Powdery Alligator-flag organisms.

The International Organization for Standardization 14040 and 14044 offer recommendations for performing LCAs. ISO 14040 outlines the general principles and framework for conducting an LCA, while ISO 14044:2006 outlines detailed guidelines and requirements for performing an LCA, covering phases such as establishing goals and scope, performing inventory analysis, evaluating impacts, and analyzing results. It is a comprehensive guideline for performing systematic and accurate life cycle assessments to identify and mitigate environmental consequences (BIS 2006). These standards help ensure consistency and reliability in LCA studies, aiding industries, policymakers, and environmentalists in making informed sustainability and resource management decisions.

The LCA study involves defining objectives and boundaries, including system boundaries, functional units, and impact categories.

- The functional unit serves as the basis for comparison, defining the product or service's quantity and function.
- The system boundary outlines the limits for assessing inputs, outputs, and impacts.
- Measures the inputs and outputs throughout the life cycle stages, encompassing raw materials, energy, emissions, and waste.
- Life cycle impact assessment (LCIA) examines environmental impacts through various analyzing categories.

Fig. 12.5 Stages of LCA

- Finally, the interpretation phase analyzes results to conclude, identify improvements, and support decision-making.

12.4 Literature Survey

A literature review on the LCA of CWs for wastewater treatment explores different system boundaries, functional units, and software tools for impact analysis, as detailed in Table 12.2. It reveals the environmental benefits, sustainability, and efficiency of CWs, highlighting their reduced pollutants, lower energy consumption, and optimized design practices compared to conventional wastewater treatment systems.

The literature review on LCAs of CWs for wastewater treatment provides a comprehensive global perspective, incorporating studies from various geographical regions using diverse methodologies and software tools. CWs emerge as environmentally beneficial solutions, offering reduced greenhouse gas emissions, lower energy consumption, and cost-effectiveness compared to conventional treatment methods.

Studies from Iceland to China and beyond examine CWs' impacts across cradle-to-gate and cradle-to-grave boundaries, focusing on functional units like 1 cubic meter of treated water or equivalent population equivalents. For instance, research in China reveals significant reductions in environmental impact and costs with enhanced CW designs incorporating bioelectrochemical processes or microalgae integration. Similarly, European studies highlight the effectiveness of CWs, especially vertical flow and hybrid systems, in reducing pollutants like nitrogen and phosphorus while promoting ecosystem services.

Software tools such as SimaPro and Open LCA facilitate rigorous evaluations of CWs' life cycle impacts, aiding comparisons with conventional technologies like activated sludge systems. These tools enable insights into CWs' environmental performance, including factors like acidification, eutrophication, and resource depletion. Overall, the literature underscores CWs' role as sustainable, adaptable solutions for decentralized wastewater treatment, supporting global efforts toward water resource management and environmental conservation.

12.5 Comparative LCA Analysis of Two CWs

12.5.1 Description of Case Study

The case studies from BlueDrop Enviro, a Hyderabad-based company, comparing the environmental footprints of their construction and operation phases using LCA methodology are now presented. The two case studies are as follows:

Table 12.2 Literature review

References	Year	Geography	System boundary	Functional unit	Life span	Result	Impact categories	Software
Hube et al. (2023)	2023	Iceland	Cradle-to-Gate	1 m^3 of treated water	20 years	LCA shows CW waste materials reduce impact; post-septic tank wetlands halve eutrophication cost-effectively	GWP, EP	Open LCA
Fang et al. (2023)	2023	China	Cradle-to-Gate	1 m^3 of treated water	20 years	The study finds bioelectrochemical CW 49.2% less impactful and 61.8% cheaper than conventional CW, despite higher initial construction impacts	GWP, ADP, AP, HT, FAEP, MAETP, POP, EP	SimaPro
Lakho et al. (2022)	2022	Belgium	Cradle-to-Grave	1 m^3 of treated water	20 years	LCA comparison of Belgian decentralized water treatment shows mobile and vertical flow CWs have lower impact, especially at shorter distances	GWP, EP, HHCP	OpenLCA
Cao et al. (2021)	2021	China	Cradle-to-Gate	1 m^3 of treated water	15 years	A study in China compares gravel and recycled concrete aggregate CWs, finding the latter more beneficial environmentally and economically, especially in phosphorus removal	GWP, PED, ADP, AP, EP, RI, POFP	CLCD (Chinese Life Cycle Database)

(continued)

Table 12.2 (continued)

References	Year	Geography	System boundary	Functional unit	Life span	Result	Impact categories	Software
Alatrista Gongora et al. (2021)	2021	Australia	Cradle-to-Grave	1 m^3 of treated water	20 years	LCA compared on-site wastewater treatment (ATUs and CWs with biochar), showing CWs have lower environmental impacts	GWP, HTT, ODP, EP, AP, POFP, ADP	SimaPro
Resende et al. (2019)	2019	Brazil	Data not available	1 m^3 of treated water	20 years	Aerated systems exhibit 1.8 times lower life cycle cost/ m^3 for treated sewage, demonstrating CWs' cost-effectiveness in small-scale treatment	GWP, FEP, FAEHT, TAP, POCP	Open LCA
Lutterbeck et al. (2017)	2017	Brazil	Cradle-to-Gate	Data not available	10 years	The study evaluates rural wastewater treatment with LCA, combining anaerobic and CW, achieving high pollutant removal; construction and anaerobic phases dominate environmental impacts	All impact categories	SimaPro

(continued)

Table 12.2 (continued)

References	Year	Geography	System boundary	Functional unit	Life span	Result	Impact categories	Software
Garfí et al. (2017)	2017	Spain	Cradle-to-Gate	1 m^3 of treated water	20 years	The study contrasts the environmental impacts and costs of conventional (activated sludge) and nature-based (CW, algal pond) wastewater treatment for small communities, favoring nature-based options	MDP, FFD, CC, ODP, TAP, FEP, MEP	SimaPro
Corbella et al. (2017)	2017	Spain	Cradle-to-Gate	1 m^3 of treated water	20 years	The study examines the environmental impact of microbial fuel cells (MFCs) in CWs, finding that graphite-based anode MFCs are the most eco-friendly option, even though they come with higher costs	GWP, ADP, ODP, AP, POP, EP	SimaPro
Fuchs et al. (2011)	2011	USA	Cradle-to-Grave	400 p.e	50 years	LCA compares VFCW and HFCW with conventional treatment, indicating VFCW has a lower environmental impact, particularly in nitrogen removal	CC, EP, FF	SimaPro

(continued)

Table 12.2 (continued)

References	Year	Geography	System boundary	Functional unit	Life span	Result	Impact categories	Software
López et al. (2010)	2010	Spain	Cradle-to-Gate	1 m^3 of treated water	30 years	CWs with Phragmites australis reduce pig slurry parameters by 80%. LCA reveals diverse environmental impacts, underscoring wetlands' role in acidification and climate change	RDP, GWP, HT, FADP, MATE, POFP	SimaPro
Machado et al. (2007)	2007	Spain and Portugal	Cradle-to-Grave	1 person. equitant	10 years	The study uses LCA to compare wastewater treatment, focusing on energy-saving systems like CW, revealing low environmental impact and offering improvement insights	GWP, AD, ODP, AP, PO, EP	SimaPro

(continued)

Table 12.2 (continued)

References	Year	Geography	System boundary	Functional unit	Life span	Result	Impact categories	Software
Fang et al. (2023)	2023	China	Cradle-to-Grave	1 m^3 of treated water	20 years	LCA compared conventional CW with bioelectrochemical-CW, revealing 49.2% lower environmental impact and 61.8% cost reduction, favoring ECW's feasibility	AD, ADF, AP, EP, GWP, ODP, HT, TE, FWAE, MAE, PO	SimaPro
Zhao et al. (2023)	2023	China	Cradle-to-Grave	1 m^3 of treated water	20 years	The study demonstrates microalgae pond-CW integration for wastewater treatment excels, enhancing pollutant removal and carbon reduction through microbial synergy, outperforming other configurations	GWP, EP, AP, PED, RD	Open LCA

12.5.1.1 Site-1: Effluent Treatment Plant, 1.2 MLD

It is designed for a cement factory in Hyderabad. It treats 1200 m^3 of wastewater per day. There are two air blowers (used for aeration) that run for eight hours a day. The treatment plant receives wastewater from residential complexes as well as effluent from the cement industry.

12.5.1.2 Site-2: Sewage Treatment Plant, 150 KLD

This system is designed for a residential housing society in Hyderabad. The STP system continuously operates a single air blower for 24 h, treating 150 m^3 of wastewater per day. The treated water is used on-site for irrigation, vehicle washing, and flushing.

Both treatment systems are designed as VFCWs with a lifespan of 25 years and undergo forced bed aeration to enhance treatment performance. The wastewater treatment systems employ facultative microbes, with a prescribed daily dosage not exceeding 75 g per liter. This dosing occurs daily at a fixed time in either the wet well or the first chamber of the holding tank. To prepare, 75 g of microbes are mixed into 1 L of freshwater, stirred for at least 5 min until completely dissolved, and then the resulting solution is applied in a single application. Electromechanical equipment integral to the CW includes an air blower, raw sewage transfer pump, pressure gauge, level sensor, flow meter, and control panel. Operation and maintenance of the aerated wetlands involve regular pruning of plants every 120 days to manage plant litter and ensure unobstructed hydrology. Water flow depth is maintained at 80–100 mm below the filter medium surface, with adjustable standpipes regulating flow. Routine removal of non-biodegradable waste from the media surface is performed for disposal. Biannual plant pruning supports plant recovery and enhances toxin absorption capability. Weekly inspections monitor overall system performance, ensuring efficient operation from water inflow through treated water outflow processes. Water quality tests for both the inlet wastewater and the final treated water are conducted at approximately 3-month intervals. Figure 12.6 illustrates the site location and includes images for both systems.

12.5.2 Methodology

12.5.2.1 Goal and Scope

In this comparative study, the environmental performance of two CW systems is evaluated: one with a capacity of 150 kL/day and the other with 1200 kL/day. The functional unit for normalization and comparison is defined as 1 m^3 of influent wastewater. The system boundary is cradle-to-gate, covering both the construction

Fig. 12.6 Site images

and operational phases. The construction phase considers materials used in infrastructure, while the operational phase focuses on electricity consumption. End-of-life considerations for infrastructure and equipment are excluded from the LCA scope for both systems. This approach aims to assess and compare the environmental impacts associated with these two CW systems based on their different capacities and operational requirements (Fig. 12.7).

12.5.2.2 Life Cycle Inventory

The LCA for both systems includes input and output data gathered from the CWs during their routine operation. They include materials, energy, electricity requirements, and daily production. The hourly electricity consumption for the two systems is approximately 5.41 kW for the ETP and about 2.91 kW for the STP. The presented LCA comparison focuses on the cradle-to-gate stage for both systems. A detailed inventory of the material and energy used during the construction and operation phase of both CWs is listed in Table 12.3 to conduct LCA. Table 12.4 shows water quality results tested at approximately 3-month intervals.

12.5.2.3 Life Cycle Impact Assessment

The various impact categories considered for this comparative LCA analysis are climate change, fossil depletion, freshwater ecotoxicity, freshwater eutrophication, terrestrial ecotoxicity, land use, human toxicity, cancer, stratospheric ozone depletion, photochemical ozone formation, ecosystem, and marine eutrophication. These impact categories provide insights into the overall environmental performance of each system across its life cycle stages.

12.5.3 Result and Discussion

The ETP system exhibits the highest total environmental impact emissions. The LCA results, depicted in Fig. 12.8, indicate that the highest emissions in the climate change category are 0.523 kg CO_2 eq/m^3 of treated wastewater for the ETP, compared to 0.372 kg CO_2 eq/m^3 of treated wastewater for the STP. Following closely are impacts in the terrestrial ecotoxicity and fossil depletion categories. The primary emissions are found to be from the operation phase in most of the impact categories represented in both systems. The studied ETP and STP systems demonstrate effective pollutant removal individually (Fig. 12.9).

The impact of different used materials for both systems under different categories is shown in Fig. 12.10. The result shows higher total emissions for the ETP system compared to the STP one. The electromechanical equipment in the ETP system resulted in more energy consumption, leading to a higher impact from the operation

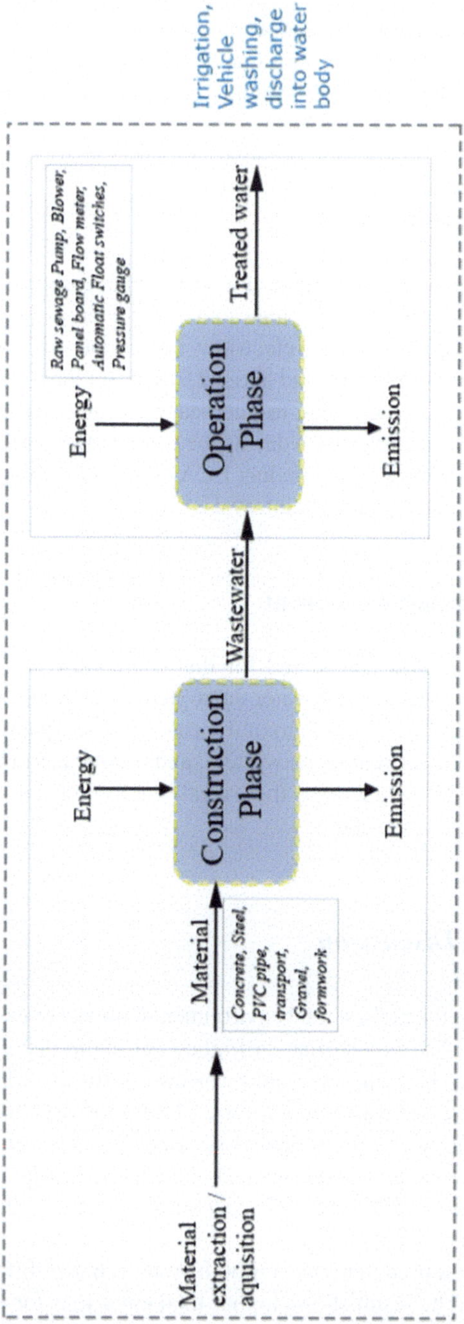

Fig. 12.7 System boundary

Table 12.3 Overview of the life cycle inventory for both systems

Phase	Input	1.2 MLD	150 KLD	Use description
Construction	Ready-mix concrete	742.02 m^3	141.07 m^3	For durable infrastructure
	Steel	53,568.68 kg	10,867.73 kg	Reinforcement steel
	Bitumen emulsion	3957.10 m^2	744.58 m^2	Waterproofing
	Gravel	2656.50 m^3	330 m^3	CW substrate
	Formwork (wood)	2814.4 m^2	889.49 m^2	To support and shape concrete structures till they gain their strength
	Electricity	263.99 kWh	32.99kWh	Used during construction
	PVC pipes	1200 m	200 m	Pipeline in CW
Operation	Electricity	260 kWh d^{-1}	70 kWh d^{-1}	To run different electromechanical equipment

Table 12.4 Summary of the concentrations of pollutants in the inlet and outlet of both CWs

Pollutants	150 KLD				1.2 MLD			
	Test-1		Test-2		Test-1		Test-2	
	Inlet	Outlet	Inlet	Outlet	Inlet	Outlet	Inlet	Outlet
pH	7.66	7.4	7.4	7.39	7.54	7.52	7.53	7.52
TSS	73	< 10	11.3	< 10	16	< 10	26	< 10
TDS	1157	1108	1130	995	1960	1950	1750	1200
COD	192	76	104	48	140	28	56	48
BOD	57.7	14	20	9.3	27.3	5.3	10.8	8.3
Oil and grease	9.6	< 5	6.7	< 5	7.6	< 5		

phase. The other system, being small in size, consumes less energy in the operation phase. Electricity used by both systems has the highest impact on all the ten impact categories.

The construction phase of each system has varying environmental impacts. For treating 1 m^3 of wastewater, the 1.2 MLD system exhibits a significant impact breakdown for the climate change category: 42% from ready-mix concrete, 29% from steel reinforcement, 12% from gravel used in wetland cells, and 17% from transport. Similarly, the 150 KLD plant shows major impacts in the climate change category, with 39% attributed to ready-mix, 30% to reinforcing steel, 12% to formwork, 11% to transport, 7% to gravel, and 1% to PVC pipes. To mitigate higher construction emissions for the STP system compared to the ETP, solutions can include sustainable material usage, energy-efficient construction, optimized system design, optimal utilization of materials, and carbon offset initiatives.

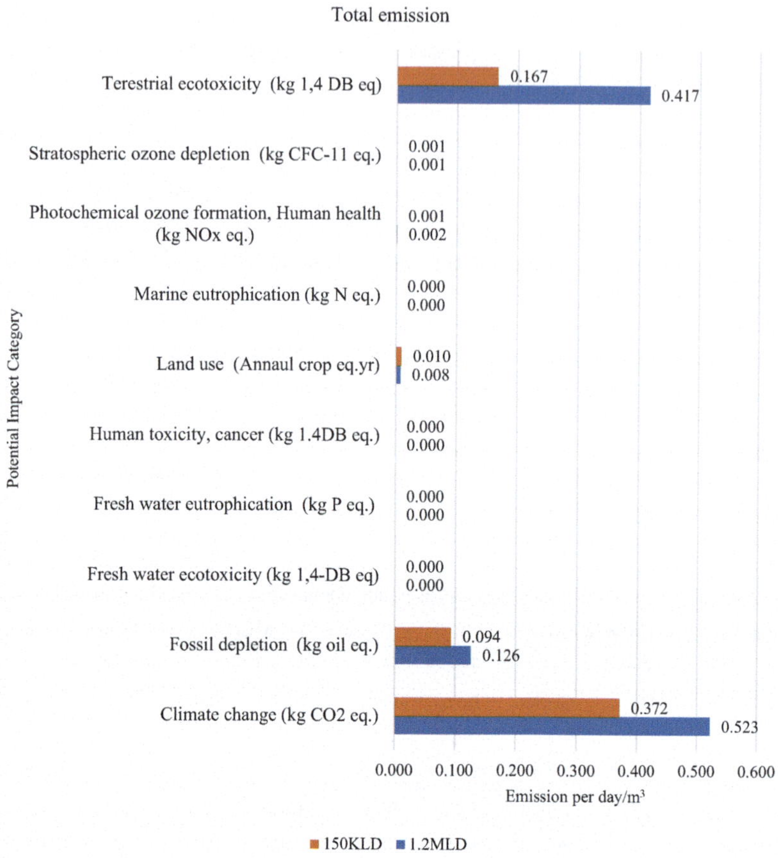

Fig. 12.8 Total emissions per day/m^3 from both the treatment systems

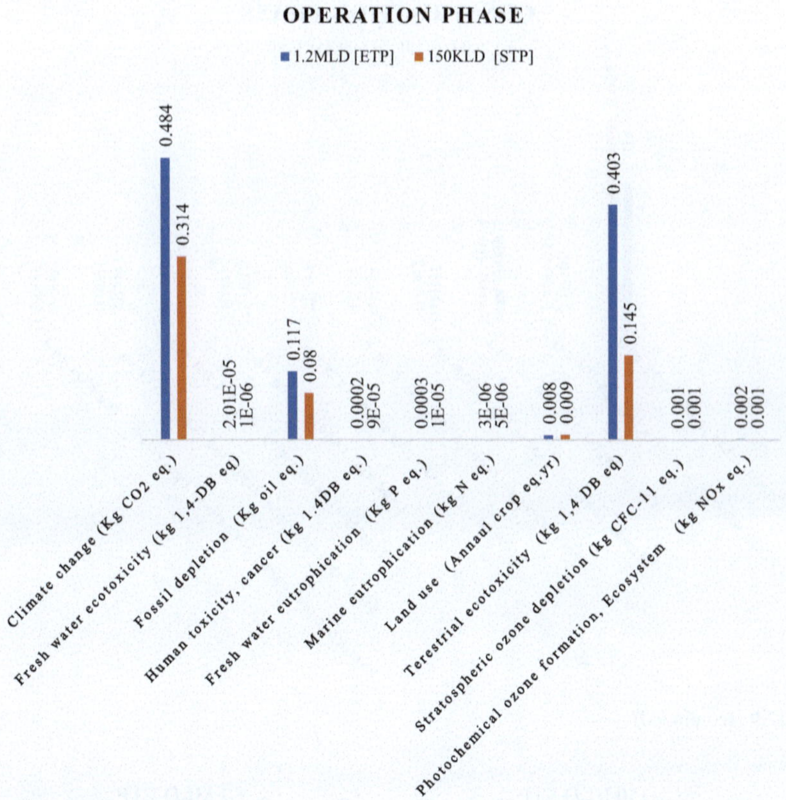

Fig. 12.9 Potential impact from the operation and construction phase of both systems

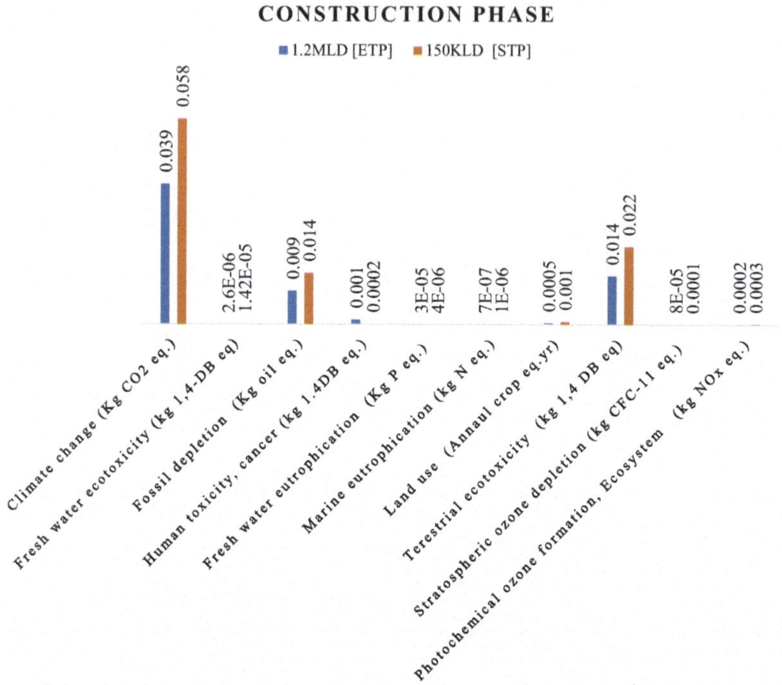

Fig. 12.9 (continued)

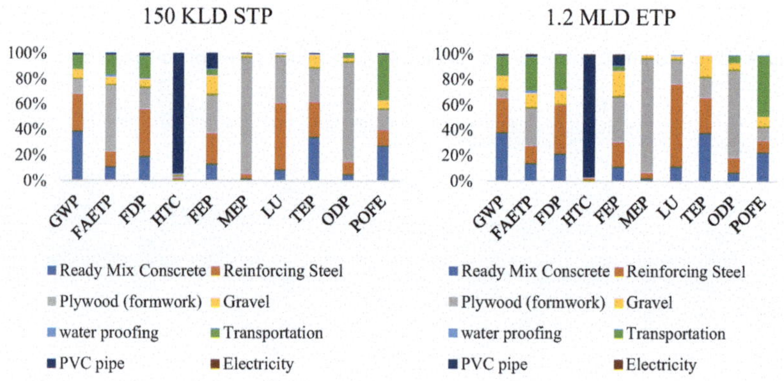

Fig. 12.10 Relative potential impact of the construction phase for both systems

12.6 Conclusion

The literature review highlights CWs as sustainable wastewater treatment solutions, showing reduced greenhouse gas emissions, energy use, and costs compared to conventional methods, supported by global studies and software tools. The comparative analysis reveals that the ETP-constructed wetland system has a higher total environmental impact compared to the STP-constructed wetland system due to greater energy consumption and emissions, primarily from its electromechanical equipment. The STP system, being smaller and more energy-efficient, results in a lower impact. Electricity usage is the major contributor to the environmental footprint of both systems. Constructed wetlands are found to be valuable for treating effluent wastewater as well. The water quality check shows that both systems effectively remove pollutants. Construction materials also significantly affect emissions. Optimize transportation, utilize renewable energy, and engage in carbon offset programs. Participating in green certification initiatives and complying with environmental standards further reduces impact, promoting sustainable wastewater treatment solutions.

References

Alatrista Gongora GR, Lu RH, El Hanandeh A (2021) Comparative life cycle assessment of aerobic treatment units and constructed wetlands as onsite wastewater treatment systems in Australia. Water Sci Technol 84:1527–1540. https://doi.org/10.2166/wst.2021.316

Almuktar SAAAN, Abed SN, Scholz M (2018) Wetlands for wastewater treatment and subsequent recycling of treated effluent: a review. Environ Sci Pollut Res 25:23595–23623

Andreo-Martínez P, Ortiz-Martínez VM, Muñoz A et al (2021) A web application to estimate the carbon footprint of constructed wetlands. Environ Modell Softw 135. https://doi.org/10.1016/j.envsoft.2020.104898

Bassi N, Gupta S, Chaturvedi K, Schmidt G (2023) Reuse of treated wastewater in India market potential and recommendations for strengthening governance

Breida M, Younssi SA, Ouammou M et al (2019) Pollution of water sources from agricultural and industrial effluents: special attention to NO_3^-, Cr(VI), and Cu(II). Water Chemistry. https://doi.org/10.5772/INTECHOPEN.86921

Bureau of Indian Standards BIS/ISO 14044 (2006): Environmental management-life cycle assessment-requirements and guidelines

Cao Z, Zhou L, Gao Z et al (2021) Comprehensive benefits assessment of using recycled concrete aggregates as the substrate in constructed wetland polishing effluent from wastewater treatment plant. J Clean Prod 288. https://doi.org/10.1016/j.jclepro.2020.125551

Chateau J, Dang G, Macdonald M et al (2023) A framework for climate change mitigation in India, WP/23/218, October 2023

Clean drinking water | UNICEF India. https://www.unicef.org/india/what-we-do/clean-drinking-water. Accessed 12 Apr 2024

Corbella C, Puigagut J, Garfí M (2017) Life cycle assessment of constructed wetland systems for wastewater treatment coupled with microbial fuel cells. Sci Total Environ 584–585:355–362. https://doi.org/10.1016/j.scitotenv.2016.12.186

Fang YK, Wang HC, Fang PH et al (2023) Life cycle assessment of integrated bioelectrochemical-constructed wetland system: environmental sustainability and economic feasibility evaluation. Resour Conserv Recycl 189. https://doi.org/10.1016/j.resconrec.2022.106740

Fuchs VJ, Mihelcic JR, Gierke JS (2011) Life cycle assessment of vertical and horizontal flow constructed wetlands for wastewater treatment considering nitrogen and carbon greenhouse gas emissions. Water Res 45:2073–2081. https://doi.org/10.1016/j.watres.2010.12.021

Garfí M, Flores L, Ferrer I (2017) Life Cycle Assessment of wastewater treatment systems for small communities: activated sludge, constructed wetlands and high rate algal ponds. J Clean Prod 161:211–219. https://doi.org/10.1016/j.jclepro.2017.05.116

How is India addressing its water needs? https://www.worldbank.org/en/country/india/brief/world-water-day-2022-how-india-is-addressing-its-water-needs. Accessed 14 Apr 2024

Hube S, Zaqout T, Ögmundarson Ó et al (2023) Constructed wetlands with recycled concrete for wastewater treatment in cold climate: performance and life cycle assessment. Sci Tot Environ 904. https://doi.org/10.1016/j.scitotenv.2023.166778

Lakho FH, Qureshi A, Igodt W et al (2022) Life cycle assessment of two decentralized water treatment systems combining a constructed wetland and a membrane based drinking water production system. Resour Conserv Recycl 178. https://doi.org/10.1016/j.resconrec.2021.106104

López G, Bayo J, Gómez-López MD et al (2010) Life cycle assessment application for slurry management in constructed wetlands

Lutterbeck CA, Kist LT, Lopez DR et al (2017) Life cycle assessment of integrated wastewater treatment systems with constructed wetlands in rural areas. J Clean Prod 148:527–536. https://doi.org/10.1016/j.jclepro.2017.02.024

Machado AP, Urbano L, Brito AG et al (2007) Life cycle assessment of wastewater treatment options for small and decentralized communities. In: Water science and technology, pp 15–22

Negi R, Chandel MK (2021) Analysing water-energy-GHG nexus in a wastewater treatment plant of Mumbai Metropolitan Region, India. Environ Res 196. https://doi.org/10.1016/j.envres.2021.110931

Ombretta T, Somalia Urban Development Programme, United Nations Human Settlements Programme (2007) First steps towards strategic urban planning. United Nations Human Settlements Programme

Population growth—Our World in Data. https://ourworldindata.org/population-growth. Accessed 5 Feb 2024

Population growth, resource consumption, and the environment | Population Summit of the World's Scientific Academies | The National Academies Press. https://nap.nationalacademies.org/read/9148/chapter/5. Accessed 5 Feb 2024

Resende JD, Nolasco MA, Pacca SA (2019) Life cycle assessment and costing of wastewater treatment systems coupled to constructed wetlands. Resour Conserv Recycl 148:170–177. https://doi.org/10.1016/j.resconrec.2019.04.034

Tsangas M, Papamichael I, Banti D et al (2023) LCA of municipal wastewater treatment. Chemosphere 341. https://doi.org/10.1016/j.chemosphere.2023.139952

United Nations. Department of Economic and Social Affairs. Population Division Global population growth and sustainable development

Water and the global climate crisis: 10 things you should know | UNICEF. https://www.unicef.org/stories/water-and-climate-change-10-things-you-should-know. Accessed 5 Feb 2024

Water: a scarce and valuable resource—resources and information on agriculture and biodiversity in the Indian Ocean. http://www.agriculture-biodiversite-oi.org/en/Nature-agriculture/News-from-the-field/Environment-and-Biodiversity/Preserving-natural-resources/Water-a-scarce-and-valuable-resource. Accessed 5 Feb 2024

Zhao X, Zhang T, Dang B et al (2023) Microalgae-based constructed wetland system enhances nitrogen removal and reduce carbon emissions: performance and mechanisms. Sci Tot Environ 877. https://doi.org/10.1016/j.scitotenv.2023.162883

Chapter 13
Assessment of Effluent Quality of Sewage Treatment Plant Using Machine Learning-Based Models

Tejash Singh, Pratham Arora, and S. K. Singal

Abstract Machine learning-based (ML) models have gained significant prominence in the environmental areas, especially in the wastewater treatment processes. Wastewater treatment plants (WWTPs) are complicated facilities that involve various biological, chemical, and physical processes and are highly subject to uncertainty. Traditional models struggle to capture the nonlinearities in these systems, but ML-based models can model these interactions effectively, which results in the rise in the use of ML models for WWTP modeling and optimization and to provide fast results. ML-based models offer numerous advantages over traditional physical models in predicting output parameters in sewage treatment plants. They can adapt and learn from data, making them particularly valuable when system dynamics are subject to change. Although there are numerous benefits, the utilization of machine learning for modeling sewage treatment plants remains largely unexplored within the Indian subcontinent, especially the comparison of two node-based models, i.e., Random Forest (RF) and artificial neural networks (ANN) in terms of output prediction of the WWTPs. Thus, this study presents ML-based approach for forecasting WWTP's output variables as a black box by explaining the relationships between an existing WWTP's different influent and effluent pollution variables for Indian tropical conditions. The study focuses on applying RF and ANN models to predict the outflow variables of wastewater treatment plants. ANN and RF models are developed to deal with number of issues of traditional mathematical models. The developed models concentrate on offering a flexible, useful, and a different approach to model the WWTP's performance. They are based on data collected from two WWTPs, which include biochemical and chemical oxygen demands, phosphates (PO_4^{-3}), and nitrates (NO_3^-) over a year and three years, respectively. The results indicate high predictive accuracy and generalization capabilities of ML-based models, with ANN achieving the highest correlation coefficient (R^2) of up to 0.9025 and RF up to 0.947 for output BOD and demonstrating the efficacy of these machine learning techniques in modeling and managing WWTP processes.

T. Singh · P. Arora (✉) · S. K. Singal
Department of Hydro and Renewable Energy, Indian Institute of Technology Roorkee, Roorkee, India
e-mail: pratham.arora@hre.iitr.ac.in

S. Souabi et al. (eds.), *Wastewater Treatment Plants*, Water Science and Technology Library 130, https://doi.org/10.1007/978-3-031-87461-1_13

Keywords Artificial neural networks · Random forest · Wastewater treatment plant · Modeling · Forecasting · Machine learning

Abbreviations

AI	Artificial intelligence
ANFIS	Adaptive neuro-fuzzy inference system
ANN	Artificial neural network
BOD	Biochemical oxygen demand
CNN	Convolutional neural network
COD	Chemical oxygen demand
DNN	Deep neural network
GBM	Gradient boosting machine
LSTM	Long-short term memory
ML	Machine Learning
RF	Random Forest
SARIMAX	Seasonal autoregressive integrated moving average
SNN	Shallow Neural Network
SVM	Support Vector Machine
TSS	Total suspended solid
WWTP	Wastewater treatment plant

13.1 Introduction

Although rapid community development has brought about unique advantages, it has also led to various environmental pollution. The wastewater generated by residential areas and the operations of commercial units is one of these problems. The sewer system is an essential element of urban water management capable of gathering sewage. Additionally, sewage systems are capable of collecting both rainwater and residential wastewater, transporting it to a centrally located wastewater treatment plant (WWTP), where it is treated before being released into different surface waterbodies (Najafzadeh and Zeinolabedini 2019).

Wastewater disposal is a major global concern, and its effective management of treatment plants is crucial in the management of wastewater. Therefore, optimal management practices should be implemented to ensure efficient treatment of wastewater (Jafar et al. 2022). Wastewater treatment plants (WWTPs) are crucial for the betterment of society and the environment, as well as for the development of circular economy and energy efficiency. Their primary purpose is to eliminate harmful pollutants from the wastewater generated by human activities that could endanger the environment. WWTPs aim to convert wastewater into reusable water that can be safely discharged into natural water resources (Salles et al. 2022).

Treatment plants are intricate systems that experience considerable variations like inflow and outflow rates, pollutant load hydraulic conditions, etc., which makes it challenging to model the treatment plants. Many mechanistic models have been used in the past to mimic treatment plant processes and forecast the behavior of specific parameters, but for the mechanistic model to function and be computationally feasible, many assumptions and simplifications must be made. Additionally, it is difficult to do the coupling of the different mechanistic models due to their difference in approaches to computing certain variables (Wang et al. 2021). Other drawbacks of physics-based models include not adequately addressing dynamic and highly nonlinear behaviors of processes affected by different evident and enigmatic factors, inability to thoroughly simulate different operations, and expensive and poor generalization performance because of which these mechanistic models are quite insufficient in modeling the treatment plants (Wang et al. 2021).

The recent advancement in computer technology has made the use of artificial intelligence (AI) more popular among researchers for monitoring and managing WWTPs (Yang et al. 2022). Technology based on AI is primarily used to forecast natural or artificial processes across the fields of study. And as a branch of AI, machine learning (ML) is used for identifying a unique pattern based on the provided data for categorization and prediction (Bagherzadeh et al. 2021). Compared to traditional models, which are based on physical approaches, data-based ML algorithms have a number of advantages. Firstly, they require fewer assumptions and do not require specialized expertise. Secondly, these models are capable of learning from large datasets more efficiently, using less processing power and computation time. As a result, they are well-suited to dealing with the complex, nonlinear, and interdisciplinary characteristics of water quality parameters (Ly et al. 2022). The primary advantage of using AI models is its data-driven approach that eliminates the necessity of specifying the system's complete physics, which makes it better to handle the irregularity and intricacy of a system, eliminating the drawbacks of using numerical models (Nadiri et al. 2018). Results of similar studies are shown in Table 13.1.

Among all the ML-based models, random forest (RF) and artificial neural network (ANN) are among the most popularly utilized by researchers because of their less complex algorithm structure and accuracy in the prediction of required data. ANN model outperformed the various other models because of less requirement of computational resources and lesser training and time to give a good and accurate forecast of the effluent parameters (Salles et al. 2022). Even when compared with support vector machine (SVM) models for predicting effluent biochemical oxygen demand (BOD), ANN is considered as better choice (Rizal and Hayder 2023). However, some research found that the accuracy of the ANN model drops significantly when more and more insignificant variables are added into the system; in that condition, RF and gradient boosting machines (GBM) are better options with relatively stable efficiency and accuracy (Bagherzadeh et al. 2021). RF model gives results comparable to deep neural network (DNN) for effluent prediction of WWTP (Wang et al. 2021). Even

Table 13.1 Broad overview of the recent similar studies in the literature

Parameters	Models	Performance metrics					Finding	References
		MAE	MSE	RMSE	R	R^2		
TN	ANN	–	–	–	–	0.55	SVM shows the better accuracy in prediction when compared to other one	Guo et al. (2015)
	SVM	–	–	–	–	1.00		
BOD, COD	MPR	–	–	–	–	0.91	The performances of models were better in estimating the parameters like BOD and COD	Singh et al. (2010)
	ANN	–	–	–	–	0.90		
COD, SS	CSWLSTM	2.05	–	2.68	–	0.83	A novel CSWLSTM-GPR prediction model outperforms other models in accuracy, suitability, performance, and reliability	Wan et al. (2022)
	SWLSTM	4.04	–	5.34	–	0.33		
	CLSTM	2.91	–	3.79	–	0.66		
	CGRU	2.31	–	3.17	–	0.76		
Flow rate	FFBP-NN	1063.7	–	1445.9	0.909	–	The performance of SVM was better than the others	Najafzadeh and Zeinolabedini (2019)
	RBF-NN	1048.7	–	1501	0.903	–		
	ANFIS	1075.4	–	1515.6	0.901	–		
	SVM	1013.1	–	1435.4	0.912	–		
COD, TN	PCA-NARX	–	–	2.9	–	0.85	Model was able to predict COD and TN accurately	Yang et al. (2022)
BOD, COD, TSS	CFL-SA	–	–	3.29	–	0.91	The prediction capabilities of SCFL were better than the other algorithms used for modeling	Nadiri et al. (2018)
	CFL-WA	–	–	1.72	–	0.94		
	SCFL	–	–	1.1	–	0.94		
TN	ANN	22E−3	–	95E−3	–	0.55	RF and GBM performed better than neural networks, and GBM has the highest accuracy	Bagherzadeh et al. (2021)
	GBM	17E−3	–	92E−3	–	0.58		
	RF	19E−3	–	95E−3	–	0.55		

(continued)

Table 13.1 (continued)

Parameters	Models	Performance metrics					Finding	References
		MAE	MSE	RMSE	R	R^2		
MLSS, Cyanide	Mechanistic	–	–	–	–	0.729	Parallel hybrid neural network outperforms the other models	Lee et al. (2002)
	Neural	–	–	–	–	0.929		
	Serial hybrid neural	–	–	–	–	0.862		
	Parallel hybrid neural	–	–	–	–	0.957		
BOD, COD, TSS	ANN	–	0.023	–	0.924	–	The model signifies good prediction capabilities	Mjalli et al. (2007)
COD	MLP	52.48	–	72.7	–	0.69	The daily concentration of COD was estimated better using the K-means-MLP model	Ay and Kisi (2014)
	RBNN	53.23	–	72.57	–	0.7		
	MLR	58.05	–	87.57	–	0.68		
	ANFIS	54.73	–	74.36	–	0.67		
	K-means-MLP	48.93	–	65.77	–	0.86		

when models based on different types of neural networks are compared, i.e., traditional feed-forward, deep feed-forward backpropagation, and deep cascade-forward backpropagation (DCB) networks then, DCB is better performed (El-Rawy et al. 2021). And in the forecasting of the intensity of odor, ANN also provided satisfactory results compared to decision trees (Byliński et al. 2019). When shallow neural networks (SNN) and deep neural networks are compared for forecasting the effluent quality, the former gives a better performance (Jafar et al. 2022). After ANN, long short-term memory (LSTM) and SVM are also widely in use for modeling WWTPs and mimicking the various processes of treatment plants. When SVM is compared with feed-forward back propagation neural network, radial basis function neural network, and adaptive neuro-fuzzy inference system (ANFIS), then SVM gives better performance than other available options (Najafzadeh and Zeinolabedini 2019).

Bagherzadeh et al. (2021) conducted a study aimed to compare several models and identify the most effective one. Researchers developed machine learning models using algorithms such as GBM, ANN, and RF. The results indicated that RF and GBM were better than ANN with respect to performance. Furthermore, it was found that ANN-based models are significantly more sensitive to duplicate variables, with their effectiveness sharply declining as the number of redundant features increases, unlike the more robust RF models. Another study goes a step further by employing variable importance measure and partial dependence plot analysis to identify the most crucial factors influencing the output in addition to comparing the models (Wang et al. 2021). The framework developed comprises RF models and DNN models aimed at improving the control on effluent quality of WWTPs. The overarching goal is to strike a balance that allows for efficient resource management while adhering to the prescribed effluent quality standards, ultimately aiming to optimization of energy usage and chemical utilization within WWTPs. Total Suspended Solids (TSS) in the effluent and phosphate (PO_4) in the effluent are two effluent parameters that are investigated. DNN is used to create and validate RF models. For effluent TSS, validation set, and test set, the performance efficiency of the RF and DNN model was [0.934, 0.862, 0.920 and 0.905, 0.870, 0.886], respectively, and [0.935, 0.892,0.920 and 0.904, 0.908, 0.872], respectively, for effluent phosphate concentration.

Model developed using LSTM gives almost 99% accuracy in predicting the ammonia concentration, which makes it a good choice for developing ML-based models (Farhi et al. 2021). Hybrid models were also developed by combining two or more than two algorithms to predict the effluent concentration and their accuracy and efficiency are compared to the individual contributing ML algorithms' performance such as convolutional neural network (CNN) and LSTM combined to develop CNN-LSTM and compared with the two standalone algorithms and perform better than both of them (Wang et al. 2019), principal component analysis-nonlinear autoregressive with external input (PCA-NARX) hybrid model is developed to give the better performance than both of the standalone ML algorithms (Yang et al. 2022) and the CSWLSTM-GPR model is innovatively crafted through the fusion of distinct algorithms.

This hybridized approach utilizes CNN for spatial feature extraction, SWLSTM for capturing temporal patterns, and GPR for introducing a probabilistic framework.

The resulting model offers a comprehensive and synergistic solution, harnessing the unique advantages of spatial, temporal, and probabilistic features for enhanced performance and robust predictions and gives better performance than all three of them (Wan et al. 2022). Model based on extreme learning machine (ELM) and seasonal autoregressive integrated moving average (SARIMAX) outperformed the LSTM, ANFIS, SVM, and gradient tree boosting (GTB) models (Ly et al. 2022; Pham et al. 2020).

The utilization of distinguish algorithms of machine learning for modeling Sewage Treatment Plants (STPs) has advanced significantly globally, although this domain is still unexplored in the context of the Indian subcontinent. Therefore, the main goal of this study is to build ML-based models that can forecast the STP's outflow variables. Two machine learning algorithms, ANN and RF, are chosen for this purpose since previous research conducted in other regions of the world has found them to be the most effective.

13.2 Methodology

13.2.1 Description of Data and WWTP

The successful development of a predictive model depends on the precise acquisition of essential data, meticulous elimination of noise, and careful selection of appropriate input and output variables. These steps are critical to ensuring the reliability and accuracy of the model. In this study, data was systematically collected from two distinct wastewater treatment facilities, each characterized by unique operational conditions and environmental contexts.

The first location is in Ramnagar, Nainital (Fig. 13.1), a Sewage Treatment Plant (STP) with a capacity of 1.5 million liters per day (MLD). At this site, critical parameters of qualities of water, including biochemical oxygen demand (BOD_5), phosphate (PO_4^{3-}), chemical oxygen demand (COD), and nitrate (NO_3^-), were diligently monitored at both the inlet and outlet points over a one year period. The second facility, located near the ITI in Srinagar, Tehsil-Srinagar, District-Pauri (Fig. 13.2), has a treatment capacity of 1.0 MLD and monitored for over 2.5 years. Similar to the Ramnagar STP, the same set of parameters, i.e., COD, BOD_5, PO_4^{3-}, and NO_3^-, was monitored at both the inlet and outlet points. However, this facility presented a different set of challenges due to the extended period of data collection and the inherent fluctuations in the wastewater composition at influent.

The monthly variation of each of the parameters at both the inlet and outlet can be seen (Fig. 13.3). At Ramnagar's STP, it can be seen by the chart that the pollutant concentration has shown an increasing trend with the passage of time at the inlet of the STP. This trend indicates an increase in strain on the treatment plant, most likely due to changes in population density, industrial activity, or other different factors within the catchment area. Despite this upward tendency, the STP is capable enough

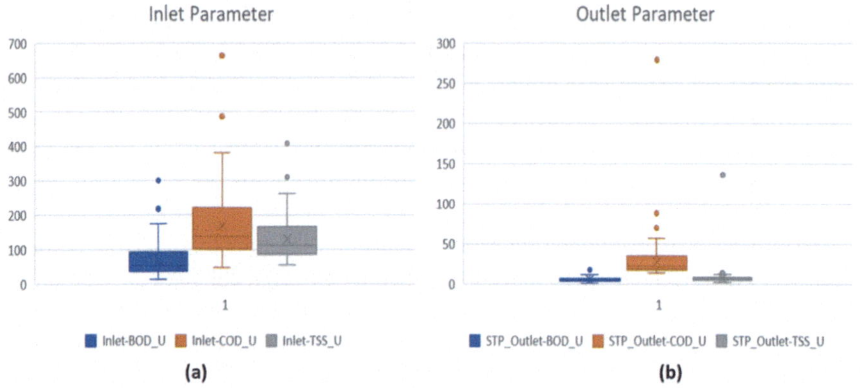

Fig. 13.1 Box and whisker plots of the **a** influent variable, and **b** effluent variable of the Ramnagar STP

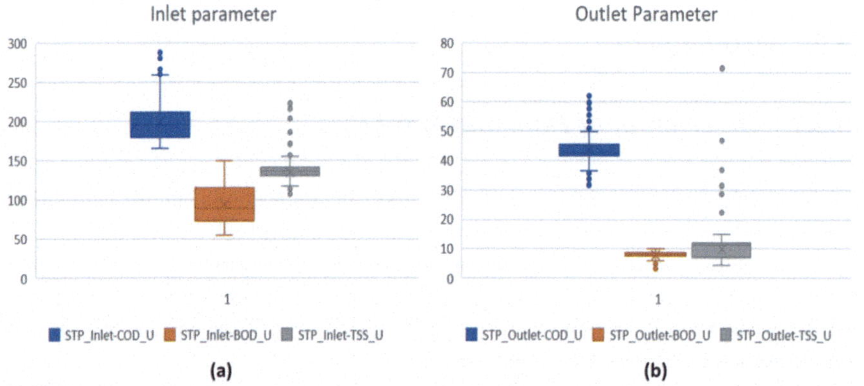

Fig. 13.2 Box and whisker plots of the **a** influent variables, and **b** effluent variables of the Srinagar STP

to effectively treat the wastewater to levels of effluent parameters under the threshold regulatory standards defined by regulatory authorities for safe discharge into surface water bodies.

In contrast, the Srinagar STP shows very large fluctuations in BOD, COD, and Total Suspended Solids (TSS) concentrations at the inlet. This signifies the variation in the quality and composition of the wastewater coming into the STP. This can occur due to seasonal variations or may also be changes in the qualities of the effluent coming out of the local industries in that area, etc. Despite these fluctuations, the STP is capable enough to efficiently reduce these under safe threshold levels to discharge the treated water in the nearby surface water body safely without harming the aquatic ecosystem. To prepare the collected data for modeling, a thorough cleaning process was undertaken. This involved the removal of outliers and highly fluctuating data points that could skew the model's accuracy. The resulting dataset was then subjected

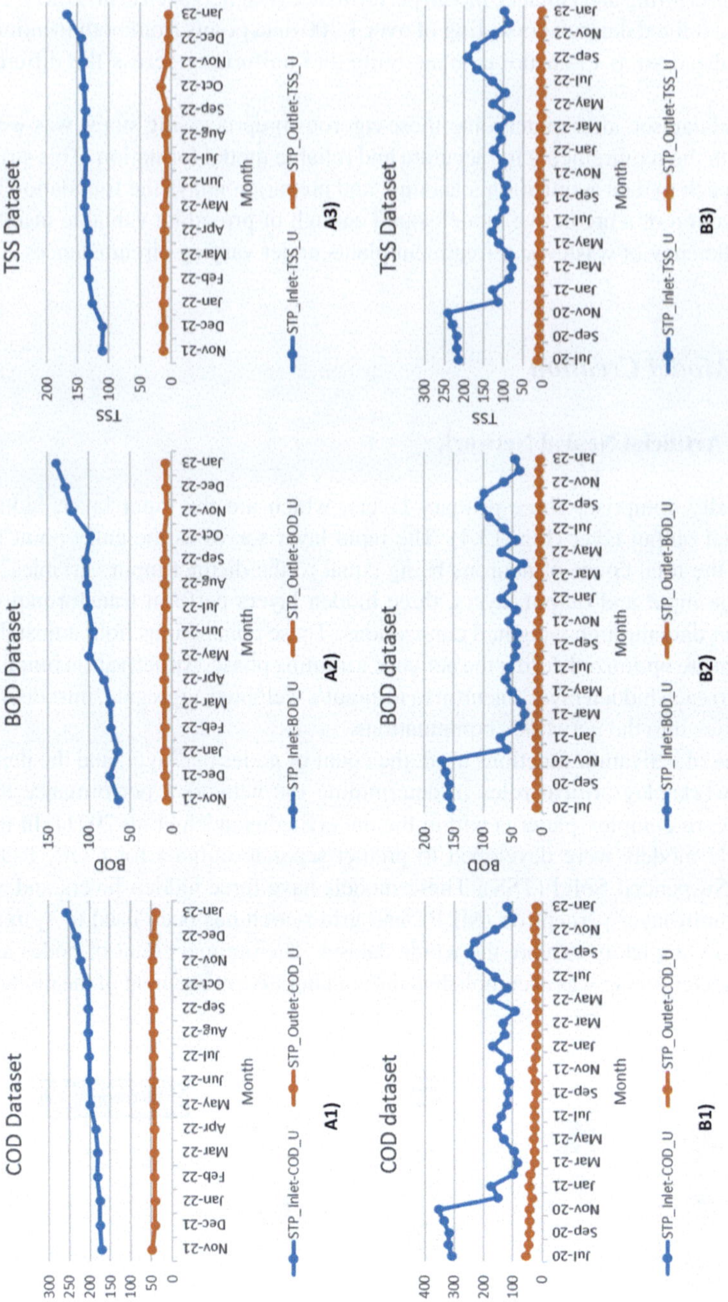

Fig. 13.3 Monthly variation of datapoints of **a** Ramnagar's STP, **b** Srinagar's STP

to feature selection, where redundant or non-contributory variables were eliminated to prevent overfitting and enhance model performance (Bagherzadeh et al. 2021). The cleaned and refined dataset, consisting of over 1,700 data points from both locations, was normalized so as to ensure comparability and uniformity across the different variables.

The final dataset, after undergoing these rigorous preprocessing steps, was well-aligned with the requirements for accurate and reliable model formation. This structured approach to data acquisition, cleaning, and preparation lays the foundation for the construction of a predictive model sound enough of providing valuable insights into the efficiency of wastewater treatment plants under varying circumstances.

13.2.2 Model Creation

13.2.2.1 Artificial Neural Network

This typically comprises three primary layers, which are the input layer, hidden layer(s), and output layer (Fig. 13.4). The input layer serves as the entry point for data, with the total count of neurons being equal to the distinct input variables. In between the input and output layers, three hidden layers perform transformations on the input data through weighted connections. These connections hold adjustable weights that are optimized during the network's training phase. An activation function is applied to each hidden layer's neuron to its input's weighted aggregate, introducing nonlinearities into the network's computations.

The type of activation functions used, the count of nodes per layer, and the depth of the network play critical roles in determining the network's performance and ability to learn complex patterns within the data (Bagherzadeh et al. 2021). In this study, ANN models were developed to predict separate outputs for COD, BOD, and Total Suspended Solid (TSS). These models have three hidden layers and are based on multi-layer perceptron (MLP), and grid search has been used to systematically and completely explore the whole dataset. The optimal count of nodes and hyperparameter tuning was accomplished through iterative refinement of the models.

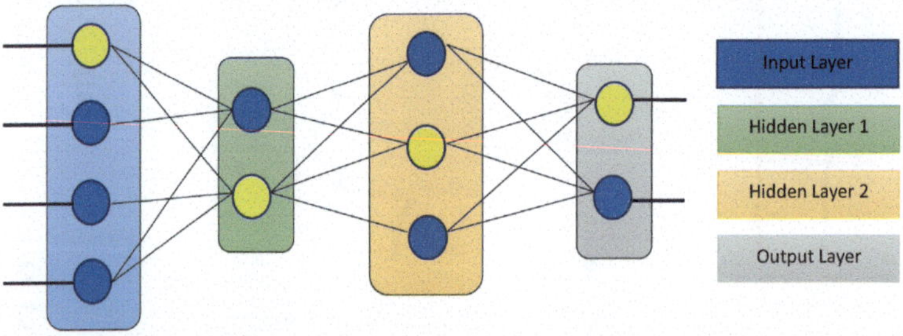

Fig. 13.4 Artificial neural network

The ANN model involves the utilization of the MLP regressor, specifically designed for linear regression tasks on the provided dataset. Here, grid search is used instead of random search, which systematically evaluates every possible combination of variables coming in and going out, facilitating the identification of the most effective model architecture. The input layer is configured with a node count equal to the number of input variables set at 6. For the hidden layer, the model undergoes training with two distinct configurations: (32, 64, 64) and (64, 128, 128) nodes distributed across three hidden layers. The output layer consists of a single node dedicated to predicting the outlet parameter. Optimization of model weights is achieved using the Adam solver, and to address potential overfitting, an L2 regularization term alpha (α) is introduced.

13.2.2.2 Random Forest

The preferred algorithm for machine learning, an aspect of the supervised learning approach, is random forest (Fig. 13.5). It could potentially be used for ML problems requiring classification as well as regression. This technique involves merging multiple classifiers to enhance the accuracy of predictions. In particular, random forest relies on a collection of decision trees applied to various input data subsets to improve prediction accuracy. Rather than depending on a single decision tree, the algorithm generates a final forecast based on the greatest number of votes of the predictions by integrating the outputs of each decision tree. This strategy improves model performance by taking advantage of the collective knowledge of multiple decision trees. The more of them there are, the more precise the results are and the less overfitting there is.

The single-output models developed based on the RF algorithm in this study have maximum tree depths of 10, 20, and 20 for BOD_{out}, COD_{out}, and TSS_{out}, and

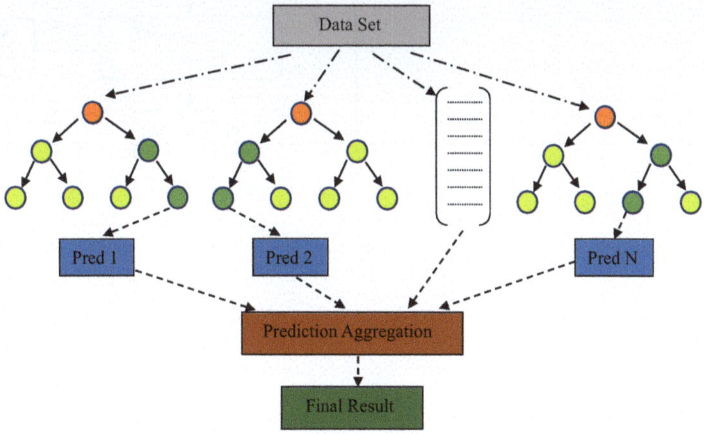

Fig. 13.5 Random forest

the random state of 100 is provided to the regressor. For the construction multiple-output model, the random forest regression model was configured with 200 trees (n_estimators = 200) and a fixed random state of 45 to ensure reproducibility. The features were standardized to ensure a consistent scale, which is beneficial for the performance of the model.

13.2.2.3 Model Construction

The entire dataset was divided into two distinct groups: the training data, comprising nearly 80% of the total dataset, and the test data, representing the remaining 20%, which was reserved as unseen data. Subsequently, preprocessing and feature selection techniques were applied to both sets. Two single-output predictive algorithms were chosen for predicting the target variable: one utilizing a fully connected ANN and the other employing RF. These models were applied to all sub-datasets. Additionally, a multiple-output model based on RF regressor was constructed to simultaneously predict all target variables, i.e., BOD, COD, TSS. All models were trained and tested using normalized input data. Using the chosen models, a number of events and structures for models were developed before the target variable was predicted. Then, using a variety of assessment indicators, the expected values were juxtaposed to the real data to determine how accurate the models were (Fig. 13.6).

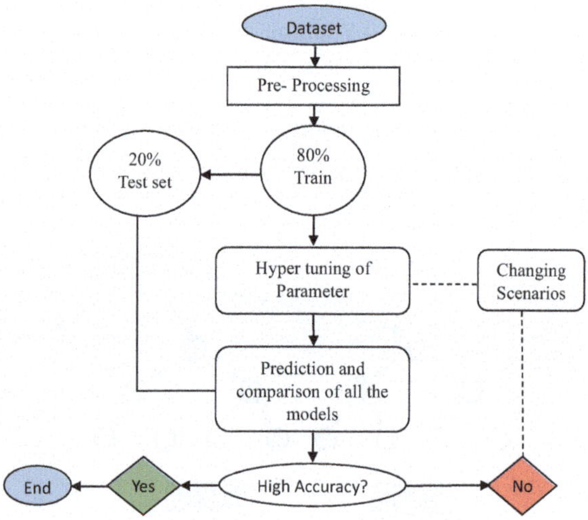

Fig. 13.6 Modeling and prediction structure

13.2.3 Evaluation Criteria

Selecting appropriate criteria for evaluation of each ML model is crucial. It has been observed that relying solely on one efficiency criterion may not provide a proper and full picture of the model performance, as each criterion has its own strengths and limitations. Therefore, in this study, we utilized two commonly used criteria in the water science field, coefficient of determination (R^2), and mean square error (MSE), for both the training and validation phases (Krause et al. 2005).

The following equation can be used for finding out R^2,

$$R^2 = \frac{\sum_i \left(Q_m(i) - \overline{Q}_m(i)\right)\left(Q_0(i) - \overline{Q}_0(i)\right)}{\sum_i \left(Q_m(i) - \overline{Q}_m(i)\right)^2 \sum_i \left(Q_0(i) - \overline{Q}_0(i)\right)^2}. \tag{13.1}$$

It informs us about the extent to which the model's prediction can account for the dispersion in the measured values. A perfect match to no association at all is indicated by a coefficient with a value between 0 and 1.

MSE calculates the average squared disparity between the forecasted values and the actual target values across a dataset. Its fundamental purpose is to measure how close a model's predictions are to the actual values in order to assess how accurate the model's forecasts are. Mathematically, MSE can be expressed as follows:

$$\text{MSE} = \frac{1}{n}\sum_{i=1}^{n}\left(Y_i - \hat{Y}_i\right)^2. \tag{13.2}$$

The numerical range of MSE varies depending on the scale and nature of the data being analyzed. However, in general, the MSE values typically range from 0 to positive infinity. Better model performance is shown by a lower MSE value, which denotes less differences between the actual and predicted values.

13.3 Results and Discussion

The information regarding the outcomes of performance capabilities of the various ML algorithms used in this study is presented in Table 13.2. Based on the results, it can be observed that in terms of R^2 and MSE values, all the machine learning techniques exhibit relatively similar performance and a very good performance capability of these algorithms in understanding the variation of the datasets.

Table 13.2 MSE and R^2 values of the models

Inflow variable	Target variable	ANN		RF	
		MSE	R^2	MSE	R^2
BOD_{in}, COD_{in}, TSS_{in}, pH, TN, TP	BOD_{out}	0.177	0.903	0.096	0.947
BOD_{in}, COD_{in}, TSS_{in}, pH, TN, TP	COD_{out}	1.596	0.895	1.714	0.901
BOD_{in}, COD_{in}, TSS_{in}, pH, TN, TP	TSS_{out}	0.377	0.901	0.677	0.946
BOD_{in}, COD_{in}, TSS_{in}, pH, TN, TP	BOD_{out}, COD_{out}, TSS_{out}	–	–	1.308	0.913

13.3.1 Single-Output Models

For all the models, irrespective of their target variable, the inflow variable taken into consideration along with the target variable to train the models are BOD_{in}, COD_{in}, TSS_{in}, pH, TN, and TP. As previously outlined, the data points were allocated for training and testing in the ratio of 8:2. Both the artificial neural network and random forest algorithms have demonstrated remarkable generalization capabilities and predictive accuracy concerning effluent variables.

For the ANN model, the R^2 for predicting BOD_{out}, COD_{out}, and TSS_{out} stands at 0.903, 0.895, and 0.901, respectively (Fig. 13.7). Meanwhile, the RF model yields R^2 values of 0.947, 0.901, and 0.946 for BOD_{out}, COD_{out}, and TSS_{out}, respectively (Fig. 13.8). These R^2 values align closely with prior research and signify the superiority of the RF model, particularly in predicting BOD_{out} and TSS_{out}. In the case of COD_{out}, the R^2 values of both RF and ANN models are comparable. Overall, it can be said that machine learning-based models are capable of finding the underlying trends of the data.

13.3.2 Multiple-Output Models

The creation of multiple-output models involved utilizing the RF regressor with input variables including BOD_{in}, COD_{in}, TSS_{in}, pH, TN, TP, BOD_{out}, COD_{out}, and TSS_{out} aimed at predicting the target variables BOD_{out}, COD_{out}, and TSS_{out}. The model yielded R^2 of 0.913, indicating a strong capability to generalize the dataset effectively. Additionally, the MSE of 1.308 signifies the model's high accuracy in prediction (Fig. 13.9).

For the RF model for BOD_{out}, COD_{out}, and TSS_{out}, the R^2 values of predictions are 0.947, 0.901, and 0.946, respectively. This shows that these models were successful in capturing most of the variation of the dataset. These models have achieved excellent

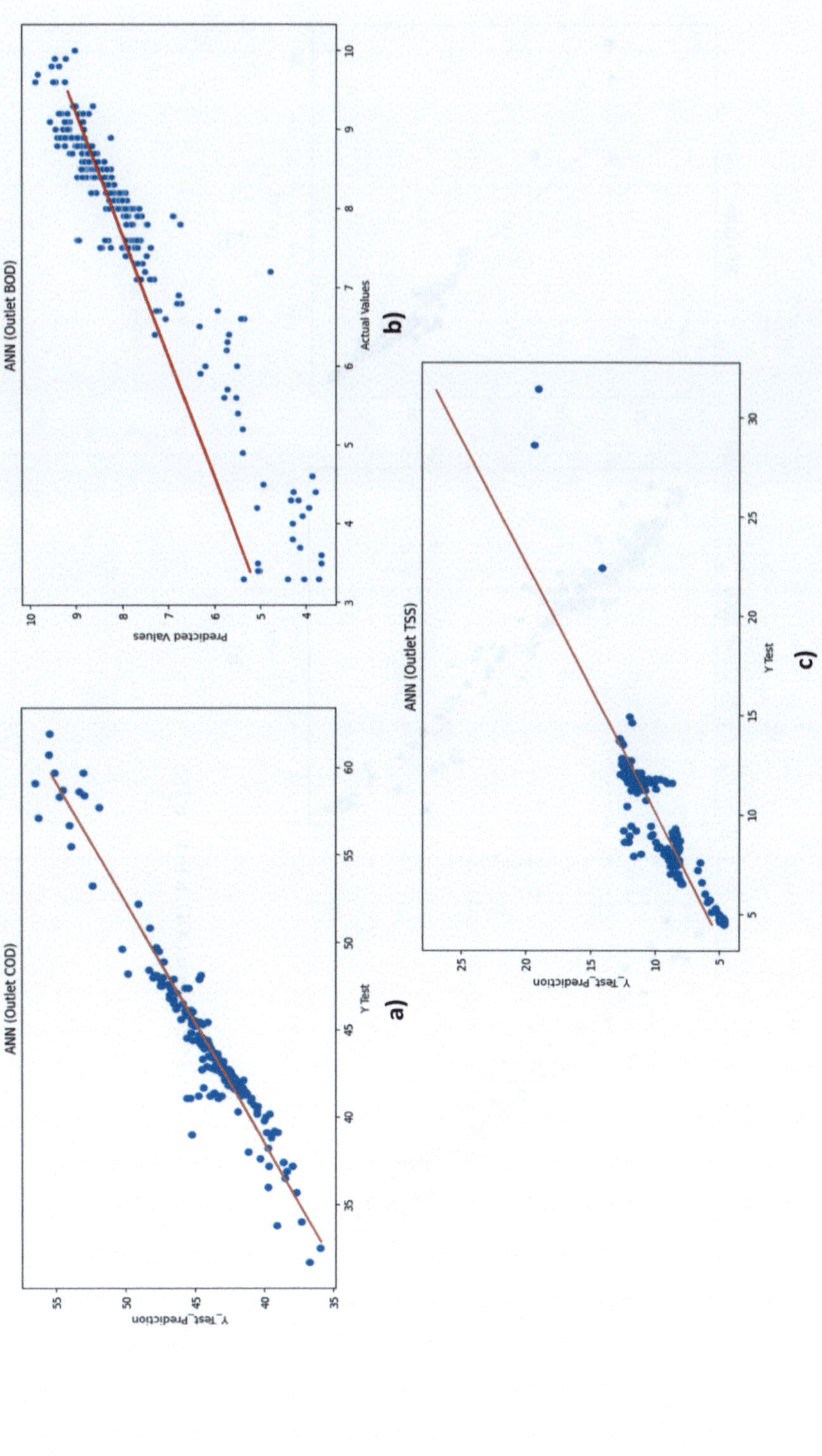

Fig. 13.7 Scatter plot of ANN model for outlet **a** COD, **b** BOD, **c** TSS

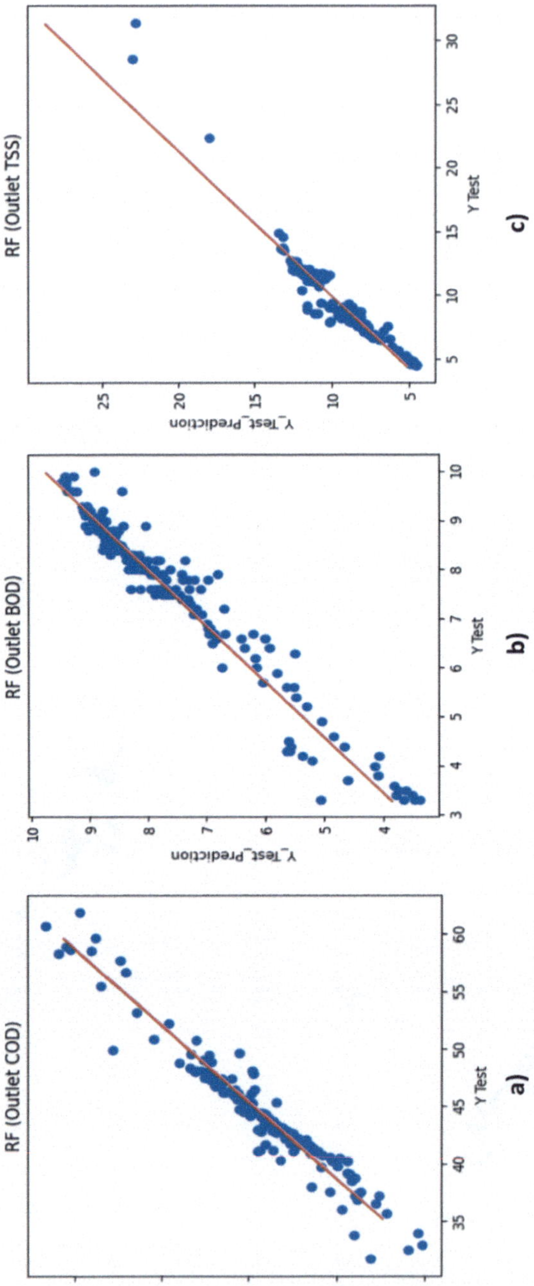

Fig. 13.8 Scatter plot of RF model for outlet **a** COD, **b** BOD, **c** TSS

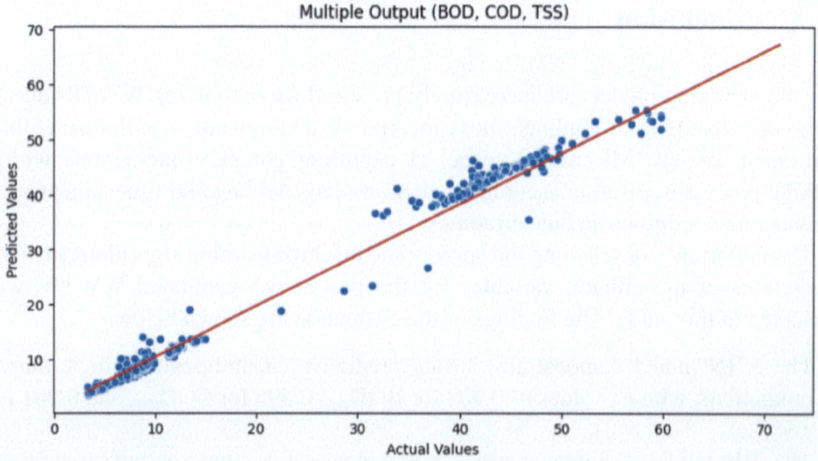

Fig. 13.9 Scatter plot of RF regressor model of multiple output

generalization performance and are capable of finding underlying patterns in the dataset.

The performance of single-output models based on the ANN algorithm is also considerable, as it has given R^2 values for predicting BOD_{out}, COD_{out}, and TSS_{out}, which stand at 0.903, 0.895, and 0.901, respectively. All these models capture approximately 90% of the variation in the provided dataset. The result of ANN-based models on this dataset shows that these models have excellent generalization performance.

Here, it can be seen that models based on RF are better performing than those based on ANN algorithms. However, for the prediction models for the COD_{out}, both models have given comparable results with the R^2 value of RF and ANN at 0.901 and 0.895, respectively. For the models created for the forecasting of the other two parameters, the performance of RF models is slightly superior to the ANN models. Because of this superior performance of the RF model, for the development of a multiple-output model, RF regressor is used. The model's strong generalization capability is evidenced by an R^2 value of 0.913, and its high predictive accuracy is underscored by MSE of 1.308.

The results of our models align with findings from similar research conducted globally in this domain. Numerous studies have yielded comparable outcomes using their predictive models, both single and multiple outputs. Our literature review indicates that the coefficient of determination for these models ranges from 0.61 to 0.96. Notably, our models demonstrate prediction capabilities near the upper end of this range, with a maximum coefficient of determination of 0.94 for single-output models, highlighting their strong predictive performance.

13.4 Conclusion

Machine learning models are increasingly preferred for optimizing WWTPs due to their adept handling of nonlinearities, uncertainty, and dynamic conditions. Unlike traditional models, ML models excel at capturing complex interactions within WWTP processes, offering accurate predictions and enabling real-time adjustments to changing conditions and uncertainties.

The importance of selecting the appropriate machine learning algorithms to make predictions of the effluent variables for the two above-mentioned WWTPs was evaluated in this study. The findings of the evaluation are shown below:

- The ANN model demonstrates strong predictive capabilities for single-output predictions, with R^2 values of 0.903 for BOD_{out}, 0.895 for COD_{out}, and 0.901 for TSS_{out}.
- The RF model exhibits superior performance for single-output predictions compared to the ANN model, with higher R^2 values of 0.947 for BOD_{out}, 0.901 for COD_{out}, and 0.946 for TSS_{out}.
- For COD_{out}, the RF and ANN models both have equivalent precision in forecasting, with R^2 values of 0.901 and 0.895, respectively.
- Overall, the RF model is particularly more effective in predicting BOD_{out} and TSS_{out}, showing its robustness in capturing the underneath trends in the data.
- The multiple-output model using the RF regressor, incorporating input variables such as BOD_{in}, COD_{in}, TSS_{in}, pH, TN, TP, BOD_{out}, COD_{out}, and TSS_{out}, achieved a commendable R^2 value of 0.913.
- The multiple-output RF model's MSE of 1.308 suggests that its predictions are very accurate.
- The better performance of the RF algorithm might be attributed to the need for substantial datasets required to train neural network models effectively.
- ML models excel at learning and adapting to changing conditions based on historical data, rendering them well-suited for modeling complicated systems such as WWTPs.

Authorship Contribution **Tejash Singh**: Writing—review and editing, writing—original draft, visualization, validation, supervision, methodology, formal analysis, data curation, conceptualization.

Pratham Arora: Writing—review and editing, writing—original draft, visualization, validation, supervision, methodology, formal analysis, data curation, conceptualization.

Sunil Kumar Singal: Writing—review and editing, writing—original draft, visualization, validation, supervision, methodology, formal analysis, data curation, conceptualization.

References

Ay M, Kisi O (2014) Modelling of chemical oxygen demand by using ANNs, ANFIS and k-means clustering techniques. J Hydrol 511:279–289. https://doi.org/10.1016/J.JHYDROL.2014.01.054

Bagherzadeh F, Mehrani MJ, Basirifard M, Roostaei J (2021) Comparative study on total nitrogen prediction in wastewater treatment plant and effect of various feature selection methods on machine learning algorithms performance. J Water Process Eng 41. https://doi.org/10.1016/j.jwpe.2021.102033

Byliński H, Sobecki A, Gebicki J (2019) The use of artificial neural networks and decision trees to predict the degree of odor nuisance of post-digestion sludge in the sewage treatment plant process. Sustainability (Switzerland) 11(16). https://doi.org/10.3390/su11164407

El-Rawy M, Abd-Ellah MK, Fathi H, Ahmed AKA (2021) Forecasting effluent and performance of wastewater treatment plant using different machine learning techniques. J Water Process Eng 44. https://doi.org/10.1016/j.jwpe.2021.102380

Farhi N, Kohen E, Mamane H, Shavitt Y (2021) Prediction of wastewater treatment quality using LSTM neural network. Environ Technol Innov 23. https://doi.org/10.1016/j.eti.2021.101632

Guo H, Jeong K, Lim J, Jo J, Kim YM, Park JP, Kim JH, Cho KH (2015) Prediction of effluent concentration in a wastewater treatment plant using machine learning models. J Environ Sci (China) 32:90–101. https://doi.org/10.1016/j.jes.2015.01.007

Jafar R, Awad A, Jafar K, Shahrour I (2022) Predicting effluent quality in full-scale wastewater treatment plants using shallow and deep artificial neural networks. Sustainability (Switzerland) 14(23). https://doi.org/10.3390/su142315598

Krause P, Boyle DP, Bäse F (2005) Comparison of different efficiency criteria for hydrological model assessment. Adv Geosci 5:89–97. https://doi.org/10.5194/adgeo-5-89-2005

Lee DS, Jeon CO, Park JM, Chang KS (2002) Hybrid neural network modeling of a full-scale industrial wastewater treatment process. Biotechnol Bioeng 78(6):670–682. https://doi.org/10.1002/bit.10247

Ly QV, Truong VH, Ji B, Nguyen XC, Cho KH, Ngo HH, Zhang Z (2022) Exploring potential machine learning application based on big data for prediction of wastewater quality from different full-scale wastewater treatment plants. Sci Tot Environ 832. https://doi.org/10.1016/j.scitotenv.2022.154930

Mjalli FS, Al-Asheh S, Alfadala HE (2007) Use of artificial neural network black-box modeling for the prediction of wastewater treatment plants performance. J Environ Manage 83(3):329–338. https://doi.org/10.1016/J.JENVMAN.2006.03.004

Nadiri AA, Shokri S, Tsai FTC, Asghari Moghaddam A (2018) Prediction of effluent quality parameters of a wastewater treatment plant using a supervised committee fuzzy logic model. J Clean Prod 180:539–549. https://doi.org/10.1016/j.jclepro.2018.01.139

Najafzadeh M, Zeinolabedini M (2019) Prognostication of waste water treatment plant performance using efficient soft computing models: an environmental evaluation. Meas: J Int Meas Confederation 138:690–701. https://doi.org/10.1016/j.measurement.2019.02.014

Pham QB, Gaya MS, Abba SI, Abdulkadir RA, Esmaili P, Linh NTT, Sharma C, Malik A, Khoi DN, Dung TD, Linh DQ (2020) Modeling of bunus regional sewage treatment plant using machine learning approaches. Desalin Water Treat 203:80–90. https://doi.org/10.5004/dwt.2020.26160

Rizal NNM, Hayder G (2023) Forecasting effluent biochemical oxygen demand in sewage treatment plants using machine learning and user-friendly interface. Int J Environ Res 17(1). https://doi.org/10.1007/s41742-022-00493-8

Salles R, Mendes J, Araujo R, Melo C, Moura P (2022) Prediction of key variables in wastewater treatment plants using machine learning models. In: Proceedings of the international joint conference on neural networks, 2022-July. https://doi.org/10.1109/IJCNN55064.2022.9892661

Singh KP, Basant N, Malik A, Jain G (2010) Modeling the performance of "up-flow anaerobic sludge blanket" reactor based wastewater treatment plant using linear and nonlinear approaches-a case study. Anal Chim Acta 658(1):1–11. https://doi.org/10.1016/j.aca.2009.11.001

Wan X, Li X, Wang X, Yi X, Zhao Y, He X, Wu R, Huang M (2022) Water quality prediction model using Gaussian process regression based on deep learning for carbon neutrality in papermaking wastewater treatment system. Environ Res 211. https://doi.org/10.1016/j.envres.2022.112942

Wang Z, Man Y, Hu Y, Li J, Hong M, Cui P (2019) A deep learning based dynamic COD prediction model for urban sewage. Environ Sci: Water Res Technol 5(12):2210–2218. https://doi.org/10.1039/c9ew00505f

Wang D, Thunéll S, Lindberg U, Jiang L, Trygg J, Tysklind M, Souihi N (2021) A machine learning framework to improve effluent quality control in wastewater treatment plants. Sci Tot Environ 784. https://doi.org/10.1016/j.scitotenv.2021.147138

Yang Y, Kim KR, Kou R, Li Y, Fu J, Zhao L, Liu H (2022) Prediction of effluent quality in a wastewater treatment plant by dynamic neural network modeling. Process Saf Environ Prot 158:515–524. https://doi.org/10.1016/j.psep.2021.12.034

Part III
Wastewater Treatment Processes

Chapter 14
Wastewater Treatment Techniques Adapted to Developing Countries

Tahira Maqsood and Abid Mahmood

Abstract Effective and affordable wastewater treatment techniques are critical issue in disease prevention and public health around the world, particularly in developing countries which often lack technical resources and finances for proper treatment. Due to the excessive use of pesticides and chemicals in today's modern agriculture and industrial sector, pesticide and chemical levels in water have increased significantly. Based on the type of pesticide and chemical and the effectiveness of the treatment procedure, an appropriate water treatment technique may be adopted for its removal. This study covers numerous chemical, physical, and biological methods including membrane bioreactors, activated sludge, adsorption, advanced oxidation processes, and membrane technologies for a critical examination and comprehensive description of pesticide and chemical removal from water. There is a detailed description and a critical discussion of both single-treatment and hybrid techniques. Potential chances to create novel pesticide removal strategies exist with the usage of hybrid removal approaches. Hybrid or integrated technologies have been determined to be an effective way to decrease the wastewater contamination level, in combination with biological methods and constructed wetlands.

Keywords Wastewater · Treatment · Physical methods · Biological methods · Chemical methods · Hybrid methods

14.1 Introduction

Water is one of the planet's most vital resources which is needed by all living things to survive in a number of environmental conditions. Even though water makes up about 71% of the earth's surface, freshwater ecosystems only make up 2.5% of it. The majority of developing countries continue to struggle with the problem of contaminated water and its sanitation due to modern urbanization and the expansion of industries (Marshall et al. 2007). Globally, freshwater shortages are rapidly

T. Maqsood · A. Mahmood (✉)
Department of Environmental Sciences, Government College University, Faisalabad, Pakistan
e-mail: drabid@gcuf.edu.pk

S. Souabi et al. (eds.), *Wastewater Treatment Plants*, Water Science and Technology Library 130, https://doi.org/10.1007/978-3-031-87461-1_14

becoming a problem. Water resources are frequently in limited supply due to the increase of human activities and population (Raouf et al. 2019). Water quality parameters measure the water quality index along with pH, temperature, biological and chemical oxygen demand, dissolved oxygen demand, total dissolved solids, color, taste, nitrogen content, phosphate, turbidity, and so on (Sathya et al. 2023).

Wastewater is categorized and defined according to where it's generated. It generally falls into two groups: first, runs that are discharged primarily from homes, For example, those released during food preparation, cleaning, washing, and personal maintenance are known as "domestic wastewater". Second, commercial/industrial wastewater is the flow discharged by industrial and commercial operations, such as food, pharmaceutical, printing, and beverage manufacturing and processing.

Wastewater that has recently been widely discharged into the environment and contains heavy metals. The most hazardous pollutants, such as copper, chromium, nickel, lead, and cadmium, are found in freshwater systems. Because of their teratogenic and carcinogenic properties, heavy metals create a major health danger and can accumulate in human bodies through the food chain (Lee et al. 2012a, b).

Not only is wastewater treatment an issue in nations that are developing, but it is also the most fundamental hygiene requirement to safeguard the surroundings and the bodies of water which provide consuming water to people all over the world. Wastewater treatment is either physio-chemically or biologically, converting the wastewater's complex organic material into more stable, odorless, and less complex substances (natural remediation). When untreated wastewater is released into surface or groundwater bodies, or onto land, it can have the following adverse impacts on the environment: As organic materials in wastewater decompose, large volumes of unpleasant gases can be released. If large amounts of organic matter-containing untreated wastewater (sewage) are dumped into water bodies, they will consume the dissolved oxygen required to fulfil the Biochemical Oxygen Demand of the wastewater that will reduce the stream's dissolved oxygen level and result in fish kills and other unfavorable outcomes. Nutrients in wastewater have the potential to encourage the formation of algal blooms and aquatic vegetation, eutrophication lakes and streams. Numerous pathogenic, or disease-causing, bacteria and hazardous substances that exist in the human digestive system or may be present in some industrial waste are typically found in untreated wastewater. These might pollute the area where the sewage is dumped, either on land or in water.

According to water management estimates, a portion of people all over the world will reside in areas affected by water in 2025, and flooding causes $14 billion in economic damage annually. It is estimated that polluted water is consumed by more than 1.1 billion people worldwide due to poor water quality. Based on current estimates and the rapidly growing global population, the world's estimated water demand and supply will be varied by 40% till 2030.

An estimated 21 (%) of transmissible diseases have occurred recently. The World Health Organization (WHO) reports that water pollution causes about 80% of human diseases, and it is responsible for 3.1% of deaths (Pawari and Gawande 2015). Water management's purpose is to provide an adequate quantity of drinking water for drinking, food preparation, sanitation, energy conservation, transportation, regeneration, and water-related dangers (pollution, drought, and flooding). Water management requires measuring a lot of different things, like optimizing cooling towers, lavatory fixtures, eliminating cooling with single pass, using smart water irrigation or landscaping, lowering the temperature in the steam sterilizers, regulating the reverse osmosis process, and utilizing less water for lab culture.

14.2 Wastewater

Pesticides, fertilizers, heavy metals, pathogens, organic waste, and novel contaminants are all contributing to the global contamination of water resources (Yadav and Pandita 2019). The diluting capacity of rivers is being reduced by the organic and inorganic waste produced from point and non-point water polluting sources (Water 2020). Grey Water is the term used to describe the wastewater that is discharged from homes after washing, drying, bathing, or showering. Black Water is the name given to the domestic wastewater that is expelled from urinals and toilets (Widyarani et al. 2022). Since it is mostly biodegradable and rich in organic content, black wastewater becomes septic and dangerous because it serves as a feeding medium for the growth of harmful microorganisms in underground pipelines (Choi et al. 2017). An aqueous discharge that industries release is known as industrial wastewater. Water is necessary for the production, processing, chilling, dilution, washing, and transportation of detergents and surfactants in domestic waste industries. The dissolved or suspended materials released during industrial processing are present in the utilized water from industries (Dash et al. 2018). The type of industry and its products determine the composition and quality of industrial wastewater (Cisneros 2011). Many sectors produce different amounts of wastewater depending on their processes, and developing countries emit significantly more wastewater than developed ones (Ranade and Bhandari 2014).

Mining, construction, and food processing industries are among the main sectors that contribute significantly to pollution (Adegoke et al. 2017). It may contain biodegradable (Titchou et al. 2021), non-biodegradable (Rawat 2021), heavy metal, or synthetic matter (Abdel-Raouf and Abdul-Raheim 2017). The situation facing developing nations is dangerous: about 70% of industrial raw waste and over 90% of sewage waste end up in aquatic bodies (Zorn 2018). Disposing of industrial trash becomes more difficult in developing nations due to high waste production for each unit area and declining land space accessible for disposal. The majority of companies release raw effluent into the groundwater, which pollutes it with insoluble and soluble salts, heavy metals, and inorganic and organic materials (Kumar et al. 2021).

Adequate wastewater treatment facilities are found in very few industries. This makes freshwater unsuitable for living organisms' consumption. The wastewater consists of waste; agricultural and industrial waste is released numerous millions of cubic meters every day, worldwide. Consequently, pesticides, fertilizers, heavy metals, waste from agriculture, organic waste, and novel pollutants enter into water

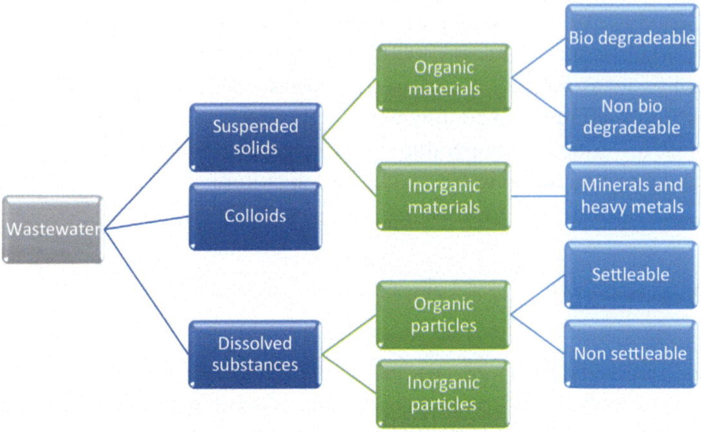

Fig. 14.1 Pollutants in wastewater

bodies (Malik et al. 2019). Morbidity and mortality from acute pesticide poisoning are mostly found in less developed countries (WWAP 2018). Substances and medications used in modern lifestyles for example hormones, personal care products, endocrine-disrupting chemicals, cosmetics, and pesticides are examples of new contaminants that have recently been identified as becoming potential hazards to human health (UN Water 2022) (Daughton and Ternes 1999). Wastewater includes in following contaminants as shown in Fig. 14.1.

14.3 Causes and Sources of Wastewater

Human activity is mainly responsible for the production of wastewater. Point and non-point sources are the two main contributors to the contamination of water. A point source is an identified source of pollution that is emitted through pipe, channel, or tunnel. It includes houses, businesses, hospitals, and enterprises that emit wastewater through pipes. On the other hand, a pollutant is considered non-point source when it reaches water resources through sources that cannot be identified (Jamwal et al. 2011). Runoffs from mines, woods, stormwater, and agriculture are some examples of non-point sources. One of the main issues facing the scientific community and authorities is monitoring and controlling non-point water contamination sources (Rane et al. 2022). Both developed and developing countries manage the non-point source of pollution of water (Leon et al. 2001). Unfortunately, because of outdated technology, unplanned and unsustainable urbanization, and inadequate infrastructure, it is more prominent in less developed countries (Kuai et al. 2015). Sources of wastewater are shown in Fig. 14.2.

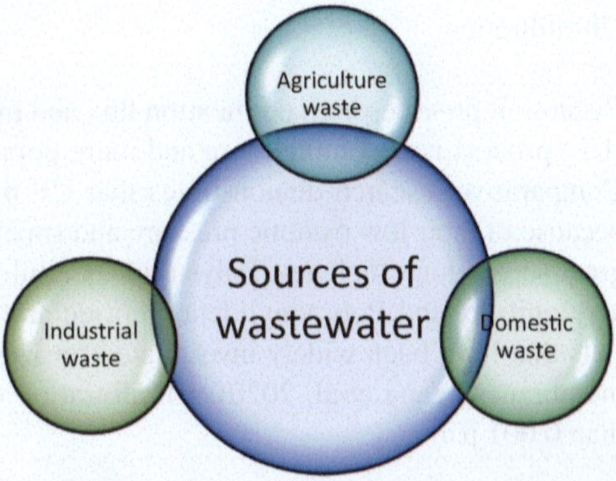

Fig. 14.2 Sources of wastewater

14.4 Wastewater Methods of Wastewater Treatment Techniques

14.4.1 Physical Methods

14.4.1.1 Membrane Filtration

One of the most effective and cost-effective ways to remediate wastewater used by many sectors is the separation or elimination of colors and organic materials by membrane filtering. Designing membranes with sufficient thermal resistance and improved performance is still challenging. The permeability, antifouling effectiveness, and rejection of membrane can be influenced by its hydrophilicity and surface charge (Jijingi et al. 2001).

Nano Filtration

It has been discovered that wastewater may be successfully treated by using Nano filtration technology for separation and purification. However, concentration polymerization and fouling may significantly limit the NF membrane's performance. Because fouling produces a coating of organic molecules that blocks membrane pores, it reduces the permeability of the membrane. Furthermore, excessive salinity inhibits membrane flow because it raises osmotic pressure because of salt rejection. The hydrophilic surface of the nano filtration membrane offers the best solution for treating textile wastewater at large throughputs (Zhu et al. 2020). Nano filtration (NF) uses pores with a size range of 1–10 nm.

Ultrafiltration

Because it provides high permeation flux and high salt absorption, the ultrafiltration (UF) process is becoming more and more popular in dyeing and salt fractionation. Comparative research demonstrates that UF membranes sustain high throughputs because of their low osmotic pressure and superior salt permeability. This results in great separation efficiency. Polysulfide, Cellulose acetate, Polyether sulfone, Polyacrylonitrile, and Poly vinylidene fluoride are among the several polymeric materials that have been widely investigated for use in the construction of effective UF membranes (Yang et al. 2020). Ultrafiltration uses pores with a size range of less than 0.001 μm.

Microfiltration

Microfiltration has pores with a size range of 0.001–0.1 μm.

14.4.1.2 Coagulation/flocculation

During coagulation, dye solution systems become unstable, releasing agglomerates and flocs.

In flocculation, suspended particles are removed and collected flocs are combined into bigger agglomerates that settle by gravity (Lee et al. 2012a, b). The textile industry often makes use of coagulation-flocculation to treat wastewater because of its low detection time, ease of usage, and economic sustainability. Using sorption, bridging, electrostatic force, and coagulating agents like ferric sulphate and chloride, sulphate of aluminum and lime bind with the pollutants and other dispersed dyes to remove them. Because of the protonated amine groups and high molecular weight polymer, respectively, sorption and bridging help in the coagulation of contaminants from wastewater and colors. The wastewater effluent's suspended and dissolved materials, settleable substances, and dyeing particles are all reduced through coagulation-flocculation (Yogalakshmi et al. 2020). As shown in Fig. 14.3.

14.4.1.3 Screening

Screening is the earliest method of physical treatment, which involves removing big, floating solid things that could hinder later stages of treatment (Naidoo and Olaniran 2014). In wastewater treatment, screening is essential to ensuring that different techniques operate as efficiently as possible by preventing potential damage, obstruction, and process failure. There are two forms of screening: coarse and fine screening, reliant on the kinds of impurities removed. Large particles, rubbish, and rags greater than 6 mm are removed through coarse screening, which can be done manually or

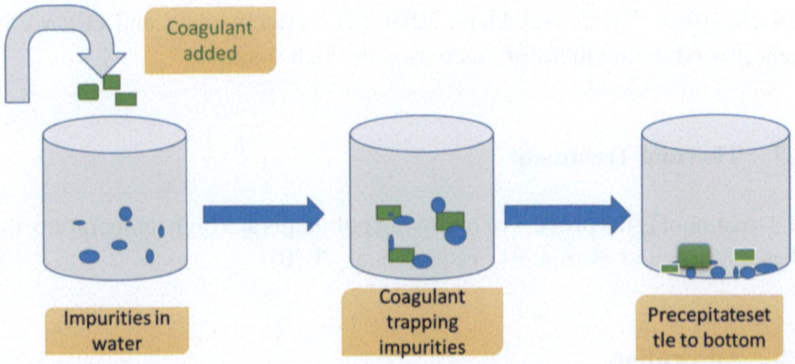

Fig. 14.3 Process of coagulation/flocculation

automatically. Fine screening eliminates suspended solids smaller than 0.001–6 mm (Bhargava 2016).

14.4.1.4 Sedimentation

The gravitational settling of particles is known as sedimentation. The technique is often used in treatment facilities for sewage and wastewater treatment plants before the filtration and disinfection processes of the main treatment stage (Yang et al. 2017).

14.4.1.5 Skimming

The term "skimming" describes the procedures used to eliminate oil, grease, and fat, as well as substances that resemble oil. Skimmer chambers are used in several large-scale WWTPs to remove oil and oil-like particles that float on top of wastewater. Skimmer types may differ according to the kind and amount of potential impurities that resemble oil for example disc, wire, and brush (Mintenig et al. 2017). The only particles that can be removed by skimming are those that are lighter than water.

14.4.1.6 Flotation

In this process, tiny gas particles are used to separate liquid or solid contaminants from wastewater. Different flotation techniques are used depending on the kind of mechanism. These include: Whereas the suction side of the pumps or an aeration tank introduces air in vacuum flotation, a spinning impeller introduces gas compounds in air flotation (Bhargava 2016). By using this method, wastewater may be treated with a wider range of impurities, including liquid, oil, gaseous, and suspended particles. Gas bubbles ensure that the pollutants of interest are removed later on in the process by trapping them. Biological sludge and suspended materials are the main uses for these mechanisms (Kyzas and Matis 2019). The effectiveness and efficacy of the treatment procedure are therefore increased through flotation.

14.4.1.7 Thermal Treatment

Thermal treatment is the process of removing pollutants at a high temperature, mainly microbes or particular chemicals (Yaqoob et al. 2020).

14.4.1.8 Adsorption

This process describes how soluble particles are eliminated when solid substrates adhere to them (Samer 2015). An adsorption process may also be used to remove multiple kinds of hazardous, inorganic, and organic contaminants. Adsorption capacities are influenced by temperature, molecular mass, concentration, size of the contaminants, and other chemical properties. Activated carbons are more effective than other adsorbents like charcoal and carbon nanotubes at removing specific pollutants (Lima 2018).

14.4.2 Biological Methods

14.4.2.1 Bioremediation

The biological process of contaminant degradation in the environment that uses microorganisms' metabolic capacity to break down a number of organic substances is known as "bioremediation" (Sisinthy and Gundlapally 2020). The most widely used biological technology that is environmentally sound for enhancing the natural decomposition process is bioremediation (Sattar et al. 2022). Phytoremediation is the ideal solution for areas with huge and intermediate concentration levels as compared to regions with elevated concentrations in small ability as heavy metals contamination of water and soil is a significant issue in countries that are developing (Timalsina et al. 2022). Because of their toxicity, bioaccumulation tendency, and non-biodegradability, heavy metals pose a risk to public conditions and environmental quality (Qing et al. 2015). By using a process known as bioremediation, which involves the employment of fungus, yeasts, bacteria, algae, cyanobacteria, and plants, heavy metals may be eliminated from both aqueous and soil media through the process of the metabolism of living organisms (Massoud et al. 2019).

One type of bioremediation is called phytoremediation, and it involves using plants to remove contaminants. Specifically, water aerophytes may be used to remove heavy metals and other pollutants from the environment by acting as organic filters in areas that are contaminated (Ansari et al. 2020).

The ability of several kinds of plant species to eliminate, become stable, or absorb inorganic or organic substances is the basis of the phytoremediation technique. Additionally, it is among the most effective methods to lessen heavy metals environmental pollution (Geng et al. 2022).

The characteristics of aquatic aerophytes enable the filtering of different contaminants within the medium, which makes them suitable for use in their removal. The tissues of these plants' developing sections store nitrogen, which is used in the synthesis of protein, amino acids, and chlorophyll as well as phosphorus, which is used in many hormonal processes in plants (Ali et al. 2019).

14.4.2.2 Biofilters

Biological filters, also referred to as bio filters, were used in the bio filtration process. Bio filters are made of filter media on which microorganisms adhere and colonies. These microorganisms are in charge of oxidizing and eliminating pollutants found in air and water because they have a tendency to catch and break down both organic and inorganic pollutants (Ralebitso-Senior et al. 2012). To create biofilm, microbes such as fungus, protozoa, bacteria, and invertebrates are used. For the microbial populations in the filter beds, fungus, and bacteria are often taken into consideration. The whole process of decomposing contaminants is carried out by these immobilized heterotrophic microbial cells. The cells first attach them to the reactor's bio filter bed's surface.

Microbial cells then begin to colonize the surface, forming an active layer that will eventually absorb the pollutant. This was accomplished through the extracellular component of polysaccharide secretion and the activation of covalent bonds with multiple surface interactions (Cohen 2001). An alternative method involves synthetically attaching the microbial cells to the bio filter bed. Micro entrapment, membrane, cross-linking, carrier bonding, and micro capsulation were used to artificially immobilize cells (Cohen 2001).

14.4.2.3 Biological Nitrification-Denitrification

Aquatic habitats depend on the nitrification process, which converts ammonia into nitrate which is the most oxygenated form of nitrogen. Nitrate prevents eutrophication and is mostly used to lower nitrogenous compounds in aquaculture systems that are circulated. Because of nitrification, physiological research is prompted by ecological observations, leading to the discovery of the special characteristics of nitrifying bacteria.

Two sequential oxidative phases carry out this chemolitho autotrophic oxidation: nitrite oxidizers, which convert nitrite to nitrate, and ammonia oxidizers, which convert ammonia to nitrite. Nitrite oxidoreductase (NOR) completes the last nitrite oxidation, whereas hydroxylamine oxidoreductase and ammonia monooxygenase catalyze the first oxidation of ammonia. Among the autotrophic nitrifies, the enzyme nitrifying genes are widely distributed.

Anaerobic nitrification is not an adequate process that may remove every nitrogenous substance found in aquaculture; denitrification is the only option. Nitrogen molecules from leftover feed and fish waste, such as nitrate and nitrite, serve as acceptors of electrons in microbial metabolism in both freshwater and marine RAS, facilitating assimilatory and dissimilatory denitrification.

Denitrification is mainly facultative, heterotrophic, and takes place in environments with low oxygen levels and is widespread across more than 50 taxa, including certain fungi, bacteria, and archaea. The denitrifying property is widely distributed among bacteria with unrelated systematic affiliations due to lateral gene transfer during evolution. The presence of denitrifying genes in plasmids is associated with the potential for horizontal gene transfer, as evidenced by studies conducted on Pseudomonadaceae and Alcaligenaceae (Preena et al. 2021). The denitrification process involves the use of enzymes like nitric oxide reductase, nitrous oxide reductase, and nitrate reductase, nitrite reductase. With the help of these enzymes, nitrate may be reduced anaerobically to nitrite, nitrite to nitrous oxide, and finally to dinitrogen gas (Lu et al. 2020).

14.4.2.4 Bio Sorption

Heavy metals are taken out of industrial trash using new biotechnological techniques including bioaccumulation and bio sorption. This is an interesting technology because of its many qualities, including little chemical or sludge production, high efficiency, high precision, no extra fertilizer needs, bio sorbent regeneration, and environmental sustainability (Bankar and Nagaraja 2018). It's an alternate approach to the conventional techniques for extracting heavy metals from wastewater. The method treats heavy metal-polluted water sources by treating biological materials (microbial biomass) as bio sorbents (Kim et al. 2006). The temperature, starting dye absorption, pH, and solution dose all affect bio sorption (Mullai et al. 2017). Many microorganisms, including fungus, bacteria, protozoa, yeast, and algae are useful for extracting heavy metals including arsenic, mercury, lead, copper, cadmium, and their cyanide complexes, because of their variety, adaptability, and flexibility.

The advancement of green technologies help regulate heavy metal emissions, heavy metal bio sorption utilizing Actinobacteria biomass is a potential tool. Microbial metal bio sorbents that fall under this category include Trichoderma, Aspergillus, Streptomyces, Rhizopus, Yarrow IA lipolytica, Bacillus, and Penicillium (El Baz et al. 2015). Actinomycetes are distinct in the bioremediation process because of a number of characteristics, including metal detoxification processes that are resistant to elevated concentrations of heavy metals. They are capable of producing a range of physiologically vigorous chemicals and change how poisonous heavy metals are.

Metal-binding proteins found on the cell wall bond with metal ions during the bio sorption process. This process involves the accumulation of heavy metals in wastewater through a variety of mechanisms, including the use of ATP, spontaneous physiochemical processes, or the presence of different inactive or non-living biomass of microbes that bind to metals (Shamim 2018).

14.4.2.5 Membrane Bioreactor

The membrane bioreactor uses the traditional activated sludge process by replacing gravity settling with interconnected membrane technologies like microfiltration or ultrafiltration for final accuracy. Based on where the membrane component is submerged (immersed) and where it is cross-flow (external/side stream), there are two main designs of membrane bioreactors that are often employed. The membrane configurations that are now in use include pleated filter cartridge, tubular type, plate and frame, spiral wound, hollow fiber, and so on. Anaerobic micro bioreactors generate high-quality wastewater that is devoid of particles and pathogens, rich in nutrients, and have a smaller environmental impact than other membrane bioreactors (Dvořák et al. 2016).

14.4.2.6 Microbial Fuel Cells

It has been determined that the best way to reduce the extra sludge and dynamism problems in traditional wastewater treatment systems is to directly convert waste into clean power (Logan and Rabaey 2012). The biological processes known as bio-electrochemical systems convert chemical energy that is found in wastewater's organic substrate into electrical power or other expensive products (Kumar et al. 2012). Bio-electrochemical systems use naturally occurring exoelectrogenic microorganisms to extract clean energy from organic waste sources (Lovley 2006). In Microbial fuel cells, this energy is harvested as bioelectricity; in microbial electrolysis cells, it is extracted as useful biofuels including hydrogen, hydrogen peroxide, ethanol, and methane. A cation exchange membrane, also known as a proton exchange membrane, separates the anode and cathode compartments.

14.4.2.7 Aerobic, Anaerobic, and Facultative Treatment

Anaerobic digestion is a commonly used biological treatment method for organic waste, sludge, and high-strength industrial effluent. In the absence of oxygen, various kinds of microbes break down complex organic compounds including bacteria and archaea. Various bacterial activity rates are the primary determinant of the degradation rate (Kondusamy and Kalamdhad 2014). As a result, suitable environments must be fulfilled within an anaerobic digester to produce the ultimate end product, which can be hydrogen sulphide, water, ammonium, carbon dioxide, and methane. One species' final output could serve as the substrate that another species requires (Sreekrishnan et al. 2004). In the presence of oxygen, aerobic digestion is facilitated by a mixed culture of autotrophic and heterotrophic bacteria. Whereas autotrophic microorganisms eat inorganic carbon like CO2, heterotrophic microbes use organic substances as their carbon source. In addition to producing a significant amount of new cells, the mixed aerobic consortium breaks down organic materials and ammonia into less toxic chemicals such as carbon dioxide, water nitrate, and nitrite. In general, the two primary steps processes in aerobic degradation are endogenous respiration and oxidation with synthesis. Generally, aerobic processes are installed after anaerobic processes because of the higher oxygen demand and longer treatment time required (Bustillo-Lecompte and Mehrvar 2017).

14.4.3 Chemical Methods

14.4.3.1 Chlorination

Chlorination is a chemical water treatment technique that helps manage taste and odor by limiting the development of microorganisms in wastewater. Advantages of chlorination is that it's less expensive and easy to use. Prechlorination, however, has the potential to oxidase micropollutants into intermediate byproducts that may be less removable and more harmful than the parent substances (Ahmed et al. 2017).

14.4.3.2 Conventional Oxidation Approaches

Photolysis

Hydroxyl free radicals are created during the process of photolysis, which is the breakdown of water molecules using energy from electromagnetic radiation. Multiple sources of radiation, such as ultraviolet and radiation from the sun can cause photolysis (Choi et al. 2020). However, in terms of wastewater treatment, each radiation has a specific function. It has been demonstrated that UV photolysis is a practical method for eliminating color from wastewater, despite its limited ability to lower the water's COD levels. However, when hydrogen peroxide was added, hydroxyl radicals were formed, which expedited the removal of chemical oxygen demand by directly absorbing ultraviolet energy (Johnson et al. 2019).

Ozonation

Strong oxidizing agent ozone is well-known for its ability to react chemically with a variety of inorganic and organic compounds (Coca et al. 2016). For the last 10 years, the use of this compound has been crucial in the treatment of wastewater because it prevents the formation of any hazardous byproducts in the reactions (Cuerda-Correa et al. 2019). Adding ozone causes a complex oxidation process called "ozonation," which significantly enhances wastewater's biodegradability. Ozonation is a very efficient way to remove nearly every kind of pharmaceutical and personal care chemicals, according to multiple research. Because of its short half-life, ozone is considered harmful if its concentration increases over a threshold, which is about 23% (Shen et al. 2019).

Fenton Oxidation Process

The Fenton oxidation procedure is another common oxidation therapy in which hydrogen peroxide and ferrous ions combine to produce hydroxyl radicals. It is commonly recognized that this procedure works effectively to eliminate the toxicity of organic chemicals found in wastewater. Nonetheless, the narrow pH range and the expense and danger of handling, storing, and transporting the chemicals are some of the process's evident drawbacks (Zhang et al. 2019). Another popular method for breaking down and eliminating persistent organic contaminants from wastewater is the Fenton process (Borah et al. 2020). Furthermore, the Fenton method's filtering, dewatering, and conditioning of wastewater has been seen as an environmentally benign procedure. Numerous studies have also attested to the Fenton approach's efficacy in lowering the heavy metals concentration including copper, chromium, aluminum, and a few more.

14.4.3.3 Advanced Oxidation Process

The latest technique for purifying wastewater that contains rigid organic compounds is the advanced oxidation process. Advanced oxidation techniques are used in wastewater treatment include photocatalysis and photo-Fenton.

Photo Catalysis

To facilitate a photon's energy transfer to a water molecule, photo catalysis is an AOP that needs the usage of catalysts. Studies have shown that the energy necessary for UV photocatalysis to reduce the total carbon content (TOC) was three times greater than that necessary for color reduction (Johnson et al. 2019). The most studied advanced oxidation technique to date that has shown interest in eliminating contaminants and microbes from wastewater is photo catalysis, which uses TiO_2 as a catalyst (Ren and Zhang 2019). Reduced costs, the ability to reuse the catalyst, room pressure, and temperature operation, and the potential to use rays of sun to release the catalyst are the main benefits of this advanced oxidation process. Furthermore, this action totally mineralizes a lot of substances. However, there are a few significant drawbacks to the photo catalysis process, such as the difficulty of attaining uniform radiation over the catalyst surface on a wider scale and the costs associated with post-use separation treatment of the catalyst.

Photo-Fenton Process

Fe(III) is often utilized in photo-Fenton treatment as a catalytic agent to create hydroxyl radicals via a reaction with two molecules of water and two in the existence of ultraviolet light (Clarizia et al. 2017). The photo-Fenton process is a hybrid of the Fenton process with UV light (Barrera-Salgado et al. 2016). The production of secondary chlorinated compounds is restricted by the formation of hydroxyl radicals. This is regarded as an excellent technique for water purification. Research has shown that this kind of wastewater treatment is low-cost, low-tech, and very effective in removing pollutants especially antibiotics from wastewater. Since the application of UV light accelerates the production of hydroxyl radicals, photo-Fenton has a lower hydrogen peroxide consumption than the conventional Fenton process (Ahmed et al. 2020).

14.4.4 Hybrid Methods

The removal of contaminants by the integration of two or more treatment approaches is known as hybrid wastewater treatment. In terms of treatment effectiveness and energy savings, this method is sturdy and long-lasting. Hybrid systems have the potential to produce bioenergy and lower system running costs. Both traditional and sophisticated oxidation methods have gained significant attention among wastewater treatment strategies in recent decades.

14.4.4.1 Adsorbents, Membrane Filtration, and Activated Sludge Process

Zinc and other hazardous substances may be removed from industrial effluents using hybrid biological absorbents. The hybrid treatment using biosorbent may effectively remove a zinc concentration of 84 to 99%, according to an experiment (Jagaba et al. 2020). Ejraei et al. (2019) studied the Chook wastewater treatment method, which combines photocatalytic degradation, membrane filtration, and adsorption. The results indicated that a higher percentage of polyvinyl pyrrolidone (PVP) would improve the hydrophilicity of the membrane. Membrane filtration worked well in isolation, but when all three processes were carried out in sequence photocatalytic degradation first, membrane filtration and last, the efficacy of pollutant removal was superior. A membrane that may be used for both membrane filtration and adsorption successfully eliminated heavy cationic metals and anionic contaminants including phosphates and nitrates. A traditional method that is frequently employed due to its efficiency in mineralizing substances from various wastewater sources is activated sludge. Still, due to its low ability to degrade pharmaceutical-related contaminants, persistent organic compounds, and anionic surfactants, a more promising alternative has been developed, which combines membrane filtration with the activated sludge process. The activated sludge method is used in conjunction with membrane filtration techniques such as ultrafiltration, microfiltration, nano filtration, etc., to treat wastewater (Ezugbe and Rathilal 2020).

14.4.4.2 Constructed Waste Stabilization Wetlands and Ponds

Naturally occurring waste maintenance wetlands and ponds are seen to be an efficient way to treat wastewater because they offer the chance to recover resources through the creation of biomass, which may then be utilized as animal feed. Latest hybrid technologies for eliminating heavy metals from wastewater include waste maintenance ponds and constructed swamplands.

In contrast, waste maintenance ponds contain less microorganisms, which prevents them from reducing shock loadings. As a result, the majority of research demonstrates that constructed wetlands are more effective at eliminating heavy metal levels (Acheampong and Ansa 2017). Constructed wetlands are used to clean wastewater from mines, tanneries, chemical factories, paper, oil refineries, and textile businesses, and overflow from roadways (Muñoz et al. 2016).

14.4.4.3 Membrane Filtration and Membrane Bioreactors

A very recent hybrid treatment technique for polluted water is membrane bioreactors. This specific hybrid technology eliminates pollutants using several methods all at once. For example, coagulants are used to flocculate large hydrophobic natural organic matters, absorbents are used to remove small to medium-sized natural organic matters, and the membrane acts as a physical barrier to separate sludge contained in chemicals and other substances (Jalilnejad et al. 2020). However, membrane technologies, which include a number of membrane filtering techniques, have also become well-liked choices for extracting water from diverse wastewater sources. Hydraulic pressure is necessary for membrane filtration methods such as ultrafiltration, nano filtration, reverse osmosis, and microfiltration to guarantee the departure of pollutants during process wastewater treatment. The size of pores is another characteristic that affects departure of pollutants (Ezugbe and Rathilal 2020).

14.4.4.4 Miscellaneous Systems

Advanced configuration to fulfil the specific requirements of the refining industries, osmotic membrane bioreactors were developed. Compared to traditional membrane bioreactors, this technique offers several advantages, including a lower fouling propensity because of increased fouling reversibility, better water quality, and the lack of hydraulic power needed for water penetration (Molinari et al. 2020). This indicates that the compounds with lower molecular weights are more effective at being removed. Reverse osmosis, electro dialysis, and membrane distillation are typical extra procedures that are combined with osmotic membrane bioreactors to generate clean water and create a hybrid system The accumulation of salt in the bioreactor and resource recovery are two further problems that this hybrid system is well-suited to solving them (Zhao et al. 2018).

14.5 Conclusion

Site-specific management plans should take into consideration the target area's social, cultural, environmental, and economic circumstances. The technology that is socially acceptable, ecologically sustainable, and reasonably priced is referred to as the "Most Appropriate Technology." The system's installation, upkeep, and operation—including any future capital improvements and required long-term repairs and replacements—should be financially feasible for the community. Selecting the right technology requires an understanding of the receiving environment, which can only be achieved through a thorough site inspection procedure. Making decisions might be made easier by creating standards for the selection of small community wastewater treatment systems. To guarantee that the decentralized wastewater treatment systems are routinely examined and maintained, centralized management is necessary. Sustainable wastewater management requires giving the local population access to the tools, knowledge, and training they need to make an impact on environmental issues that concern them. It is necessary to expand the body of knowledge on environmental issues and their remedies in emerging nations, taking into account both national practical experience and scientific theory. Employee training programs are crucial to the efficient use, upkeep, and monitoring of buildings and equipment, including wastewater quality monitoring. Even though wastewater management in developing nations has several obstacles and difficulties, they may be addressed with careful planning and the execution of policies.

References

Abdel-Raouf MS, Abdul-Raheim ARM (2017) Removal of heavy metals from industrial waste water by biomass-based materials: a review. Mercury 8:15–19

Acheampong MA, Ansa EDO (2017) Low-cost technologies for mining wastewater treatment. J Environ Sci Eng 6:391–405

Adegoke KA, Oyewole RO, Lasisi BM, Bello OS (2017) Abatement of organic pollutants using fly ash based adsorbents. Water Sci Technol 76(10):2580–2592

Ahmed MB, Zhou JL, Ngo HH, Guo W, Thomaidis NS, Xu J (2017) Progress in the biological and chemical treatment technologies for emerging contaminant removal from wastewater: a critical review. J Hazard Mater 323:274–298

Ahmed Y, Lu J, Yuan Z, Bond PL, Guo J (2020) Efficient inactivation of antibiotic resistant bacteria and antibiotic resistance genes by photo-Fenton process under visible LED light and neutral pH. Water Res 179:115878

Ali F, Jilani G, Fahim R, Bai L, Wang C, Tian L, Jiang H (2019) Functional and structural roles of wiry and sturdy rooted emerged macrophytes root functional traits in the abatement of nutrients and metals. J Environ Manage 249:109330

Ansari AA, Naeem M, Gill SS, AlZuaibr FM (2020) Phytoremediation of contaminated waters: an eco-friendly technology based on aquatic macrophytes application. Egypt J Aquat Res 46(4):371–376

Bankar A, Nagaraja G (2018) Recent trends in biosorption of heavy metals by actinobacteria. In: New and future developments in microbial biotechnology and bioengineering. Elsevier, pp 257–275

Barrera-Salgado KE, Ramírez-Robledo G, Álvarez-Gallegos A, Pineda-Arellano CA, Sierra-Espinosa FZ, Hernández-Pérez JA, Silva-Martínez S (2016) Fenton process coupled to ultrasound and UV light irradiation for the oxidation of a model pollutant. J Chem 2016(1):4262530

Bhargava A (2016) Physico-chemical waste water treatment technologies: an overview. Int J Sci Res Educ 4(5):5308–5319

Borah P, Kumar M, Devi P (2020) Recent trends in the detection and degradation of organic pollutants. In: Abatement of environmental pollutants. Elsevier, pp 67–79

Bustillo-Lecompte C, Mehrvar M (2017) Slaughterhouse wastewater: treatment, management and resource recovery. Physico-Chemical Wastewater Treatment and Resource Recovery, 153–174

Choi Y-Y, Baek S-R, Kim J-I, Choi J-W, Hur J, Lee T-U, Park C-J, Lee BJ (2017) Characteristics and biodegradability of wastewater organic matter in municipal wastewater treatment plants collecting domestic wastewater and industrial discharge. Water 9(6):409

Choi SW, Shahbaz HM, Kim JU, Kim D-H, Yoon S, Jeong SH, Park J, Lee D-U (2020) Photolysis and TiO_2 photocatalytic treatment under UVC/VUV irradiation for simultaneous degradation of pesticides and microorganisms. Appl Sci 10(13):4493

Cisneros BJ (2011) Safe sanitation in low economic development areas. Treatise Water Sci 147

Clarizia L, Russo D, Di Somma I, Marotta R, Andreozzi R (2017) Homogeneous photo-Fenton processes at near neutral pH: a review. Appl Catal B 209:358–371

Coca M, González-Benito G, García-Cubero MT (2016) Chemical oxidation with ozone as an efficient pretreatment of lignocellulosic materials. Biomass fractionation technologies for a lignocellulosic feedstock based biorefinery, pp 409–429

Cohen Y (2001) Biofiltration–the treatment of fluids by microorganisms immobilized into the filter bedding material: a review. Biores Technol 77(3):257–274

Cuerda-Correa EM, Alexandre-Franco MF, Fernández-González C (2019) Advanced oxidation processes for the removal of antibiotics from water. an overview. Water 12(1):102

Dash S, Chaudhuri H, Gupta R, Nair UG (2018) Adsorption study of modified coal fly ash with sulfonic acid as a potential adsorbent for the removal of toxic reactive dyes from aqueous solution: kinetics and thermodynamics. J Environ Chem Eng 6(5):5897–5905

Daughton CG, Ternes TA (1999) Pharmaceuticals and personal care products in the environment: agents of subtle change? Environ Health Perspect 107(suppl 6):907–938

Dvořák L, Gómez M, Dolina J, Černín A (2016) Anaerobic membrane bioreactors—a mini review with emphasis on industrial wastewater treatment: applications, limitations and perspectives. Desalin Water Treat 57(41):19062–19076

Ejraei A, Aroon MA, Saravani AZ (2019) Wastewater treatment using a hybrid system combining adsorption, photocatalytic degradation and membrane filtration processes. J Water Process Eng 28:45–53

El Baz S, Baz M, Barakate M, Hassani L, El Gharmali A, Imziln B (2015) Resistance to and accumulation of heavy metals by actinobacteria isolated from abandoned mining areas. Sci World J 2015(1):761834

Ezugbe EO, Rathilal S (2020) Membrane technologies in wastewater treatment: a review. Membr (Basel) 10:89

Geng N, Xia Y, Lu D, Bai Y, Zhao Y, Wang H, Ren L, Xu C, Hua E, Sun G (2022) The bacterial community structure in epiphytic biofilm on submerged macrophyte *Potamogetom crispus* L.

and its contribution to heavy metal accumulation in an urban industrial area in Hangzhou. J Hazard Mater 430:128455

Jagaba AH, Kutty SRM, Khaw SG, Lai CL, Isa MH, Baloo L, Lawal IM, Abubakar S, Umaru I, Zango ZU (2020) Derived hybrid biosorbent for zinc (II) removal from aqueous solution by continuous-flow activated sludge system. J Water Process Eng 34:101152

Jalilnejad E, Sadeghpour P, Ghasemzadeh K (2020) Achievements in membrane bioreactors for wastewater and water treatment. In: Current trends and future developments on (bio-) membranes. Elsevier, pp 187–220

Jamwal P, Mittal AK, Mouchel J-M (2011) Point and non-point microbial source pollution: a case study of Delhi. Phys Chem Earth Parts a/b/C 36(12):490–499

Jijingi HE, Sara Kazemi Yazdi SK, Yousif Abdalla Abakar Y, Etim E (2001) Evaluation of membrane bioreactor (MBR) technology for industrial wastewater treatment and its application in developing countries: a review. Case Stud Chem Environ Eng. https://doi.org/10.1016/j.cscee.2024.100886

Johnson I, Ali MAS, Kumar M (2019) Cyanobacteria/microalgae for distillery wastewater treatment-past, present and the future. In: Microbial wastewater treatment. Elsevier, pp 195–236

Kim SJ, Chung JH, Kim TY, Cho SY (2006) Biosorption of heavy metals and cyanide complexes on biomass. Stud Surface Sci Catal 141–144

Kondusamy D, Kalamdhad AS (2014) Pre-treatment and anaerobic digestion of food waste for high rate methane production–A review. J Environ Chem Eng 2(3):1821–1830

Kuai P, Li W, Liu N (2015) Evaluating the effects of land use planning for non-point source pollution based on a system dynamics approach in China. PLoS ONE 10(8):e0135572

Kumar AK, Reddy MV, Chandrasekhar K, Srikanth S, Mohan SV (2012) Endocrine disruptive estrogens role in electron transfer: bio-electrochemical remediation with microbial mediated electrogenesis. Biores Technol 104:547–556

Kumar P, Inwati GK, Mathpal MC, Ghosh S, Roos WD, Swart HC (2021) Defects induced enhancement of antifungal activities of Zn doped CuO nanostructures. Appl Surf Sci 560:150026

Kyzas GZ, Matis KA (2019) The flotation process can go green. Processes 7(3):138

Lee J-C, Son Y-O, Pratheeshkumar P, Shi X (2012) Oxidative stress and metal carcinogenesis. Free Radical Biol Med 53(4):742–757

Lee KE, Morad N, Teng TT, Poh BT (2012) Development, characterization and the application of hybrid materials in coagulation/flocculation of wastewater: a review. Chem Eng J 203:370–386

Leon LF, Soulis ED, Kouwen N, Farquhar GJ (2001) Nonpoint source pollution: a distributed water quality modeling approach. Water Res 35(4):997–1007

Lima EC (2018) Removal of emerging contaminants from the environment by adsorption. Ecotoxicol Environ Saf 150:1–17

Logan BE, Rabaey K (2012) Conversion of wastes into bioelectricity and chemicals by using microbial electrochemical technologies. Science 337(6095):686–690

Lovley DR (2006) Bug juice: harvesting electricity with microorganisms. Nat Rev Microbiol 4(7):497–508

Lu J, Zhang Y, Wu J, Wang J (2020) Nitrogen removal in recirculating aquaculture water with high dissolved oxygen conditions using the simultaneous partial nitrification, anammox and denitrification system. Biores Technol 305:123037

Malik P, Inwati GK, Mukherjee TK, Singh S, Singh M (2019) Green silver nanoparticle and Tween-20 modulated pro-oxidant to antioxidant curcumin transformation in aqueous CTAB stabilized peanut oil emulsions. J Mol Liq 291:111252

Marshall FM, Holden J, Ghose C, Chisala B, Kapungwe E, Volk J, Agrawal M, Agrawal R, Sharma RK, Singh RP (2007) Contaminated irrigation water and food safety for the urban and peri-urban poor: appropriate measures for monitoring and control from field research in India and Zambia. Incpetion Report DFID Enkar 8160(3):1–12

Massoud R, Hadiani MR, Hamzehlou P, Khosravi-Darani K (2019) Bioremediation of heavy metals in food industry: application of *Saccharomyces cerevisiae*. Electron J Biotechnol 37:56–60

Mintenig SM, Int-Veen I, Löder MGJ, Primpke S, Gerdts G (2017) Identification of microplastic in effluents of waste water treatment plants using focal plane array-based micro-Fourier-transform infrared imaging. Water Res 108:365–372

Molinari R, Lavorato C, Argurio P (2020) Application of hybrid membrane processes coupling separation and biological or chemical reaction in advanced wastewater treatment. Membranes 10(10):281

Mullai P, Yogeswari MK, Vishali S, Namboodiri MMT, Gebrewold BD, Rene ER, Pakshirajan K (2017) Aerobic treatment of effluents from textile industry. In: Current developments in biotechnology and bioengineering. Elsevier, pp 3–34

Muñoz MA, Rosales RM, Gabarrón M, Faz A, Acosta JA (2016) Effects of the hydraulic retention time on pig slurry purification by constructed wetlands and stabilization ponds. Water Air Soil Pollut 227:1–13

Naidoo S, Olaniran AO (2014) Treated wastewater effluent as a source of microbial pollution of surface water resources. Int J Environ Res Public Health 11(1):249–270

Pawari MJ, Gawande S (2015) Ground water pollution & its consequence. Int J Eng Res General Sci 3(4):773–776

Preena PG, Rejish Kumar VJ, Singh ISB (2021) Nitrification and denitrification in recirculating aquaculture systems: the processes and players. Rev Aquac 13(4):2053–2075

Qing X, Yutong Z, Shenggao L (2015) Assessment of heavy metal pollution and human health risk in urban soils of steel industrial city (Anshan), Liaoning, Northeast China. Ecotoxicol Environ Saf 120:377–385

Ralebitso-Senior TK, Senior E, Di Felice R, Jarvis K (2012) Waste gas biofiltration: advances and limitations of current approaches in microbiology. Environ Sci Technol 46(16):8542–8573

Ranade VV, Bhandari VM (2014) Industrial wastewater treatment, recycling and reuse. Butterworth-Heinemann

Rane NR, Tapase S, Kanojia A, Watharkar A, Salama E-S, Jang M, Yadav KK, Amin MA, Cabral-Pinto MMS, Jadhav JP (2022) Molecular insights into plant–microbe interactions for sustainable remediation of contaminated environment. Biores Technol 344:126246

Raouf MEA, Maysour NE, Farag RK, Abdul-Raheim AM (2019) Wastewater treatment methodologies, review article. Int J Environ & Agri Sci 3:18

Rawat K (2021) Organic pollutants in wastewater and its remediation approaches using graphene adsorbent. Int J Agric Appl Sci 2:8–13

Ren H, Zhang X (2019) High-risk pollutants in wastewater. Elsevier

Samer M (2015) Biological and chemical wastewater treatment processes. Wastewater Treat Eng 150(10.5772):61250

Sathya R, Arasu MV, Al-Dhabi NA, Vijayaraghavan P, Ilavenil S, Rejiniemon TS (2023) Towards sustainable wastewater treatment by biological methods–A challenges and advantages of recent technologies. Urban Clim 47:101378

Sattar S, Hussain R, Shah SM, Bibi S, Ahmad SR, Shahzad A, Zamir A, Rauf Z, Noshad A, Ahmad L (2022) Composition, impacts, and removal of liquid petroleum waste through bioremediation as an alternative clean-up technology: a review. Heliyon 8(10)

Shamim S (2018) Biosorption of heavy metals. Biosorption 2:21–49

Shen J, Ding T, Zhang M (2019) Analytical techniques and challenges for removal of pharmaceuticals and personal care products in water. In: Pharmaceuticals and personal care products: waste management and treatment technology. Elsevier, pp 239–257

Sisinthy S, Gundlapally SR (2020) Mitsuaria chitinivorans sp. nov. a potential candidate for bioremediation: emended description of the genera Mitsuaria, Roseateles and Pelomonas. Arch Microbiol 202(7):1839–1848

Sreekrishnan TR, Kohli S, Rana V (2004) Enhancement of biogas production from solid substrates using different techniques—a review. Biores Technol 95(1):1–10

Timalsina H, Gyawali T, Ghimire S, Paudel SR (2022) Potential application of enhanced phytoremediation for heavy metals treatment in Nepal. Chemosphere 306:135581

Titchou FE, Zazou H, Afanga H, El Gaayda J, Akbour RA, Hamdani M (2021) Removal of persistent organic pollutants (POPs) from water and wastewater by adsorption and electrocoagulation process. Groundw Sustain Dev 13:100575

UN Water (2022) UN world water development report 2022. UN Water, Geneva, Switzerland

Widyarani W, DR., Hamidah U, Komarulzaman A, Rosmalina RT, Sintawardani N (2022) Domestic wastewater in Indonesia: generation, characteristics and treatment. Environ Sci Pollut Res 29(22):32397–32414

WWAP (2018) The united nations world water development report 2018. Nature-Based Solutions for Water. UNESCO, Paris, France

Yadav VK, Pandita PR (2019) Fly ash properties and their applications as a soil ameliorant. In: Amelioration technology for soil sustainability. IGI Global, pp 59–89

Yang C, Xu W, Nan Y, Wang Y, Hu Y, Gao C, Chen X (2020) Fabrication and characterization of a high performance polyimide ultrafiltration membrane for dye removal. J Colloid Interface Sci 562:589–597

Yang Y, Ok YS, Kim K-H, Kwon EE, Tsang YF (2017) Occurrences and removal of pharmaceuticals and personal care products (PPCPs) in drinking water and water/sewage treatment plants: a review. Sci Total Environ 596:303–320

Yaqoob AA, Khatoon A, Mohd Setapar SH, Umar K, Parveen T, Mohamad Ibrahim MN, Ahmad A, Rafatullah M (2020) Outlook on the role of microbial fuel cells in remediation of environmental pollutants with electricity generation. Catalysts 10(8):819

Yogalakshmi KN, Das A, Rani G, Jaswal V, Randhawa JS (2020) Nano-bioremediation: a new age technology for the treatment of dyes in textile effluents. Bioremediation of industrial waste for environmental safety: Volume I: industrial waste and its management, pp 313–347

Zhang M, Dong H, Zhao L, Wang D, Meng D (2019) A review on Fenton process for organic wastewater treatment based on optimization perspective. Sci Total Environ 670:110–121

Zhao L, Deng J, Sun P, Liu J, Ji Y, Nakada N, Qiao Z, Tanaka H, Yang Y (2018) Nanomaterials for treating emerging contaminants in water by adsorption and photocatalysis: systematic review and bibliometric analysis. Sci Total Environ 627:1253–1263

Zhu Z, Liu D, Cai S, Tan Y, Liao J, Fang Y (2020) Dyes removal by composite membrane of sepiolite impregnated polysulfone coated by chemical deposition of tea polyphenols. Chem Eng Res des 156:289–299

Zorn M (2018) Natural disasters and less developed countries. Nature, tourism and ethnicity as drivers of (de) marginalization: insights to marginality from perspective of sustainability and development, pp 59–78

Chapter 15
Role of Microbial Electrochemical Technologies in the Removal of Micropollutants from Wastewater

Koran Barman, Nehaun Zargar, and Sovik Das

Abstract The ubiquitous detection of micropollutants in the environment, mainly stemming from anthropogenic activities, poses significant risks to human health as well as to the ecosystem. The conventional WWTPs are not designed to effectively remove these trace contaminants from wastewater, leading to their persistence in treated effluents and subsequent entry into the human body and ecosystem. In this context, the present chapter highlights the role of microbial electrochemical technologies (METs) for the removal of these biorefractory micropollutants revealing that METs, such as microbial fuel cells, bioelectro-Fenton, microbial electrolysis cells, and microbial remediation cells, can effectively degrade these micropollutants with simultaneous power production and value-added product recovery. This critical assessment presents a detailed mechanism of factors such as the biocatalyst, electrode material, initial concentration of micropollutants, and the nature of micropollutants, which significantly influence the degradation efficiency. Even though METs offer promising alternative to conventional treatment methods for micropollutant removal, scalability, operational stability, and economic viability hinder the commercialization of METs. Thereby, future research should focus on optimizing METs for large-scale applications, improving their economic feasibility, and addressing operational challenges to enhance the practical implementation of METs in wastewater treatment systems.

Keywords Bioelectro-Fenton · Bioelectrochemical technologies · Emerging contaminants · Microbial fuel cell · Microbial remediation cell

K. Barman · N. Zargar · S. Das (✉)
Department of Civil Engineering, Indian Institute of Technology Delhi, New Delhi, India
e-mail: dassovik@iitd.ac.in

S. Souabi et al. (eds.), *Wastewater Treatment Plants*, Water Science and Technology Library 130, https://doi.org/10.1007/978-3-031-87461-1_15

15.1 Introduction

Over the last century, the explosion in human population, advancements in human lifestyle, and rampant industrialization have resulted in the incremental release of biorefractory pollutants, also known as micropollutants, into the environment (Ahmad et al. 2022). Micropollutants are also referred to as contaminants of emerging concerns (CECs), emerging contaminants (ECs), and microconstituents (Behera et al. 2023; Sikarwar et al. 2024). Micropollutants are a particular class of unregulated substances originating from both natural and anthropogenic activities and typically detected in trace concentration in aquatic environments ranging from μg/L to ng/L, posing risks to humans and the environment (Raj et al. 2024b). Micropollutants include pharmaceuticals, plasticisers, detergents, pesticides, insulating foams, waterproofing agents, flame retardants, endocrine-disrupting chemicals, and microplastics (MPs) to name a few, majorly released from anthropogenic sources, namely, domestic, municipal, industrial, and hospital wastewater, livestock, landfill leachate, aquaculture, and agricultural runoff (Ansari and Khan 2024; Kim and Zoh 2016). Currently, CECs are unregulated substances and lack well-defined regulations and legal frameworks. Additionally, due to their recalcitrant nature, complex diversity, and presence at trace level, micropollutants are difficult to detect and degrade from wastewater through conventional wastewater treatment plants (WWTPs), as the conventional WWTPs are not specifically designed for the removal of micropollutants (Ahmad et al. 2022; Priyadarshini et al. 2022a, b). As a result, micropollutants remain unaltered during treatment process in conventional WWTPs and lead to release of micropollutants into the water bodies, creating a menace to aquatic organisms and adversely affecting human health (Bhatt et al. 2022; Luo et al. 2014).

Micropollutant exposure can lead to various detrimental effects in humans and other flora and faunas, including acute or chronic toxicity, impairment of the reproductive system, physical abnormalities, antimicrobial resistance, feminization of some aquatic species, disruption of endocrine systems, bioaccumulation in different trophic levels of the food chain (Bhatt et al. 2022; Raj et al. 2024a; Sharma et al. 2024). Over the past few decades, researchers have extensively investigated the adverse effects of micropollutants on human health and other organisms (Barceló et al. 2023; Vélez et al. 2015). For instance, Vélez et al. (2015) reported decreased fecundability and infertility in human females due to increased concentrations of perfluorooctanoic acid and perfluorohexane sulfonate in maternal plasma. Also, recent investigations have detected MPs in human blood, urine, and stool primarily due to ingestion of contaminated potable water, and food (Barceló et al. 2023). Moreover, a recent investigation by Barceló et al. (2023) reported the presence of MPs, including polypropylene, polyethylene, and polyvinylchloride, in the range of 2–12 μm in human breast milk. Along with human health, micropollutants such as pharmaceuticals are widely detected in aquatic environments as pharmaceuticals are released in their native form or as metabolites from the human body. In this regard, a higher concentration of antibiotics has been detected in the aquatic environment, with some rivers containing antibiotics 300 times higher than the safe limit, which may lead to antibiotic resistance (Bhatt et al. 2022).

Therefore, considering the harmful effects of micropollutants on human and the ecosystem, it is crucial to eliminate micropollutants from wastewater to avert their incremental accumulation in the environment (Ahmad et al. 2022). In this regard, advanced wastewater treatment technologies have emerged as an effective solution for the effective degradation of pollutants from wastewater rather than conventional wastewater technologies due to the inefficiency of conventional wastewater technologies in the degradation of micropollutants from wastewater, as these technologies are specifically designed for the degradation of nutrients and organic portion of the wastewater (Ahmad et al. 2022, 2023). Also, conventional technologies are resource-intensive as they need high energy input and possess higher land footprint and emit greenhouse gases. Considering that in recent times, various advanced wastewater treatment technologies including, ultraviolet-based treatment, membrane bioreactor (MBR), membrane filtration, electrochemical and bio-electrochemical systems, ozonation, plasma and other treatment technologies have been extensively employed to remove micropollutants from contaminated water (Sharma et al. 2024). However, multiple drawbacks are also associated with these technologies, such as fouling of membrane in MBR and reverse osmosis technology, high operational cost in electrochemical oxidation and electrocoagulation due to utilization of high external energy, generation of chemical sludge in coagulation, high cost of treatment in ozonation and adsorption, and generation of secondary sludge in adsorption to highlight a few.

Thereby, the development and utilization of effective and sustainable wastewater treatment technologies for efficacious removal of micropollutants is the need of the hour (Ahmad et al. 2022). In this regard, microbial electrochemical technologies (METs) namely, bioelectro-Fenton (BEF), microbial electrolysis cells (MECs), microbial fuel cells (MFCs), microbial remediation cells (MRCs) and microbial carbon capture cells (MCCs) have emerged as a promising alternative to conventional WWTPs due to the utilization of robust exoelectrogenic microorganism for effective degrdation of micropollutants from wastewater, concurrently generating bioelectricity and value-added bioresources (Ahmad et al. 2022; Priyadarshini et al. 2022a; Rasheed et al. 2023). Thus, the present chapter provides an overview of different types of METs and removal of micropollutants from wastewater via METs, factors influencing micropollutant degradation in METs, limitations of METs, and future perspectives on efficacious degradation of micropollutants from wastewater.

15.2 Microbial Electrochemical Technologies

The METs are a group of advanced technology employed for wastewater treatment and concurrently generating electrical energy and produce a wide range of by-products, including green CH_4 and H_2, H_2O_2, and other valuable inorganic and organic compounds (Logan and Rabaey 2012; Priyadarshini et al. 2022b). In METs, the exoelectrogenic microorganisms are employed as biocatalysts for effective degradation of biodegradable organic compounds via the breakdown of the chemical bonds of the organic compounds and directly transform the chemical energy of these bonds into bioenergy or other resources (Ahmad et al. 2022; Gude 2016). Based on the configuration of the setup, valuables produced, and reactions occurring on the electrodes, METs are classified into MFCs, MECs, microbial electrosynthesis cell (MES), microbial desalination cells (MDC), and many more (Priyadarshini et al. 2022a; Rasheed et al. 2023).

15.2.1 Microbial Fuel Cell

The MFC is a type of microbial electrochemical technology that utilizes exoelectrogenic microorganisms in the anodic chamber, to effectively degrade biodegradable organic compounds present in the wastewater and transforming the stored chemical energy of the organic compounds into electrical energy (Ali et al. 2019; Logan et al. 2006; Silveira et al. 2020). The utilization of microbes as a catalyst for electricity generation was first reported by Michael Cresse Potter in 1911, where Potter observed the disintegration of biodegradable organic compounds through the fermentative activity of *Saccharomyces cerevisia* and *Escherichia coli* producing a voltage between 0.3 and 0.5 V, and the study of bioelectrochemistry officially began with this ground-breaking discovery (Potter 1911). The initial experiments conducted by Potter and subsequent researchers relied on electron shuttles or chemical mediators to transfer the metabolically produced electrons within the bacteria cells to the exogenous electrodes (Potter 1911). However, these mediators had various disadvantages, including short lifetimes, high cost, and toxicity to both exoelectrogenic microbes and environment (Kim et al. 1999). Therefore, to address these drawbacks, a significant breakthrough in MFC technology took place when Kim et al. (1999) invented a mediator-less biofuel cell, where exoelectrogens or electroactive bacteria were employed to facilitate the direct transfer of electrons from bacterial cell to the external anode in the absence of a mediator. Along with that, exoelectrogens reduce the sludge quantity by consuming the reducing equivalents, which are generated from the substrate oxidation reaction (Ahmadi et al. 2023). This innovation led to utilizing wastewater as a suitable substrate in the MFCs, while accomplishing biological wastewater treatment (Kim et al. 1999).

Based on the design, MFCs are categorized into single and dual-chamber systems: a single-chamber system houses both the cathode and anode electrodes in a single compartment, whereas a dual-chamber MFC consists of two distinct chambers, namely an aerobic cathodic chamber and an anaerobic anodic chamber, physically separated by a proton exchange membrane (PEM) or a salt bridge (Fig. 15.1) (Ahmadi et al. 2023; Priyadarshini et al. 2022a, b). The anaerobic anodic chamber houses the electrochemical active microorganisms, also referred to as "exoelectrogens," that act as biocatalysts to oxidize the organic matters present in the anodic chamber and, in the process, generate electrons, protons, and CO_2 (Ahmadi et al. 2023;

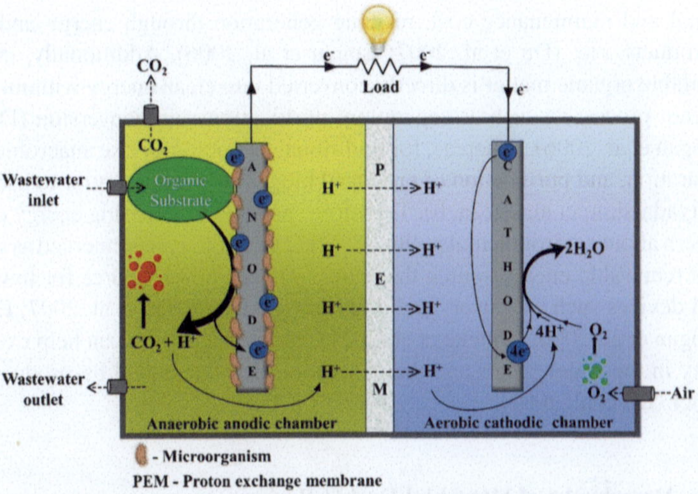

Fig. 15.1 Schematic of a double-chamber microbial fuel cell reactor

Priyadarshini et al. 2022a). Under the anoxic environment, these exoelectrogenic microbes transfer the produced electrons from the cell to the anode surface by direct membrane-associated electron transfer mechanism, by electron mediator or shuttles, i.e. cysteine, phenazines, and flavins, or by electrically conductive proteinaceous nanowires mechanism (Logan et al. 2006). Subsequently, the generated electrons travel to the cathode through a peripheral electrical circuit, thereby producing an electrical current (Ahmadi et al. 2023; Priyadarshini et al. 2022a). Simultaneously, through a PEM or a salt bridge, the protons travel to the cathodic chamber and react with terminal electron acceptors such as oxygen, potassium dichromate, potassium ferricyanide, ferric chloride, etc., to generate H_2O_2 via two-electron pathway and water via four-electron pathway (Du et al. 2007; Logan et al. 2006; Obileke et al. 2021; Sharma et al. 2024).

15.2.1.1 Benefits of Microbial Fuel Cell

In the last few decades, the successive advancement of MFC technology has provided multiple advantages to this novel technology as compared to traditional wastewater treatment technologies, including activated sludge process (ASP), sequencing batch reactor (SBR), anaerobic digestion, etc. (Ahmadi et al. 2023; Priyadarshini et al. 2022a). As compared to these traditional technologies, MFC has various environmental benefits including low carbon footprint, direct electricity generation, ability of water reclamation, low sludge production, and less odour; operational advantages, such as no aeration requirements, use of self-generating microorganisms, better resistance to external environmental stress; and various economic benefits such as low operational and maintenance cost, revenue generation through energy and value-added products, etc. (Du et al. 2007; Logan et al. 2006). Additionally, in MFC, biodegradable organic matter is directly converted into clean energy without further downstream processes, such as separation, purification, and conversion (Du et al. 2007; Logan et al. 2006). Whereas, for traditional technologies like anaerobic digestion, separation, and purification of produced biogas are required prior to use (Gude 2016; Priyadarshini et al. 2022a, b). Therefore, with ever-increasing energy demand and concern about environmental pollution, MFC technology has emerged as a potential clean renewable energy source that can serve as a power source for low-power electrical devices such as sensors and monitoring devices (Du et al. 2007; He et al. 2017; Logan et al. 2006; Obileke et al. 2021). Moreover, MFCs can help exemplify circularity in the water sector and the generation of value-added by-products from wastewater (Du et al. 2007).

15.2.1.2 Drawbacks of Microbial Fuel Cell

As discussed in the previous section, although MFC provides various advantages, several challenges are also associated with MFC, for example, the low energy efficiency of MFC resulting from various energy losses, including activation, over potential, ohmic, and mass transfer losses (Logan et al. 2006). Also, the voltage output of MFC is unstable, inconsistent, comparatively low, and has long-term performance limitations (Logan et al. 2006). Therefore, because of these drawbacks, MFCs are currently unable to attain their theoretical energy generation potential. In addition to low-power yield and slow start-up, high fabrication cost remains a significant limitation of this novel technology, which limits its application at a pragmatic scale (Du et al. 2007; Obileke et al. 2021). Moreover, at lower temperatures, microbial activity reduces, leading to poor performance of MFC. Additionally, utilization of MFC for wastewater treatment is majorly limited by the biofouling of membranes and electrodes (Du et al. 2007; Obileke et al. 2021). Biofouling of electrodes and membranes reduces MFC's performance by hindering the transportation of protons and competition for substrate utilization (Do et al. 2018; Priyadarshini et al. 2022a, b). Therefore, substantial technical advancement such as increasing coulombic efficiency (CE), energy output, etc.; along with reduction of the fabrication cost is essential for the swift transition of MFC technology from laboratory to field-scale practical application.

15.2.2 Bioelectro-Fenton

The idea of the BEF was first introduced in 2009 by Zhu and Ni (2009) while coupling MFC with Fenton technology to achieve complete mineralization of persistent organic pollutants (POPs), including p-nitrophenol, and approximately 85% degradation of total organic carbon was obtained. The BEF process integrates MET with the electro-Fenton oxidation process, featuring two chambers: an anaerobic anodic chamber, which houses the exoelectrogenic microbes, and an aerobic cathodic chamber for H_2O_2 production, with a PEM separating the chambers (Olvera-Vargas et al. 2018). In the anodic chamber of BEF, the exoelectrogenic microbes utilize the organic substrate of the wastewater to produce H^+, CO_2 and electrons (Fig. 15.2) (Olvera-Vargas et al. 2018). These generated H^+ and electrons transfer spontaneously, without any external energy input, across the bulk solution and the external circuit, respectively, to the cathodic chamber to react with terminal electron acceptor O_2 and produce H_2O_2 through a two-electron reduction pathway (Li et al. 2018; Olvera-Vargas et al. 2018). Furthermore, as shown in Eqs. 15.1–15.3, $^{\cdot}OH$ radicals are produced when H_2O_2 reacts with Fe^{2+} ion sources such as FeOOH, $FeSO_4$, etc., facilitating efficacious removal of the complex organic compounds at the cathodic chamber through Fenton reaction, simultaneously producing Fe^{2+} ions (Priyadarshini et al. 2022a). Various applications of BEFs include the remediation of several pharmaceuticals, micropollutants, dyes, and industrial pollutants such as p-nitrophenol, and the treatment of various kinds of wastewater including landfill leachate, swine wastewater, etc. (Li et al. 2018).

$$Fe^{2+} + H_2O_2 \rightarrow Fe^{3+} + {}^{\cdot}OH + OH^- \tag{15.1}$$

$$O_2 + 2H^+ + e^- \rightarrow H_2O_2 \tag{15.2}$$

$$Fe^{3+} + e^- \rightarrow Fe^{2+} \tag{15.3}$$

The BEF offers various advantages compared to conventional wastewater technologies, as well as advanced wastewater treatment technologies namely advanced oxidation processes (AOPs) such as efficient removal of contaminants, high degradation rate and energy conservation (Li et al. 2018). Also, the in-situ synthesis of H_2O_2 at the cathode reduces the operating cost of BEF systems compared to traditional

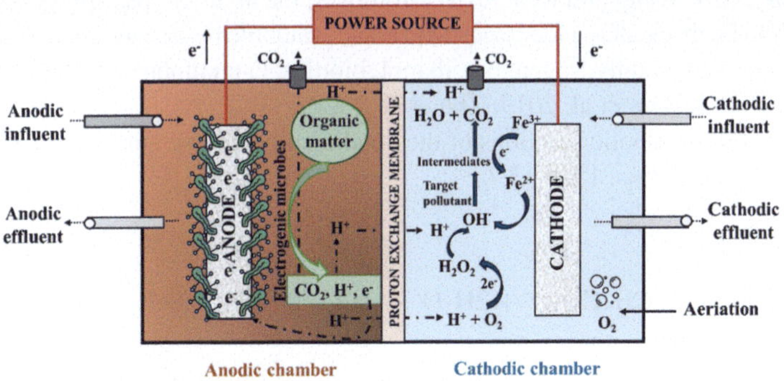

Fig. 15.2 Schematic of a Bioelectro-Fenton reactor

electro-Fenton oxidation, as the spontaneous mechanism of BEF system simultaneously treats wastewater by oxidizing the organics via Fenton oxidation requiring minimum energy input (Li et al. 2018; Olvera-Vargas et al. 2018). However, the bottlenecks associated with BEFs are scavenging of H^+ by methanogens, high cost of electrodes and membranes used, scalability and commercialization of the technology. Wastewater treatment using BEFs has been comprehensively studied at lab scale, thus more research should be carried out at the pilot-scale for the scaling up of this technology at site conditions.

15.2.3 Microbial Carbon Capture Cell

The MCC system is a bioelectrochemical technology primarily utilized to capture and sequester atmospheric CO_2 and produce electrical energy during wastewater treatment (Das et al. 2019). A MCC is a modified MFC, wherein the cathodic chamber houses the photoautotrophic microorganisms that facilitate the sequestration of CO_2 via the photosynthesis process and supplies O_2 for the cathodic reduction compared to conventional MFC, where external aeration is provided for cathodic oxygen reduction reaction (ORR) (Kumar and Jujjavarappu 2023). Typically, MCC consists of an aerobic cathodic and anaerobic anodic chamber physically separated by a PEM or salt bridge (Kumar and Jujjavarappu 2023; Wang and Ren 2013). In MCC, wastewater is introduced to the anaerobic anodic chamber, where anaerobic electrogenic microorganisms oxidize the organic matter and, in the process, generate protons, electrons, and CO_2 (Fig. 15.3) (Das et al. 2019; Wang and Ren 2013). The CO_2 generated during the oxidative degradation of complex organic matter in the anaerobic anodic chamber is transported to the aerobic cathodic chamber for sequestration and biomass synthesis by photosynthetic organisms such as microalgae and cyanobacteria through photosynthesis (Das et al. 2019; Wang and Ren 2013). Additionally, in MCC, O_2 required for the cathodic reduction is generated during the photosynthesis process by algae or cyanobacteria, instead of external aeration, thus reducing the operating cost (Das et al. 2019; Wang and Ren 2013). Moreover, the algal biomass produced in the MCC could be utilized as a feedstock to produce valuable by-products such as antibiotics, biofuels like biohydrogen, bioethanol, biodiesel and biobutanol, biofertilizers, and amino acids (Das et al. 2019; Zhu et al. 2022).

The overall metabolic reactions of the anodic and cathodic chamber of MCC are expressed through Eqs. 15.4–15.7

In the anodic chamber:

$$CH_3COO^- + 2H_2O \rightarrow 2CO_2 + 7H^+ + 8e^- \tag{15.4}$$

In the cathodic chamber, light-dependent reaction:

$$nCO_2 + nH_2O \rightarrow (CH_2O)_n + nO_2 \tag{15.5}$$

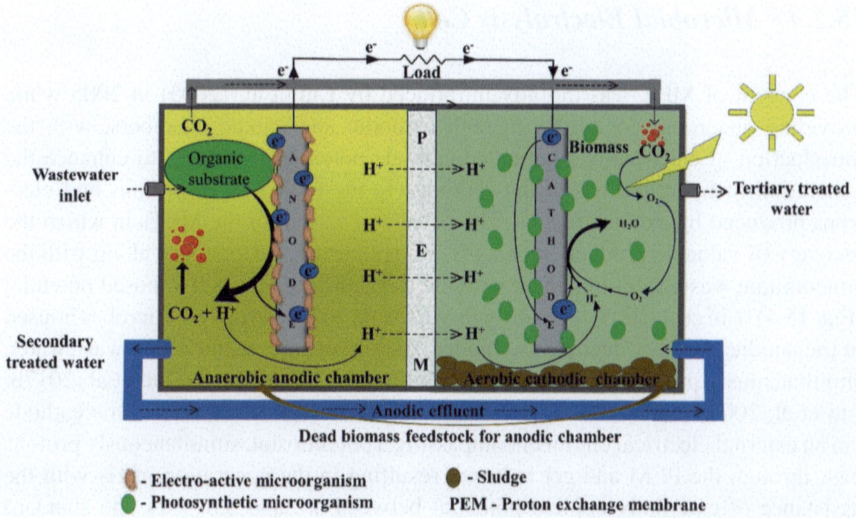

Fig. 15.3 Schematic of a microbial carbon capture cell

$$2O_2 + 8e^- + 8H^+ \rightarrow 4H_2O \tag{15.6}$$

In the cathodic chamber, light-independent reaction:

$$C_2H_4O_2 + 2O_2 \rightarrow 2CO_2 + 2H_2O \tag{15.7}$$

In recent times, MCC has emerged as an environmentally sustainable alternative to conventional MFC for their concurrent electricity generation, CO_2 sequestration, and generation of valuable by-products such as hydrogen, amino acids, ethanol, and biofertilizer during wastewater treatment (Das et al. 2019; Wang and Ren 2013). Furthermore, other significant advantages of MCC are reduction in CO_2 emission and non-requirement of aeration (Das et al. 2019). Although significant progress has been made on MCC, there are still many significant limitations associated with this technology including, low carbon sequestration rate, energy efficiency, and low yield of by-products, etc. that confines its field-scale application (Kumar and Jujjavarappu 2023). Moreover, high fabrication cost of MCCs is an important parameter that hinders the commercialization of MCCs. Thereby for successful field-scale implementation of MCCs, it is very essential to reduce capital costs along with the optimization of various operating factors that affect the performance and productivity of MCCs (Kumar and Jujjavarappu 2023; Yaashikaa et al. 2023).

15.2.4 *Microbial Electrolysis Cell*

The concept of MEC was initially introduced by Liu et al. (2005) in 2005 while providing anaerobic conditions to both cathodic and anodic chambers, with the introduction of electrically augmented cathode potential in a MFC to enhance the generation of hydrogen gas at the cathode via the reduction of protons and electrons produced by exoelectrogens. The MEC is a variant of the METs in which the recovery of valuable products such as H_2, CH_4, acetate, etc., is made along with the concomitant wastewater treatment with the assistance of minimal imposed potential (Fig. 15.4) (Liu et al. 2005). Primarily in MECs, the exoelectrogenic microbes housed in the anodic chamber degrade the biodegradable organic matter of the wastewater, simultaneously producing protons, electrons and CO_2, electrons (Guo et al. 2017b; Liu et al. 2005). These electrons subsequently flow from the anode to the cathode via an external electrical circuit encompassing a potentiostat, simultaneously protons pass through the PEM and get reduced, resulting in the generation of H_2 with the assistance of externally applied potential between 0.2 and 0.8 V vs. the standard hydrogen electrode (SHE) (Guo et al. 2017b). Additionally, CH_4 can be generated in the cathodic chamber of MEC either through the proliferation of hydrogenotrophic methanogens or through the, particularly when the cathodic chamber is operated under biotic conditions with organic substrate (Priyadarshini et al. 2022a). Apart from the production of CH_4 and H_2 at the cathodic chamber, the diverse applications of MECs include remediation of dyes, micropollutants, aromatic and chlorinated compounds, endocrine-disrupting compounds, xenobiotics, and heavy metals present in wastewater (Chakraborty et al. 2020b).

The MECs offer several significant advantages compared to conventional wastewater remediation technologies, including adsorption, MBR, AOPs, ozonation, electrocoagulation, and ASP, including the recovery of a range of value-added chemicals, including methane, hydrogen, acetate, ethanol, formic acid, and hydrogen peroxide, etc., and utilization of self-regenerating biocatalysts, reduced sludge generation, and

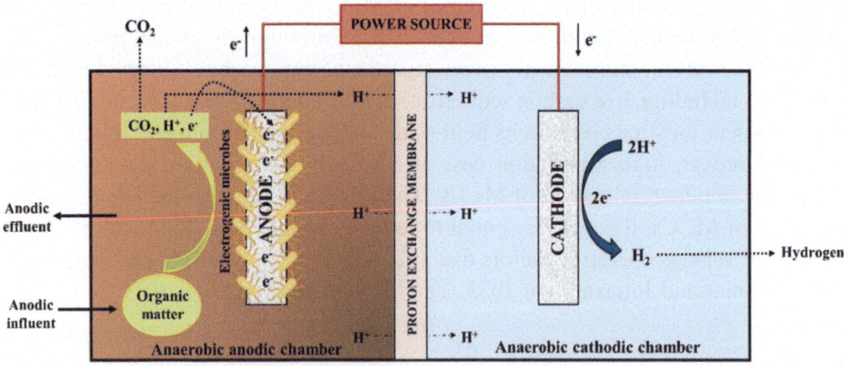

Fig. 15.4 Schematic of a microbial electrolysis cell reactor

low operational costs (Zhang and Angelidaki 2014). The bottlenecks associated with this technology include the low yield of value-added by-products and the necessity for externally applied power, which pose obstacles to scalability and replicability of MECs (Zhang and Angelidaki 2014). The technology of MECs holds the promise of providing long-term sustainability towards wastewater treatment. Therefore, to address these challenges, future research should focus on the optimization of this technology in terms of operational parameters like electrode material, initial contaminants of pollutants, applied potential, etc., which should be explored for mitigation of the above-mentioned challenges (Chakraborty et al. 2020b; Zhang and Angelidaki 2014).

15.2.5 Microbial Remediation Cell

The MRC is a green and eco-friendly derivate of MET that utilizes the biocatalytic activity of exoelectrogenic microbes in the anodic chamber of a MET for the enhanced degradation of pollutants from different environmental matrices including soil, groundwater, and wastewater and simultaneously generating electricity (Ahmad et al. 2022, 2023). Based on the reactor configuration, MRC can consist of two enclosed chambers, the anodic and cathodic chambers, separated by a PEM or a salt bridge (Ahmad et al. 2023). Further, the distinctive feature of MRC is that it utilizes single or multiple electrodes within the reactor setup for enhanced mineralization of recalcitrant pollutants in the anodic chamber via stimulating the electroactive microbes and increased electron transfer rate (Dummi Mahadevan and Zhao 2017; Wang and Ren 2013). In the anodic chamber of MRC, organic pollutants are oxidized via electroactive microbes, simultaneously generating protons, electrons, and CO_2 (Chakraborty et al. 2020b). Consequently, protons and electrons are then transferred to the cathodic through PEM and external circuit, respectively (Fig. 15.5) (Chakraborty et al. 2020b). Simultaneously, in the cathodic chamber of the MET, electrons and protons oxidize pollutants such as U(VI), Cr(VI), perchlorate, chlorinated solvents, etc. leading to their decontamination (Ahmad et al. 2023; Dummi Mahadevan and Zhao 2017). Additionally, microorganism of the cathodic chamber plays an essential role in reducing various pollutants such as metals and chlorinated solvents into less toxic forms (Ahmad et al. 2023).

Over recent decades, MRC has gained significant attention as a sustainable alternative to conventional wastewater treatment technologies such as ASP, trickling filter, SBR, etc. due to low treatment cost, low to minimal sludge generation, low emission of greenhouse gases and simultaneous bioelectricity production (Ahmad et al. 2023). Also, MRC can effectively remediate a wide range of pollutants, including, hydrocarbons, azo dyes, petroleum waste, perchlorates, radionuclides, and chlorinated carbons, to name a few. Additionally, a wide range of pollutants can be degraded in the both anodic and cathodic compartments of MRC (Ahmad et al. 2023). Although MRC provides various advantages as mentioned previously, still there are several limitations associated with MRC, for instance, long-term performance limitations

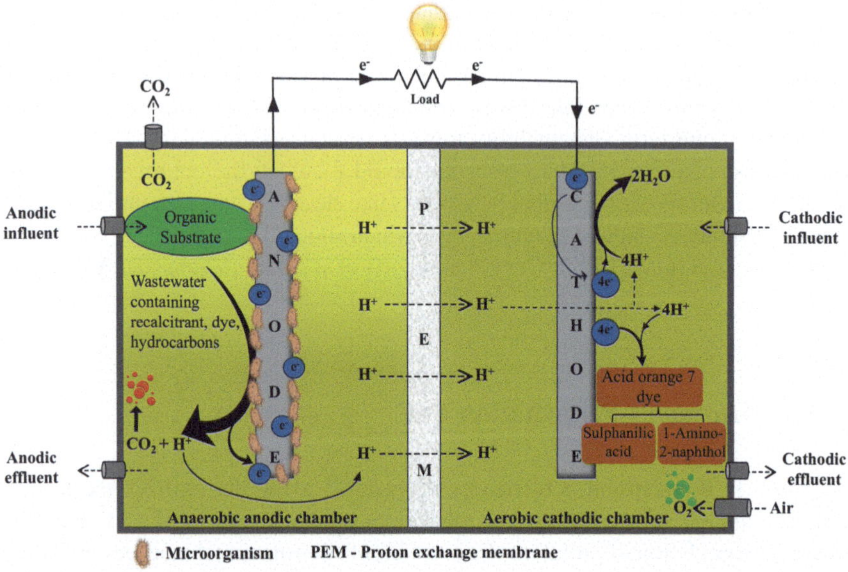

Fig. 15.5 Schematic of a microbial remediation cell

due to inconsistency, instability, and comparatively low voltage output (Do et al. 2018). Also, the high capital cost of MRC system is an important factor that restricts their field-scale application. Additionally, in some cases, the cathodic degradation of certain pollutants can produce toxic by-products, which further need downstream treatment for the effective mineralization of toxic by-products, thus increasing the treatment cost. Moreover, the anodic microbial community is vulnerable to high concentrations of toxic metals, which may lead to a change in the microbial dynamics and, ultimately, a reduction in degradation efficiency. Therefore, the successful transition of MRC from the lab to the field requires a reduction of fabrication cost by utilization of economically viable electrodes, catalysts, and membranes. Additionally, further research and development of MRC is required to enhance the voltage output by reducing various overpotential losses. Moreover, pre-treatment of effluents containing high concentrations of toxic metals is required prior to its treatment in MRC. Furthermore, investigation on microbial reactions at the genetic level could help to enhance the micropollutant removal efficiency of MRC by identifying the optimal operating conditions.

15.3 Factors Affecting the Removal Efficiency of Micropollutants in Microbial Electrochemical Technologies

The degradation efficiency of micropollutants in METs significantly depends on factors such as the initial concentration and nature of the micropollutants and operational parameters such as electrode material, microbes present as biocatalysts in an anodic chamber, and applied voltage (Du et al. 2007; Logan et al. 2006). Optimization of these factors is crucial for enhancing the performance and sustainability of METs. Thereby, this section provides a detailed discussion on the effects of the aforementioned parameters on the removal of micropollutants through METs.

15.3.1 Effect of Electrode Material on the Removal Efficacy of Micropollutants

The electrode material significantly influences the performance of MET, as it directly affects microbial adhesion to the electrode, electron transfer and electrochemical efficacy of the system (Mier et al. 2021). Generally, stainless steel is predominantly used as an anode material due to the uniform biofilm attachment on its surface and performance efficiency in removing organic pollutants (Köktaş and Gökkuş 2022; Mier et al. 2021). Furthermore, various carbon-based electrodes, including carbon felt, carbon paper, carbon cloth, carbon fiber, reticulated verified carbon, and carbon mesh, have also been explored as potential electrode materials due to their abundance, low toxicity, affordability, ease of processing, versatility in forming various shapes, high electrical conductivity, electrocatalytic activity, stability with microbial inoculums, biocompatibility, functioning in wide potential range and adjustable porosity (Ahmadi et al. 2023; Mier et al. 2021; Paul et al. 2018). Moreover, few researchers have explored the application of modified electrodes for enhanced electrochemical behaviour and enriched anodic microbial community by infusing carbon electrodes with metal or metal oxide, or by modifying graphite anodes with the coating of active polymers, electron mediators, quinone, and polyaniline groups (Gajda et al. 2018; Rhee et al. 2004). In this context, Xu et al. (2018) investigated the effect of metal and metal oxide modified anodes for removing pharmaceutical compounds with simultaneous power generation through MFCs. The study demonstrated that the modified electrodes generated higher power density and illustrated better removal efficacy for pharmaceuticals such as diclofenac, ibuprofen, and carbamazepine, with a notable increase in removal efficiency of 75%, 33.2% and 10.7%, respectively, as compared to non-modified control anode (Xu et al. 2018). This enhanced degradation of pharmaceuticals could be attributed to the enriched microbial communities, unique metal and metal oxide catalytic functions, minimized internal resistance, and ameliorated electron transfer (Xu et al. 2018). Additionally, augmenting the available surface area of electrodes provides additional space for microbial growth and attachment, enhancing electron transfer rate and further enhancing the pollutant degradation and energy generation from METs (Zhang et al. 2012). Therefore, it can be inferred that modified electrodes contribute to improved electron transfer efficiency by enhancing the active surface area and conductivity (Xu et al. 2018). These modified electrodes also aid in redox reactions and, hence, mineralizing micropollutants into harmless by-products (Xu et al. 2018; Zhang et al. 2012).

15.3.2 *Effect of Biocatalyst on the Removal Efficacy of Micropollutants*

The exoelectrogenic microbes of the anodic chamber of the MET act as biocatalysts, significantly influencing the degradation efficiency of the micropollutants (Ahmad et al. 2022). These exoelectrogenic microbes transform the chemical energy of biodegradable organic substrates of the wastewater into electrical energy and valuable resources by utilising both pure and mixed culture inoculums, which facilitates the mineralization of ECs and other organic contaminants from wastewater, respectively (Ahmad et al. 2022; Chatterjee et al. 2019). In a study, Feng et al. (2016) explored the effect of microbial diversity, especially electroactive microbial growth, on METs start-up time. The study showed that adding 0.001 M and 0.01 M concentrations of quaternary ammonium compound (QAC) in anolyte while treating real wastewater accelerated METs start-up time with a reduction of 20% and 29% in the start-up time, respectively compared to the control. This reduction in start-up time is likely due to the strong hydrogen and covalent bonding interactions between the anode surface and microbes, as the addition of QAC in the system transforms the hydrophobic anodic surface into hydrophilic (Feng et al. 2016). Furthermore, in an investigation, Arkatkar et al. (2020) explored the bioelectrochemical behaviour of a sequentially introduced biocatalytic co-culture comprising *Alcaligenes faecalis*, *E. coli* and *Pseudomonas aeruginosa* in an anodic chamber of a MFC with two distinct combinations of: *A. faecalis* and P. *aeruginosa* (b) *E. coli* and *A. faecalis*. The study showed that the biocatalytic co-culture exhibited a syntrophic relationship, increasing the power generation to $207.593 \pm 1.705\ \mu W/m^2$ with the combination of *E. coli* and *P. aeruginosa*. Hence, this study proves that adding an exoelectrogen to any non-exoelectrogenic bacterial culture enhances the power output of MFCs (Arkatkar et al. 2020). Thereby, these results illustrated that for the optimization of METs start-up time, power production and substrate degradation efficiency, various parameters such as inoculum selection, elimination of competing microbes, diversity of microbes and operational parameters for enrichment of electroactive bacteria's need to be explored further (Arkatkar et al. 2020; Feng et al. 2016).

15.3.3 *Effect of Type of Substrate on the Removal Efficacy of Micropollutants*

The removal of micropollutants in METs is significantly influenced by the type of substrate because in METs, substrates serve as the primary source of nutrients and energy for the microbes, and thereby, it is considered one of the most important biological factors for power generation (Pant et al. 2010). A wide range of organic substrates, including simple and pure organic compounds such as glucose and acetate to a complex amalgam of different organic compounds in wastewater, such as sewage, landfill leachate, brewery wastewater, and swine wastewater, have been globally used in METs due to their biodegradability (Pant et al. 2010). In a study, Feng et al. (2008) compared the power production in an air–cathode MFCs, utilizing domestic and brewery wastewater as substrates, with COD values ranging from 200 to 350 mg/L. The power output achieved with brewery wastewater as a substrate (an average power density of 58 mW/m^2) was approximately 60% lower compared to that of domestic wastewater (146 mW/m^2) (Feng et al. 2008). This can be attributed to the complexity of brewery wastewater, which contains diverse organic compounds and potential inhibitors such as ethanol and extracts of *Humulus lupulus*, affecting the microbial growth and electron transfer kinetics in MFC (Feng et al. 2008). Additionally, imbalances in nutrient availability and differences in microbial community composition may further contribute to reduced power generation compared to domestic wastewater (Feng et al. 2008). Similarly, Chae et al. (2009) investigated the impact of different substrates, namely, glucose, acetate, propionate, and butyrate on overall MFCs performance. The study reported that the acetate-fed MFC was most efficient for electricity generation, achieving the maximum CE of 72.3%, followed by butyrate, propionate, and glucose with CE of 43.0%, 36.0%, and 15.0%, respectively (Chae et al. 2009). Glucose yielded the lowest CE owing to its fermentable nature, which resulted in the utilisation of glucose by various non-electrogenic bacteria. Thereby, these results highlight the significance of choosing the right substrate in METs as it directly impacts microbial metabolism and electron transfer.

15.3.4 *Effect of Externally Applied Voltage on the Removal Efficacy of Micropollutants*

Externally applied voltage is an essential parameter that significantly affects the degradation efficiency of micropollutants in METs (Guo et al. 2017b; Priyadarshini et al. 2022b). Considering the same, Yang et al. (2018) explored the effect of applied potential on the removal efficiency of 20 mg/L sulfadiazine (SDZ) in MEC, employing graphite as an anode, active carbon fiber, and a stainless steel ring with conductive adhesive as a cathode. The investigation elucidates that the removal of SDZ significantly improved from 79.3% to 91.9% as the applied cell voltage increased from 0 V to 1.5 V, respectively, within 36 h of treatment duration (Yang et al. 2018). Similarly, in an investigation, Wen et al. (2013) reported an enhanced degradation efficiency of 1 mg/L 4-chlorophenol from 61.8% to 92.5% with an increase in the applied potential from 0.2 to 0.7 V within 24 h of treatment in an MEC, utilizing carbon felt as anode and Pd catalyst containing graphite sheet, as cathode (Wen et al. 2013). Further, in a research, Guo et al. (2017a, b) reported an enhanced degradation efficiency of 50 mg/L of chloramphenicol (CAP) from 70 to 96% with the application of higher negative cathode potential from −0.5 V to −1.25 V versus SHE, within 72 h of treatment in a MEC, utilizing carbon-graphite as electrodes (Guo et al. 2017a). Furthermore, an enhanced mineralization efficiency of aniline from 66.8 ± 3.1% to 93.1 ± 1.2% was achieved by increasing the applied cell voltage from 0.3 to 0.5 V in a bipolar membrane MEC-Fenton system, employing carbon fiber brush and graphite plate as anode and cathode, respectively, within six days of treatment (Li et al. 2017). Additionally, the enhanced mineralization rate at 0.5 V of applied cell voltage could be ascribed to increased production of H_2O_2 at the cathodic chamber (Li et al. 2017). Thus, externally applied voltage can significantly influence the degradation efficiency of micropollutants in METs via various possible ways such as in MECs, utilization of suitable applied cell voltage can trigger bacterial activity and enhance biofilm formation (Li et al. 2017; Yang et al. 2018). Additionally, in MEC, application of more negative cathode potential can produce ample electrons and change the microbial community for enhanced degradation of pollutants in the system. Moreover, in the MEC-Fenton system, the mineralization efficiency of pollutants is controlled by applied voltage since the generation of hydroxyl radicals is controlled by the applied voltage (Ahmad et al. 2022; Guo et al. 2017a; Yang et al. 2018).

15.3.5 Effect of Initial Concentration and Characteristics of the Pollutant on the Removal Efficacy of Micropollutants

The degradation of ECs in METs greatly depends on various factors, including the initial concentration of the pollutant and its properties (Miran et al. 2017). Based on the properties of the pollutant, METs are more capable of treating wastewater containing readily degradable organic matter compared to wastewater containing recalcitrant, xenobiotic, and biorefractory contaminants mainly due to their toxicity to microorganisms and poor biodegradability (Raj et al. 2023a; Wang et al. 2020). Considering that, Liu et al. (2013) explored the impact of the initial concentration of *p*-nitrophenol (PNP) on its removal efficiency in a dual-chamber MFC, utilizing kryptol as electrodes. The study elucidates that an increase in the concentration of PNP from 50 to 300 mg/L decreases the PNP degradation percentage from 64.69 to 33.67% (Liu et al. 2013). This reduction in the degradation efficiency of PNP is attributed to its toxicity, as PNP inhibits the proliferation of exoelectrogenic microbes in an anodic chamber and ultimately reduces the microbial activity (Liu et al. 2013). Similarly, Miran et al. (2017) studied the effect of initial concentrations of 4-chlorophenols (4-CP) on its degradation efficiency in a dual-chamber MFC, utilizing a carbon felt as anode and carbon cloth as cathode, respectively. According to the investigation, with the initial concentration of 4-CP from 10 to 100 mg/L decreases its degradation rate from 74.2 to 50% within 48 h of operation (Miran et al. 2017). This decrease in the removal efficiency of 4-CP is primarily due to its toxicity to the anodic exoelectrogenic microorganisms, which adversely affects the biodegradation ability of these microorganisms (Miran et al. 2017). In another exploration, Ondon et al. reported a decrease in the removal efficiency of antibiotic norfloxacin (NFLX) from 65.5 to 48.4% while increasing the initial concentration of NFLX from 4 mg/L to 128 mg/L in a single-chamber MFC, utilizing carbon cloth and Pt/C as anode and cathode, respectively (Ondon et al. 2020). Therefore, from all the above results it is evident that the degradation rate of micropollutants in METs decreases with an increase in influent concentration due to an increase in toxicity to the microbes (Liu et al. 2013; Miran et al. 2017).

15.4 Application of Microbial Electrochemical Technologies for the Removal of Micropollutants from Wastewater: Mechanism of Pollutants Removal

In recent times, various kinds of micropollutants have been effectively removed from wastewater via the utilization of METs and the same has been discussed below (Table 15.1).

15.4.1 Removal of Micropollutants Through Microbial Fuel Cell

Over the last few decades, both single-chamber and dual-chamber MFCs have been utilized by several researchers for the elimination of various types of micropollutants from wastewater with considerable efficiency (Ahmad et al. 2022; Priyadarshini et al. 2022a). In this regard, a study by Thung et al. (2015) demonstrated 80.6% degradation efficiency of 50 mg/L acid orange 7 (AO7) while achieving 8.57 $\pm$ 0.2 mW/m^2 of power generation in a single-chamber up-flow membrane-less MFC, utilizing carbon felt and carbon plate as anode and cathode, respectively. Additionally, it was also observed that an increase in the initial dye concentration from 50 to 75 mg/L leads to reduced power production of 7.07 $\pm$ 1.4 mW/m^2 (Thung et al. 2015). This indicates that both azo dye molecules and anode compete for electrons generated from the degradation of the organic substrate (Thung et al. 2015). Similarly, in another investigation, Oon et al. (2018) achieved 83% decolourization efficiency of 25 mg/L

Table 15.1 Degradation of micropollutants via different microbial electrochemical technologies

Treatment technology	Pollutant	Initial concentration (mg/L)	Removal efficiency for the treatment period	Observation	References
MFC	RR120	100	100% in 72 h	*Staphylococcus equoruma RAP2* bacterial strain was isolated from a wastewater treatment plant and utilized in the anodic chamber of MFC for degradation of RR120	(Shahi et al. 2021)
	Fipronil	Approximately 30	79% in 12 h	Apart from fipronil, MFC was also utilized for the efficacious degradation of other refractory aromatic compounds, namely, 4 chloronitrobenzene, fluoroglycofen, sulfanilamide, and azoxystrobin with a degradation efficiency of 93%, 84%, 74% and 65%, respectively	(Zhang et al. 2019)
	Oxyfluorfen	50	77% in 24 h	Microbial community analysis of anodic biofilm identified the prominent functional bacteria genera namely, *Azonexus, Arcobacter, Comamonas Azospirillum,* and *Acinetobacter* were the prominent functional bacteria genera found in the anodic biofilm	(Zhang et al. 2018)

(continued)

Table 15.1 (continued)

Treatment technology	Pollutant	Initial concentration (mg/L)	Removal efficiency for the treatment period	Observation	References
BEF	Metoprolol	0.01	95% in 6 h	Maximum removal efficiency was achieved with an applied voltage of 0.2 V, pH 3, and 11.168 mg/L of Fe^{2+}	(Yang et al. 2021)
	MP	10	85.70 ± 3.30% in 4 h	MP was degraded from real wastewater	(Raj et al. 2023b)
	TC	50	71.4 ± 0.4% in 2.5 h	Maximum removal of TC from stimulated wastewater was achieved at a pH of 2.0 with 16.75 mg/L of Fe^{2+}. Also, it was found that acidic condition is favourable for the efficacious removal of TC, as acidic condition promotes the H_2O_2 production, and subsequently, it reacts with Fe^{2+} to produce hydroxyl radicals, which ultimately degraded the TC	(Song et al. 2022)
MEC	Sulfamethoxazole	30	92.53% ± 1.5% in 72 h	The end product of sulfamethoxazole degradation was identified as methane and aniline	(Xue et al. 2019)

(continued)

Table 15.1 (continued)

Treatment technology	Pollutant	Initial concentration (mg/L)	Removal efficiency for the treatment period	Observation	References
	Erythromycin	20	99.02% in 48 h	ermG was the prominent antibiotic resistance gene found in the anodic biofilm and effluent	(Hua et al. 2019)
	2C4NP	30	100% in 72 h	Enhanced degradation of 2C4NP was observed with an applied voltage of 0.5–0.75 V	(Yu et al. 2020)
MRC	Diclofenac	5	75.59% in 48 h	Ru-Fe-modified anode was utilized for Diclofenac removal	(Qiu et al. 2020)
	1,2-DCA	102	95% in 14 days	Anodophilic microbial consortia were utilized for the effective removal of 1,2-DCA	(Pham et al. 2009)
CW-MFC	PCL	5	97.1% in 48 h	Maximum removal of PCL and DCF was achieved through utilization of duckweed plant (*Lemna minor*) in CW-MFC	(Jain et al. 2023)
	DCF	5	87.5% in 48 h		
	SMX	0.06	98.6% in 48 h	Higher degradation efficiency of SMX and TC was achieved via the utilization of 200 mg/L of glucose as a co-substrate	(Wen et al. 2021)
	TC	0.025	99.2% in 48 h		
	CR	50 ± 10	89.99 ± 0.03% in 24 h	Low-cost earthen separator was used as a proton exchange membrane for the removal of CR	(Saket et al. 2022)

* MFC—Microbial fuel cell; RR120—Reactive Red 120; BEF—Bio-electro-Fenton; MP—Methylparaben; TC—Tetracycline; MEC—Microbial electrolysis cell; 2C4NP—2-chloro-4-nitrophenol; MRC—Microbial remediation cell; 1,2-DCA—1,2-dichloroethane; CW-MFC—Constructed wetland coupled microbial fuel cell; PCL—Paracetamol; DCF—Diclofenac; SMX—Sulfamethoxazole; TC—Tetracycline; CR—Congo red

New Coccine within a treatment duration of one day while achieving a power output of 20.13 ± 0.37 mW/m^3 in a double-chamber MFC, employing anaerobically and aerobically inoculated activated carbon flakes as cathode and anode.

In addition to the above-discussed azo dyes, POPs, such as phenol, are frequently detected in industrial wastewater due to its widespread use in the petrochemical, pharmaceutical, chemical, coking, and other sectors. Moreover, the bioaccumulation and biomagnification of phenol in the food chain results in increased concentration of phenol in higher trophic levels of the food chain, consequently posing a significant health risk to aquatic and human life and hence, the removal of phenol from wastewater is essential and the same has been demonstrated through MFCs (Luo et al. 2009). Considering that, in an investigation, Luo et al. (2009) reported phenol degradation efficiency greater than 95% within 60 h of treatment through the utilization of 400 mg/L of phenol as substrate in a dual-chamber MFC, employing carbon paper and platinum-coated carbon paper as anode and air–cathode, respectively. Similarly, other major micropollutants, such as pesticides and pharmaceuticals, have also been effectively removed through MFCs, for instance, in an investigation, Cao et al. (2015) reported 71.15% removal of hexachlorobenzene (HCB) pesticide along with 77.5 mW/m^2 of power production with an applied external resistance of 1000 Ω in a soil MFC, utilizing granular activated carbon layer as electrodes. Furthermore, in recent years, various pharmaceuticals, such as anti-inflammatory drugs, beta-blockers, antibiotics, and antidiabetics, were reported to be present in wastewater, and considering the same, several investigations were carried out exploring the degradation of these pharmaceuticals in MFCs. In this regard, a study by Ondon et al. (2020) reported 65.5% removal of 4 mg/L norfloxacin (NFX) in a single-chamber MFC utilizing carbon mesh and Pt/C as anode and cathode, respectively. Thus, the above-discussed example elucidates that MFC can be utilized for the degradation of micropollutants with simultaneous energy harvesting (Cao et al. 2015; Luo et al. 2009).

15.4.2 Removal of Micropollutants Through Microbial Electrolysis Cell

In recent years, the global focus on developing sustainable technologies has led to significant advancements in alternative energy resources and the "waste to energy" concept. One such advancement is MEC, a cutting-edge technology that enables the simultaneous production of biohydrogen and wastewater treatment (Chakraborty et al. 2020b). Derived from MFCs, MECs feature an anaerobic cathode and operate with an externally applied voltage, facilitating hydrogen production as well as offering diverse applications such as the production of CH_4, remediation of micropollutants, dyes, aromatic and chlorinated compounds, endocrine-disrupting compounds, xenobiotics, and heavy metals present in wastewater (Chakraborty et al. 2020b). In this regard, Zhao et al. (2024) explored the remediation of antibiotic namely sulfaquinoxaline through MEC using sodium acetate as co-substrate, reporting 94.2% degradation efficiency of sulfaquinoxaline. The analysis of biodegradation products was performed in the study, which proved that the degradation mechanism primarily involved hydroxylation, with sodium acetate likely facilitating the breaking of C-N bonding in nitrogen-containing heterocycles, thereby initiating hydroxylation and accelerating sulfaquinoxaline degradation. Similarly, Min et al. (2024) reported 94.5 ± 3.2% removal of polycyclic aromatic hydrocarbon with methane yield of 227 ± 2 mL/g COD using MEC. Furthermore, Hua et al. (2019) investigated the removal of erythromycin through MEC using carbon cloth anodes, reporting approximately 99% remediation of erythromycin within 48 h. Thereby, the aforementioned studies demonstrated that the MEC stands as a promising technique capable of efficiently remediating micropollutants and recovering value-added products from wastewater.

15.4.3 Removal of Micropollutants Through Microbial Remediation Cell

Over the last few decades, increment in water pollution has resulted in accumulation of various kinds of pollutants such as nutrients, heavy metals, organic matter, micropollutants, and many more in natural water resources (Ahmad et al. 2023). Additionally, conventional wastewater treatment technologies such as ASP, sequencing batch reactor, trickling filter, are inefficient in the removal of micropollutants from wastewater, which can be attributed to the recalcitrant nature of micropollutants. Thereby, over the recent decades, several researchers have utilized MRCs for efficacious degradation of micropollutants from wastewater. For instance, in an investigation, Zhong and co-workers explored the degradation of 30 mg/L of methyl orange dye (MO) in the anodic chamber of MRC, utilizing NiO/MnO_2-carbon felt and carbon felt as anode and cathode, respectively (Zhong et al. 2018). The results revealed MO degradation efficiency of 92.5% within a treatment time of 48 h, along with generating a maximum power density of 0.628 W/m^2 (Zhong et al. 2018). Similarly, in an investigation, Cao et al. (2010) achieved 98% decolourization of 300 mg/L of Congo red dye while utilizing glucose as co-substrate in the anodic chamber of an air–cathode single-chamber MRC, employing carbon paper as anode during 36 h of retention time. In addition to azo dyes, various organic pollutants such as aromatic amines are also effectively removed through MRCs. For instance, in a study, Cheng et al. (2015) achieved 91.2 ± 2.2% degradation of aniline with an initial concentration of 260.76 ± 9.31 mg/L fed in the anodic chamber of an MRC employing carbon brushes and carbon fibers as anode and cathode, respectively, during 144 h of retention time. Furthermore, cathodic chamber is also simultaneously utilized by researchers for the efficacious degradation of pollutants such as U(VI), Cr(VI), perchlorate, chlorinated solvents, etc. from wastewater. For instance, in an investigation, Shea et al.

(2008) achieved 96% removal of perchlorate with influent perchlorate concentration of 10 mg/L in the denitrifying bio-cathodic chamber of MRC, utilizing graphite granules as electrodes. Thereby, from the above-discussed studies, it is evident that both the anodic and cathodic chambers of MRC can be employed for effective degradation of micropollutants from various wastewater streams (Ahmad et al. 2023).

15.4.4 Removal of Micropollutants Through Hybrid Microbial Electrochemical Systems

The METs offer a novel approach for treating wastewater containing persistent micropollutants with concurrent resource recovery. However, field-scale applications face challenges due to the variety of micropollutants present in wastewater, making standalone treatment difficult. To address these limitations, researchers have integrated METs with other technologies, such as constructed wetlands, electro-Fenton processes, adsorption, and MBR, resulting in more cost-effective remediation of micropollutants and performance enhancement of the system (Ahmad et al. 2022). For instance, MECs require external power for pollutant remediation, which can be addressed by utilizing the power generated from MFCs (Okabe 2021). Furthermore, in an investigation, Xu et al. (2024) explored the removal of an antibiotic namely, sulfamethoxazole through a constructed wetland coupled with MFC (CW-MFC) using Mn ore as the bedding media near the electrode. In the investigation approximately 91.33% degradation of 0.26 mg/L of sulfamethoxazole was achieved in the CW-MFC, utilizing Mn ore as an electron acceptor, enhancing the growth of specific exoelectrogenic bacteria capable of transferring electrons from inside of their cell to the external anode in the MFC (Xu et al. 2024). Thereby, CW-MFCs demonstrate immense potential in enhancing the degradation of micropollutants through controlled electron transfer to conductive electrodes. In addition to antibiotics, surfactants are also efficiently removed through hybrid METs. For instance, Sathe et al. (2020) explored the removal of sodium dodecyl sulfate through a photo-cathode MFC and laterite-based adsorption hybrid treatment system. The study showed that the hybrid system achieved approximately 96.3% removal of sodium dodecyl sulfate with a treatment time of 24 h and an initial concentration of 10 mg/L, offering an effective solution for the degradation of micropollutants in wastewater, contributing to the mitigation of surfactant-related adverse effects in the water bodies (Sathe et al. 2020). Thereby, these results illustrated that hybrid METs present a promising and cost-effective solution for sustainable wastewater management with efficient enhancement in the degradation of micropollutants.

15.5 Limitations in the Removal of Micropollutants Through Microbial Electrochemical Technologies

The METs are a group of noble and environment-friendly technologies primarily utilized for wastewater treatment along with concurrent bioresource production. In recent decades, several researchers have utilized METs for the effective degradation of various micropollutants from wastewater; however, these technologies are majorly limited to lab-scale due to various limitations such as high capital costs, attributed to the utilization of expensive membranes, electrodes, and catalysts in the setups (Khandelwal et al. 2022). As a result, the average capital cost of MFC utilized for domestic wastewater treatment is 30 times higher than conventional ASP technology, which adversely affects the economic feasibility of MFCs (He et al. 2017). Hence to circumvent this drawback, researchers have employed cost-effective carbon-based electrodes namely, carbon sheet, carbon fiber, carbon cloth, graphite plate, and graphite rod, whereas these carbon-based electrodes lead to decreased power output and wastewater treatment efficiency due to hydrophobicity, poor conductivity and inefficient physicochemical stability of carbon-based electrodes (He et al. 2017; Khandelwal et al. 2022). In addition to high capital cost, poor power generation in MFC due to the poor ionic conductivity of the wastewater and higher internal resistance of the reactor components is a major hurdle for the upscaling of MFC (He et al. 2017). In METs such as MFC, biofouling of PEM due to the formation of biofilm at the PEM surface is a major constraint that significantly increases the resistance offered to mass transfer and ion transport and eventually reduces the life span of PEM subsequently decreasing the degradation efficiency of pollutants (Koók et al. 2019; Priyadarshini et al. 2022a). Additionally, the conventional polymer-based membranes that are applied in lab-scale reactors can't withstand high hydrostatic pressure at large-scale MECs, thereby limiting their field-scale application. Moreover, the degradation of pollutants in METs is significantly influenced by external environmental factors such as temperature, since the bacterial metabolic activities, growth, and biochemical reactions associated with micropollutant removal are influenced by temperature (Yan et al. 2019).

In the past few decades, several investigations have demonstrated that the degradation efficiency of micropollutants in METs is significantly influenced by the initial concentration and properties of the pollutants (Ahmad et al. 2022). Increased influent concentration in METs partially reduces the removal efficiency of micropollutants, as increased influent pollutant concentration in the anodic chamber of METs leads to inhibition of microbial growth, consequently decreasing the removal efficiency (Ahmad et al. 2022). Additionally, METs are more capable of degrading readily biodegradable organic matter from wastewater compared to biorefractory, xenobiotic and recalcitrant micropollutants from wastewater due to low-biodegradability and toxicity to electroactive microorganisms (Raj et al. 2023a; Wang et al. 2020). Thereby, the commercialization and upscaling of METs for the effective degradation of micropollutants from wastewater need further research and development addressing the above-discussed bottlenecks, such as the development of cost-efficient, durable, and stable electrodes, catalysts, and membranes. Moreover, future research should emphasize enhancing power generation by minimizing the ohmic, kinetic, and transportation resistance (Mahmoud et al. 2022). Furthermore, the development of hybrid wastewater treatment technology via utilizing other wastewater treatment technologies, such as AOPs along with METs could overcome the limitations associated with the degradation of micropollutants through METs.

15.6 Future Perspectives

The role of METs in removing micropollutants from wastewater has exhibited promising potential; whereas, scalability and commercialization of these technologies are still challenging owing to the requirement of expensive electrode material, electrocatalyst, PEM, high energy consumption and lower yield of value-added products that need to be addressed (Priyadarshini et al. 2022a). Therefore, future research should aim to overcome these challenges by developing durable, stable, and low-cost materials for electrodes, membranes, and electrocatalysts. For instance, utilising porous ceramic membranes and waste-derived or sulfonated polymer membranes significantly reduces capital as well as operational cost of METs (Chakraborty et al. 2020a; Li et al. 2018). Additionally, the advancements in electrode materials, particularly waste-derived cathode materials facilitating in-situ production of H_2O_2 in METs can also be beneficial in curtailing the capital and operational cost of the METs (Noori et al. 2020; Priyadarshini et al. 2022a). Moreover, modifications of reactor configuration such as use of 3D electrodes over 2D electrodes and exploration of up-flow mode single-chamber designs address the mass transfer limitation of METs, further enhancing their efficacy (Ferreira 2008). Additionally, integrating METs with other wastewater treatment technologies, such as AOPs, constructed wetlands, and algal ponds, can provide comprehensive solutions for the complete mineralization of micropollutants (Olvera-Vargas et al. 2018; Wen et al. 2020). Along with this, further research is needed to focus on understanding the degradation pathways of the micropollutants and the toxicity profile of the intermediate by-products and the end products formed during micropollutant degradation via METs (Ahmad et al. 2022, 2023). Furthermore, understanding the microbial dynamics and their role in the biodegradation mechanisms of micropollutants is also essential for elucidating the performance of METs. Moreover, the development of sophisticated models to predict MET behaviour under various operating conditions, such as varying initial concentration of the pollutant, pH, anolyte, and operational time will aid in expediting field-scale applications of METs. Therefore, future research should aim to overcome these existing technological barriers, develop economically viable materials, optimize reactor configurations, and integrate METs with other treatment technologies for sustainable wastewater management.

15.7 Conclusion

Micropollutants, being toxic compounds, pose a significant challenge due to the ineffectiveness of conventional wastewater treatment technologies for their removal. However, as detailed in previous sections, METs have proved to be feasible, cost and energy-efficient solution for addressing this issue. In this regard, the present chapter explored different METs and their application for the removal of micropollutants present in wastewater. The METs exhibit dual advantages of micropollutant removal as well as valuable resource recovery of H_2, CH_4, acetate, etc., with added benefit of minimising the sludge production. However, the efficacy of degradation of micropollutants in METs is significantly influenced by various process parameters such as pollutant concentration, nature of microbes, and electrolyte. Furthermore, the removal of micropollutants in METs is also dependent on reaction kinetics of the system, degradation mechanisms, the associated pathway, and the formation of by-products. Therefore, future research should prioritize optimizing these parameters and reactor design, along with comprehensive life cycle and techno-economic assessments, to advance the field-scale applicability and sustainability of METs for treating wastewater containing micropollutants (Ahmad et al. 2022, 2023).

References

Ahmad A, Priyadarshani M, Das S, Ghangrekar MM (2022) Role of bioelectrochemical systems for the remediation of emerging contaminants from wastewater: a review. J Basic Microbiol 62(3–4):201–222

Ahmad A, Priyadarshini M, Ghangrekar MM (2023) Microbial remediation cell. In: Ghangrekar MM, Duteanu NM, Surampalli RY, Zhang TC (eds) Microbial electrochemical technologies: fundamentals and applications, vol 2. Wiley-VCH. New Jersey, pp 445–465

Ahmadi S, Rezae A, Ghosh S, Malloum A, Banach A (2023) A review on bioelectrochemical systems for emerging pollutants remediation: a computational approaches. J Environ Chem Eng 11(3):110021

Ali J, Wang L, Waseem H, Sharif HMA, Djellabi R, Zhang C, Pan G (2019) Bioelectrochemical recovery of silver from wastewater with sustainable power generation and its reuse for biofouling mitigation. J Clean Prod 235:1425–1437

Ansari K, Khan MA (2024) Introduction to micropollutants, 1 edn. In: Khalid M, Park Y, Karri RR, Walvekar R (eds) Advanced oxidation processes for micropollutant remediation. CRC Press. Boca Raton, pp 1–11

Arkatkar A, Mungray AK, Sharma P (2020) Bioelectrochemical behaviour of a sequentially added biocatalytic coculture in a microbial fuel cell. J Basic Microbiol 60(7):562–573

Barceló D, Picó Y, Alfarhan AH (2023) Microplastics: detection in human samples, cell line studies, and health impacts. Environ Toxicol Pharmacol 101:104204

Behera M, Rout PR, Bhunia P, Surampalli RY, Zhang TC, Kao CM, Ghangrekar MM (2023) Introduction to Microconstituents. In: Surampalli RY, Zhang TC, Kao CM, Ghangrekar MM, Bhunia P, Behera M, Rout PR (eds) Microconstituents in the environment: occurrence, fate, removal and management. John Wiley & Sons Ltd., New Jersey, pp 1–35

Bhatt P, Bhandari G, Bilal M (2022) Occurrence, toxicity impacts and mitigation of emerging micropollutants in the aquatic environments: recent tendencies and perspectives. J Environ Chem Eng 10(3):107598

Cao Y, Hu Y, Sun J, Hou B (2010) Explore various co-substrates for simultaneous electricity generation and Congo red degradation in air-cathode single-chamber microbial fuel cell. Bioelectrochemistry 79(1):71–76

Cao X, Song H-L, Yu C-Y, Li X-N (2015) Simultaneous degradation of toxic refractory organic pesticide and bioelectricity generation using a soil microbial fuel cell. Biores Technol 189:87–93

Chae K-J, Choi M-J, Lee J-W, Kim K-Y, Kim IS (2009) Effect of different substrates on the performance, bacterial diversity, and bacterial viability in microbial fuel cells. Biores Technol 100(14):3518–3525

Chakraborty I, Das S, Dubey B, Ghangrekar M (2020a) Novel low cost proton exchange membrane made from sulphonated biochar for application in microbial fuel cells. Mater Chem Phys 239:122025

Chakraborty I, Sathe S, Khuman C, Ghangrekar M (2020b) Bioelectrochemically powered remediation of xenobiotic compounds and heavy metal toxicity using microbial fuel cell and microbial electrolysis cell. Mater Sci Energy Technol 3:104–115

Chatterjee P, Dessì P, Kokko M, Lakaniemi A-M, Lens P (2019) Selective enrichment of biocatalysts for bioelectrochemical systems: a critical review. Renew Sustain Energy Rev 109:10–23

Cheng H-Y, Liang B, Mu Y, Cui M-H, Li K, Wu W-M, Wang A-J (2015) Stimulation of oxygen to bioanode for energy recovery from recalcitrant organic matter aniline in microbial fuel cells (MFCs). Water Res 81:72–83

Das S, Das S, Das I, Ghangrekar M (2019) Application of bioelectrochemical systems for carbon dioxide sequestration and concomitant valuable recovery: a review. Mater Sci Energy Technol 2(3):687–696

Do M, Ngo H, Guo W, Liu Y, Chang S, Nguyen D, Nghiem L, Ni B (2018) Challenges in the application of microbial fuel cells to wastewater treatment and energy production: a mini review. Sci Total Environ 639:910–920

Du Z, Li H, Gu T (2007) A state of the art review on microbial fuel cells: a promising technology for wastewater treatment and bioenergy. Biotechnol Adv 25(5):464–482

Dummi Mahadevan G, Zhao F (2017) A concise review on microbial remediation cells (MRCs) in soil and groundwater radionuclides remediation. J Radioanal Nucl Chem 314:1477–1485

Feng Y, Wang X, Logan BE, Lee H (2008) Brewery wastewater treatment using air-cathode microbial fuel cells. Appl Microbiol Biotechnol 78:873–880

Feng C, Liu Y, Li Q, Che Y, Li N, Wang X (2016) Quaternary ammonium compound in anolyte without functionalization accelerates the startup of bioelectrochemical systems using real wastewater. Electrochim Acta 188:801–808

Ferreira B (2008) Three-dimensional electrodes for the removal of metals from dilute solutions: a review. Miner Process Extr Metall Rev 29(4):330–371

Gajda I, Greenman J, Ieropoulos IA (2018) Recent advancements in real-world microbial fuel cell applications. Curr Opin Electrochem 11:78–83

Gude VG (2016) Wastewater treatment in microbial fuel cells–an overview. J Clean Prod 122:287–307

Guo N, Wang Y, Yan L, Wang X, Wang M, Xu H, Wang S (2017a) Effect of bio-electrochemical system on the fate and proliferation of chloramphenicol resistance genes during the treatment of chloramphenicol wastewater. Water Res 117:95–101

Guo Z, Thangavel S, Wang L, He Z, Cai W, Wang A, Liu W (2017b) Efficient methane production from beer wastewater in a membraneless microbial electrolysis cell with a stacked cathode: the effect of the cathode/anode ratio on bioenergy recovery. Energy Fuels 31(1):615–620

He L, Du P, Chen Y, Lu H, Cheng X, Chang B, Wang Z (2017) Advances in microbial fuel cells for wastewater treatment. Renew Sustain Energy Rev 71:388–403

Hua T, Li S, Li F, Ondon BS, Liu Y, Wang H (2019) Degradation performance and microbial community analysis of microbial electrolysis cells for erythromycin wastewater treatment. Biochem Eng J 146:1–9

Jain M, Kiran PS, Ghosal PS, Gupta AK (2023) Development of microbial fuel cell integrated constructed wetland (CMFC) for removal of paracetamol and diclofenac in hospital wastewater. J Environ Manage 344:118686

Khandelwal A, Vijay A, Jadhav D, Lens P, Swaminathan J, Ghosh PC, Chendake A (2022) Genesis and recent advancement in microbial fuel cells: Wastewater treatment and resource recovery perspectives, 1 edn. In: Mungray A, Mungray A, Sonawane S, Sonawane S (eds) Novel approaches towards wastewater treatment and resource recovery technologies. Elsevier, Amsterdam, pp 23–36

Kim M-K, Zoh K-D (2016) Occurrence and removals of micropollutants in water environment. Environ Eng Res 21(4):319–332

Kim BH, Park DH, Shin PK, Chang IS, Kim HJ (1999) Mediator-less biofuel cell, (ed.) U.S. Patent, Vol. USOO5976719A, Korea Institute of Science and Technology, Seoul, Rep. of Korea. United State, p 18

Köktaş İY, Gökkuş Ö (2022) Removal of salicylic acid by electrochemical processes using stainless steel and platinum anodes. Chemosphere 293:133566

Koók L, Bakonyi P, Harnisch F, Kretzschmar J, Chae K-J, Zhen G, Kumar G, Rózsenberszki T, Tóth G, Nemestóthy N (2019) Biofouling of membranes in microbial electrochemical technologies: causes, characterization methods and mitigation strategies. Biores Technol 279:327–338

Kumar T, Jujjavarappu SE (2023) A critical review on an advanced bio-electrochemical system for carbon dioxide sequestration and wastewater treatment. Total Environ Res Themes 5:100023

Li X, Jin X, Zhao N, Angelidaki I, Zhang Y (2017) Efficient treatment of aniline containing wastewater in bipolar membrane microbial electrolysis cell-Fenton system. Water Res 119:67–72

Li X, Chen S, Angelidaki I, Zhang Y (2018) Bio-electro-Fenton processes for wastewater treatment: advances and prospects. Chem Eng J 354:492–506

Liu H, Grot S, Logan BE (2005) Electrochemically assisted microbial production of hydrogen from acetate. Environ Sci Technol 39(11):4317–4320

Liu H, Hu T, Zeng G, Yuan X, Wu J, Shen Y, Yin L (2013) Electricity generation using p-nitrophenol as substrate in microbial fuel cell. Int Biodeterior Biodegradation 76:108–111

Logan BE, Rabaey K (2012) Conversion of wastes into bioelectricity and chemicals by using microbial electrochemical technologies. Science 337(6095):686–690

Logan BE, Hamelers B, Rozendal R, Schröder U, Keller J, Freguia S, Aelterman P, Verstraete W, Rabaey K (2006) Microbial fuel cells: methodology and technology. Environ Sci Technol 40(17):5181–5192

Luo H, Liu G, Zhang R, Jin S (2009) Phenol degradation in microbial fuel cells. Chem Eng J 147(2–3):259–264

Luo Y, Guo W, Ngo HH, Nghiem LD, Hai FI, Zhang J, Liang S, Wang XC (2014) A review on the occurrence of micropollutants in the aquatic environment and their fate and removal during wastewater treatment. Sci Total Environ 473:619–641

Mahmoud RH, Gomaa OM, Hassan RY (2022) Bio-electrochemical frameworks governing microbial fuel cell performance: technical bottlenecks and proposed solutions. RSC Adv 12(10):5749–5764

Mier AA, Olvera-Vargas H, Mejía-López M, Longoria A, Verea L, Sebastian P, Arias DM (2021) A review of recent advances in electrode materials for emerging bioelectrochemical systems: from biofilm-bearing anodes to specialized cathodes. Chemosphere 283:131138

Min Z, Rui T, Yu L (2024) A combination of microbial electrolysis cells and bioaugmentation can effectively treat synthetic wastewater containing polycyclic aromatic hydrocarbon. Water Sci Technol 89(10):2716–2731

Miran W, Nawaz M, Jang J, Lee DS (2017) Chlorinated phenol treatment and in situ hydrogen peroxide production in a sulfate-reducing bacteria enriched bioelectrochemical system. Water Res 117:198–206

Noori MT, Vu MT, Ali RB, Min B (2020) Recent advances in cathode materials and configurations for upgrading methane in bioelectrochemical systems integrated with anaerobic digestion. Chem Eng J 392:123689

Obileke K, Onyeaka H, Meyer EL, Nwokolo N (2021) Microbial fuel cells, a renewable energy technology for bio-electricity generation: a mini-review. Electrochem Commun 125:107003
Okabe S (2021) Bioelectrochemical anoxic ammonium nitrogen removal by an MFC driven single chamber microbial electrolysis cell. Chemosphere 274:129715
Olvera-Vargas H, Trellu C, Oturan N, Oturan MA (2018) Bio-electro-Fenton: a new combined process–principles and applications, 1 edn. In: Zhou M, Oturan MA, Sirés I (eds) Electro-Fenton process: new trends and scale-up. Springer Singapore, Singapore, pp 29–56
Ondon BS, Li S, Zhou Q, Li F (2020) Simultaneous removal and high tolerance of norfloxacin with electricity generation in microbial fuel cell and its antibiotic resistance genes quantification. Biores Technol 304:122984
Oon Y-S, Ong S-A, Ho L-N, Wong Y-S, Oon Y-L, Lehl HK, Thung W-E, Nordin N (2018) Disclosing the synergistic mechanisms of azo dye degradation and bioelectricity generation in a microbial fuel cell. Chem Eng J 344:236–245
Pant D, Van Bogaert G, Diels L, Vanbroekhoven K (2010) A review of the substrates used in microbial fuel cells (MFCs) for sustainable energy production. Biores Technol 101(6):1533–1543
Paul D, Noori M, Rajesh P, Ghangrekar M, Mitra A (2018) Modification of carbon felt anode with graphene oxide-zeolite composite for enhancing the performance of microbial fuel cell. Sustain Energy Technol Assess 26:77–82
Pham H, Boon N, Marzorati M, Verstraete W (2009) Enhanced removal of 1, 2-dichloroethane by anodophilic microbial consortia. Water Res 43(11):2936–2946
Potter MC (1911) Electrical effects accompanying the decomposition of organic compounds. Proc Royal Soc London. Series B, Containing Papers Biol Char 84(571):260–276
Priyadarshini M, Ahmad A, Das S, Ghangrekar MM (2022a) Application of innovative electrochemical and microbial electrochemical technologies for the efficacious removal of emerging contaminants from wastewater: a review. J Environ Chem Eng 10(5):108230
Priyadarshini M, Ahmad A, Das S, Ghangrekar MM (2022b) Application of microbial electrochemical technologies for the treatment of petrochemical wastewater with concomitant valuable recovery: a review. Environ Sci Pollut Res 29(41):61783–61802
Qiu B, Hu Y, Liang C, Wang L, Shu Y, Chen Y, Cheng J (2020) Enhanced degradation of diclofenac with Ru/Fe modified anode microbial fuel cell: kinetics, pathways and mechanisms. Biores Technol 300:122703
Raj R, Das S, Behera M, Ghangrekar MM (2023a) Bio-Electrochemical Systems for Micropollutant Removal. In: Surampalli RY, Zhang TC, Kao CM, Ghangrekar MM, Bhunia P, Behera M, Rout PR (eds) Microconstituents in the environment: occurrence, fate, removal and management. John Wiley & Sons Ltd., New Jersey, pp 455–490
Raj R, Sathe S, Das S, Ghangrekar M (2023b) Nickel–iron-driven heterogenous bio-electro-fenton process for the degradation of methylparaben. Chemosphere 341:139989
Raj R, Dhanda A, Kumar S, Das S, Ghangrekar MM (2024a) Repurposing waste-iron electrocoagulated algal biomass as effective heterogenous (bio) electro-fenton catalyst for phthalate removal from wastewater. Sci Rep 14(1):23932
Raj R, Tripathi A, Das S, Ghangrekar M (2024b) Is waste-derived catalyst mediated electro-Fenton a sustainable option for mitigating emerging contaminants from wastewater? Curr Opin Environ Sci Health 37:100523
Rasheed A, Sharma N, Surampalli RY, Das S (2023) Evaluating treatment solutions: critical review on technologies employed for microplastic removal from water matrices. Curr Opin Environ Sci Health, 100516
Rhee S-K, Liu X, Wu L, Chong SC, Wan X, Zhou J (2004) Detection of genes involved in biodegradation and biotransformation in microbial communities by using 50-mer oligonucleotide microarrays. Appl Environ Microbiol 70(7):4303–4317
Saket P, Mittal Y, Bala K, Joshi A, Yadav AK (2022) Innovative constructed wetland coupled with microbial fuel cell for enhancing diazo dye degradation with simultaneous electricity generation. Biores Technol 345:126490

Sathe S, Bhowmick G, Dubey B, Ghangrekar M (2020) Surfactant removal from wastewater using photo-cathode microbial fuel cell and laterite-based hybrid treatment system. Bioprocess Biosyst Eng 43:2075–2084

Shahi A, Chellam PV, Singh RS, Verma A (2021) Biodegradation of reactive red 120 in microbial fuel cell by Staphylococcus equoruma RAP2: statistical modelling and process optimization. J Water Process Eng 40:101913

Sharma N, Rasheed A, Zargar N, Das S (2024) A critical review on mitigation strategies for per- and polyfluoroalkyl substances from water matrices with special emphasis on techno-economic and life cycle assessment: current status, knowledge gaps and future perspectives. J Environ Chem Eng 114863

Shea C, Clauwaert P, Verstraete W, Nerenberg R (2008) Adapting a denitrifying biocathode for perchlorate reduction. Water Sci Technol 58(10):1941–1946

Sikarwar D, Bashir Y, Sharma N, Rasheed A, Das S (2024) Critical perspective on the elimination of emerging contaminants from industrial wastewater via microbial electrochemical technologies. CLEAN–Soil, Air, Water, 2300346

Silveira G, de Aquino Neto S, Schneedorf JM (2020) Development, characterization and application of a low-cost single chamber microbial fuel cell based on hydraulic couplers. Energy 208:118395

Song X, Jo C, Zhou M (2022) Enhanced electricity generation and tetracycline removal of bioelectro-Fenton with electroactive biofilm induced by multi external resistance. Chemosphere 289:133070

Thung W-E, Ong S-A, Ho L-N, Wong Y-S, Ridwan F, Oon Y-L, Oon Y-S, Lehl HK (2015) A highly efficient single chambered up-flow membrane-less microbial fuel cell for treatment of azo dye acid orange 7-containing wastewater. Biores Technol 197:284–288

Vélez M, Arbuckle T, Fraser W (2015) Maternal exposure to perfluorinated chemicals and reduced fecundity: the MIREC study. Hum Reprod 30(3):701–709

Wang H, Ren ZJ (2013) A comprehensive review of microbial electrochemical systems as a platform technology. Biotechnol Adv 31(8):1796–1807

Wang Y, Jiang X, Yang H, Li W, Mu Y, Shen J (2020) Degradation and mineralization of recalcitrant compounds in bioelectrochemical systems, 1 edn. In: Tiquia-Arashiro SM, Pant D (eds) Microbial electrochemical technologies. CRC Press, Boca Raton, pp 331–342

Wen Q, Yang T, Wang S, Chen Y, Cong L, Qu Y (2013) Dechlorination of 4-chlorophenol to phenol in bioelectrochemical systems. J Hazard Mater 244:743–749

Wen H, Zhu H, Yan B, Xu Y, Shutes B (2020) Treatment of typical antibiotics in constructed wetlands integrated with microbial fuel cells: roles of plant and circuit operation mode. Chemosphere 250:126252

Wen H, Zhu H, Xu Y, Yan B, Shutes B, Bañuelos G, Wang X (2021) Removal of sulfamethoxazole and tetracycline in constructed wetlands integrated with microbial fuel cells influenced by influent and operational conditions. Environ Pollut 272:115988

Xu H, Quan X, Xiao Z, Chen L (2018) Effect of anodes decoration with metal and metal oxides nanoparticles on pharmaceutically active compounds removal and power generation in microbial fuel cells. Chem Eng J 335:539–547

Xu W, Yang B, Wang H, Zhang L, Zhang Y, Zhang M, Shi J (2024) Microbial functional insights into antibiotics and nitrogen removal in constructed wetland-microbial fuel cells packed with mine waste substrate. J Water Process Eng 64:105709

Xue W, Zhou Q, Li F (2019) Bacterial community changes and antibiotic resistance gene quantification in microbial electrolysis cells during long-term sulfamethoxazole treatment. Biores Technol 294:122170

Yaashikaa P, Saravanan A, Kumar PS, Thamarai P, Rangasamy G (2023) Role of microbial carbon capture cells in carbon sequestration and energy generation during wastewater treatment: a sustainable solution for cleaner environment. Int J Hydrogen Energy 52:799–820

Yan W, Xiao Y, Yan W, Ding R, Wang S, Zhao F (2019) The effect of bioelectrochemical systems on antibiotics removal and antibiotic resistance genes: a review. Chem Eng J 358:1421–1437

Yang X-L, Zhang S, Li H, Zhang L-M, Song H-L, Wang Y-W (2018) Effects of voltage on sulfadiazine degradation and the response of sul genes and microbial communities in biofilm-electrode reactors. Ecotoxicol Environ Saf 151:272–278

Yang X, Zou R, Tang K, Andersen HR, Angelidaki I, Zhang Y (2021) Degradation of metoprolol from wastewater in a bio-electro-Fenton system. Sci Total Environ 771:145385

Yu Q, Xiong W, Huang D, Luo C, Yang Q, Guo T, Wei Q (2020) Cathodic reduction characteristics of 2-chloro-4-nitrophenol in microbial electrolysis cell. Environ Eng Res 25(6):854–861

Zhang Y, Angelidaki I (2014) Microbial electrolysis cells turning to be versatile technology: recent advances and future challenges. Water Res 56:11–25

Zhang L-J, Tao H-C, Wei X-Y, Lei T, Li J-B, Wang A-J, Wu W-M (2012) Bioelectrochemical recovery of ammonia–copper (II) complexes from wastewater using a dual chamber microbial fuel cell. Chemosphere 89(10):1177–1182

Zhang Q, Zhang L, Wang H, Jiang Q, Zhu X (2018) Simultaneous efficient removal of oxyfluorfen with electricity generation in a microbial fuel cell and its microbial community analysis. Biores Technol 250:658–665

Zhang Q, Zhang L, Li Z, Zhang L, Li D (2019) Enhancement of fipronil degradation with eliminating its toxicity in a microbial fuel cell and the catabolic versatility of anodic biofilm. Biores Technol 290:121723

Zhao Y, Zheng J, Li X, Wang S, Zhou A, Li H, Zhao B, Yue X (2024) Co-metabolism driven sulfaquinoxaline removal in microbial electrolysis cells: a mechanistic analysis based on DFT calculation, metabolic pathway and functional enzyme activity. J Environ Chem Eng 12(2):112153

Zhong D, Liu Y, Liao X, Zhong N, Xu Y (2018) Facile preparation of binder-free NiO/MnO_2-carbon felt anode to enhance electricity generation and dye wastewater degradation performances of microbial fuel cell. Int J Hydrogen Energy 43(51):23014–23026

Zhu X, Ni J (2009) Simultaneous processes of electricity generation and p-nitrophenol degradation in a microbial fuel cell. Electrochem Commun 11(2):274–277

Zhu X, Lei C, Qi J, Zhen G, Lu X, Xu S, Zhang J, Liu H, Zhang X, Wu Z (2022) The role of microbiome in carbon sequestration and environment security during wastewater treatment. Sci Total Environ 837:155793

Chapter 16
Wastewater Treatment Solutions for Small Towns: A Practical Guide

Zain Muhammad and Abid Mahmood

Abstract Small towns face unique challenges in treating wastewater due to limited resources, expertise, and land availability. This chapter explores sustainable treatment solutions tailored to their specific needs. We introduce three main categories of suitable technologies: Natural systems (constructed wetlands, lagoons, infiltration trenches), biological treatment systems (activated sludge, anaerobic digestion), and physical–chemical treatment systems (screening, sedimentation, filtration). Each method is explained, focusing on its benefits and applicability for small towns. A comparison table highlights key factors for choosing the most suitable technology, including treatment efficiency, land requirements, operational costs, maintenance needs, and environmental impact. Additionally, the chapter showcases real-world success stories through 3 diverse case studies of small towns implementing sustainable wastewater treatment solutions. By analyzing these case studies and considering local factors, small towns can make informed decisions. The chapter concludes by offering practical recommendations for selecting the right technology and exploring future trends in sustainable wastewater treatment for small towns.

Keywords Wastewater · Small towns · Treatment · Technologies · Environmental impact

16.1 Introduction

16.1.1 What Are Small Towns? A Brief Look

The basic knowledge about a small town should be in your mind before implementing a wastewater treatment solution. While the exact definition of "**small town**" can vary by country, they generally share some characteristics that contribute to their distinct

Z. Muhammad · A. Mahmood (✉)
Department of Environmental Sciences, Government College University, Faisalabad, Pakistan
e-mail: drabid@gcuf.edu.pk

S. Souabi et al. (eds.), *Wastewater Treatment Plants*, Water Science and Technology Library 130, https://doi.org/10.1007/978-3-031-87461-1_16

character. They have close-knit communities, a slower pace of life, and reliance on local resources.

Big cities around the world have gotten much better at treating wastewater than small towns and rural areas. These smaller areas are way behind. This is a big problem because dumping lots of untreated or partially treated sewage into water sources like rivers and oceans is very harmful, especially in developing countries (Massoud et al. 2009; Ozimek and Pijanowska 2024).

16.1.2 Unique Challenges in Small Towns

While small towns offer a unique and cherished way of life, they also face distinct challenges. These challenges can limit their growth and development, requiring innovative solutions to ensure their continued vibrancy. Sanitation projects in poor countries, especially small ones, often try to fix things with cheap technology (Maqbool et al. 2020). They mainly concentrate on low-cost solutions (Tilley et al. 2014). Here's a closer look at some of the key hurdles faced by small towns:

- **Limited Economic Opportunities**: Small towns often struggle to attract and retain businesses, leading to limited job options for residents. This can result in population decline, as younger generations seek out opportunities in larger cities.
- **Infrastructure Deficiencies**: Aging infrastructure, including roads, water systems, and wastewater treatment facilities, can be a significant burden for small towns with limited budgets.
- **Regulatory Compliance**: Regulations governing wastewater discharge are becoming increasingly stringent. Small towns that fail to meet these standards can face hefty fines or even enforcement actions that could further limit their resources.
- **Public Health Risks**: Improper wastewater treatment can expose residents to harmful pathogens and contribute to the spread of water-borne diseases. This is a major concern, especially for small towns with limited access to advanced healthcare facilities (Gholipour et al. 2021).
- **Lower Wastewater Volumes**: The smaller population size in small towns translates to lower wastewater volumes than larger cities. This can make some large-scale treatment technologies less economical.

16.1.3 Opportunities for Small Towns

Importantly, it is equally true that small towns' wastewater treatment presents numerous difficulties. Yet, these communities have the strengths which can be employed in order to create effective and sustainable solutions. Here's how small towns can transform their limitations into opportunities:

- **Land Availability**: For those towns, one of the biggest strengths is always the availability and flexibility of space, which is limited in comparison to cities. This land is available and would enable the establishment of treatment systems that require large areas such as the constructed wetlands.
- **Simpler Wastewater Characteristics**: Generally, industries are not as prevalent in the smaller towns as they are in the big cities. This can lead to wastewater with a more uniform composition, which could mean that some processes that simplify the task of the treatment plant have been completed. Areas of a species might perform better in biological treatment systems with a less composite influent. What people can understand is that large city sewage treatment plants failed in small towns. For this reason, they require special methods of handling waste in such small towns and cities. This implies either modifying the solutions developed for the big cities or coming up with fresh methods of dealing with waste (Massoud et al. 2009).

Nevertheless, small towns can make use of these peculiarities and search for new treatment approaches, turning wastewater from a problem into an opportunity to construct environmental initiatives and develop the community.

16.1.4 Sustainable Solutions Overview

Thus, the way to maintain a clean water supply in small towns is to focus on more efficient and eco-friendly approaches to wastewater treatment. These solutions are far from simple wastewater treatment since they are aimed at reducing the negative effects on the environment and maximizing the utility of the resources in the overall community's sustainable welfare. Here's an overview of some key principles that guide the development of sustainable wastewater treatment options for small towns:

- **Energy Efficiency**: As a result of sustainable solutions, the focus is made on the use of energy-saving technologies for reduction of the overall energy demand for treatment. This may include the use of natural treatment systems such as constructed wetlands, improving operational processes to increase efficiency, and reduce energy consumption, or generation of energy through biogas and utilizing for power the treatment plant among others (Metcalf and Eddy, Inc. 2014).
- **Resource Recovery**: Sewage treatment technologies can still get other important elements or compounds for instance phosphorus and nitrogen back from the wastewater. It can produce some fertilizer opportunities and mitigate the options of using chemical fertilizers (Singh et al. 2022). Also, COD and BOD wastewaters can create a new revenue stream for producing electricity and other gases through the process of anaerobic digestion.
- **Minimal Environmental Impact**: Sustainable strategies are mainly concerned with the reduction of the environmental impact in the case of wastewater treatment. These mechanisms consist of reducing the utilization of toxic chemicals,

properly disposing of the byproducts of the treatment processes, and preserving the adjoining environment from pollution (Apgar and Moore 2010).

- **Community Engagement**: In the sustainable discharge of wastewater, the community needs to be water users who are active participants. As with any program, raising the awareness of the people, especially the residents regarding the use of water and making them feel that the system is their own is very important for the provision of a properly functioning treatment system (Langston 2008).

Exploring Diverse Technologies:

The following content of this chapter will go deeper into various wastewater treatment technologies available for small towns, each offering unique advantages in terms of sustainability:

Natural Treatment Systems: These systems employ natural processes in filtration sedimentation and biodegradation to filter pollutants from the wastewater. Some natural treatment systems include constructed wetlands and lagoons. These are convenient to implement in small communities since they are cheaper and consume little power.

Biological Treatment Systems: These systems use the capability of small organisms—microbes to decompose pollutants found in the wastewater. Activated sludge and sequencing batch reactors are some of the most effective and employed types of wastewater treatments. Still, managing their energy usage has to be done effectively for sustainability purposes.

Physical–Chemical Treatment Systems: These systems are not the complete solution in and of themselves, however, they can be designed and used to accommodate particular requirements, for example, when the pollutants which need to be removed are particularly unpleasant, or where the process needs to be pretreated before biological treatment. The eco-friendly aspect covers the concept of sparingness of chemicals and their proper utilization in the treatment process.

Learning about these principles and familiarizing oneself with the vast variety of sustainable treatment technologies allows small towns to set a specific course of development that leads to efficient and sustainable WWT practices, which contributes to the general population's safety and well-being, as well as the preservation of the natural environment and future prosperity of the community.

16.2 Suitable Treatment Technologies for Small Towns

16.2.1 Natural Systems: Nature's Power for Wastewater Treatment

Natural wastewater treatment systems rely on natural resources to eliminate pollutants within the wastewater. They have been cited to be cheap and environmentally friendly relative to the standard methods of mechanical and chemical treatment (Tchobanoglous et al. 2003). Here's a closer look at the two main categories of natural treatment systems (Fig. 16.1):

A. Constructed Wetlands: Constructed wetlands are human-made systems that mimic the natural filtering function of **marshes** and **swamps**. Wastewater flows through a shallow, lined basin filled with wetland plants, soil, and microbial communities. These components work together to remove pollutants through various mechanisms (Wallace and Reddy 2000):

- **Physical Filtration**: Plant roots and the wetland media physically trap suspended solids and debris from the wastewater (Wallace and Reddy 2000).
- **Sedimentation**: Heavier particles settle out as the flow slows through the wetland (Wallace and Reddy 2000).
- **Biological Treatment**: Microbial communities attached to plant roots and living in the soil biodegrade organic matter in the wastewater. This process is similar to what happens in a natural stream or river (Vymazal 2009).
- **Plant Uptake**: Wetland plants absorb nutrients like nitrogen and phosphorus from the wastewater for their growth (Vymazal 2008).

Types of Constructed Wetlands: There are three main types of constructed wetlands, each with its own design and operational characteristics:

- **Free Water Surface Wetlands (FWWs)**: These resemble natural marshes with shallow open water areas and emergent plants like reeds or bulrushes. Wastewater

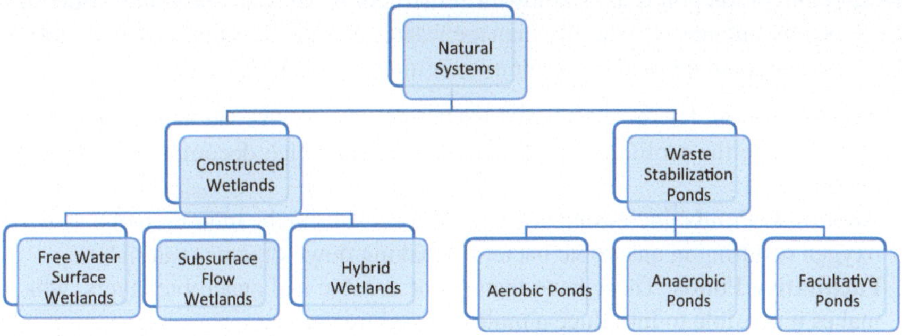

Fig. 16.1 Different methods in natural systems to treat wastewater

flows freely through the surface of the wetland. FWWs are efficient in removing organic matter, nutrients, and some pathogens (Vymazal 2009).

- **Subsurface Flow Wetlands (SSWs)**: These systems have a gravel or rock bed below the surface planted with wetland vegetation. Wastewater is evenly distributed below the surface, allowing for efficient treatment with minimal odor or mosquito breeding concerns. SSFs are particularly effective in removing nitrogen and pathogens (Vymazal 2009).
- **Hybrid Wetlands**: As the name suggests, hybrid systems combine features of both FWWs and SSFs. This allows for customization to address specific wastewater treatment needs (Vymazal 2008).

Benefits of Constructed Wetlands

- **Cost-Effective**: Natural systems require lower capital investment compared to conventional treatment plants and have minimal operation and maintenance costs (Tchobanoglous et al. 2003).
- **Sustainable**: They utilize natural processes and require minimal energy input (Wallace and Reddy 2000).
- **Habitat Creation**: Constructed wetlands provide habitat for diverse bird, insect, and amphibian species, promoting biodiversity (Vymazal 2009).
- **Odor Control**: Properly designed and maintained natural systems generate minimal odors (Wallace and Reddy 2000).

Limitations of Constructed Wetlands

- **Land Availability**: Constructed wetlands require more land compared to some mechanical treatment options (Tchobanoglous et al. 2003).
- **Treatment Efficiency**: Natural systems might not always achieve the same level of effluent quality as advanced treatment technologies (Vymazal 2009).
- **Climate Dependence**: Performance can be affected by seasonal temperature variations and extreme weather events (Vymazal 2009).

B. Waste Stabilization Ponds (WSPs)

Waste stabilization ponds are shallow in construction. They are hydraulic structures developed by humans to naturally manage wastewater (Tchobanoglous et al. 2003). They are categorized based on the dominant microbial activity zone:

- **Aerobic Ponds**: These ponds have high dissolved oxygen content near the surface to encourage the proliferation of aerobic bacteria which decompose the organic materials.
- **Anaerobic Ponds**: Increasing depth in WSPs results in the limitation of dissolved oxygen resulting in anaerobic bacteria breaking down organic materials.
- **Facultative Ponds**: These have features of aerobic and anaerobic layers, which makes it possible to introduce a multi-layered treatment.

Treatment Mechanisms in WSPs

Waste stabilization ponds rely on a combination of physical, biological, and chemical processes to treat wastewater (Tchobanoglous et al. 2003):

- **Sedimentation**: As in constructed wetlands, this occurs in WSPs tributaries because of low flow velocity that leads to the settling of suspended solids.
- **Biodegradation**: The organic matter in the pond is decomposed by aerobic and anaerobic bacteria that inhabit various zones of the pond depending on the degree of oxygenation used (Tchobanoglous et al. 2003).
- **Algal Activity**: Photosynthesizing algae found in the photic zone are responsible for the production of oxygen. Algae also assist in polluting substances' elimination from water as they convert it into cellular mass (Vymazal 2009).

Benefits of Waste Stabilization Ponds

- **Simple and Robust Technology**: WSPs are easy to construct and operate, and there is a small necessity for technology expertise (Tchobanoglous et al. 2003).
- **Low Maintenance**: Of less energy demand and low utility expenses because the natural processes are incorporated in their operations.
- **Land Availability**: WSPs occupy relatively more land than advanced treatments. Comparing it to the constructed wetland though, WSPs have a relatively smaller area of construction (Tchobanoglous et al. 2003).

Limitations of Waste Stabilization Ponds

- **Land Availability**: This may still be true in matters to do with the availability of land especially in the developing world where population pressures precede economic development (Tchobanoglous et al. 2003).
- **Odor Control**: Abandonment of proper full aerobics in WSPs produces bad odor due to the anaerobic activity. This can be a worry to people in the neighboring areas of the treatment facility.
- **Climate Dependence**: It is important to note that the rate of treatment mentioned can be affected by fluctuations in temperature depending on which season of the year it belongs to. Low temperature can inhibit the biological actions which can be a failure in the treatment capability.
- **Effluent Quality**: The WSP effluent likely needed to be treated further to achieve the chain discharge criteria, especially for nutrients.

Additional Considerations While Applying Natural Systems in a Small Town

While natural treatment systems offer numerous advantages for small towns, there are some additional factors to consider before implementation:

- **Pre-treatment Requirements**: Based on the nature of the influent wastewater some degree of pre-treatment may be required before the wastewater is discharged into the natural environment. This could include straining to exclude large objects or sedimentation to settle out gross particles. Treatment procedures enhance the effectiveness and service delivery of the natural treatment system before the treatment.

- **Integration with Other Systems**: Natural systems can be integrated easily with the other treatment technologies which means that it is a unit process in forming the treatment train. For example, constructed wetlands can fit into polishing the effluent from the septic tank system. WSPs could be used as the first point in the treatment process and then polishing with a natural system like the reed bed.
- **Public Perception and Community Engagement**: Natural wastewater treatment systems might have issues as far as aesthetic value or possible stench. Public participation remains important during the planning and execution phases of social marketing programs. That is why increasing awareness of the benefits and correct practices in the population can help to avoid rejection and discuss the issue (Cornwell et al. 2016).
- **Regulatory Compliance**: Make certain that the selected natural treatment system complies with all the local and national laws concerning the quality of the effluents to be discharged. Which one should be coordinated with the authorities during the pre-planning to avert violation of the set laws?
- **Long-Term Operation and Maintenance**: The natural systems too need maintenance activities in the form of seasons of cleaning of inlet and outlet structures, vegetative control in constructed wetlands, pond depth, and algae accumulation in WSPs (Vymazal 2009).

Thus, by paying more attention to these extra factors, small towns can make the most of the natural wastewater treatment systems and guarantee their successful utilization of them toward a sustainable future.

16.2.2 Biological Treatment Systems: Harnessing Microbes for Wastewater Purification

Conventional systems of wastewater treatment are mainly supported by biological treatment systems. They use bacteria, such as those that are found in sewage such as protozoa, to decompose and eliminate the organic pollutant from the wastewater (Metcalf and Eddy, Inc. 2014). Such systems therefore offer the best option for attaining a high-quality effluent.

Here's a closer look at the two main categories of biological treatment systems (Fig. 16.2):

Here's a closer look at the two main categories of biological treatment systems:

A. Activated Sludge Process

Tchobanoglous et al. documented in their work that the activated sludge process is currently the most popular bio-treatment system in the world. It works by developing a densely concentrated population of microorganisms known as activated sludge in a tank that has to be supplied with oxygen at all times. They are thoroughly mixed

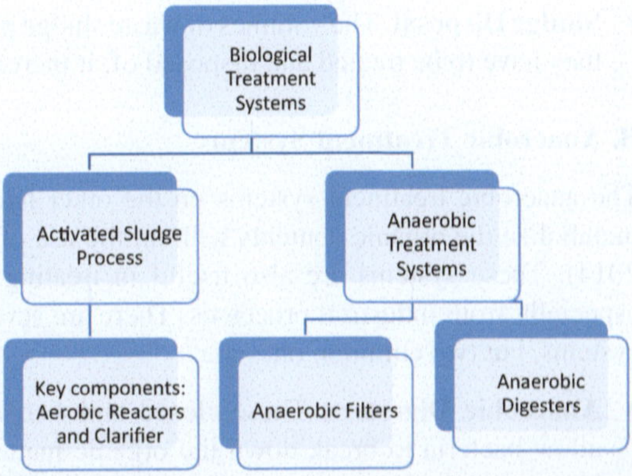

Fig. 16.2 Review of types of biological treatment systems

with the wastewater permitting easy access for microorganisms to degrade organic compounds (Metcalf and Eddy, Inc. 2014).

Key Components of an Activated Sludge System

- **Aerobic Reactors**: These are the tanks where activated sludge is stored and the flow of air in the tanks must be constant to foster aerobic conditions for the microbes to decompose organic matter (Tchobanoglous et al. 2003).
- **Clarifier**: The mixed liquor that leaves the aeration tank which is the wastewater together with the activated sludge then goes into a settling tank commonly known as a clarifier. In this section, the activated sludge descends and the effluent that has been treated, and clear water is separated from the sludge. The settled sludge is thereafter returned to the aeration tank as it helps in breaking down the organic matter efficiently (Metcalf and Eddy, Inc. 2014).

Benefits of Activated Sludge

- **Highly Efficient**: This process helps in the elimination of a large number of organic pollutants making the quality of the effluent to be high.
- **Versatility**: The activated sludge process is flexible in the sense that it can be used to manage fluctuating conditions of the influent as regards its composition and flow (Metcalf and Eddy, Inc. 2014).
- **Widely Used Technology**: A great deal of information and expertise is available concerning the design, management, and troubleshooting of activated sludge systems.

Limitations of Activated Sludge

- **Energy Intensive**: In continuous aeration, there is constant circulation of air which may take a lot of energy hence adding on the expenses incurred.
- **Bulking and Foaming**: Some of the typical operational problems include bulking, and foaming and this affects the efficiency of the treatment plant and correct measures have to be taken.

- **Sludge Disposal**: The volumes of waste sludge produced are high hence whenever they have to be treated and disposed of, it increases operational cost.

B. Anaerobic Treatment Systems

The anaerobic treatment systems on the other hand employ anaerobic bacteria to metabolize the organic contents without the use of oxygen (Metcalf and Eddy, Inc. 2014). These systems are also useful in treating wastewater with high strength, especially from industrial processes. There are several types of anaerobic treatment systems, but two common ones are:

- **Anaerobic Digesters**: These closed tanks create an anaerobic environment to allow bacteria to break down the organic matter. It results in the generation of biogas which if harnessed can be used as a source of energy, (Lettinga 1995).
- **Anaerobic Filters**: It is a stream of wastewater in direct contact with a packed media bed with anaerobic microorganisms on the surface of the media. The biofilm breaks down the organic content of the material and can moreover offer comparable treatment with lesser space than digesters (McCarty and McKinney 1961).

Benefits of Anaerobic Treatment

- **Energy Production**: The resulting biogas can be reclaimed for cogeneration or CHP which reduces the use of conventional energy resources (Lettinga 1995).
- **Reduced Sludge Production**: Anaerobic processes produce less sludge compared to the aerobic treatments this was established by McCarty and McKinney in 1961.
- **Suitable for High-Strength Wastewater**: They are efficient in dealing especially with wastewater that has a high-biochemical oxygen demand (Tchobanoglous et al. 2003).

Limitations of Anaerobic Treatment

- **Sensitivity to Compounds**: Anaerobic bacteria can be sensitive to certain chemicals or heavy metals present in wastewater, affecting treatment performance (Lettinga 1995).
- **Lower Treatment Efficiency**: Compared to activated sludge, anaerobic systems might require additional treatment stages to achieve the same level of effluent quality for certain pollutants.
- **Odor Control**: Strict management practices are necessary to control potential odor emissions from anaerobic treatment systems.

16.2.3 Physical–Chemical Treatment Systems: A Targeted Approach for Wastewater Purification

Physical–chemical, also referred to as physicochemical, treatment systems use both physical and chemical methods to eliminate pollution and other undesirable contents from wastewater. These systems provide considerable means of solution to certain issues or certain pollutants that are not adequately dealt with in biological treatments. Here's a jump into the key components and functionalities of physicochemical treatment:

16.2.3.1 Unit Processes

Physicochemical treatment involves the application of several unit processes where each focusing on particular issues in the wastewater. Here are some commonly employed processes:

- **Screening and Filtration**: Rag and debris screens reduce substantially bigger objects such as rags and debris, while screens or filters, finer, retain smaller approving solids. This does not allow the buildup of materials that would hamper the treatment processes thus enhancing the efficiency of the downstream processes.
- **Coagulation and Flocculation**: Commonly, coagulants which have positive charges are discharged into the water; these charges then attract those on the colloidal particles in the wastewater and lead to the coagulation process. Flocculants then bind onto these destabilized particles to create big flocs (clumps) that settle faster (Tchobanoglous et al. 2003).
- **Sedimentation**: In gravity settling tanks (clarifiers), the resultant flocs which occurred due to coagulation and flocculation process will sink from the wastewater. The clear and treated supernatant is also collected in this stage while the settled sludge on the other hand is treated or discharged.
- **Flotation**: For wastewater with lighter flocs that resist settling, dissolved air flotation (DAF) can be employed. Air bubbles are introduced into the wastewater, attaching to the flocs and causing them to float to the surface, where they can be skimmed off (Metcalf and Eddy, Inc. 2014). Where flocs generated are lighter and are not easily settled, dissolved air flotation (DAF) can be used. The air is then diffused in the water and it sticks on the flocs which float on the water's surface and can easily be removed by skimming (Metcalf and Eddy, Inc. 2014).
- **Neutralization**: If the wastewater is very acidic or very alkaline, the pH might also be an issue and chemicals can be dosed to bring the pH to an appropriate level for the biological treatments or for meeting the discharge standards.
- **Oxidation**: Substances like chlorine or the ozone can be used to disinfect wastewater to kill bacteria and other pathogen within wastewater.
- **Adsorption**: Certain materials, like activated carbon, have a high affinity for specific pollutants. Wastewater is passed through a bed of this adsorbent material, where the target pollutants adhere to the surface. (Metcalf and Eddy, Inc. 2014).
- **Membrane Filtration**: Membranes with tiny pores allow water to pass through while rejecting larger molecules and suspended solids. This technology can be used for various applications, such as microfiltration (removal of bacteria and

protozoa), ultrafiltration (removal of macromolecules and colloids), and reverse osmosis (removal of dissolved salts and ions) (Tchobanoglous et al. 2003).

16.2.3.2 Applications of Physicochemical Treatment

Physicochemical treatment systems find application in various wastewater treatment scenarios, including:

- **Pre-treatment for Biological Systems**: Some matter such as solids, organics, and certain pollutants should be taken out before reaching the biological treatment because they can further improve the efficiency of biological treatment and also protect the useful biological organisms.
- **Phosphorus Removal**: There are possibility to treat phosphorus, inflow, which can cause algae in the receiving water bodies, by the chemical precipitation or adsorption processes (Metcalf and Eddy, Inc. 2014).
- **Color and Odor Removal**: Specific adsorption media or oxidation processes can address color or unpleasant odors present in wastewater.
- **Industrial Wastewater Treatment**: Some industries have specific pollutants and a high degree of purification and this is where the physicochemical processes will be useful in the treatment of wastewater.
- **Tertiary Treatment**: After the biological treatment, there are physicochemical processes by which effluent quality can be brought to even higher standards.

16.2.3.3 Advantages of Physicochemical Treatment

- **Targeted Treatment**: These systems can deal with particular pollutants or contaminants which cannot be effectively dealt by the biological treatment process alone.
- **Flexibility**: It was found that several unit processes can be aligned to fit the general needs of the wastewater treatment plant.
- **Reliable Performance**: Physicochemical systems, as a rule, are not as sensitive to changes in waste characteristics as biological ones.

16.3 Comparing Treatment Options for Small Towns

See the Table 16.1.

Factors Influencing Technology Selection: A Decision-Making Framework

Selection of an efficient wastewater treatment system for a small town is not an easy task of merely selecting between a set of entities. Here are some key factors to consider that will influence technology selection:

Table 16.1 Brief comparison of different treatment options

Wastewater treatment option	Description	Advantages	Disadvantages
Natural treatment systems	Mimic natural processes like wetlands or soil infiltration	Cost-effective, low energy consumption, environmentally friendly	Large land requirement, dependent on climate, lower treatment capacity
Biological treatment systems	Utilize microorganisms to break down organic matter	Efficient treatment, relatively small footprint, adaptable to various influent conditions	Higher operational costs, requires sludge management
Physical–chemical treatment systems	Employ physical and chemical processes to remove specific pollutants	Targeted treatment of specific contaminants, rapid results	High operational costs, requires proper residual management

- **Community Size and Wastewater Volume**: Ideally, where there are larger towns with higher volumes of wastewater it may be advantageous to provide a more sophisticated system of biological treatment. The use of a natural treatment system might be cheaper for the smaller communities as established by USEPA in 2018.
- **Wastewater Characteristics**: The method and the type and concentration of pollutants in the wastewater flow will determine the treatment method. Organic matters are easily removed through biological methods while certain pollutants such as heavy metals, may need physical–chemical methods (Tchobanoglous et al. 2003).
- **Land Availability**: Natural treatment systems entail the use of additional land for the construction of wetlands or land application. Lack of space could lead to going for solutions that take up less space such as BIO TUBE biological treatment units (USEPA 2018).
- **Financial Resources**: Life cycle costs of the building construction and the cost of operating the building also have to be considered. Indeed, though the natural treatment systems may start with less financial investment, these may require more acreage. The long-term energy, personnel, and maintenance costs should also be considered (Metcalf and Eddy, Inc. 2014).
- **Regulatory Requirements**: Treatment systems have to meet the local and national requirements concerning the quality of effluent (USEPA 2018). Knowledge of the specific discharge standards will then help in choosing the technology that is capable of attaining the needed level of treatment.
- **Technical Expertise**: The management and maintenance of some of the established treatment systems demand expertise. Where the health project will be based and whether there are enough competent staff within the area or by collaboration with neighboring institutions will determine whether the project will adopt complicated technologies (Langston 2008).

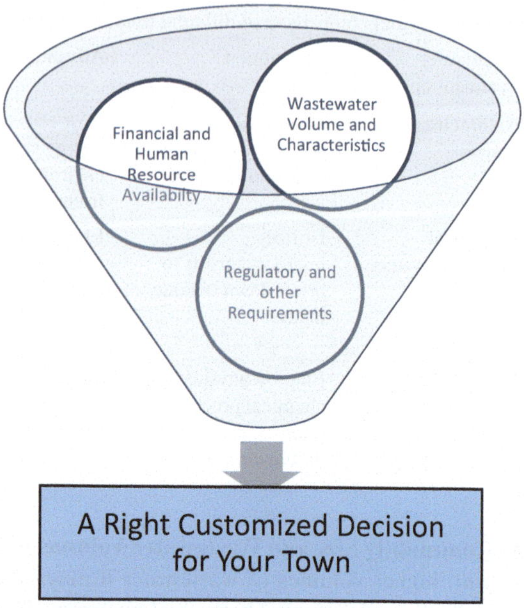

Fig. 16.3 Right decision-making process includes a lot of factors

- **Community Needs and Sustainability**: In addition to the core solution, communities may want to think about aspects like the collection and reuse of nutrients and/or biogas potential or ability to add on in the future (Singh et al. 2022). To encourage environmental preservation and assure future health of the community the following goals should be achieved: Use sustainable solutions that will affect the environment least of all (Apgar and Moore 2010) (Fig. 16.3).

Combining Technologies: Integrated Wastewater Treatment Systems

- **Natural Systems as Pre- or Post-treatment**: Wetland systems can also be used as initial treatment steps, removing the solid and organic material from the wastewater prior to the application of a biological purification treatment system. This can enhance the efficiency and the life cycle of the biological treatment process (Tchobanoglous et al. 2003). Also, natural systems can be employed in polishing the effluent generated from a biological treatment process to further increase the quality of the water before discharge (Vymazal 2009).
- **Natural Systems with On-Site Treatment**: Constructed wetlands could also be used in small communities or even individual households involving on-site wastewater treatment such as septic tanks where constructed wetlands can be incorporated as a polishing step to enhance the effluent quality and consequently minimize impacts on the environment (Wallace and Reddy 2000).
- **Combining Biological Processes**: The biological treatment systems can themselves be configured in series or parallel (Meng et al. 2024). For instance, an

anaerobic digester followed by the first stage treatment; higher strength wastewater treatment and activated sludge method or nitrification, which is conversion of ammonia to nitrate (Metcalf and Eddy, Inc. 2014).

16.4 Case Studies: Success Stories in Small Towns

16.4.1 Introduction

However, there are several difficulties nevertheless, many small towns across the globe are using efficient wastewater treatment technologies. Let's explore some inspiring case studies:

- **Brattleboro, Vermont, USA**: The specific case study is a town with a population of around 12,000 inhabitants that has developed an interesting system of using and treating wastewater, including the system of anaerobic digestion and constructed wetland. The organic matter in the waste is decomposed by the anaerobic digester to produce the biogas which in turn can power the facility hence eliminating the hitches sometimes encountered while relying on external energy sources (Lowe et al. 2004).
- **Fraser, Colorado, USA**: Lacking much local treatment capacity and getting increasing pressure to remove nutrients from the effluent of its wastewater treatment plant, the town of Fraser, with a population of some 1300, developed a system that incorporates SBRs and a constructed wetland. It thus effectively takes out such nutrients as phosphorus and nitrogen, thus safeguarding the Fraser River (McCarty et al. 2011).
- **Saanen, Switzerland**: This is a small alpine village in Switzerland with a population of slightly over 800 people showing dedication to sustainable wastewater treatment. In their treatment plant, they use both biological processes and sand filtration and yet, they produce high effluent quality while at the same time reducing impact on the environment (Ahring et al. 2007).

16.4.2 Diverse Case Studies

16.4.2.1 A Town in a Developing Country: Ecosan in San Pedro La Laguna, Guatemala

San Pedro La Laguna, a Mayan village nestled on the shores of Lake Atitlán in Guatemala, faced a significant challenge: Poor state of sanitation facilities thus causing water pollution and posing health risks. The population was estimated to be about 16,000 but the community was without an efficient sever system, thus most of the houses in the community used pit latrines which were a threat to the quality of the groundwater (Mara and Drangert 2014).

As a response, the municipality of San Pedro La Laguna began a first-of-its-kind effort following the principles of a sustainable ecosan-based approach for the adequate management of wastewater (Lüderitz et al. 2019). The ecosan toilets were put in the people's homes taking the place of pit toilets. These toilets divide human waste into two fractions; liquid and solids. The liquid waste is processed in several biodigesters and is thereby handled in a way that degrades and is finally discharged. Solid waste is composted thus being recycled and used as fertilizer for the nearby crops (Montalvo et al. 2009).

It has become quite apparent that the ecosan approach in San Pedro La Laguna has been rather effective. There has been a progressive enhancement of water quality in the lake thus lowering the occurrences of water-borne diseases. The composted fertilizer has been responsive in feeding local farmers as well as encouraging sustainable agriculture, and increasing productivity (Lüderitz et al. 2019). In addition, it has also brought about an enhancement of the operational and maintenance of the systems by involving the locals in the project (Montalvo et al. 2009).

Thus, with effective waste treatment and resource recovery in ecosan systems, public health is enhanced, the environment is protected, and local economies are boosted.

16.4.2.2 A Town in a Developed Country: Sequencing Batch Reactors and Wetlands in Scandinavia

A few small towns in the whole of Scandinavia have achieved cost-effective and excellent wastewater treatment through the use of SBR and constructed wetlands. Here's a closer look:

- **Example 1: Östersund, Sweden**: This is a small town in Sweden with a population of about 50,000 people which recently rehabilitated its WWTP in 2010 (Jönsson et al. 2015). Due to the flash in flow and concentrations of various pollutants, the SBRs offer a versatile and effective biological treatment system. The second constructed wetland works as a polishing filter where most of the nutrients and pollutants are removed before discharge (Vymazal 2011). The said combination enables Östersund to obtain high effluent quality, utilizing less energy and chemicals (Jönsson et al. 2015).
- **Example 2: Hvalba, Faroe Islands**: This even smaller place with an approximate population of 600 people had problems with its lagoon-based wastewater treatment system which was overburdened. A new treatment plant was installed in 2008 to include both SBRs and a compact reed bed wetland as described by (Christensen et al. 2013). This was found to be useful in treating the wastewater since the SBRs successfully do this while the polishing step in the reed bed wetland is compact and efficient. Therefore, vast improvements have been observed in the quality of effluent and the environmental manifestation in Hvalba (Christensen et al. 2013).

The outlined examples of the Scandinavian countries demonstrate the opportunities for the integrated application of SBRs and constructed wetlands in small towns. SBRs are flexible and provide effective biological treatment, constructed wetlands are low-cost and friendly to the environment and are added as a polishing stage.

16.4.2.3 A Town with a Unique Geographic Location: Combining Nature and Technology in a Mountain Town—Lake Tahoe, USA

Lake Tahoe situated in California and Nevada, among the Sierra Nevada Mountains faces the challenges of wastewater treatment. The town currently has about 65,000 residents, although this number is subject to seasonal changes regarding tourists thus the town requires a system that can follow the high but also protect the alpine lakes ecosystem (Tahoe Regional Planning Agency 2023).

Lake Tahoe adopted a multi-stage treatment approach that leverages both natural processes and advanced technology:

1. **Primary Treatment**: The process of wastewater includes screening and grit removal to filter off any large solid and other interrelated debris.
2. **Advanced Biological Treatment**: The used biological treatment process is called sequencing batch reactors (SBRs) and it is characterized by high flexibility and efficacy regarding the degradation of organic substances in the investigated wastewater (Jönsson et al. 2015).
3. **Membrane Filtration**: Any living microorganisms and other impurities are filtered out using the most sophisticated membrane technologies so that the quality of the effluent before its discharge is optimal (IWA Publishing 2007).
4. **Land Application with Disinfection**: The treated wastewater is disinfected with ultraviolet light to have a complete pathogen die-off. Of this amount, a part is applied on the managed land in the specific areas where it percolates through the ground and again undergoes natural purification (USEPA).

16.5 Conclusion and Recommendations

16.5.1 Summary

This chapter examined the prospects for the treatment of wastewater in small towns. It recognized the difficulties and underscored the need to choose a treatment system based on several factors ranging from the number of residents within a community, types of wastewater, available land, available capital, current laws, expertise, and users' requirements. The document then jumped into the three main categories of wastewater treatment systems. Realizing that a single technology might not address all needs, this chapter emphasized the importance of combining elements from different categories to create a customized treatment train. This allows for targeted and efficient multi-stage treatment. We concluded by showcasing inspiring case studies from around the world.

16.5.2 Choosing the Right Technology for Your Small Town

There is a set of criteria that have to be met to suggest the most appropriate wastewater treatment technology for a given small town (Fig. 16.4).

Here's a breakdown of key steps to guide you:

1. **Data Collection and Analysis**
   - **Wastewater Characteristics**: Investigate the flow and contents of the effluent discharged in your town. Such as the content present in organic matter; the nutrient status of the water; the amount of nitrogen, and phosphorus and the degree of present industrial contaminants (Metcalf and Eddy, Inc. 2014).
   - **Community and Regulatory Requirements**: Think of the population and growth estimate for those certain areas. Acquaintance with the local and/or national standards regarding the effluent discharge quality, and possible reuse (WSAA 2023).
2. **Technology Assessment**
   - Looks at the three significant treatment divisions, namely Natural, Biological, and Physicochemical, and their advantages as well as weaknesses.
   - Assess possible synergistic interactions of the applied technologies that when integrated in a specific manner will form the optimal treatment train. Some of the subcategory factors include the rate of production and how efficiently the factory can produce the expected use; this is together with costs such as capital and material cost, pollution to the environment, frequently required services and replacements, and availability of enough land to accommodate the factory (Jimenez and Ferrer 2006).
3. **Feasibility Studies and Cost Analysis**
   - Thus, feasibility studies of short-listed treatments have to be carried out. This might entail, for example, pilot testing or discussion with engineers working on wastewater treatment in small towns.

Fig. 16.4 Steps to choose the right technology for your small town

4. **Develop a Comprehensive Cost Analysis, Including**
   - Upfront capital costs for construction and equipment.
   - Ongoing operational expenses for energy, personnel, and maintenance.
   - Potential costs associated with residual management (sludge disposal for biological systems) (Jimenez and Ferrer 2006).
5. **Community Engagement and Decision-Making**
   - Let the community or the governed has a say in what is going to be done to them. Show available information about treatments and further treatment possibilities tourists might undergo alongside their approximate cost, usefulness for the tourist, and impact on the environment (Cornwell et al. 2016).
   - Obtain the views and also hear what the community has to say about the subject. Here, the main sources are Cornwell et al. (2016).

Additional Resources

- USEPA's Small Community Wastewater Systems
- The World Bank's Wastewater Treatment and Reuse Guide for Small Towns

16.5.3 Recommendations for Small Towns: A Sustainable Path Forward

Here are some key recommendations to ensure a sustainable and successful journey:

1. **Embrace a Long-Term Vision**: In planning and designing the system future growth projections must be considered to avoid early compromise (Jimenez and Ferrer 2006).
2. **Operation and Maintenance (O&M) is Key**: Provide sufficient funds for consistent running and sustaining of the treatment system. Conduct adequate training of the employees so that they can run it efficiently without the occurrence of expensive breakages (Jimenez and Ferrer 2006). It is recommended to cooperate and collaborate with local players if specific knowledge is not available in the community (WSAA 2023).
3. **Embrace Continuous Improvement**: The firms should embrace changes in new technologies and practices that may come up in the advance of time to reduce performance and environmental effects (Jimenez and Ferrer 2006).

16.5.4 Future Trends in Small-Town Wastewater Treatment

The advancement of wastewater treatment in small towns is currently with new technologies and positive attention to sustainable solutions. Here are some emerging trends to watch:

1. **Decentralized Treatment and Resource Recovery**: Instead of large centralized centers, individual or compact and constructions that are spread out are becoming more popular, as are so-called distributed facilities that are closer to the sources of wastewater (Jimenez and Ferrer 2006). This cuts down on dependence on the establishment of massive structures.
2. **Advanced Technologies for Compact and Efficient Treatment**: MBRs are highly mobile and effective for treating water and hence can be used in small cramped towns (Judd 2008). Also, there is an improved knowledge in the breakdown of organic wastes and biogas generation through anaerobic digestion.
3. **Nature-Based Solutions Gaining Ground Modern**: Constructed wetlands are more advanced in their designs and the types of plants used for water treatment. Moreover, applications like rain gardens and bioswales for wastewater management should also be incorporated into green infrastructure solutions which in turn saves the stress of natural filtration to treatment plants (Davis et al. 2009).

These trends hold immense promise for small towns to achieve sustainable and cost-effective wastewater treatment.

References

Ahring BK, Sandberg M, Angelidaki I, Johansen KW, Rasmussen P (2007) Improving biogas production by thermophilic digestion of primary sludge and household organic waste. Bioresour Technol 98(5):1008–1017

Apgar WC, Moore KM (2010) Small towns and rural communities: a handbook for local leaders, 1st ed. Santa Barbara, ABC-CLIO

Christensen MH, Nielsen A, Jørgensen PE (2013) Small-scale wastewater treatment plants: experiences from the Faroe Islands with special reference to nitrogen removal. J Environ Eng 139(6):795–802

Cornwell DA, Spear CJ, Merriman KL (2016) Engaging the public in wastewater decision-making: a case study of community perceptions of water reuse. J Environ Plann Manage. 59(12):2220–2240

Davis AP, Traver D, Hunt WF (2009) Improving urban stormwater quality using bioretention areas. North Carolina State University

Metcalf and Eddy, Inc. (2014) Wastewater engineering: treatment and reuse, 5th ed. McGraw-Hill Education

Gholipour S, Mohammadi F, Nikaeen M, Shamsizadeh Z, Khazeni A, Sahbaei Z, Mousavi S.M, Ghobadian M, Mirhendi H (2021) COVID-19 infection risk from exposure to aerosols of wastewater treatment plants. Chemosphere 273. https://doi.org/10.1016/j.chemosphere.2021.129701

Jimenez JA, Ferrer JS (2006) Wastewater treatment plants: planning, design, and operation. CRC Press

Jönsson S, Åström S, Svensson K, Carlsson M (2015) Five years' experience of a large-scale municipal sequencing batch reactor (SBR) for treatment of municipal wastewater. Water Sci Technol 71(7):1022–1028

Judd S (2008) The MBR book: principles and applications of membrane bioreactors in water and wastewater treatment. Elsevier

Langston CS (2008) The heartbeat of a nation: How small towns shape America. University of Kansas Press, Kansas

Lettinga G (1995) Anaerobic wastewater treatment with UASB reactors. Wageningen Academic Publishers, Wageningen

Lowe E, Milner S, Pryor S (2004) Anaerobic digestion of biosolids and municipal solid waste: potential for co-digestion. Waste Manag 24(7):713–722

Lüderitz C, Lehnig M, Amezquita A, Kroiß H (2019) Decentralized wastewater management in a developing country: a case study on the implementation of ecological sanitation in San Pedro La Laguna, Guatemala. J Water Sanit Hyg Dev. 9(2):313–324

Maqbool T, Zhang J, Qin Y, Ly QV, Asif MB, Zhang X, Zhang Z (2020) Seasonal occurrence of N-nitrosamines and their association with dissolved organic matter in full-scale drinking water systems: determination by LC-MS and EEM-PARAFAC. Water Res 183.https://doi.org/10.1016/j.watres.2020.116096

Mara D, Drangert B (2014) Sanitation and health. Routledge, London

Massoud MA, Tarhini A, Nasr JA (2009) Decentralized approaches to wastewater treatment and management: applicability in developing countries. J Environ Manag 90(1):652–659

McCarty PL, McKinney RE (1961) The role of research in the development of anaerobic waste treatment processes. J Water Pollut Control Fed 33(7):761–776

McCarty PL, Bae J, Kim JW (2011) The future of anaerobic treatment technology. Bioresour Technol 102(10):7688–7702

Meng X, Huang Z, Ge G (2024) Upgrade and reconstruction of biological processes in municipal wastewater treatment plants. Desalination Water Treatment 317. https://doi.org/10.1016/j.dwt.2024.100299

Montalvo FF, Jimenez JA, Ferrer JS (2009) Experience with the ecosan approach in San Pedro La Laguna, Guatemala. Ecol Eng 35(8):1207–1214

Ozimek T, Pijanowska J (2024) Social perception of the need and efficiency of ecological sewage treatment plants in small towns in Poland. Ecohydrol Hydrobiol 24(2):367–373. https://doi.org/10.1016/j.ecohyd.2023.08.004

Publishing IWA (2007) Membrane technology for water and wastewater treatment. IWA Publishing, London

Singh R, Singh P, Kaushik R, Verma Y (2022) An overview on the utilization of wastewater sludge as a source of nutrients and soil amendment. Bioresour Technol 342:122278

Tahoe Regional Planning Agency. Wastewater Treatment. Published 11 Jan 2023. Accessed https://www.trpa.gov/

Tchobanoglous G, Burton FL, Stensel HD (2003) Wastewater treatment process, 5th ed. Metcalf & Eddy, Inc.

Tilley E, Ulrich L, Lüthi C, Reymond P, Zurbrügg C (2014) Compendium of sanitation systems and technologies, 2nd revised ed. Swiss Federal Institute of Aquatic Science and Technology (Eawag)

U.S. Environmental Protection Agency (USEPA). Small community wastewater systems. Published 29 Aug 2018. Accessed https://www.epa.gov/water-research/small-wastewater-systems-research

Vymazal J (2008) Constructed wetlands for wastewater treatment. Ecol Eng 61(6):888–895

Vymazal J (2009) Constructed wetlands for wastewater treatment. CRC Press

Vymazal J (2011) Horizontal flow constructed wetlands. CRC Press

Wallace S, Reddy KR (2000) Treatment of wastewater in constructed wetlands. Lewis Publishers

WSAA (Water Services Association of Australia). Wastewater treatment plants. Accessed https://www.wsaa.asn.au/

Chapter 17
A Critical Understanding of Biological Treatment Techniques of Sewage Wastewater

Sukhendu Dey, Megha Santra, Apurba Ratan Ghosh, Prem Rajak, Abhratanu Ganguly, and Palas Samanta

Abstract One of the most difficult problems influencing wastewater treatment in the developed world is the treatment and disposal of sewage sludge. Sewage sludge is often disposed of either spreading it over land used for agriculture or by making a cake that may be deposited in a landfill or burned. Water needs to be taken out in order to make a sludge cake. The manner in which water is bonded to the flocs or sludge particles limits the amount of water that may be removed by pressure-based dewatering technologies. In order to drive water under an electrical gradient from the positive (anode) electrode to the negative (cathode) electrode, a DC voltage must be applied across the sludge. The typical metallic anode dissolves quickly in the acidic environment produced by the electrolysis of water, which is one of the reasons this technology has not been used in practice. Currently, membrane separation, ion exchange, electrodialysis, and adsorption are the chemical and physical treatment techniques utilized to dispose of wastewater. But these treatment techniques always result in some issues, such excessive expense, and secondary pollution, hence these technologies are only used under specific circumstances. On the other hand, biological treatments—particularly the activated sludge process—are recommended due to their superior economic performance and ability to prevent secondary pollution. Accordingly, this chapter will discuss the different biological treatment technologies of sewage/wastes, also called secondary treatment methods, in a very comprehensive way.

Keywords Activated sludge process · Biological treatments · Sewage sludge · Secondary treatment methods · Wastewater

S. Dey · M. Santra · A. R. Ghosh
Department of Environmental Science, The University of Burdwan, Burdwan, West Bengal, India

P. Rajak · A. Ganguly
Department of Animal Science, Kazi Nazrul University, Asansol, West Bengal, India

P. Samanta (✉)
Department of Environmental Science, Sukanta Mahavidyalaya, University of North Bengal, Dhupguri, West Bengal, India
e-mail: samanta.palas2010@gmail.com

S. Souabi et al. (eds.), *Wastewater Treatment Plants*, Water Science and Technology Library 130, https://doi.org/10.1007/978-3-031-87461-1_17

17.1 Introduction

Growing demand-driven agricultural evolution, smart life, and industrial modernization are the main causes of the growing pollution in water. Water bodies are receiving contaminated effluents from industries on a constant basis (Babuponnusami et al. 2023). According to Kinigopoulou et al. (2022), the effluents comprise a variety of organic and inorganic materials, such as microplastics, polyaromatic hydrocarbons, heavy metals, and medications. Unsuitable farming methods lead to the intermingling of household and electronic waste, as well as the deposition of heavy metals such as cadmium and copper in water bodies. Additionally, storm runoff and atmospheric deposition raise the concentration of organic and inorganic contaminants, including heavy metals, in water bodies, contaminating freshwater (Keskin et al. 2007). Accordingly, wastewater treatment and proper handling of the resulting sludge are essential from an environmental safety perspective due to the harmful effects of municipal, industrial, and healthcare wastewater on water, land, air, and agricultural commodities (Choudri et al. 2018).

In this scenario, reclaiming or recycling wastewater is most crucial requirements, despite the freshwater supply being rapidly depleted and rising. According to Alsharhan et al. (2020), rivers and subterranean water sources provide over 70% freshwater utilized for irrigation. Water used for agriculture accounts for 92% of total water use worldwide. There is a water problem for irrigation because 40% of the world's population lives in significant water-stressed regions (Grogan et al. 2017). Accordingly, to overcome this problem treating wastewater could be a viable solution and as a result of that treating of wastewater and their subsequent uses growing rapidly across the globe. Among Asian countries, China has an approximately 1.3 million hectares of reused wastewater, while only 37.6% of India's sewage from cities is thought to be processed at the moment (Kesari et al. 2021).

Numerous contaminants like pesticides, pharmaceuticals, heavy metals, microplastics, polyaromatic hydrocarbons, etc., including nutrients, including phosphorus, nitrogen, potassium, and sulfur, are present in both sewage water and treated sewage (Schweitzer and Noblet 2018). In one hand, this treated sewage enhances the fertility of soil, boosts crop yield, lowers the need for fertilizers, and may even lower agricultural production costs but in other hand reclaimed wastewater revealed a number of global health issues connected to public health and the environment (Ungureanu et al. 2020). This is basically due to risk factors associated with contaminants like pesticides, pharmaceuticals, heavy metals, microplastics, polyaromatic hydrocarbons, etc., as well as pathogen exposure. Thus, it is crucial to evaluate the health risks associated with reclaimed wastewater. For the aforementioned reason, prior to releasing wastewater into rivers, farmland, or soils, a sophisticated wastewater treatment procedure needs to be employed (Barancheshme and Munir 2018). As a result, the present study also suggested an advanced wastewater management approach. Currently, wastewater is treated using a variety of chemical and physical techniques, such as membrane separation, ion exchange, electrodialysis, and adsorption (Peng and Guo 2020). But these treatment techniques always result in some issues, such as excessive expense and secondary pollution, hence these technologies are only used under specific circumstances. However, because of their superior economic performance and ability to prevent secondary contamination, biological treatments—particularly the activated sludge process—are recommended (Ahmed et al. 2021). Accordingly, this chapter will discuss the different biological treatment technologies of sewage/wastes, also called secondary treatment methods, in a very comprehensive way.

17.2 Toxic Pollutants in Sewage Wastewater

The toxic pollutants that accumulate in wastewater can have grave consequences for the environment and human health if left untreated. In order to mitigate these risks, various treatment techniques have been developed to effectively remove these pollutants from our waterways. Two of the most widely utilized methods are activated sludges and lagoons (Karri et al. 2021; Rezai and Allahkarami 2021). Activated sludge involves the use of bacteria to break down and transform organic matter in wastewater. The bacteria consume these pollutants and convert them into sludge, which is then treated and removed from the water. This process is both safe and efficient, producing clean, pollutant-free water (Saravanan et al. 2021). Lagoons are large, open bodies of water that are designed to naturally purify wastewater through a combination of natural processes, including aeration, sedimentation, and skillful management of algae and other aquatic life (Shammas et al. 2009). By allowing the water to be exposed to natural elements, pollutants are broken down and removed from the water, leaving it safe for consumption. Both activated sludges and lagoons can be effective methods for treating wastewater, depending on the specific situation. While activated sludges are usually preferred for larger, more complex wastewater treatments, lagoons often prove to be a suitable solution for smaller, simpler operations (de Gisi et al. 2015; Mohammadi and Fataei 2019; Bachi et al. 2022). Regardless of the technique employed, however, the ultimate goal is always the same: to ensure that our rivers, lakes, and oceans are kept safe and free from toxic pollutants.

17.2.1 Pharmaceuticals

Basically, pharmaceuticals compounds are the man-made chemicals in our environment. Large volumes of fresh water are used for the production of pharmaceutical chemicals, and enormous amounts of wastewater are produced. Because of its large volume, complexity, and risk, this pharmaceutical manufacturing activity is classified as a "red category" and has high levels of solids content, chemical oxygen demand (COD), biological oxygen demand (BOD), and supplemental chemicals (Adishkumar et al. 2012; Changotra et al. 2017; Gadipelly et al. 2014; Martinez et al. 2018). Comparison with domestic waste pharmaceutical compounds production generates 500–6500 m^3 wastewater to produce one ton of antibiotics along with huge organic pollutants (Zhang et al. 2018). To treat pharmaceutical wastewater many procedures are taken such as submerged hollow fiber membrane bioreactor (Wen et al. 2004), activated carbon filter, flotation, sedimentation, coagulation and membrane separation (Guo et al. 2017; Choi et al. 2008), advanced oxidation processes (Changotra et al. 2017; Pal 2018; Klavarioti et al. 2009), coupled advanced oxidation processes (AOPs), and biological treatments (Martinez et al. 2018; Changotra et al. 2018; Oller et al. 2011). From the different pharmaceutical industry different kinds of wastewater are generated at the time of production scale and process. Biopharmaceutical wastewater is generated due to highly concentrated antibiotic wastewater which has characteristics as high supersaturation concentration, sulfate concentration, low carbon/nitrogen (C/N) ratio, and complicated composition, strong fluctuation in quantity, high chroma, and biological toxicity. On the other hand chemical pharmacy also contains high concentrated salt content, lack of nutrients and all the compounds are difficult to biodegrade and toxicity to microbiology. The wastewater also contains sugar, glycosides, cellulose, lignin, organic pigment, tannins, anthraquinone, alkali content, and other organic matter. Biological treatment technology is divided into anaerobic and aerobic process and combination process (Guo et al. 2017).

17.2.2 Personal Care Products

The use of personal care products such as shampoos, conditioners, and styling products are an integral part of daily life for many people. As such, proper disposal techniques are of utmost importance, in order to preserve the environment and ensure safe and healthy habitats for plants and animals alike (Khalid and Abdollahi 2021). In this section, we will explore two such treatment methods: activated sludges and lagoons. Wastewater treatment techniques known as activated sludges make it easier to remove organics, sediments, and other pollutants from wastewater. In these systems, bacteria and other microorganisms are used to provide biological cleansing in addition to mechanical aeration. Compared to other approaches, the process of treating wastewater using activated sludge is comparatively efficient and economical because it uses less energy and land (Salgado et al. 2012; Ahansazan et al. 2014). Conversely, lagoons are open-air, shallow ponds used for treating or storing wastewater. These waterways can be utilized alone or in combination with other treatment techniques, including activated sludges, to treat wastewater. Lagoons are often used to extract pollutants such as phosphorus and nitrogen, as well as to provide a habitat for species of plants and animals (Randrianarison and Ashraf 2017; Rodrigues-Filho et al. 2023). Both activated sludges and lagoons are important treatments techniques for personal care product wastewater (Meyer et al. 2019), helping to ensure that the environment is not harmed by the disposal of such products. By understanding the benefits and drawbacks of both methods, individuals can make an informed decision about which method works best for them.

17.2.3 Pesticides

Pesticides are made up of several chemical compounds that are mostly employed in horticulture to control insect populations, undesired organisms, and plant growth. Varied pesticides have varied chemical compositions, such as amides, carbamates, triazoles, organophosphates, neonicotinoids, and organochlorines. Additionally, pesticides are utilized in homes, particularly in gardens and lawns, where runoff from these areas may expose chemicals to nearby water bodies or the sewage system. The nature of the influent, the chemical structure of the pesticide, and the pH value of the pesticides poisoning the water, which ranges from 0.5 to 14, make the removal of pesticides from wastewater exceedingly complex and hard (Bracamontes-Ruelas et al. 2022).

17.2.4 Heavy Metals and Microplastics

Sewage effluent contains heavy metals and microplastics, which is a serious environmental risk. Microplastics and heavy metals have the potential to negatively impact human health and ecosystems. Microplastics are minuscule particles of plastic, often less than 5 μm, which arise from the disintegration of larger plastic objects or are purposefully produced for different uses (microbeads in personal care products, for example) (Karbalaei et al. 2018; Khalid et al. 2021). There are several ways that microplastics might end up in sewage effluent. For example, plastic particles from clothing, home objects, and industrial processes can find their way into both domestic and industrial wastewater (Ngo et al. 2019). Microplastics have the ability to linger in the environment for a very long period, build up in aquatic ecosystems, and perhaps endanger marine life. They might also make their way into the food chain, endangering people's health (Yuan et al. 2022).

On the other hand, heavy metals that can find their way into sewage systems include lead, mercury, cadmium, and chromium. These metals can enter through stormwater runoff, industrial discharges, and domestic sewage from homes that use goods that contain those (Hwang et al. 2016). Heavy metals are typically present in sewage discharge due to industrial operations. The final effluent may still contain certain heavy metals even though treatment systems are built to eliminate impurities (Dutta et al. 2021). Fish, invertebrates, and other aquatic life may be harmed by heavy metal poisoning. Long-term hazards may result from their accumulation in the sediments of aquatic bodies (Al-Sarraj et al. 2022). Furthermore, heavy metals can have a negative impact on human health and cause a variety of diseases and problems when they enter the food chain (Sonone et al. 2020).

Reducing the use of plastics and products containing heavy metals, increasing public awareness, and implementing technological breakthroughs in wastewater treatment are all necessary to address the problem of microplastics and heavy metals in sewage effluent (Xu et al. 2022). The reduction of these pollutants' negative effects on the environment and public health necessitates multidisciplinary cooperation and integrated water management strategies (Pirali zefrehei et al. 2022).

17.3 Different Treatment Options for Wastewater Treatment

Sewage/wastewater treatment is the answer that safeguards both the environment and people. It constitutes a collection of methods that lessen and/or remove contaminants from the water to ensure it can be used again for farming or released into the natural environment without causing harm (Sembuil et al. 2022). Different types of advanced processes are used for destruction of inorganic and organic contaminants in wastewater through inactivation of pathogens and pathogenic indicators (Ikai et al. 2010; Cho et al. 2005; Glaze et al. 1987; Glaze 1987). Details about different techniques like supercritical water oxidation, wet air oxidation, fenton reagent, ultrasound oxidation, photocatalytic oxidation, electrochemical oxidation and ozonation, etc., are summarized in Table 17.1. In this chapter, we mainly address the different aerobic treatment options for treating sewage effluents.

Biological process is a treatment technique used for wastewater treatment through the action of bacteria and other microorganisms to degrade organic matter in the wastewater. The process is based on the principle of biological oxygen demand (BOD) and includes the use of mechanical aerators to provide oxygen for the microorganisms in the water (Chan et al. 2009; Skouteris et al. 2020). Activated sludges are one popular method of aerobic wastewater treatment. The process involves the maintenance of a mix of microorganisms, bacteria, and protozoa in a tank containing anaerobically treated wastewater (Tocchi et al. 2012). The organisms are given a steady supply of oxygen to break down the wastewater into simpler organic compounds for removal. Sludge is then removed from the tank and sent to a settling pond or lagoon, where it is allowed to settle and solidify before being sent on for further treatment (Kodali 2017). Lagoons are open-air facilities used for storing and treating wastewater. In this process, wastewater is pumped into a series of lagoons where natural processes like evaporation, biological degradation, and sedimentation take place (Pishgar et al. 2020). Depending on the size of the lagoons and the amount of wastewater being treated, the lagoons can be either aerobic or anaerobic in nature (Chan et al. 2009). Mechanical aerators are used in aerobic lagoons to supply oxygen to the microorganisms that decompose the organic content and particles in the wastewater. In anaerobic lagoons, the microbes rely on a natural supply of oxygen from the atmosphere. The end result of either process is a cleaner, clearer wastewater

Table 17.1 Advantages and disadvantages of different wastewater techniques

Treatment technology	Advantages	Disadvantages	References
Biological treatment process	At high loading rates and low temperature good removal effectiveness can participate in system, this is comparatively simple reactors due to construction and process of operation, bioenergy produced as methane, this process is used as high performance to remove emergent contaminants from pharmaceutical wastewater	Nutrients and pathogens are removed but post-treatment is required, where hydrogen sulfide also generated and start-up process takes time due to low progress rate of microbes to be active so higher settling time and area is required	(Cecconet et al. 2017; Couto et al. 2019; Dogan et al. 2020; Tiwari et al. 2017)
Constructed wetlands process	Low energy consumption, operational and maintenance costs observed and worked as high pharmaceutical contaminants removal	Some of pharmaceuticals cannot remove completely	(Ingrao et al. 2020; Li et al. 2014a, b; Liu et al. 2019; Verlicchi and Zambello 2014)
Adsorption process	Lower energy consumption, little amount of sludge production and simple operating conditions observed where large volume of wastewater treated and higher organic load also. This process has very good ionic and colloidal matter removal efficiency	Low operating cost making and effective process is difficult where residual management and recycling is serious concern. Electrode replacement is needed	(Garba et al. 2019; Kyzas et al. 2015; Quesada et al. 2019; Siyal et al. 2020)
Electrocoagulation process	It is a moderately faster treatment process than the others and huge amount can be treated. Using electrode cost is relatively low where particles electroflotation occurred by hydrogen bubbles	Only contaminants can separate	(Feng et al. 2013; Sires and Brillas 2012; Song et al. 2017)
Membrane filtration process	It is moderately quick and has finest removal efficiency of organic matter and salt	Membrane fouling occurred and here also only contaminants can separate	(Fonseca Couto et al. 2018; Gu et al. 2018; Janssens et al. 2017; Tan et al. 2019)

(continued)

Table 17.1 (continued)

Treatment technology	Advantages	Disadvantages	References
Advanced oxidation process	Through this process heavy metals can also be removed in the forms of precipitated of M(OH)*x*. High-energy conversion is required due direct chemical energy conversion within substrate to electricity. Due to involvement of hydroxyl radical it virtually reacts with every aqueous pollutant without discriminating	Very much costly and requires OH radical and other reagents proportional to contaminants. It is not cost-effective to use AOPs to handle huge amount of wastewater	(Anjali and Shanthakumar 2019; Fatta-Kassinos et al. 2011; Michael-Kordatou et al. 2018; Serpone et al. 2017)
Microbial fuel cell process	Excellent capacity to remove pharmaceutical pollutants at low temperature and gas treatment lacking and rich in CO_2 with no useful energy. To reduce the operating cost cathode is aerated passively	Depending on power output and treatment efficiency this system is not fully developed and ready for application. Utilization of MFC is the low-power density which hinders the scaling up of the system	(Gude 2016; Ismail and Habeeb 2017; Nayak and Ghosh 2019; Xia et al. 2018; Xu et al. 2018)
Ozonation process	In ozonation process oxidant compound assisting in sterilization and disinfection properties. Organic contaminants removed efficiently	Little employment in pharmaceutical is required to remove contaminants and radical scavenger is disrupted	(Gomes et al. 2017; Mehrjouei et al. 2015; Wang and Chen 2020)
Ion exchange process	It is efficient and very much useful technique for water softening where no perforation of substances into soft water. Treated wastewater also used for water treatment and most of the heavy metal can reuse	The machines used for softening the water is known as ion exchangers which are cleaned for high saturation level and require high operational cost. To make water soft sodium ions are involved which is not safe	(Ahmed and Hameed 2019; Krstić et al. 2018; Mazur et al. 2018)

that is safe to discharge into the environment (Gavrilescu and Macoveanu 1999; Al-Amshawee et al. 2020). The details of different biological treatment processes of sewage wastewater are described below:

17.3.1 Activate Sludge Process

Activated sludge process (ASP) is the most widely used and widely accepted wastewater treatment method. Microorganisms' metabolism is employed in a complicated biochemical reaction process to eliminate organic contaminants from wastewater. The temperature of the wastewater is one example of an external factor affecting this process. A variety of processes, including nitrification and denitrification, biological phosphorus removal, and the removal of organic carbon, are involved in ASP (Sheik and Ambati 2023). In biochemical process of ASP wastewater passed through different tanks, where in the presence of oxygen it goes through artificial agitation and forms many nitrifying organisms stuck to the sludge, which eventually settled to the tank's bottom (Abbasi and Abbasi 2010). The first component of ASP's infrastructure is a single bioreactor that runs constantly and is used by suspended microorganisms to eat dissolved organic materials. Because aerobic biodegradation requires dissolved oxygen (DO), the reactor is aerated. Second, treated wastewater is separated from flocculated biomass or activated sludge in a settling tank, also known as a secondary settler or clarifier. Sludge recycling lines, thirdly, facilitate the return of settled sludge to the bioreactor. Fourth, a line for sludge waste at the bottom of the clarifier, from which a little quantity of sludge is taken out to keep the bioreactor's biomass concentration at a constant level (Hreiz et al. 2015). The benefits and drawbacks of the activated sludge process are as follows:

Advantages:

- Lower space is needed; shorter footprint
- Quicker procedure
- Greater effectiveness of BOD elimination
- Higher concentration of germs
- Able to handle more complex waste routes.

Disadvantages:

- Greater initial funding
- Increased operating costs for disposing of sediments and dewatering substances
- Needs should be effectively handled in order to more successfully manage upheavals
- Vulnerable to depositing sludge problems (bulking)
- Needs greater focus and more tests to regulate; highly skilled operators are needed.

17.3.2 Lagoons

Lagoons had always been a subject of intrigue for scientists, and the treatment techniques of lagoons in particular had been garnering increasing attention in recent times. It was believed that lagoons could be used for the treatment of a variety of wastewater sources due to their natural characteristics. Activated sludge processes, on the other hand, had been used for decades to improve the quality of wastewater before they are discharged into receiving waters (Bachi et al. 2022). The activated sludge processes used oxygen as an essential nutrient that encouraged the growth of microorganisms. The growth of these microbial populations transformed the organic matter present in the wastewater into carbon dioxide, water, and other products (Yeasmin et al. 2023). The combination of both treatment techniques, activated sludge and lagoons, had proved to be a highly effective way of removing pollutants from wastewater sources (Ahmed et al. 2017; Kumar et al. 2023). In many cases, their combined use had resolved problems that had previously been difficult to solve. An earthenware reservoir is used for wastewater treatment; a lagoon uses both biological and natural methods for treating sewage wastewater. Certain remote locales utilize them because they need lesser upkeep and consume less energy than other treatment alternatives. As part of their own wastewater treatment facilities, companies and factories utilize sewage treatment lagoons (Mancuso et al. 2023).

Lagoons could be either natural or constructed. Constructed lagoons were typically permanent, and could be designed to be either aerobic or anaerobic (Shammas et al. 2009). The natural lagoons, on the other hand, were often seasonal and required careful monitoring and treatment procedures to ensure effective wastewater treatment. The benefits of combining both the activated sludge and lagoon techniques were numerous (Teh et al. 2016). Not only did the combination create a more effective wastewater treatment system, but it also reduced the amount of sludge that would otherwise be generated in the process. Additionally, the combination allowed for the reuse of treated wastewater, providing a valuable resource for agricultural and other uses. Furthermore, lagoons were relatively easy to construct and maintain, making them an affordable wastewater treatment solution (Yahya et al. 2020).

Lagoon treatment plants generally encompass a range of easily distinguishable approaches that vary in terms of mixing intensity and aeration technique (Joshiba et al. 2019). The oxidizing pond represents an extreme case where oxygen supply is nearly solely a result of photosynthesis and mingling is entirely reliant on external factors, primarily winds. The pond's highest load and how it reacts with it are primarily determined by the surrounding environmental variables. The suspended mixed aerated lagoon represents the opposite extremity; here, mingling and oxygen delivery are accomplished using technical methods that offer the highest-level system regulation (Tong et al. 2020). Based on these two parameters, five different types of lagoon systems are generally employed for treating sewage wastewater, which are ranked according to how much technological support is used to achieve the necessary mixing and oxygenation:

(1) Oxidation ponds

(2) Mechanically assisted oxidation ponds
(3) Aerated oxidation ponds
(4) Facultative lagoons
(5) Suspension mixed aerated lagoons.

According to the kind of pollutants lagoons are intended to handle, sewage treatment lagoons may be either aerobic or anaerobic. Animal manure from farms is primarily treated in anaerobic lagoons (Agga et al. 2022). But more prevalent are aerobic lagoons, which can be completely or only partially aerating (this is also referred to as facultative lagoons) (Zema et al. 2016). Municipal and industrial wastewater with strength of low to medium is treated in facultative lagoons. Algae represent a necessary component of facultative and aerated lagoons for the treatment of wastewater (Mahapatra et al. 2013). In 2022, 4657 discharging lagoon systems have been identified by the EPA, among which 83% were found to be owned by public treatment plants. Rural communities with fewer than 3000 residents are served by these lagoon systems (84%).

17.3.2.1 Oxidation Ponds

Lagoons and oxidation ponds are common names for stabilization ponds. Using those, wastewater effluents undergo rudimentary secondary treatment (Shammas et al. 2009). Heterotrophic bacteria in an oxidation pond break down organic waste in the sewage to create minerals and cellular components. They help algae to flourish (Sharma and Arivalagan 2021). By generating oxygen, algal population growth promotes the breakdown of organic materials. The oxygen utilized by heterotrophic bacteria is replenished by this oxygen's generation. Generally, the algal development in oxidation ponds requires a depth of fewer than ten feet. Furthermore, since these ponds are highly affected by seasonal temperature fluctuations, their application is primarily limited to areas with warmer climates. Because of settlement of bacterial and algal cells during sewage's breakdown, ponds also have a tendency to fill. Oxidation ponds need big storage capacity and lengthy retention periods, and they are often ineffective. Because of the sluggish rate of decomposition, the ponds' wastewater carrying oxidized byproducts must be routinely emptied. Oxidation ponds are generally four kinds, namely aerobic (high-rate), anaerobic, facultative, and maturation oxidation ponds (Kadri et al. 2021) which are described below in detail.

Aerobic (High-Rate) Ponds

Since algae are actively engaged in photosynthetic processes, aerobic ponds, also called high-rate algae ponds, are able to sustain dissolved oxygen levels across their 30- to 45-cm depth (Ranjan et al. 2019). In daytime, oxygen is supplied by photosynthetic activity, and at midnight, oxygenation is produced by the wind since the pond is shallower (Baxa et al. 2021). Biochemical oxygen demand (BOD) elimination capacity of aerobic ponds is widely recognized, making them a great option for locations with low land costs. Some additional attributes are a BOD loading rate of

112–225 kg/1000 m^3 day, a holding period of 2–6 days, and a 95% BOD elimination efficiency (Ali et al. 2020).

Anaerobic ponds

This is very small set up of oxidation pond system. An oxygen-free environment is required for anaerobic ponds to function (Uddin and Wright 2022). The ponds are usually intended to be between two and five meters deep, with a retention period of one to one and a half days, an ideal pH of below 6.2, temperature higher than fifteen degrees Celsius, and an organic load capacity of three thousand kg per day. The lagoon experiences severe anaerobic state (no dissolved oxygen) due to excessive organic loading. Anaerobic ponds are very effective in hot climates and often operate similarly to open septic tanks. About 60% of BOD (at 20 °C) could be eliminated using this pond (Turker et al. 2009). This efficacy varies with climate. Sedimentation is the primary process that drives cleaning. Bacteria and viruses disappear from the pond by adhering themselves to sinking solids or by dying due to predatory activity or feed depletion. Helminths sink into the bottom of the pond. Practically speaking, facultative ponds and anaerobic ponds are combined together (Martinez et al. 2014). Sewage that has a BOD up to 300 mg/l and temperature greater than 20 °C can be retained throughout a day using hydraulic retention time. The smell that a design could produce has long been a concern for engineers (Ashadullah et al. 2021). In addition, the level of SO_4^{2-} in sewage might be below 500 mg/l, which will minimize odor issues in well-constructed ponds. Anaerobic reactors of all kinds use identical mechanisms for elimination of organic materials from their systems, which in ponds (del Castillo et al. 2022).

Facultative ponds

An anaerobic and aerobic mixed treatment system is called a facultative pond. A typical pond has three primary sections: an anaerobic bottom area with anaerobic microbes, an aerobic outside area with bacteria and algae, and an area that is halfway between the two habitats where microbes can thrive in both (Chan et al. 2009). Water that comes from a pre-treatment origin reaches the pond if it is utilized in sequence. Facultative ponds eliminate 95% of the BOD, usually in the range of 100–400 kg BOD/ha/day (Mulligan and Gibbs 2003). The period of treatment might be as long as two or three weeks due to photosynthesis reactions that take place within the unit that utilize algae as decomposing organisms. The typical depth of a facultative pond is between one and two meters (Abdel-Raouf et al. 2012).

Maturation Ponds

Mature ponds employ algae for their main source of energy, much like facultative ponds do. They have a far more varied algal community than facultative ponds. From pond to pond throughout the series, the algal diversity rises as a result (Kaid et al. 2022). Together with photo-oxidation, algal activity controls the primary removal processes, particularly for pathogens and fecal coliforms. These ponds accept wastewater from facultative ponds, and their quantity and size are determined by the final effluent's necessary bacteriological quality. Still, maturation ponds eliminate fecal coliform, infectious agents, and nutrients, whereas facultative ponds usually treat BOD (Dias et al. 2017). A shorter depth range of 1–1.15 m distinguishes it from the other pond types. This is the only pond type not included in aerobics pond's characteristics.

Following are the advantages and disadvantages of oxidation pond process for treating sewage wastewater:

Advantages:

- Little cost, simple to build, and little maintenance
- It is not necessary for disinfecting wastewater
- Ability to manage a wide range of hydraulic workloads
- Perfect for tropical regions and tiny settlements
- Finishes the cleanup of sludge
- Treats several kinds of wastewater (municipal or industrial), with higher removal rates of helminth, fecal coliform, and BOD than other treatment techniques such biological filters, activated sludge, and rotating biological contactors.

Disadvantages:

- Need a sizable amount of land
- Algal levels with elevated BOD and TSS
- Lower removal efficiency, especially during the winter
- Greater footprint.

17.3.2.2 Mechanically-Assisted Oxidation Ponds

By deliberately boosting mixing behavior, the pond might be helped to keep the non-motile algae in suspended form. Therefore reducing the negative effects of (i) stratification on oxygen production and (ii) increased loading of the pond supernatant due to sludge feedback during hot weather (Owusu-Twum and Sharara 2020). Placing a recirculation pump that turns the pond's level either once or twice daily may assist to achieve it in tiny pools. Installing a floating stirrer is recommended for big ponds (>10 ha). Though the blades are positioned deep beneath the water's surface of the pond and revolve slowly, they're comparable to floating aerators. The goal is for the flow of a lot of water slowly since this will enable mixing to spread out beyond the stirrer somewhat. Just about 0.1 W/m^3 of pool capacity is needed in terms of energy. The pond's fundamental physical and biological activities remain unchanged when stirring is installed. Only in situations where the natural surroundings are conducive to stratification is going to happen in the pond, does it offer additional protection for healthy non-motile algae development by preserving a minimal mixing level. There is no impact on the sludge layer's existence or activity (Kadlec 2003; Ho and Goethals 2020).

17.3.2.3 Aerated Oxidation Ponds

This technique installs air tubes with diffusion devices at set times along the pond bottom to increase pond's natural oxygenation capability. The conduits are occasionally elevated over the bottom. Photosynthesis from algae is still the prime producer of oxygen within the pond, but bursting bubbling combine the contents and increases the amount of oxygen present (Boyd 1998; Boyd et al. 2018). A sludge layer builds, similar to what happens in an oxidation pond, since there is not enough mixing power to stop settleable organic substances from the influent from settling. The mud layer remains unaltered so that fermentation can continue, unlike in oxidation pond, thanks to the aeration pipes being raised over the pond bottom (Ghangrekar 2022).

17.3.2.4 Facultative Lagoons

Facultative lagoons are another prime option for treating wastewater from industries and municipalities. These oxygenated and manually mixed clay lagoons typically range in depth from 1.2 to 2.4 m (4–8 feet) (Yahya et al. 2020). The oxygen exists due to result of algal breathing and atmospheric reaeration, which fosters the growth of facultative and aerobic species. Anaerobic microorganisms are found in the lagoon's bottom layer, which also contains sewage deposits (Owusu-Twum and Sharara 2020). Known as the facultative zone, an intermediate oxygen-deprived layer is anaerobic at the bottom and aerobic at the top. Temperature-induced fluctuations in water density could cause such layers to remain on the surface for extended periods of time. Because of temperature variations, inversions can happen in the spring and fall whenever the water's surface might possess greater density compared to lower levels. During these unsteady times, this denser water sinks, causing turbulence and unpleasant smells (Buchanan et al. 2023; Mahapatra et al. 2022).

There are two distinct varieties of such ponds: primary facultative ponds, which collect raw sewage/wastewater (Kadri et al. 2021), and secondary facultative ponds, which get settled wastewater from first *stage*, typically anaerobic ponds effluent (Adhikari and Fedler 2020). In order to support the algal growth, facultative ponds have been designed for removing BOD based on a minimal organic substrate loading. In this manner, algae provide the oxygen that's required to eliminate soluble BOD. Deep green water color indicates healthy populations of algae, although sometimes these communities might become red or pink because of purple sulfide-oxidizing photosynthetic activity. There is a small overloading that causes this ecological shift (Almasi et al. 2015). As a result, a qualitative sign of a cleaning mechanism operating at peak efficiency is the shift in color in facultative ponds. A facultative pond operating at peak efficiency will have varying concentrations of algae, typically 500–2000 μg chlorophyll/liter, depending on temperatures and organic load (Babuponnusami et al. 2023). Dissolved oxygen content and pH levels vary throughout the day because of algae's photosynthetic activities. The way facultative ponds behave is significantly influenced by factors like wind speed, which causes the pond water to mingle. The optimal wastewater stabilization results from a higher mixing degree, which guarantees a consistent distribution of BOD, dissolved oxygen, microbes, and algae. Currently, near about 7000 facultative lagoons in use in the United States, where the technology behind them has been widely used for at least 90 years (Li et al. 2014a, b). Following are the advantages and disadvantages of facultative lagoons process:

Advantages:

- Uses little energy (gravity feeding)
- Sturdy over load variations
- Doesn't need harvesting biomass
- Cheaper to build than subterranean treating methods
- Carbon-free operation (because of photosynthesis against respiration)
- Easy operation
- Ammonia, bacteria, BOD, fecal coliform, and settleable matter are all removed with a moderate degree of effectiveness.

Disadvantages:

- Possible environment for mosquitoes
- Elevated levels of algae within the wastewater
- Algae absorb most nitrogen, released into the atmosphere as ammonia
- Larger space required
- Periodically, inert substances and accumulated sludges need to be removed
- Ammonia levels in wastewater are unpredictable and difficult to regulate
- Because of decreased microbial activity in colder regions, sludge formation is expected to be greater.

17.3.2.5 Suspension Mixed Aerated Lagoons

Because of system's extremely higher oxygenation energy input, suspended solids do not settle. With the exception of (i) not having a tank for settling and (ii) not having sewage recycle, the setup is the same that of a typical AS system as long as the lagoons are kept in an aerobic condition. Accordingly, the sludge age (Rs) and the hydraulic retention time (Rh) are equal (dos Santos et al. 2016). The endogenous (XE), active OHO (XBH), and unbiodegradable particle organics from the influent (XI) constitute the majority of the organic particulate matter in the effluent, which is AS generated from the influent organics. The decline in COD between the effluent and unfiltered influent COD corresponds to the carbon-based need for oxygen per unit influent flow for endogenous respiration and growth, as determined by the COD balance. In overall, the COD decrease is unsatisfactory since the effluent's COD levels are too higher to be discharged unrestrictedly into recipient waterways (Abdel-Shafy et al. 2023). To obtain a somewhat good (albeit not nitrified) discharge quality, further processing is required having the explicit goal of eliminating the settleable particulates from wastewater flow without requiring for settling tanks (Orhon 2015).

On average, a number of facultative lagoons or oxidation ponds are used to accomplish further treatment. A wastewater that is mostly devoid of solids is produced in second lagoons when solids sink out and create a sludge layer (Ghaitidak and Yadav 2013). As the waste material inside this paper ferments, methane gas is released, accounting for a significant energy input. There exists an oxygen requirement in supernatant as a result of fermentation's recycling of electricity (COD) emanating from sludge layer to supernatant layers (Molino et al. 2013). Consequently, the behavior of second facultative lagoon doesn't initially appear different from the setup in which the wastewater is released straight into a facultative lagoon and the suspending mixed lagoon is removed. There's a significant distinction, though. Through physical/biological procedures like growth, adsorption, and flocculation, the water-soluble and particle organics in suspension mixed lagoon are converted to settleable solids (Fakhru'l-Razi et al. 2009). Because of this, the bulk of organic material settles out much more quickly and completely in the second pond. Thus, in next facultative pond, the suspending mixed lagoon serves to be a biological "flocculator," encouraging the segregation of solid and liquid (Opitz et al. 2021).

17.3.3 Trickling Filters

Trickling filter (TF) technology is a popular technique to remove nitrogen and chemical oxygen demand (COD) from surface and ground water (Ji et al. 2013; Van den Akker et al. 2008, 2011). Details Trickling Filters was a well-known environmental engineering firm specializing in water treatment techniques. It had developed a reputation for being the go-to source of advice and guidance on the most effective, cost-efficient procedures for cleaner water and downstream ecosystems (Ali et al. 2017; Sathya et al. 2022). Because microbial communities have a natural metabolic and physiological character that allows them to remove large amounts of pollutants, trickling filters are a preferred approach for treating biological wastewater where microorganisms play a vital role to limit pollutant strength (Hamza et al. 2016). The microbes' characteristic behavior is to adhere to moist surfaces, proliferate, and embed themselves in a slick environment to create biofilms made of extracellular polymeric compounds (Wang et al. 2015). Protozoa, fungus, algae, and bacteria make up the majority of the 0.1–0.2 mm thick biofilm. At the lower level of the filter, nitrifying bacteria are present for nitrification; fungi help to reduce pollutants at low pH conditions, while protozoa help to reduce effluent turbidity and maintain the biofilm. Wastewater flows over the biofilm where organic matter degrades into water and carbon dioxide due to the nature of microbes (Rangaswamy and Singh 2022). The under-drain system, filter medium, and rotary distributor are the three main parts of the TFs. Comparing with other technique TFs contain high process stability, low hydraulic retention time (HRT), and sludge processing. TFs established that 80–85% COD removal and 85–90% BOD removal is possible (Ali et al. 2017).

17.3.4 Rotating Biological Contactor

For the purpose of treating municipal and industrial wastewater, one form of biological wastewater treatment method is the rotating biological contactor (RBC) (Waqas et al. 2023). The principal wastewater in this wastewater treatment process is partially immersed in a number of huge plastic discs that are installed on a horizontal shaft (Rezai and Allahkarami 2021). Identical to biofilters, rotating biological contactors (RBCs) are fixed-film reactors where microorganisms are affixed to support medium. The supporting media in RBC are gradually spinning discs which are partially buried in sewage which is flowing through the reactor (Hassard et al. 2015). Since air is transported to sewage by exterior agitation caused by the spinning of discs, the adhering biofilm receives oxygen via the liquid while immersed as well as from the atmosphere while the layer is not immersed. Sloughed biofilm fragments are eliminated using the identical process as biofilters (Maurya et al. 2023). Simply, the spinning discs are periodically subjected to both air and sewage, which enables a coating of bacteria to develop on the surfaces of discs and break down the organic materials in sewage (Hassard et al. 2015).

When used in conjunction with primary sedimentation, high-rate biological treatments can remove up to 85% of heavy metals and BOD_5 that were initially present in raw sewage (Murshid et al. 2023). When compared to biofilters or RBCs, activated sludge typically yields an effluent with a marginally greater grade of these components. These procedures can remove a significant amount of germs and viruses, but not all of them, when combined with a disinfection phase. But they barely eliminate any dispersed minerals, non-biodegradable organics, nitrogen, or phosphorus (Ifelebuegu 2011; Inyinbor et al. 2021).

17.3.5 Combined Biofilm and Activated Sludge Technology

The treatment technique of "combined biofilm and activated sludge technology" is among the most efficient and cost-effective methods of wastewater treatment available (Babuponnusami et al. 2023). In short, the process involves the combination of both biological and mechanical treatment processes, which allows for the removal of pollutants at a fraction of the cost of traditional activated sludge processes (Ahmed et al. 2021). The effectiveness of this technology relies on the combination of both biological and mechanical components. The biological component, or the "biofilm," is a thin layer of microorganisms that live on the surface of the wastewater and absorbs pollutants from the water (Patel et al. 2022). The mechanical component, or "activated sludge," is composed of bacteria and other microorganisms that consume the pollutants in the wastewater, breaking them down into harmless byproducts (Saravanan et al. 2021). The combined biofilm and activated sludge process is one of the most efficient and cost-effective methods of wastewater treatment available today. It has been demonstrated to lower costs and enhance effluent quality in both municipal and industrial wastewater treatment facilities (Waqas et al. 2020). Additionally, this process is relatively simple to maintain and does not require any costly or complicated equipment.

17.3.6 Aerobic Photo-Bio Reactors Anaerobic Process

Miles of pipes, networks of tanks, and various aerobic and anaerobic processes coexisted in harmony within the treatment facility. It was a complex system, but the staff at the facility had a detailed understanding of its various components (Borgaonkar and Marhaba 2021). The most common treatments were activated sludges and lagoons, both of which relied on the use of aerobic photo-bio reactors (Silva et al. 2013). In the case of activated sludges, these reactors provided a home for bacterial cultures that helped to break down the pollutants in wastewater, while lagoons offered a more benign environment for the same purpose (Haq and Raj 2020). Of course, the treatment process also relied on a variety of other techniques, from chemical additions to physical separations. But the aerobic photo-bio reactors remained at the core of the treatment facility's efficacy.

17.4 Economic Aspect of the Treatment Technique

The treatment technique of activated sludges is one that has been a mainstay of water treatment systems for many years. Used in conjunction with aerated lagoons, it is often the most cost-effective method when a water treatment system is required. Activated sludges are generally used in water treatment systems when a large volume of wastewater needs to be treated, such as in a municipal or industrial wastewater treatment plant (Adonadaga 2014). Activated sludge is a biological process that requires a mixture of aerobic bacteria that break down organic matter in the wastewater. As the bacteria break down the organic material, they use up the oxygen in the wastewater, and the sludge is then removed from the water (Sabliy et al. 2019). Lagoons are another form of water treatment that can be used in conjunction with activated sludges. Lagoons are typically used in smaller volumes of wastewater where the flow rate is much slower and a high volume of bacteria can be present. These lagoons can also utilize algae and other aquatic plants to further remove organic matter from the water (Musa and Idrus 2021). The economic aspect of using activated sludges and lagoons as a treatment technique is an important consideration (Kamali et al. 2019). While activated sludges and lagoons may be more expensive to install and maintain than other treatment techniques, they are usually the most cost-effective over the long term (Rezai and Allahkarami 2021). Additionally, they are less labor-intensive than some other treatment methods, meaning that they can require fewer personnel to operate them. Furthermore, activated sludges and lagoons can often offer higher levels of wastewater treatment than some other systems, providing a better value for the money spent (Jóźwiakowska and Marzec 2020).

17.5 Conclusions

The conclusion from the research team's investigation into various treatment techniques was that activated sludges were the most effective. The team found that the sludge, when activated, was able to break down organic compounds and other contaminants in wastewater, resulting in a higher quality effluent. Lagoons were found to be the second most effective treatment technique, providing some degree of treatment. However, due to the lack of oxygenation in lagoons, the team concluded that they were not as effective at breaking down organic compounds as activated sludges. Overall, the research team concluded that activated sludges were the most effective treatment technique for wastewater. They found that the sludge was able to break down organic compounds and other contaminants in wastewater, resulting in a higher quality effluent. Lagoons were also found to be an effective treatment technique, though not as effective as activated sludges.

References

Abbasi T, Abbasi SA (2010) Enhancement in the efficiency of existing oxidation ponds by using aquatic weeds at little or no extra cost—the macrophyte-upgraded oxidation pond (MUOP). Bioremediat J 14(2):67–80. https://doi.org/10.1080/10889861003767076

Abdel-Raouf N, Al-Homaidan AA, Ibraheem I (2012) Microalgae and wastewater treatment. Saudi J Biol Sci 19(3):257–275. https://doi.org/10.1016/j.sjbs.2012.04.005

Abdel-Shafy HI, Ibrahim AM, Al-Sulaiman AM, Okasha RA (2023) Landfill leachate: sources, nature, organic composition, and treatment: an environmental overview. Ain Shams Eng J 102293. https://doi.org/10.1016/j.asej.2023.102293

Adhikari K, Fedler CB (2020) Pond-In-Pond: an alternative system for wastewater treatment for reuse. J Environ Chem Eng 8(2):103523. https://doi.org/10.1016/j.jece.2019.103523

Adishkumar S, Sivajothi S, Rajesh Banu J (2012) Coupled solar photo-fenton process with aerobic sequential batch reactor for treatment of pharmaceutical wastewater. Desalin Water Treat 48(1–3):89–95. https://doi.org/10.1080/19443994.2012.698799

Adonadaga MG (2014) Nutrient removal efficiency of activated sludge plants treating industrial and municipal wastewater in Ghana. J Environ Pollut Hum Health 2(3):58–62. https://doi.org/10.12691/jephh-2-3-1

Agga GE, Couch M, Parekh RR, Mahmoudi F, Appala K, Kasumba J, Loughrin JH, Conte ED (2022) Lagoon, anaerobic digestion, and composting of animal manure treatments impact on tetracycline resistance genes. Antibiotics 11(3):391. https://doi.org/10.3390/antibiotics11030391

Ahansazan B, Afrashteh H, Ahansazan N, Ahansazan Z (2014) Activated sludge process overview. Int J Environ Sci Dev 5(1):81. https://doi.org/10.7763/IJESD.2014.V5.455

Ahmed MJ, Hameed BH (2019) Insights into the isotherm and kinetic models for the coadsorption of pharmaceuticals in the absence and presence of metal ions: a review. J Environ Manage 252:109617. https://doi.org/10.1016/j.jenvman.2019.109617

Ahmed MB, Zhou JL, Ngo HH, Guo W, Thomaidis NS, Xu J (2017) Progress in the biological and chemical treatment technologies for emerging contaminant removal from wastewater: a critical review. J Hazard Mater 323:274–298. https://doi.org/10.1016/j.jhazmat.2016.04.045

Ahmed SF, Mofijur M, Nuzhat S, Chowdhury AT, Rafa N, Uddin MA, Inayat A, Mahila TMI, Ong HC, Chia WI, Show PL (2021) Recent developments in physical, biological, chemical, and hybrid treatment techniques for removing emerging contaminants from wastewater. J Hazard Mater 416:125912. https://doi.org/10.1016/j.jhazmat.2021.125912

Al-Amshawee SK, Yunus MY, Azoddein AA (2020) A review on aerobic biological processes for palm oil mill effluent: Possible approaches. In: IOP conference series: materials science and engineering, vol 736, No. 2. IOP Publishing, p 022035. https://doi.org/10.1088/1757-899X/736/2/022035

Ali I, Khan Z, Peng C, Naz I, Sultan M, Ali M, Mahmood, Niaz Y (2017) Identification and elucidation of the designing and operational issues of trickling filter systems for wastewater treatment. Polish J Environ Stud 26(6). https://doi.org/10.15244/pjoes/70627

Ali AE, Salem WM, Younes SM, Kaid M (2020) Modeling climatic effect on physiochemical parameters and microorganisms of Stabilization Pond Performance. Heliyon 6(5). https://doi.org/10.1016/j.heliyon.2020.e04005

Almasi A, Sharafi K, Hazrati S, Fazlzadehdavil M (2015) A survey on the ratio of effluent algal BOD concentration in primary and secondary facultative ponds to influent raw BOD concentration. Desalin Water Treat 53(13):3475–3481. https://doi.org/10.1080/19443994.2013.875945

Al-Sarraj ES, Eskandera MZ, Al-Taee SK (2022) Heavy metal pollution in Iraqi rivers and impact on human and fish health: a review. Biol App Environ Res 6(2):95–112. https://doi.org/10.51304/baer.2022.6.2.95

Alsharhan AS, Rizk ZE, Alsharhan AS, Rizk ZE (2020) Overview on global water resources. Water Resourc Integrat Manage United Arab Emirates 17–61. https://doi.org/10.1007/978-3-030-31684-6_2

Anjali R, Shanthakumar S (2019) Insights on the current status of occurrence and removal of antibiotics in wastewater by advanced oxidation processes. J Environ Manage 246:51–62. https://doi.org/10.1016/j.jenvman.2019.05.090

Ashadullah AKM, Shafiquzzaman M, Haider H, Alresheedi M, Azam MS, Ghumman AR (2021) Wastewater treatment by microalgal membrane bioreactor: evaluating the effect of organic loading rate and hydraulic residence time. J Environ Manage 278:111548. https://doi.org/10.1016/j.jenvman.2020.111548

Babuponnusami A, Sinha S, Ashokan H, Paul MV, Hariharan SP, Arun J, Gopinath KP, Le QH, Pugazhendhi A (2023) Advanced oxidation process (AOP) combined biological process for wastewater treatment: a review on advancements, feasibility and practicability of combined techniques. Environ Res 116944. https://doi.org/10.1016/j.envres.2023.116944

Bachi OE, Halilat MT, Bissati S, Al-Ansari N, Saggai S, Kouadri S, Najm HM (2022) Wastewater treatment performance of aerated lagoons, activated sludge and constructed wetlands under an arid algerian climate. Sustainability 14(24):16503. https://doi.org/10.3390/su142416503

Barancheshme F, Munir M (2018) Strategies to combat antibiotic resistance in the wastewater treatment plants. Front Microbiol 8:2603. https://doi.org/10.3389/fmicb.2017.02603

Baxa M, Musil M, Kummel M, Hanzlík P, Tesařová B, Pechar L (2021) Dissolved oxygen deficits in a shallow eutrophic aquatic ecosystem (fishpond)–Sediment oxygen demand and water column respiration alternately drive the oxygen regime. Sci Total Environ 766:142647. https://doi.org/10.1016/j.scitotenv.2020.142647

Borgaonkar AD, Marhaba TF (2021) Evaluation of sustainability strategies—a water quantity and quality perspective. In: Handbook of water purity and quality. Academic Press, pp 409–454. https://doi.org/10.1016/B978-0-12-821057-4.00011-2

Boyd CE (1998) Pond water aeration systems. Aquacultural Eng 18(1):9–40. https://doi.org/10.1016/S0144-8609(98)00019-3

Boyd CE, Torrans EL, Tucker CS (2018) Dissolved oxygen and aeration in ictalurid catfish aquaculture. J World Aquaculture Soc 49(1):7–70. https://doi.org/10.1111/jwas.12469

Bracamontes-Ruelas AR, Ordaz-Díaz LA, Bailón-Salas AM, Ríos-Saucedo JC, Reyes-Vidal Y, Reynoso-Cuevas L (2022) Emerging pollutants in wastewater, advanced oxidation processes as an alternative treatment and perspectives. Processes 10(5):1041. https://doi.org/10.3390/pr1005 1041

Buchanan PJ, Sun X, Weissman JL, Zakem E (2023) Oxygen intrusions sustain aerobic nitrite oxidation in anoxic marine zones. bioRxiv 2023–02. https://doi.org/10.1101/2023.02.22.529547

Cecconet D, Molognoni D, Callegari A, Capodaglio AG (2017) Biological combination processes for efficient removal of pharmaceutically active compounds from wastewater: a review and future perspectives. J Environ Chem Eng 5(4):3590–3603. https://doi.org/10.1016/j.jece.2017. 07.020

Chan YJ, Chong MF, Law CL, Hassell DG (2009) A review on anaerobic–aerobic treatment of industrial and municipal wastewater. Chem Eng J 155(1–2):1–18. https://doi.org/10.1016/j.cej. 2009.06.041

Changotra R, Rajput H, Dhir A (2017) Natural soil mediated photo Fenton-like processes in treatment of pharmaceuticals: batch and continuous approach. Chemosphere 188:345–353. https:// doi.org/10.1016/j.chemosphere.2017.09.016

Changotra R, Varshney L, Paul Guin J, Dhir A (2018) Performance of hematite particles as an Iron source for the degradation of ornidazole in photo-fenton process. J Sol-Gel Sci Technol 85:203–212. https://doi.org/10.1007/s10971-017-4499-4

Cho M, Chung H, Choi W, Yoon J (2005) Different inactivation behaviors of MS-2 phage and Escherichia coli in TiO_2 photocatalytic disinfection. Appl Environ Microbiol 71(1):270–275. https://doi.org/10.1128/AEM.71.1.270-275.2005

Choi KJ, Kim SG, Kim SH (2008) Removal of antibiotics by coagulation and granular activated carbon filtration. J Hazard Mater 151(1):38–43. https://doi.org/10.1016/J.JHAZMAT. 2007.05.059

Choudri BS, Charabi Y, Ahmed M (2018) Health effects associated with wastewater treatment, Reuse and disposal. Water Environ Res 90(10):1759–1776. https://doi.org/10.2175/106143018 X15289915807425

Couto CF, Lange LC, Amaral MCS (2018) A critical review on membrane separation processes applied to remove pharmaceutically active compounds from water and wastewater. J Water Process Eng 26:156–175. https://doi.org/10.1016/j.jwpe.2018.10.010

Couto CF, Lange LC, Amaral MC (2019) Occurrence, fate and removal of pharmaceutically active compounds (PhACs) in water and wastewater treatment plants—a review. J Water Process Eng 32:100927. https://doi.org/10.1016/j.jwpe.2019.100927

De Gisi S, Sabia G, Casella P, Farina R (2015) An integrated approach for monitoring efficiency and investments of activated sludge-based wastewater treatment plants at large spatial scale. Sci Total Environ 523:201–218. https://doi.org/10.1016/j.scitotenv.2015.03.106

del Castillo AF, Garibay MV, Senés-Guerrero C, Orozco-Nunnelly DA, de Anda J, Gradilla-Hernández MS (2022) A review of the sustainability of anaerobic reactors combined with constructed wetlands for decentralized wastewater treatment. J Clean Prod 133428. https:// doi.org/10.1016/j.jclepro.2022.133428

Dias DFC, Passos RG, von Sperling M (2017) A review of bacterial indicator disinfection mechanisms in waste stabilisation ponds. Rev Environ Sci Biotechnol 16:517–539. https://doi.org/10. 1007/s11157-017-9433-2

Dogan A, Płotka-Wasylka J, Kempińska-Kupczyk D, Namieśnik J, Kot-Wasik A (2020) Detection, identification and determination of chiral pharmaceutical residues in wastewater: problems and challenges. TrAC, Trends Anal Chem 122:115710. https://doi.org/10.1016/j.trac.2019.115710

dos Santos SL, Chaves SRM, Van Haandel A (2016) Influence of phase separator design on the performance of UASB reactors treating municipal wastewater. Water Sa 42(2):176–182. https:// doi.org/10.4314/wsa.v42i2.01

Dutta D, Arya S, Kumar S (2021) Industrial wastewater treatment: current trends, bottlenecks, and best practices. Chemosphere 285:131245. https://doi.org/10.1016/j.chemosphere.2021.131245

Fakhru'l-Razi A, Pendashteh A, Abdullah LC, Biak DRA, Madaeni SS, Abidin ZZ (2009) Review of technologies for oil and gas produced water treatment. J Hazardous Mater 170(2–3), 530–551. https://doi.org/10.1016/j.jhazmat.2009.05.044

Fatta-Kassinos D, Vasquez MI, Kümmerer K (2011) Transformation products of pharmaceuticals in surface waters and wastewater formed during photolysis and advanced oxidation processes–degradation, elucidation of byproducts and assessment of their biological potency. Chemosphere 85(5):693–709. https://doi.org/10.1016/j.chemosphere.2011.06.082

Feng L, van Hullebusch ED, Rodrigo MA, Esposito G, Oturan MA (2013) Removal of residual anti-inflammatory and analgesic pharmaceuticals from aqueous systems by electrochemical advanced oxidation processes. A review. Chem Eng J 228:944–964. https://doi.org/10.1016/j.cej.2013.05.061

Gadipelly C, Pérez-González A, Yadav GD, Ortiz I, Ibáñez R, Rathod VK, Marathe KV (2014) Pharmaceutical industry wastewater: review of the technologies for water treatment and reuse. Ind Eng Chem Res 53(29):11571–11592. https://doi.org/10.1021/ie501210j

Garba ZN, Zhou W, Lawan I, Xiao W, Zhang M, Wang L, Chen L, Yuan Z (2019) An overview of chlorophenols as contaminants and their removal from wastewater by adsorption: a review. J Environ Manage 241:59–75. https://doi.org/10.1016/j.jenvman.2019.04.004

Gavrilescu M, Macoveanu M (1999) Process engineering in biological aerobic waste-water treatment. Acta Biotechnol 19(2):111–145. https://doi.org/10.1002/abio.370190205

Ghaitidak DM, Yadav KD (2013) Characteristics and treatment of greywater—a review. Environ Sci Pollut Res 20:2795–2809. https://doi.org/10.1007/s11356-013-1533-0

Ghangrekar MM (2022) Ponds and wetlands for treatment of wastewater. In: Wastewater to water: principles, technologies and engineering design. Springer Nature Singapore, Singapore, pp 555–592. https://doi.org/10.1007/978-981-19-4048-4_13

Glaze WH (1987) Drinking-water treatment with ozone. Environ Sci Technol 21(3):224–230. https://doi.org/10.1021/es00157a001

Glaze WH, Kang JW, Chapin DH (1987) The chemistry of water treatment processes involving ozone, hydrogen peroxide and ultraviolet radiation. https://doi.org/10.1080/01919518708552148

Gomes J, Costa R, Quinta-Ferreira RM, Martins RC (2017) Application of ozonation for pharmaceuticals and personal care products removal from water. Sci Total Environ 586:265–283. https://doi.org/10.1016/j.scitotenv.2017.01.216

Grogan DS, Wisser D, Prusevich A, Lammers RB, Frolking S (2017) The use and re-use of unsustainable groundwater for irrigation: a global budget. Environ Res Lett 12(3):034017. https://doi.org/10.1088/1748-9326/aa5fb2

Gu Y, Huang J, Zeng G, Shi L, Shi Y, Yi K (2018) Fate of pharmaceuticals during membrane bioreactor treatment: status and perspectives. Biores Technol 268:733–748. https://doi.org/10.1016/j.biortech.2018.08.029

Gude VG (2016) Wastewater treatment in microbial fuel cells–an overview. J Clean Prod 122:287–307. https://doi.org/10.1016/j.jclepro.2016.02.022

Guo Y, Qi PS, Liu YZ (2017) A review on advanced treatment of pharmaceutical wastewater. In: IOP conference series: earth and environmental science, vol 63, No. 1). IOP Publishing, p 012025. https://doi.org/10.1088/1755-1315/63/1/012025

Hamza RA, Iorhemen OT, Tay JH (2016) Advances in biological systems for the treatment of high-strength wastewater. J Water Process Eng 10:128–142. https://doi.org/10.1016/j.jwpe.2016.02.008

Haq I, Raj A (2020) Pulp and paper mill wastewater: ecotoxicological effects and bioremediation approaches for environmental safety. Bioremed Ind Waste Environ Saf: Volume II: Biol Agents Methods Industrial Waste Manage 333–356. https://doi.org/10.1007/978-981-13-3426-9_14

Hassard F, Biddle J, Cartmell E, Jefferson B, Tyrrel S, Stephenson T (2015) Rotating biological contactors for wastewater treatment–a review. Process Saf Environ Prot 94:285–306. https://doi.org/10.1016/j.psep.2014.07.003

Ho L, Goethals PL (2020) Municipal wastewater treatment with pond technology: historical review and future outlook. Ecol Eng 148:105791. https://doi.org/10.1016/j.ecoleng.2020.105791

Hreiz R, Latifi MA, Roche N (2015) Optimal design and operation of activated sludge processes: state-of-the-art. Chem Eng J 281:900–920. https://doi.org/10.1016/j.cej.2015.06.125

Hwang HM, Fiala MJ, Park D, Wade TL (2016) Review of pollutants in urban road dust and stormwater runoff: part 1. Heavy metals released from vehicles. Int J Urban Sci 20(3):334–360. https://doi.org/10.1080/12265934.2016.1193041

Ifelebuegu AO (2011) The fate and behavior of selected endocrine disrupting chemicals in full scale wastewater and sludge treatment unit processes. Int J Environ Sci Technol 8:245–254. https://doi.org/10.1007/BF03326213

Ikai H, Nakamura K, Shirato M, Kanno T, Iwasawa A, Sasaki K, Niwano Y, Kohno M (2010) Photolysis of hydrogen peroxide, an effective disinfection system via hydroxyl radical formation. Antimicrob Agents Chemother 54(12):5086–5091. https://doi.org/10.1128/aac.00751-10

Ingrao C, Failla S, Arcidiacono C (2020) A comprehensive review of environmental and operational issues of constructed wetland systems. Curr Opin Environ Sci Health 13:35–45. https://doi.org/10.1016/j.coesh.2019.10.007

Inyinbor AA, Bello OS, Dada OA, Oreofe TA (2021) Emerging water pollutants and wastewater treatments. In: Two-Dimensional (2D) nanomaterials in separation science. Springer International Publishing, Cham, pp 13–42. https://doi.org/10.1007/978-3-030-72457-3_2

Ismail ZZ, Habeeb AA (2017) Experimental and modeling study of simultaneous power generation and pharmaceutical wastewater treatment in microbial fuel cell based on mobilized biofilm bearers. Renewable Energy 101:1256–1265. https://doi.org/10.1016/j.renene.2016.10.008

Janssens R, Mandal MK, Dubey KK, Luis P (2017) Slurry photocatalytic membrane reactor technology for removal of pharmaceutical compounds from wastewater: towards cytostatic drug elimination. Sci Total Environ 599:612–626. https://doi.org/10.1016/j.scitotenv.2017.03.253

Ji G, He C, Tan Y (2013) The spatial distribution of nitrogen removal functional genes in multimedia biofilters for sewage treatment. Ecol Eng 55:35–42. https://doi.org/10.1016/j.ecoleng.2013.02.009

Joshiba GJ, Kumar PS, Femina CC, Jayashree E, Racchana R, Sivanesan S (2019) Critical review on biological treatment strategies of dairy wastewater. Desalin Water Treat 160:94–109. https://doi.org/10.5004/dwt.2019.24194

Jóźwiakowska K, Marzec M (2020) Efficiency and reliability of sewage purification in long-term exploitation of the municipal wastewater treatment plant with activated sludge and hydroponic system. Archiv Environ Protect 46(3). https://doi.org/10.24425/aep.2020.134533

Kadlec RH (2003) Pond and wetland treatment. Water Sci Technol 48(5):1–8. https://doi.org/10.2166/wst.2003.0266

Kadri SUT, Tavanappanavar AN, Nagesh Babu R, Bilal M, Singh B, Gupta SK, Bharagava RN, Govarthanan M, Savanur MA, Mulla SI (2021) Overview of waste stabilization ponds in developing countries. In: Cost-efficient wastewater treatment technologies: natural systems. Springer International Publishing, Cham, pp 153–175. https://doi.org/10.1007/698_2021_790

Kaid M, Ali A, Shamsan A, Younes S, Abdel-Raheem S, Abdul-Malik M, Salem W (2022) Efficiency of maturation oxidation ponds as a post-treatment technique of wastewater. Curr Chem Lett 11(4):415–422. https://doi.org/10.5267/j.ccl.2022.4.005

Kamali M, Costa ME, Aminabhavi TM, Capela I (2019) Sustainability of treatment technologies for industrial biowastes effluents. Chem Eng J 370:1511–1521. https://doi.org/10.1016/j.cej.2019.04.010

Karbalaei S, Hanachi P, Walker TR, Cole M (2018) Occurrence, sources, human health impacts and mitigation of microplastic pollution. Environ Sci Pollut Res 25:36046–36063. https://doi.org/10.1007/s11356-018-3508-7

Karri RR, Ravindran G, Dehghani MH (2021) Wastewater—sources, toxicity, and their consequences to human health. In: Soft computing techniques in solid waste and wastewater management. Elsevier, pp 3–33. https://doi.org/10.1016/B978-0-12-824463-0.00001-X

Kesari KK, Soni R, Jamal QMS, Tripathi P, Lal JA, Jha NK, Ruokolainen J et al (2021) Wastewater treatment and reuse: a review of its applications and health implications. Water, Air, Soil Pollut 232:1–28. https://doi.org/10.1007/s11270-021-05154-8

Keskin S, Kayrak-Talay D, Akman U, Hortaçsu Ö (2007) A review of ionic liquids towards supercritical fluid applications. J Supercritical Fluids 43(1):150–180. https://doi.org/10.1016/j.supflu.2007.05.013

Khalid M, Abdollahi M (2021) Environmental distribution of personal care products and their effects on human health. Iranian J Pharmaceutical Res: IJPR 20(1):216. https://doi.org/10.22037/ijpr.2021.114891.15088

Khalid N, Aqeel M, Noman A, Khan SM, Akhter N (2021) Interactions and effects of microplastics with heavy metals in aquatic and terrestrial environments. Environ Pollut 290:118104. https://doi.org/10.1016/j.envpol.2021.118104

Kinigopoulou V, Pashalidis I, Kalderis D, Anastopoulos I (2022) Microplastics as carriers of inorganic and organic contaminants in the environment: a review of recent progress. J Mol Liq 350:118580. https://doi.org/10.1016/j.molliq.2022.118580

Klavarioti M, Mantzavinos D, Kassinos D (2009) Removal of residual pharmaceuticals from aqueous systems by advanced oxidation processes. Environ Int 35(2):402–417. https://doi.org/10.1016/J.ENVINT.2008.07.009

Kodali RK (2017) Smart waste water treatment. In: 2017 IEEE region 10 symposium (TENSYMP). IEEE, pp 1–5. https://doi.org/10.1109/TENCONSpring.2017.8070092

Krstić V, Urošević T, Pešovski B (2018) A review on adsorbents for treatment of water and wastewaters containing copper ions. Chem Eng Sci 192:273–287. https://doi.org/10.1016/j.ces.2018.07.022

Kumar V, Sehgal R, Gupta R (2023) Microbes and wastewater treatment. In: Development in wastewater treatment research and processes. Elsevier, pp 239–255. https://doi.org/10.1016/B978-0-323-88505-8.00010-3

Kyzas GZ, Fu J, Lazaridis NK, Bikiaris DN, Matis KA (2015) New approaches on the removal of pharmaceuticals from wastewaters with adsorbent materials. J Mol Liq 209:87–93. https://doi.org/10.1016/j.molliq.2015.05.025

Li Y, Zhu G, Ng WJ, Tan SK (2014) A review on removing pharmaceutical contaminants from wastewater by constructed wetlands: design, performance and mechanism. Sci Total Environ 468:908–932. https://doi.org/10.1016/j.scitotenv.2013.09.018

Li H, Zhu J, Agnew S, Devine J, Bengoechea J, Kulick FM, Faeth M, Rothermel B (2014) A pilot study for treating facultative lagoon effluent utilizing a fixed bed biofilm reactor. In: WEFTEC 2014. Water Environment Federation. https://doi.org/10.2175/193864714815941144

Liu X, Guo X, Liu Y, Lu S, Xi B, Zhang J, Wang Z, Bi B (2019) A review on removing antibiotics and antibiotic resistance genes from wastewater by constructed wetlands: performance and microbial response. Environ Pollut 254:112996. https://doi.org/10.1016/j.envpol.2019.112996

Mahapatra DM, Chanakya HN, Ramachandra TV (2013) Treatment efficacy of algae-based sewage treatment plants. Environ Monit Assess 185:7145–7164. https://doi.org/10.1007/s10661-013-3090-x

Mahapatra S, Samal K, Dash RR (2022) Waste Stabilization Pond (WSP) for wastewater treatment: a review on factors, modelling and cost analysis. J Environ Manage 308:114668. https://doi.org/10.1016/j.jenvman.2022.114668

Mancuso G, Lavrnić S, Canet-Martí A, Zaheer A, Avolio F, Langergraber G, Toscano A (2023) Performance of lagoon and constructed wetland systems for tertiary wastewater treatment and potential of reclaimed water in agricultural irrigation. J Environ Manage 348:119278. https://doi.org/10.1016/j.jenvman.2023.119278

Martínez F, Molina R, Rodríguez I, Pariente MI, Segura Y, Melero JA (2018) Techno-economical assessment of coupling Fenton/biological processes for the treatment of a pharmaceutical wastewater. J Environ Chem Eng 6(1):485–494. https://doi.org/10.1016/J.JECE.2017.12.008

Maurya A, Kumar R, Raj A (2023) Biofilm-based technology for industrial wastewater treatment: current technology, applications and future perspectives. World J Microbiol Biotechnol 39(5):112. https://doi.org/10.1007/s11274-023-03567-7
Mazur LP, Cechinel MA, de Souza SMGU, Boaventura RA, Vilar VJ (2018) Brown marine macroalgae as natural cation exchangers for toxic metal removal from industrial wastewaters: a review. J Environ Manage 223:215–253. https://doi.org/10.1016/j.jenvman.2018.05.086
Mehrjouei M, Müller S, Möller D (2015) A review on photocatalytic ozonation used for the treatment of water and wastewater. Chem Eng J 263:209–219. https://doi.org/10.1016/j.cej.2014.10.112
Meyer MF, Powers SM, Hampton SE (2019) An evidence synthesis of pharmaceuticals and personal care products (PPCPs) in the environment: imbalances among compounds, sewage treatment techniques, and ecosystem types. Environ Sci Technol 53(22):12961–12973. https://doi.org/10.1021/acs.est.9b02966
Michael-Kordatou I, Karaolia P, Fatta-Kassinos D (2018) The role of operating parameters and oxidative damage mechanisms of advanced chemical oxidation processes in the combat against antibiotic-resistant bacteria and resistance genes present in urban wastewater. Water Research 129:208–230. https://doi.org/10.1016/j.watres.2017.10.007
Mohammadi M, Fataei E (2019) Comparative life cycle assessment of municipal wastewater treatment systems: lagoon and activated sludge. Caspian J Environ Sci 17(4):327–336. https://doi.org/10.22124/cjes.2019.3806
Molino A, Nanna F, Ding Y, Bikson B, Braccio G (2013) Biomethane production by anaerobic digestion of organic waste. Fuel 103:1003–1009. https://doi.org/10.1016/j.fuel.2012.07.070
Mulligan CN, Gibbs BF (2003) Innovative biological treatment processes for wastewater in Canada. Water Qual Res J 38(2):243–265. https://doi.org/10.2166/wqrj.2003.018
Murshid S, Antonysamy A, Dhakshinamoorthy G, Jayaseelan A, Arivalagan P (2023) A review on biofilm-based reactors for wastewater treatment: recent advancements in biofilm carriers, kinetics, reactors, economics, and future perspectives. Sci Total Environ 164796. https://doi.org/10.1016/j.scitotenv.2023.164796
Musa MA, Idrus S (2021) Physical and biological treatment technologies of slaughterhouse wastewater: a review. Sustainability 13(9):4656. https://doi.org/10.3390/su13094656
Nayak JK, Ghosh UK (2019) Post treatment of microalgae treated pharmaceutical wastewater in photosynthetic microbial fuel cell (PMFC) and biodiesel production. Biomass Bioenerg 131:105415. https://doi.org/10.1016/j.biombioe.2019.105415
Ngo PL, Pramanik BK, Shah K, Roychand R (2019) Pathway, classification and removal efficiency of microplastics in wastewater treatment plants. Environ Pollut 255:113326. https://doi.org/10.1016/j.envpol.2019.113326
Oller I, Malato S, Sánchez-Pérez J (2011) Combination of advanced oxidation processes and biological treatments for wastewater decontamination—a review. Sci Total Environ 409(20):4141–4166. https://doi.org/10.1016/J.SCITOTENV.2010.08.061
Opitz J, Alte M, Bauer M, Peiffer S (2021) The role of macrophytes in constructed surface-flow wetlands for mine water treatment: a review. Mine Water Environ 40(3):587–605. https://doi.org/10.1007/s10230-021-00779-x
Orhon D (2015) Evolution of the activated sludge process: the first 50 years. J Chem Technol Biotechnol 90(4):608–640. https://doi.org/10.1002/jctb.4565
Owusu-Twum MY, Sharara MA (2020) Sludge management in anaerobic swine lagoons: a review. J Environ Manage 271:110949. https://doi.org/10.1016/j.jenvman.2020.110949
Pal P (2018) Treatment and disposal of pharmaceutical wastewater: toward the sustainable strategy. Sep Purif Rev 47(3):179–198. https://doi.org/10.1080/15422119.2017.1354888
Patel HK, Kalaria RK, More BS, Jokhakar PH, Khimani MR (2022) Microbial ecology of biofiltration. In: An Innovative Role of Biofiltration in Wastewater Treatment Plants (WWTPs). Elsevier, pp 235–266. https://doi.org/10.1016/B978-0-12-823946-9.00008-5

Peng H, Guo J (2020) Removal of chromium from wastewater by membrane filtration, chemical precipitation, ion exchange, adsorption electrocoagulation, electrochemical reduction, electrodialysis, electrodeionization, photocatalysis and nanotechnology: a review. Environ Chem Lett 18:2055–2068. https://doi.org/10.1007/s10311-020-01058-x

Pirali zefrehei AR, Kolahi M, Fisher J (2022) Ecological-environmental challenges and restoration of aquatic ecosystems of the Middle-Eastern. Sci Rep 12(1), 17229. https://doi.org/10.1038/s41598-022-21465-0

Pishgar R, Lee J, Dominic JA, Hosseini S, Tay JH, Chu A (2020) Augmentation of biogranules for enhanced performance of full-scale lagoon-based municipal wastewater treatment plants. Appl Biochem Biotechnol 191:426–443. https://doi.org/10.1007/s12010-020-03256-3

Quesada HB, Baptista ATA, Cusioli LF, Seibert D, de Oliveira Bezerra C, Bergamasco R (2019) Surface water pollution by pharmaceuticals and an alternative of removal by low-cost adsorbents: a review. Chemosphere 222:766–780. https://doi.org/10.1016/j.chemosphere.2019.02.009

Randrianarison G, Ashraf MA (2017) Microalgae: a potential plant for energy production. Geol Ecol Landscapes 1(2):104–120. https://doi.org/10.1080/24749508.2017.1332853

Rangaswamy B, Singh IB (2022) PCR based fingerprinting techniques to transpire the microbial populations diversity of biofilm based nitrifying packed bed bioreactor in the bioremediation of aquaculture wastewater. Green Technol Resilien Sustain 2(1):5. https://doi.org/10.1007/s44173-022-00005-z

Ranjan S, Gupta PK, Gupta SK (2019) Comprehensive evaluation of high-rate algal ponds: wastewater treatment and biomass production. Application of Microalgae in Wastewater Treatment: Volume 2: Biorefinery Approaches of Wastewater Treatment, 531–548. https://doi.org/10.1007/978-3-030-13909-4_22

Rezai B, Allahkarami E (2021) Wastewater treatment processes—techniques, technologies, challenges faced, and alternative solutions. In: Soft computing techniques in solid waste and wastewater management. Elsevier, pp 35–53. https://doi.org/10.1016/B978-0-12-824463-0.00004-5

Rodrigues-Filho JL, Macêdo RL, Sarmento H, Pimenta VR, Alonso C, Teixeira CR, Pagliosa PR, Netto SA, Santos NCL, Daura-Jorge FG, Rocha O, Horta P, Branco JO, Sartor R, Muller J, Cionek VM (2023) From ecological functions to ecosystem services: linking coastal lagoons biodiversity with human well-being. Hydrobiologia 1–43. https://doi.org/10.1007/s10750-023-05171-0

Sabliy L, Kuzminskiy Y, Zhukova V, Kozar M, Sobczuk H (2019) New approaches in biological wastewater treatment aimed at removal of organic matter and nutrients. Ecol Chem Eng S 26(2):331–343. https://doi.org/10.1515/eces-2019-0023

Salgado R, Marques R, Noronha JP, Carvalho G, Oehmen A, Reis MAM (2012) Assessing the removal of pharmaceuticals and personal care products in a full-scale activated sludge plant. Environ Sci Pollut Res 19:1818–1827. https://doi.org/10.1007/s11356-011-0693-z

Saravanan A, Kumar PS, Jeevanantham S, Karishma S, Tajsabreen B, Yaashikaa PR, Reshma B (2021) Effective water/wastewater treatment methodologies for toxic pollutants removal: processes and applications towards sustainable development. Chemosphere 280:130595. https://doi.org/10.1016/j.chemosphere.2021.130595

Sathya K, Nagarajan K, Carlin Geor Malar G, Rajalakshmi S, Raja Lakshmi P (2022) A comprehensive review on comparison among effluent treatment methods and modern methods of treatment of industrial wastewater effluent from different sources. Appl Water Sci 12(4):70. https://doi.org/10.1007/s13201-022-01594-7

Schweitzer L, Noblet J (2018) Water contamination and pollution. In: Green chemistry. Elsevier, pp 261–290. https://doi.org/10.1016/B978-0-12-809270-5.00011-X

Sembuil A, Nor A, A-Halim MH (2022) Sustainable indicators for designing the wastewater treatment plant. In: IOP conference series: earth and environmental science (vol. 1091, No. 1). IOP Publishing, p 012057. https://doi.org/10.1088/1755-1315/1091/1/012057

Serpone N, Artemev YM, Ryabchuk VK, Emeline AV, Horikoshi S (2017) Light-driven advanced oxidation processes in the disposal of emerging pharmaceutical contaminants in aqueous media:

a brief review. Curr Opin Green Sustain Chem 6:18–33. https://doi.org/10.1016/j.cogsc.2017.05.003

Shammas NK, Wang LK, Wu Z (2009) Waste stabilization ponds and lagoons. Biological Treatment Processes 315–370. https://doi.org/10.1007/978-1-60327-156-1_8

Sharma NK, Arivalagan AR (2021) Algae or bacteria—the future of biological wastewater treatment. In: Handbook of advanced approaches towards pollution prevention and control. Elsevier, pp 217–247. https://doi.org/10.1016/B978-0-12-822134-1.00008-7

Sheik AG, Ambati SR (2023) Design of supervisory controllers to reduce energy consumption in plant-wide biological wastewater treatment plants with feedback signals from non-ideal sensors. Chem Eng Commun, 1–17. https://doi.org/10.1080/00986445.2023.2218290

Silva TF, Silva MEF, Cunha-Queda AC, Fonseca A, Saraiva I, Boaventura RA, Vilar VJ (2013) Sanitary landfill leachate treatment using combined solar photo-Fenton and biological oxidation processes at pre-industrial scale. Chem Eng J 228:850–866. https://doi.org/10.1016/j.cej.2013.05.060

Sirés I, Brillas E (2012) Remediation of water pollution caused by pharmaceutical residues based on electrochemical separation and degradation technologies: a review. Environ Int 40:212–229. https://doi.org/10.1016/j.envint.2011.07.012

Siyal AA, Shamsuddin MR, Low A, Rabat NE (2020) A review on recent developments in the adsorption of surfactants from wastewater. J Environ Manage 254:109797. https://doi.org/10.1016/j.jenvman.2019.109797

Skouteris G, Rodriguez-Garcia G, Reinecke SF, Hampel U (2020) The use of pure oxygen for aeration in aerobic wastewater treatment: a review of its potential and limitations. Biores Technol 312:123595. https://doi.org/10.1016/j.biortech.2020.123595

Song P, Yang Z, Zeng G, Yang X, Xu H, Wang L, Xu R, Xiong, Ahmad K (2017) Electrocoagulation treatment of arsenic in wastewaters: a comprehensive review. Chem Eng J 317:707–725. https://doi.org/10.1016/j.cej.2017.02.086

Sonone SS, Jadhav S, Sankhla MS, Kumar R (2020) Water contamination by heavy metals and their toxic effect on aquaculture and human health through food Chain. Lett ApplNanobioscience 10(2):2148–2166. https://doi.org/10.33263/LIANBS102.21482166

Tan X, Acquah I, Liu H, Li W, Tan S (2019) A critical review on saline wastewater treatment by membrane bioreactor (MBR) from a microbial perspective. Chemosphere 220:1150–1162. https://doi.org/10.1016/j.chemosphere.2019.01.027

Teh CY, Budiman PM, Shak KPY, Wu TY (2016) Recent advancement of coagulation–flocculation and its application in wastewater treatment. Ind Eng Chem Res 55(16):4363–4389. https://doi.org/10.1021/acs.iecr.5b04703

Tiwari B, Sellamuthu B, Ouarda Y, Drogui P, Tyagi RD, Buelna G (2017) Review on fate and mechanism of removal of pharmaceutical pollutants from wastewater using biological approach. Biores Technol 224:1–12. https://doi.org/10.1016/j.biortech.2016.11.042

Tocchi C, Federici E, Fidati L, Manzi R, Vincigurerra V, Petruccioli M (2012) Aerobic treatment of dairy wastewater in an industrial three-reactor plant: Effect of aeration regime on performances and on protozoan and bacterial communities. Water Res 46(10):3334–3344. https://doi.org/10.1016/j.watres.2012.03.032

Tong L, Qin L, Guan C, Wilson ME, Li X, Cheng D, Ma J, Liu H, Gong F (2020) Antibiotic resistance gene profiling in response to antibiotic usage and environmental factors in the surface water and groundwater of Honghu Lake, China. Environ Sci Pollut Res 27:31995–32005. https://doi.org/10.1007/s11356-020-09487-5

Türker U, Okaygün M, Almaqadma SJ (2009) Impact of anaerobic lagoons on the performance of BOD and TSS removals at the Haspolat (Mia Milia) Wastewater Treatment Plant. Desalination 249(1):403–410. https://doi.org/10.1016/j.desal.2009.06.044

Uddin MM, Wright MM (2022) Anaerobic digestion fundamentals, challenges, and technological advances. Phys Sci Rev. https://doi.org/10.1515/psr-2021-0068

Ungureanu N, Vlăduț V, Voicu G (2020) Water scarcity and wastewater reuse in crop irrigation. Sustainability 12(21):9055. https://doi.org/10.3390/su12219055

Van den Akker B, Holmes M, Pearce P, Cromar NJ, Fallowfield HJ (2011) Structure of nitrifying biofilms in a high-rate trickling filter designed for potable water pre-treatment. Water Res 45(11):3489–3498. https://doi.org/10.1016/j.watres.2011.04.017

van Den Akker B, Holmes M, Cromar N, Fallowfield H (2008) Application of high rate nitrifying trickling filters for potable water treatment. Water Res 42(17):4514–4524. https://doi.org/10.1016/j.watres.2008.07.038

Verlicchi P, Zambello E (2014) How efficient are constructed wetlands in removing pharmaceuticals from untreated and treated urban wastewaters? A review. Sci Total Environ 470:1281–1306. https://doi.org/10.1016/j.scitotenv.2013.10.085

Wang J, Chen H (2020) Catalytic ozonation for water and wastewater treatment: recent advances and perspective. Sci Total Environ 704:135249. https://doi.org/10.1016/j.scitotenv.2019.135249

Wang H, Ji G, Bai X, He C (2015) Assessing nitrogen transformation processes in a trickling filter under hydraulic loading rate constraints using nitrogen functional gene abundances. Biores Technol 177:217–223. https://doi.org/10.1016/j.biortech.2014.11.094

Waqas S, Bilad MR, Man Z, Wibisono Y, Jaafar J, Mahlia TMI, Aslam M et al (2020) Recent progress in integrated fixed-film activated sludge process for wastewater treatment: a review. J Environ Manage 268, 110718. https://doi.org/10.1016/j.jenvman.2020.110718

Waqas S, Harun NY, Sambudi NS, Bilad MR, Abioye KJ, Ali A, Abdulrahman A (2023) A review of rotating biological contactors for wastewater treatment. Water 15(10):1913. https://doi.org/10.3390/w15101913

Wen X, Ding H, Huang X, Liu R (2004) Treatment of hospital wastewater using a submerged membrane bioreactor. Process Biochem 39(11):1427–1431. https://doi.org/10.1016/S0032-9592(03)00277-2

Xia C, Zhang D, Pedrycz W, Zhu Y, Guo Y (2018) Models for microbial fuel cells: a critical review. J Power Sources 373:119–131. https://doi.org/10.1016/j.jpowsour.2017.11.001

Xu H, Quan X, Xiao Z, Chen L (2018) Effect of anodes decoration with metal and metal oxides nanoparticles on pharmaceutically active compounds removal and power generation in microbial fuel cells. Chem Eng J 335:539–547. https://doi.org/10.1016/j.cej.2017.10.159

Xu R, Yang Z, Niu Y, Xu D, Wang J, Han J, Wang H (2022) Removal of microplastics and attached heavy metals from secondary effluent of wastewater treatment plant using interpenetrating bipolar plate electrocoagulation. Sep Purif Technol 290:120905. https://doi.org/10.1016/j.seppur.2022.120905

Yahya MN, Gökçekuş H, Ozsahin DU (2020) Comparative analysis of wastewater treatment technologies. Jurnal Kejuruteraan 32(2):221–230. https://doi.org/10.17576/jkukm-2020-32(2)-06

Yeasmin F, Rasheduzzaman M, Manik M, Hasan MM (2023) Activated sludge process for wastewater treatment. In: Advanced and innovative approaches of environmental biotechnology in industrial wastewater treatment. Springer Nature Singapore, Singapore, pp 23–50. https://doi.org/10.1007/978-981-99-2598-8_2

Yuan Z, Nag R, Cummins E (2022) Human health concerns regarding microplastics in the aquatic environment—from marine to food systems. Sci Total Environ 823:153730. https://doi.org/10.1016/j.scitotenv.2022.153730

Zema DA, Andiloro S, Bombino G, Caridi A, Sidari R, Tamburino V (2016) Comparing different schemes of agricultural wastewater lagooning: depuration performance and microbiological characteristics. Water Air Soil Pollut 227:1–9. https://doi.org/10.1007/s11270-016-3132-4

Zhang L, Ji X, Ren X, Ma Y, Shi X, Tian Z, Asiri AM, Chen L, Tang B, Sun X (2018) Electrochemical ammonia synthesis via nitrogen reduction reaction on a MoS_2 catalyst: theoretical and experimental studies. Adv Mater 30(28):1800191. https://doi.org/10.1002/adma.201800191

Chapter 18
Electroplating Industry Wastewater Treatment: Advanced Electrochemical Oxidation Process

Sonal Rajoria, Manish Vashishtha, and Vikas K. Sangal

Abstract Water is an exceptionally essential natural source for all living things. Due to rapid industrialization and population increase, water quality has been adversely affected. Globally, the volume of wastewater generated and contaminant loads are increasing due to various factors. Especially, water pollution is growing rapidly due to the wastewater discharge from industries. The untreated discharge of electroplating wastewater may lead to the formation of hazardous metal–organic complexes that are harmful to the environment. Therefore, it is crucial to develop economically and environmentally friendly treatment procedures to protect the environment and remove electroplating effluents from the aquatic system. The advanced electrochemical oxidation process (AEOP), which is low-cost and highly efficient, may be applicable in electroplating wastewater treatment with heavy metals and organic acids. Research on AEOP has recently become an interesting topic as it has been shown to oxidize most organic contaminants to inorganic carbon through mineralization effectively. Hence, this study reviews the currently most promising AEOP, introduces its basic principles, mechanisms, and describes the properties of various electrode materials. At the same time, the effects of operating conditions such as reactor configuration, supporting electrolytes, and electrode materials on the degradation effect are discussed.

Keywords Advanced electrochemical oxidation process · Electrode materials · Electroplating industry · Wastewater treatment methods

S. Rajoria · M. Vashishtha · V. K. Sangal (✉)
Department of Chemical Engineering, Malaviya National Institute of Technology, Jaipur, Rajasthan, India
e-mail: vksangal.chem@mnit.ac.in

S. Souabi et al. (eds.), *Wastewater Treatment Plants*, Water Science and Technology Library 130, https://doi.org/10.1007/978-3-031-87461-1_18

18.1 Introduction

The global scientific community widely recognizes environmental pollution as a significant issue. Providing clean water, safe food, and renewable energy remains a major challenge in today's era (Krishna et al. 2023). Water is a fundamental resource crucial for sustaining life on Earth. It plays a vital role in supporting ecosystems, agriculture, industry, and human health. The value of water to the ecosystem is now widely acknowledged on a global scale. Despite being important for life, freshwater is an extremely rare resource on Earth. Freshwater makes up less than 3% of the water on Earth; the other 97% is salt water from seas or oceans. Most of the world's freshwater is not available to humans (Gusain et al. 2020). Around 69% of the freshwater on Earth is frozen in the form of glaciers and polar ice caps, and the remaining 30% is found in groundwater beneath the Earth's surface. Thus, only over 1% of the freshwater on Earth is still easily accessible to humans. Unfortunately, the available surface water is not equally distributed throughout the world. Roughly 25% of the global population lives in regions with limited water resources, where an individual typically obtains less than 1000 m^3 of water annually. The lack of water significantly affects people's access to clean, usable water and impacts the geopolitics and economic growth of many regions. Access to clean and safe drinking water is essential for human health, while aquatic ecosystems depend on water for habitat support and biodiversity. From facilitating biochemical reactions within cells to serving as a medium for transportation and irrigation, water is indispensable for functioning biological, chemical, and physical processes. Water resources are becoming more and more in demand due to the growth of cities, businesses, and natural demands. Therefore, it's critical to seek solutions that allow water to be effectively cleaned so that it can be reused in a variety of ways.

Many industries contribute to the generation of wastewater as a byproduct of their manufacturing processes. Numerous sectors, such as food processing, chemical manufacturing, textile manufacturing, and mining, are responsible for a wide range of pollutants. For example, chemical industries discharge wastewater containing heavy metals, solvents, and other toxic substances, posing significant risks to water quality and ecosystems. Similarly, food processing facilities generate wastewater containing organic matter, nutrients, and suspended solids, which can reduce oxygen levels in water bodies and harm aquatic life. Textile industries release wastewater containing dyes, bleaches, and other chemicals, leading to contamination of water sources and soil. Additionally, mining operations, electronics, and electroplating industries produce wastewater containing heavy metals, acids, and other pollutants, which can have long-lasting impacts on water quality and ecosystem health. The electroplating sector has been found to be one of the fastest growing of all. This rapid expansion also results in the generation of substantial volumes of wastewater containing high metal content. According to reports, the wastewater discharged from the electroplating industry contains considerably heavy metals and cyanide. These substances are deemed hazardous due to their potential toxicity to both humans and the environment. In the electroplating industry, electroplating is the primary process responsible for producing hazardous compounds. This process involves altering the characteristics of an object by applying an electrolytic coating. Addressing this challenge requires advanced decontamination methods that are both effective and environmentally friendly. In recent years, several techniques have been developed and widely utilized to mitigate environmental pollutants from industrial wastewater. These include coagulation, flocculation, membrane filtration, adsorption, ion exchange, and advanced oxidation processes (AOPs), each with its own set of advantages and limitations. However, advanced electrochemical treatment methods, particularly electro-oxidation (EO), have gained significant attention for their efficiency and versatility in treating complex industrial wastewater. As a promising electrochemical technique, EO offers several benefits for electroplating wastewater treatment. Several factors, including the type of electrode materials, electrolyte types, and operating conditions influence the effectiveness of EO. In this chapter, our aim is to highlight the main aspects of advanced electrochemical oxidation treatment method and discuss its significant benefits for electroplating wastewater treatment. This chapter also explores the critical factors affecting EO process performance and provides insights into the latest advancements and applications in this field.

18.2 Electroplating Industry

The electroplating industry is one such sector that generates wastewater because of its manufacturing processes. The electroplating process involves depositing a thin layer of metal onto a surface using an electric current, typically to enhance the appearance, corrosion resistance, or conductivity of the substrate material. While essential for enhancing product durability and aesthetics, the electroplating process poses significant environmental and safety concerns. This process often produces wastewater that contains various pollutants, including heavy metals such as copper, nickel, chromium, and zinc, as well as acids, cyanides, and other chemicals (Fig. 18.1) used in the plating baths. Complexing agents, brighteners, stabilizers, and other functional additives, including malic acid, citric acid, ethylene diamine tetra acetic acid, and dihydroxy succinic acid, serve as organic compounds to electroplating effluents (Kong et al. 2020). Electroplating wastewater also contains turbidity, total suspended and dissolved solids, chemical oxygen demand, biochemical oxygen demand, phosphate, sulfate, and phosphate. The industry generates waste, contributes to air and water pollution, and utilizes hazardous chemicals, necessitating careful management to mitigate adverse impacts. Wastewater from industries like electroplating must be treated due to its high concentration of pollutants, which pose significant risks to the environment and human health if left untreated. These pollutants, including heavy metals, cyanides, acids, and organic compounds, can severely degrade water quality in receiving bodies such as rivers and lakes, rendering them unsuitable for various purposes such as drinking, irrigation, and supporting aquatic life. Exposure to these substances can cause many health issues, including disruption of the

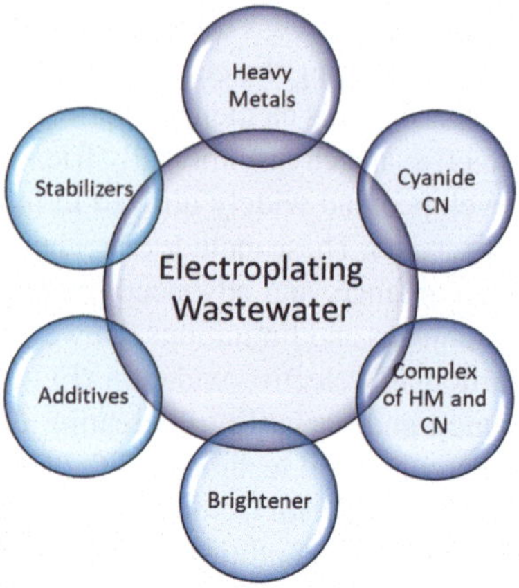

Fig. 18.1 Composition of electroplating industry wastewater (Rajoria et al. 2021, 2022)

endocrine system, respiratory problems, skin irritation, and even more severe conditions like kidney and liver damage or cancer. The hazardous nature of the wastewater persists when metals dissolve or remain suspended in the water. Even after treatment, this contaminated water is still discharged into the sewage system. Therefore, it is essential to employ an effective treatment method before discharging such highly hazardous wastewater into open water channels. We are currently in the 'Pollutant Removal Age,' where the focus is on effectively removing contaminants from wastewater to protect the environment and public health. Industry wastewater treatment has evolved from a voluntary and optional process to a crucial multistage procedure essential for significantly improving the quality of potable water and wastewater released into the environment. The demand for affordable treatment options has increased due to strict regulations on allowable pollutant levels in discharged effluent, which have led to higher disposal costs (Gitis and Hankins 2018).

18.3 Electroplating Industry Wastewater Treatment Methods

Addressing wastewater from the electroplating industry, which typically contains harmful substances such as heavy metals and organic compounds, involves a combination of physical, chemical, and biological techniques. These methods aim to remove or neutralize contaminants to meet regulatory standards and protect the environment. Figure 18.2 shows the classification of wastewater treatment methods for the electroplating industry, categorized into physicochemical and advanced techniques.

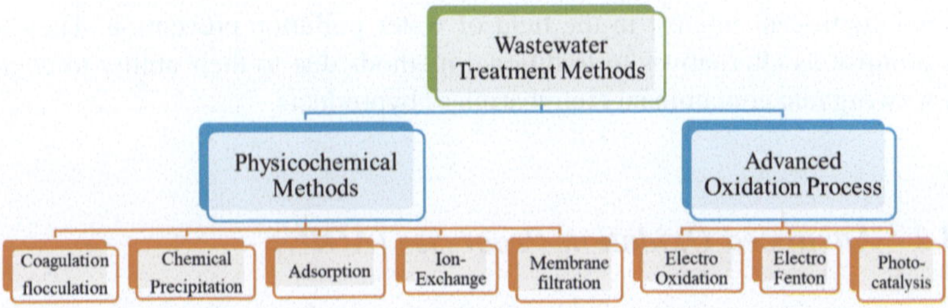

Fig. 18.2 Classification of electroplating industry wastewater treatment methods

18.3.1 Conventional Methods

The physical and chemical conventional techniques that have been employed for decent amounts of removal of heavy metals include coagulation/flocculation, adsorption, chemical precipitation, membrane filtration, and ion exchange.

Chemical precipitation, involving the addition of chemicals to precipitate heavy metals, and coagulation/flocculation, which promotes the formation of larger particles for easier removal, are commonly used initial steps. Ion exchange, adsorption onto surfaces like activated carbon, and membrane filtration techniques like reverse osmosis and nano-filtration are also employed to target specific contaminants. Furthermore, biological treatments employ microorganisms to decompose organic substances and decrease the levels of heavy metals, which is another conventional approach. Nevertheless, these conventional approaches encounter many difficulties that can reduce their efficiency. Chemical precipitation and coagulation/flocculation processes may have difficulty dealing with complicated wastewater mixtures, necessitating extensive pre-treatment or resulting in incomplete removal of pollutants. Ion exchange and adsorption can be costly and may require frequent regeneration or replacement of adsorbents. Filtration methods that involve membranes can consume a lot of energy and get clogged easily, which lowers their performance over time. However, membrane processes can be capital-intensive, requiring significant upfront investment, and may incur high operational costs for maintenance and replacement of membranes. The biological treatments may be slow and less effective for removing certain heavy metals or persistent organic pollutants. In addition, fluctuations in wastewater composition and operating conditions can affect the performance of biological treatment systems, necessitating careful monitoring and management. Furthermore, the existence of cyanide compounds in electroplating wastewater can create further difficulties, as traditional approaches may not fully degrade or remove these extremely harmful substances. Moreover, the scale and variability of electroplating operations can complicate the design and operation of treatment systems, requiring careful optimization and monitoring to ensure consistent performance. As industries need to find more efficient and sustainable solutions for electroplating wastewater treatment, various modern methods have emerged to address these challenges caused by heavy metals and organic compounds. Among these, AOPs have garnered significant interest in the field of water pollution prevention. They hold great promise as alternatives to traditional methods due to their ability to degrade refractory organic contaminants into harmless byproducts.

18.3.2 Advanced Oxidation Processes (AOPs)

According to Glaze et al. (1987), AOPs are characterized as water treatment methods conducted under ambient temperature and pressure conditions. These processes are based on the in situ generation of highly reactive oxidizing agents, such as hydroxyl radicals ($^{\cdot}OH$), which are present in sufficient concentrations to efficiently degrade contaminants and purify the water. AOPs can be classified into various categories according to different criteria, including (i) the method used to produce the oxidizing radicals ($^{\cdot}OH$, $^{\cdot}O$, $^{\cdot}HO_2$, etc.); (ii) whether or not the oxidation processes involve UV or/and Vis light irradiation; and (iii) whether or not the oxidation processes involve a single phase or heterogeneous catalysts (Table 18.1).

Regardless of the methodology employed, all AOPs have the ability to generate extremely reactive species, also referred to as reactive oxygen species (ROS), such as $^{\cdot}OH$, which subsequently attack the organic substrates found in contaminated aquatic matrices through the general reaction described in Eq. (18.1) (Vagi and Petsas 2020):

$$\begin{aligned} &\text{AOPs} \rightarrow \text{Production of unselective radicals} \\ &\quad \rightarrow \text{Attack the organic pesticide} \rightarrow CO_2 \rightarrow H_2O + \text{inorganic ions} \end{aligned} \tag{18.1}$$

AOPs have been effectively applied to a variety of wastewater streams, including agricultural, municipal, and industrial sources. Research on the sustainability of

Table 18.1 Three groups of AOPs technologies among different criteria

GROUP I: Tested methodology for the generation of the oxidizing radicals ($^{\cdot}OH$), $^{\cdot}HO_2$, $^{\cdot}O$)	GROUP II: Oxidation processes occur with UV/Vis light irradiation or not	GROUP III: Oxidation processes occur in a single phase
Chemical AOPs: • Peroxonation (O_3/H_2O_2) • Fenton's reagent (Fe^{2+}/H_2O_2) **Photo-chemical AOPs:** • Photo-Fenton process (UV/H_2O_2/Fe^{2+}) • Photolysis of H_2O_2 (UV/H_2O_2)	**Photo-chemical AOPs:** • UV/H_2O_2 • Photo-Fenton process (UV/H_2O_2/Fe^{2+}) •Photo-electro-Fenton technologies	**Homogenous AOPs:** •Persulfate-based technologies
Electrochemical AOPs: •Electro-Fenton •Electrochemical oxidation	**Non-photo-chemical AOPs:** • Advanced electro-oxidation processes (AEOP)	**Heterogenous AOPs:** • Heterogeneous Fenton-like • Catalytic ozonation

AOPs has primarily focused on four key areas: reducing energy consumption, advancing circular economy practices, minimizing waste, and improving the performance of catalytic AOPs. By optimizing AOPs for energy efficiency, employing catalysts, and breaking down complex organic molecules into simpler, less harmful byproducts, these processes can significantly lessen the environmental impact of water treatment. AOP degradation effectiveness is highly dependent on a number of reaction parameters, including temperature, irradiation wavelength, oxidant concentration, current density, and pH of the treated solution. AOPs can be classified into several processes, including ozone-based AOPs, photocatalysis, non-thermal plasma, Fenton and Fenton-like oxidation, and EO. These methods offer high removal efficiencies but can be energy-intensive and produce harmful byproducts if not properly controlled. However, AOPs treatment, particularly electrochemical oxidation (EO), has gained significant attention for its efficiency and versatility in treating complex electroplating industrial wastewater.

18.4 Electrochemical Oxidation (EO)

The advanced electrochemical oxidation process (AEOP) has been identified as a promising alternative water treatment technology to address this environmental concern. AEOP has garnered growing interest due to their ability to completely mineralize highly refractory organic pollutants, including antibiotics (Vinayagam et al. 2024), dyes (Chen et al. 2015), pharmaceuticals (Öztürk et al. 2021), phenols (Zhang et al. 2022), and even organic matter (Tran et al. 2015). The AEOP not only effectively removes recalcitrant pollutants but also has a number of other environmentally significant features, including (1) complete automation; (2) fewer hazardous byproducts; (3) simultaneous removal of organic and metal contaminants; (4) no sludge production; (5) rapid, well-regulated operation with high removal efficiency; (6) easy integration with other traditional water treatment technologies; (7) no additional chemical requirements; and (8) no complicated dosing system (Rajoria et al. 2022). These key characteristics make the electrochemical method both efficient and environmentally friendly. As research and development efforts continue to advance in this field, the EO process holds great promise for promoting the transition toward more effective and sustainable solutions for electroplating wastewater treatment. EO has been used for wastewater treatment since the nineteenth century when studies were conducted on the electrochemical breakdown of cyanide. This technology has become the subject of extensive research since the late 1970s (Kuhn 1971). Over the past two decades, research has focused on improving the efficiency of oxidizing various pollutants with different electrodes and enhancing the electrocatalytic activity and stability of electrode materials. Researchers have also investigated factors that influence process performance and explored the mechanisms and kinetics involved in pollutant degradation.

18.4.1 Fundamentals

An EO process generally involves a reduction reaction at the cathode and an oxidation reaction at the anode (Botte 2017). The basic principle of EO processes is to remove pollutants by taking advantage of the redox reactions that happen at both the anode (pollutant oxidation) and cathode (heavy metal reduction), which have been widely used as a heavy metal remediation solution (Asaithambi et al. 2022; Nabgan et al. 2022). Meanwhile, the use of EO processes for removing organic pollutants lies in their ability to achieve either partial degradation or complete mineralization through anodic oxidation reactions. Thus, the electrocatalytic properties of the anodic materials used are crucial to the efficiency of organic pollutant removal in EO technology. In an electrolytic cell, organic pollutants are oxidized through two distinct processes: (i) direct anodic oxidation and (ii) indirect oxidation (Särkkä et al. 2015; Garcia-Segura et al. 2018; Yan et al. 2024).

18.4.2 Oxidation Mechanisms

EOPs involve two main processes for the oxidation of organic pollutants: (i) direct electron transfer to the surface of anode (R); (ii) production of reactive oxygen species as oxygen intermediates of water oxidation (Eqs. 18.2–18.4); additional, weaker oxidizing agents are produced electrochemically from bulk ions.

$$R + H_2O \rightarrow R(^{\cdot}OH) + H^+ + e^- \tag{18.2}$$

$$2R(^{\cdot}OH) \rightarrow 2RO + H_2O_2 \tag{18.3}$$

$$3H_2O \rightarrow O_3 + 6H^+ + 6e^- \tag{18.4}$$

Depending on the variety of electrode (anode) materials, the subsequent products are produced when $R(^{\cdot}OH)$ interacts with the anode surface: (i) As demonstrated by Eq. (18.5), $R(^{\cdot}OH)$ is transformed into RO superoxide for an active anode with low O_2^- overpotentials. In this case, R functions acts as a unique facilitator in the oxidation process of organic compound when it interacts with the anode surface. For a non-active anode with minimal O_2^- overpotentials, $R(^{\cdot}OH)$ is weakly adsorbed on the anode surface. This weak adsorption enables $R(^{\cdot}OH)$ to effectively interact with organic substances, thereby facilitating their mineralization through the oxidation process.

$$R(^{\cdot}OH) \rightarrow RO + H^+ + e^- \tag{18.5}$$

Lower access of R($^{\cdot}$OH) with the surface of anode, along with enhanced chemical reactivity is linked to a higher potential for O_2 evolution from the anode materials. Active chlorine species is one of the primary indirect oxidation agents used in treatment of industrial wastewater. The direct oxidation of chloride ions at the anode surface is the basis of chlorine-mediated oxidation. It generates chlorine (Cl_2), as indicated by Eq. (18.6). The chlorine then diffuses away from the anode and undergoes disproportionation, resulting in the formation of hypochlorous acid (HClO) and chloride ions, as shown in Eq. (18.7) (Zhu et al. 2019).

$$2Cl^- \rightarrow Cl_2 + 2e^- \quad (18.6)$$

$$Cl_2 + H_2O \rightarrow HClO + Cl^- + H^+ \quad (18.7)$$

The hypochlorite (ClO^-) ion with a pKa of 7.5 is in equilibrium with HClO in bulk solution, as demonstrated by Eq. (18.8) (Kapałka et al. 2010).

$$HClO \rightarrow ClO^- + H^+ \quad (18.8)$$

The electrochemical conversion of hypochlorous acid (HClO) to chlorate ions ($ClO_3{}^-$) can reduce the HClO content, which in turn lowers the rate of inorganic/organic compounds reduction at ambient temperature (Eq. (18.9)) (Cotillas et al. 2018).

$$6HClO + 3H_2O \rightarrow 2ClO_3^- + 4Cl^- + 12H^+ + 1.5O_2 + 6e^- \quad (18.9)$$

As described in Eq. (18.10) with a standard electrode potential (E°) of 0.68 V/SHE, two electrons are involved in the oxygen reduction reaction at the cathode surface that can be reduced to generate H_2O_2 in bulk in an acidic or neutral medium (Li et al. 2018).

$$O_2 + 2H^+ + 2e^- \rightarrow H_2O_2 \quad (18.10)$$

The reduction of four electrons for oxygen in H_2O, with a standard electrode potential (E°) of 1.23 V/SHE, is more complex than described in Eq. (18.10). The stability and production of H_2O_2 depend on several variables, including the operating circumstances, cathode characteristics, and cell configuration (Torres-Pinto et al. 2019). Parasitic reactions can occur in both divided and undivided electrochemical cells, potentially decreasing H_2O_2 accumulation in the system. These include the electrochemical reduction of H_2O_2 at the cathode surface (Eq. (18.11)) and the disproportionation of H_2O_2 in the bulk solution (Eq. 18.12), the latter occurring to a lesser extent.

$$H_2O_2 + 2H^+ + 2e^- \rightarrow 2H_2O \quad (18.11)$$

$$2H_2O_2 \rightarrow O_{2(g)} + 2H_2O \tag{18.12}$$

In addition, H_2O_2 in an undivided cell can oxidize to oxygen at the anode (Eqs. 18.13 and 18.14), producing weakly bound hydroperoxyl radicals (HO_2^-) as intermediates (Ridruejo et al. 2018).

$$H_2O_2 \rightarrow HO_2^{\cdot} + H^+ \rightarrow e^- \tag{18.13}$$

$$HO_2^{\cdot} \rightarrow O_2 + H^+ + e^- \tag{18.14}$$

In AOPs, the organic pollutants are oxidized into CO_2, water, and mineral acid. The AOP has garnered significant interest due to its higher removal efficiency, versatility, and adaptability (Guo et al. 2022). The AOPs constitute a direct and clean way to electrochemically generate $^{\cdot}OH$ radicals without additional chemical reagents or secondary contaminants. The electroplating effluents were quickly and effectively degraded by indirect oxidation using $^{\cdot}OH$ produced during the EO process (Eq. 18.15) (Moradi et al. 2020; Rajoria et al. 2023a).

$$R[^{\cdot}OH]_{ads} + \text{Heavy Metals} \rightarrow \text{oxidation products} + nCO_2 + yH_2O \tag{18.15}$$

Consequently, a variety of reactive species, including atomic hydrogen, O_2, $^{\cdot}OH$, and H_2O_2, are continuously produced in situ by the electrochemical process and the effective operation of the anode. These reactive species effectively remove both organic and inorganic contaminants from electroplating wastewater. A potential mechanism for the electroplating effluent degradation throughout the EO treatment method is depicted in Fig. 18.3.

18.4.3 Affecting Factors

Two factors that affect the electrocatalytic oxidation reaction's efficiency and activity are the rate of electron transfer and the stimulation and production of active oxidizing groups. Reactor designs, elements pertaining to water quality, operating circumstances, supporting electrolytes, and electrode materials are some of the variables that affect this intended outcome. Considering all these contributing factors is essential for precisely controlling and optimizing the electrocatalytic oxidation reaction process through rational regulation.

18.4.3.1 Typical Reactor Designs

Among EO reactor designs, the bipolar cell is the most basic. In addition to plane electrodes, cylindrical electrodes can also be used. An EO system must be designed

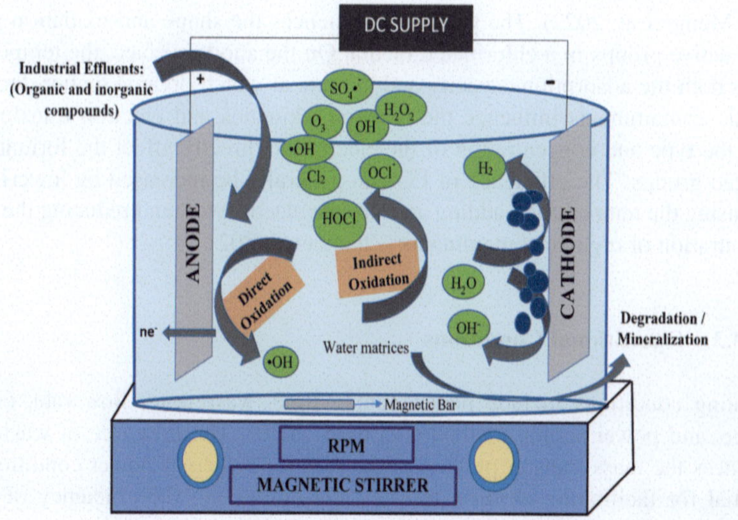

Fig. 18.3 Schematic representation of oxidation mechanisms (direct and indirect oxidation)

considering various electrode materials, cell configurations, and operating conditions. Since electrode surfaces are the primary sites of reaction in electrochemical processes, maintaining high mass transfer rates is a crucial consideration in cell configuration design. Electrodes come in two varieties, primarily two- and three-dimensional structures. The latter ensures that the electrode surface-to-cell volume ratio is high. Static electrodes and moving electrodes are two categories for both kinds of electrodes. The use of moving electrodes enhances the mass-transport coefficient by promoting turbulence. Recent investigations of EO for industrial wastewater use static parallel and cylindrical electrode cells as the main types of two-dimensional electrodes in reactor designs. The parallel plate geometry in a filter press arrangement is a commonly used cell design because it is easy to scale up to bigger electrode sizes. Reactor design must consider not only the geometry and configuration of the electrodes but also the cell configuration, whether divided or undivided. An ion-conducting membrane or a porous diaphragm divides the anolyte and catholyte in divided cells. Choosing suitable electrode materials is just as important as selecting the diaphragm or membrane to maintain the proper operation of the electrolytic process in divided cells. Overall, the use of divided cells should be avoided whenever possible, as separators are costly and assembling a divided cell is complex, often leading to mechanical and corrosion issues.

18.4.3.2 Water Quality Factors

Factors influencing water quality include temperature, pH, the kind and amount of organic pollutants present, and the electrolytes in the reaction solution (Xie et al.

2021; Meng et al. 2022). The pH value influences the shape and oxidation potential of active groups in a chlorinated media. On the anode surface, the temperature affects both the adsorption capacity and the rate at which organic matter degrades. Organic contaminants influence the reaction's kinetics and electron transfer rate, while the type and concentration of the electrolyte directly affect the formation of oxidized groups. The efficiency of EO can generally be increased by lowering the pH, raising the temperature, adding appropriate electrolytes, and reducing the initial concentration of organic contaminants (Zhang et al. 2021).

18.4.3.3 Operational Conditions

Operating conditions include power supply time, wastewater flow rate, current, voltage, and power supply method (Lei et al. 2020). The flow rate of wastewater influences the mass transfer process of the reaction, whereas power conditions are essential for facilitating electron transfer and optimizing the efficiency of active group generation. Therefore, to achieve high efficiency and low energy consumption, selecting the right flow rate, current/voltage, power supply time, and method is essential.

18.4.3.4 Supporting Electrolytes

Various supportive electrolytes have been used for the EO of organic pollutants with different electrode materials to increase solution conductivity. Significant research has focused on examining the impact of various types and concentrations of supporting electrolytes on the EO of organic pollutants. In general, salts such as reducing salts ($NaSO_3$ and $NaNO_2$) and neutral salts (Na_2SO_4, $NaClO_4$, NaCl, and $NaNO_3$) were used to prepare supportive electrolytes (Niu et al. 2013). $NaClO_4$ and $NaNO_3$ are two of the most stable compounds among them. They can prevent the production of any oxidizing species that could react with organic compounds, which make them suitable for studying direct anodic oxidation at the anode surface. However, An et al. (2012) found that adding chloride ions mostly increased the decolorization efficiency but did not decrease the TOC, suggesting a more successful attack by active chlorine on the chromophoric groups of the dye compound.

18.4.3.5 Electrode Materials

The performance of the electrode materials is a critical factor in EO and directly influences the variations in electrocatalytic oxidation technology (Baddouh et al. 2018). Therefore, selecting the right electrode material is crucial since it influences the process's efficiency and selectivity (Rajoria et al. 2023b). Several characteristics must be present in the electrode material:

- Excellent electrical conductivity; resistance to erosion, corrosion, and the production of passivation layers; high physical and chemical stability.
- Selectivity and catalytic activity.
- Low lifetime cost ratio. Cost-effective and long-lasting electrode materials have to be preferred.
- Long service life, strong catalytic activity, easy manufacture, and a broad operating potential window for the hydrogen and oxygen evolution reaction (OER) (Salazar-Banda et al. 2021).

Different electrode materials were employed to examine their oxidation power for the removal of contaminants in wastewater (Guvenc et al. 2022). Table 18.2 lists the types and oxidation power of the anode materials used in wastewater treatment. Complete combustion can be achieved by non-active anodes like SnO_2, PbO_2, or BDD, whereas active anodes like IrO_2, RuO_2, or Pt promote partial and selective oxidation of contaminants. Therefore, non-active anodes are considered the ideal electrodes to achieve complete oxidation of organics to CO_2 during wastewater treatment. Several studies have shown that a broad range of bio-refractory organic compounds can be oxidized with platinum (Pt) electrodes at relatively low OEP (1.6 V against SHE in 0.5 M H_2SO_4).On the other hand, there has been a lot of interest recently in the use of BDD electrodes for chemically destroying contaminants. The BDD electrode has drawn a lot of interest as a unique electrode material in recent decades and is thought to be the best electrode material for EO of numerous pollutants from various industries' wastewater. Table 18.3 lists the merits and demerits of different electrode materials for EO treatment.

18.5 Conclusions and Future Perspectives

All living things need water, which is one of the most important natural resources. The rising health problems related to environmental issues are increasingly attributed to insufficient decontamination of industrial wastewater. This challenge has inspired scientists and engineers to develop and implement advanced technologies that efficiently remove organic pollutants while maintaining cost-effectiveness. Therefore, there is a need to develop novel methods for degrading electroplating wastewater by applying a new class of advanced oxidation processes. The application of advanced oxidation processes (AOPs) in the treatment of industrial wastewater is becoming more widespread. This chapter demonstrates that the electrochemical oxidation (EO) process, as one of the AOPs, is highly effective for degrading wastewater containing hazardous contaminants. The rapid decontamination and complete oxidation of electroplating effluents are largely driven by the substantial production of reactive oxygen species (ROS) and other oxidants during EO treatments. To maximize these advantages, optimizing operating conditions, reactor design, and electrode materials is essential. EO treatment approaches are less maintenance-intensive, more

Table 18.2 Classification of anode materials based on oxidation power and potential for OER (Patel et al. 2013)

Anode type	Active component	Electrode	Electrocatalytic ability for OER	Oxidation potential (V)	Overpotential (V)	Adsorption enthalpy of M-˙OH	Oxidation power of anode
Active	RuO_2	RuO_2-TiO_2 (DSA-Cl_2)	Good	1.4–1.7	0.18	Chemisorption of ˙OH radical	Low
	IrO_2	IrO_2-Ta_2O_2 (DSA-O_2)	Good	1.5–1.8	0.25		
	Pt	Ti/Pt	Good	1.7–1.9	0.30		
	Carbon and graphite	Ti/GO	Good	1.7	–		
Non-active	PbO_2	Ti/PbO_2	Poor	1.8–2.0	0.50		
	SnO_2	Ti/SnO_2-Sb_2O_5	Poor	1.9–2.2	0.70		
	BDD	p-Si/BDD	Poor	2.2–2.6	1.3	Physisorption of ˙OH radical	High

Table 18.3 Comparison of different electrode materials in EO treatment (Rajoria et al. 2021)

Electrode	Merits	Demerits	Comparison with the other electrodes
Ti	Stable	Expensive	–
Pt	Resistant to corrosion, excellent repeatability qualities, low oxygen evolution overpotential	Expensive, low efficiency to mineralize organics	Low performance in anodic oxidation
PbO_2	Cost-effective, simple to assemble, relatively high overpotential toward OER, good current efficiency, relatively high mineralization efficiency	Corrosive Pb^{2+} ions could be released, inadequate use in the treatment of industrial wastewater	–
DSA electrodes	Supports indirect oxidation, good conductivity properties, cost-effective, easy availability, good current efficiency	Short lifespan, poor electrochemical stability	In comparison with BDD, the degradation rate is minimal
BDD	High current efficiency, inert in tough conditions, electrochemical stability, high OEP, excellent electrochemical stability	Very expensive, efficiency reduced in diluted solutions	Higher activity

cost-effective, and exhibit a reservoir effect compared to alternative methods. Additionally, EO has the greatest potential for indirect oxidation by chlorine/hypochlorite formed during treatment and hydroxyl radicals generated at the anode to disinfect water. However, the formation of toxic byproducts can be a concern in certain cases and requires careful examination prior to advancing to pilot or industrial-scale implementations. The effectiveness of the EO method relies significantly on the choice of electrode materials. Further investigation is necessary to consider fundamental issues such as a variety of electrode materials, electrode service life, an overall economic analysis of the process for industrial wastewater treatment, and optimization of electrolytic reactor.

Future developments are expected to focus on optimizing electrochemical reactors at the pilot scale upon their implementation. Additionally, combining various electrochemical techniques could be explored to enhance wastewater treatment in industrial applications. These technologies have a bright future. Future advancements will depend on the collaborative efforts of analytical chemists, electrochemists, and engineers to effectively deploy and enhance electrochemical technologies. This collaboration will be essential for identifying innovative pre- and post-treatment methods. Thus, it is essential to highlight the importance of advanced electrochemical approaches as a key technical measure for effective environmental protection.

References

An H et al (2012) Fabrication and electrochemical treatment application of a microstructured TiO_2-NTs/Sb-SnO_2/PbO_2 anode in the degradation of C.I. Reactive Blue 194 (RB 194). Chem Eng J 209:86–93. https://doi.org/10.1016/j.cej.2012.07.089

Asaithambi P et al (2022) A review of hybrid process development based on electrochemical and advanced oxidation processes for the treatment of industrial wastewater. Int J Chem Eng 2022:1–15. https://doi.org/10.1155/2022/1105376

Baddouh A et al (2018) Electrochemical decolorization of Rhodamine B dye: influence of anode material, chloride concentration and current density. J Environ Chem Eng 6:2041–2047. https://doi.org/10.1016/j.jece.2018.03.007

Botte GG (2017) Electrochemical technologies for water treatment, management, and efficiency. Electrochem Soc Interface 26(2):53–61. https://doi.org/10.1149/2.F04172if

Chen C et al (2015) Electrochemical decolorization of dye wastewater by surface-activated boron-doped nanocrystalline diamond electrode. J Environ Sci 45:100–107. https://doi.org/10.1016/j.jes.2016.01.004

Cotillas S et al (2018) Removal of Procion Red MX-5B dye from wastewater by conductive-diamond electrochemical oxidation. Electrochim Acta 263:1–7. https://doi.org/10.1016/j.electacta.2018.01.052

Garcia-Segura S, Ocon JD, Chong MN (2018) Electrochemical oxidation remediation of real wastewater effluents—a review. Process Saf Environ Prot 113:48–67. https://doi.org/10.1016/j.psep.2017.09.014

Gitis V, Hankins N (2018) Water treatment chemicals: trends and challenges. J Water Process Eng 25:34–38. https://doi.org/10.1016/j.jwpe.2018.06.003

Glaze WH, Kang J, Douglas H (1987) The chemistry of water treatment processes involving ozone hydrogen peroxide and ultraviolet radiation. Ozone: Sci Eng J Int Ozone Assoc Eng 9:335–352. https://doi.org/10.1080/01919518708552148

Guo H et al (2022) How to improve lead dioxide anodes performance in organic wastewater treatment: review and prospect. Process Saf Environ Prot 164:189–207. https://doi.org/10.1016/j.psep.2022.06.012

Gusain R, Kumar N, Ray SS (2020) Recent advances in carbon nanomaterial-based adsorbents for water purification. Coord Chem Rev 405:213111. https://doi.org/10.1016/j.ccr.2019.213111

Guvenc SY et al (2022) Pre-coagulated landfill leachate treatment by electro-oxidation using MMO/Ti, Pt/Ti, and graphite anodes. Environ Eng Res 28(1):210419. https://doi.org/10.4491/eer.2021.419

Kapałka A et al (2010) Direct and mediated electrochemical oxidation of ammonia on boron-doped diamond electrode. Electrochem Commun 12(12):1714–1717. https://doi.org/10.1016/j.elecom.2010.10.004

Kong X et al (2020) A novel technique of COD removal from electroplating wastewater by Fenton—alternating current electrocoagulation. Environ Sci Pollut Res 27(13):15198–15210. https://doi.org/10.1007/s11356-020-07804-6

Krishna RH et al (2023) Carbon nanotubes and graphene-based materials for adsorptive removal of metal ions—a review on surface functionalization and related adsorption mechanism. Appl Surface Sci Adv 16:100431. https://doi.org/10.1016/j.apsadv.2023.100431

Kuhn AT (1971) Electrolytic decomposition of cyanides, phenols and thiocyanates in effluent streams—a literature review. J Appl Biotechnol 21:29–34

Lei J et al (2020) Linear attenuation current input mode: a novel power supply mode for electrochemical oxidation process. J Water Process Eng 36:101305. https://doi.org/10.1016/j.jwpe.2020.101305

Li H et al (2018) NH_3 inhibits mercury oxidation over low-temperature MnO_x/TiO_2 SCR catalyst. Fuel Process Technol 176:124–130. https://doi.org/10.1016/j.fuproc.2018.03.022

Meng C et al (2022) Efficient electrochemical oxidation of COVID-19 treatment drugs favipiravir by a novel flow-through Ti/TiO_2-NTA/Ti_4O_7 anode. Electrochim Acta 430:141055. https://doi.org/10.1016/j.electacta.2022.141055

Moradi M et al (2020) Service life and stability of electrodes applied in electrochemical advanced oxidation processes: a comprehensive review. J Ind Eng Chem 87:18–39. https://doi.org/10.1016/j.jiec.2020.03.038

Nabgan W et al (2022) A state of the art review on electrochemical technique for the remediation of pharmaceuticals containing wastewater. Environ Res 210:112975. https://doi.org/10.1016/j.envres.2022.112975

Niu J et al (2013) Electrochemical mineralization of pentachlorophenol (PCP) by Ti/SnO_2-Sb electrodes. Chemosphere 92:1571–1577. https://doi.org/10.1016/j.chemosphere.2013.04.035

Öztürk H, Barışçı S, Turkay O (2021) Paracetamol degradation and kinetics by advanced oxidation processes (Aops): electro-peroxone, ozonation, goethite catalyzed electro-fenton and electro-oxidation. Environ Eng Res 26(2):180332. https://doi.org/10.4491/eer.2018.332

Patel PS et al (2013) Electro-catalytic materials (electrode materials) in electrochemical wastewater treatment. Procedia Eng 51:430–435. https://doi.org/10.1016/j.proeng.2013.01.060

Rajoria S, Vashishtha M, Sangal VK (2021) Review on the treatment of electroplating industry wastewater by electrochemical methods. Mater Today: Proc 47:1472–1479. https://doi.org/10.1016/j.matpr.2021.04.165

Rajoria S, Vashishtha M, Sangal VK (2022) Treatment of electroplating industry wastewater: a review on the various techniques. Environ Sci Pollut Res 29:72196–72246. https://doi.org/10.1007/s11356-022-18643-y

Rajoria S, Vashishtha M, Sangal VK (2023a) Electrochemical treatment of electroplating wastewater using synthesized GO/TiO_2 nanotube electrode. Environ Sci Pollut Res 30:71226–71251. https://doi.org/10.1007/s11356-023-27366-7

Rajoria S, Vashishtha M, Sangal VK (2023b) Preparation and characterization of novel SnO_2-GO/TiO_2NTs electrode for the degradation of electroplating effluents. J Environ Chem Eng 11:110679. https://doi.org/10.1016/j.jece.2023.110679

Ridruejo C et al (2018) On-site H_2O_2 electrogeneration at a CoS_2-based air-diffusion cathode for the electrochemical degradation of organic pollutants. J Electroanal Chem 808:364–371. https://doi.org/10.1016/j.jelechem.2017.09.010

Salazar-Banda GR et al (2021) Developments in electrode materials for wastewater treatment. Curr Opin Electrochem 26:100663. https://doi.org/10.1016/j.coelec.2020.100663

Särkkä H, Bhatnagar A, Sillanpää M (2015) Recent developments of electro-oxidation in water treatment—a review. J Electroanal Chem 754:46–56. https://doi.org/10.1016/j.jelechem.2015.06.016

Torres-Pinto A et al (2019) Metal-free carbon nitride photocatalysis with in situ hydrogen peroxide generation for the degradation of aromatic compounds. Appl Catal B: Environ 252:128–137. https://doi.org/10.1016/j.apcatb.2019.03.040

Tran NH et al (2015) A critical review on characterization strategies of organic matter for wastewater and water treatment processes. Bioresource Technol 193:523–533. https://doi.org/10.1016/j.biortech.2015.06.091

Vagi MC, Petsas AS (2020) Recent advances on the removal of priority organochlorine and organophosphorus biorecalcitrant pesticides defined by Directive 2013/39/EU from environmental matrices by using advanced oxidation processes: an overview (2007–2018). J Environ Chem Eng 8(1):102940. https://doi.org/10.1016/j.jece.2019.102940

Vinayagam V et al (2024) Recent developments on advanced oxidation processes for degradation of pollutants from wastewater with focus on antibiotics and organic dyes. Environ Res 240:117500. https://doi.org/10.1016/j.envres.2023.117500

Xie W et al (2021) Electrospun iron/cobalt alloy nanoparticles on carbon nanofibers towards exhaustive electrocatalytic degradation of tetracycline in wastewater. Chem Eng J 405:126585. https://doi.org/10.1016/j.cej.2020.126585

Yan Y et al (2024) Electrochemical oxidation processes based on renewable energy towards carbon neutrality: oxidation fundamentals, catalysts, challenges and prospects. Chem Eng J 487:150447. https://doi.org/10.1016/j.cej.2024.150447

Zhang J et al (2021) Current progress in electrochemical anodic-oxidation of pharmaceuticals: mechanisms, influencing factors, and new technique. J Hazard Mater 418:126313. https://doi.org/10.1016/j.jhazmat.2021.126313

Zhang Z et al (2022) The fabrication of Ti_4O_7 particle composite modified PbO_2 coating electrode and its application in the electrochemical oxidation degradation of organic wastewater. J Alloy Compd 897:162742. https://doi.org/10.1016/j.jallcom.2021.162742

Zhu K et al (2019) Comparative degradation of atrazine by anodic oxidation at graphite and platinum electrodes and insights into electrochemical behavior of graphite anode. Electrocatalysis 10(1):35–44. https://doi.org/10.1007/s12678-018-0493-z

Chapter 19
Exploring the Utility of Membrane Bioreactors (MBRs) in Wastewater Treatment

Shyam Kodape

Abstract The membrane process is a widely used separation process in wastewater treatment technology, which leads to an efficient treatment technology along with excellent performance and process economics. Generally, membrane-based technologies primarily rely on secondary treatment to produce high-quality effluent suitable for recycling and various applications. However, over the years, membrane-based reactors (MBRs) have proven to be quite effective in secondary treatment technology; typically, the membranes' range of application is between microfiltration (MF) and ultrafiltration (UF). Researchers have developed numerous membrane processes and are actively working on improving others. Many researchers proved that the MBR process offers excellent performance in removing inorganic contaminants, organic compounds, and biological entities from wastewater. Except for the separation of suspended microorganisms and treated wastewater, the MBR technology is like the activated sludge process. The MBR type of system separates suspended growth and treated water simultaneously, whereas the separation occurs in the conventional system's second stage. MBR has numerous benefits, comprising better control over biochemical activity, excellent effluent that is free of microbes, a smaller footprint, and the ability to handle variable organic load rates. Aside from municipal wastewater treatment, the MBR method makes it simple to treat a variety of industrial wastewater, from olive oil mill wastewater containing phenolic chemicals to refinery wastewater containing oil.

S. Kodape (✉)
Department of Chemical Engineering, Visvesvaraya National Institute of Technology, Nagpur, India
e-mail: smkodape@che.vnit.ac.in

S. Souabi et al. (eds.), *Wastewater Treatment Plants*, Water Science and Technology Library 130, https://doi.org/10.1007/978-3-031-87461-1_19

Graphical Abstract

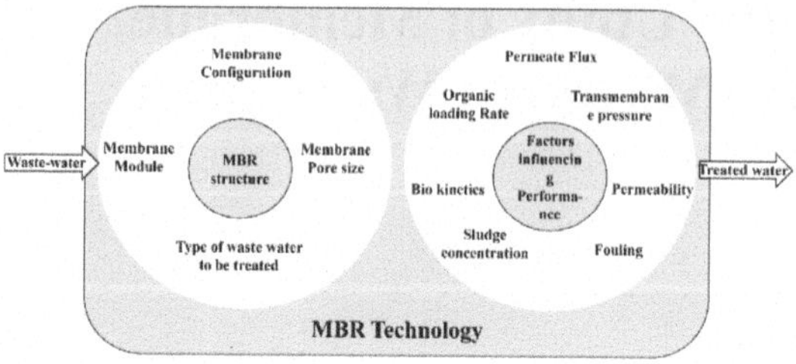

Keywords Side-stream MBR · Submerged MBR · Biodegradation kinetic · Fouling · Respirometric technique

19.1 Introduction

The percentage of available freshwater on Earth is extremely low, because a lack of freshwater is becoming an increasingly serious problem in several countries, therefore, attempts should be made to treat wastewater (Sartor et al. 2008; Sgroi et al. 2018; Oliveira Filho et al. 2021; Liu et al. 2021). There are different types of membrane separation processes such as reverse osmosis (RO), nanofiltration (NF), microfiltration (MF), electrodialysis (ED), ultrafiltration (UF), and electro deionization (EDI), out of all these RO, NF, MF, UF processes give permeate and concentrate (Zhang et al. 2021a). Wastewater treatment is one of the fields that has been fascinated by membrane separation. In general, the membrane process is feed stream that is sent from one of the sides of the membrane and permeate means allowing particles and it allows from another side of the membrane, and retentate means rejected materials from the membrane (Lu and Li 2021). Membrane bioreactor integration is a promising new development as a replacement for the conventional secondary clarifier and includes a suspended activated sludge growth system that gives microporous membranes for separation of solids and liquids (Brepols et al. 2008; Tian et al. 2008; Innocenti et al. 2002; Badani et al. 2005).

Membrane bioreactor (MBR) technology offers significant potential for wastewater reuse due to its small footprint, high-quality effluent, complete solid–liquid separation, reduced sludge production, and better process reliability (Sartor et al. 2008; Sgroi et al. 2018). It saves plant space and reduces processing hold-up time, producing minimal sludge and removing microorganisms and nutrients (Chon et al. 2013; Choi et al. 2009). MBR is increasingly popular for decentralized, recovery, reclamation, and reuse applications (Chon et al. 2013; Melin et al. 2006; Keffala et al. 2024). Dorr-Oliver Inc. was the first to introduce the technique, which combined a biologically activated sludge process with a crossflow membrane filtration circuit (Merz et al. 2007). In the 1960s, an ultrafiltration membrane was employed to remove activated sludge from a biological wastewater treatment system's final effluent, and the mixed liquor was recycled back into the aeration tank. Since then, the MBR process has grown, and MBR technology research has exploded, notably in the twentieth century (Arévalo et al. 2014). This chapter investigates the potential of MBR in water treatment by studying its capacity for pollutant removal as well as its operational and biological performance. We also discuss the configuration of MBRs and the control of fouling.

19.2 MBR Configuration

Microbial behavior classifies treatment methods in membrane bioreactors into aerobic and anaerobic categories, whereas the configuration of the membrane bioreactor is determined by the filtration driving power. In general, it is comprised of two units; an activated sludge bioreactor for the removal of biodegradable pollutants and an ultrafiltration (UF) or microfiltration (MF) membrane module for solid/liquid separation (Melin et al. 2006; Buer and Cumin 2010). Membrane separation can be accomplished using either pressure-driven filtration, in which membranes are attached to the bioreactor via an external loop (recirculated or side stream), or vacuum-driven filtration, in which membranes are submerged in the bioreactor, as in dead-end mode (Ng and Kim 2007). So, there are two types of MBR configurations based on where the membrane is located: side-stream membrane unit is located outside the bioreactor known as recirculated or external loop MBR and the membrane separation unit is submerged in the bioreactor in the submerged MBR or immersed MBR as shown in Fig. 19.1.

In a side-stream MBR system, the influent wastewater is directly exposed to microorganisms. Wastewater and biomass are channeled through membranes within the recirculation loop. The biomass that is retained is returned to the reactor, whereas the treated water is discharged. A recirculation pump provides a crossflow velocity

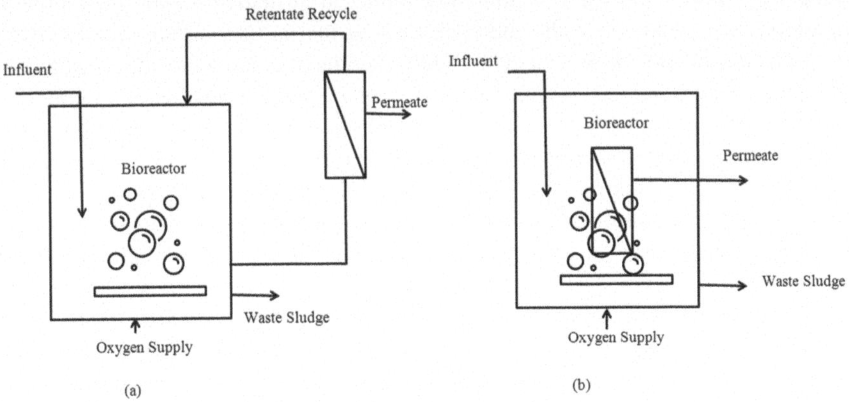

Fig. 19.1 MBR system, **a** side-stream MBR, **b** submerged MBR

(CFV) at the surface of the membrane, reducing suspended solids deposition, but it comes at a high operational cost due to the recirculation loop's installation (Le-Clech et al. 2006; Rahman and Al-Malack 2006; Cicek 2003; Nguyen et al. 2016). External membrane assemblies utilizing tubular, spiral wound, hollow fiber (HF), or crossflow plate-frame membranes necessitate substantial flow recirculation (Merz et al. 2007; Mutamim et al. 2013; Gkotsis et al. 2014). The side-stream MBR has the longest history and is mostly utilized in industrial wastewater treatment, where it provides direct hydrodynamic fouling control at a high energy cost. Table 19.1 is the summary of the range of operational and performing parameters in side-stream MBR. MBR is a mix of high-performance biochemical activities and membrane separation, which is an effective, accurate, and compact biological wastewater treatment that is small and easy to use (Mehmood et al. 2021). The side-stream MBR can tolerate higher loading rates at various temperatures than that of activated sludge (Messaoudi et al. 2021). It has the capability to achieve a water quality meeting reuse criteria (Lübbecke et al. 1995). Adaptation of advanced add-on processes such as airlift, jet loop mechanism appears to be performance-enhancing tools (Abeynayaka and Visvanathan 2011). Researchers use ultrafiltration, polyvinylidene fluoride, tubular, hydraulic, tubular crossflow membrane module, microstructure fibers, polysulfone (PS), ceramic membrane, and polypropylene membrane in MBR setup (Badani et al. 2005; Lübbecke et al. 1995; Abeynayaka and Visvanathan 2011; Brik et al. 2006; Farizoglu and Uzuner 2011; Chakrabarty et al. 2010).

Submerged MBR design is possible with plate and frame/hollow fiber membranes, where biological degradation occurs around the membrane, keeping the biomass contained within the reactor (Çulfaz et al. 2011). Submerged MBRs have been extensively studied by researchers. In industrial wastewater treatment, submerged membrane bioreactors are commonly used. Individual homes, municipal treatment plants, hotels, airports, golf courses, sports clubs, and even chemical industries have demonstrated that submerged MBR can be used effectively for wastewater treatment in a number of studies published in scientific literature (see Table 19.2). Most of the time, submerged MBRs use vertical flat sheets or HF membranes that are mounted vertically or horizontally. HF membrane modules are usually cheaper, have a lot of membrane area, and can withstand backwashing that is very strong (Kader 2015; Defrance and Jaffrin 1999). Gui et al. (2003) reported that the most important factor in sludge deposition on the membrane surface is membrane flux; aeration intensity only becomes relevant when the MLSS concentration is high, and intermittent aeration operations, such as suction time and non-suction time, can have both positive and negative effects (Chandrasekhar et al. 2022). Different SRTs have made it possible for a pilot ultrafiltration MBR to remove nutrients and micropollutants from water, as well as to cut down on waste sludge production (Innocenti et al. 2002; Turano et al. 2002).

In comparison with submerged systems, side-stream MBR requires significantly more power (Dhaouadi and Marrot 2008) shown in Table 19.3. Pumping accounts for 60–80% of total energy consumption in side stream aerobic MBRs, with aeration accounting for only 20–40% (Yeh and Wu 1997). External plate-frame or hollow fiber membranes, on the other hand, offer promise in terms of lowering aeration and

Table 19.1 Summary of range of operational and performing parameters in side-stream MBR

Ref	Wastewater source	Membrane module	Influent COD range	Removal COD	Temperature	CFV	TMP	MLSS/MLVSS
Sartor et al. (2008)	Kitchen wastewater	Nanofiltration (NF) membrane made of hydrophilized polyamide (HPA), spiral wound, effective surface area of 1.2 m^2, pore size = nanopores	2180 mg/lit	> 97%	NA	NA	0.098 kPa	230 mgTSS/lit
Sgroi et al. (2018)	Synthetic	Ceramic, pore size = 0.45 μm, MF and 150 kDa, UF dia = 10 mm, length = 250 mm, area = 0.0125 m^2	11,225 mg/lit	91%	47 °C, 60 °C and 30 °C	2.28 m/s (combined air and mixed liquor)	20–90 kPa	7000–9000 mgMLSS/lit
Oliveira Filho et al. (2021)	Textile	PVDF, UF (tubular), pore size = 25 nm	1280–5600 mg/lit	97%	–	2 m/s	60–100 kPa	5–15 mgMLSS/lit
Liu et al. (2021)	Polyester finishing factory (textile)	PVDF, tubular cross flow UF, area = 0.28 m^2, pore size = 15 kDa, dia = 12.7 mm	1380–6033 mg/lit	60–95%	32–34	2 m/s	40–80 kPa	

(continued)

Table 19.1 (continued)

Ref	Wastewater source	Membrane module	Influent COD range	Removal COD	Temperature	CFV	TMP	MLSS/MLVSS
Zhang et al. (2021)	Dairy industry	Ceramic, pore size = 50 nm, 0.24 m^2, no. of channel = 37	921–9004 mg/lit	98%	22 °C	1.32–2.65 m/s	60–400 kPa	2312–38,684 mg/lit
Lu and Li (2021)	Petroleum industry (oily water)	PS, flat sheet, 3.21 nm	–	–	–	Cross flow rate = 3.1×10^{-6} m^3/s	Constant 103 kPa and 172.4 kPa	–
Brepols et al. (2008)	Colloidal yeast	PES microstructure, round fiber, and twisted, HF, pore size = 12 nm	–	–	–	0.09 m/s for structured and twisted, 0.08 m/s for round fiber	–	0.025 wt% Baker's yeast
Tian et al. (2008)	Wastewater treatment plant	Ceramic, tubular, area = 0.25 m^2, length = 120 cm, no. of channels = 19, pore size = 0.1 μm	–	–	–	1–5 m/s at fixed TMP,1,2, 3, 4 m/s at fixed permeate flux	–	–
Innocenti et al. (2002)	Olive mill	PS, flat sheet, pore size = 17 kDa	–	90%	30 °C	Re = 7500 and 25,000	50 and 300 kPa	

(continued)

Table 19.1 (continued)

Ref	Wastewater source	Membrane module	Influent COD range	Removal COD	Temperature	CFV	TMP	MLSS/MLVSS
Badani et al. (2005)	Olive mill	Ceramic (MF), length = 1.2 m, id/od = 6 mm/ 10 mm, pore size = 0.14 μm	1500–5300 mg/lit	81%	–	–	100–150 kPa	12 × 10^3 mgMLSS/lit
Fazal et al. (2015)	Polyvinylepyrrolidone PVP 360 aqueous solution	PP, HF, fibers = 250, area = 0.06 m^2	Different concentration of PVP 0.1, 0.5, 1.0, and 2.0 wt%		25 °C	0.0723, 0.1209, 0.1684, and 0.2195 m/s	19, 38, 57, 77, 96, and 115 kPa	–
Radjenović et al. (2008)	SWW, biological suspension (BS), SWW + BS	PVDF, UF, pore size = 0.3 and 0.7 μm. area = 0.0072 m^2	–	–	–	0.1 and 3.5 m/s	100 kPa	2000–10,000 (mgMLSS/lit)
Le-Clech et al. (2006)	Synthetic	PP, hollow fiber, MF pore = 0.23–0.27 μm Area = 0.087 m^2	312–584 mg/lit	> 96.5 (without PAC) and > 99 (with powered activated carbon PAC)	24–34	0.9–1.9 m/s	70–140 kPa	2100 × 10^3–3600 × 10^3 mgMLSS/lit

(continued)

Table 19.1 (continued)

Ref	Wastewater source	Membrane module	Influent COD range	Removal COD	Temperature	CFV	TMP	MLSS/MLVSS
Kader (2015)	Filtered primary clarifier effluent	PS, HF, Area = 7.9 m^2, id = 0.762 mm, pore size = 0.01 μm	BOD = 74 mg/lit	> 51%	–	1.01 and 2.04 m/s	22–34 kPa	TS = 29 mg/lit
Chon et al. (2013)	Skim milk	ZrO_2 alpha alumina supported, MF rig Tubular, pore size = 0.1 μm, dia = 6.8 mm	Case in micelles	95%	50 °C	3–7.2 m/s	5–300 kPa	–
Choi et al. (2009)	SWW	PS, UF tubular, dia = 11.5 mm, area = 0.1 m^2, pore size = 0.01 μm	8500–17,600 mg/lit	> 99%	27 °C	3.5–6.6 m/s	135–260 kPa	6×10^3 to 40×10^3 mgMLSS/lit
Melin et al. (2006)	Petroleum refinery (oily waste water)	Ceramic, HF, tubular, pore size = 0.1 μm, 0.019 m^2	160–2300 gm/lit	> 93%	–	3.24, 2.69, 2.21 m/s	79.29–165.47 kPa	3000 and 5000 mg/lit const two run
Keffala et al. (2024)	Ice cream and dairy product factory	Ceramic, tubular, pore size = 0.2 μm, area = 0.06 m^2, dia = 0.02 m	11,000–13,300 mg/lit	> 83%	15, 25 and 35 °C	3.53 m/s	–	Bacterial count cell = 10^9 cfu/ml–2.4 × 10^8 viable cells/ml

(continued)

Table 19.1 (continued)

Ref	Wastewater source	Membrane module	Influent COD range	Removal COD	Temperature	CFV	TMP	MLSS/MLVSS
Merz et al. (2007)	Glass industry	Mixed cellulose ester, MF dead end, area = 0.00287 m^2 and plate Frame area = 0.00301 m^2	< 20%	–	–	1–5 m/s	49–250 kPa	–
Arévalo et al. (2014)	Urban waste water	Skin Zirconia (ZrO_2) + support alumina (gamma AL_2O_3), dia = 4.5 mm, area = 0.04 m^2, pore size = 0.45 μm	30–1424 mg/lit	95	–	2–3 m/s	–	1500×10^3–$15{,}100 \times 10^3$ mgMLSS/lit

Abbreviations TS—Total solid; TSS—Total suspended solid; CFV—Cross flow velocity; COD—Chemical oxygen demand; MLSS—Mixed liquor suspended solid; MLVSS—Mixed liquor volatile suspended solid; TMP—Transmembrane pressure membrane cleaning costs (Choi et al. 2005; Cao et al. 2005). The lower amount of membrane surface shear generated by submerged MBRs results in less energy consumption but lower permeate flow rates when operating at reduced pressure. As a result, a larger membrane surface area is required for this application (Bourgeous et al. 2001). The debate over whether to use submerged aerobic MBRs or side stream layouts appears to have been settled in favor of submerged aerobic MBRs. In fact, because of the lower associated energy requirements, the submerged configuration is currently used in the majority of commercial applications (Berre and Daufin 1996).

19.3 Membrane Types in MBR System

Over the previous years, MBRs have been quite effective in secondary treatment technology; normally the membranes' applicable range is MF and UF (Dhaouadi and Marrot 2008). Membrane types can be classified based on either cylindrical or planar geometry. Industries use five different types of membrane configurations, including plate-and-frame (i.e., flat sheet (FS), HF, spiral-wound, tubular, and plated filter cartridges), for efficient separation; membrane surface modification is widely used in these configurations (Scott and Smith 1997). Plate-and-frame type membrane modules consist of flat sheet membranes along with support and/or membrane separators. Plates are affixed to the membrane, using a vacuum-controlled procedure where water passes over the membrane and treated water is collected through pipelines (Kang and Choo 2003). In the HF type of module, multiple HF membranes are arranged, and an epoxy block connected from the external side seals the ends of the fibers together. Two configurations exist: one involves the flow of water from the membrane's interior to its exterior under pressure, while the other involves the flow from the external to the internal section when operating under vacuum. Various manufacturers produce these configurations in different ways (Xing et al. 2001a). In reverse osmosis and nanofiltration operations, the spiral-wound shape of membranes is widely employed. These membranes are tightly sealed up against the perforated tube through which the permeate exits. Major manufacturers universally manufacture spiral-wound modules with uniform dimensions, which simplifies and reduces the cost of membrane production (Bazrafshan et al. 2021). Plants that have a high capacity can arrange various types of membrane modules either in parallel or in series. Typically, pressure vessels are used to connect tubular modules and plated filter cartridges, allowing the pumping of mixed liquor suspended solids into them. These configurations mainly apply to side-stream systems (Al-Amri et al. 2010). By immersing hollow fiber and flat sheet membrane modules directly in a mixed liquor, vacuum pumps typically drawn-out effluent from these modules. To ensure the safety of membranes and avoid excessive cleaning frequencies resulting from clogging by suspended particles and other fibrous materials, surface modification techniques such as applying a thin screen on the membranes are employed (Choi et al. 2005). A comparison of tubular, hollow, flat sheet membranes is shown in Table 19.4.

Table 19.2 Summary of range of operational and performing parameters in submerged MBR

Reference	Wastewater source	Membrane module	Influent COD range	% Removal COD	Temp	TMP (final or mean)	MLSS or MLVSS
Buer and Cumin (2010)	Municipal wastewater	Polyvinylidene fluoride (PVDF) flat sheet, pore size of 0.1 μm and surface area of 0.1 m^2	600 mg/lit	94%	Room temperature	< 30 kPa	2600–2800 mg MLSS/lit
Ng and Kim (2007)	Petrochemical wastewater	HF, polytetrafluoroethylene (PTFE), nominal pore size = 0.1 μm, surface area = 0.1 m^2	128 ± 74 mg/lit	> 87%	27 ± 1 °C	< 5 kPa	3000–4000 mgMLSS/lit
Rahman and Al-Malack (2006)	Synthetic municipal wastewater	PES, HF, id = 0.55 mm, pore size = 0.2 μm, area = 0.065 m^2	584 ± 175 mg/lit	95% at 25 °C, 85% at 35 °C, 50% at 45 °C	25, 35 and 45 °C	Increased in TMP 5.3 kPa at 25 °C, 7 kPa at 35 °C and 38.4 kPa at 45 °C	8000–10,000 mg MLSS/lit
Cicek (2003)	Volatile fatty acid mixture (anaerobic treatment)	Polysulfone, microfiltration, pore size = 0.2 μm, area = 0.042 m^2	10,000 mg/lit	> 98%	30 °C	10–30 kPa	40,000 mgMLSS/lit, 35,000 mgMLVSS/lit
Nguyen et al. (2016)	Tannery wastewater	Hollow fiber, microfiltration, pore size = 0.4 μm	3690 mg/lit	78%	18–26 °C	20 kPa	15,100 mgMLSS/lit
Mutamim et al. (2013)	Synthetic	PVDF, hollow fiber, pore size = 0.1 μm, area = 0.02 m^2	2000 mg/lit	> 99%	40 °C	12 kPa	1000–6000 mgMLSS/lit

(continued)

Table 19.2 (continued)

Reference	Wastewater source	Membrane module	Influent COD range	% Removal COD	Temp	TMP (final or mean)	MLSS or MLVSS
Gkotsis et al. (2014)	Mixed sewage (industrial + domestic source)	PVDF, hollow fiber, pore size = 0.1 μm, area = 10.2 m^2/module, id = 650 μm	87–800 mg/lit	95%	28–32 °C	13 kPa	10,000–12,000 mgMLSS/lit and 8500–10,000 mgMLVSS/lit
Mehmood et al. (2021)	Real wastewater	Hollow fiber	295 ± 116 mg/lit	> 95%	–	–	4000–16,000 mgMLSS/lit
Messaoudi et al. (2021)	Piggery	Polyethylene, hollow fiber, microfiltration, pore size = 0.4 μm, area = 0.04 m^2	4685–5390 mg/lit	> 95%	–	23–25 kPa	9382–36,100 mgMLSS/lit
Lübbecke et al. (1995)	Distillery	Nylon mesh, opening = 30 μm, area = 0.05 m^2	30,200–40,000 mg/lit	50%	–	3.5 kPa	11,000–12,000 mgMLSS/lit
Abeynayaka and Visvanathan (2011)	Textile effluent	Thin-film composite, base polypropylene, hollow fiber, nanofiltration, pore size = 0.2 μm, area = 0.023 m^2	180–230 mg/lit	> 85%	25 °C	50–90 kPa	–
Brik et al. (2006)	Synthetic	Synthetic resin, tubular, pore size = 0.25 μm, area = 0.05 m^2	220 mg/lit	> 96%	–	33.3 kPa	4000–5000 mg/lit
Farizoglu and Uzuner (2011)	Paper mill wastewater	Hydrophilic PES, hollow fiber, microfiltration, pore size = 0.1 μm, area = 15 m^2	≤ 1500 mg/lit	92%	25–34 °C	30 kPa	8000 mgMLSS/lit

Table 19.3 Filtration conditions comparison for side-stream and immersed MBRs (Radjenović et al. (2008)

Parameter	Submerged Membrane	Side-Stream Membrane
Surface area (m^2)	46	2
Permeate flux (lit/m^2/s)	0.33–0.83	0.83–1.66
Pressure (kPa)	20–50	400
Airflow rate (m^3/s)	0.66	-
Energy For filtration (kWh/m^3)	0.3–0.6	4–12

Table 19.4 Comparison of hollow fiber, tubular, and flat sheet membranes (Dohare and Trivedi 2014)

Membrane type	Arrangement	Packing density	Energy demand	Cleaning	Replacement	Specific fluxes m^3/m^2/s
Tubular membranes	External recycling	Low	High (turbulent flow)	Efficient + physical cleaning possible	Tubes or element	40–60
Flat sheet membranes	External/ submerged	Moderate	Low-moderate (laminar flow)	Moderate	Sheet	115
Hollow fiber membranes	External/ submerged	High	Low	Back washing possible	Element	50–65

Materials for the construction of membranes are made up of different organic and inorganic substances, but a metallic type of membrane also exists (Qin et al. 2007). Generally, the most commonly used substances are polyamides, celluloses, charged polysulfone, polysulfone, polyvinylidene difluoride (PVDF), polyacrylonitrile (PAN), polyethylene (PE), polyethylsulfone (PES), and polypropylene (PP), etc. (Satyawali and Balakrishnan 2008). When it comes to physical and chemical resistance, polymeric substances outperform all other types of substances by a wide margin. However, hydrophobic polymeric membranes foul more easily than hydrophilic membranes. Most foulant and membrane interactions occur when the membrane is hydrophobic (Dhaouadi and Marrot 2008). Commercially available membranes are altered to increase their hydrophilicity using techniques like grafting, plasma treatment, organic chemical reactions, and chemical oxidation. The modification process, combined with the membrane module fabrication process, typically differentiates one membrane from another (Gander et al. 2000; Tran et al. 2021).

As an alternative to metal oxide-based ceramic membranes, nanomaterials such as zeolites, metal–organic frameworks, and carbon-based materials have attracted considerable interest in the preparation of inorganic membranes due to their highly configurable nanoscopic structural features that can make them extremely effective at either blocking or allowing water to pass through (Shin and Bae 2018). Inorganic membranes outperform polymeric membranes in terms of temperature and chemical stability, corrosive resistance, cleaning ease, and lifespan (Xiao et al. 2021). The surface coating of membranes is an upcoming approach for getting new properties for the membranes, varied compared with the original type of material (McAdam and Judd 2006; Faridizad et al. 2022). Morphology of membranes and surface chemistry plays a major part in the proficiency of the membrane modification process. Generally, the aim of the modification process is not only for improving the flux and/or selectivity but also to enhance the chemical resistance (Huang et al. 2020b). In this context, several methods have been used to differ the chemistry of membranes, these include pore-filled grafting, dip coating, and surface grafting (Jeison and Lier 2007; Munz et al. 2008b; Tay et al. 2007). The photo-induced graft type of polymerization of a monomer is given in the polymer matrix in an acceptable range for changing the surface, it differs the properties such as wettability or permeability (Kornboonraksa and Lee 2009). The grafting technique provides a high degree of control over the degree of change by enabling the deliberate choice of reaction conditions and irradiation (Zheng et al. 2013). In addition, the graft polymerization is initiated without prior modification of a surface. The procedure is comparatively simple, cost-effective, energy efficient, and well-suited for integration with other technologies (Jung et al. 2006; Zhang et al. 2009).

19.4 Selection Criteria of Membrane Material for Water Treatment

The main reason for selecting a good polymer is the one that takes out the contaminants more effectively. Secondarily, it is useful for obtaining the highest flux (permeate rate per unit area of membrane) (Beuscher et al. 2021). Efficient flux can be achieved by effectively managing fouling. Polyampholyte amphiphilic copolymer is highly effective in filtering a wide range of foulants, including oil, organics, proteins, and complex mixtures (Rabiller-baudry et al. 2006). Cellulose acetate (CA) has a very low protein binding membrane and is used for aqueous-based samples and biological samples. CA is a mechanically strong membrane that is also hydrophilic (Huang et al. 2021). Polyethylsulfone is a hydrophilic membrane, it has high flow characteristics, and it can be used for high temperature liquids (Goon et al. 2021). The functionalized PES membrane exhibits chemical robustness and is effective for various organic and aqueous materials, attributed to its hydrophilicity. This property results in a higher water permeation flux compared to the pristine membrane (Puron 2008; Zuo et al. 2021; Nunes et al. 2020). Polyvinylidene difluoride (PVDF) is a hydrophilic membrane and is useful for most general biological sample filtration (Zhu et al. 2017). Many of the researchers reported the usage of these membrane materials in wastewater treatment for the good performance of membrane bioreactors. Zhiwei WANG et al. reported that a PVDF membrane with a pore size of

0.2 μm can be used for the hybrid membrane process (Wu et al. 2021). Tay et al. (2007) proved that PVDF with a different pore size can be used in membrane bioreactors (Devendra Dohare and Trivedi 2014). The ceramic membrane can be used for a crossflow membrane bioreactor with a high conversion rate that can sustain high temperatures with different types of pore sizes in the treatment of wastewater (Messaoudi et al. 2021; Goh and Ismail 2018; Rani and Kumar 2021; Yin and Zhang 2021; Ramakrishnan et al. 2004; Lee et al. 2016). Nowadays mostly used membranes in MBR are PVDF and ceramic. But in these two types of membrane material, preferable is PVDF because of good performance and low cost of membrane compared with ceramic (Chavan et al. 2018; Yang et al. 2021).

19.5 Range of Wastewater Treated by MBR System

The removal of micropollutants to zero discharge levels poses a significant challenge for wastewater treatment (Nadizadeh and Mahdavi 2021; Huang et al. 2020a). For this purpose, a membrane bioreactor (MBR) is one of the most effective techniques available (Seman et al. 2012). MBRs can treat a wide range of wastewater types, including textile and olive mill wastewaters, as well as municipal and petroleum refinery wastewater, tannery wastewater, piggery wastewater, paper making, and other types of wastewater (Bhalani et al. 2021; Deng et al. 2021; Yadav et al. 2021; Mazzaferro et al. 2025; Andrade et al. 2021; Mokarizadeh and Raisi 2021). Reuse of treated water allows the lowering of the making cost. Badani et al. (2005) demonstrated that treatment of wastewater from the textile industry by MBR gives the high effluent quality and high removal of ammonia–nitrogen, and color for the treated effluent (Badani et al. 2005). They have also reported that the performance of MBR at low HRT (1–2 days) was adequate to obtain a reduction of chemical oxygen demand. Nanofiltration was suggested as a way to improve the effluent from MBRs (Lübbecke et al. 1995). During the process of making olive oil, a lot of black waste called olive mill wastewater is made. It is distinguished by the wide range of contaminants such as phenol. A membrane bioreactor (MBR) configured with an external ceramic ultrafiltration membrane can deliver a consistent permeate flow devoid of suspended particles and phenolic compounds. The implementation of the back pulse mechanism facilitates the uninterrupted operation of the MBR (Zhang et al. 2021c, d). Pharmaceutical, endocrine disrupting substances, and personal care items in municipal waste streams can be efficiently eliminated at longer sludge retention time because bacteria that break down those compounds may multiply slowly, requiring more time for biomass accumulation (Deng et al. 2021; Ahmadi et al. 2023). Chemical oxygen demand removal can be enhanced up to 96% by the addition of a powder form of activated carbon or/and aluminum sulfate into a recirculated membrane bioreactor to treat municipal wastewater (Wang et al. 2021a). Dissolved and suspended oil substances, minerals, heavy metals, salt, oil and grease, total organic carbon, chemical components, and dissolved gases, among other things, are commonly found in petroleum refinery wastewater. The MBR method, applied in surface refining processes within the metalworking industry, can greatly enhance the quality of degreasing solutions. The permeate shows no presence of solid particles, and this method can achieve a reduction in hydrocarbon concentrations by as much as 90% (Buer and Cumin 2010; Yadav et al. 2021). Pig farming has resulted in an overabundance of piggery waste. Piggery effluent is rich in organic materials, suspended particles, phosphate, and nitrogen (Wang et al. 2010). The COD, biological oxygen demand, suspended solids, organic matter, and ammonia–nitrogen extraction efficiency of a hybrid system comprises of chemical precipitation and membrane bioreactor that are above 90% (Xing et al. 2001b). MBRs are capable of effectively eliminating surplus pollutants, in terms of COD, BOD, and ammonia–nitrogen, such as the pulp and paper sector, which discharged nearly half of all effluent into rivers, according to studies (Fuchs et al. 2003). With solid retention time (SRT) increment, ultrafiltration hollow fiber MBR is also possible to boost the removal of metals such as Al, Ag, Ba, and Se (Innocenti et al. 2002). In the dairy industry, cheese-making operations lead to producing wastewater that contains a high amount of organic compounds that leads to a major environmental problem. Cheese whey contains a high amount of concentrated wastewater with respect to phosphorous and nitrous matter content. The anaerobic fluidized bed membrane bioreactor and jet loop membrane bioreactor systems demonstrate effective performance in treatment applications (Rani and Kumar 2021; Amin et al. 2022). Conversely, an increase in the concentration of mixed liquid suspended solids (MLSS) enhances the treatment efficiency of high-strength wastewater (Turano et al. 2002). The effect of different sludge retention times (SRTs) shows that the quality of effluent quality improved as SRT increased along with the reduction in biomass production (Innocenti et al. 2002).

19.5.1 Nitrogen and Phosphorous Removal

Usage of pesticides and agricultural fertilizers and the release of partially treated municipal and industrial types of wastewater contain a high amount of nitrogen and phosphorous concentration in groundwater and surface water (Kumari et al. 2021). The aggregation of nitrate in closed marine systems has a problem for both the environment and marine life. This usually leads to an extreme amount of growth of algae of lakes and other water bodies, providing an enormous growth of certain weak species (Seo et al. 2024; Carletti et al. 2003). Both nitrogen and phosphorous are important rate-limiters for eutrophication, so getting rid of them is very important for successful wastewater treatment. Studies showed that at thermophilic temperatures, the removal efficiencies of soluble organic nitrogen decreased, and the removal efficiencies of total nitrogen increased with temperature (Messaoudi et al. 2021). The fusion of other technology with MBR proved the successful removal of nitrogen and phosphorous (Kumar et al. 2023). Integrated technology such as tangential-flow membrane bioreactor, ion exchange, anaerobic/aerobic type of membrane bioreactor showed good performance of total nitrogen and organic removal (Badani et al. 2005; Goh and Ismail 2018; Pathak et al. 2020; Sari Erkan et al. 2020; Isik et al. 2021;

Razavi and Miri 2015). Burhanettin Farizoglu et al. (2007) used a membrane bioreactor which consists of a jet loop type of bioreactor (aeration rate is high) and another unit consisting of a microfiltration membrane separation unit (Rani and Kumar 2021).

19.5.2 Removal of Trace Organic Contaminants

Pharmaceutically active chemicals and persistent chemical residues are transferred to wastewater treatment plants (WWTPs) through urban sewage pipe network systems due to the high rate of medication use in human healthcare and inefficient drug metabolism (Munz et al. 2008a). Quantitative structure–activity relationship analysis serves as an effective qualitative method for the removal of trace organic pollutants, including bisphenol A and sulfamethoxazole. The extraction of trace organic contaminants is conceivable utilizing a submerged type of MBR system; nevertheless, the results show that the removal efficiencies are dependent on the physicochemical features of the trace organic contaminants (Prado et al. 2009; Zhu et al. 2025; Song et al. 2022).

19.5.3 Removal of Heavy Metals

MBR is able to enhance the removal of metal like Ag, Al, Cd, Ba, Fe, Co, Mn, Hg, V, Zn, and Ni; B, As, Pb, Se, and Cu removal (Chon et al. 2013; Nadizadeh and Mahdavi 2021; Lee and Han 2016; Farizoglu et al. 2007; Parihar et al. 2024). Membrane bioreactors have the ability to effectively remove a significant amount of heavy metals from a variety of wastewaters under a variety of operating conditions and flow patterns. Biological treatment procedures achieve this through the adsorption and complex formation of metals with microorganisms (Alidadi et al. 2016). The interaction of metal ions with the negatively charged surfaces of microorganism results in the mixing of microorganisms with the metal ions, leading to the adsorption of these ions onto the cell surfaces of the microorganisms. Metals can also form complexes with carboxyl groups existing in microbial polysaccharides as well as other polymers, or they can be absorbed into biological cells through protein components (Kakar 1981).

19.6 Factors Influencing Performance of Membrane System

19.6.1 Permeate Flux

The decline in permeate flux within membrane bioreactors is predominantly attributed to the presence of mixed liquor suspended solids, rather than the characteristics of the wastewater itself (Hota et al. 2021). Flux decline can be categorized into four distinct mechanisms: adsorption, concentration polarization, reversible fouling, and irreversible fouling. The filtration experiments indicated that the permeate flux decreased more rapidly with higher feed concentration and larger membrane pore size, along with a reduction in tangential flow within the side-stream MBR membrane module (Tomar et al. 2023). Several analyses support the use of a reduced flow rate to decelerate the initial fouling and produce loose fouling layers on membranes. Membrane flux decline shows pollutant accumulation during operation, causing pore obstruction and transmembrane pressure variations (Matos et al. 2009). Thin-film nanofibrous composite-cellulose nanofibril (TFNC-CNF)-coated membrane flux reduction has been studied and compared with PVDF membranes for the treatment of domestic wastewater, as illustrated in Fig. 19.2. When fouling developed, both membranes experienced an abrupt reduction in flux of greater than 25% for a period of time before gradually reaching a plateau over a period of time (Sun et al. 2011).

According to Darcy's law the membrane flux is measured in liters per hour per square meter (lit/hr/m^2) (Hasar 2009; Corsino et al. 2020).

$$J = \frac{\Delta P}{\mu R_t}, \tag{19.1}$$

where

J Permeate flux, lit/hr/m^2
ΔP Transmembrane pressure, kPa
μ Dynamic permeate viscosity, kPa s
R_t Total intrinsic membrane resistance, 1/m.

Fig. 19.2 Flux reduction in TFCN-CNF-coated membrane and PVDF membranes in MBR (Lotfikatouli et al. 2021)

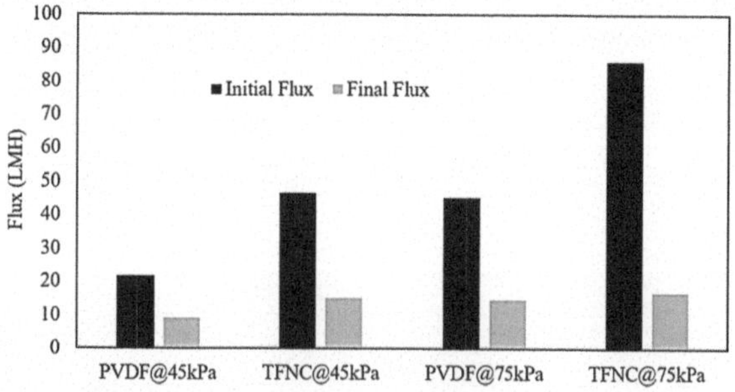

19.6.2 Transmembrane Pressure (TMP)

The liquid flow from the tangential flow membrane module is related to the driving force, a pressure drops. Two types of pressure points, one is the static pressure at which permeate flow is zero and the other is dynamics pressure with the flow of permeate. From these two types of pressures, the TMP can be determined using Eq. 19.2:

$$\text{TMP(kPa)} = \text{Static pressure} - \text{dynamic pressure} \tag{19.2}$$

The initial flux drop observed during operation at a constant TMP could be due to the permeate drag force causing bacterial floc attachment to membrane surfaces. The flux profile at the filtration pressures can be divided into two stages: a sharp flux decrease followed by a gradual flux decline until the pressure drops to the reversibility pressure. At higher filtration pressures, the rapid flux decline in the initial period becomes more pronounced. This decline typically starts high at the beginning of a filtration cycle and gradually decreases as fouling progresses (Vieno and Sillanpää 2014). A typical pressure profile displayed in Fig. 19.3a is achieved in a typical tangential flow membrane module when the filtrate ports are fully open. Because the filtrate pressure is zero, the transmembrane pressure is non-uniform throughout the entire membrane surface, as seen in Fig. 19.3b, resulting in a decrease in flux as the gel layer forms (Nghiem et al. 2009). Because permeation flux is proportional to TMP, membrane filtration can be done in either constant flux or constant (TMP) modes (Zhang et al. 2021b). TMP is managed in constant flux mode to keep flux constant at the target flux. The flux changes in constant TMP mode as fouling and cleaning progress (Gao et al. 2022).

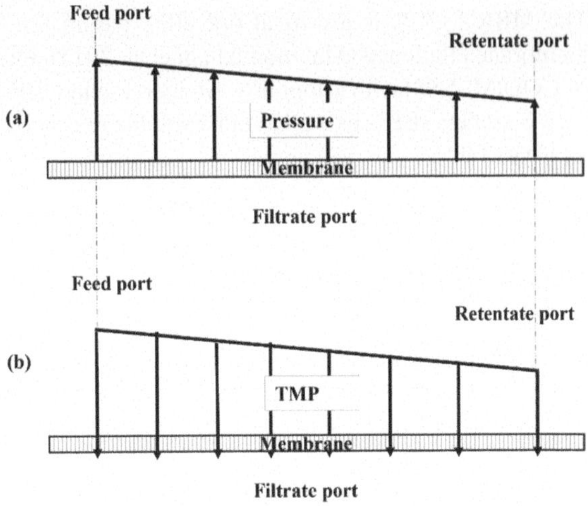

Fig. 19.3 **a** Pressure profile and **b** TMP profile in tangential flow membrane module with fully open feed, retentate, and filtrate ports (adapted from Defrance and Jaffrin 1999)

19.6.3 Permeability

The flow permeability (K) of a fluid is a major mechanical property that affects the membrane's selectivity (Pettinato et al. 2015). Excessive biofilm growth on membrane can greatly limit permeability. The permeate flux and porosity of the supporting material are critical in determining the thickness of the biofilm layer. It is possible to preserve the membrane's high permeability and TMP by applying the necessary negative filtration pressure and giving it enough time to recover (Metals et al. 2020). Permeability is the measure of permeate sending through unit area of membrane per unit time in unit transmembrane pressure, Eq. (19.3).

$$K = \frac{J}{\mathrm{TMP}}, \tag{19.3}$$

where J is measured flux (lit/hr/m^2) and TMP is transmembrane pressure (kPa).

In the performance of the membrane, the temperature of the bioreactor content plays a major role because of its change in permeate and concentrate viscosity. TMP or the driving force required for flux increases as the temperature drops, which lower membrane permeability by impacting viscosity and intrinsic membrane resistance (Kader 2015; Elangovan and Philip 2009). The maintenance of membrane permeability can be achieved through a combination of physical cleaning methods, such as relaxation or back flushing, along with periodic chemical cleaning in situ, often using chemically enhanced back flushing (Bai et al. 2023).

19.6.4 Organic Loading Rate (OLR)

The MBR's OLR is a critical operating parameter. The rise in OLR lowers the membrane efficiency (Mahmoudkhani et al. 2014). OLR can be calculated as a ratio of COD/MLSS/day in kilograms. An increase in OLR led to an increase in extracellular polymers (EPS) as well as other soluble microorganism products (SMPs) (Field and Wu 2022).

19.6.5 Sludge Concentration

When sensitive microorganisms are exposed to high concentrations of total dissolved solid (TDS), total suspended solid (TSS), and toxicity substances, among other pollutant such as nitrogen compounds, nanometal oxides, microplastics, all have varied effects on microbial structure, metabolic processes (Liu et al. 2024). TDS concentrations have been found to be especially effective in increasing osmotic pressure, separating microbes, destroying bacterial enzyme configurations, and limiting the rate at which microbes can grow in such conditions. Low MLVSS concentrations are caused by unfavorable wastewater conditions, which inhibited biomass growth (Lotfikatouli et al. 2021). As a result of these factors, the reaction time and operating costs for organic degradation via biological treatment systems have increased significantly. High TDS (salinity) levels are one of the most significant influences on the performance of biological treatment systems because it contains a high proportion of non-biodegradable materials derived from the tanning industry. Tannery wastewater must be treated before discharge, which can be done through gravity settling or coagulation. This could help with waste biological degradation and the overall quality of treated wastewater (Vo et al. 2021).

19.7 Biodegradation Kinetic

Cell growth occurs simultaneously with the oxidation of organic compounds. Due to the high proportion of organic material in biomass, the increase in biomass can be quantified by estimating the MLSS or MLVSS, COD consumption. As shown in Fig. 19.4, in order to design an efficient submerged MBR, it is very important to use both experimental along with rational parameters that come from biochemical expressions to figure out the specific growth rate of biomass (μ), maximum specific growth rate (μ_m), the half velocity constant (k_s), maximum cell yield (Y), and endogenous decay coefficient (k_d) that are highly recommended (Deschamps et al. 2021).

The relationship between microbial growth and the rate of substrate consumption, as given by Monod, is expressed as

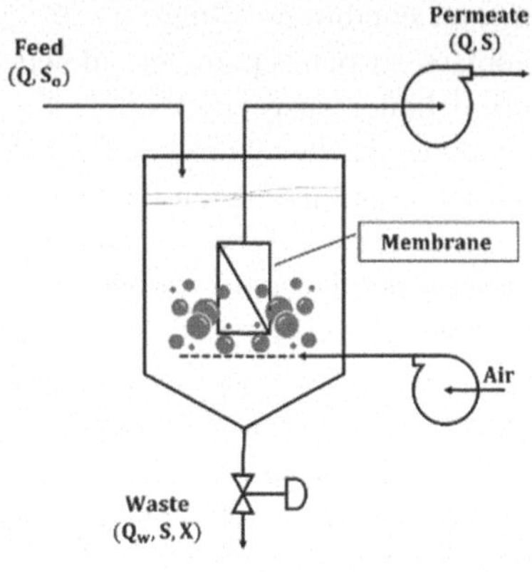

Fig. 19.4 Material balance over submerged MBR

$$\mu = \frac{\mu_m S}{k_s + S}, \tag{19.4}$$

where S is the concentration of growth limiting substrate (mg/lit), the k_s and μ_m parameters of the Monod model represent the concept of a growth limiting substrate, which is illustrated in Eq. 19.5. Calculating the biokinetic coefficients requires a thorough analysis of the biomass and substrate mass balances, which must accurately reflect the performance of the MBR in detail.

After applying steady-state conditions and balancing biomass concentration and substrate balance over the membrane bioreactor, the following equations will occur, which may be mathematically stated as (Chang et al. 2002).

$$\frac{Q}{VX}(S_o - S) = \frac{1}{Y}\frac{1}{\text{SRT}} + \frac{k_d}{Y}, \tag{19.5}$$

$$\frac{\text{SRT}}{1 + (\text{SRT}k_d)} = \frac{K_s}{\mu_m}\left(\frac{1}{S}\right) + \frac{1}{\mu_m}, \tag{19.6}$$

where V is reactor volume (lit); X is concentration of biomass in the reactor (mg/lit); k_d is decay coefficient of biomass (1/day); Q is flow rate of wastage (1/s); t is time (s); and SRT is V/Q, sludge retention time.

According to the Eq. (19.5), when plotting $Q(S_o - S)/VX$ versus $1/\text{SRT}$, k_d and Y can be determined using intersection and slope of the line respectively and according Eq. (19.6) on plotting $\text{SRT}/(1 + (\text{SRT}k_d))$ versus $1/S$, μ_m and k_s can be determined.

19.8 Respirometric Measurements of Activated Sludge

Respirometry involves the examination of the metabolic consumption rate of an inorganic electron acceptor by microbes, conducted under precisely controlled experimental conditions (Bugge et al. 2012). Aeration provides dissolved oxygen to biomass, suspends particles, allowing for rapid treatment rates and a relatively high mixed liquor suspended solids concentration, and improves membrane cleaning in submerged membrane bioreactors. Instead, it's the main power input for MBR systems (Hall and Gyepi-Garbrah 2001). The inclusion of suspended biomass and biofilm, particularly when compared to processes involving only pure suspended or attached biomass, may influence the kinetics of both types of biomasses. This is attributed to the potential competition for substrate and oxygen availability (Hasan et al. 2012). The mass transfer coefficient (k_{La}) was used to figure out how well the system used oxygen. As a result, the physical, chemical, and biological characteristics of the wastewater must be determined (Hong et al. 2019; Song et al. 2021). The respirometric technique involves aerating the mixed liquor in a bioreactor until oxygen saturation is achieved. Based on the results of phase I (see Fig. 19.5) and after adding a certain amount of substrate, the concentration of dissolved oxygen

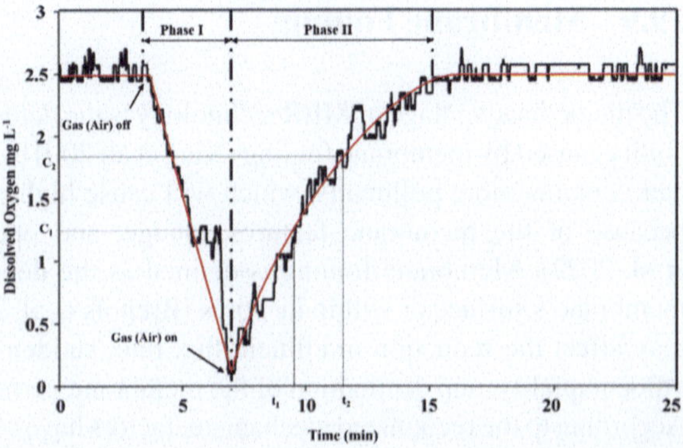

Fig. 19.5 Respirometric technique for estimation of OUR (adapted from Badani et al. 2005; Arévalo et al. 2014)

in the bulk of the mixed fluid drops rapidly, indicating a high rate of biomass oxygen consumption, as shown in Fig. 19.5. By switching to gassing-on, the oxygen level is restored to its original setting (see phase II) (Goh and Ismail 2018; Zhang et al. 2021d).

The difference between the transfer of oxygen into solution and oxygen uptake rate by the respiring culture, as stated by Eq. (19.7), accounts for the observed increase in concentration of dissolved oxygen during phase II.

$$\frac{dC_L}{dt} = k_l a\left(C^* - C_L\right) - xQ_{O_2}, \tag{19.7}$$

where

C^*	saturated dissolved oxygen concentration (mg/lit)
C_L	Concentration of dissolved oxygen in the fermentation broth (mg/lit)
Q_{O_2}	specific respiration rate, (mg/lit/min)
x	biomass concentration (MLSS), (mg/lit)
xQ_{O_2}	uptake rate of oxygen.

Indeed, the oxygen uptake rate (OUR) is commonly acknowledged as a key indicator of biomass viability. Additionally, OUR can be used to determine the biodegradability of influent as well as effluent wastewater (Pandey et al. 2023; Tao et al. 2022; Buzatu et al. 2012). The slope of the line in phase I can be used to calculate OUR. Using phase II data, the entire oxygen transfer coefficient, $k_L a$, can be computed. MBR sludge is reaerated at the end of phase I, and the progression of DO versus time is depicted in phase II in Fig. 19.5. Equation (19.8) may be used to get the overall oxygen transfer coefficient $k_L a$.

$$k_L a = \frac{\ln\left(\frac{C^* - C_1}{C^* - C_2}\right)}{t_2 - t_1} \tag{19.8}$$

19.9 Membrane Fouling

The major disadvantage of MBR technology is the deterioration of membrane permeability caused by membrane fouling (Xing et al. 2001b). Using high-strength wastewater contains more pollutants, which will cause high obstruction for the membrane because of the membrane features, sludge, and operating parameters (Bokhary et al. 2022). Membrane fouling is defined as the deposition of compounds on the membrane's surface or within its pores (Brepols et al. 2008; Vo et al. 2021). Fouling may affect the reduction in effluent flux rate; sudden increases in TMP, effects on effluent quality, and declination of the membrane permeability (Yoneda et al. 2024). According to the recognized mechanism, factors have a role in the fouling phenomena and are shown in Fig. 19.6. Standard blockage occurs when particles smaller than the pore size of the membrane are trapped within the membrane's pore walls, reducing the size of the membrane's pore openings. Intermittent blocking occurs when pores from one particle are blocked by pores from another particle in superimposition, resulting in intermediate blocking. Particles bigger than the pore size of membrane adhere to the membrane surface during cake filtration. Pores are only partially blocked when a single particle is present and no other particles are superimposed on it. Organic fouling, colloidal fouling, biofouling, and scaling are all types of fouling that can be classified according to the foulants they contain (Pophali et al. 2003; Alizad Oghyanous et al. 2020).

Reversible fouling caused by suspended solids is the most common type, whereas irreversible fouling is most often caused by certain dissolved or colloidal organics. Consequently, membrane fouling is complicated in MBR technology because the processes of membrane filtration and bio-reaction are combined in a single reactor. Membrane fouling involves physicochemical relationships among components of the biochemical reactor, membrane module configuration, and system operating parameters, among other factors (Devendra Dohare and Trivedi 2014; Al-Malack 2006). Furthermore, a high MLSS value may result in an increase in the apparent viscosity of the bioreactor content, which may result in a reduction in the oxygen transfer in a suspension of liquid. Thus the insufficient supply of oxygen to microorganisms causes low purification yields (Zhang et al. 2021d), which makes the MBR expensive as the investment and operational costs are high (Melin et al. 2006). Fouling tendency was calculated from detailed analysis of resistances and continuous TMP

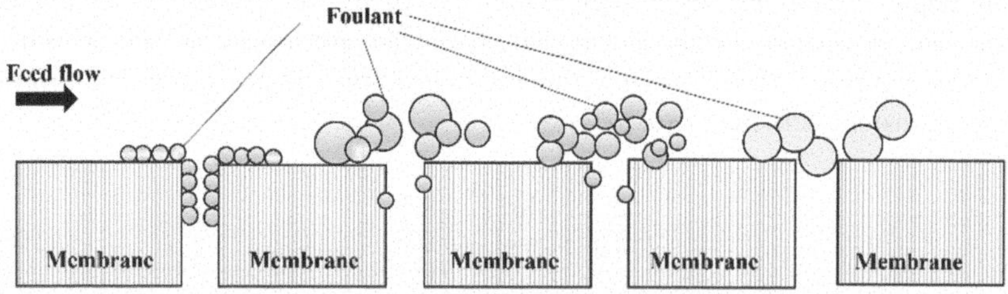

Fig. 19.6 Mechanisms of fouling

value. Total filtration resistance can be divided into three components: resistance provided by the fouling layer on the surface of membrane, resistance provided by the membrane-solution interface, and membrane resistance itself (19.9). For MBRs, membrane resistance (1/*m*) is often given as clean water permeability.

$$R_t = R_c + R_g + R_f + R_m, \tag{19.9}$$

where

R_t is the total membrane resistance
R_c reversible resistce from the cake layer
R_g gel resistance
R_f resistance that is irreversible due to pore blocking that can be recovered through chemical cleaning, and
R_m intrinsic membrane resistance.

19.9.1 Organic Fouling

Organic fouling occurs because of the deposition of organic compounds on the surface of the membrane or in its pores because of the intermolecular interactions between organic compounds and the membrane. The most commonly occurring fouling in drinking water filtration processes is natural organic matter (NOM) fouling. For the separation of wastewater and activated sludge generally fouling occurs because of soluble microbial products (SMP) (Moya-Llamas et al. 2023).

19.9.2 Colloidal Fouling

Small particles can aggregate on the surface of the membrane; as a result, the formation of the filter cake is called colloidal fouling. The particulates can either be microorganisms, suspended solids, and colloids. Colloidal fouling is the most common type of fouling in ultrafiltration (UF) and microfiltration (MF) systems. Most of the type of MBRs occurs normally colloidal and organic fouling by using MF and UF (Germain and Stephenson 2005; Majewsky et al. 2011; Rodríguez et al. 2011).

19.9.3 Biofouling

The adhesion and development of microorganisms on the surface of the membrane because of the organization of a layer is known as biofouling, which leads to a declination of membrane performance (Corsino et al. 2021). Generally, the formation

Table 19.5 Typical values of resistance in MBR system

MBR Type	Resistance (1/m)	Pore size (μm)	Membrane Material	References
Sub- HF	R_m=1.05×10^{11}, R_c=13.84×10^{11}, R_t=17.43×10^{11}	0.1	Ceramic	Corsino et al. (2020)
Submerged	R_m=1.24×10^{11}, R_f=1.77×10^{13}	0.2	PVDF	Corsino et al. (2021)
Sub- HF	R_m=0.67×10^{12}, R_c=1.09×10^{12}, R_t=2.7×10^{12}	0.1	Ceramic	Çulfaz et al. (2011)
Sub- HF	R_m=1.2×10^{12}	0.1	Polysulfone	Defrance and Jaffrin (1999)

of a layer can occur on all kinds of surfaces like synthetic and natural, because of the fact that the development of bacteria can elaborate adhesion mechanisms. Because of their low flux rate and limited cleaning types of membranes, nanofiltration (NF) and reverse osmosis (RO) processes suffer more biofouling (Trapani et al. 2018).

19.9.4 Scaling

Scaling means the deposition of particles on the surface of the membrane. Generally, over-saturation of $CaCO_3$, $CaSO_4$, $BaSO_4$, $SrSO_4$, $MgCO_3$, and SiO_2 in operation forms scaling, with consider concerning in nanofiltration (NF) and reverse osmosis (RO) (Martín-Pascual et al. 2015). The scaling effect is not that much in MBR fouling. Precipitation of calcium or iron will happen in some cases. For restoring the membrane permeability, if oxidant cleaning is not enough then acid cleaning must be considered. In the given below Table 19.5 shows a different resistances value from the literature. It shows that each type of resistance value is strongly related to experimental conditions like pore size and membrane material, operating conditions, and biomass characteristics (Pajoum Shariati et al. 2013).

19.10 Fouling Control Methods

Membranes may require regular cleaning either by physical or chemical means to maintain an efficient process (Chon et al. 2013). There are several strategies that have been investigated to control the fouling that occurs during MBR operation. Some important strategies are reduction of concentration polarization, chemical, and physical cleaning techniques, preliminary treatment of influent wastewater, aeration intensity, change in mixed liquor concentration, electric field-assisted MBR, addition of adsorbent, and surface modification (Nguyen et al. 2016; Hosseinzadeh et al. 2015;

Zarragoitia-González et al. 2008; Zhao et al. 2020). Fouling mainly relates with concentration polarization that can be decreased by reducing flux. The main physical cleaning techniques are backwashing (where effluent is sent in the opposite direction) and membrane relaxation (where filtration is paused). Mainly used chemicals for membrane cleaning are NaOCl, citric acid, and oxalic acid (Le-Clech et al. 2006). Backwashing can be done by NaOCl solution followed by soaking and flushing out with permeate for calculated time (Chandrasekhar et al. 2022). Mineral deposits are typically removed by acidic solutions, whereas organic compounds are removed by alkaline solutions. Muhammad Muhitur Rahaman et al. used three different chemicals to clean a ceramic membrane: Clorox (5.25% sodium hypochlorite), Persil (washing detergent), and superclean (acidic detergent with a pH range of 1.5) (Buer and Cumin 2010). To reduce fouling, HF and FS membrane modules require wastewater pre-treatment (Lin et al. 2019). Figure 19.7 depicts a typical scheme for calculating resistances step by step. Physical methods such as water wash and sponge rubbing are used to remove cake layer and gel blocking, respectively, whereas chemical washing is required to remove foulants from the inside pore membrane.

It is commonly in use of different types of aeration intensities for biological treatment and for control in membrane fouling, in order to get more efficient use of energy for system demands. Aeration intensity has more impact on permeation through the membrane and the removal of the cake layer (Lindamulla et al. 2021). But the adverse effect of the large intensity in aeration gives in high pore blocking of the membrane as the high shear could release more polymeric substances, impact the sludge stabilities (Guo et al. 2020). Both types of MBR configurations require shear force at the membrane's surface to control fouling caused by the membrane's components and the mixed liquor. Side-stream type of MBRs gives a shear force by sending sludge into the reactor. The other type of membrane processes like submerged processes gives shear force by aerating in the bioreactor (Lu et al. 2021). While shear enrichment is critical for permeate flux and fouling reduction in a membrane, promoting shear requires energy, which is the primary reason for the submerged configuration's predominance. Also, in an immersed type of MBR module fouling is stronger, due to its flux rate being high. Allowing activated sludge leads to shear stress for microorganisms, causing them to separate, this shows a reduction in the size of particles and discharge of foulant substance from the microorganisms. This may greatly support the fouling rate of a membrane (Dhaouadi and Marrot 2008). Mixed liquor properties impact the propensity of fouling and can also change for minimization of fouling (Sim et al. 2011). To reduce the extra soluble products or soluble microbial product concentration, mixed liquor is generally changed by the addition of chemicals (Filloux et al. 2016). The surface modifications of membranes are a good approach to get new properties to the membranes, different from those of the original material. Morphology of membranes and surface chemistry plays a major role in the efficiency of the membrane modification process. From this surface modification process, fouling can also be controlled (Zheng et al. 2018).

Recent advances such as encapsulated self-forming dynamic bio-membranes (Jeong et al. 2007), metal–organic framework (MOF) modified membranes (Çulfaz et al. 2011), electro algae-activated sludge membrane bioreactor (Yan et al. 2019),

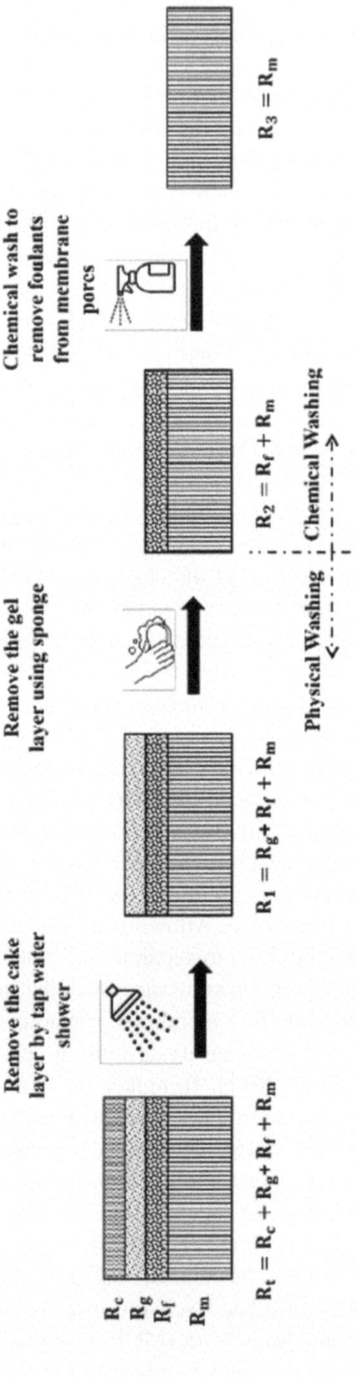

Fig. 19.7 Stages to calculate filtration resistances

laser energy (Tay et al. 2020), and osmotic membrane bioreactor have also been offered technologies to increase the value of reclaimed water from the wastewater stream. The change in influent OLR impacts the conventional MBR more widely than the moving bed MBR. Membrane fouling behavior is high in the moving bed MBR compared with conventional MBR because the filamentous bacteria in the moving bed MBR resulted in a dense and thick cake layer formation on the surface of the membrane (Metzger et al. 2007). Physical relaxation and backwash, in conjunction with shear-enhanced techniques such as air sparging, the addition of fluidized particles, and mechanical vibration, are becoming increasingly common for in situ fouling mitigation in MBRs that operate continuously for a long period of time (Karamouz et al. 2024; Qrenawi and Rabah 2023).

19.11 Conclusion

Several studies have demonstrated that the science behind MBR technology holds great promise as a key component of wastewater reclamation/reuse schemes. Wastewater reclamation and reuse can benefit greatly from MBR technology. There are several advantages to MBR technology over conventional processes in terms of space occupancy and effluent quality, to name a few. Next-generation membrane fouling studies should be the focus of future research efforts. This has been demonstrated by treating everything from olive oil mill wastewater, which contains phenolic compounds, to refinery wastewater, which contains oil content, and everything in between. While reviewing the research to date and evaluating the advantages and limitations associated with MBR technology for wastewater treatment, the paper also identifies areas that require additional research to be completed. MBR technology has several advantages over conventional processes in terms of space occupancy and effluent quality, to name a couple of examples. Over the last several years, the application of MBR technology on a large scale has grown in importance. Especially, future research efforts should emphasize advancements in membrane fouling study that reduce energy requirements. MBR has excellent performance in the removal of COD, TN, ammonia–nitrogen, suspended particles, and organic matter from a wide range of wastewaters, including municipal and industrial waste. The treatment that takes place in the membrane bioreactor (MBR) type of system results in a significant advancement in the removal of organic compounds and suspended solids. In addition, the MBR process can remove nitrogen compounds in a nitrification/denitrification mode, and it can be combined with the use of a coagulant to remove phosphorus compounds from wastewater.

Fouling is a frequent and serious issue in MBR applications. It is a significant issue in membrane-type processes but becomes even more challenging to control and predict in an MBR due to the highly heterogeneous nature of the mixed liquor in the bioreactor, and particularly the effect of the active sludge on the process. Future research efforts should emphasize advancements in membrane fouling studies that reduce energy requirements. Numerous researchers concluded that the degree of aeration had a greater effect on membrane permeation and cake layer removal than any other factor. The researchers discussed, however, the possibility that a high aeration intensity could result in a significant amount of membrane pore blockage. The filtration resistance of all membranes increased in a similar manner as they were cleaned physically and chemically, implying that chemical cleaning influenced MBR fouling in a subsequent operation. Recent advancements in technology, such as laser technology, are being used to modify ultrafiltration polyethersulfone membranes to reduce biofouling. The focus of future research should be on next-generation membrane fouling studies.

References

Abeynayaka A, Visvanathan C (2011) Performance comparison of mesophilic and thermophilic aerobic sidestream membrane bioreactors treating high strength wastewater. Bioresour Technol 102:5345–5352. https://doi.org/10.1016/j.biortech.2010.11.079

Ahmadi M, Davallo M, Kiarostami V et al (2023) Exploring the application of the hybrid nano-bioreactor technology based on the developed polyethersulfone mixed-matrix membrane for industrial effluent treatment. Arab J Chem 16:105189. https://doi.org/10.1016/j.arabjc.2023.105189

Al-Amri A, Salim MR, Aris A (2010) The effect of different temperatures and fluxes on the performance of membrane bioreactor treating synthetic-municipal wastewater. Desalination 259:111–119. https://doi.org/10.1016/j.desal.2010.04.023

Alidadi H, Najafpoor AA, Hosseinzadeh A et al (2016) Waste recycling by vermicomposting: maturity and quality assessment via dehydrogenase enzyme activity, lignin, water soluble carbon, nitrogen, phosphorous and other indicators. J Environ Manage 182:134–140. https://doi.org/10.1016/j.jenvman.2016.07.025

Alizad Oghyanous F, Etemadi H, Yegani R (2020) Foaming control and determination of biokinetic coefficients in membrane bioreactor system under various organic loading rate and sludge retention time. Biochem Eng J 157:107491. https://doi.org/10.1016/j.bej.2020.107491

Al-Malack MH (2006) Determination of biokinetic coefficients of an immersed membrane bioreactor. J Memb Sci 271:47–58. https://doi.org/10.1016/j.memsci.2005.07.008

Amin MSA, Stüber F, Giralt J et al (2022) Ceramic-supported graphene oxide membrane bioreactor for the anaerobic decolorization of azo dyes. J Water Process Eng 45:102499. https://doi.org/10.1016/j.jwpe.2021.102499

Andrade MC, Pereira JC, de Almeida N et al (2021) Improving hydraulic permeability, mechanical properties, and chemical functionality of cellulose acetate-based membranes by co-polymerization with tetraethyl orthosilicate and 3-(aminopropyl)triethoxysilane. Carbohydr Polym 261.https://doi.org/10.1016/j.carbpol.2021.117813

Arévalo J, Ruiz LM, Pérez J, Gómez MA (2014) Effect of temperature on membrane bioreactor performance working with high hydraulic and sludge retention time. Biochem Eng J 88:42–49. https://doi.org/10.1016/j.bej.2014.03.006

Badani Z, Ait-Amar H, Si-Salah A et al (2005) Treatment of textile waste water by membrane bioreactor and reuse. Desalination 185:411–417. https://doi.org/10.1016/j.desal.2005.03.088

Bai F, Liu S, Ma J, Zhang Y (2023) Biodegradation of sulfate and elimination of heavy metals by immobilized-microbial bioaugmentation coupled with anaerobic membrane bioreactor. Chem Eng J 473:145196. https://doi.org/10.1016/j.cej.2023.145196

Bazrafshan N, Dadashi Firouzjaei M, Elliott M et al (2021) Preparation and modification of low-fouling ultrafiltration membranes for cheese whey treatment by membrane bioreactor. Case Stud Chem Environ Eng 4:100137. https://doi.org/10.1016/j.cscee.2021.100137

Beuscher U, Kappert EJ, Wijmans JG (2021) Membrane research beyond materials science. J Memb Sci 119902.https://doi.org/10.1016/j.memsci.2021.119902

Bhalani DV, Trivedi JS, Jewrajka SK (2021) Selective grafting of morphologically modified poly(vinylidene fluoride) ultrafiltration membrane by poly(acrylic acid) for inducing antifouling property. Appl Surf Sci 544:148905. https://doi.org/10.1016/j.apsusc.2020.148905

Bokhary A, Leitch M, Liao BQ (2022) Effect of organic loading rates on the membrane performance of a thermophilic submerged anaerobic membrane bioreactor for primary sludge treatment from a pulp and paper mill. J Environ Chem Eng 10:107523. https://doi.org/10.1016/j.jece.2022.107523

Bourgeous KN, Darby JL, Tchobanoglous G (2001) Ultrafiltration of wastewater: effects of particles, mode of operation, and backwash effectiveness. Water Res 35:77–90. https://doi.org/10.1016/S0043-1354(00)00225-6

Brepols C, Dorgeloh E, Frechen FB et al (2008) Upgrading and retrofitting of municipal wastewater treatment plants by means of membrane bioreactor (MBR) technology. Desalination 231:20–26. https://doi.org/10.1016/j.desal.2007.11.035

Brik M, Schoeberl P, Chamam B et al (2006) Advanced treatment of textile wastewater towards reuse using a membrane bioreactor. Process Biochem 41:1751–1757. https://doi.org/10.1016/j.procbio.2006.03.019

Buer T, Cumin J (2010) MBR module design and operation. Desalination 250:1073–1077. https://doi.org/10.1016/j.desal.2009.09.111

Bugge TV, Jørgensen MK, Christensen ML, Keiding K (2012) Modeling cake buildup under TMP-step filtration in a membrane bioreactor: cake compressibility is significant. Water Res 46:4330–4338. https://doi.org/10.1016/j.watres.2012.06.015

Buzatu P, Zsirai T, Aerts P, Judd SJ (2012) Permeability and clogging in an immersed hollow fibre membrane bioreactor. J Memb Sci 421–422:342–348. https://doi.org/10.1016/j.memsci.2012.07.039

Cao JH, Zhu BK, Lu H, Xu YY (2005) Study on polypropylene hollow fiber based recirculated membrane bioreactor for treatment of municipal wastewater. Desalination 183:431–438. https://doi.org/10.1016/j.desal.2005.02.056

Carletti G, Pavan P, Bolzonella D, Cavinato C (2003) Occurrence and removal of heavy metals from industrial and municipal wastewater: a comparison between MBR and conventional activated sludge processes (CAS)

Chakrabarty B, Ghoshal AK, Purkait MK (2010) Cross-flow ultrafiltration of stable oil-in-water emulsion using polysulfone membranes. Chem Eng J 165:447–456

Chandrasekhar SS, Sekhar SC, Kalyani S, Sridhar S (2022) Performance assessment of a side-stream membrane bioreactor for the treatment of kitchen wastewater. Biochem Eng J 180:108366. https://doi.org/10.1016/j.bej.2022.108366

Chang I-S, Le Clech P, Jefferson B, Judd S (2002) Membrane fouling in membrane bioreactors for wastewater treatment. J Environ Eng 128:1018–1029. https://doi.org/10.1061/(asce)0733-9372(2002)128:11(1018)

Chavan V, Agarwal C, Adya VC, Pandey AK (2018) Hybrid organic-inorganic anion-exchange pore-filled membranes for the recovery of nitric acid from highly acidic aqueous waste streams. Water Res 133:87–98. https://doi.org/10.1016/j.watres.2018.01.023

Choi H, Zhang K, Dionysiou DD et al (2005) Effect of permeate flux and tangential flow on membrane fouling for wastewater treatment. Sep Purif Technol 45:68–78. https://doi.org/10.1016/j.seppur.2005.02.010

Choi JH, Park SK, Ng HY (2009) Membrane fouling in a submerged membrane bioreactor using track-etched and phase-inversed porous membranes. Sep Purif Technol 65:184–192. https://doi.org/10.1016/j.seppur.2008.10.019

Chon K, Cho J, Shon HK (2013) Fouling characteristics of a membrane bioreactor and nanofiltration hybrid system for municipal wastewater reclamation. Bioresour Technol 130:239–247. https://doi.org/10.1016/j.biortech.2012.12.007

Cicek N (2003) A review of membrane bioreactors and their potential application in the treatment of agricultural wastewater. Can Biosyst Eng/Genie Biosyst Canada 45:37–49

Corsino SF, de Oliveira TS, Di Trapani D et al (2020) Simultaneous sludge minimization, biological phosphorous removal and membrane fouling mitigation in a novel plant layout for MBR. J Environ Manage 259:109826. https://doi.org/10.1016/j.jenvman.2019.109826

Corsino SF, Di Trapani D, Torregrossa N, Piazzese D (2021) Preliminary evaluation of biopolymers production by mixed microbial culture from citrus wastewater in a MBR system using respirometric techniques. J Water Process Eng 41:102003. https://doi.org/10.1016/j.jwpe.2021.102003

Çulfaz PZ, Haddad M, Wessling M, Lammertink RGH (2011) Fouling behavior of microstructured hollow fibers in cross-flow filtrations: critical flux determination and direct visual observation of particle deposition. J Memb Sci 372:210–218. https://doi.org/10.1016/j.memsci.2011.02.002

Defrance L, Jaffrin MY (1999) Comparison between filtrations at fixed transmembrane pressure and fixed permeate flux: application to a membrane bioreactor used for wastewater treatment. J Memb Sci 152:203–210. https://doi.org/10.1016/S0376-7388(98)00220-8

Deng L, Li S, Qin Y et al (2021) Fabrication of antifouling thin-film composite nanofiltration membrane via surface grafting of polyethyleneimine followed by zwitterionic modification. J Memb Sci 619:118564. https://doi.org/10.1016/j.memsci.2020.118564

Deschamps L, Merlet D, Lemaire J et al (2021) Excellent performance of anaerobic membrane bioreactor in treatment of distillery wastewater at pilot scale. J Water Process Eng 41.https://doi.org/10.1016/j.jwpe.2021.102061

Devendra Dohare E, Trivedi ER (2014) A Review on membrane bioreactors: an emerging technology for industrial wastewater treatment. Int J Emerg Technol Adv Eng 4:226–236. Website www.ijetae.com

Dhaouadi H, Marrot B (2008) Olive mill wastewater treatment in a membrane bioreactor: process feasibility and performances. Chem Eng J 145:225–231. https://doi.org/10.1016/j.cej.2008.04.017

Di Trapani D, Mannina G, Viviani G (2018) Membrane bioreactors for wastewater reuse: respirometric assessment of biomass activity during a two year survey. J Clean Prod 202:311–320. https://doi.org/10.1016/j.jclepro.2018.08.014

Elangovan R, Philip L (2009) Performance evaluation of various bioreactors for the removal of Cr(VI) and organic matter from industrial effluent. Biochem Eng J 44:174–186. https://doi.org/10.1016/j.bej.2008.11.014

Faridizad G, Abdollahzadeh E, Bonakdarpour B (2022) The use of membrane bioreactors in high rate activated sludge processes: how and why sludge retention time affects membrane fouling. J Water Process Eng 47:102807. https://doi.org/10.1016/j.jwpe.2022.102807

Farizoglu B, Uzuner S (2011) The investigation of dairy industry wastewater treatment in a biological high performance membrane system. Biochem Eng J 57:46–54. https://doi.org/10.1016/j.bej.2011.08.007

Farizoglu B, Keskinler B, Yildiz E, Nuhoglu A (2007) Simultaneous removal of C, N, P from cheese whey by jet loop membrane bioreactor (JLMBR). J Hazard Mater 146:399–407. https://doi.org/10.1016/j.jhazmat.2006.12.051

Fazal S, Zhang B, Zhong Z et al (2015) Industrial wastewater treatment by using MBR (membrane bioreactor) review study. J Environ Prot (Irvine, Calif) 06:584–598. https://doi.org/10.4236/jep.2015.66053

Field RW, Wu JJ (2022) Permeate flux in ultrafiltration processes—understandings and misunderstandings. Membranes (Basel) 12:1–18. https://doi.org/10.3390/membranes12020187

Filloux E, Gernjak W, Gallard H, Croue JP (2016) Investigating the relative contribution of colloidal and soluble fractions of secondary effluent organic matter to the irreversible fouling of MF and UF hollow fibre membranes. Sep Purif Technol 170:109–115. https://doi.org/10.1016/j.seppur.2016.06.034

Fuchs W, Binder H, Mavrias G, Braun R (2003) Anaerobic treatment of wastewater with high organic content using a stirred tank reactor coupled with a membrane filtration unit. Water Res 37:902–908. https://doi.org/10.1016/S0043-1354(02)00246-4

Gander M, Jefferson B, Judd S (2000) Aerobic MBRs for domestic wastewater treatment: a review with cost considerations. Sep Purif Technol 18:119–130. https://doi.org/10.1016/S1383-5866(99)00056-8

Gao J, Yue Z, Yan Y et al (2022) Application of the membrane bioreactor process in upgrading a norfloxacin pharmaceutical wastewater treatment system. Desalin Water Treat 248:49–60. https://doi.org/10.5004/dwt.2022.28130

Germain E, Stephenson T (2005) Biomass characteristics, aeration and oxygen transfer in membrane bioreactors: their interrelations explained by a review of aerobic biological processes. Rev Environ Sci Biotechnol 4:223–233. https://doi.org/10.1007/s11157-005-2097-3

Gkotsis PK, Banti DC, Peleka EN et al (2014) Fouling issues in Membrane Bioreactors (MBRs) for wastewater treatment: major mechanisms, prevention and control strategies. Processes 2:795–866. https://doi.org/10.3390/pr2040795

Goh PS, Ismail AF (2018) A review on inorganic membranes for desalination and wastewater treatment. Desalination 434:60–80. https://doi.org/10.1016/j.desal.2017.07.023

Goon GSS, Labban O, Hao Z et al (2021) Deformation-induced cleaning of organically fouled membranes : fundamentals and techno-economic assessment for spiral-wound membranes. J Memb Sci 626:119169. https://doi.org/10.1016/j.memsci.2021.119169

Gui P, Huang X, Chen Y, Qian Y (2003) Effect of operational parameters on sludge accumulation on membrane surfaces in a submerged membrane bioreactor. Desalination 151:185–194. https://doi.org/10.1016/S0011-9164(02)00997-9

Guo Y, Li T, Xiao K et al (2020) Key foulants and their interactive effect in organic fouling of nanofiltration membranes. J Memb Sci 610:118252. https://doi.org/10.1016/j.memsci.2020.118252

Hall C, Gyepi-Garbrah I (2001) Critical issues for the use of tangential flow filtration systems for cell harvesting applications critical issues for the use of tangential flow filtration systems for cell harvesting applications. In: Animal cell technology: from target to market. Springer, Dordrecht, pp 482–484

Hasan SW, Elektorowicz M, Oleszkiewicz JA (2012) Correlations between trans-membrane pressure (TMP) and sludge properties in submerged membrane electro-bioreactor (SMEBR) and conventional membrane bioreactor (MBR). Bioresour Technol 120:199–205. https://doi.org/10.1016/j.biortech.2012.06.043

Hasar H (2009) Simultaneous removal of organic matter and nitrogen compounds by combining a membrane bioreactor and a membrane biofilm reactor. Bioresour Technol 100:2699–2705. https://doi.org/10.1016/j.biortech.2008.12.024

Hong PN, Noguchi M, Matsuura N, Honda R (2019) Mechanism of biofouling enhancement in a membrane bioreactor under constant trans-membrane pressure operation. J Memb Sci 592:117391. https://doi.org/10.1016/j.memsci.2019.117391

Hosseinzadeh M, Bidhendi GN, Torabian A et al (2015) A new flat sheet membrane bioreactor hybrid system for advanced treatment of effluent, reverse osmosis pretreatment and fouling mitigation. Bioresour Technol 192:177–184. https://doi.org/10.1016/j.biortech.2015.05.066

Hota S, Kar BB, Mishra M (2021) Removal of phosphorus from contaminated water sources using composite matrix fabricated from agro-based waste materials. Mater Today Proc. https://doi.org/10.1016/j.matpr.2020.11.608

Huang T, Li J, Chen Y et al (2020a) Improving permeability and antifouling performance of poly (ether ether ketone) membranes by photo-induced graft polymerization. Mater Today Commun 23:100945. https://doi.org/10.1016/j.mtcomm.2020.100945

Huang S, Pooi CK, Shi X et al (2020b) Performance and process simulation of membrane bioreactor (MBR) treating petrochemical wastewater. Sci Total Environ 747:141311. https://doi.org/10.1016/j.scitotenv.2020.141311

Huang Y, Xiao C, Huang Q et al (2021) Progress on polymeric hollow fiber membrane preparation technique from the perspective of green and sustainable development. Chem Eng J 403:126295. https://doi.org/10.1016/j.cej.2020.126295

Innocenti L, Bolzonella D, Pavan P, Cecchi F (2002) Effect of sludge age on the performance of a membrane bioreactor: influence on nutrient and metals removal. Desalination 146:467–474. https://doi.org/10.1016/S0011-9164(02)00551-9

Isik O, Batyrow M, Abdelrahman AM et al (2021) Dynamic membrane bioreactor performance for treatment of municipal wastewaters at different sludge concentrations. Environ Technol Innov 22:101452. https://doi.org/10.1016/j.eti.2021.101452

Jeison D, van Lier JB (2007) Cake formation and consolidation: main factors governing the applicable flux in anaerobic submerged membrane bioreactors (AnSMBR) treating acidified wastewaters. Sep Purif Technol 56:71–78. https://doi.org/10.1016/j.seppur.2007.01.022

Jeong TY, Cha GC, Yoo IK, Kim DJ (2007) Characteristics of bio-fouling in a submerged MBR. Desalination 207:107–113. https://doi.org/10.1016/j.desal.2006.07.006

Jung YJ, Koh HW, Shin WT, Sung NC (2006) A novel approach to an advanced tertiary wastewater treatment: combination of a membrane bioreactor and an oyster-zeolite column. Desalination 190:243–255. https://doi.org/10.1016/j.desal.2005.02.088

Kader AM (2015) A review of membrane bioreactor (MBR) technology and their applications in the wastewater treatment systems. Desalin Water Treat 32:111–119

Kakar YP (1981) Nitrate pollution of ground water in Southern and South-Western Haryana, India. In: Quality of groundwater, proceedings of an international symposium, Noordwijkerhout, Netherlands. 17:125–129. https://doi.org/10.1016/S0166-1116(08)71894-6

Kang SK, Choo KH (2003) Use of MF and UF membranes for reclamation of glass industry wastewater containing colloidal clay and glass particles. J Memb Sci 223:89–103. https://doi.org/10.1016/S0376-7388(03)00311-9

Karamouz F, Hazrati H, Khoshfetrat AB (2024) Mitigation of fouling in membrane bioreactor using the integration of electrical field and nano chitosan adsorbent. Environ Technol Innov 36:103774. https://doi.org/10.1016/j.eti.2024.103774

Keffala C, Jupsin H, Zouhir F et al (2024) Effect of biomass concentration in membrane bioreactor design on alpha factor for two large pilot systems: case study of mechanical surface aeration and air insufflation system. Desalin Water Treat 320:100724. https://doi.org/10.1016/j.dwt.2024.100724

Kornboonraksa T, Lee SH (2009) Factors affecting the performance of membrane bioreactor for piggery wastewater treatment. Bioresour Technol 100:2926–2932. https://doi.org/10.1016/j.biortech.2009.01.048

Kumar V, Singh E, Singh S et al (2023) Micro- and nano-plastics (MNPs) as emerging pollutant in ground water: environmental impact, potential risks, limitations and way forward towards sustainable management. Chem Eng J 459:141568. https://doi.org/10.1016/j.cej.2023.141568

Kumari R, Ankit H, Basu S (2021) Reclamation of water from dairy wastewater using membrane bioreactor (MBR)—membrane filtration processes. Mater Today Proc 47:1452–1456. https://doi.org/10.1016/j.matpr.2021.03.452

Le Berre O, Daufin G (1996) Skimmilk crossflow microfiltration performance versus permeation flux to wall shear stress ratio. J Memb Sci 117:261–270. https://doi.org/10.1016/0376-7388(96)00076-2

Le-Clech P, Chen V, Fane TAG (2006) Fouling in membrane bioreactors used in wastewater treatment. J Memb Sci 284:17–53. https://doi.org/10.1016/j.memsci.2006.08.019

Lee YS, Han GB (2016) Complete reduction of highly concentrated contaminants in piggery waste by a novel process scheme with an algal-bacterial symbiotic photobioreactor. J Environ Manage 177:202–212. https://doi.org/10.1016/j.jenvman.2016.04.025

Lee JS, Heo SA, Jo HJ, Min BR (2016) Preparation and characteristics of cross-linked cellulose acetate ultrafiltration membranes with high chemical resistance and mechanical strength. React Funct Polym 99:114–121. https://doi.org/10.1016/j.reactfunctpolym.2015.12.014

Lin W, Li M, Wang Y et al (2019) Quantifying the dynamic evolution of organic, inorganic and biological synergistic fouling during nanofiltration using statistical approaches. Environ Int 133:105201. https://doi.org/10.1016/j.envint.2019.105201

Lindamulla LMLKB, Jegatheesan V, Jinadasa KBSN et al (2021) Integrated mathematical model to simulate the performance of a membrane bioreactor. Chemosphere 284:131319. https://doi.org/10.1016/j.chemosphere.2021.131319

Liu X, Ren Z, Ngo HH et al (2021) Membrane technology for rainwater treatment and reuse: a mini review. Water Cycle 2:51–63. https://doi.org/10.1016/j.watcyc.2021.08.001

Liu A, Dai K, Wang F et al (2024) Performance of anaerobic dynamic membrane bioreactor treatment of wastewater from food waste pretreatment: dissolved organic matter changes and overloading effects. Environ Technol Innov 35:103737. https://doi.org/10.1016/j.eti.2024.103737

Lotfikatouli S, Hadi P, Yang M et al (2021) Enhanced anti-fouling performance in membrane bioreactors using a novel cellulose nanofiber-coated membrane. Sep Purif Technol 275:119145. https://doi.org/10.1016/j.seppur.2021.119145

Lu Q, Li N (2021) Preparation of hydrophilic polyvinylidene fluoride/polyvinyl alcohol ultrafiltration membrane via polymer/non-solvent co-induced phase separation method towards enhance anti-fouling performance. J Environ Chem Eng 9:106431. https://doi.org/10.1016/j.jece.2021.106431

Lu X, Zheng C, Zhen G et al (2021) Roles of colloidal particles and soluble biopolymers in long-term performance and fouling behaviors of submerged anaerobic membrane bioreactor treating methanolic wastewater. J Clean Prod 290:125816. https://doi.org/10.1016/j.jclepro.2021.125816

Lübbecke S, Vogelpohl A, Dewjanin W (1995) Wastewater treatment in a biological high-performance system with high biomass concentration. Water Res 29:793–802. https://doi.org/10.1016/0043-1354(94)00215-S

Mahmoudkhani R, Torabian A, Hassani AH, Mahmoudkhani R (2014) Copper, cadmium and ferrous removal by membrane bioreactor. APCBEE Proc 10:79–83. https://doi.org/10.1016/j.apcbee.2014.10.020

Majewsky M, Gallé T, Yargeau V, Fischer K (2011) Active heterotrophic biomass and sludge retention time (SRT) as determining factors for biodegradation kinetics of pharmaceuticals in activated sludge. Bioresour Technol 102:7415–7421. https://doi.org/10.1016/j.biortech.2011.05.032

Martín-Pascual J, Reboleiro-Rivas P, López-López C et al (2015) Effect of the filling ratio, MLSS, hydraulic retention time, and temperature on the behavior of the hybrid biomass in a hybrid moving bed membrane bioreactor plant to treat urban wastewater. J Environ Eng 141:04015007. https://doi.org/10.1061/(asce)ee.1943-7870.0000939

Matos CT, Sequeira AM, Velizarov S et al (2009) Nitrate removal in a closed marine system through the ion exchange membrane bioreactor. J Hazard Mater 166:428–434. https://doi.org/10.1016/j.jhazmat.2008.11.038

Mazzaferro L, Trinidad K, Kaplan DL, Asatekin A (2025) Fouling resistance of novel amphiphilic polyampholyte membranes with varying hydrophobicity towards common foulants and cell culture media. Sep Purif Technol 354:129090. https://doi.org/10.1016/j.seppur.2024.129090

McAdam EJ, Judd SJ (2006) A review of membrane bioreactor potential for nitrate removal from drinking water. Desalination 196:135–148. https://doi.org/10.1016/j.desal.2006.03.008

Mehmood CT, Waheed H, Tan W, Xiao Y (2021) Photocatalytic quorum quenching: a new antifouling and in-situ membrane cleaning strategy for an external membrane bioreactor coupled to UASB. J Environ Chem Eng 9:105470. https://doi.org/10.1016/j.jece.2021.105470

Melin T, Jefferson B, Bixio D et al (2006) Membrane bioreactor technology for wastewater treatment and reuse. Desalination 187:271–282. https://doi.org/10.1016/j.desal.2005.04.086

Merz C, Scheumann R, El Hamouri B, Kraume M (2007) Membrane bioreactor technology for the treatment of greywater from a sports and leisure club. Desalination 215:37–43. https://doi.org/10.1016/j.desal.2006.10.026

Messaoudi M, Douma M, Tijani N, Messaoudi L (2021) Study of the permeability of tubular mineral membranes: application to wastewater treatment. Heliyon 7:e06837. https://doi.org/10.1016/j.heliyon.2021.e06837

Metals H, Yazdi SH, Vosoogh A, Bazargan A (2020) The application of membrane bioreactors (MBR) for the removal of organic matter, nutrients, and heavy metals from landfill leachate. Handb Environ Mater Manag. https://doi.org/10.1007/978-3-319-58538-3

Metzger U, Le-Clech P, Stuetz RM et al (2007) Characterisation of polymeric fouling in membrane bioreactors and the effect of different filtration modes. J Memb Sci 301:180–189. https://doi.org/10.1016/j.memsci.2007.06.016

Mokarizadeh H, Raisi A (2021) Industrial wastewater treatment using PES UF membranes containing hydrophilic additives: experimental and modeling of fouling mechanism. Environ Technol Innov 23:101701. https://doi.org/10.1016/j.eti.2021.101701

Moya-Llamas MJ, Pacazocchi MG, Trapote A (2023) Respirometric tests in a combined UASB-MBR system treating wastewater containing emerging contaminants at different OLRs and temperatures: biokinetic analysis. J Environ Manage 345:118643. https://doi.org/10.1016/j.jenvman.2023.118643

Munz G, Gualtiero M, Salvadori L et al (2008a) Process efficiency and microbial monitoring in MBR (membrane bioreactor) and CASP (conventional activated sludge process) treatment of tannery wastewater. Bioresour Technol 99:8559–8564

Munz G, Gori R, Cammilli L, Lubello C (2008b) Characterization of tannery wastewater and biomass in a membrane bioreactor using respirometric analysis. Bioresour Technol 99:8612–8618. https://doi.org/10.1016/j.biortech.2008.04.004

Mutamim NSA, Noor ZZ, Hassan MAA et al (2013) Membrane bioreactor: applications and limitations in treating high strength industrial wastewater. Chem Eng J 225:109–119. https://doi.org/10.1016/j.cej.2013.02.131

Nadizadeh Z, Mahdavi H (2021) Grafting of zwitterion polymer on polyamide nanofiltration membranes via surface-initiated RAFT polymerization with improved antifouling properties as a new strategy. Sep Purif Technol 254:117605. https://doi.org/10.1016/j.seppur.2020.117605

Ng ANL, Kim AS (2007) A mini-review of modeling studies on membrane bioreactor (MBR) treatment for municipal wastewaters. Desalination 212:261–281. https://doi.org/10.1016/j.desal.2006.10.013

Nghiem LD, Tadkaew N, Sivakumar M (2009) Removal of trace organic contaminants by submerged membrane bioreactors. Desalination 236:127–134. https://doi.org/10.1016/j.desal.2007.10.059

Nguyen NC, Nguyen HT, Chen SS et al (2016) A novel osmosis membrane bioreactor-membrane distillation hybrid system for wastewater treatment and reuse. Bioresour Technol 209:8–15. https://doi.org/10.1016/j.biortech.2016.02.102

Nunes SP, Culfaz-Emecen PZ, Ramon GZ et al (2020) Thinking the future of membranes: perspectives for advanced and new membrane materials and manufacturing processes. J Memb Sci 598:117761. https://doi.org/10.1016/j.memsci.2019.117761

Oliveira Filho M, Mailler R, Rocher V et al (2021) Comprehensive study of supported PVDF membrane ageing in MBR: a direct comparison between changes at bench scale and full scale. Sep Purif Technol 279:119695. https://doi.org/10.1016/j.seppur.2021.119695

Pajoum Shariati F, Heran M, Sarrafzadeh MH et al (2013) Biomass characterization by dielectric monitoring of viability and oxygen uptake rate measurements in a novel membrane bioreactor. Bioresour Technol 140:357–362. https://doi.org/10.1016/j.biortech.2013.04.099

Pandey AK, Kim SM, Yoon YH et al (2023) Influence of the permeate flux on continuous biohydrogen production, permeability, and granulation in a dynamic membrane bioreactor. Chem Eng J 475:145684. https://doi.org/10.1016/j.cej.2023.145684

Parihar RK, Bhandari K, Burnwal PK et al (2024) Advancing dairy wastewater treatment: exploring two-stage fluidized bed anaerobic membrane bioreactor for enhanced performance, fouling, and microbial community analysis. J Water Process Eng 58:104917. https://doi.org/10.1016/j.jwpe.2024.104917

Pathak N, Van Tran HH, Phuntsho S, Shon HY (2020) Membrane bioreactors for the removal of micro-pollutants. Elsevier BV

Pettinato M, Chakraborty S, Arafat HA, Calabro' V (2015) Eggshell: a green adsorbent for heavy metal removal in an MBR system. Ecotoxicol Environ Saf 121:57–62.https://doi.org/10.1016/j.ecoenv.2015.05.046

Pophali GR, Kaul SN, Mathur S (2003) Influence of hydraulic shock loads and TDS on the performance of large-scale CETPs treating textile effluents in India. Water Res 37:353–361. https://doi.org/10.1016/S0043-1354(02)00268-3

Prado N, Ochoa J, Amrane A (2009) Zero nuisance piggeries: long-term performance of MBR (membrane bioreactor) for dilute swine wastewater treatment using submerged membrane bioreactor in semi-industrial scale. Water Res 43:1549–1558. https://doi.org/10.1016/j.watres.2008.12.043

Puron T (2008) Introduction to membranes—MBRs manufacturers' comparison: part 3—supplier review. 23–25

Qin JJ, Wai MN, Tao G et al (2007) Membrane bioreactor study for reclamation of mixed sewage mostly from industrial sources. Sep Purif Technol 53:296–300. https://doi.org/10.1016/j.seppur.2006.08.002

Qrenawi LI, Rabah FKJ (2023) Fouling in membrane bioreactors: the influence of some parameters and the effectiveness of some control strategies. Desalin Water Treat 315:62–80. https://doi.org/10.5004/dwt.2023.30099

Rabiller-baudry M, Paugam L, Bégoin L et al (2006) Alkaline cleaning of PES membranes used in skimmed milk ultrafiltration: from reactor to spiral-wound module via a plate-and-frame module. 191:334–343.https://doi.org/10.1016/j.desal.2005.07.028

Radjenović J, Matošić M, Mijatović I et al (2008) Membrane bioreactor (MBR) as an advanced wastewater treatment technology. In: Emerging contaminants from industrial and municipal waste, vol 5. pp 37–101. https://doi.org/10.1007/698_5_093

Rahman MM, Al-Malack MH (2006) Performance of a crossflow membrane bioreactor (CF-MBR) when treating refinery wastewater. Desalination 191:16–26. https://doi.org/10.1016/j.desal.2005.05.022

Ramakrishnan S, McDonald CJ, Prud'homme RK, Carbeck JD (2004) Latex composite membranes: structure and properties of the discriminating layer. J Memb Sci 231:57–70.https://doi.org/10.1016/j.memsci.2003.10.036

Rani SLS, Kumar RV (2021) Insights on applications of low-cost ceramic membranes in wastewater treatment: a mini-review. Case Stud Chem Environ Eng 4:100149. https://doi.org/10.1016/j.cscee.2021.100149

Razavi SMR, Miri T (2015) A real petroleum refinery wastewater treatment using hollow fiber membrane bioreactor (HF-MBR). J Water Process Eng 8:136–141. https://doi.org/10.1016/j.jwpe.2015.09.011

Rodríguez FA, Poyatos JM, Reboleiro-Rivas P et al (2011) Kinetic study and oxygen transfer efficiency evaluation using respirometric methods in a submerged membrane bioreactor using pure oxygen to supply the aerobic conditions. Bioresour Technol 102:6013–6018. https://doi.org/10.1016/j.biortech.2011.02.083

Sari Erkan H, Çağlak A, Soysaloglu A et al (2020) Performance evaluation of conventional membrane bioreactor and moving bed membrane bioreactor for synthetic textile wastewater treatment. J Water Process Eng 38.https://doi.org/10.1016/j.jwpe.2020.101631

Sartor M, Kaschek M, Mavrov V (2008) Feasibility study for evaluating the client application of membrane bioreactor (MBR) technology for decentralised municipal wastewater treatment in Vietnam. Desalination 224:172–177. https://doi.org/10.1016/j.desal.2007.04.086

Satyawali Y, Balakrishnan M (2008) Treatment of distillery effluent in a membrane bioreactor (MBR) equipped with mesh filter. Sep Purif Technol 63:278–286. https://doi.org/10.1016/j.seppur.2008.05.008

Scott JA, Smith KL (1997) A bioreactor coupled to a membrane to provide aeration and filtration in ice-cream factory wastewater remediation. Water Res 31:69–74. https://doi.org/10.1016/S0043-1354(96)00234-5

Seman MNA, Khayet M, Hilal N (2012) Comparison of two different UV-grafted nanofiltration membranes prepared for reduction of humic acid fouling using acrylic acid and N-vinylpyrrolidone. Desalination 287:19–29. https://doi.org/10.1016/j.desal.2010.10.031

Seo Y, Pooi CK, Ng HY (2024) Recommendation on the selection of powdered activated carbon as carrier to enhance performance of polymeric UF membrane. J Memb Sci 123226.https://doi.org/10.1016/j.memsci.2024.123226

Sgroi M, Vagliasindi FGA, Roccaro P (2018) Feasibility, sustainability and circular economy concepts in water reuse. Curr Opin Environ Sci Heal 2:20–25. https://doi.org/10.1016/j.coesh.2018.01.004

Shin C, Bae J (2018) Bioresource technology current status of the pilot-scale anaerobic membrane bioreactor treatments of domestic wastewaters: a critical review. Bioresour Technol 247:1038–1046. https://doi.org/10.1016/j.biortech.2017.09.002

Sim LN, Ye Y, Chen V, Fane AG (2011) Investigations of the coupled effect of cake-enhanced osmotic pressure and colloidal fouling in RO using crossflow sampler-modified fouling index ultrafiltration. Desalination 273:184–196. https://doi.org/10.1016/j.desal.2011.01.059

Song S, Rong L, Dong K et al (2021) Pore-scale numerical study of intrinsic permeability for fluid flow through asymmetric ceramic microfiltration membranes. J Memb Sci 642:119920. https://doi.org/10.1016/j.memsci.2021.119920

Song J, Wu H, Yuan H et al (2022) Reclamation of pharmaceutical wastewater by UAF, contact oxidation, and auto brush-cleaning aiding nano ZrO_2 coated ceramic membrane MBR—an industrial-scale practice and functional microbial diversity. J Water Process Eng 46:102563. https://doi.org/10.1016/j.jwpe.2022.102563

Sun J, Wang X, Li R et al (2011) Hyperhaline municipal wastewater treatment of a processing zone through pilot-scale A/O MBR, part II: nitrogen and phosphorous removal. Procedia Environ Sci 8:781–788. https://doi.org/10.1016/j.proenv.2011.10.118

Tao C, Parker W, Bérubé P (2022) Assessing the role of cold temperatures on irreversible membrane permeability of tertiary ultrafiltration treating municipal wastewater. Sep Purif Technol 278.https://doi.org/10.1016/j.seppur.2021.119556

Tay JH, Yang P, Zhuang WQ et al (2007) Reactor performance and membrane filtration in aerobic granular sludge membrane bioreactor. J Memb Sci 304:24–32. https://doi.org/10.1016/j.memsci.2007.05.028

Tay MF, Lee S, Xu H et al (2020) Impact of salt accumulation in the bioreactor on the performance of nanofiltration membrane bioreactor (NF-MBR)+Reverse osmosis (RO) process for water reclamation. Water Res 170:115352. https://doi.org/10.1016/j.watres.2019.115352

Tian J, Liang H, Yang Y et al (2008) Membrane adsorption bioreactor (MABR) for treating slightly polluted surface water supplies: as compared to membrane bioreactor (MBR). J Memb Sci 325:262–270.https://doi.org/10.1016/j.memsci.2008.07.047

Tomar SK, Padhi SK, Dikshit PK et al (2023) Effect of hydraulic retention time on biological nutrients removal in an anaerobic membrane bioreactor treating low-strength drain wastewater. Bioresour Technol Rep 24:101599. https://doi.org/10.1016/j.biteb.2023.101599

Tran VH, Lim S, Jungwon P et al (2021) Submerged versus side-stream osmotic membrane bioreactors using an outer-selective hollow fiber osmotic membrane for desalination. Desalination 515:115196. https://doi.org/10.1016/j.desal.2021.115196

Turano E, Curcio S, De Paola MG et al (2002) An integrated centrifugation–ultrafiltration system in the treatment of olive mill wastewater. J Memb Sci 209:519–531. https://doi.org/10.1016/S0376-7388(02)00369-1

Vieno N, Sillanpää M (2014) Fate of diclofenac in municipal wastewater treatment plant—a review. Environ Int 69:28–39. https://doi.org/10.1016/j.envint.2014.03.021

Vo TDH, Bui XT, Dang BT et al (2021) Influence of organic loading rates on treatment performance of membrane bioreactor treating tannery wastewater. Environ Technol Innov 24:101810. https://doi.org/10.1016/j.eti.2021.101810

Wang Z, Wang Q, Wu Z, Wang X (2010) A hybrid membrane process for simultaneous thickening and digestion of waste activated sludge. 4:272–279.https://doi.org/10.1007/s11783-010-0234-x

Wang A, Shao M, Yang F et al (2021a) Preparation and properties of antibacterial PVDF composite thin films. Eur Polym J 160:110803. https://doi.org/10.1016/j.eurpolymj.2021.110803

Wu L, Liu Y, Hu J et al (2021) Preparation of polyvinylidene fluoride composite ultrafiltration membrane for micro-polluted surface water treatment. Chemosphere 284:131294. https://doi.org/10.1016/j.chemosphere.2021.131294

Xiao X, You S, Guo H et al (2021) Mycelial pellets for alleviation of membrane fouling in membrane bioreactor. J Memb Sci 635:119545. https://doi.org/10.1016/j.memsci.2021.119545

Xing CH, Qian Y, Wen XH et al (2001a) Physical and biological characteristics of a tangential-flow MBR for municipal wastewater treatment. J Memb Sci 191:31–42. https://doi.org/10.1016/S0376-7388(01)00472-0

Xing CH, Wen XH, Qian Y, Tardieu E (2001b) Microfiltration-membrane-coupled bioreactor for urban wastewater reclamation. Desalination 141:63–73. https://doi.org/10.1016/S0011-9164(01)00389-7

Yadav P, Ismail N, Essalhi M et al (2021) Assessment of the environmental impact of polymeric membrane production. J Memb Sci 622:118987. https://doi.org/10.1016/j.memsci.2020.118987

Yan Z, Yang H, Qu F et al (2019) Application of membrane distillation to anaerobic digestion effluent treatment: identifying culprits of membrane fouling and scaling. Sci Total Environ 688:880–889. https://doi.org/10.1016/j.scitotenv.2019.06.307

Yang Y, Fu W, Chen L et al (2021) One-step dip-coating method for preparation of ceramic nanofiber membrane with high permeability and low cost. J Eur Ceram Soc. https://doi.org/10.1016/j.jeurceramsoc.2021.09.049

Yeh HM, Wu HH (1997) Membrane ultrafiltration in combined hollow-fiber module systems. J Memb Sci 124:93–105

Yin J, Zhang H (2021) A combined physical blending and surface grafting strategy for hydrophilic modification of polyethersulfone membrane toward oil/water separation. Polymer (Guildf) 233:124177. https://doi.org/10.1016/j.polymer.2021.124177

Yoneda I, Nishiyama M, Watanabe T (2024) Comparative experiment to select water quality parameters for modelling the survival of *Escherichia coli* in lakes. Environ Pollut 357:124423. https://doi.org/10.1016/j.envpol.2024.124423

Zarragoitia-González A, Schetrite S, Alliet M et al (2008) Modelling of submerged membrane bioreactor: conceptual study about link between activated slugde biokinetics, aeration and fouling process. J Memb Sci 325:612–624. https://doi.org/10.1016/j.memsci.2008.08.037

Zhang Y, Ma C, Ye F et al (2009) The treatment of wastewater of paper mill with integrated membrane process. Desalination 236:349–356. https://doi.org/10.1016/j.desal.2007.10.086

Zhang B, Zhang S, Wu Y et al (2021a) Process Intensification Enhanced separation performance of microfiltration carbon membranes for oily wastewater treatment by an air oxidation strategy. Chem Eng Process—Process Intensif 169:108620. https://doi.org/10.1016/j.cep.2021.108620

Zhang S, Lei Z, Dzakpasu M et al (2021b) Removal of trace organic contaminants in municipal wastewater by anaerobic membrane bioreactor: efficiencies, fates and impact factors. J Water Process Eng 40:101953. https://doi.org/10.1016/j.jwpe.2021.101953

Zhang J, Wang L, Zhang C et al (2021c) MnO_x-mineralized oxidized-polypropylene membranes for highly efficient oil/water separation. Sep Purif Technol 276:119343. https://doi.org/10.1016/j.seppur.2021.119343

Zhang Z, Yu D, Xu X et al (2021d) A polypropylene melt-blown strategy for the facile and efficient membrane separation of oil–water mixtures. Chinese J Chem Eng 29:383–390. https://doi.org/10.1016/j.cjche.2020.03.033

Zhao C, Song T, Yu Y et al (2020) Insight into the influence of humic acid and sodium alginate fractions on membrane fouling in coagulation-ultrafiltration combined system. Environ Res 191:110228. https://doi.org/10.1016/j.envres.2020.110228

Zheng Y, Yu S, Shuai S et al (2013) Color removal and COD reduction of biologically treated textile effluent through submerged filtration using hollow fiber nanofiltration membrane. Desalination 314:89–95

Zheng Y, Tang B, Ye J et al (2018) A crucial factor towards a sustainable process for municipal wastewater treatment: fouling effects of different statuses of biomass in the membrane bioreactors with no sludge discharge. J Clean Prod 192:877–886. https://doi.org/10.1016/j.jclepro.2018.05.048

Zuo K, Wang K, DuChanois RM et al (2021) Selective membranes in water and wastewater treatment: role of advanced materials. Mater Today. https://doi.org/10.1016/j.mattod.2021.06.013

Zhu Z, Jiang J, Wang X et al (2017) Improving the hydrophilic and antifouling properties of polyvinylidene fluoride membrane by incorporation of novel nanohybrid GO@SiO_2 particles. Chem Eng J 314:266–276. https://doi.org/10.1016/j.cej.2016.12.038

Zhu MC, Wang JW, Lu YZ et al (2025) Enhancement of membrane filtration performance and mitigation of sludge calcification in an intermittent anaerobic membrane bioreactor treating high-calcium wastewater. Sep Purif Technol 354.https://doi.org/10.1016/j.seppur.2024.129028

Chapter 20
Aerobic Suspended Processes for Wastewater Treatment

Karthik Balasundaram

Abstract Worldwide, aerobic suspended growth process has been widely used for the biological treatment of wastewater, the most common being the *activated sludge process* (ASP). The conventional ASP has several advantages like relatively low capital cost, high treatment efficiency, good process, and operational control. The limitations of the conventional ASP include requirement of large land area, high operating costs, and generation of large sludge volume resulting in higher disposal costs. To overcome these limitations, several modifications in the design of conventional ASP have been made. One of the widely used modification of the conventional ASP is the sequencing batch reactor (SBR). The SBR is an activated sludge system operating in "batch" mode. This chapter aims to cover the fundamentals of the ASP including the microbial ecology, growth kinetics, oxygen and nutrient requirements, and the classic reactor configuration. The chapter discusses the design and operating criteria for the conventional ASP in detail. It discusses the various operational problems associated with the conventional ASP along with the remedies. The chapter also provides an overview of various modified processes of the conventional ASP like extended aeration, tapered aeration, step aeration, and contact stabilisation. Next, the fundamentals of design of SBR process are discussed in detail. Finally, to connect the fundamentals with practical design, a design example is provided for SBR.

Keywords Aerobic suspended growth · Activated sludge process · Sequential batch reactor

20.1 Introduction

The primary aim in the biological treatment of wastewater is the degradation and subsequent removal of colloidal and dissolved organic matter by using microorganisms. The biological treatment can be achieved either by the aerobic or anaerobic process. Aerobic treatment processes are those in which the degradation of organic

K. Balasundaram (✉)
Department of Civil Engineering, Visvesvaraya National Institute of Technology, Nagpur, India
e-mail: karthik@civ.vnit.ac.in

S. Souabi et al. (eds.), *Wastewater Treatment Plants*, Water Science and Technology Library 130, https://doi.org/10.1007/978-3-031-87461-1_20

matter happens by using dissolved oxygen as an electron acceptor. The aerobic treatment is further classified based on the physical state of the microorganisms in the reactor during the treatment process, i.e. either suspended or attached. In the suspended type processes the microorganisms bring about the degradation of organic matter while remaining in suspension in the reactor and in the attached processes, they remain attached to a media during the degradation process. Globally, the activated sludge process (ASP) is the extensively used aerobic suspended process for the treatment of wastewater (Orhon 2015). Invented in the early twentieth century, the ASP has become synonymous with the aerobic suspended treatment in present day. The earliest study can be traced back to 1882, where, in a study on the mud samples collected from the Thames Estuary, it was observed that sewage was destroyed by microorganisms (Orhon 2015). Based on this observation, it was argued that the aerobic bacteria oxidised the sewage and that this mechanism could be adopted to treat sewage. Studies during this period highlighted the important role of oxygen and the microbial community in the degradation and removal of organic matter. These studies gave a totally new perspective to the hitherto understanding of sewage treatment and paved the way for a new biological treatment system that would degrade organic matter in sewage. The studies were focused only on microbial degradation without forced aeration. In fact, many studies even concluded that dissolved air was not effective in oxidising the sewage (Orhon 2015).

The first study on application of aeration to sewage was performed by Angus Smith in the year 1882 (Orhon 2015). The subsequent studies also reported that the flocculent suspended solids generated during the treatment had an impact on the oxidation of organic matter. It was also observed that the solids that settled down in the aeration tank improved efficiency of the treatment. The credit for discovering the ASP as used in the present day is given to Ardern and Lockett (Metcalf and Eddy 2014; Rittmann and McCarty 2020). They observed that flocculent suspended particles were formed by the aeration of sewage. These flocculent particles when retained in the aeration tank significantly decreased the time required to remove organic matter from the sewage. They referred to the flocculent suspended particles, as being *'activated'* and hence the process came to be known as *activated sludge process*. The early twentieth century witnessed path breaking discoveries which helped in understanding the basic fundamentals of the ASP. The outcome of these studies resulted in (1) design and operation of a sewage treatment process completely based on aerobic microorganisms in a suspended process, (2) understanding the role of 'activated sludge' in achieving the desired process efficiency, and (3) understanding the crucial role of oxygen in the performance of the system.

The first ASP for sewage treatment was installed in England in the year 1914 (Orhon 2015). The plant had a capacity of 300 m^3/day with a fill and draw type configuration. Since then, the ASP has come a long way. In years to come, the ASP went on to become the first choice for treatment of sewage worldwide.

20.2 Conventional Activated Sludge Process

The treatment units commonly used in a conventional ASP is as shown in the flow diagram in Fig. 20.1. The conventional ASP operates in a plug flow regime. To ensure the plug flow regime, the aeration tank of conventional ASP has a length to width ratio more than 5 (CPHEEO 2013). The main units of the ASP are the aeration tank, secondary sedimentation tank (SST) and a separate sludge recirculation line between the SST and inlet of aeration tank for recycling the settled sludge from the SST. The settled sludge in the SST mainly contains a mix of aerobic microorganisms and is known as 'activated sludge'. The preliminary and primary treatment for the removal of grit and large solid particles always precede the aeration tank process when treating low-strength wastewater. However, a recent study (Latif 2022) has shown that removing primary settling from the conventional ASP treatment process for low strength wastewater did not have any effect on the BOD and COD removal. There was a decrease in the removal of total suspended solids and increase in NH_4-N removal. The main objective of the ASP is to remove the colloidal and dissolved organic matter from the wastewater. The wastewater after the primary treatment is mixed with the activated sludge from the SST at the inlet of the aeration tank. This mixture is now referred as mixed liquor. Aerobic condition is maintained in the aeration tank with forced supply of air and the mixture is aerated for approximately 4–8 h. The aerobic microorganisms oxidise the organic matter partly to synthesis new cells and partly to derive energy. The microorganisms produce extracellular polymeric substances (EPS) during the degradation process. The polysaccharides, lipids, nucleic acids, proteins found on the interior and exterior of the microorganism cells are known as EPS (Minnie et al. 2022). However, the complete details about the molecular structure, compositional, and structural properties of the EPS are not available till date (Yu 2020). This produced EPS acts as a glue that causes the suspended and colloidal solids to agglomerate and form flocs. The formed flocs tend to remain in suspension in the aeration tank. The aeration tank effluent containing the flocs is taken to the SST where the flocs settle under gravity at the bottom of the tank and form sludge. The clearer effluent from the top of SST is either sent for final disposal or is subjected to further tertiary treatment. A portion of the bottom sludge from the SST is recycled and mixed with the primary treated sewage in the aeration tank. The remaining portion, termed as excess sludge is taken to the sludge thickener and then to sludge drying beds. Alternatively, the excess sludge can also be removed from the aeration tank as mixed liquor. The size of the flocs in the activated sludge ranges from 2.20 to 2.21 μm, in thickened sludge from 0.35 to 0.65 μm (Wiercik et al. 2022). Plants up to 300 MLD capacities have been built in India (CPHEEO 2013).

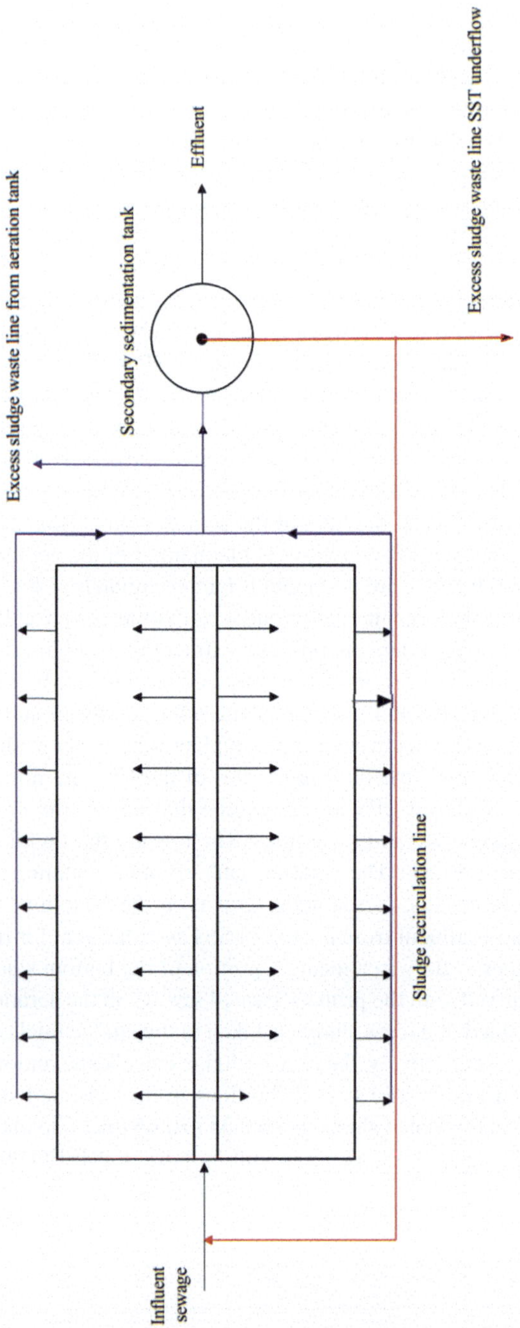

Fig. 20.1 Conventional activated sludge process

20.3 Microbial Ecology

The workforce of the ASP is the aerobic microorganisms that oxidise the organic matter in sewage by using them as the electron donor. The wastewater treatment in the ASP is brought about by a diverse group of microorganisms. This group of microorganisms also have the capacity to carry out the oxidation while remaining in a suspended mode. The major group include the prokaryotes (bacteria) and eukaryotes (protozoa, crustacea nematodes, and rotifers) (Rittmann and McCarty 2020; Sobczyk et al. 2021). Fungi are usually not present in significant proportion. The heterotrophic bacteria are the primary microorganisms responsible for the oxidation of organic matter. Some bacterial species oxidise a variety of organic compounds whereas some species are selective and oxidise only a particular type of the organic compound. The common species found in an ASP system include *Pseudomonas, Arthrobacter, Zoogloea, Comamonas Lophomonas, Flavobacterium, Bacillus, Dokdonella, Sphaerotilus, Azotobacter, Chromobacterrum, Achromobacter, Arcobacter and Nocardia* (Xia et al. 2018; Chen et al. 2020; Rittmann and McCarty 2020; Lu et al. 2023). As multiple bacterial species are involved in the ASP, even if the population of a particular species is affected, another species can replace it quickly so that the overall treatment process is not affected significantly. This ensures an inherent redundancy in the microbial ecosystem and is a major reason for the reliability of ASP systems.

20.4 Bacterial Growth Kinetics in ASP

In the ASP, the organic matter present in wastewater is oxidised by bacteria under aerobic conditions to gain energy and form new bacterial cell mass. In this process, the organic matter and the oxygen act as electron donor and electron acceptor, respectively. The exact stoichiometry of the biological oxidation of the organic matter is never known. However, as a practise the organic matter is represented as $C_6H_{12}O_6$ (glucose) and the new cells are represented as $C_5H_7NO_2$ (Metcalf and Eddy 2014). Thus, the oxidation of organic matter and the formation of new bacterial cells can be shown as in Eq. (20.1).

$$3C_6H_{12}O_6 + 8O_2 + 2NH_3 \quad 8CO_2 + 14H_2O + 2C_5H_7O_2N \tag{20.1}$$

After oxidation of the entire organic matter present in wastewater, the bacteria then undergo endogenous respiration. During the endogenous respiration the bacteria oxidise the cell mass to carbon-di-oxide and water as per Eq. (20.2)

$$C_5H_7O_2N + 5O_2 \rightarrow 5CO_2 + 2H_2O + NH_3 \tag{20.2}$$

Based on the stoichiometric relationship as in Eq. (20.2), it can be observed that 160 g of O_2 is consumed during the oxidation of 113 g of cell mass, i.e. 1.42 mgO_2/mg of cellular mass. It should be noted that it is difficult to directly measure the concentration of microorganisms that are involved in the oxidation of organic matter as indicated in Eq. (20.1). Therefore, the concentration of microorganisms present in the system is measured indirectly by measuring the concentration of suspended solids present in the mixed liquor. The suspended solids are typically biologically active solids and represent the biomass produced in the reactor. These biologically active suspended solids are known as mixed liquor suspended solids (MLSS). The MLSS indicates the concentration of microorganisms and the solids (both inorganic and organic) in the mixed liquor. However, the MLSS as an indicator has certain limitations. The MLSS contains inert organic solids (dead cells and inert cellular matter) and inorganic solids like silt, clay, and sand that escaped in the primary treatment. The MLSS also contains microorganisms that do not oxidise the organic matter present in sewage and thus do not participate in the reaction shown in Eq. (20.1). Therefore, to have a better measure of the microorganisms actually responsible for oxidation, the volatile fraction of MLSS, the mixed liquor volatile suspended solids (MLVSS) is used as an indicator. The MLVSS measures the concentration of volatile suspended solids thus indicating the active microorganisms and biodegradable biomass in the mixed liquor. The MLVSS concentration at steady state in a complete mix process is given as:

$$X_v = \frac{Y \cdot Q(S_0 - S)\theta_c}{(1 + k_d\theta_c)t} \tag{20.3}$$

where,

X_v = MLVSS concentration in the aeration tank (mg/l)
Y = yield coefficient (0.4–0.8 mg VSS/mg BOD_5 removed, typical value = 0.6)
Q = influent flowrate, (m^3/d)
S_0 = influent substrate concentration (soluble BOD) (mg/l)
S = effluent substrate concentration (soluble BOD) (mg/l)
k_d = decay coefficient of microorganisms (0.04–0.075 day^{-1} typical value = 0.06)
θ_c = Solids retention time (days)
t = Hydraulic retention time (days).

Generally, the MLVSS is approximately 65–75% of MLSS (Rittmann and McCarty 2020).

20.5 Fundamentals of Process Design

When the ASP system was first invented in 1914, very little was known about the fundamentals of microbial ecology and bacterial growth kinetics happening inside the aeration tank. The various design parameters were based upon empirical observations. As time progressed, several investigations and studies resulted in a better understanding of the various factors affecting the process. These studies led to the emergence of the process parameters founded on the fundamentals of the microbial ecology and bacterial growth kinetics. The process parameters currently used for the design and operation of ASP are derived both empirically as well as using the fundamental principles of microbial kinetics. The important parameters used in the design of an ASP include:

1. Hydraulic retention time (HRT)
2. Organic loading/volumetric loading
3. Food to microorganism ratio (F/M ratio) and
4. Sludge age.

20.5.1 Hydraulic Retention Time (HRT)

The HRT is an empirical parameter and is defined as *the rate at which wastewater is applied to the aeration tank*. It is given as:

$$\text{HRT}(\text{t, hours}) = \frac{V}{Q} \times 24 \qquad (20.4)$$

where,

V = Volume of aeration tank, m^3
Q = Sewage flow, m^3/d.

20.5.2 Organic Loading Rate

The organic loading rate is another empirical parameter which is defined as *the BOD load applied per unit volume of the aeration tank per day*. This is also known as volumetric loading rate and is given as:

$$\text{Volumetric load}\left(\text{kg BOD/m}^3 \cdot d\right) = \frac{Q \cdot S_0}{V} \qquad (20.5)$$

20.5.3 Food-to-Microorganisms (F/M) Ratio

The (F/M) ratio developed in 1960s is the widely used process parameter. It is known as a process parameter as it is used to measure or control the performance of biological treatment process. It is simple, easy to understand conceptually, and the value can be obtained from the routine measurements made at most wastewater treatment plants. It is defined as *the ratio of food supplied to the microorganisms present in the aeration tank*. The food supplied is measured in terms of BOD of the wastewater coming into the aeration tank and the microorganisms are measured in terms of MLVSS in the aeration tank.

$$\frac{F}{M} = \frac{\text{food applied to the reactor in kg}}{\text{Biomass in the system in kg}}$$

$$\frac{F}{M} = \frac{Q}{V} \cdot \frac{S_0}{X_t} \tag{20.6}$$

where,

X_t = MLSS mg/l.

The F/M ratio values for ASP are in the range 0.3–0.5. When F/M ratio is high, the system will be loaded with food resulting in production of excessive biomass. The excessive biomass will be dispersed in the entire volume of reactor. Thus, effluent will have high concentration of suspended solids. When the F/M ratio is low, the BOD removal will be higher in the plant. The system will operate in the endogenous growth phase of biomass and pin-pointed, non-settleable biological floc will be produced. Effluent will contain undesirable suspended solids of high concentration. Optimum F/M ratio has been found to produce settleable flocs and effluent with low suspended solids concentration. The F/M ratio can be maintained in a desired range by adjusting the MLSS concentration.

20.5.4 Sludge Age

The sludge age (θ_c) is defined as *the average time for which the suspended solids remain in the aeration tank*. It is also known as the mean cell residence time (MCRT) or solids retention time (SRT). The wastewater passes only once through the aeration tank and remains in the aeration tank only for the duration of HRT. However, the solids are recycled from the SST back to the aeration tank. Therefore, the solids pass through the aeration tank multiple times resulting in an increased retention time of the solids within the aeration tank. This duration is known as sludge age or SRT or MCRT and is given as:

$$\theta_c = \frac{\text{Total microbial mass in reactor}}{\text{Total mass of solids withdrawn daily+ solids leaving with the effluent}}$$

$$\theta_c = \frac{V \cdot X_t}{(Q_w \cdot X_r) + (Q_e \cdot X_e)} \tag{20.7}$$

For a system working efficiently under steady state, the solids concentration in the effluent is much less as compared to the MLSS concentration and hence it is neglected. Therefore:

$$\theta_c = \frac{V \cdot X_t}{Q_w \cdot X_r} \tag{20.8}$$

where,

Q_w = waste sludge flow rate (m^3/d)
X_r = MLSS concentration in the waste /return sludge (mg/l)
X_e = MLSS concentration in the effluent (mg/l)
Q_e = effluent flowrate (m^3/d).

The θ_c for ASP is in the range 5–8 days. The ASP plants are typically designed to keep θ_c values between 4 and 10 days (CPHEEO 2013).

20.6 Design Consideration of Activated Sludge Process

The design of ASP involves calculating the capacity of aeration tank and HRT, oxygen requirement, design of SST, sludge production, sludge recycling rate, and excess sludge wasting. The design parameters for the ASP are given in Table 20.1. The following assumptions are made in the development of equations for design of different components:

(i) The ASP is operating in complete mix regime. The complete mix ASP system consists of an aeration tank followed by SST unit and a sludge recycle line. Figure 20.2 shows a flow diagram for a completely mixed activated sludge system.
(ii) There are no microorganisms in the influent. Although the waste water contains microorganisms, there concentration is much less than the concentration maintained in the aeration tank. Hence, the microorganisms in the incoming sewage are neglected.
(iii) There may be some oxidation of waste in the secondary clarifier. However, it is assumed that the oxidation of organic matter by microorganisms occurs only in the aeration tank.
(iv) The θ_c for the system is calculated considering volume of aeration tank only.
(v) The system is operating under steady state.

Table 20.1 Design parameters for activated sludge system for sewage

Process parameter	Unit	Flow regime		
		Conventional	Complete mix	Extended aeration
MLSS	mg/l	1500–3000	3000–4000	3000–5000
MLSS/MLVSS	Ratio	0.8	0.8	0.6
F/M	day^{-1}	0.3–0.4	0.3–0.6	0.1–0.18
HRT	Hours	4–6	4–6	12–24
θ_c	days	5–8	5–8	10–26
Q_r/Q	Ratio	0.25–0.5	0.25–0.8	0.25–1.0
BOD removal	%	85–92	85–92	95–98
kg O_2/kg BOD removed	Ratio	0.8–1.0	0.8–1.0	1.0–1.2
*Air required/kg BOD removed	m^3	40–100	50–75	100–135

*Adopted from (Tirthakar 2022)

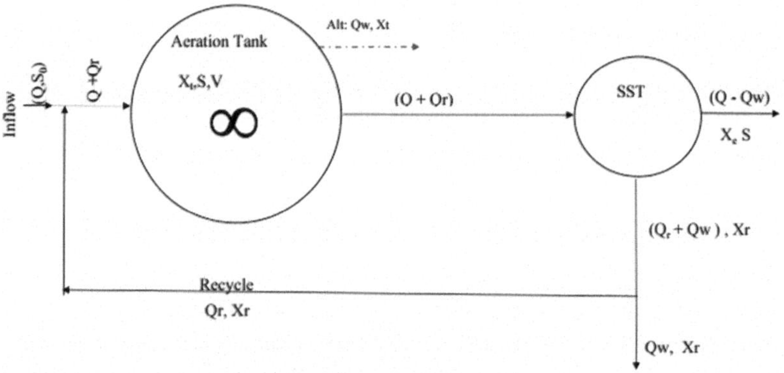

Fig. 20.2 Complete mix activated sludge process

(vi) The BOD of the incoming wastewater is assumed constant.

20.6.1 Aeration Tank

The volume of aeration tank can be calculated from Eq. (20.3) by replacing t with V/Q as:

$$V \cdot X_v = \frac{Y \cdot Q(S_0 - S)\theta_c}{(1 + k_d\theta_c)} \tag{20.9}$$

The volume can be calculated using the above equation by assuming suitable values for Y, k_d and θ_c. The value of θ_c depends on the expected temperature of mixed liquor during operation, settleability of the sludge and the requirement of nitrification. A minimum SRT of 3 days is required for bacteria to form flocs. Generally, a design value of 5 days and 10 days is adopted for warmer and temperate climate, respectively (Rittmann and McCarty 2020). The other design parameters are HRT and MLSS concentration. The aeration tank volume can be minimised by adopting a higher value of MLSS concentration. However, a high MLSS concentration in aeration tank requires a large SST for effective settling of flocs and thickening of the sludge. The maximum value of the MLSS concentration is governed by the limitations on the removal of flocs in the SST. Generally, the value of MLSS lies between 1000 and 4000 mg/l (CPHEEO 2013). Other design factors in calculating the volume of aeration tank include: Oxygen transfer rate by the aeration equipment for a given reactor volume, minimum HRT required for stable operation under varying flow conditions and the solids loading on SST. After determining the volume, the geometric configuration of the aeration tank can be fixed by considering the values of dispersion number (D/UL) (Arceivala and Asolekar 2009). The dispersion number is defined by the ratio D/UL where, U—is mean flow velocity along reactor length, D—axial or longitudinal dispersion coefficient and L—length of axial travel in reactor. The dispersion number represents the mixing conditions in a reactor. A value of (D/UL) between 3.0 and 4.0 represents a completely mixed process and a value of (D/UL) ≤ 0.2 represents a conventional plug flow process (Arceivala and Asolekar 2009). The length of the aeration tank is determined by the value of dispersion number which in turn indicates the type of ASP system (completely mixed or plug flow). The aeration tanks are generally constructed using reinforced concrete. Important factors in the design of aeration tanks are (CPHEEO 2013):

(i) The depth of aeration tank is in the range of 3–4.5 m with 0.3–0.6 m free board. The capacity of aerators is proportional to the depth of the aeration tank. More the depth higher capacity aerators are provided.
(ii) The width of aeration tank is usually between 5 and 10 m.
(iii) The width-depth ratio must be in the range of 1.2–2.2.
(iv) Multiple tank units should be provided when the volume of aeration tank exceeds 70 MLD.
(v) Dead spots should be avoided by providing baffles and fillets in the corner.
(vi) When multiple units are provided, the inlet and outlet structures should be so designed that an individual unit can be isolated for service or maintenance.
(vii) Froth control system in the form of effluent spray nozzle with provision of adding antifoaming agent into the spray water should be provided along the length of tank.

20.6.2 Oxygen Requirement of Aeration Tank

In the ASP, oxygen is required for oxidation of the organic matter, development of new cell and endogenous respiration of the microorganisms. Thus, the total oxygen required is:

$$\text{Total } O_2 \text{ required (gm/day)} = \frac{Q(S_0 - S)}{f} \quad (20.10)$$

where,

$f = \frac{BOD_5}{BOD_u}$ the ratio of BOD_5 to ultimate BOD.

However, the entire oxidised organic matter is not utilised for energy. A part of the oxidised matter is used for the synthesis of new cells. At steady state, the amount of cell mass produced is equal to the amount of cell mass removed from the system. Therefore, the new cells formed are removed from the system without undergoing oxidation and hence there is no oxygen demand from the organic matter converted to new cell mass. From Eq. (20.2), the amount of oxygen required to oxidise the cell mass is 1.42 mgO_2/mg of cellular mass. Therefore, this oxygen demand should be deducted from Eq. (20.10) to get the net oxygen requirement for the oxidation of organic matter to energy.

$$\text{Net } O_2 \text{ required (gm/day)} = \frac{Q(S_0 - S)}{f} - 1.42 \cdot Q_w \cdot X_r \quad (20.11)$$

The above equation gives the oxygen requirement for degrading the carbonaceous BOD only. The equation does not include the additional oxygen required for nitrification. The additional oxygen required for nitrification is 4.56 kgO_2/kg NH_3–N oxidised to NO_3–N (CPHEEO 2013).

20.6.3 Secondary Sedimentation Tank

An efficient SST is necessary for effective solid–liquid separation and removal of the flocs from the effluent of aeration tank. The SST is provided to fulfil two major requirements: (i) separation of MLSS (flocs) from the effluent of aeration tank to get a clarified effluent and (ii) thickening of settled solids at the bottom of SST to remove excess water and obtain the required concentration of solids. The MLSS concentration in aeration tank is entirely dependent on the concentration of the solids in the underflow of the SST. Any failure on the part of the SST in the production of the solids of desired concentration in the underflow will lead to an unsatisfactory performance of the entire ASP. The operational problems of activated sludge process are actually due to poor SST design rather than biological factors. The SST may be either rectangular or circular in shape. The circular tanks have diameter ranging from 3 to 60 m (CPHEEO 2013). The ratio of tank diameter to the side water depth should be less than 5 (Tirthakar 2022). Circular tanks may have either central feed or periphery-feed arrangement. The tanks are provided with scrappers to remove the settled sludge from the bottom. The scrapper rotates at 1–2 revolutions per hour with peripheral speed of 2.5–4 cm/sec (CPHEEO 2013). The rotating scrapper scrap and push the sludge at the bottom of the tank towards the sludge sump from where it is pumped out. In circular SSTs provided with centre-feed system, the diameter of the inlet pipe is about 15–20% of the tank diameter (Tirthakar 2022). Also, to avoid scouring of the deposited sludge, the inlet pipe should not extend more than 1 m below the surface (Tirthakar 2022). The SLR and the surface overflow rate (SOR) are the main parameters considered in the design of SST. The SLR is defined as *the MLSS applied per unit surface area of the SST per unit time.* The SOR is defined as *the sewage flow per unit surface area of the SST*. The SST design is governed by the solids loading rate when the MLSS concentration values are greater than 2000 mg/l. As the performance of the SST is affected by the variations in flow rate, the units are generally designed both for average and peak overflow rates. The surface area of SST is determined at both average and peak flow rates and the higher surface area is considered for design. The other design parameters are depth and weir loading rate. The SST is to be designed considering the dual role of clarification and thickening of sludge. Generally, it is expected that an SST under efficient operation will concentrate the MLSS by approximately 3 times (CPHEEO 2013). The depth of tank should be adequate to satisfy the volume required for both the functions. The depth of SST is in the range of 2.4–4.2 m with a detention time of 2.2–4.2 h (CPHEEO 2013). Longer detention period in SST may cause denitrification which results in poor settling of MLSS. Another issue is that during the longer detention period, the live MLSS in the SST end up consuming the residual dissolved oxygen in the treated wastewater. Subsequently, these MLSS will face deficiency of oxygen until they reach the aeration tank. The oxygen deficiency hinders their metabolism within the aeration tank which affects the process efficiency. The weir loading or the weir overflow rate represents the wastewater flow per unit length of the weir. High weir overflow rates cause high velocities at the outlet. High exit velocities tend to cause turbulence in the settling zone which causes the flocs to be drawn into the effluent. The loading should be such that the clarified effluent is uniformly withdrawn over the entire perimeter of the SST. The design guidelines for SST for ASP are given in Table 20.2.

Table 20.2 Design guidelines for SST for activated sludge process

Overflow rate m^3/m^2/day		Solid loading kg/day/m^2		Side water depth m	Weir loading m^3/m^2/day
Average	Peak	Average	Peak	Average	Average
15–35	40–50	70–140	210	≥ 3.0–3.5	185

20.6.4 *Sludge Production*

The oxidation of organic matter by microorganism results in the generation of new cells which are separated out as sludge in the SST. The rate of sludge produced is given as:

$$P_x = \frac{Y \cdot Q(S_0 - S)}{1 + k_d \theta_c} \quad (20.12)$$

where,

P_x = sludge production $\left(\frac{\text{kg}}{\text{day}}\right)$

20.6.4.1 Sludge Volume Index

The sludge volume index (SVI) is determined experimentally and is defined as *the volume occupied in ml by one gram of solids in the mixed liquor after settling for 30 min.* The SVI is given as:

$$\text{SVI(ml/g)} = \frac{1000 \cdot V_s}{X_t} \quad (20.13)$$

where,

V_s = Volume of settled sludge (ml/l).

SVI indicates the settling characteristics of the MLSS in the system. A value of SVI < 80 ml/g indicates dense sludge which will settle quickly. An SVI value between 100 and 150 ml/g indicates a relatively slow settling sludge removing more solids during the settling process (CPHEEO 2013). An SVI > 150 ml/g is typically linked with growth of filamentous microorganisms resulting in sludge with poor settling characteristics (CPHEEO 2013). A value over 200 definitely indicates sludge bulking (Rittmann and McCarty 2020). It is recommended that the SVI value be determined daily to check the state of the sludge. Based on the obtained SVI value, the plant operator should decide whether to increase or decrease the SVI value. An increase in the SVI value is achieved by decreasing the return sludge rate. On the other hand, a decrease in the SVI value is achieved by increasing the return sludge rate.

20.6.4.2 Sludge Recycle

One of the vital components of the conventional ASP is the sludge recycling from the SST. The activated sludge that is returned to the aeration tank is known as return activated sludge (RAS). The RAS is required (i) to maintain the required MLSS concentration in the aeration tank and (ii) to have θ_c value greater than HRT. The sludge recycling is done by pumping the sludge through a separate sludge recycle line. The amount of sludge to be recycled depends on the sludge settleability in SST. The higher the concentration of solids in the RAS, the lower recycling rate is required to maintain the desired MLSS concentration. The recycling rate can be calculated by doing a mass balance on the complete mix ASP system (Fig. 20.2) under steady state and neglecting the solids concentration in the incoming sewage. The mass balance gives:

$$Q_r \cdot X_r = (Q + Q_r) \cdot X_t$$

Rearranging the above equation,

$$R = \frac{Q_r}{Q} = \frac{X_t}{X_r - X_t}$$

Assuming that there is no difference in the settling nature of solids in laboratory and the settling of solids on field in SST, the value of X_r can be obtained from the SVI value calculated in laboratory as below:

$$X_r = \frac{10^6}{\text{SVI}}(\text{mg/l})$$

$$R = \frac{Q_r}{Q} = \frac{X_t}{\left(\frac{10^6}{\text{SVI}} - X_t\right)} \tag{20.14}$$

where,

R = Recirculation ratio
Q_r = RAS recycle rate (m³/d)
X_r = MLSS in the RAS or excess sludge (mg/l).

20.6.4.3 Excess Sludge Wasting

The remaining portion of the settled sludge apart from the RAS is known as excess sludge. The excess sludge is wasted so that the quantity of biomass produced due to bacterial growth is balanced by wasting an equivalent quantity of sludge and the system operates in steady state. For a conventional ASP, the excess sludge to be wasted will be about 0.35–0.5 kg/kg BOD removed for domestic wastewater (CPHEEO 2013). If the excess sludge is not wasted, the concentration of solids continuously increases in the SST. This results in overloading of solids in the SST. The SST becomes inefficient in settling the solids causing an increase in the thickness of the sludge layer. As the thickness of the sludge layer increase beyond a certain level, the solids are directly drawn into the effluent resulting in a poor-quality effluent. The interval between sludge removals should be less than 4 h and never exceed 12 h (CPHEEO 2013). There are two methods to waste the excess sludge depending on the location of removal (i) from the underflow of SST or (ii) as mixed liquor from the aeration tank. The excess sludge is sent to a sludge thickener. The thickened sludge is then sent to a sludge digester. The excess sludge concentration and flow can be determined as:

(i) **Removal of the Excess Sludge from the Aeration Tank**

This method is known as the hydraulic control of sludge age (Chen et al. 2020). The solids concentration in excess sludge is equal to the MLSS concentration (i.e. $X_r = X_t$). Thus, for a constant θ_c, Q_w is given as:

$$Q_w = \frac{V}{\theta_c} \tag{20.15}$$

(ii) **Removal of the Excess Sludge from the SST Underflow**: In this method, the solids are removed from the SST underflow via the sludge recirculation line. The solids concentration in the excess sludge is same as the solids concentration in the recycled sludge (i.e. X_r). Thus, for a constant θ_c, Q_w is given as

$$Q_w = \frac{V \cdot X_t}{\theta_c \cdot X_r} \tag{20.16}$$

20.7 Operational Problems in ASP

20.7.1 Bulking Sludge

Activated sludge showing poor settling property is known as bulking sludge. Bulking refers to the flocs that settles slowly and compacts poorly. As the bulking sludge tend to remain in suspension, the concentration of solids in the settled sludge is very low. Therefore, a higher sludge recycle rate is required to maintain a steady MLSS concentration in the aeration tank. If the sludge is not removed quickly, then the size of the sludge layer will continue to increase until it occupies the entire SST. In this situation, the MLSS is discharged directly into the effluent without settling. This results in rapid decrease in the concentration of MLSS from aeration tank causing decrease in the SRT. The decrease in SRT will increase the concentration of SS and BOD in the effluent resulting in poor effluent quality. Bulking sludge if not managed in time can lead to failure of the entire ASP system. The bulking of sludge can be either filamentous of non-filamentous (viscous bulking) (Minnie et al. 2022). The excessive growth of filamentous bacteria results in the filamentous bulking. The growth of the filamentous bacteria outside of the flocs forms a bridge between the nearby flocs which hinders the flocculation causing poor settling. Bulking sludge can be identified by SVI and SS concentrations in the returned sludge. The SVI has a direct correlation to sludge bulking. An SVI value > 200 ml/g indicates sludge bulking (Rittmann and McCarty 2020). Similarly, low value of SS concentration in the return sludge is an indicator of bulking sludge. Filamentous bulking of sludge is generally due to low dissolved oxygen (DO), high F/M ratio, and presence of reduced sulphur compounds in the effluent from aeration tank. The remedy to mitigate the different types of bulking depends on the primary cause. Therefore, it is necessary to identify the primary reason for bulking before applying any remedy. Bulking due to low DO is caused by filamentous bacteria such as Type 021N, *Sphaerotilus natans*, and Type 1701 (Rittmann and McCarty 2020). The remedy is to increase the DO concentration in the effluent from aeration tank. Higher DO concentrations can be achieved by enhancing the capacity of the aerators or by reducing the BOD load on the aeration tank. The bulking due to low-F/M ratio occurs due to long SRT. The filamentous species responsible for bulking due to low F/M ratio include Type 0581, *Microthrix paravicella,* Type 0092, *Haliscomenobacter hydrosis*, and Type 0041 (Rittmann and McCarty 2020). This type of bulking can be manged by adjusting the F/M ratio and reducing the SRT to less than 6 days. The third type of bulking occurs when the influent to ASP has high concentrations of reduced sulphur compounds (mostly sulphides). This type of bulking is caused primarily by sulphur-oxidising species such as *Thiothrix* and 021N (Rittmann and McCarty 2020). Exclusion of influent water rich in sulphide compounds helps avoid this type of bulking. Similar to reduced sulphur, carbohydrate rich waters are also prone to give rise to filamentous microorganisms. Exclusion of such wastewaters therefore help in controlling filamentous sludge bulking. Viscous (non-filamentous) bulking is caused by the excessive growth of EPS by Zoogloea species (Minnie et al. 2022). The high concentration of EPS results in electrostatic repulsion amongst the flocs due to the presence of negative charge on the EPS. This repulsion hinders the flocculation process producing a poor settling sludge. A recent study (Li et al. 2020) has shown the role of EPS in damaging the floc structure and stability resulting in bulking. The excessive growth of filamentous bacteria brought changes in the EPS structure resulting in decrease in protein content and increase in polysaccharides content. The altered EPS structure resulted in flocs with poor settling ability. An effective method to control the growth of filamentous microorganisms is addition of chlorine. The addition of chlorine at an appropriate dose selectively kills the filamentous microorganisms. Attention should be paid to the dose of chlorine. Too little chlorine will be ineffective whereas too much chlorine will end up killing the beneficial floc forming microorganisms causing poor settling. As a rule of thumb 5 kg as Cl_2 per 1000 kg MLSS is added per day till the bulking problem persists (Rittmann and McCarty 2020). The chlorination should be discontinued once the bulking problem is over.

20.7.2 Rising Sludge

Rising sludge refers to a state where the sludge in SST tend to float or rise to the surface. Rising sludge can be identified by observing the occurrence of tiny gas bubbles along with the floating solids. The problem of rising sludge is observed when denitrification occurs in the SST. During denitrification, the nitrites and nitrates in the wastewater are converted to nitrogen gas. The nitrogen gas bubbles are entrapped within the sludge mass. Prolonged denitrification causes formation of sufficient nitrogen gas to make the sludge mass buoyant. This buoyant sludge mass rises to the surface thereby deteriorating the quality of the effluent from the SST. Denitrification in the SST occurs when the influent to the SST has nitrate concentrations of 6–8 mg/l. High temperatures (above 20 °C) do accelerate the rate of denitrification. Therefore, nitrates concentration more than 6–8 mg/l in the influent of SST along with high temperatures can result in rising sludge even in the normal detention period. Treatment plants located in warm climate areas are more vulnerable to this problem. The viable solution to this problem is to denitrify the effluent of the aeration tank before allowing its entry into the SST.

20.7.3 Foaming

Foaming refers to the formation of foam or scum on the surface of aeration tanks. The microorganisms responsible for causing foaming include (i) *Nocardia sp.,* (ii) *Microthrix,* (iii) type 1863 (Tirthakar 2022). Rapid growth of *Nocardia* is caused by warm temperature, presence of grease, oil, and fat in the wastewater and longer SRTs (generally greater than 9 days).

Foaming causes reduction in oxygen transfer, poor quality of effluent and requires additional maintenance. As the foaming occurs with a long SRT, it indicates that the microorganisms responsible for foaming grow slowly. Therefore, foaming can be controlled by reducing the SRT to less than 6 days. Another practise is to chlorinate the RAS. Since the foam and scum gets accumulated on the surfaces of the SST, an effective remedy to control foaming is to frequent removal of foam and scum from the surface of the tanks.

20.8 Modifications of Conventional ASP System

In addition to the operational problems, the other disadvantage of conventional ASP is related to the oxygen requirement and aeration. In the conventional ASP, the organic load is higher at the inlet and lower at the outlet whereas the aeration is more or less uniform over the length of the tank. Therefore, there is a possibility that the available oxygen at the inlet may be insufficient for the degradation of organic matter. The oxidation reaction may be limited by the availability of oxygen. The deficiency of oxygen may adversely affect the process efficiency. To improve the overall efficiency of the system and to achieve economy in the operation, various modifications to the conventional ASP were developed. The modified processes are listed as under:

(i) Step aeration
(ii) Tapered aeration
(iii) Complete mix process
(iv) Extended aeration
(v) Contact stabilisation
(vi) Aerobic granular sludge (AGS)
(vii) Sequencing batch reactor (SBR).

20.8.1 Step Aeration

In step aeration process (Fig. 20.3), the returned sludge from the SST enters at the head of aeration tank and is mixed with only a part of the influent. The remaining influent is distributed along the length of the aeration tank in steps. This configuration enables uniform oxygen demand over the length of the aeration tank and efficient use of supplied air. This configuration has significant capacity to absorb shock organic loadings. The configuration results in significant reduction in the volume of aeration tank for achieving similar BOD removal efficiency.

20.8.2 Complete Mix Process

The complete mix ASP (Fig. 20.2) operates as a completely mixed flow reactor. Mechanical aerators are used to achieve complete mixing in the aeration tanks. The completely mixed process allows the operation of the system at a high MLSS

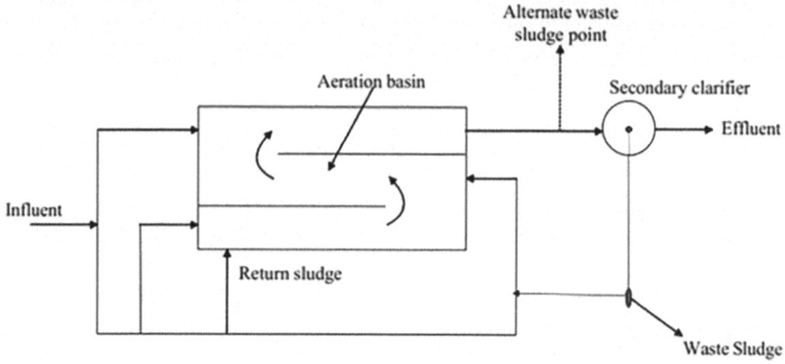

Fig. 20.3 Step aeration

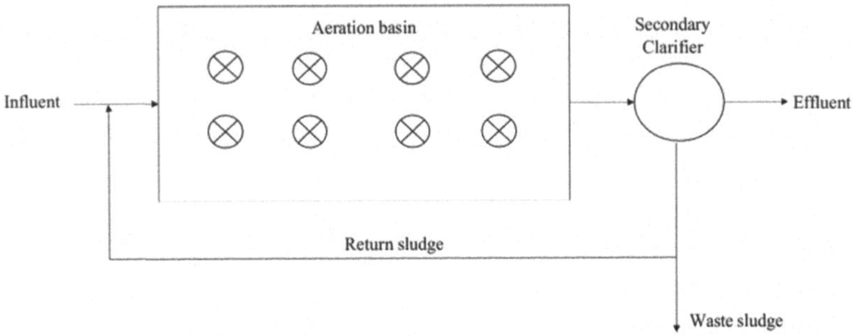

Fig. 20.4 Tapered aeration

concentration resulting in the decrease in volume of aeration tank. The complete mix process remains stable under shock loadings and has increased capacity to treat toxic biodegradable wastes like phenols (CPHEEO 2013). Also, the use of micro bubble and nano bubble aeration has shown to improve the oxygen transfer efficiency with lower energy consumption as compared to the conventional aeration (Zhou et al. 2022).

20.8.3 Tapered Aeration

In conventional ASP systems, the air is distributed uniformly over the entire volume of the aeration tank. It is observed that in the conventional system, as the MLSS pass through the aeration tank, the DO demand is high at the inlet and tapers out near the exit. The tapered aeration system (Fig. 20.4) was developed to match the DO requirement. In this system, the quantity of air supplied is proportional to the demand exerted by the microorganisms as the MLSS pass through the aeration tank. To meet the high oxygen demand near the inlet of the aeration tank, air is supplied at a high rate. The air supply rate is reduced gradually to match the reduced oxygen demand. The advantage of this configuration is that supply of air is optimised with respect to the demand thus avoiding excessive aeration. Optimum supply of air will prove to be economical and will also inhibit the growth of nitrifying organisms in the system.

20.8.4 Extended Aeration

Extended aeration system (Fig. 20.5) is provided for treatment of low-volume wastewaters where conveyance of the wastewater to a sewage treatment plant may not be economical and practically feasible. One major advantage of this configuration is that primary sedimentation is not required here. The extended aeration configuration

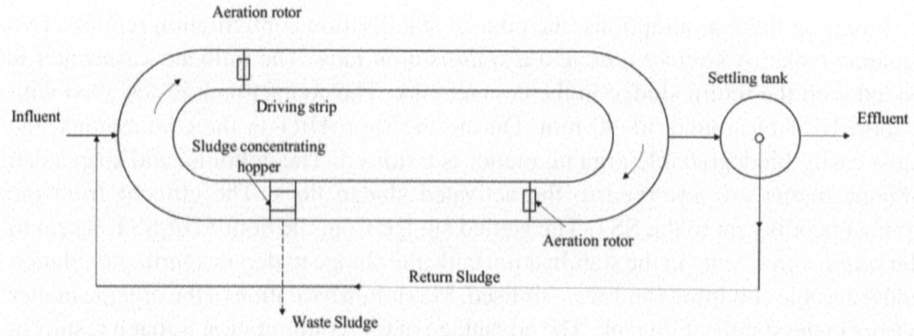

Fig. 20.5 Extended aeration

requires large aeration tanks. The configuration is operated as a completely mixed system. Key features of the extended aeration process include, long HRT (around 24 h), high sludge age (θc), high MLSS concentration (around 3000–6000 mg/l), and low F/M ratio (CPHEEO 2013). Due to long HRT, the MLSS in the aeration tank go through endogenous respiration resulting in a stabilised sludge. The endogenous respiration results in minimum production of excess sludge. Thus, the process does not require the elaborate sludge digestion system as in conventional ASP. The major disadvantage is large land requirement for the aeration tank.

20.8.5 Contact Stabilisation

In this configuration (Fig. 20.6), it is assumed that the oxidation of organic matter is a two-step process:

(i) The organic matter is first adsorbed by the microorganisms.
(ii) The adsorbed organic matter is then oxidised by the microorganisms.

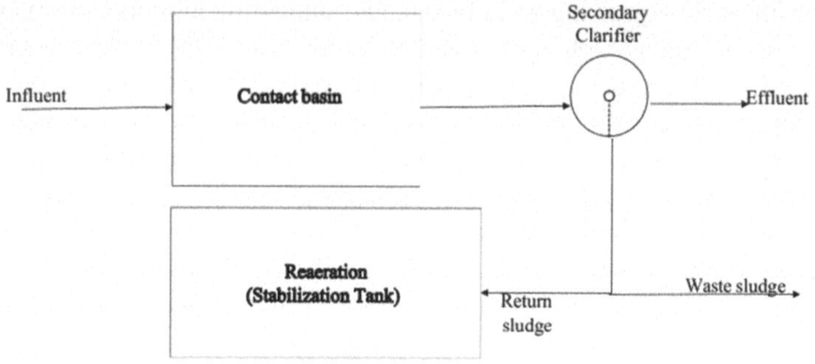

Fig. 20.6 Contact stabilization process

Based on these assumptions, the contact stabilisation configuration requires two separate tanks: A *contact tank* and a *stabilisation tank*. The influent wastewater is mixed with the return sludge in the contact tank. The contact tank is designed with a short HRT of around 30–90 min. During the short HRT in the contact tank, the most easily biodegradable organic matter is oxidised. The colloidal and suspended organic matter are adsorbed to the activated sludge flocs. The effluent from the contact tank is sent to the SST. The settled sludge from the bottom of SST is sent to the *stabilisation tank*. In the stabilisation tank, the sludge undergoes further oxidation under aerobic conditions and are stabilised. Maximum oxidation of the organic matter occurs in the stabilisation tank. The advantage of this configuration is that it results in approximately 50% saving in tank volume (Tirthakar 2022). The contact stabilisation process removes the colloidal and suspended organic matter effectively. A major disadvantage of this process is that it the operation of the process requires highly skilled labour and continuous monitoring as there two mixed liquors that need to be monitored in two different tanks (Metcalf and Eddy 2014). Also, the process is vulnerable to sudden increase in loading.

20.8.6 Aerobic Granular Sludge

The AGS technology is a biofilm-based biotechnology that does not require any media for attachment (Ekholm et al. 2022). The AGS was first reported by Morgenroth et al. (1997), in which granules of 2–8 mm in diameter were formed from aerobic activated sludge. Thereafter, several researchers studied and added valuable insights to the AGS and subsequently the AGS technology was adopted for full-scale application for waste water treatment under the commercial name Nereda® (Semaha et al. 2023). The key feature of the AGS is the high-density and large diameter ($\geq$ 200 μm) granules which results in excellent solid–liquid separation (Semaha et al. 2023). The AGS is formed from the activated sludge flocs in a sequencing batch reactor (SBR) (de Kreuk et al. 2005; Rashid et al. 2020). The SBR is operated with anaerobic, aerobic, and short settling phases in the cycle. These operating conditions result in the growth of slow-growing and glycogen accumulating microorganisms such as nitrifying bacteria and polyphosphate accumulating bacteria (Nancharaiah and Kiran Kumar Reddy, 2018; Nancharaiah and Sarvajith, 2019). These microorganisms form dense granules and have a high settling velocity than that of the activated sludge flocs. The dense microbial structure, large granule size, and aerobic respiration in the exterior of the granule results in the presence of aerobic, anoxic and anaerobic conditions in a single granule even under external aeration. The presence of the different micro environments makes it possible to remove nitrogen, organic carbon, and phosphorus simultaneously from wastewater. The exact mechanism of the AGS formation is not clearly known. A proposed mechanism for the AGS granulation process involves four distinct stages: (Nancharaiah and Kiran Kumar Reddy 2018; Ali et al. 2021; Setianingsih et al. 2024a).

Stage 1: The hydrodynamic and thermodynamic forces result in the microbe-to-microbe contact forming aggregates.
Stage 2: The chemical bonding, physical forces such as van der Waal forces, and biochemical factors produce aggregates.
Stage 3: The secreted EPS act as a glue for the bonding of aggregates.
Stage 4: The 3D structure of the granule is sustained by the hydrodynamic shear forces.

There are several advantages of the AGS like 25–75% reduction in land area requirement, 20–50% reduction in energy consumption, decreased sludge handling, enhanced nutrient removal, improved ability to sustain shock loadings (Bengtsson et al. 2019; Nancharaiah et al. 2019; We et al. 2020; Hamza et al. 2022; Miyake et al. 2023; Setianingsih et al. 2024b; Vydehi et al. 2024; Yu and Wang 2024). Also, there is possibility of resource recovery from AGS in the form of phosphorus compounds, alginate-like exopolymers, biodegradable plastics, and energy (De Sanctis et al. 2020).

20.8.7 Sequencing Batch Reactor (SBR)

The SBR is a variation of the conventional ASP. The unit processes in the SBR are similar to the conventional ASP systems. The SBR technology is based on a 'fill and-draw type' treatment process as against the continuous process in conventional ASP. The primary difference amongst the two systems is that the SBR performs the oxidation of organic matter and solid–liquid separation in a single tank in batch mode in a time-controlled sequence whereas in a conventional ASP all the process happens simultaneously in different tanks in continuous mode. Thus, the SBR performs the same treatment in time what the conventional ASP does in space. The U.S. EPA (1999) has aptly stated that 'the SBR is no more than an activated sludge system which operates in time rather than in space'. It is a misconception that the SBR is a recent technology. In fact, the 'fill and draw processes' similar to SBR were popular in the early twentieth century (Dutta and Sarkar 2015). The use of the term SBR can be attributed to R.L. Irvine (1979) who first introduced this term. As the technology related to aeration and process control improved in the late 1950s, significant increase in the use of SBR technology occurred during this period. The SBR-based treatment systems are capable to remove higher proportion of BOD, COD, ammonium, and phosphorous from the wastewater as compared to conventional ASP (Alagha et al. 2020; Bărbulescu and Barbeş 2021; Sánchez et al. 2021).

20.8.7.1 Operating Principles

The SBR is a batch reactor that operates in five steps: Fill, react, settle, decant, and idle (von Sperling 2007; Askari et al. 2024).

Fill: In this first step, the wastewater is added in the tank and mixed with the biomass (activated sludge) remaining in the tank after the preceding cycle. Depending on the operation of mechanical mixing and aeration the fill phase can be operated in the following three modes:

Static Fill—In static fill, the influent wastewater enters the tank without mixing or aeration. This type of fill is used during the initial start-up phase of a facility and at plants designed for nutrient removal. As there is no mixing or aeration involved, the static fill results in reduction in energy saving.
Mixed Fill—In mixed fill, as the influent wastewater enters the tank it is subjected to mixing by mechanical mixers. However, the aerators are kept off. The mechanical mixing result in a uniform mixing of the influent wastewater and biomass. As there is no aeration, an anoxic or anaerobic condition is maintained in the tank.
Aerated Fill—In aerated fill, the influent wastewater is subjected to both the mechanical mixing and aeration. The contents of the tank are aerated and complete aerobic condition is maintained in the tank.

The system can alternate between the three modes during the fill cycle. The fill period operates for approximately 25% of the entire duration.

React: The influent wastewater flow into the tank is restricted during the react stage. The organic matter oxidation reaction that began during fill stage is completed during this stage. Aerobic, anoxic, or anaerobic conditions can be achieved in react stage by controlling the aeration. The desired level of treatment can be attained by varying the duration of aeration and mixing. The react stage operates for more than 50% of the entire cycle duration.

Settle: The objective of this stage is to achieve solid–liquid separation. In this stage, there is no inflow and outflow of wastewater from the tank. The aeration tank operates similar to the SST of the conventional ASP. The settle stage lasts for about 30–90 min.

Draw (decant): In this stage, the clarified supernatant is disposed as effluent from the aeration tank. There are several mechanisms available for discharging the effluent. The draw stage lasts typically for about 45 min which is 5 to 30% of the entire cycle duration.

Idle: This is final phase of the cycle. The idle stage is used only when there are multiple SBR tanks in use. The idle stage is the period between the draw and fill stages. When multiple SBRs are in operation, the idle phase is required to balance the working cycle of one tank with the working cycle of another tank. The excess sludge is also wasted during this stage.

The SBR systems are generally automated so that all five steps are controlled automatically. Recent study (Kwon et al. 2023) has applied the IoT system to the conventional SBR processes. The use of IoT system has reduced the cycle time by approximately 22%. The SBRs are so designed that each aeration tank can be shut off separately for maintenance purposes without affecting the operation of other aeration tanks.

Table 20.3 Design parameters for SBR

Process parameter	Unit	Continuous flow and intermittent decant	Intermittent flow and intermittent decant
MLSS	mg/l	3000–4000	3500–5000
F/M	day^{-1}	0.05–0.08	0.05–0.3
θ_c	days	15–20	4–20
Sludge yield	kg dry solids/kg BOD	0.75–0.85	0.75–1.0
Cycle time	Hours	4–8	2.5–6
Settling time	Hours	> 0.5	> 0.5
Decant depth	m	1.5	2.5
Fill volume base	–	Peak flow	Peak flow
kg O_2/kg BOD removed	Ratio	1.1	1.1
kg O_2/kg TN removed	Ratio	4.6	4.6

Table 20.4 Design parameters for conventional SBR

Process parameter	Unit	Municipal	Industrial
MLSS	mg/l	2000–2500	2000–4000
F/M	day^{-1}	0.15–0.4	0.15–0.6
Cycle time	Hours	4	4–24
HRT	Hours	6–14	Varies

20.8.7.2 Design Criteria

The general design guidelines for the SBR as per CPHEEO (2013) are given in Table 20.3. Also, the key parameters for the design of SBR for municipal and industrial wastewater as per U.S. EPA (1999) are given in Table 20.4.

Example: Design a SBR of 50 MLD capacity to achieve the effluent quantity as per given details. Assume 4 SBR to be provided.

Parameter	Influent (mg/l)	Effluent (mg/l)
BOD	200	≤ 10
COD	450	≤ 50
MLSS	3200	≤ 10
TKN	40	≤ 5
TN	5	≤ 1

Solution:

S. No	Detail	Value	Unit	Remarks
1	**Flow calculation**			
1	Given flow	50.00	MLD	
		50,000.00	m^3/day	
		2083.33	m^3/hr	
2	Wastewater flow to each tank	520.83	m^3 /hr	Assuming 4 SBR
2	**SBR operating cycle**			
1	Fill and aeration period	2.00	hrs	Assumption
2	Settle	1.00	hr	Assumption
3	Decant	1.00	hr	Assumption
4	Total cycle time	4.00	hrs	Assumption
5	Idle	20.00	min	Assumption
6	Total number of tanks provided	4.00		Given
7	No. of tanks receiving flow simultaneously	2.00		Assumption
8	No. of tanks aerated simultaneously	2.00		Assumption
9	MLSS in aeration tank (X_t)	3200.00	mg/l	Given
10	MLVSS in aeration tank	2560	mg/l	Assumption: 80% of MLSS
11	Decant depth	0.30		Assumption: 30% of tank depth
12	SVI	150.00	ml/g	Assumption
13	No. of cycles/tank/d	6.00	cycles	
14	Total no. of cycles in a day	24.00	cycles	
15	Fill volume per cycle	2083.33	m^3/fill	
3	**Volume determination**			
	Using F/M ratio (Method 1)			
1	F/M ratio	15		Assumption
2	Volume	26,041.67	m^3	Using $\frac{F}{M} = \frac{Q}{V}.\frac{S_0}{X_t}$
3	Volume required per tank	6510.42	m^3	
	Method 2			
4	Volume exchange ratio	0.3		Assumption
5	Volume per tank	6944.44	m^3	V = Fill volume per Cycle/0.3
6	Choosing higher value amongst both volume (V_T)	6944.44	m^3	
4	**HRT calculation**			
1	Full liquid depth	5.50	m	Assumption
2	Decant depth	1.65	m	0.3 * 5.5

(continued)

(continued)

S. No	Detail	Value	Unit	Remarks
3	V_T	6944.44	m^3	From Sr no 3(6)
4	HRT	13.33	hrs	(4 * V_T)/50,000
5	**Dimensions of each SBR tank**			
1	V_T	6944.44	m^3	
2	Side water depth	5.50	m	Assumption
3	Area of each tank	1262.63	m^2	
4	Length of tank	40.00	m	Assumption
5	Width of the tank	31.57	m	
6	Free board	0.80	m	Assumption
7	Depth of tank	6.30	m	5.50 + 0.80
8	Provide four SBR tanks each of size	**40 m x 32 m x 6.3 m**		
6	**Oxygen requirement**			
	Oxygen required for BOD removal			
1	S_0	200.00	mg/l	Given
2	S	10.00	mg/l	Given
3	BOD to be removed	190.00	mg/l	
4	Total BOD to be removed	9500.00	kg/day	
5	Assuming theoretical O_2 required	1.20	kg O_2/kg of BOD removed	
6	Oxygen required for BOD removal	11,400.00	kg/day	**A**
	Oxygen required for nitrogen removal			
7	Nitrogen assimilation in bacterial cell produced	5.00	Assumption %/kg BOD removed in bacterial cell	
8	Nitrogen assimilation	475.00	kg	
9	Inlet nitrogen	2000.00	kg	
10	Outlet nitrogen	250.00	kg	
11	Nitrogen to be removed	1275.00	kg/day	
12	Theoretical O_2 required for N_2 removal	4.60	kg O_2/kg NH_3–N removed	
14	Total oxygen requirement	17,265.00	kg	**A + B**
7	**Sludge wasting and pump design**			
1	Sludge generation rate	0.40	kg/kg BOD removal assumption	
2	Excess sludge generated	3800.00	kg/d	
3	SVI	150	ml/g	
4	Total volume of sludge wasted	570.00	m^3/day	SVI*excess sludge generated
5	Sludge wasted per tank	142.50	m^3/day	

(continued)

(continued)

S. No	Detail	Value	Unit	Remarks
6	Sludge wasted per tank per cycle	23.75	m^3/day	Sludge wasted/no. of cycles/tank/d
7	Pump running period per cycle	15.00	mins	Assumption
8	Pump capacity required	95.00	m^3/day	
9	Solids retention time (SRT)	18.71	days	(V_T xMLVSS)/(sludge wasted × 1000)
8	**Decant rate**			
1	Time for decanting	0.5	hr	
2	Decant rate	6250.00	m^3/hr	(Q × 1.5)/0.5
		104.17	m^3/min	

20.9 Conclusion

Aerobic suspended treatment processes continue to be the mainstay of domestic wastewater treatment. Low chemical and energy consumption, easy operation process, potential for resource recovery are some of the key features of these processes. However, these processes do have some drawbacks like large land footprint, large volumes of sludge handling, higher operation costs, and ineffective biological nutrient removal. Continuous research in recent years have led to the development of several new modifications to the traditional aerobic wastewater treatment processes. Most of the research has focused on developing different process modifications with an objective to improve the overall treatment efficiency. Technological advance in sample characterisation techniques has allowed the use of computer modelling, genetic, and microbiological studies to further understand the inherent processes and microbial ecology. The new findings are also transferred to the field resulting in the upgradation and retrofitting of the conventional wastewater treatment plants and construction of new plants based on the modified technologies like SBR and AGS. However, some technological gaps still remain to be addressed and more research is required to resolve the various operational problems. Also, the future research needs to focus on the new methods for the reduction in sludge volume and valorisation of the waste sludge to recover value added products.

References

Alagha O et al (2020) Suitability of SBR for wastewater treatment and reuse: pilot-scale reactor operated in different anoxic conditions. Int J Environ Res Public Health 17(5).https://doi.org/10.3390/ijerph17051617

Ali NSA et al (2021) Initialization, enhancement and mechanisms of aerobic granulation in wastewater treatment. Sep Purif Technol. Available at: https://doi.org/10.1016/j.seppur.2020.118220

Arceivala SJ, Asolekar SR (2009) Wastewater treatment for pollution control and reuse, 3rd edn. McGraw-Hill

Askari SS et al (2024) Enhancing sequencing batch reactors for efficient wastewater treatment across diverse applications: a comprehensive review. Environ Res. Available at: https://doi.org/10.1016/j.envres.2024.119656

Bărbulescu A, Barbeş L (2021) Statistical methods for assessing water quality after treatment on a sequencing batch reactor. Sci Total Environ 752. https://doi.org/10.1016/j.scitotenv.2020.141991

Bengtsson S et al (2019) A comparison of aerobic granular sludge with conventional and compact biological treatment technologies. Environ Technol (UK) 40(21):2769–2778. Available at: https://doi.org/10.1080/09593330.2018.1452985

Chen GH et al (2020) Biological wastewater treatment. IWA Publishing

CPHEEO (2013) Manual on sewerage and sewage treatment systems, 3rd edn. Central Public Health & Environmental Engineering Organisation, Ministry of Housing and Urban Affairs, Government of India

de Kreuk M, de Bruin LMM, van Loosdrecht M (2005) Aerobic granular sludge—from idea to pilot plant. In: Bathe S et al (eds) 1st IWA-workshop aerobic granular sludge, 1st edn. IWA Publishing, pp 111–117

De Sanctis M et al (2020) Aerobic granular-based technology for water and energy recovery from municipal wastewater. New Biotechnol 56:71–78. https://doi.org/10.1016/j.nbt.2019.12.002

Dutta A, Sarkar S (2015) Sequencing batch reactor for wastewater treatment: recent advances. Curr Pollution Rep 177–190. https://doi.org/10.1007/s40726-015-0016-y

Ekholm J et al (2022) Full-scale aerobic granular sludge for municipal wastewater treatment—granule formation, microbial succession, and process performance. Environ Sci: Water Res Technol 8(12):3138–3154. https://doi.org/10.1039/d2ew00653g

EPA (1999) Wastewater technology fact sheet sequencing batch reactors. Washington, D.C.

Hamza R et al (2022) A review of the state of development of aerobic granular sludge technology over the last 20 years: full-scale applications and resource recovery. Case Stud Chem Environ Eng 5.https://doi.org/10.1016/j.cscee.2021.100173

Irvine RL (1979) Sequencing batch biological reactors—an overview. J-Water Pollut Control Fed. http://www.jstor.org/page/info/about/policies/terms.jsp

Kwon H et al (2023) Optimization of a sequencing batch reactor with the application of the Internet of Things. Water Res 229. https://doi.org/10.1016/j.watres.2022.119511

Latif EF (2022) Applying novel methods in conventional activated sludge plants to treat low-strength wastewater. Environ Monit Assess 194(4).https://doi.org/10.1007/s10661-022-09968-9

Li WM et al (2020) New insights into filamentous sludge bulking: the potential role of extracellular polymeric substances in sludge bulking in the activated sludge process. Chemosphere 248.https://doi.org/10.1016/j.chemosphere.2020.126012

Lu X et al (2023) A review of filamentous sludge bulking controls from conventional methods to emerging quorum quenching strategies. Water Res. https://doi.org/10.1016/j.watres.2023.119922

Metcalf W, Eddy C (2014) Wastewater engineering treatment and resource recovery, 5th edn

Minnie J et al (2022) Modelling filamentous bacteria in activated sludge systems and the advancements of secondary settling tank models: a review. Biochem Eng J. https://doi.org/10.1016/j.bej.2022.108598

Miyake M et al (2023) Pilot-scale demonstration of aerobic granular sludge augmentation applied to continuous-flow activated sludge process for the treatment of low-strength municipal wastewater. J Water Process Eng 51.https://doi.org/10.1016/j.jwpe.2022.103392

Morgenroth E et al (1997) Aerobic granular sludge in a sequencing batch reactor. Wat Res 31(12):3191–3194

Nancharaiah YV, Kiran Kumar Reddy G (2018) Aerobic granular sludge technology: mechanisms of granulation and biotechnological applications. Bioresource Technol 1128–1143. https://doi.org/10.1016/j.biortech.2017.09.131

Nancharaiah YV, Sarvajith M (2019) Aerobic granular sludge process: a fast growing biological treatment for sustainable wastewater treatment. Curr Opn Environ Sci Health 57–65. https://doi.org/10.1016/j.coesh.2019.09.011

Nancharaiah YV, Sarvajith M, Mohan TVK (2019) Aerobic granular sludge: the future of wastewater treatment. Curr Sci 3(3):395–404

Orhon D (2015) Evolution of the activated sludge process: the first 50 years. J Chem Technol Biotechnol 90(4):608–640. https://doi.org/10.1002/jctb.4565

Rashid SS, Liu YQ, Zhang C (2020) Upgrading a large and centralised municipal wastewater treatment plant with sequencing batch reactor technology for integrated nutrient removal and phosphorus recovery: environmental and economic life cycle performance. Sci Total Environ 749. https://doi.org/10.1016/j.scitotenv.2020.141465

Rittmann B, McCarty P (2020) Environmental biotechnology principles and applications,2nd edn. McGraw-Hill

Sánchez JB, Vuono M, Dionisi D (2021) Model-based comparison of sequencing batch reactors and continuous-flow activated sludge processes for biological wastewater treatment. Comput Chem Eng 144.https://doi.org/10.1016/j.compchemeng.2020.107127

Semaha P et al (2023) Transition of biological wastewater treatment from flocculent activated sludge to granular sludge systems towards circular economy. Bioresource Technol Rep 21.https://doi.org/10.1016/j.biteb.2022.101294

Setianingsih NI, Hadiyanto, Budihardjo MA, Yuliasni R, Vistanty H et al (2024a) Effective sludge activation in a sequencing batch reactor (SBR) towards aerobic granular sludge (AGS) technology. Desalin Water Treat 320.https://doi.org/10.1016/j.dwt.2024.100774

Setianingsih NI, Hadiyanto, Budihardjo MA, Yuliasni R, Malik RA et al (2024b) Potential application and strategies of aerobic granular sludge (AGS) technology for wastewater treatment in Indonesia: a review. Desalin Water Treat 320. https://doi.org/10.1016/j.dwt.2024.100756

Sobczyk M et al (2021) Multivariate analysis of activated sludge community in full-scale wastewater treatment plants. Environ Sci Pollut Res 28(3):3579–3589. https://doi.org/10.1007/s11356-020-10684-5

Tirthakar SN (2022) Process and hydraulic design of wastewater treatment plants, 1st edn. Standard Book House, New Delhi

von Sperling M (2007) Activated sludge and aerobic biofilm reactors, 1st edn

Vydehi P et al (2024) Aerobic granular sludge-based sustainable wastewater treatment: process, bottlenecks, and knowledge gap through scientometric perspective. J Hazard Mater Adv. https://doi.org/10.1016/j.hazadv.2024.100462

We ACE, Aris A, Mohd Zain NA (2020) A review of the treatment of low-medium strength domestic wastewater using aerobic granulation technology. Environ Sci: Water Res Technol 6(3):464–490. https://doi.org/10.1039/c9ew00606k

Wiercik P et al (2022) Evaluation of changes in activated sludge and sewage sludge quality by FTIR analysis and laser diffraction. Desalin Water Treat 273:114–125. https://doi.org/10.5004/dwt.2022.28855

Xia Y et al (2018) Diversity and assembly patterns of activated sludge microbial communities: a review. Biotechnol Adv 1038–1047. https://doi.org/10.1016/j.biotechadv.2018.03.005

Yu HQ (2020) Molecular insights into extracellular polymeric substances in activated sludge. Environ Sci Technol 54(13):7742–7750. https://doi.org/10.1021/acs.est.0c00850

Yu C, Wang K (2024) Continuous flow aerobic granular sludge: recent developments and applications. Water Sci Technol 89(5):1155–1178. https://doi.org/10.2166/wst.2024.055

Zhou S et al (2022) Microbubble- and nanobubble-aeration for upgrading conventional activated sludge process: a review. Bioresource Technol. https://doi.org/10.1016/j.biortech.2022.127826

Part IV
Purification Satiations and Valorization of Treated Wastewater, Sludge, Biogas

Chapter 21
Photosynthetic Biogas Upgrading with Microalgae Coupled with Nutrient Recovery from Digestate

Edwin G. Hoyos, Laura Vargas-Estrada, Raúl Muñoz, and María del Rosario Rodero

Abstract Biogas is generated during the anaerobic digestion of wastewater or biodegradable solid wastes and has been recently considered an important energy vector to achieve the energy transition towards the use of clean energy to limit global warming and meet the long-term decarbonization goals of the world economy. Biogas is mainly composed of methane (40–75%), and other compounds considered as contaminants such as carbon dioxide (15–60%), hydrogen sulphide (0.005–2%), water (3–10%), oxygen (0–1%) and nitrogen (0–2%), which significantly reduce its heating value and/or generate corrosion in engines. Thereby, biogas must be upgraded to biomethane in compliance with international standards (methane $\geq$ 90%, carbon dioxide $\leq$ 2–4% and oxygen $\leq$ 1%), to be marketed as natural gas and ensure its techno-economic viability. In this sense, photosynthetic biogas purification using microalgal-bacterial consortium has positioned as a promising technology due to its environmental benefits and low operation costs, especially when integrated with digestate treatment since nutrients, such as nitrogen and phosphorous, can be recovered as biofertilizers and biostimulants. However, the low aqueous CO_2 solubility and limited microalgae productivity in the photobioreactors are the major drawbacks of this biotechnological platform since they entail the requirement of extensive areas for CO_2 uptake by microalgae, hindering its application on an industrial scale. Recently, the use of nanoparticles during photosynthetic biogas upgrading has become an attractive strategy to significantly boost process performance based on their ability to improve CO_2 reduction and nutrient assimilation by microalgae. This chapter reviews the state of the art of the different methods applied for biogas purification to date. Special interest is devoted to providing a detailed description of the fundamentals, recent progresses and research results on the photosynthetic biogas purification method combined with nutrient valorisation from digestates. Finally, the evaluation

E. G. Hoyos · L. Vargas-Estrada · R. Muñoz · M. R. Rodero (✉)
Institute of Sustainable Processes, University of Valladolid, Valladolid, Spain
e-mail: mariarosario.rodero@uva.es

Department of Chemical Engineering and Environmental Technology, University of Valladolid, Valladolid, Spain

S. Souabi et al. (eds.), *Wastewater Treatment Plants*, Water Science and Technology Library 130, https://doi.org/10.1007/978-3-031-87461-1_21

of the impact of nanoparticles on the yield of photosynthetic biogas purification is discussed.

Keywords Biomethane · Microalgal-bacterial symbiosis · Nanoparticles · Wastewater treatment

21.1 Introduction

Fossil fuels remain to perform a major position in worldwide energy production. In 2019, ~ 84% of the world's primary energy came from coal, oil and natural gas (Ritchie and Rosado 2022). According to McKinsey & Company (2023), global fossil fuel consumption is expected to peak in 2030, but the outlook remains uncertain. The main issue with the fossil fuel consumption is the massive release of CO_2, which is one of the main reasons of global warming. In this regard, the planet's average temperature has increased at a rate of 0.15–0.20 °C per decade since 1975 mainly due to anthropogenic greenhouse gas releases from the fossil fuel consumption (NASA Earth Observatory 2023). In this context, the International Paris Agreement was signed in 2015 with the central objective of keeping the rise in the planet's average temperature < 2 °C during this century and redoubling efforts to limit it to 1.5 °C. The European Union has set the goals of decreasing greenhouse gas productions by at least 55% by 2030 and reaching climate neutrality by 2050, by promoting and accelerating the use of renewable energy (European Council 2023).

Over the past decade, biogas has received increasing attention as a renewable and sustainable fuel (Ferreira et al. 2022). Biogas is normally produced from the anaerobic biodegradation of organic substances from both solid waste and wastewater in the lack of O_2 and in the presence of anaerobic microorganisms. Raw biogas is composed of CH_4 (40–75%), CO_2 (15–60%) and minor levels of other compounds like H_2O (3–10%), hydrogen sulphide (0.005–2%), O_2 (0–1%) and N_2 (0–2%) (Yang et al. 2017; Ryckebosch et al. 2011). Biogas can be employed for heat and electricity generation in industries and, after a strict purification, as a substitute of natural gas or as vehicular fuel (Ferreira et al. 2022; Thiruselvi et al. 2021). The latter represents an economic benefit since the treated biogas can be marketed at the same price or higher than natural gas. However, this route for biogas valorisation has been limited in recent years due to the high chemical and energy demand of conventional biogas upgrading processes, as well as technical and regulatory limitations for the transport and storage of biomethane. This scenario has often resulted in biogas flaring or venting into the environment in small-medium industries, thus contributing to uncontrolled greenhouse gas discharges (Cantera et al. 2020). This fact has encouraged the search for more profitable and sustainable strategies for biogas upgrading.

Microalgae-based photosynthetic biogas upgrading is considered an innovative and very attractive platform technology for biogas upgrading (Meier et al. 2015). This biotechnological platform presents significant economic and environmental benefits compared to traditional physical–chemical processes of biogas upgrading (Rodero et al. 2023; Wu et al. 2020). In addition, this technology can be easily combined with wastewater treatment, as microalgae can uptake N and P from wastewater streams during their growth (Meier et al. 2015; Garlapati et al. 2021). During photosynthetic biogas upgrading through microalgal-bacterial symbiosis, microalgae fix CO_2 from biogas photosynthetically and produce oxygen, which can be utilized by bacteria to eliminate other pollutants such as H_2S from biogas via oxidation. Furthermore, the microalgal-bacterial biomass generated in the system can be employed to produce pharmaceutical, cosmetic, medical, food and energy compounds, improving the profitability and sustainability of this technology (Rodero et al. 2023; Garlapati et al. 2021).

The limited CO_2 gas–liquid mass transfer is a major bottleneck in photosynthetic biogas purification technology (Nagarajan et al. 2019). The low microalgae productivity due to poor CO_2 fixation is one of the factors influencing this latter, which makes the implementation of this biotechnological platform on a large scale difficult due to the need for a large area of land to obtain an effective process performance (Rodero et al. 2023). Key environmental and operational parameters affecting CO_2 mass transfer are the pH and alkalinity of the algal–bacterial culture medium, as well as the liquid–biogas (*L/G*) ratio in the biogas purification column (Nagarajan et al. 2019). Unfortunately, increasing the pH and alkalinity of the scrubbing solution to improve CO_2 absorption entails significant chemical consumption. Several studies on the influence of pH, alkalinity and *L/G* ratio on photosynthetic biogas purification coupled with digestate treatment have been performed over the last decade (Marín et al. 2019, Méndez et al. 2022, Posadas et al. 2017; Rodero et al. 2018). Nevertheless, novel strategies to improve biomethane quality in algal–bacterial photobioreactors without external alkalinity addition are required to make this biotechnology more sustainable and cost-effective (Marín et al. 2021). In this context, nanoparticle supplementation in the photobioreactor culture broth has recently emerged as a potential technique to significantly improve photosynthetic biogas upgrading due to enhanced CO_2 capture by microalgae, while boosting nutrient recovery from wastewater (Hoyos et al. 2024a, b; Vargas-Estrada et al. 2022, 2023a, b). This book chapter will summarize the latest developments in the field of biogas and biomethane, with particular attention on photosynthetic biogas purification boosted by nanoparticles.

21.2 State of the Art of Biogas Purification Techniques

The transformation of raw biogas into biomethane consists mainly of three stages: (1) a cleaning process to remove a major fraction of trace contaminants like H_2S, particles, siloxanes and H_2O that can deteriorate the CO_2 removal equipment, (2) an upgrading process to remove CO_2 and (3) a final polishing step to remove most trace contaminants (H_2S, VOCs, siloxanes, mercaptans, NH_3 and so on) in order to produce a biomethane in accordance with the required Standard specifications (Ryckebosch et al. 2011).

21.2.1 H_2S Removal

H_2S must be removed from biogas based on occupational safety reasons, equipment and pipeline protection, and biogas quality (Cattaneo et al. 2023). European regulations require negligible levels of H_2S in biomethane ($H_2S + COS < 15$ mg/Nm^3) prior addition into natural gas pipelines or used as autogas (UNE-EN16723 2018). Today, physicochemical techniques are still the preferred methods for the elimination of hydrogen sulphide from biogas, despite being energy-intensive processes that may produce hazardous waste requiring an expensive treatment (Cattaneo et al. 2023). Table 21.1 shows the conventional physicochemical techniques for H_2S removal from gas streams that dominate the market.

On the other hand, biological gas desulphurization methods, such as biofilters, bioscrubbers and biotrickling filters, have recently received much attention, as they are more environmentally friendly and more economically sustainable compared to physicochemical techniques (Khoshnevisan et al. 2017; Arıman and Koyuncu 2022). Biological technologies can be carried out under either aerobic or anoxic conditions. Nevertheless, the aerobic process dilutes the biogas as a result of the air injection. This is a major drawback, especially when the desulphurized biogas is intended to be converted into biomethane, since polishing efforts to eliminate oxygen and nitrogen are increased. In this sense, anoxic H_2S oxidation arises as a promising alternative to biologically remove H_2S from biogas, since it avoids biogas dilution

Table 21.1 Traditional physical–chemical methods for H_2S removal from gas streams

Addition of iron salts/oxides to the digester slurry	
Adsorption	– **Activated carbon** – Molecular sieve – **Iron oxides (iron sponge, SulfaTreat, Sulphur-Rite)** – Zinc oxides – Alkaline solids
Absorption/Scrubbing	– **Water** – **No-water physical solvents (Selexol)** – Alkaline solutions – Zinc oxides slurries – Iron oxide slurries – **Iron salts, chelated and no chelated (Lo-Cat, SulFerox, Sulfothane)** – Quinone and vanadium salts – Chemical oxidants: hypochlorite, H_2O_2, $KMnO_4$ – Amines
Membrane purification	
Claus process (+ SCOT/+ Superclaus)	
Incineration	

The most suitable methods for removal of H_2S from biogas are highlighted in bold (Allegue and Hinge 2014)

and explosion hazards as a result of the absence of O_2 in the process (Almenglo et al. 2023). During biological anoxic desulphurization of biogas, chemolithotrophic bacteria oxidize H_2S using NO_3^- or NO_2^- as the final electron acceptor instead of O_2, and producing N_2 instead of H_2O as a by-product (Almenglo et al. 2023). Phototrophic sulphur bacteria (green sulphur bacteria and purple sulphur bacteria) can also oxidize hydrogen sulphide in the absence of O_2. These microorganisms carry out anoxygenic photosynthesis and use reduced sulphur substances as electron donors for photosynthetic CO_2 reduction (Struk et al. 2020). Finally, H_2S can be removed from biogas in situ employing microorganisms of the *Thiobacillus* genus, which transform hydrogen sulphide to elemental sulphur using carbon dioxide from the anaerobic digester as their carbon source. In this process, in situ aeration is needed in the digester at an air concentration of 2–6% (depending on the proportion of H_2S in the biogas) to perform the H_2S oxidation to sulphur (Ghimire et al. 2021). This approach avoids the costs of manufacturing and implementation of a new bioreactor module for desulphurization. In contrast, this method slightly reduces the volatile suspended solids available for anaerobic metabolism and decreases the biogas CH_4 concentration because of the addition of N_2 and an additional electron acceptor such as O_2 during aeration (Cattaneo et al. 2023).

21.2.2 CO_2 Removal

Carbon dioxide is the main pollutant of biogas (15–60% v/v of biogas), which decreases its heating value, flammability limit and flame speed (Aghel et al. 2022). According to most international standards, the CO_2 content in biomethane must be less than 2–4% to be added biomethane into natural gas networks or utilized as automotive fuel (Marín et al. 2021).

Conventional methods for biogas decarbonization are based on physicochemical techniques, some of them at industrial scale such as absorption, adsorption and membrane separation and others still in the development and evaluation stage such as cryogenic process and chemical hydrogenation (Angelidaki et al. 2018; Halim et al. 2022). Table 21.2 lists the advantages and drawbacks of conventional physicochemical techniques for the removal of CO_2 from biogas. Physicochemical processes still dominate the market for biogas upgrading despite its high capital, maintenance and operational expenses, high power consumption and environmental impact. However, during the last two decades, research efforts have intensified the search for new alternatives for biogas upgrading more cost-effective as well as more environmentally friendly. In this context, biological methods can be a good alternative since they transform CO_2 into biomass that can be utilized to obtain high value added products under conditions of moderate temperatures and atmospheric pressure.

Nevertheless, most of them have recently been demonstrated on a laboratory scale and are in an early pilot or large-scale phase (Angelidaki et al. 2018).

In this context, carbon dioxide from biogas can also be separated by in situ desorption. CH_4, H_2S and CO_2 dissolved in the anaerobic mixed sludge are continuously

Table 21.2 Advantages and disadvantages of physical–chemical technologies for biogas decarbonization

Technology	Advantages	Drawbacks	References
Water scrubbing	Widespread use in industry and simplicity of design $CH_4 > 97\%$ and $CO_2 < 2\%$ (for a treatment capacity of 10–10,000 $Nm^3\ h^{-1}$) 1–2% CH_4 losses	It is highly recommended a previous stage for H_2S removal High capital and operating costs due to the need for high pressure, antifoaming agents and low energy efficiency Environmental problems related to the high carbon footprint as well as the acidification and human toxicity of the water generated in the process	Struk et al. (2020), Golmakani et al. (2022)
Organic solvent scrubbing	Improved efficiency compared to water scrubbing in terms of CO_2 separation which implies lower footprint and volumes of the upgrading unit	Prior removal of H_2S and H_2O is recommended More energy is required for the solvent regeneration compared to water scrubbing The cost of the organic solvents is significantly higher than that of water	Angelidaki et al. (2018), Ullah Khan et al. (2017)
Amine absorption	High efficiency can be achieved (> 99% CH_4) with very low CH_4 slip (< 0.1%) Low operational cost Regenerative	High energy required for absorbent regeneration High investment cost H_2S removal prior is highly recommended Corrosion issues, possible foaming and need to treat chemical waste and accumulation of contaminants	Aghel et al. (2022), Ward et al. (2021)
Pressure swing adsorption	Biomethane purity of 96–98% Low energy demand and emissions Removal of N_2 and O_2 Small capacities (< 500 Nm^3/h) can be treated	Adsorbents are damaged by hydrogen sulphide Methane slip up to 4% Requires a complex cycle control High investment and operational costs	Struk et al. (2020), Ullah Khan, et al. (2017), Zhao et al. (2010), Bahrun et al. (2022)

(continued)

Table 21.2 (continued)

Technology	Advantages	Drawbacks	References
Membrane separation	High energy efficiency No hazardous secondary effluents and small footprint The ability to scale down the process without large efficiency losses Easy retrofitting to higher upgrading capacities	Need of a previous stage to remove minor contaminants such as H_2S, particles, siloxanes and H_2O CH_4 purity < 96% High methane loss (7–10%) Membranes are fragile and expensive	Angelidaki et al. (2018), Golmakani et al. (2022), Baena-Moreno et al. (2020), Baena-Moreno et al. (2020), Bauer et al. (2013)
Cryogenic process	High CH_4 purity (up to 99%) No chemical solvents are required Production of marketable CO_2	High investment and operating costs Use of many process equipment Pretreatment needs for minor impurities such as H_2S, water, siloxanes and halogens Substantial CH_4 slip	Struk et al. (2020), Aghel et al. (2022), Zhao et al. (2010), Andriani et al. (2014)

stripped out by air in a desorption unit, which decreases the carbon dioxide and hydrogen sulphide concentration in the biogas (Muñoz et al. 2015). Nonetheless, the main challenges of this technology are CH_4 losses, which range between 2 and 8%, and pH control, since CO_2 desorption causes the pH in the anaerobic digester to increase and pH levels above 8.5 can inhibit methanogenic activity in the bioreactor (Ahmed et al. 2021).

Hot potassium carbonate (HPC) has recently been suggested as a novel approach and potential substitute to chemical scrubbing with amines for CO_2/CH_4 separation during biogas upgrading (Paolini et al. 2019; Barbera et al. 2019). This technology basically consists of the absorption of CO_2 using K_2CO_3 solutions with subsequent regeneration of the resulting $KHCO_3$ solution by decreasing the pressure and rising the temperature. The advantages of HPC are the very low CH_4 slip (< 0.1%), low power consumption (low operating costs), small equipment dimensions and absence of harmful chemicals during operation. On the other hand, heat consumption is high, but a large part of it (over 80%) can be recovered via heat integration. This converts HPC into a competitive technique for biomethane production, both economically and environmentally (Paolini et al. 2019; Barbera et al. 2019).

Finally, absorption–desorption processes based on Fe-EDTA-carbonate solutions represent a novel physicochemical method for the removal of carbon dioxide and hydrogen sulphide from biogas in an individual stage. On a laboratory scale, this innovative process provided contents of methane > 97%, carbon dioxide < 2%, oxygen < 1% and a complete removal of hydrogen sulphide under atmospheric pressure and temperature conditions, without external chemical supplementation and at an power consumption lower compared to its physicochemical competitors (Marín et al. 2020). However, this technology still requires additional efforts and research to be validated on a pilot and industrial scale.

An innovative biological strategy for biogas upgrading is based on hydrogenotrophic methanogenesis, in which carbon dioxide and hydrogen are biologically transformed into methane via autochthonous hydrogenotrophic methanogens without any extra energy demand (apart from that required for bioreactor operation) (Ullah Khan, et al. 2017). The electricity used to obtain hydrogen through electrolysis must come from renewable energy like wind or solar to make this method sustainable (Aghel et al. 2022). Nonetheless, hydrogen storage is one of the major issues of this technology due to its very low volumetric energy density and high flammability. Additionally, this process is limited to small installations in which the required CH_4 concentration is < 95% (Aghel et al. 2022).

The photosynthetic potential of microalgae and cyanobacteria to fix CO_2 provides the basis of another biological alternative for biogas upgrading. This technique consists of CO_2 sequestration into microalgal biomass via photosynthesis. In addition, H_2S can be removed from biogas because of the existence of some sulphur-oxidizing bacteria in the microalgal-bacterial broth. These bacteria oxidize H_2S into sulphate using the dissolved oxygen generated by microalgae, which prevents the inhibitory levels of H_2S for microalgae and contributes to reduce the O_2 proportion in the biomethane. Photosynthetic biogas purification is considered one of the most environment-friendly technologies for biogas purification since, in addition to its carbon dioxide capture potential, algal biomass allows recovering the dissolved nutrients present in digestate and this biomass can be further valorised as a feedstock for obtaining of commercial bioproducts (biostimulants, biofertilizers and so on) (Ahmed et al. 2021). One of the technical limitation of this technology is still the O_2 (0.5–2%) and N_2 concentration in the biomethane (6–9%), which still requires further development (Ahmed et al. 2021).

21.2.3 Removal of Other Impurities from Biogas

Siloxanes in biogas cause equipment abrasion when biogas is utilized as an energy resource (in combustion). This issue only affects biogas generated in landfills and sewage sludge, where the concentration of siloxanes is typically > 50 mg m^{-3}, which causes a rapid deterioration of equipment, and therefore, siloxane removal from biogas is required before its use as energy carrier (Abatzoglou and Boivin 2009; Golmakani et al. 2022). Siloxanes can be removed from biogas by several methods: absorption, adsorption, cryogenic, membrane and biological. However, the most widely used strategy is adsorption on activate carbon, which presents moderate/high operating costs due to the high working pressure and maintenance of the absorbent (Muñoz et al. 2015; Ullah Khan et al. 2017). On the other hand, biological techniques represent a potential emergent substitute for the elimination of siloxanes from biogas because of their greater cost-effectiveness and environmental-friendliness, especially biotrickling filtration, which exhibits the lowest operating costs compared to physical–chemical methods (Pascual et al. 2021). However, biodegradation of siloxanes is still in the research and development phase and its application is still limited. Furthermore, understanding the specific microorganisms as well as the degradation processes carried out in biological removal of siloxanes remains a challenge (Pascual et al. 2023).

The presence of water in biogas generates corrosion in equipment due to the formation of acids as a consequence of its reaction with CO_2, H_2S and NH_3, and hence its elimination is essential for any biogas application except on-site heat and electricity production (Ullah Khan et al. 2017). Today, biogas moisture can only be removed by physicochemical techniques like absorption, adsorption and refrigeration (Muñoz et al. 2015; Golmakani et al. 2022). However, these technologies imply high capital investment and operating expenses, energy demand and environmental impact (Muñoz et al. 2015).

NH_3 in biogas can cause corrosion of equipment and can be transformed into NO_x, which are greenhouse gases. International standards specify an ammonia concentration $\leq$ 3–20 ppm in the biomethane for its use as natural gas substitute or vehicle fuel (Awe et al. 2017). Halogenated compounds are also impurities present in biogas that can produce corrosion in equipment and whose limit concentration depends on each country, but under no circumstances should be above 50 mg m^{-3} (Awe et al. 2017). Halocarbons soluble in water and ammonia can be removed via water scrubbing (Ullah Khan et al. 2017). Adsorption with activated carbon is also utilized for the elimination of halogenated compounds and VOCs (Muñoz et al. 2015). So far, biological technologies have not yet been tested for the elimination of ammonia and halogenated impurities in the biogas upgrading process.

21.3 Photosynthetic Biogas Purification and Nutrient Recovery from Wastewater

Photosynthetic biogas purification is a biological technology that simultaneously removes CO_2, H_2S and other compounds at trace levels such as NH_3 and volatile organic compounds (VOCs) from biogas (Franco-morgado et al. 2018). This process utilizes the principles of scrubbing, where gas compounds are typically absorbed in a bubble column. The low solubility of methane in water, based on its water/gas Henry's law constant in dimensionless form (0.03 at 25 °C), compared to other compounds (0.83, 2.44 and 1462 for CO_2, H_2S and NH_3 at 25 °C), supports biogas upgrading via scrubbing (Sander 2015). Therefore, the other undesirable compounds can be absorbed while CH_4 losses from biogas remained low (< 5%) (Posadas et al. 2017). The liquid stream containing the absorbed gases is then directed to a photobioreactor where a symbiotic community of microalgae and/or cyanobacteria (jointly denoted as microalgae), along with heterotrophic/autotrophic bacteria, consume these unwanted compounds from biogas. Consequently, this biological step eliminates the need for absorbent regeneration while producing valuable bioproducts (valuable algal–bacterial biomass) and limiting greenhouse gas emissions. Thus, this process is considered a sustainable approach from both environmental and economic perspectives (Toledo-cervantes et al. 2017a).

One of the bottlenecks of this process is the CO_2 mass transfer from the raw biogas to the liquid phase (culture media) in the absorption column. Despite being greater than that of methane, the aqueous solubility of carbon dioxide is relatively low, which affects its removal efficiency from biogas (Rodero et al. 2019b). Therefore, improving the hydrodynamics to ensure a homogeneous flow regime is essential for achieving high CO_2 removal efficiencies. For this reason, the design of the absorption column, including the height to diameter (*H/D*) ratio, gas–liquid flow configuration and an appropriate biogas diffuser, is of paramount relevance. High *H/D* ratio values require large absorption columns, which limits the scale-up and increases oxygen/nitrogen stripping, while low *H/D* ratios shorten the interaction period between the gas and liquid. High *H/D* ratios (> 22) are typically reported for photosynthetic biogas purification units at laboratory scale (Bose et al. 2021). However, a 150 L bubble column with a *H/D* ratio of ~ 7 was used by Rodero et al. (2019a) during the scale-up of the process up to a 32 m^2 HRAP, in agreement with the common values applied (3–10) in the design of large-scale absorption columns (Jakobsen 2014). Counter-current flow enhances CO_2 gas–liquid mass transfer rates in absorption units but leads to incomplete H_2S oxidation and increases O_2/N_2 stripping during photosynthetic biogas upgrading (Toledo-Cervantes et al. 2017b). Therefore, co-current flow operation is preferred in this process. On the other hand, small pore-size diffusers made of polymers or stainless steel are applied in biogas upgrading to promote small gas bubbles with large gas–liquid contact area (Marín et al. 2020).

The optimization of the operating parameters like the biogas retention time, *L/G* ratio in the absorption column is also a must. Gas residence times ranging between 20 and 120 min are typically reported during photosynthetic biogas upgrading (Table 21.3). Nevertheless, further optimization of the CO_2 absorption is still needed to decrease these values down to 3–15 min as in biotrickling filters for biogas desulphurization (Fernández et al. 2013; Fortuny et al. 2008). The *L/G* ratio in the scrubbing column is a crucial operational parameter. Increasing this ratio enhances the elimination of CO_2 and H_2S but also promotes the stripping of oxygen and nitrogen from the liquid to the biomethane, worsening the quality of biomethane and causing risks of explosion due to a higher O_2/CH_4 ratio (Bose et al. 2022; Meier et al. 2019). *L/G* ratios in the range of 0.5–2 have been reported as ideal (Rodero et al. 2019a; Bose et al. 2022; Marín et al. 2019). However, optimization of *L/G* ratio over time is necessary, as the optimal value may vary with changes in environmental conditions and the properties of the cultivation medium (such as pH, alkalinity and temperature) (Rodero et al. 2019b).

The pH and alkalinity of the culture broth influence the efficiency of CO_2 and H_2S elimination because of the acidic nature of these biogas pollutants. Operating at higher pH levels enhances gas–liquid mass transfer in the scrubbing column but may also negatively impact biological activity. Typically, pH values between 9 and 10 are recommended for this process (Bahr et al. 2014). High alkalinity or inorganic

Table 21.3 Process efficiency in biogas upgrading combined with wastewater treatment in algal–bacterial systems

Unit design	WW type	HRT (d)	Gas residence time (h)	CO_2-RE (%)	H_2S-RE (%)	CH_4 concentration (%)	WW treatment-RE (%)	References
9.2 L HRAP interconnected to a 0.7 L AC	Synthetic sewage	2.5–4.0	0.77	75	99	–	TN: 77–84; TP:64–98; TOC:12–26; IC:40–50	Guenka et al. (2021)
9.6 m^3 HRAP interconnected to a 150 L AC	Domestic WW	3.5–8	0.33–0.55	59–89	86–98	79–89	COD: 87–91; TN: 76–86; P-PO_4^{3-}: 81–87	Rodero et al. (2019a)
300 L HRAP interconnected to a 3.2 L AC	MSM and leachate (15:1 v/v)	138	0.38	80	100	70–90	TN:35–62	Velasco et al. (2023)
0.25 L stirred conical flasks	Swine WW	7	–	53–68	–	–	COD: 72–83; TN:73–84; TP: 75–84	Cheng et al. (2024)
0.16 L serum bottles	Synthetic swine manure digestate	5	72	90	100	–	COD: 68; TN:10; P-PO_4^{3-}:7	Xie et al. (2023)
180 L HRAP interconnected to a 2.5 L AC	Mixture synthetic wastewater/ digestate (3.1:1)	23–58	0.7–1.2	85–90	–	81–94	TN: 100; TP:100	Marín et al. (2022)
180 L HRAP interconnected to a 2.5 L AC	Digestate	45	0.7–2.0	90–98	–	85–91	TN: 100; TP:100	Méndez et al. (2022)

(continued)

Table 21.3 (continued)

Unit design	WW type	HRT (d)	Gas residence time (h)	CO_2-RE (%)	H_2S-RE (%)	CH_4 concentration (%)	WW treatment-RE (%)	References
9.6 m^3 HRAP interconnected to a 150 L AC	Simulated digestate	73	0.33–0.55	78–99	96–100	83–91	COD: 74; TN: 86; P-PO_4^{3-}: 68	Rodero et al. (2019a)
180 L HRAP interconnected to a 2.2 L AC	Digestate	180	0.87	98–100	99–100	98–99	COD: 81–93; TN: 97–99; TP: 90–99	Franco-morgado et al. (2018)
180 L HRAP interconnected to a 2.5 L AC	Digestate	52	0.8	64–96	100	85–98	TC: 47–100; TN: 58–99; P-PO_4^{3-}: 99–100	Marín et al. (2018)

WW wastewater, *HRT* hydraulic retention time, *RE* removal efficiency, *TN* total nitrogen; *TP* total phosphorous, *TOC* total organic carbon, *IC* inorganic carbon, *COD* chemical oxygen demand carbon concentrations are necessary to keep the pH of the medium elevated along the scrubbing column. Inorganic carbon concentrations around 1.5–2.1 g L^{-1} are optimum to ensure nearly complete removal of CO_2 and H_2S at low *L/G* ratios (which minimize O_2/N_2 stripping) (Rodero et al. 2018; Bose et al. 2022). However, higher concentrations can increase CO_2 emissions to the atmosphere and jeopardize long-term biological activity, diminishing environmental benefits (Rodero et al. 2020a).

Another crucial step in this process is the subsequent assimilation of biogas pollutants by microalgae and bacteria. Microalgae fix dissolved CO_2 in the cultivation broth using light energy in a process called oxygenic photosynthesis. During this process, electrons released by light-mediated photolysis of water are used to reduce CO_2, resulting in the generation of O_2 and algal biomass from CO_2, H_2O, nutrients, and light energy (Fig. 21.1). The most typical photosynthetic microorganisms in these systems are *Spirulina, Chlorella, Anabaena, Scenedesmus* and *Chlorococcum*, as they can grow at high pH levels (9–10) and elevated concentrations of inorganic carbon (Bose et al. 2019). Microalgae exhibit higher photosynthetic efficiencies compared to plants, fixing approximately 1.83 g of CO_2 per gram of microalgal biomass produced (Shahid et al. 2020). Nevertheless, microalgal biomass productivities, mainly determined by the system configuration and environmental conditions, still require extensive surface areas (621 m^2 for 1 Nm^3/h of treated biogas) for effective carbon dioxide capture (Toledo-cervantes et al. 2017a). This extensive land requirement, which lead to high investment costs, is one of the current limitations for the scale-up of this biotechnology. These costs can be partially reduced by coupling the process with the treatment of wastewater or anaerobic effluents, from which nutrients and H_2O needed for algal–bacterial growth can be obtained. The use of the biomass as a biofertilizer, biostimulant or feedstock for biofuel production can also outbalance the high investment costs (Arif et al. 2021; Torres-Franco et al. 2021).

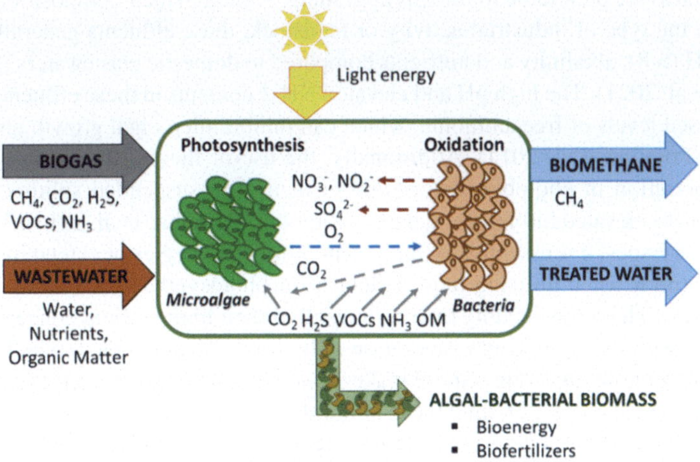

Fig. 21.1 Schematic diagram of photosynthetic biogas upgrading coupled with wastewater treatment

On the other hand, bacteria in the photobioreactors use the oxygen generated through photosynthesis, along with CO_2 or organic carbon and nutrients, to grow and oxidize H_2S, NH_3, VOCs, and organic matter when using wastewaters as a nutrient source (Fig. 21.1). H_2S oxidation can be either chemical or biological (carried out by sulphur-oxidizing bacteria). Due to the excess of oxygen in microalgal systems, sulphate (SO_4^{2-}) is the main product of this oxidation, although elemental sulphur (S^0) can also be generated. Despite H_2S concentrations in the biogas can be detrimental to microalgae growth, the rapid oxidation of H_2S in the scrubbing column prevents reduction of algae activity (Meier et al. 2018). NH_3 oxidation occurs in two steps: ammonia-oxidizing bacteria first convert ammonia to NO_2^-, which is then further oxidized of NO_2^- to NO_3^- by nitrite-oxidizing bacteria (Winkler and Straka 2019). The low NH_3 levels in biogas requires an extra nitrogen source to promote algal–bacterial activity. Heterotrophic bacteria in photobioreactors can effectively remove minor levels of VOCs found in biogas and/or organic matter during wastewater treatment (Franco-Morgado et al. 2018).

Wastewaters and anaerobic effluents are rich in carbon and nutrients. These compounds must be reduced to acceptable limits before being released into natural water systems (Abdur et al. 2017). The characteristics of the wastewater vary depending on its source. Domestic wastewater typically has a chemical oxygen demand of 200–1200 mg O_2/L, total nitrogen concentrations of 15–170 mg/L, primarily in the form of NH_4^+, and phosphorus levels between 5 and 50 mg/L (Ángeles et al. 2019). Algal–bacterial systems achieve high nutrient recovery when treating these wastewaters (De Cassia et al. 2023). The main limitation of treating household wastewaters simultaneously with biogas purification in microalgal reactors is their low alkalinity (inorganic carbon content < 180 mg/L) and pH values (~ 7). Consequently, CO_2 removal efficiencies with these wastewaters are low (Table 21.3).

The use of agro-industrial wastewaters or anaerobic effluents (the liquid fraction of digestates) are preferable for biogas upgrading. Although their composition varies based on the type of industrial activity or feedstock, these effluents generally have higher pH (> 8), alkalinity and nutrients compared to domestic wastewaters (Torres-Franco et al. 2021). The high pH and elevated NH_4^+ contents in these effluents result in increased levels of free ammonia, which can inhibit microalgal growth and nitrification (Posadas et al. 2014). Additionally, the use of these effluents can reduce light penetration in photobioreactors due to their high suspended solids content, which causes elevated turbidity and dark coloration (Marcilhac et al. 2014). To overcome these issues, it is essential to implement dilution strategies or extend hydraulic retention times when utilizing these effluents for photosynthetic biogas upgrading (Table 21.3). These low feeding rates as a result of their high nutrient concentration, combined with the significant evaporation in high-rate algal ponds, typically result in zero effluent output and in require make-up water to address elevated evaporation rates (Marín et al. 2018). Despite these minor drawbacks, nearly complete removals of CO_2, H_2S, and nutrients have been reported using these high-strength effluents, particularly digestates, as a H_2O and nutrient source (Table 21.3).

21.4 Impact of Nanoparticles on Photosynthetic Biogas Purification

The purification of biogas with microalgae has demonstrated to be a promising platform to treat biogas with final CO_2 concentrations < 2% at pilot-lab scale and demo scale (Rodero et al. 2020a, b). Nonetheless, the scalability of the process to industrial levels is mainly limited by the CO_2 mass transfer to the cultivation medium, the limited photosynthetic microalgal activity and the dependence of the upgraded biomethane quality to the variations of alkalinity and pH of the cultivation media, *L/G* ratio and environmental conditions.

CO_2 mass transfer to the cultivation media has been a challenge in biogas purification techniques regardless of the physical, chemical or biological technology used. Thus, different techniques have been implemented to improve the mass transfer of CO_2, and recently the use of mesoporous nanomaterials has caught the attention in the research field. Nanoparticles (NPs) have demonstrated a better performance as a result of their specific physicochemical properties that capable of enhancing CO_2 mass transfer. For instance, the high number of active sites can increase their adsorption capacity, which along with their larger surface area to volume ratio, can improve CO_2 capture (Anwar et al. 2018). In this context, Kluytmans et al. (2003) observed that the supplementation of carbon NPs increased the specific gas–liquid interfacial area in a sparged stirred tank reactor and in a bubble column. Thereby, the incorporation of NPs to microalgae cultures dedicated to purification of biogas could be a potential technique to enhance the transfer of CO_2 to the culture broth.

The addition of NPs in microalgae cultures is a polemical topic since most of literature was dedicated to their toxicity and little was known on their beneficial effect on microalgae metabolism. For instance, the supplementation of metal oxide NPs such as silicon dioxide (SiO_2), hematite (Fe_2O_3) and magnesium sulphate has demonstrated to enhance the biomass productivity and the accumulation of macromolecules. Jeon et al. (2017) demonstrated that the supplementation of silica NPs improved the biomass concentration and lipid production of *Chlorella vulgaris* mainly mediated by an increase of 31% of the CO_2 mass transfer to the cultivation media. Similarly, the accumulation of lipid in *C. vulgaris* cultures was improved by up to 110% when 1 g L^{-1} $MgSO_4$ NPs were added to the cultures (Sarma et al. 2014). On the other hand, the addition of polymeric nanofibers covered with Fe_2O_3 NPs have demonstrated to improve both the CO_2 biofixation rate and lipid accumulation of *Chlorella fusca* by 45 and 11%, respectively (da Silva Vaz et al. 2019). Fe_2O_3 NPs have received particular attention since beneficial effects on different microalgae strains (i.e. *Scenedesmus obliquus* and *Chlorella* spp.) have been obtained at dosages lower than 50 mg L^{-1}. For instance, He et al. (He et al. 2017) demonstrated that the addition of 2–20 mg L^{-1} of hematite NPs increased the biomass concentration of *S. obliquus*, while lipid production was increased by 40% with the supplementation of 5 mg L^{-1} of Fe_2O_3 NPs. Similarly, the supplementation of 20 mg L^{-1} Fe_2O_3 NPs to *Chlorella pyrenoidosa* cultures improved the biomass concentration and lipid accumulation by 34 and 17%, respectively (Rana et al. 2020). In addition, the supplementation of Fe_2O_3 at concentrations ranging between 50 and 100 mg L^{-1} improved the lipid concentration of *C. vulgaris* between 25 and 40% (Bibi et al. 2021). Thus, the supplementation of NPs to microalgae cultures dedicated to biogas purification has the potential to increase CO_2 absorption while producing high-added compounds.

In 2022, a hybrid system of microalgae-NPs dedicated to biogas purification was initially proposed by Vargas-Estrada et al. (2022). The authors evaluated the effect of three different NPs, namely SiO_2, Fe_2O_3 and carbon-coated zero-valent iron (from the company CALPECH), on the performance of a mixed microalgae-bacteria consortium during biogas upgrading. These three NPs exhibited specific physicochemical properties to enhance CO_2 adsorption. However, only the carbon-coated zero-valent iron NPs from CALPECH significantly increased microalgae growth and carbohydrate accumulation when 70 mg L^{-1} were added. Then, Vargas-Estrada et al. (2023a) investigated the behaviour of *Chlorella sorokiniana* cultures dedicated to biogas purification supplemented with iron-based NPs: Fe_2O_3, CALPECH and carbon-coated zero-valent NPs with different physicochemical properties (different pore size and iron content). These authors confirmed the beneficial impact of CALPECH NPs on *C. sorokiniana* growth and metabolites accumulation, especially when 70 mg L^{-1} were added. This concentration mediated increases in biomass productivities between 40 and 45% in *C. sorokiniana* and the mixed consortium, as well as in carbohydrate accumulation between 100 and 150%, respectively. Vargas-Estrada et al. (2023b) also validated the effect of the supplementation of 70 mg L^{-1} of CALPECH NPs in the operation of a continuous lab-scale pilot algal pond devoted to biogas purification combined with nutrient recovery under indoor conditions. The system consisted of an open-high-rate algal pond interconnected to a biogas scrubbing column by means of a settler as previously described in Méndez et al. (2022). Interestingly, the biomass concentration inside the HRAP increased from 1.56 to 3.26 g L^{-1} while the CO_2 removal efficiency increased from 86 to 92% after the addition of the NPs. Even if the system operated at a low inorganic carbon (IC) concentration (< 600 mg L^{-1}), the supplementation of the NPs mediated a raise in the pH of the system from 8.55 to 9.08. However, this enhanced microalgae growth decreased the IC content of the culture broth, resulting in a loss of the buffer capacity, which ultimately limited the transfer of CO_2. It is important to stress that the carbon-coated zero-valent NPs did not negatively impact nutrient removal in the system and an intense nutrient consumption was recorded throughout the study. Similarly, the boosted microalgal activity increased the dissolved oxygen concentration (DO) in the system, which significantly enhanced the oxidation of H_2S to SO_4^{2-} and of NH_4^+ to NO_3^- by the end of their study.

Recently, Hoyos et al. (2024) assessed the effect of increased concentrations of CALPECH NPs (70, 140 and 280 mg L^{-1}) on the operation of an open algal pond upgrading biogas to biomethane. An outstanding process performance was recorded after the supplementation of 140 mg L^{-1} of CALPECH NPs, since the biomass content increased from 1.38 to 3.10 g VSS L^{-1}, which entailed an increase in biomass productivity from 22.5 to 48.2 g m^{-2} d^{-1} to prevent biomass accumulation in the HRAP. It is important to stress that this intense microalgal activity required the addition of 1.7 g L^{-1} Na_2CO_3 to maintain the buffer capacity in the system.

Under these particular conditions, CO_2 removal efficiencies prevailed around 95% and the CH_4 content in the purified biogas remained at 94%. Similarly to the previous work conducted by Vargas-Estrada et al. (2023b), this increased microalgae activity mediated an intense nutrient assimilation in the system, especially of $N–NH_4^+$ and $P–PO_4^{3-}$. On the other hand, $S–SO_4^{2-}$ concentration doubled after the addition of the NPs, which was attributed to the higher oxidation of H_2S induced by the high DO in the system. Finally, Hoyos et al. (2024) recently tested the potential of liquid NPs, more specifically CALPECH liquid NPs, in a similar system configuration. The liquid CALPECH NPs performance, similarly to their solid counterparts, significantly increased the pH of the system from 8.7 to 9.3, which mediated an improved CO_2 mass transfer to the cultivation media and an increased IC content in the culture medium. This resulted in a raise in the biomass content from 1.2 to 3.5 g VSS L^{-1} and a CO_2 removal of 94%.

Lastly, it is essential to indicate that at the early stages of this technology, the mechanism of interaction between NPs and microalgae was not well identified. Vargas-Estrada et al. (2023b) initially attributed the boosted performance in the system to the physical properties of the NPs, which could enhance the CO_2 availability to the microalgae culture. However, in an abiotic study, Hoyos et al. (2024) confirmed that the supplementation of NPs likely induced an improvement in the metabolism of microalgae and the physical properties of the NPs only mediated the reduction in nitrogen and oxygen proportion in the purified biogas. Overall, the supplementation of algal cultures with NPs has emerged as a potential strategy to enhance one of the main drawbacks of photosynthetic biogas purification, since high biomass contents (> 3 g VSS L^{-1}) and CO_2 removals (> 90%) can be obtained at relatively low IC concentrations (400–800 mg L^{-1}) (Fig. 21.2). Thereby, this rise in biomass productivity could pave the way to the widespread implementations of photobiorefineries, where high-added-value products can be recovered from the microalgal biomass, which will substantially enhance the economy and sustainability of this biotechnology.

21.5 Conclusion

Nowadays, global warming is one of the largest environmental challenges. In this context, there is an urgent need to quickly decrease the high need for fossil fuels and increase the deployment of renewable energies in the coming years. Biogas represents a renewable and sustainable resource with promising potential to help partially replacing the fossil fuel consumption. One of the potential routes to valorise biogas is through its purification to biomethane. However, conventional physicochemical technologies for biogas purification require a high energy demand and chemical consumption and entail high environmental impacts. Hence, innovative and sustainable technologies are required, which has been the research focus during the last decades in the biogas purification field.

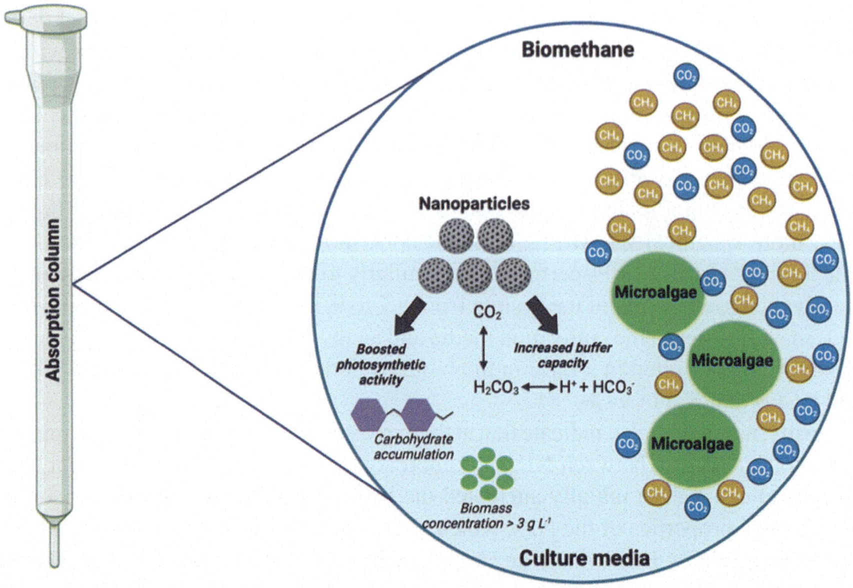

Fig. 21.2 Effect of nanoparticles on microalgae cultures devoted to photosynthetic biogas upgrading

Microalgal-bacterial systems are a cost-effective and sustainable biotechnology for the concomitant treatment of biogas and high-strength effluents like agro-industrial wastes and digestates. The main bottlenecks of this biotechnology are the CO_2 gas–liquid mass transfer and its subsequent photosynthetic fixation, which require large surface areas. Improving gas–liquid mass transfer is necessary to decrease gas residence time and, consequently, the size of the absorption columns. Additionally, new photobioreactor designs and strategies to promote biomass growth should be explored to enhance productivity. Despite these challenges, this biotechnology effectively removes pollutants from biogas, providing multiple benefits, including concomitant wastewater treatment and the generation of high-added-value biomass. Furthermore, it is environmentally friendly and has a lower carbon footprint compared to physicochemical technologies.

The integration of nanotechnology to microalgae cultures has become a hot topic in recent years. Even if at the early stages of this new research field the attention was focused on the negative side of using nanomaterials in microalgae cultures recently, recent research has focused on the benefits of supplying nanomaterials as a feasible technology to increase microalgal productivity. For instance, it has been widely demonstrated that some nanomaterials, such as polymeric nanofibers, can enhance CO_2 availability to microalgae cultures. Moreover, metal NPs, mainly iron-based NPs, can induce lipid and carbohydrate accumulation in *Chlorella* species. More recently, the incorporation of nanoparticles to photosynthetic biogas upgrading process has demonstrated promising results not only to boost microalgae metabolism, but to obtain a biomethane with low CO_2 and N_2 concentrations as well. Despite limited knowledge about the mechanism of interaction between nanoparticles and microalgae, it has been demonstrated that the physicochemical composition of the nanoparticles, e.g. porosity, pore size, chemical composition, size, etc., is key parameters to support a cost-effective biogas upgrading process. Thereby, further investigations are required to comprehend and optimize this research field, in particular the understanding of the fate of nanoparticles. Thus, the loop of nanoparticle-microalgae system can be closed in a sustainable way. The addition of nanoparticles to photosynthetic processes (e.g. photosynthetic biogas upgrading) will pave the way towards a new generation of photobiorefineries that will be the cornerstone of a more sustainable economy.

References

Abatzoglou N, Boivin S (2009) A review of biogas purification processes. Biofuels Bioprod Biorefining 3(1):42–71. https://doi.org/10.1002/bbb.117

Abdur S, Aldin S, Ali M, Mozahar M, DeLasa H (2017) Biological CO_2 fixation with production of microalgae in wastewater—a review. Renew Sustain Energy Rev 76:379–390. https://doi.org/10.1016/j.rser.2017.02.038

Aghel B, Behaein S, Wongwises S, Shadloo MS (2022) A review of recent progress in biogas upgrading: with emphasis on carbon capture. Biomass Bioenerg 160:106422. https://doi.org/10.1016/j.biombioe.2022.106422

Ahmed SF et al (2021) Biogas upgrading, economy and utilization: a review. Environ Chem Lett 19:4137–4164. https://doi.org/10.1007/s10311-021-01292-x

Allegue L, Hinge J (2014) Biogas upgrading: evaluation of methods for H_2S removal (Danish Technological Institute)

Almenglo F, González-Cortés JJ, Ramírez M, Cantero D (2023) Recent advances in biological technologies for anoxic biogas desulfurization. Chemosphere 321:138084. https://doi.org/10.1016/j.chemosphere.2023.138084

Andriani D, Wresta A, Atmaja TD, Saepudin A (2014) A review on optimization production and upgrading biogas through CO_2 removal using various techniques. Appl Biochem Biotechnol 172(4):1909–1928. https://doi.org/10.1007/s12010-013-0652-x

Ángeles R, Rodero R, Carvajal A, Muñoz R, Lebrero R (2019) Potential of microalgae for wastewater treatment and its valorization into added value products. In: Application of microalgae in wastewater treatment. Springer, Cham, pp 281–215. https://doi.org/10.1007/978-3-030-13909-4_13

Angelidaki I et al (2018) Biogas upgrading and utilization: current status and perspectives. Biotechnol Adv 36(2):452–466. https://doi.org/10.1016/j.biotechadv.2018.01.011

Anwar MN et al (2018) CO_2 capture and storage: a way forward for sustainable environment. J Environ Manage 226:131–144. https://doi.org/10.1016/j.jenvman.2018.08.009

Arif M, Li Y, El-dalatony MM, Zhang C, Li X, Salama E (2021) A complete characterization of microalgal biomass through FTIR/TGA/CHNS analysis: an approach for biofuel generation and nutrients removal. Renew Energy 163:1973–1982. https://doi.org/10.1016/j.renene.2020.10.066

Arıman S, Koyuncu S (2022) Removal of hydrogen sulfide in biogas from wastewater treatment sludge by real-scale biotrickling filtration desulfurization process. Water Pract Technol 17(7):1406–1420. https://doi.org/10.2166/wpt.2022.072

Awe OW, Zhao Y, Nzihou A, Minh DP, Lyczko N (2017) A review of biogas utilisation, purification and upgrading technologies. Waste Biomass Valor 8(2):267–283. https://doi.org/10.1007/s12649-016-9826-4

Baena-Moreno FM, le Saché E, Pastor-Pérez L, Reina TR (2020) Membrane-based technologies for biogas upgrading: a review. Environ Chem Lett 18(5):1649–1658. https://doi.org/10.1007/s10311-020-01036-3

Bahr M, Ignacio D, Dominguez A, González Sánchez A, Muñoz R (2014) Microalgal-biotechnology as a platform for an integral biogas upgrading and nutrient removal from anaerobic effluents. Environ Sci Technol 48(1):573–581. https://doi.org/10.1021/es403596m

Bahrun MHV, Bono A, Othman N, Zaini MAA (2022) Carbon dioxide removal from biogas through pressure swing adsorption—a review. Chem Eng Res des 183:285–306. https://doi.org/10.1016/j.cherd.2022.05.012

Barbera E, Menegon S, Banzato D, D'Alpaos C, Bertucco A (2019) From biogas to biomethane: a process simulation-based techno-economic comparison of different upgrading technologies in the Italian context. Renew Energy 135:663–673. https://doi.org/10.1016/j.renene.2018.12.052

Bauer F, Persson T, Hulteberg C, Tamm D (2013) Biogas upgrading—technology overview, comparison and perspectives for the future. Biofuels Bioprod Biorefining 7(5):499–511. https://doi.org/10.1002/bbb.1423/pdf

Bibi M, Zhu X, Munir M, Angelidaki I (2021) Bioavailability and effect of α-Fe_2O_3 nanoparticles on growth, fatty acid composition and morphological indices of Chlorella vulgaris. Chemosphere 282:131044. https://doi.org/10.1016/j.chemosphere.2021.131044

Bose A, Lin R, Rajendran K, Shea RO, Xia A, Murphy JD (2019) How to optimise photosynthetic biogas upgrading: a perspective on system design and microalgae selection. Biotechnol Adv 37(8):107444. https://doi.org/10.1016/j.biotechadv.2019.107444

Bose A, Shea RO, Lin R, Murphy JD (2021) Design, commissioning, and performance assessment of a lab-scale bubble column reactor for photosynthetic biogas upgrading with Spirulina platensis. Ind Eng Chem Res 60:5688–5704. https://doi.org/10.1021/acs.iecr.0c05974

Bose A, O'Shea R, Lin R, Murphy JD (2022) Optimisation and performance prediction of photosynthetic biogas upgrading using a bubble column. Chem Eng J 437:134988. https://doi.org/10.1016/j.cej.2022.134988

Cantera S et al (2020) A systematic comparison of ectoine production from upgraded biogas using Methylomicrobium alcaliphilum and a mixed haloalkaliphilic consortium. Waste Manag 102:773–781. https://doi.org/10.1016/j.wasman.2019.11.043

Cattaneo CR, Muñoz R, Korshin GV, Naddeo V, Belgiorno V, Zarra T (2023) Biological desulfurization of biogas: a comprehensive review on sulfide microbial metabolism and treatment biotechnologies. Sci Total Environ 893:164689. https://doi.org/10.1016/j.scitotenv.2023.164689

Cheng P, Wang Z, Lu B, Zhao Y, Zhang H (2024) Co-culturing microalgae with endophytic bacteria from bamboo for efficient nutrient and heavy metal removal coupling with biogas upgrading. Water Enviro Res (a research publication of the Water Environment Federation) 96(1):e10977. https://doi.org/10.1002/wer.10977

da Silva Vaz B, da Silva Mastrantonio DJ, Costa JAV, de Morais MG (2019) Green alga cultivation with nanofibers as physical adsorbents of carbon dioxide: evaluation of gas biofixation and macromolecule production. Bioresour Technol 287:121406. https://doi.org/10.1016/j.biortech.2019.121406

De Cassia B et al (2023) Microalgae-based domestic wastewater treatment: a review of biological aspects, bioremediation potential, and biomass production with biotechnological high-value. Environ Monit Assess 195. https://doi.org/10.1007/s10661-023-12031-w

European Council (2023) Climate change: what the EU is doing. https://www.consilium.europa.eu/en/policies/climate-change/. Accessed 7 Dec 2023

Fernández M, Ramírez M, María R, Manuel J, Cantero D (2013) Hydrogen sulphide removal from biogas by an anoxic biotrickling filter packed with Pall rings. Chem Eng J 225:456–463. https://doi.org/10.1016/j.cej.2013.04.020

Ferreira AF, Rodero-Raya R, Muñoz R (2022) The experiences of success and failure in the pilot and real-scale photosynthetic biogas production. In: 3rd generation biofuels: disruptive technologies to enable commercial production, pp 1037–1059. https://doi.org/10.1016/B978-0-323-90971-6.00004-8

Fortuny M et al (2008) Biological sweetening of energy gases mimics in biotrickling filters. Chemosphere 71:10–17. https://doi.org/10.1016/j.chemosphere.2007.10.072

Franco-Morgado M, Toledo-Cervantes A, González-Sánchez A, Lebrero R, Muñoz R (2018) Integral (VOCs, CO_2, mercaptans and H_2S) photosynthetic biogas upgrading using innovative biogas and digestate supply strategies. Chem Eng J 354(August):363–369. https://doi.org/10.1016/j.cej.2018.08.026

Garlapati VK, Sharma S, Sevda S (2021) Photosynthetic biogas upgrading: an attractive biological technology for biogas upgrading. INC. https://doi.org/10.1016/B978-0-12-822808-1.00014-3

Ghimire A, Gyawali R, Lens PNL, Lohani SP (2021) Technologies for removal of hydrogen sulfide (H_2S) from biogas. In: Emerging technologies and biological systems for biogas upgrading. INC, pp 295–320. https://doi.org/10.1016/B978-0-12-822808-1.00011-8

Golmakani A, Ali Nabavi S, Wadi B, Manovic V (2022) Advances, challenges, and perspectives of biogas cleaning, upgrading, and utilisation. Fuel 317:123085. https://doi.org/10.1016/j.fuel.2021.123085

Guenka P et al (2021) Integration of algae-based sewage treatment with anaerobic digestion of the bacterial-algal biomass and biogas upgrading. Bioresour Technol 340:125552. https://doi.org/10.1016/j.biortech.2021.125552

Halim HNA, Rajiman V, Shariff AM (2022) A review on CO_2 absorption using chemical solvents at low and high CO_2 partial pressure conditions in a packed column. Open Chem Eng J 16(1):1–19. https://doi.org/10.2174/18741231-v16-e2204140

He M et al (2017) Improvement on lipid production by Scenedesmus obliquus triggered by low dose exposure to nanoparticles. Sci Rep 7(1):1–12. https://doi.org/10.1038/s41598-017-15667-0

Hoyos EG, Kuri R, Toda T, Muñoz R (2024a) Innovative design and operational strategies to improve CO_2 mass transfer during photosynthetic biogas upgrading. Bioresour Technol 391:129955. https://doi.org/10.1016/j.biortech.2023.129955

Hoyos EG, Amo-Duodu G, Gulsum Kiral U, Vargas-Estrada L, Lebrero R, Muñoz R (2024b) Influence of carbon-coated zero-valent iron-based nanoparticle concentration on continuous photosynthetic biogas upgrading. Fuel 356:129610. https://doi.org/10.1016/j.fuel.2023.129610

Jakobsen HA (2014) Bubble column reactors. In: Chemical reactor modeling: multiphase reactive flows, 2nd ed. Springer Cham, pp 883–927. https://doi.org/10.1007/978-3-319-05092-8

Jeon HS, Park SE, Ahn B, Kim YK (2017) Enhancement of biodiesel production in Chlorella vulgaris cultivation using silica nanoparticles. Biotechnol Bioprocess Eng 22(2):136–141. https://doi.org/10.1007/s12257-016-0657-8

Khoshnevisan B et al (2017) A review on prospects and challenges of biological H_2S removal from biogas with focus on biotrickling filtration and microaerobic desulfurization. Biofuel Res. J. 4(4):741–750. https://doi.org/10.18331/BRJ2017.4.4.6

Kluytmans JHJ, van Wachem BGM, Kuster BFM, Schouten JC (2003) Mass transfer in sparged and stirred reactors: Influence of carbon particles and electrolyte. Chem Eng Sci 58(20):4719–4728. https://doi.org/10.1016/j.ces.2003.05.004

Marcilhac C, Sialve B, Pourcher A, Ziebal C, Bernet N, Fabrice B (2014) Digestate color and light intensity affect nutrient removal and competition phenomena in a microalgal-bacterial ecosystem, vol 64, pp 278–287. https://doi.org/10.1016/j.watres.2014.07.012

Marín D et al (2018) Seasonal variation of biogas upgrading coupled with digestate treatment in an outdoors pilot scale algal-bacterial photobioreactor. Bioresour Technol 263:58–66. https://doi.org/10.1016/j.biortech.2018.04.117

Marín D, Vega M, Lebrero R, Muñoz R (2020) Optimization of a chemical scrubbing process based on a Fe-EDTA-carbonate based solvent for the simultaneous removal of CO_2 and H_2S from biogas. J Water Process Eng 37:101476. https://doi.org/10.1016/j.jwpe.2020.101476

Marín D et al (2019) Influence of liquid-to-biogas ratio and alkalinity on the biogas upgrading performance in a demo scale algal-bacterial photobioreactor. Bioresour Technol 280:112–117. https://doi.org/10.1016/j.biortech.2019.02.029

Marín D, Carmona-martínez AA, Lebrero R, Muñoz R (2020) Influence of the diffuser type and liquid-to-biogas ratio on biogas upgrading performance in an outdoor pilot scale high rate algal pond. Fuel 275:117999. https://doi.org/10.1016/j.fuel.2020.117999

Marín D, Carmona-Martínez AA, Blanco S, Lebrero R, Muñoz R (2021) Innovative operational strategies in photosynthetic biogas upgrading in an outdoors pilot scale algal-bacterial photobioreactor. Chemosphere 264:128470. https://doi.org/10.1016/j.chemosphere.2020.128470

Marín D et al (2022) Anaerobic digestion of food waste coupled with biogas upgrading in an outdoors algal-bacterial photobioreactor at pilot scale. Fuel 324:30–40. https://doi.org/10.1016/j.fuel.2022.124554

McKinsey & Company (2023) Global energy perspective 2023. https://www.mckinsey.com/industries/oil-and-gas/our-insights/global-energy-perspective-2023. Accessed 3 Dec 2023

Meier L, Pérez R, Azócar L, Rivas M, Jeison D (2015) Photosynthetic CO_2 uptake by microalgae: an attractive tool for biogas upgrading. Biomass Bioenerg 73:102–109. https://doi.org/10.1016/j.biombioe.2014.10.032

Meier L, Stará D, Bartacek J, Jeison D (2018) Removal of H_2S by a continuous microalgae-based photosynthetic biogas upgrading process. Process Saf Environ Prot 119:65–68. https://doi.org/10.1016/j.psep.2018.07.014

Meier L, Martínez C, Vílchez C, Bernard O, Jeison D (2019) Evaluation of the feasibility of photosynthetic biogas upgrading: simulation of a large-scale system. Energy 189:116313. https://doi.org/10.1016/j.energy.2019.116313

Méndez L, García D, Perez E, Blanco S, Muñoz R (2022) Photosynthetic upgrading of biogas from anaerobic digestion of mixed sludge in an outdoors algal-bacterial photobioreactor at pilot scale. J Water Process Eng 48:102891. https://doi.org/10.1016/j.jwpe.2022.102891

Muñoz R, Meier L, Diaz I, Jeison D (2015) A review on the state-of-the-art of physical/chemical and biological technologies for biogas upgrading. Rev Environ Sci Biotechnol 14(4):727–759. https://doi.org/10.1007/s11157-015-9379-1

Nagarajan D, Lee DJ, Chang JS (2019) Integration of anaerobic digestion and microalgal cultivation for digestate bioremediation and biogas upgrading. Bioresour Technol 290:121804. https://doi.org/10.1016/j.biortech.2019.121804

NASA Earth Observatory (2023) World of change: global temperatures. https://earthobservatory.nasa.gov/world-of-change/global-temperatures. Accessed 6 Dec 2023

Paolini V et al (2019) CO_2/CH_4 separation by hot potassium carbonate absorption for biogas upgrading. Int J Greenh Gas Control 83:186–194. https://doi.org/10.1016/j.ijggc.2019.02.011

Pascual C, Cantera S, Muñoz R, Lebrero R (2021) Siloxanes removal in a two-phase partitioning biotrickling filter: influence of the EBRT and the organic phase. Renew Energy 177:52–60. https://doi.org/10.1016/j.renene.2021.05.144

Pascual C, Lebrero R, Cantera S (2023) Toward a sustainable and cost-efficient biological-based platform for siloxanes removal. Crit Rev Environ Sci Technol 53(1):70–86. https://doi.org/10.1080/10643389.2022.2044254

Posadas E, Bochon S, Coca M, García-González MC, García-Encina PA, Muñoz R (2014) Microalgae-based agro-industrial wastewater treatment: a preliminary screening of biodegradability. J Appl Phycol 26:2335–2345. https://doi.org/10.1007/s10811-014-0263-0

Posadas E, Marín D, Blanco S, Lebrero R, Muñoz R (2017) Simultaneous biogas upgrading and centrate treatment in an outdoors pilot scale high rate algal pond. Bioresour Technol 232:133–141. https://doi.org/10.1016/j.biortech.2017.01.071

Rana MS, Bhushan S, Sudhakar DR, Prajapati SK (2020) Effect of iron oxide nanoparticles on growth and biofuel potential of Chlorella spp. Algal Res 49:101942. https://doi.org/10.1016/j.algal.2020.101942

Ritchie H, Rosado P (2022) Fossil fuels. OurWorldInData.org. https://ourworldindata.org/fossil-fuels#article-licence. Accessed 3 Dec 2023

Rodero MR, Posadas E, Toledo-Cervantes A, Lebrero R, Muñoz R (2018) Influence of alkalinity and temperature on photosynthetic biogas upgrading efficiency in high rate algal ponds. Algal Res 33:284–290. https://doi.org/10.1016/j.algal.2018.06.001

Rodero MR et al (2019a) Technology validation of photosynthetic biogas upgrading in a semi-industrial scale algal-bacterial photobioreactor. Bioresour Technol 279:43–49. https://doi.org/10.1016/j.biortech.2019.01.110

Rodero MR et al (2019b) Development of a control strategy to cope with biogas flowrate variations during photosynthetic biogas upgrading. Biomass Bioenerg 131:105414. https://doi.org/10.1016/j.biombioe.2019.105414

Rodero MR, Severi CA, Rocher-Rivas R, Quijano G, Muñoz R (2020a) Long-term influence of high alkalinity on the performance of photosynthetic biogas upgrading. Fuel 281:118804. https://doi.org/10.1016/j.fuel.2020.118804

Rodero MR et al (2020b) Performance evaluation of a control strategy for photosynthetic biogas upgrading in a semi-industrial scale photobioreactor. Bioresour Technol 307:123207. https://doi.org/10.1016/j.biortech.2020.123207

Rodero MR, Ángeles R, García-Depraect O, Lebrero R, Muñoz R (2023) Recent advances on photosynthetic biogas upgrading to biomethane. https://doi.org/10.1016/B978-0-443-18479-6.00010-7

Ryckebosch E, Drouillon M, Vervaeren H (2011) Techniques for transformation of biogas to biomethane. Biomass Bioenerg 35(5):1633–1645. https://doi.org/10.1016/j.biombioe.2011.02.033

Sander R (2015) Compilation of Henry's law constants (Version 4.0) for water as solvent. Atmos Chem Phys 15(8):4399–4981. https://doi.org/10.5194/acp-15-4399-2015

Sarma SJ et al (2014) Application of magnesium sulfate and its nanoparticles for enhanced lipid production by mixotrophic cultivation of algae using biodiesel waste. Energy 78:16–22. https://doi.org/10.1016/j.energy.2014.04.112

Shahid A et al (2020) Cultivating microalgae in wastewater for biomass production, pollutant removal, and atmospheric carbon mitigation: a review. Sci Total Environ 704:135303. https://doi.org/10.1016/j.scitotenv.2019.135303

Struk M, Kushkevych I, Vítězová M (2020) Biogas upgrading methods: recent advancements and emerging technologies. Rev Environ Sci Biotechnol 19(3):651–671. https://doi.org/10.1007/s11157-020-09539-9

Thiruselvi D et al (2021) A critical review on global trends in biogas scenario with its up-gradation techniques for fuel cell and future perspectives. Int J Hydrogen Energy 46(31):16734–16750. https://doi.org/10.1016/j.ijhydene.2020.10.023

Toledo-Cervantes A, Estrada JM, Lebrero R, Muñoz R (2017a) A comparative analysis of biogas upgrading technologies: photosynthetic vs physical/chemical processes. Algal Res 25:237–243. https://doi.org/10.1016/j.algal.2017.05.006

Toledo-Cervantes A, Madrid-Chirinos C, Cantera S, Lebrero R, Muñoz R (2017b) Influence of the gas-liquid flow configuration in the absorption column on photosynthetic biogas upgrading in algal-bacterial photobioreactors. Bioresour Technol 225:336–342. https://doi.org/10.1016/j.biortech.2016.11.087

Toledo-Cervantes A, Lebrero R, Cavinato C, Muñoz R (2017c) Biogas upgrading using algal-bacterial processes. In: Microalgae-based biofuels and bioproducts, pp 283–304. https://doi.org/10.1016/B978-0-08-101023-5.00012-1

Torres-Franco A, Passos F, Figueredo C, Mota C, Muñoz R (2021) Current advances in microalgae-based treatment of high-strength wastewaters: challenges and opportunities to enhance wastewater treatment performance. Rev Environ Sci Biotechnol 20(1):209–235. https://doi.org/10.1007/s11157-020-09556-8

Ullah Khan I et al (2017) Biogas as a renewable energy fuel—a review of biogas upgrading, utilisation and storage. Energy Convers Manag 150:277–294. https://doi.org/10.1016/j.enconman.2017.08.035

UNE-EN16723 (2018) Natural gas and biomethane for use in transport and biomethane for injection in the natural gas network

Vargas-Estrada L, Hoyos EG, Sebastian PJ, Muñoz R (2022) Elucidating the role of nanoparticles on photosynthetic biogas upgrading: influence of biogas type, nanoparticle concentration and light source. Algal Res 68:102899. https://doi.org/10.1016/j.algal.2022.102899

Vargas-Estrada L, Hoyos EG, Sebastian PJ, Muñoz R (2023a) Influence of mesoporous iron based nanoparticles on Chlorella sorokiniana metabolism during photosynthetic biogas upgrading. Fuel 333:126362. https://doi.org/10.1016/j.fuel.2022.126362

Vargas-Estrada L, Hoyos EG, Méndez L, Sebastian PJ, Muñoz R (2023b) Boosting photosynthetic biogas upgrading via carbon-coated zero-valent iron nanoparticle addition: a pilot proof of concept study. Sustain Chem Pharm 31:100952. https://doi.org/10.1016/j.scp.2022.100952

Velasco A et al (2023) Organic leachate and biogas utilization in outdoor microalgae cultivation under alkaline conditions at pilot-scale. Waste Biomass Valorization 15(7):4265–4277. https://doi.org/10.1007/s12649-023-02223-3

Ward AJ, Feng L, Møller HB (2021) Chemical absorption-amine absorption/stripping technology for biogas upgrading. In: Emerging technologies and biological systems for biogas upgrading, pp 29–55. https://doi.org/10.1016/B978-0-12-822808-1.00003-9

Winkler MKH, Straka L (2019) New directions in biological nitrogen removal and recovery from wastewater. Curr Opin Biotechnol 57:50–55. https://doi.org/10.1016/j.copbio.2018.12.007

Wu L, Wei W, Song L, Woźniak-Karczewska M, Chrzanowski Ł, Ni BJ (2021) Upgrading biogas produced in anaerobic digestion: biological removal and bioconversion of CO_2 in biogas. Renew Sustain Energy Rev 150(Nov 2020):111448. https://doi.org/10.1016/j.rser.2021.111448

Xie T, Herbert C, Zitomer D, Kimbell L, Stafford M, Venkiteshwaran K (2023) Biogas conditioning and digestate recycling by microalgae: acclimation of Chlorella vulgaris to H_2S-containing biogas and high NH_4-N digestate and effect of biogas: digestate ratio. Chem Eng J 453(P2):139788. https://doi.org/10.1016/j.cej.2022.139788

Yang T, Lu L, Wang S, Cao W, Lu X (2017) Biogas upgrading using single-walled carbon nanotubes by molecular simulation. Mol Simul 43(13–16):1034–1044. https://doi.org/10.1080/08927022.2017.1342120

Zhao Q, Leonhardt E, MacConnell C, Frear C, Chen S (2010) Purification technologies for biogas generated by anaerobic digestion. Climate friendly farming, compressed biomethane. In: CSNRA research report 2010-001, Chap 9, pp 1–24

Chapter 22
Treatment and Utilization of Sewage Sludge

Wafa Hassen, Bilel Hassen, Marwa El Ouaer, Solwene Kouki, and Abdennaceur Hassen

Abstract Sludges are semi-solid substances that come from various industrial processes, mainly water treatment, waste treatment, and wastewater treatment. Sludge treatment involves all the methods used to reduce or eliminate health risks, extract water from the sludge, and reduce its weight and volume, to decrease the costs of final disposal. Common sludge treatment methods reduce risks to the natural environment while respecting the water cycle. Sludge generated naturally by sewage treatment plants could now be considered a valuable resource, because proper treatment of these sludges and wastewater directly reduces the operating costs of a treatment plant or sewage treatment plant, offering an economic opportunity for some of the biowaste resulting from degradation and settling processes. Various technologies such as thickening, dehydration anaerobic digestion, thermal hydrolysis, co-digestion, and thermal drying are used to transform sludge into valuable energy, the exploitation of which depends on the energy cost in each country. Most countries around the world use part of their energy produced from wastewater treatment plants after the transformation of sewage sludge, whereas in some countries, a portion of this energy produced in large digesters is sold as green energy (Biogas). Other countries prefer to agronomically valorize these sludges after proper and well-managed

W. Hassen · M. E. Ouaer · S. Kouki · A. Hassen (✉)
Laboratory of Treatment and Wastewater Valorization, Center of Research and Water Technologies (CERTE), Soliman, Tunisia
e-mail: abdohas@gmail.com

W. Hassen
e-mail: hassen.wafa@gmail.com

M. E. Ouaer
e-mail: maroua_el@yahoo.fr

S. Kouki
e-mail: soulwene.kouki@gmail.com

B. Hassen
Laboratories of Microbiology and Pathology of Aquatic Organisms, National Institute of Navy Sciences and Technologies (INSTM), Salammbô, Tunisia
e-mail: bilel.hassen@fst.utm.tn

S. Souabi et al. (eds.), *Wastewater Treatment Plants*, Water Science and Technology Library 130, https://doi.org/10.1007/978-3-031-87461-1_22

treatments. Antibiotic is increasingly being considered as a source of valuable by-products and agricultural inputs, such as phosphorus extracted from struvite, which is a mineral that can naturally crystallize in industrial recovery processes through anaerobic degradation. These crystals are sources of phosphorus, magnesium, and nitrogen recovery from residual sludge.

Keywords Sewage sludge · Sludge processing · Industrial and agricultural valorization · Valuable energy · Valuable by-products

Sewage sludge

22.1 Definition and Origins of Sewage Sludge

The Committee for European Standardization (CES) defines sludge as a mixture of water and solid materials, separated from the types of water that have it through natural or artificial processes. It originates from the treatment of domestic or industrial wastewater. This treatment occurs in multiple stages, based on the principles of physics, chemistry, and biology. The result is purified water discharged into the natural environment and a main residue: sludge. The sludge comprises inert mineral materials, nitrogen, phosphorus, and organic matter. Depending on the treated effluents and the techniques used, the composition of the sludge may vary. Wastewater treatment allows the removal of easily degradable organic matter and various compounds present in wastewater before their return to the water cycle. It is important to distinguish between wastewater treatment and sludge treatment within wastewater treatment plants, considering the possibility of transitioning from one process to the other. Lagooning technique is a simple yet cost-effective method for purifying wastewater. This technique involves the natural treatment of wastewater through the accumulation of water in ponds or basins, also known as biological or stabilization ponds. Within these ponds, a combination of biological, biochemical, and physical processes occurs. Lagoons are stationary systems with a continuous flow, involving of multiple ponds operating in parallel where the inlet and outlet flows are equal. The depth of the ponds typically ranges from 1 to 5 m, and wastewater remains in the ponds for a period of 2–10 days for purification purposes. Thus, lagoon purification serves as a highly effective method for treating wastewater in small to medium-sized communities, as well as in communities with fluctuating populations. In some cases, lagoon plants can be constructed specifically to treat wastewater generated by seasonal food industries.

Therefore, residual sludge refers to the mixture of solids and water extracted from a settler or collected at the bottom of a farming unit. In wastewater treatment, the deposit of semi-solid material remains after the removal of most liquids from the wastewater. Although they can be treated in a septic tank, these sludges are often unusable. Residual sludge is mainly sewage. Sewage sludge eventually dries up in the last waters of a sewage treatment plant (Oleszczuk et al. 2012). Sewage sludge is defined as the waste resulting from the treatment of domestic and industrial wastewater, to reduce their pollutant load and their impact on the environment. There are mainly two types of residual sludge:

1. Primary residual sludge is produced at the beginning of the sedimentation process during wastewater treatment.
2. Secondary residual sludge is produced toward the end of wastewater treatment through microbial digestion.

The sludge from wastewater treatment plants is composed of water, organic, and mineral matter in dissolved or solid form. Even after being dehydrated or treated in vegetative beds or in another form, sludge from wastewater treatment remains. There are two types of sludge: raw sludge and treated sludge.

Raw sludge is present in wastewater treatment plants as the primary sludge during the mechanical cleaning stage or as excess sludge during the biological stage. Excess sludge mainly contains microorganisms such as protists and bacteria. Aerobic or anaerobic treatment of raw sludge results in less odorous sludge, known as treated sludge or stabilized sludge. Anaerobic treatment is carried out in large wastewater treatment plants as biogas, with a methane content of 60%.

At first, wastewater sludge is of liquid consistency and dark. Its gravity thickening reaches a solid content of around 2–5%.

By using flocculation tools, sludge can be drained, for example, through centrifuges, screw presses, or belt installations, up to 35%. It is also possible to achieve higher levels of drainage using chamber filter presses or sludge dewatering installations, which are also undergoing organic matter biodegradation. Sewage sludge is rich in nutrients because microorganisms use the contents of wastewater in the biological stage to build biomass, thus concentrating the nutrient salts present in the wastewater. Nitrates, phosphates, and other nutrients are important for agriculture.

In sewage sludge, many organic compounds with different properties and effects can be found. Some of these compounds can be carcinogenic, mutagenic, toxic, etc. Heavy metals pose particular problems. For example, chromium is essentially non-toxic as chromium (III), but toxic and carcinogenic as chromium (VI). German regulations on sewage sludge and fertilizers set limit values to reduce risks to human health and the environment. Even though these substances are present only in low concentrations in wastewater, there is a risk of accumulation after agricultural spreading ("bioaccumulation") and contamination of the food chain.

22.2 Sludge Characterization and Classification

The commonly transported pollutants in wastewater and the liquid phase by-products from any water treatment process eventually accumulate in more or less concentrated suspensions called "sludge". These sludges have in common that they are still very liquid waste, generally of low or no value. Some are chemically inert, but those derived from biological treatments are often fermentable and emit a foul odor. Therefore, in any study on residual sludge, it is essential to characterize these sludges to determine the proper treatment method and estimate the performance of the equipment to be used. Table 22.1 presents a classification of different purification sludges based on their origin, composition, and especially the nature of the purification treatment they undergo.

Type of Sewage Sludge

Sludge types can be classified according to their dryness or dry matter/water ratio. Sludge dryness is defined as the residual water content of sludge, expressed as a percentage (%) (Table 22.1). The second classification recommended in some countries is based on the dry matter/water content ratio. Thus, these distinctions are made:

- Hydrophilic organic sludge (URW: urban wastewater, FIA: agri-food industry).
- Hydrophilic oily sludge (refinery, mechanics).
- Hydrophobic oily sludge (rolling).
- Hydrophilic mineral sludge (surface water, surface treatment, tannery).
- Hydrophobic mineral sludge (drilling, surface water, gas washing).
- Fibrous sludge (paper mills).

Table 22.1 Sludge characteristics based on dryness and some selected properties

	Dryness (%)	Physical characteristics	Type of transport
Liquid sludge	1–2	Liquid	Pumping and tankers
Thickened liquid sludge	8–12	Liquid	Pumping and tankers
Pasty sludge	18–30	Pelletizable	Pumping and tipping
Solid sludge	30–90	Pelletizable, stackable, and crushable	Conveyor belt transport
Dry sludge	> 90	Powdered or granulated	Conveyor belt or pneumatic system
Compost sludge	50–60	Shovelable, stackable, and crushable	Conveyor belt transport

Table 22.2 Sewage sludge classifications

	Class A	Class B	Class C	Class D
Definition	Primary sludge, physico-chemical, and high load	Biological	Mixed A + B	Stabilized in digestion, stabilization
Parameter				
Suitability for concentration	Excellent	Medium	Good to low	Medium to low
Suitability for stabilization	Excellent	Good to low	Medium to good	–
Ability to not ferment	Medium to low	Good	Low	Good
Treatability	Very good	Medium to low	Good	Good

Considering the treatability of sludge, a classification has been developed in France based on American standards. Table 22.2 details the four proposed classes. **Likewise, sewage sludge is often classified according to the treatment it undergoes.**

22.2.1 Primary Sludge

Primary sludges are produced during the initial purification process of a wastewater treatment plant. They are derived from the treatment of effluents by settling the wastewater in primary settling tanks. These waters have already undergone a physical pretreatment, including sand removal and screening, before passing through a settling tank to remove the heaviest materials. The residues collected at the bottom of the tank contain a high concentration of mineral and inorganic matter. On the other hand, the other solid materials collected by screening and sand removal are directly directed to recovery bins to be transported to landfills. These pretreatments are essential when the wastewater arrives at the treatment plant because they carry various materials, often coarse or waste, that could disrupt the proper functioning of the subsequent treatment steps. These coarse elements could clog the different units of the facility and compromise the efficiency of the subsequent treatments. Screening is carried out using grids with varying bar spacing depending on the size of the objects to be kept. After screening, comes the sieving step, which ensures a thorough screening by filtering the raw water on fabrics, meshes, or perforated sheets, with more or less fine meshes to remove solid matter (impurities) present in the water. Sand removal follows the sieving step and allows the removal of gravel, sand, or mineral particles through sediment traps, to prevent the deposition of these particles in the facilities and to protect the treatment units against abrasion. Finally, as the final pretreatment step, oil removal includes recovering oils using scrapers.

The pretreated waters then follow their purification path toward the settling basins and then toward the aeration or biological activation basins. Thus, the sludge extracted from the primary treatments, usually the sludge from the primary clarifier, is isolated by settling during the primary treatment. This treatment includes removing the remaining dissolved pollutant load and residual suspended matter from the water that has undergone pretreatment. The coarse particles that were not trapped during the pretreatment stages, whether they are organic or mineral, finely dispersed or in a colloidal state, and with highly variable particle density, are mainly removed during the settling stage. The pretreated wastewater passes through the settling basins at a low velocity, letting the suspended matter settle. To promote the sedimentation processes and concentrate the sludge in the center of the basin, the primary clarifiers are equipped with a rotating bridge. This rotating bridge concentrates the settled matter toward the center of the basin, where it is then extracted by pumping stations to be transported to the sludge treatment facilities.

Decantation processes only allow for the removal of particles larger than 50 μ. However, the separation of finely dispersed materials and colloids, which are responsible for water discoloration, cannot be achieved directly through decantation. Colloidal species present in raw water or wastewater include clay, silica, iron, other heavy metals, and organic solids such as debris from dead organisms, with an average size smaller than 1 μ. Their settling velocity is slow under laminar flow conditions, making them difficult to decant. Therefore, it is essential to resort to secondary treatments involving biochemical reactions, physico-chemical processes, or microbial biological processes to eliminate non-settleable particles and dissolved materials. This includes secondary treatment with activated sludge, bacterial beds, etc.

This process is activated by a proper insufflation and injection of compressed air into the wastewater to reactivate the general bio-physical–chemical processes that govern the basin. The sludge generated during this stage will be transferred to another type of treatment basin, called a clarifier basin. This is the final stage of water treatment. It involves letting the water settle, so the last impurities can settle at the bottom. After this stage, the water is finally discharged into nature. However, it is not yet clear and not clean enough to be discharged into the natural environment. A final treatment stage, called tertiary treatment, is necessary for the treated wastewater to be clean enough to be safely discharged into nature without harming ecosystems. The sludge recovered from the clarifiers is called the secondary sludge.

The settling processes only allow the removal of large particles, larger than 50 μ. However, the separation of finely dispersed materials and colloids responsible for water discoloration cannot be achieved directly through settling. Colloidal species present in raw water or wastewater include clay, silica, iron, other heavy metals, and organic solids such as debris from dead organisms, with an average size smaller than 1 μ. Their settling velocity is low under laminar flow, making them difficult to settle. So it is essential to resort to secondary treatments involving biochemical reactions and physico-chemical or biological processes of microbial origin to remove non-settleable particles and dissolved materials. The secondary treatment process of activated sludge, bacterial beds, etc., is necessary. Thus, during the biological treatment with activated sludge, the wastewater to be treated will undergo aeration by forced air injection into the "activated sludge basin" to ensure the suitable conditions (O_2) for the development of microorganisms that aggregate and form the bacterial floc. Polluting organic materials will be captured by these flocs and form activated sludge that is stirred and ensures the purification of wastewater in the basin.

As for treatment by a biological bed, the principle includes letting the water be treated, previously settled during the primary treatment, to flow over a porous support made of plastic materials often and carrying purifying microorganisms. The water to be treated is evenly distributed on the filter surface as rain dispersion via a distribution grid, followed by a phase of natural aeration induced by the effect of air current to provide the oxygen throughout the biomass of the bed to maintain the proper conditions for the growth and development of the purifying microflora. The ultimate aim is not to develop biomass but to return purified water after transforming primarily dissolved materials into microbial biomass that can be removed through filtration or settling effects. A liquid-biomass separation step is required and is ensured by a clarification device. Like activated sludge, a portion of these sludges is used to reseed biological ponds while the rest is transferred to the sludge treatment system.

Following the treatment process, as previously mentioned, a clarifier (or secondary settler) allows for the isolation and separation of sludge from the treated wastewater. To maintain a constant and enough stock of bacteria in the activated sludge tank, part of the sludge extracted from the clarifier is returned to the head of the tank. The remaining part is removed from the system and directed toward the sludge treatment units.

The biological processes are often activated by a good insufflation and injection of compressed air into the mass of wastewater to reactivate the general bio-physical–chemical processes present in the basin and the sludge generated. This is the process of biological flocculation or bio-flocculation energized by the resident microbial flora. The sludge-laden wastewater will then be directed to a clarifier basin for the final stage of water purification. This stage involves letting the water settle so the last impurities fall to the bottom. After that, the water is discharged into nature, but it is not yet clear or ready to be discharged into the natural environment. A final treatment step, called ultimate treatment, involving physical disinfection by irradiation or chemical disinfection by chlorination, is then necessary. The resulting sludge separated from the treated water and recovered from the clarifiers is called the secondary sludge.

As for tertiary treatment, it is not systematic, and it would be a refining operation of the effluent before reintroducing the purified water into the natural environment by reducing suspended solids and biodegradable organic pollution through tertiary filtration which, depending on the material used, allows for primarily physical or biological purification, or by reducing non-biodegradable dissolved organic pollution such as detergents through adsorption on activated carbon. Similarly, tertiary treatment can be done by reducing phosphorus pollution, also known as de-phosphatization, since the main disadvantage of high phosphorus content in water is to promote eutrophication of water bodies after the discharge of treated wastewater into natural receiving environments, lakes, and marine environments.

Eutrophication occurs after the excessive enrichment of an aquatic environment (especially if the water is stagnant or has reduced circulation) with nutrients or organic matter, causing an overabundant growth of plant biomass. The subsequent decomposition of this biomass consumes, partially or entirely, the dissolved oxygen in the water and reduces the biodiversity of the aquatic environment. Therefore, reducing the phosphorus inputs from wastewater is often considered the key factor in combating eutrophication in rivers and lakes. Phosphorus in water can originate from three sources: diffuse inputs from agriculture (livestock, fertilizers), industrial inputs (agro-food industry, pulp and paper industry), and domestic inputs, mainly from detergents. The commonly applied techniques for phosphorus removal involve precipitation reactions (physico-chemical processes) or the use of purifying microorganisms that ensure an over-elimination of phosphorus (biological processes). Chemical phosphorus removal is achieved through reagents (such as iron and aluminum salts or lime) that form insoluble precipitates or complexes, which are then separated from the water using solid–liquid separation techniques.

Biological de-phosphatization is based on transferring phosphorus from the liquid phase (wastewater to be treated) to the solid phase (sludge) through intracellular storage. This accumulation leads to the formation of polyphosphate granules (poly-P) and results in a progressive enrichment of the sludge in phosphorus to high levels. It is then easy to ensure the removal of phosphorus from the water by simply withdrawing excess sludge after a settling step. In the final stage of the known tertiary treatment, the focus is on the elimination of pathogenic germs: disinfection. Since urban wastewater represents the liquid waste from individual and collective life in urban areas, it commonly has pathogenic microorganisms that can cause diseases such as bacteria, viruses, and parasites.

The main purpose of disinfection is to destroy pathogenic microorganisms. Various disinfection processes that can be implemented include chlorine in gaseous form or as sodium hypochlorite or bleach. This method of disinfection and sterilization is the most commonly used and well-controlled; chlorine dioxide (ClO_2), which is a highly effective reagent but requires delicate implementation and significant monitoring; ozone, which is highly effective in eliminating viruses but is underutilized because of its high implementation cost; and finally, using ultraviolet UV-C254 rays, which are a highly developed disinfection technique. This process is increasingly being used in wastewater treatment plants because of its simplicity, cleanliness, and interesting efficiency. Among the various tertiary treatment techniques mentioned alongside UV-C254 irradiation, phosphorus removal is primarily used usually.

22.2.2 *Physico-chemical Sludges*

Physico-chemical sludges are produced by adding reagents such as iron salts, aluminum, called chemical coagulant to agglomerate fine particles in water during the treatment of mainly drinking water and industrial wastewater carrying often one or more toxic, refractory, and non-biodegradable compounds that could not be neutralized during the primary treatment such as hexavalent chromium, cyanides, pesticides, and certain hydrocarbon compounds. This physico-chemical treatment aims to agglomerate fine particles and certain soluble pollutants, which have difficulties in naturally settling or do not settle, into precipitates or stable solids called "flocs" before passing through a separator such as a decanter or a floater. The coagulation process is a mechanism that allows the agglutination and agglomeration of different particles and compounds that will transform into a block, called a floc, more or less compact. The small particles in a liquid often have the same (negative) charge and repel each other. With coagulation, a chemical added to the water or coagulant eliminates the charges of the particles to help with their agglomeration. Thus, coagulation is a chemical treatment process widely used in water treatment. It is mainly used to remove suspended matter particles from water by manipulating the electrostatic charges of the particles suspended in the water. This process introduces small highly charged molecules into the water or coagulant to destabilize the charges of the particles, colloids, or oily substances in suspension. Coagulation is commonly used in drinking water treatment plants as a primary treatment and in many wastewater treatment plants in areas known for high contamination by recalcitrant and non-biodegradable products such as industrial areas for paper and textiles processing. It is an effective physico-chemical process to support conventional primary treatment. The many coagulant products or agents are simple common chemicals such as ferric chloride and aluminum sulfate. Aluminum sulfate derives its strength from its versatility. It is also the most versatile chemical product used in municipal and industrial markets, while ferric chloride is the coagulant of choice for many industrial and sanitary wastewater treatment applications, because of its high efficiency in clarification and its usefulness as a sludge dewatering agent.

Although it leaves a slight residual color, this chemical product offers excellent turbidity removal. The coagulation process is often followed by a mechanical phase called flocculation. Flocculation forces the neutralized particles to come together. This second phase is essential to ensure the effectiveness of the initial treatment phase. Therefore, flocculation is the step that follows coagulation in the treatment of drinking water and wastewater. The flocculation process is primarily a mechanical process aimed at gathering clusters of suspended matter that have already been coagulated. After flocculation, enormous masses called flocs are formed, which are clusters of interconnected matter. Eventually, these flocs will transform into precipitate and settle to the bottom to form chemical sludge.

22.3 Common Sludge Treatment Techniques

The processing of sewage plant sludge depends on its composition and involves four commonly used techniques for sludge treatment.

22.3.1 Sewage Sludge Thickening

Thickening is an essential step to increase the dry matter content of the sludge up to 6–8%. This process is important to ensure better treatment quality afterward. First, thickening helps reduce the volume of sludge from wastewater treatment plants. Its goal is to increase the dry matter content of the sludge without changing its liquid consistency, to optimize the subsequent steps of conditioning, stabilization, and dehydration for their final destination. The performance of thickening is decisive, as it directly influences the sizing and operating cost of the sludge treatment chain. The physico-chemical characteristics of the sludge vary from one wastewater treatment plant to another, each requiring specific treatment before reaching its final destination or being valorized. Thickening can be mechanically achieved through drainage (drainage table, thickening drum) or flotation (dissolved air flotation).

To get a less liquid sludge, the thickening process of sewage sludge does not involve adding a thickener or thickening substance, but removing some of the water (dehydration). The reduction of sludge volume therefore involves a decrease in water content, which helps with and reduces transportation and storage costs, especially for agricultural use. Wastewater treatment processes typically focus on separating water and constituent matter. Biological flocculation, through microbial action, or physico-chemical flocculation, through the addition of a chemical coagulant, is an important first step in the thickening process, as it allows the dry matter to aggregate into "flocs". This step aims to give a "structure" to the sludge so it can effectively undergo subsequent steps for optimal separation of water and matter. Reducing the volume of sludge from wastewater treatment plants recommended in the treatment plants mainly defines the subsequent steps by reducing the size of equipment, energy consumption, and operating costs.

Thickening is the first, often unavoidable, step in reducing the volume of sludge extracted from the wastewater treatment process. It optimizes the subsequent conditioning, stabilization, and dewatering steps by reducing the dimensions of the facilities and operating costs. The sludge, very diluted at the process outlet for water contents usually ranging from 99 to 99.8%, will be concentrated to 90–96% in water but will remain fluid enough to be pumped. Thickening represents an additional investment and a potential source of nuisances, but this is largely offset by the advantages downstream in the process: reduction in the size of anaerobic digesters or aerobic stabilization reactors; elimination of thickeners downstream of biological stabilization processes and reduction of pollution returns at the head of the plant; possibility to extract diluted sludge from primary clarifiers, thus avoiding fermentations in these installations, leading to foaming and bulking with the development of filamentous bacteria or actinomycetes in biological treatment; reduction in conditioning rates upstream of dewatering; increase in the production of dewatering equipment and reduction in energy consumption; finally, a significant decrease in pumping rates, transfer pipelines, and potential intermediate storages. The main thickening processes commonly used can be summarized:

- Gravity thickening relies on the effects of gravity. Fresh sludge feed occurs in the central zone, thickened sludge is collected in the lower part, and the supernatant remains in the upper part. This system is used for primary, physico-chemical, and mixed sludges that settle well by gravity. Biological sludges recovered from biological treatment processes such as trickling filters or activated sludge settle mainly slowly.
- Flotation thickening: The sludge concentrates and aggregates in the upper part through the attachment of micro-bubbles, usually air, to the suspended solids, which are ultimately less dense than water. This system is specified and recommended for thickening biological sludges because of their low sedimentation capacity.
- Mechanical thickening: The concentration and aggregation of sludge are achieved by increasing gravitational forces through centrifugation. Centrifugal force is applied, allowing separation. It is mainly used for biological sludges. These are mostly expensive devices that require proper maintenance measures, such as a rotary drum leading to separation through filtration across the rotary drum. It is used with biological sludges. The intervention costs are not high, they require little space, and they do not produce odors. Finally, thickening tables and separation occur by draining water through a moving porous horizontal belt. They are suitable for activated or digested sludges. They are not suitable for physico-chemical sludges.

22.3.2 *The Stabilization*

The stabilization and hygienization process involves reducing the water content in sludge significantly (up to 80%), even drying it (5–10% water content). This step helps to decrease odors, stabilize organic matter, and eliminate any organisms or bacteria that could cause diseases. One of the three fundamental processes used for sludge stabilization aims to reduce the amount of organic matter (OM) present, decrease pathogens, eliminate odors, and most importantly, reduce or eliminate the putrefaction capacity of OM. Stabilization processes are divided into.

22.3.2.1 Biological Stabilization

Aerobic Stabilization: Aerobic stabilization is a biological process where organic matter is oxidized by microorganisms with the addition of oxygen in open digesters. This helps reduce the final amount of sludge by transforming it to make it suitable for the subsequent treatment stages. This process is commonly used as a secondary treatment in a wastewater treatment plant without primary treatment. It can still be employed to mix sludges with a higher oxygen input. Factors that influence this process include retention time, temperature, and oxygen and mixing requirements.

Aerobic Thermophilic Stabilization: This self-heating aerobic digestion method was developed to meet increasingly stringent standards. It is based on the conservation of thermal energy generated during the aerobic fermentation of organic substances in the sludge.

Composting: This process involves the biological decomposition and stabilization of organic matter under controlled and aerobic conditions, resulting in the development of thermophilic temperatures and the generation of biological heat. The outcome is a stable and pathogen-free product. Organic matter decomposes into CO_2, water, minerals, and stabilized organic matter. This process can be carried out only with sludge or by mixing it with structuring agents that facilitate aerobic conditions. Typically, carbon-based woody residues (wood chips, dead leaves, etc.) are mixed with organic materials to create a suitable composting mixture. The structuring agents help adjust the moisture content or carbon/nitrogen (C/N) ratio by increasing porosity and promoting air circulation within the mixture (Hassen et al. 2001).

The main steps of the process can be summarized as follows: mixture preparation, fermentation or composting, stabilization and biological maturation, and finally mechanical refining. Composting is an effective process in the decontamination of organic pollutants such as petroleum hydrocarbons, mono-aromatic compounds, explosives, chlorophenols, certain pesticides, and polycyclic aromatic compounds. The role of microorganisms is essential as they partially or even degrade the main compounds present in the sludge through good oxygenation of the environment, either through natural aeration or air insufflation. Microorganisms can thus partially or completely transform and mineralize the main organic compounds. However, as for certain specific pollutants, such as metallic pollutants, they are not significantly eliminated during the process. Oxidation and reduction reactions of these pollutants alter their solubility, thereby decreasing their bioavailability and toxicity within the solid phase (Hachicha et al. 2006, 2009). It is essential to adequately control parameters such as temperature, pH, aeration or oxygen content, moisture, and C/N ratio during the composting process in order to avoid anaerobic conditions that lead to an increase in unpleasant anaerobic odors and limit the functioning of aerobic microorganisms, which are considered the main decomposers of matter and actors in the composting cycle (Hassen et al. 2000; Fersi et al. 2018).

Anaerobic Stabilization: It is one of the most commonly used methods for stabilizing sludge, also known as biomethanization. It involves the degradation of organic matter in the absence of oxygen, releasing energy and primarily producing methane (CH_4), carbon dioxide (CO_2), and water (H_2O) through the action of specific types of anaerobic bacteria called methanogens. This process occurs in four distinct stages: hydrolysis, acidogenesis, acetogenesis, and finally methanogenesis with the release of methane or CH_4. The implementation of this process depends on the organic load of the environment, which allows for the distinction between low-load systems, high load systems, anaerobic contact systems, and systems with gas separation. It is essential to control certain parameters such as pH, temperature, sludge feeding, retention, and residence time, as well as biogas content at the outlet during this biomethanization process (Reverdy et al. 2011; Traore et al. 2016).

22.3.2.2 Chemical Stabilization

This is an alternative to biological stabilization for sludge treatment. The objective of this stabilization is to reduce pathogens and substantially reduce odor-producing microorganisms.

Stabilization with Limestone or Lime: The predominantly applied product is limestone. It is added to the sludge in the proper dosage to maintain a pH of 12 for enough time, at least 2 h, to suppress or reduce anaerobic microorganisms responsible for odors. This system is widely used:

- In small wastewater treatment, plants with sludge incorporated into natural land or stored before transportation.
- Wastewater treatment plants requiring additional stabilization.
- Complementary stabilization system when other systems are out of service.

It is usually added before the sludge is dried, but it can also be used later on using smaller quantities of limestone. The dosage of limestone depends on the sludge, the chemical composition of the sludge, including the organic matter content, and the sludge concentration.

During the sludge treatment process with quicklime or slaking, it is necessary to maintain the pH above 12 for at least 2 h to ensure the destruction of pathogens and provide enough residual alkalinity, so the pH does not drop below 11. These conditions are necessary to ensure the complete cessation of biological activity within these sludges and allow for proper storage or disposal of the stabilized sludge. The amount of limestone required to stabilize the sludge is determined by its duration, chemical composition, and solids concentration. This quantity often varies between 6 and 51%. It should be noted that primary sludge requires less limestone than activated sludges. The sludges that can be conditioned only with lime are as follows: certain sludges from surface treatment industries, tanneries, or drinking water.

The use of lime after thickening or dewatering of the sludge increases the dry matter content of the sludge, improves its texture, stabilizes it by reducing its fermentability to mitigate or eliminate foul odors, and finally makes the sludge hygienic by reducing all pathogenic microorganisms to a non-hazardous level.

Chlorine Oxidation: A significant amount of chlorine is introduced into the sludge to be treated. This procedure is carried out in sealed reactors and requires short retention times. This method has not yet been widely implemented in the industry.

22.3.2.3 Sludge Conditioning

Gelatinous sludges can complicate drying operations, which is why pre-conditioning is recommended to enhance their characteristics for dehydration and stabilization. The most common methods include (Table 22.3):

Table 22.3 Most commonly used sludge conditioning processes

	System	Advantages	Disadvantages
Biological stabilization	Aerobic	Easy control and operation Low exploitation risks (non-flammable) Biologically harmless and stable sludge Final residue applicable to natural terrains Reduction of volatile solids Almost complete elimination of odors	Poor results with cooling; Poor results in drying units; Large reactor volume
	Thermophile	High destruction of volatile solids; Low reactor volume Cleaned sludge Lower specified oxygen requirement (30–40% less than conventional processes) Drier dewatered sludge	Pre-thickening required; high agitation Tank with good thermal insulation; Occasional odors Foams Higher polyelectrolyte consumption
	Composting	High reduction in initial sludge volume High moisture reduction (60–70%) Decomposition and stabilization of organic matter High weight reduction in the final product Cleaned sludge: removal of pathogens and parasites and inhibition of seeds Significant odor reduction Optimal end product for use as a fertilizer or organic modification	Need for large areas mainly in low-intensity technologies (composting in piles) No energy recovery Problematic management of the water generated in the process; need for structuring agents Dust and aerosol generation in certain types of composting; in open-air plants, the composting process in winter is slowed down Odor generation that requires control of critical parameters

(continued)

Table 22.3 (continued)

	System	Advantages	Disadvantages
	Anaerobic	Lower reactor volume; biomass reduction; pathogen reduction Lower energy consumption; Energy recovery (CH_4) Odor elimination	Slow process Comprehensive control of critical parameters
Chemical stabilization	Coagulation–flocculation	Strong pathogen reduction; high odor reduction; larger usable volume Free neutralization in acidic soils Backup method for other stabilization systems Farm-friendly end product (high lime and nitrogen content)	Critical pH control to prevent the reappearance of odors and pathogens pH adjustments can cause gas emissions (high ammonia pH/low sulfuric pH) Higher transportation costs due to increased volume of the final product

Chemical Conditioning: This process leads to the coagulation of solids and the release of absorbed water, and it is used before the drying process. Commonly used chemicals include ferric chloride, quicklime or limestone, aluminum sulfate, and organic polymers. The latter, grouped under the general name of organic polymers, belongs to a vast family divided into two main groups: natural polymers such as sawdust, plant fibers, and latex, and synthetic polymers including plastics such as PVC, PTFE, polyethylene, polycarbonates, and artificial rubbers. Among natural polymers, the subgroup of biopolymers can be notably mentioned, which are polymers derived from living organisms or synthesized from monomers obtained from renewable resources, such as polysaccharides like starch, cellulose, chitin, proteins like casein, wheat gluten, collagen, and finally bacterial polyesters and microbial or transgenic polymers. The first three products, namely ferric chloride, quicklime or limestone, and aluminum sulfate, enable the disinfection and stabilization of sludge, while organic polymers do not have a disinfecting effect but are easier to handle and usually more cost-effective. Finally, it is essential to emphasize that the purpose of this stabilization through chemical conditioning is to reduce pathogens and substantially reduce odor-producing microorganisms, namely anaerobic metabolism microorganisms.

Thermal Treatment: The sludge is heated to temperatures ranging between 160 and 210 °C for short periods under pressure. This process results in the coagulation of solids and a change in structure, thus reducing the water affinity for the sludge solids.

The sludge becomes sterilized, almost deodorized, and its dehydration capacity goes up significantly.

Dehydration: Dehydration is a physical operation (natural or mechanical) used to reduce the moisture content and volume of sludge. Its primary objectives are to increase the dry matter content of sludge from 3 to 40%, reduce transportation costs by decreasing volume, improve handling and transportation of sludge, prevent odors, and increase calorific value by reducing moisture. The most commonly used systems are mechanical compared to natural systems. From an economic perspective, the preferred dehydration technologies are, in decreasing order, centrifuges, belt filters, and filter presses.

22.3.3 *Mechanical Systems*

Sewage sludge can be subjected to various mechanical treatment processes. Among them we can cite.

22.3.3.1 Centrifuges

A cylindrical-conical drum with a horizontal axis is used for the separation of the solid phase from water by relying on centrifugal force. There are two types of centrifugations in the sludge dewatering process: (a) countercurrent centrifugation: solids and liquid flow in opposite directions in the drum, and (b) co-current centrifugation: the solid fraction and the liquid fraction flow in the same direction.

22.3.3.2 Filter Press

Filter presses involved a series of vertical rectangular plates arranged one behind the other on a frame. Filter clothes, usually made of synthetic fabrics, are placed on the faces of these plates. The space between two plates, in the hollow center, determines the thickness of the resulting cake, which can vary between 15 and 30 mm. The surface area of filter presses can reach up to 400 m^2, with plates of 2 m^2 each. These filters typically include over 100 plates. The filtering process can last up to 25 h, depending on different stages, such as filling, filtering, unloading, and cleaning. This process allows for a sealing efficiency of 35–45% depending on the characteristics of the sludge being treated. It is essential to have specialized and qualified staff to ensure the maintenance and operation of these filters.

22.3.3.3 Belt Filter

This system allows for continuous feeding of the sludge while performing chemical conditioning, usually with polyelectrolytes. In belt filters, gravity drainage is first carried out, followed by dehydration through the mechanical application of pressure through a porous fabric. This is an economical method as it does not require significant initial investment, maintenance, and operating costs are reduced, and the installation consumes low energy.

22.3.4 Natural Systems

22.3.4.1 Drying Period

This system relies on natural dehydration. The layers of material are vertically arranged in a container. The sludge passes over these layers of gravel or sand, allowing for filtration and dehydration of the sludge through evaporation. This evaporation will depend on the climatic conditions of the area, the number of days the sludge is exposed, and the characteristics of the sludge.

The composition of the draining material usually includes layers of sand, typically 10 cm thick, and a layer of gravel, with a depth ranging from 10 to 20 cm. At the bottom, a network of drainage pipes is positioned to gather the water. The sand layer needs to be regularly replenished, as some sand is lost during the filtering and sludge recovery process. The drawback of this process is the extensive land area required, while its major advantage is the low maintenance and upkeep cost.

The sludge from wastewater treatment, once properly treated, can be exploitable resources rather than harmful waste. Composting them lets businesses and communities become energy self-sufficient, and even sell the energy produced to a local network. The use of sludge helps combat the depletion of raw materials by resorting to methanization. This biological process transforms sludge from urban treatment plants into biogas, a form of green energy. Unlike solar or wind energy, biomass from organic waste, such as residual sludge, can be stored and used continuously and at the right time. Once treated, the sludge is ready to be integrated into a system.

As for valorization, it is possible to use certain sludges in the agricultural sector by spreading them in fields as fertilizer or compost. Other sludges, after dehydration, can produce energy in thermal furnaces. Sometimes, it is possible to reinject the sludges into the production process of the industry.

Regarding disposal, sludges with difficult-to-manage particles should be incinerated. Some sludges are also disposed of in landfills with household waste. It all depends on the treatment they have undergone and the sector from which they originate. Sludges from the nuclear and chemical sectors are normally disposed of rather than valorized. On the other hand, high-quality sludge from the agri-food sector would be beneficial if spread in fields.

As mentioned earlier, sewage sludges are residues from the treatment of liquid effluents by sewage treatment plants. They involve organic and mineral materials, and their dry matter and organic matter content depend on the effluent treatment process. Sludges undergo different treatments to reduce their water content, stabilize them, and prevent fermentation, and biological contamination risks, depending on their intended use. If the infrastructure or characteristics of the sludges do not allow for valorization, they are stored in storage facilities.

This reuse is seen from two angles: disposal and/or reuse of sludge.

The main disposal and/or reuse routes are limited to controlled landfills, incineration in furnaces, and agricultural reclamation.

22.3.4.2 Controlled Landfill

Controlled landfilling represents an average of 20–25% of the tonnage in most European countries. It involves burying sludge, often mixed with household waste, while respecting certain conditions such as compacting the residues, ensuring the site is waterproof, recovering and treating leachate, as well as equipping and managing the site (Allen 2001). These percentages usually decrease in line with the European Union directives, which foresee a gradual reduction in landfilling of biodegradable and recoverable municipal waste (nutrient source, energy), including residual sludge.

The sludge needs to undergo stabilization and dehydration prior to its use, ensuring that its moisture content does not exceed 70%.This method has gradually lost its appeal because of financial considerations and environmental issues such as unpleasant odors, mosquito proliferation, the addition of fertilizing elements (nitrates, phosphates), toxic products into surface waters and contamination of groundwater.

Landfills should only accept waste that can no longer be reasonably recovered, or that is non-recyclable or hazardous, also known as ultimate waste.

22.3.4.3 Incineration

The incineration method allows for the destruction of organic matter in waste by burning it at high temperatures (over 500 °C), which produce smoke and mineral residues called ashes. It is between 15 and 20% of the total weight of sludge, but it incurs high costs for wastewater treatment plant operators. The waste is burned in a high-temperature chamber, preferably at 1200 °C, with enough oxygen to oxidize all organic materials. This method can thus destroy 99.9% of organic waste, including chlorinated organic compounds if used correctly. However, this process is expensive and can be carried out with or without energy recovery. It is mainly used in large urban areas, communities where sludge spreading is competing with the spreading of organic waste from agriculture, and in communities with a tourist vocation. This technique is also harmful from an ecological and environmental point of view, as it not only contributes to the waste of useful organic matter for the soil but also to the release of highly toxic gases such as NO, NO_2, CO, SO, and dioxins. This contributes to air pollution and greenhouse gas emissions.

22.3.4.4 Usage of Sludge in Agriculture

The adoption of sewage sludge for fertilization and soil improvement is of great interest in agriculture, as the cost of commercial fertilizers continues to rise (Annabi et al. 2007). This valorization process is optimal for eliminating the pollution caused by wastewater treatment. Agricultural valorization involves spreading sludge to provide fertilizing elements (NPK) and organic matter (OM) to soils that are depleted in these elements (Hamdi et al. 2007, 2012, 2019; Picariello et al. 2021). The purpose is to preserve natural resources and avoid any waste of organic matter through incineration or burial of landfills (Hechmi et al., 2020a). This practice is commonly used in agricultural lands and accounts for 55–65% of the total volume of sludge. It has been the preferred method for recycling sewage sludge since 1975 because of its agronomic benefits. This practice not only applies to sewage sludge but also to hundreds of millions of tons or cubic meters of various materials. When properly applied, these materials contribute to maintaining soil fertility, thus reducing the need for commercial fertilizers. In agriculture, sludge is used as fertilizer, meaning that it is a product capable of providing crops with the needed nutrients for their growth and development (Miranda et al. 2018). Given the many treatment processes used in different plants, which generate various types of quality sludge, the composition of sludge can vary depending on the type of treatment used, the incoming effluent, or the size of the plant.

22.3.4.5 Interest in Sludge Application in Agriculture

Organic matter: The organic matter found in the soil acts as an energy source for soil microorganisms. These microorganisms decompose it, releasing essential nutrients and CO_2 for plant growth. In return, plants return organic matter to the soil as crop residues, leaves, and dead roots, thus completing the cycle. The effects of organic matter transformed into humus in the soil are diverse:

- Improvement of soil structure and stability.
- Balance of the soil humic balance.
- Ease of soil workability.
- Soil warming.
- Water retention.
- Progressive supply of nutrients.
- Stimulation of biological activity.

The soil contains between 2 and 3 kg/m^2 of biomass. The biological activity of this biomass contributes to phenomena like soil structure and porosity soil permeability, combating the leaching of nutrients or erosion, and the availability of certain nutrients such as phosphorus or potassium.

The C/N ratio, or carbon to nitrogen ratio, is an indicator that allows to assess the extent of evolution of organic matter, its ability to decompose more or less rapidly in the soil:

- C/N < 15: Nitrogen production, decomposition rate increases; it is the maximum for a C/N ratio = 10.
- 15 < C/N < 20: Additional nitrogen is needed to promote good decomposition of carbonaceous matter.
- C/N > 20: Insufficient nitrogen to allow carbon decomposition (competition between plant absorption and transformation of organic matter by soil microorganisms, the phenomenon of "nitrogen hunger"). Nitrogen is then drawn from the soil reserves. Mineralization is slow and only returns a little mineral nitrogen to the soil (Jedidi et al. 1995; Hassen et al. 1998).

It is widely accepted that the higher the C/N ratio of a product, the slower its decomposition in the soil, but the humus achieved is more stable (Hassen et al. 2023).

Fertilizing Materials: Sludges are used to improve soil fertility as organic and mineral fertilizers, providing nutrients to plants and soil microorganisms to stimulate overall biological activities. Additionally, they help reduce the use of commercial fertilizers (Hechmi et al. 2020b). Fertilizing elements play an important role in plant development. The key elements present in sewage sludge are nitrogen, phosphorus, and potassium. Some elements are less important in terms of quantity, such as calcium, magnesium, sodium, and sulfur. The composition of sludge depends on both the nature of the initial water pollution and the adopted purification processes to which this water has been subjected: physical, physico-chemical, and biological treatments. Two major characteristics support this classification: organic or mineral nature: The organic nature usually requires the stabilization of organic matter or allows for an ultimate step of thermal oxidation; and hydrophilic or hydrophobic nature: The dehydration of liquid sludge becomes more difficult when the suspended matter is strongly bound to water, thus hydrophilic.

22.4 Reuse of Sewage Sludge By-Products

Sewage sludge mainly contains nitrogen, phosphorus, and potassium, which play an essential role in many complex reactions involved in the growth and development of plants. Nitrogen is the most important element for plant production yield and is found in sludge in different forms, more readily assimilated by plants than others. Secondary elements such as residual antibiotic compounds, magnesium, calcium, sodium, and sulfur are also present in sludge and have a relative role in plant growth. Finally, trace elements and metalloids such as boron, chlorine, copper, iron, manganese, molybdenum, and zinc are present in small quantities in sludge and are entirely bound to the solid phase. All these trace elements are useful in small quantities and are always present in all metabolic chemical reactions of plants, mainly as coenzymes or growth factors (Mokni-Tlili et al. 2023).

22.4.1 Restrictions on Sludge for Agricultural Purposes

The operation of treated sewage sludge on agricultural land poses a potential risk because of pathogenic microorganisms, heavy metals (ETMs), and organic trace compounds (CTOs). Improper application over the years can lead to surface water pollution (through drainage or erosion), as well as contamination of groundwater and water sources. In addition, pollutants accumulate in the soil, leading to a long-term decrease in soil fertility, which is reflected in a reduction in diversity and microbial activity, crop quality, and yield.

Metallic Trace Elements: Sewage sludge can contain high levels of heavy metals. The transfer of heavy metals to the soil and subsequently to plants still poses potential risks, as they can enter the food chain and harm humans and livestock. High concentrations of heavy metals in the soil lead to a reduction in soil biological activity and result in decreased agricultural yields and the accumulation of these same metals in plants. Table 22.4 shows the maximum limits for metallic trace elements (MTE) in the end product resulting from the composting of sewage sludge.

Organic Trace Compounds (OTC): Undesirable organic substances are elements that can hinder the treatment of sludge or their use in agriculture. These are mainly synthetic chemicals used in everyday life and found in domestic wastewater. High concentrations of detergents and medications (especially antibiotics that affect sludge flora) are particularly observed (Mokni-Tlili et al. 2023). Two major categories of

Table 22.4 Trace metal (TME) levels and limits (AFNOR 2003)

Metal trace elements (MTE)	Limit value (mg/kg DM)	Maximum annual flux averaged over 10 years (g/ha/yr)
Arsenic (As)	18	90
Cadmium (Cd)	3	15
Chromium (Cr)	120	600
Copper (Cu)	300	1000
Mercury (Hg)	2	10
Nickel (Ni)	60	300
Lead (Pb)	180	900
Selenium (Se)	12	60
Zinc (Zn)	600	3000

OTC are distinguished: polycyclic aromatic hydrocarbons (PAHs) and polychlorinated biphenyls (PCBs). Table 22.5 summarizes the concentration limits of organic trace compounds (OTC) for the final product attained by composting sewage sludge.

Pathogenic Microorganisms: Potential pathogens such as bacteria, viruses, protozoa, and helminths, which can pose serious risks to human and animal life, are typically present in wastewater and raw sewage sludge.

It is important not to use residual sludge for agricultural purposes in direct contact with plants intended to be consumed raw. Disinfection with lime helps significantly reduce the number of pathogens and reduce the risks of contamination. Table 22.6 summarizes the survival duration of certain pathogens at 20°–30 °C and the limits of microorganisms of health interest for different crops.

Groups of experts often study health risks to assess the health risks associated with the spreading of sewage sludge from wastewater treatment plants. These studies consider various pathways of human exposure and contamination, such as ingestion, inhalation, contact, and the food chain.

They also consider different spreading periods, including duration, climate, soil nature, and cultural system. Figure 22.1 summarizes the main potential transmission and contamination pathways for humans and their immediate environment.

The results of the study have mainly shown that the risks of contamination for humans are usually considered acceptable, reflecting the absence of significant

Table 22.5 Organic trace compound levels (OTC) and thresholds (AFNOR 2003)

Organic trace compound		Value threshold (mg/kg DM)	Maximum annual flow averaged over 10 years (g/ha/y)
A total of 7 PCB	PCB 28 + 52 + 101 + 118 + 138 + 153 + 180	0.8	1.2
HPA	Fluoranthene	4	6
	Benzo (b) fluoranthene	2.5	4
	Benzo (a) pyrene	1.5	2

Table 22.6 Survival times of some typical pathogens under temperatures ranging from 20° to 30 °C (AFNOR 2003)

Pathogens	Survival time in soil	Survival time on cultivated plants
Viruses enterovirus	Between 100 and 20 days	Between 60 and 15 days
Bacteria: fecal coliforms Salmonella Vibrio cholerae	Between 70 and 20 days Between 70 and 20 days Between 20 and 10 days	Between 30 and 20 days From 30 to 15 days From 5 to 2 days
Protozoa *Entamoeba histolytica* cysts Tapeworm eggs Ascaris egg	Between 70 and 10 days Many months Many months	From 10 to 20 days From 60 to 30 days From 60 to 30 days

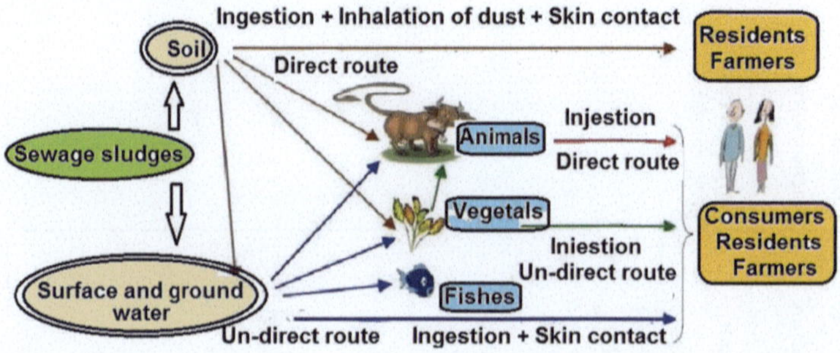

Fig. 22.1 Main potential transmission and contamination pathways for humans and their immediate environment

chronic effects on farmers and neighboring populations. So, it is necessary to detect and quantify certain pathogens in sewage sludge and to apply effective hygiene treatments, if needed, at the level of the farmer population and the crops produced. Additionally, there is often a significant decrease in pathogens after the first two months because of unfavorable conditions for their survival in the soil, such as variations in desiccation, humidity, and temperature, as well as natural competition with microbial ecology, etc. In agricultural areas, some pathogens are already present and have their usual habitat in cultural environments at the time of spreading, mainly related to livestock farming, birds, etc.

Wastewater and treatment process residues contain a variety of living microorganisms. Fortunately, most of these microorganisms are inactivated by the different bio-physico-chemical processes used in wastewater treatment. Only a small portion of them are pathogenic and/or phytopathogenic. These residual sludges may contain phytopathogenic microorganisms (which attack plants and crops), zoo pathogenic microorganisms (which attack animals), and anthropo-zoonose pathogenic microorganisms (which attack both humans and animals). Therefore, the spreading of these residual sludges, whether treated or not, requires monitoring of their quality, precautions during spreading, and hygienization treatments such as digestion, liming, and composting of the sludges.

There are typically five types of pathogens that may be present in sewage sludge: metazoan parasites such as helminths and worms, protozoa, fungi, bacteria, and viruses. The European Commission (2001) conducted a study on untreated and treated sludge, which provided a comprehensive inventory of all pathogens that may be present in urban sludge (Table 22.7).

The disinfection of sludge allows for the reduction of contamination by pathogens such as Staphylococcus aureus, salmonella, enteroviruses, and helminth eggs and prevents any reoccurrence of these pathogens (Huang et al. 2023). Total disinfection is achieved when indicator pathogens of potential contamination are not detected according to the thresholds defined by current regulations. In France, for example, Article 16 of the order of January 8, 1998, specifies that the number of salmonella must be less than 8 per 10 g of dry matter, the concentration of enteroviruses must be

Table 22.7 Pathogens in municipal sewage sludge

Bacteria	Viruses	Helminthes	Protozoa	Fungi	Yeast
Salmonella spp.	Poliovirus	*Ascaris lumbrocoides*	*Giardia lamblia*	*Aspergillus spp.*	*Candida spp.*
Pseudomonas aeruginosa	Coxsackievirus	*Trichuris sp.*	*Cyclospora cayetanensis*	*Phialophora richardsiae*	*Trichosporon*
Listeria monocytogenes	Echovirus	*Hymenolepis sp.*	*Entamoeba histolytica*	*Geotrichum candidum*	*Cryptococcus neoformans*
Campylobacter spp.	Parvovirus	*Taenia saginata*	*Toxoplasma gondii*	*Trichophyton spp.*	
Clostridium botulinum	Adenovirus	*Toxocara*	*Sarcocystis spp.*	*Epidermophyton spp.*	
Clostridium perfringens	Reovirus	*Diphyllobothrium latum*	*Cryptosporidium parvum*		
Shigella spp.	Hepatitis A, C, and E viruses	*Echinococcus granulosus*	*Encephalitozoon intestinalis*		
Mycobacterium spp.	Rotavirus	*Ancylostoma duodenale*	*Vittaforma corneae*		
Staphylococcus (coagulase-positive strains)	Astrovirus	*Necator americanus*			
Streptococcus (strains beta hemolysis)	Calicivirus				
Escherichia coli (strains enteropathogenic)	Coronavirus				
Yersinia enterocolitica	Norwalk-like Calicivirus				
Bacillus anthracis	Virus adeno-associated				
Vibrio spp.	Virus influenza				

(continued)

Table 22.7 (continued)

Bacteria	Viruses	Helminthes	Protozoa	Fungi	Yeast
Leptospira spp.	Polyomavirus (JC et al., BK)				
Aeromonas mobiles					
Arcobacter spp.					
Brucella spp.					

less than 3/10 g DM, viable pathogenic helminth eggs must be at a maximum of 3/10 g DM, and a characterization analysis of thermotolerant coliforms (resistant to thermal treatments and developing at 44 °C) must be carried out every fifteen days during the spreading period to show the absence of recontamination and the development of these opportunistic bacteria. An opportunistic pathogen is a microorganism that rarely causes disease but can become pathogenic under certain conditions, particularly when the individual's immune system and resistance are weakened, especially in immunocompromised and unhealthy individuals.

Furthermore, sewage sludge is not a conducive environment for the survival of pathogenic microorganisms because of the various constraints experienced during the treatment process, such as pressure, acidity or pH, moisture, and temperature. During storage and spreading, their destruction is often accelerated by climatic effects such as lack of oxygen, temperature, solar radiation, and moisture, as well as the effects of the sludge composition itself owing to the concentration of certain micro-pollutants and toxic metals, and the effects of the soil such as competition with other microorganisms and physico-chemical conditions. All these factors, often unfavorable for the development of these pathogens accustomed to living in luxurious conditions for organisms, also play an important role in the degradation of pathogenic elements. Burial can slow down their disappearance and loss of viability. Three treatment techniques for hygienization are commonly available in biological wastewater treatment plants: composting, pasteurization, and digestion; chemical treatment (liming, stabilization with nitrite salts); and finally, physical treatment (thermal treatment in a dryer at 90 °C).

The mesophilic anaerobic digestion only allows for partial sludge sanitization, as the treatment is carried out at temperatures below 40 °C. Although fecal bacteria are eliminated, enteroviruses are inactivated, and helminth eggs can survive. When combined with quicklime treatment, an additional sanitization action occurs on the digested product, increasing temperature and pH. The complementarity of these techniques ensures better control over the reduction of pathogenic microorganisms.

After spreading on agricultural soils, pathogens from sewage sludge can survive, depending on environmental conditions, in the soils and potentially contaminate various environmental media (soil, water, and air) based on spreading practices and weather conditions (wind, rain).

Several elements interact complexly with the survival of pathogens in soils. Natural microbial life allows for competition with pathogenic sludges naturally eliminated in soils. The major factors influencing the lifespan of pathogens are listed in Table 22.8 (Wray 1975; Smith 1996; ADEME/SYPREA/FP2E/INERIS, 2007). Sandy soil promotes oxygen diffusion in the soil mass, which is conducive to aerobic microorganisms. On the other hand, clayey soil, which is more compact, favors anaerobes and their adhesion to soil particles, thus reinforcing their physical, chemical, and biological protection within the soil matrix.

Survival in soils depends on various parameters related to sludge characteristics, spreading conditions, and weather conditions. Generally, viruses and bacteria have a relatively short lifespan in soils, while helminths and fungi can survive for several months after spreading. However, many unfavorable factors in arable land limit their

Table 22.8 Factors influencing the lifespan of pathogens in soil amended with sewage sludges

Parameters	Impact on period of survival			
	Viruses	Bacteria	Protozoa	Helminthes
Temperature	–	–	–	–
Sun exposure	–	–	–	–
Humidity	+	+	+	+
Drought	–	–	–	–
Soil clay content	+	+	?	+
Soil water holding capacity	?	+	+	+
pH 6–8	?	+	?	?
Availability of nutrients	/	+	?	?
Presence of harmful antimicrobial substances	/	–	/	/
Soil microflora	?	–	+	?
Plant cover	?	+	+	+
Depth of application	+	+	+	+

–: decreases survival; +: increases survival; ?: effect unknown; /: not applicable

survival and presence in the surface layer to a period ranging from 30 to 90 days. It is widely accepted that gram-negative bacteria have a lower survival rate compared to gram-positive ones, whereas the survival rate of sporulated bacteria will show a higher survival rate than the vegetative ones.

After spreading, most pathogens accumulate in the top few centimeters of soil. Water is the main pathway for pathogen dissemination in the environment. Groundwater contamination is unlikely given the current spreading conditions (delimitation of spreading area outside of close or immediate protection zones of drinking water sources and control of the amount spread). Given the prevailing conditions of spread, which involve restricting the quantity of spread and delineating the area of spread beyond the protection zones of nearby or immediate drinking water sources, the possibility of groundwater contamination is highly improbable.

The contamination of surface waters can theoretically occur through runoff or leaching during rainy events after spreading. However, runoff is usually very limited on crop residues, and spreading is done far away over 35 m from watercourses and water points.

The air contamination following spreading can only occur when using spray bars equipped with nozzles for liquid slurry in windy conditions. In France, for example, spreading is prohibited with air-dispersion devices. Moreover, the viability of pathogens on plants (while dispersing on grassland) rarely exceeds one month, as they become deactivated by desiccation and exposure to sunlight. No published study has also shown the transmission of pathogens from animals that have grazed on a spread plot with humans. The National Veterinary Health Surveillance Unit has never shown a direct link between slurry spreading and pathological accidents since the beginning of its operation. Furthermore, the application of livestock effluents can have to get an impact into impact on agricultural soils where spreading occurs. It is important to note that animal waste, which may harbor harmful pathogens, contributes significantly to the overall environmental contamination in soil and surface water systems.

22.5 State of Art of Sewage Sludge Spreading in Agricultural Soils

To preserve the quality of the receiving environment, it is necessary to keep, transform, and recycle the elements of residual sludge in agricultural soils. The sludge must be kept in the soil, organic compounds must be transformed into inorganic compounds assimilable by plants, and harmful elements must be removed or inactivated. However, certain parameters of the receiving site can influence these processes. The main parameters considered are the thickness and nature of the soil, the depth of the water table, drainage, slope, pH, cation exchange capacity, vegetation, and climate. Standards and recommendations exist for most of these parameters. For example, the soil must have a minimum thickness of 1.5 m, the water table must be located more than one meter from the soil surface, the soil must be well-drained, and the pH should be around 6.5. These standards and recommendations have let potential receiving sites be classified into usable, unusable, or conditionally usable zones for the spreading of residual sludge.

The residual sludges can be spread only on agricultural soils if they respect the principle of "agronomic interest" and are free from high concentrations of inorganic or organic pollutants. The spreading of sludges must follow strict rules, such as minimum distances from banks, sources, wells, and dwellings, to avoid rapid infiltration toward surface or groundwater, as well as any runoff. A distance of at least three meters from roads and ditches must be respected. However, spreading is prohibited on frozen soils, steep slopes during periods of heavy rain, and on regularly cultivated lands (such as market gardens). To reduce contamination by pathogenic microorganisms and olfactory nuisances, only sludges stabilized by aerobic or anaerobic digestion, treated with lime, composted, disinfected by pasteurization, and dried at high temperatures can be used. Besides the problems related to the hygienic quality of sludges to be spread on agricultural soils, other economic or technical problems, and obstacles to the application of spreading rules may arise, such as storage and transport, as well as the inadequacy or insufficiency of stabilization and dehydration techniques.

To better manage the agronomic use of sewage sludge by spreading, it is essential to consider and show the following elements:

1. A preliminary study on the characteristics of the sludge, the soil, and their suitability for spreading and on the technical implementation methods such as the periods and equipment for spreading and the storage conditions. This study justifies compliance with the conditions imposed by the law, which also requires that

storage capacities consider periods when spreading is prohibited, such as winter or drought. In addition, this study should provide an alternative disposal plan in case of incidents.

2. A "spreading register" that summarizes the spreading activities, including the spread plots, the characteristics of the sludge and soil, etc. A copy of this register must be submitted to the prefecture every year, and the sludge producer must keep this document for ten years.
3. An annual forecast program that, before each campaign, specifies the receiving plots for the following campaign, as well as their characteristics and the crops practiced.
4. An annual report on the spreading program, as a register, which shows continuous monitoring of the quality of the spreading and sludge, specifying, in particular, their heavy metal content, micro-pollutants, and pathogenic microorganisms, as well as their agronomic value, origin, and source. This register must also include the signed agreement with the user, the quantities of sludge involved, and the usage recommendations. It must also define the dates of use and the receiving plots.

Depending on the quality of the sludge, especially their pollutant load, the frequency of analysis may vary. It is widely accepted that this frequency should be halved after the first year. Additionally, if a sludge mixture is allowed, although this is quite rare, these analyses must be carried out on the different components and must point out the mixture proportion. Finally, it is necessary to plan an alternative solution to spread if the operator must manage a batch of sludge that does not meet the regulatory spreading requirements. Despite the increasing costs of treating sludge for agricultural use, spreading remains the most economical method. Before being spread, the sludge may have undergone one or more treatments (Fig. 22.2).

The treated sludges for spreading can be categorized as follows:

- Liquid sludges: with a dry matter content of 2–5%, produced by small plants (< 2000 PE), they have undergone thickening or storage in a silo.
- Paste-like sludges: with a dry matter content of 15–25%, resulting from biological digestion, corresponding to medium-sized WWTPs (65,000–20,000 PE). They have undergone dehydration in a belt filter press or centrifugation.

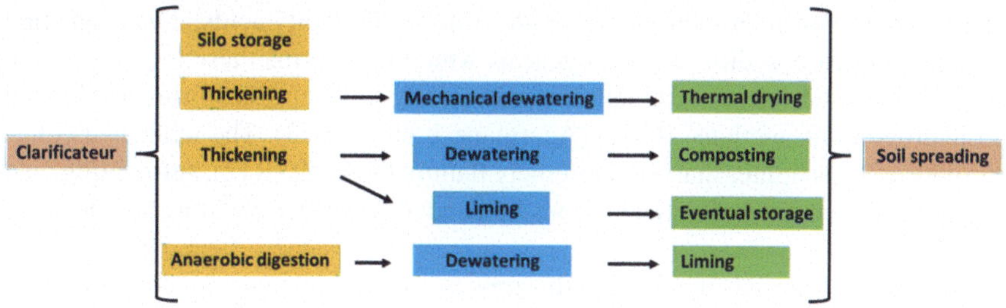

Fig. 22.2 Different stages of sludge treatment before land application

- Lime-treated sludges: with a dry matter content of 25–30%, this sludge can be paste-like or solid, produced by medium to large WWTPs (20,000–100,000 PE). They result from lime treatment after dehydration.
- Composted sludges: with a dry matter content of 35–70%, resulting from composting of dehydrated sludges.
- Lagoon sludges: with a dry matter content of 10–20%, they correspond to a particular category of liquid sludges treated extensively by macrophytes.
- Solid sludges: resulting from treatment by a filter press or thermal drying. They are mainly produced by large WWTPs (with a relatively high production cost).

The spreading of sludge offers undeniable agronomic advantages. Sludge from wastewater treatment plants is rich in fertilizing elements such as nitrogen and phosphorus. In liquid sludge, the most of nitrogen is as ammonium, which facilitates its assimilation by plants. As for pasty and solid sludge, nitrogen is present in organic form and remains available in the long-term. Furthermore, the contribution of phosphorus from sludge is significant, almost equivalent to chemical fertilizers. Despite these advantages, it is important to emphasize that this valorization of sludge has certain limitations. It is often poorly perceived by the common public because of heavy metals, toxic organic compounds, and pathogenic germs in sludge.

22.6 Biomethanization

Among the potential uses of residual sludge, anaerobic digestion or biomethanization is another area to consider. Biomethanization involves mainly the synthesis of methane gas, CO_2, and some residues such as SO_2, following a biological transformation under the action of anaerobic microorganisms called methanogens, in specific conditions of pressure, pH, temperature, and in a reactor called a biomethanizer. This process allows the degradation of simple organic matter into complex organic matter through anaerobic digestion, thus releasing methane gas, which is often used as an energy source. It is inspired by the natural digestion of ruminants and takes place in four essential steps involving five different enzymatic groups: hydrolysis and fermentation which break down large molecular complexes (fats, starch, proteins, etc.) into simpler and smaller molecules (sugars, alcohols, amino acids, etc.); acidogenesis which transforms simple molecules into volatile fatty acids, thus producing hydrogen and carbon dioxide; acetogenesis which transforms these molecules into acetic acid or hydrogen; and finally methanogenesis, which transforms acetic acid and hydrogen into methane (CH_4) (Azarmanesh et al. 2023). The latter is a colorless, odorless gas lighter than air, extremely flammable in air (R12), suffocating, and explosive. It is a highly potential gaseous fuel under normal temperature and pressure conditions.

The organic materials involved in this process are those that do not contain fibers, meaning indigestible lignin compounds.

In a bioreactor, the industrial parameters to monitor are mainly the absence of oxygen (anoxia), pH, and temperature. The acidity of the pH involves other enzymatic processes that do not lead to methane production, resulting in a decrease in efficiency. Temperature is also an important factor, as it determines the choice of inoculum and operating methods. Temperature levels can be mesophilic (37–40 °C) or thermophilic (55–60 °C). The carbon/nitrogen ratio should be around 30. It is also important to consider toxic substances for the microbial inoculum operator of biomethanization, such as heavy metals, antibiotics, and PAHs, which can inhibit different enzymatic reactions and prevent material transformations.

The technological choice will affect civil engineering aspects, such as tank size, storage locations, and storage form.

There are several biomass valorization techniques, depending on the dry matter concentration (DM). The distinction between liquid and dry biomethanization is made at a threshold of 20% DM. Below this threshold, it is called liquid biomethanization, which includes other technical subdivisions. Above 20% is looked at as dry biomethanization or high-rate biomethanization.

Before the digesters, it is possible to find a phase separator that separates the fibers (solids) from the liquid phase, as these are not digestible in the liquid phase, hence the different units of liquid biomethanization. The Infinitely Mixed unit includes a bag containing the materials to be digested, continuously mixed with the microbial mixture (bacteria and yeast) responsible for digestion for 40 days. The digestate comes out accompanied by the microbial bed, which will be renewed in the bag. Digestion allows the release of biogas. The fixed-bed unit involves a slightly original system: The slurries are pumped into a container containing the fixed bacterial mixture on canvases. The slurry progresses along these canvases and is digested as they advance, transforming into biogas and digesting for 6–8 days. A bag downstream of the container (optional in these installations) allows for "post-digestion" of digestates that have not been completely digested in the container, which has two advantages: a greater recovery of biogas and partial purification of the biogas from sulfur gases. Loading the materials into a "silo" is part of the Fixed Bed and Batch Unit, where they are hydrolyzed and transformed into acetate. The bacterial mixture is fixed on tubes as a biofilm in which the materials are digested. The second tank allows for methanization. Both are connected and allow the passage of materials being digested from one tank to another. The biogas is recovered from the top of the tanks. Upstream of the actual digestion, the materials are hydrolyzed, and downstream, the post-digester finishes degrading the slower-to-digest organic materials that may not have been completely transformed into the liquefier.

Finally, dry anaerobic digestion can be carried out in batch or continuous mode, with or without mixing, combined with a liquid phase or not, and depending on the level of methanization.

The general principle is comparable to that of wet anaerobic digestion, except that the fermentation juices are regularly sprinkled on the materials, percolate through them while fermenting, and are recovered by gravity. These juices are re-injected to sprinkle the materials, and so on, until the substrates are fully fermented. Depending on the complexity of the system, the digestion time varies from 15 to 30 days, and the biogas is recovered in the upper part of the silos.

The digestate corresponds to the organic matter residues after digestion. It can be either liquid, as it is recovered at the end of the process or dried to take the form of pellets (Fig. 22.3).

The technologies and processes of biomethanization typically vary depending on the incoming raw materials (or inputs) and are distinguished by their dry matter content (% DM) and their biodegradability. These two parameters influence how the inputs develop in the digester. This also affects the choice of technologies to be implemented. The choice of technique is based on the average dry matter content of the total ration and its "pump ability" during digestion.

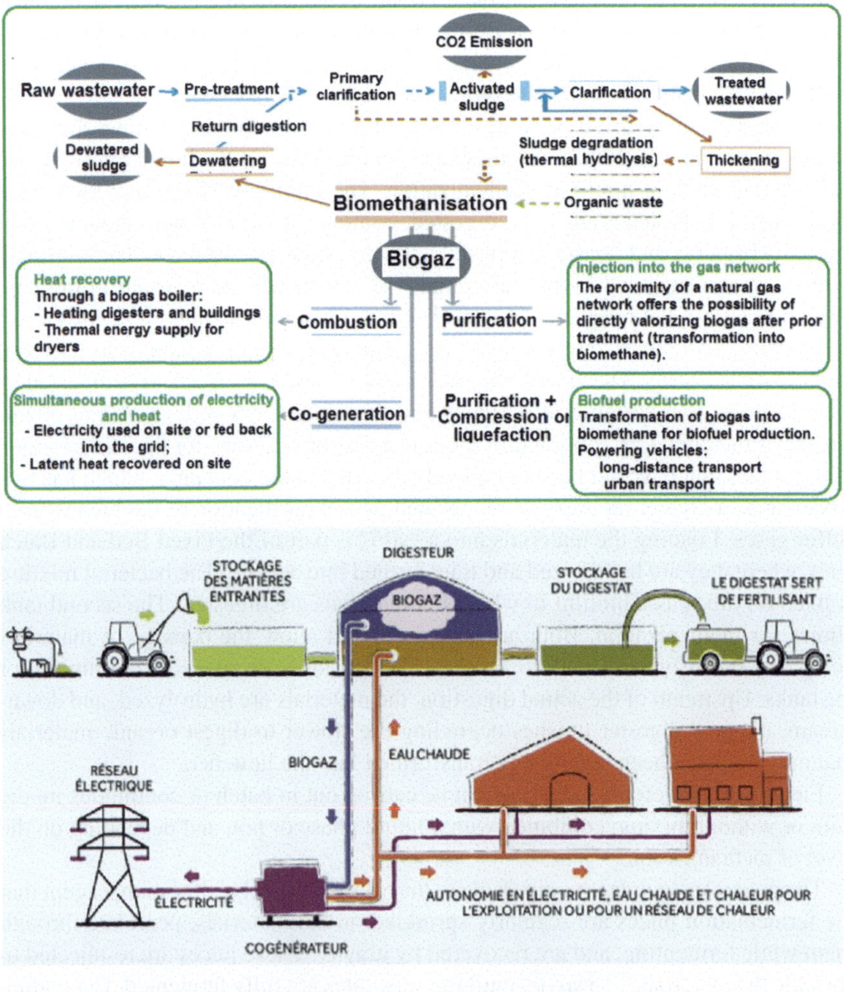

Fig. 22.3 Biomethanization and its technologies: general principle

It is also important to note that the materials require variable residence times in the digester, depending on their degradation kinetics and the intended technique. The materials can be solid or liquid at the inlet; they liquefy in the digester. If required, the liquid is added or recirculated from the digestate to dilute the content to facilitate mixing and pumping (water, rainwater, liquid fraction of the digestate, etc.). Considering all the mentioned parameters, residual sludge can be a preferred and favorable input for the practice of biomethanization, as this technology simultaneously guarantees the bioremediation of a residue and its valorization by producing energy and fertilizing material. The various methods of biogas valorization primarily focus on the production of heat or steam through combustion in a boiler, the production of electricity through combustion in an engine driving an alternator, the production of steam through a boiler driving a turbine that also powers an alternator, the production of electricity and heat (cogeneration) through the same processes as electricity production but with a heat recovery system, and finally, the production of biomethane (CH_4) through the purification of its carbon dioxide and compression of the biogas (Fig. 22.3). This biomethane can be used on-site as a biofuel or injected into the natural gas network. The use of injected biomethane occurs outside the production site and can be varied: in boilers, electricity production (with or without heat valorization), or as a biofuel. The estimates of the theoretical yields of biomethanization commonly found, depending on the type of conversion, are summarized in Table 22.9.

The performance displayed by industrial manufacturers of biomethanization units does not always correspond to the actual performance achieved, which is often lower. Part of the energy produced is used by the biomethanization process itself to heat the tanks, operate the various engines (agitation, pump), electronic systems, etc. This self-consumption varies depending on the technique used, the type of inputs, the chosen temperature, the influence of the climate, etc. As for the purification of biogas into biomethane, the valorization of biomethane into final energy does not depend on the producer. The energy efficiency of using injected biomethane is then linked to that of natural gas and it is stored either in an internal or external gasometer.

Table 22.9 Estimation of theoretical yields as a function of conversion type

Conversion type	Efficiency
Boiler	80–100%
Electricity generation via motor	30–40% (depending on power)
Cogeneration engine	– 80–90% overall efficiency – Can vary from ~ 20% electrical and ~ 60% thermal – 42% electrical and ~ 43% thermal
Biogas purification (for injection or biofuel)	– 85 to 99.9% methane recovery, depending on the purification technique – The biomethane is then used, depending on the efficiency of the user device

22.7 Biomethanization Digestate Recovery Techniques

After the end of the sludge digestion process, the remaining material is called the digestate. Just like biogas, digestate has great agronomic value. It is the residue of the decomposition of incoming materials during anaerobic digestion. Through proper management of the anaerobic digestion process, digestate can be an amendment and fertilizer, replacing agricultural effluents and synthetic fertilizers. As an amendment, digestate is as beneficial as manure, as it provides stable carbon that promotes humus production and improves soil structure. Only the quickly biodegradable organic carbon is converted into biogas, while the carbon that contributes to humus formation is preserved. Besides, through the digestion of volatile organic compounds, the digestate has little odor and keeps its fertilizing properties in terms of NPK.

Regarding its properties as a fertilizer, both micro- and macro-nutrients are fully preserved. Digestion has partially mineralized these elements, making them directly assimilable by plants, like chemical fertilizers. Therefore, in the short term, the liquid fraction of digestate acts as a synthetic fertilizer, where the elements, especially nitrogen and potassium, are directly assimilable.

In the long-term, the solid fraction containing nutrients in organic form takes over, gradually degrading in the soil and gradually releasing the nutrients, similar to a traditional farm fertilizer. The status of digestate is linked to anaerobic digestion and closed-loop spreading on the farm, and the administrative management of digestate is relatively simple when adhering to spreading standards. The digestate is seen as waste if imported inputs are waste or if it leaves the production site.

Its valorization on agricultural land is still possible, as it originates from natural organic materials. However, it is necessary to follow waste legislation to ensure soil quality preservation. Digestate treatments can include raw spreading or undergo various other treatments to facilitate spreading, transportation, and even produce higher value-added fractions for agriculture. Currently, at the European level, digestate is valorized through various methods: it can be used in its raw form, separated into a liquid fraction rich in mineral nitrogen and a solid fraction with higher humus content, composted often in combination with green waste or subjected to drying or evaporation processes to reduce transport volumes.

Besides the energy and digestate parts as soil fertilizers, anaerobic digestion offers other services. Anaerobic digestion plays a role at multiple levels in reducing greenhouse gas emissions and fossil energy consumption. By replacing fossil resources with renewable and local biomass (neutral CO_2 cycle), it directly reduces greenhouse gas emissions (unused fossil energy) and indirectly (fossil energy extraction, transportation to consumption sites, etc.). It avoids the spreading of chemical fertilizers by using digestate, reducing dependence on energy intensive synthetic fertilizers, such as nitrogen production, or those derived from mining activities with limited stocks, such as phosphorus and potassium (Fig. 22.4). This is significant energy savings for the planet and helps many countries become more independent from imported resources, such as natural gas, oil, and mineral resources. Additionally, anaerobic

Fig. 22.4 Services provided by anaerobic digestion sludge

digestion practices help maintain and create local jobs, making farms self-sufficient in energy and fertilizers. This lets farmers control costs and not be subject to international price volatility. Anaerobic digestion also offers solutions for the valorization of agricultural co-products, waste, and household or agri-food industry by-products that currently have no other means of valorization. By maintaining and creating local jobs, anaerobic digestion contributes to population retention in their region.

Therefore, a biomethanization unit requires between 4 and 10 individuals, depending on its size and site activities. It also contributes to generating indirect and local employment opportunities, such as biological monitoring (by laboratories), equipment maintenance (electromechanical, automation), development of new activities on-site, and scientific research in the field, etc.

This activity focused on biomethanization has a significant potential in terms of research and innovation. It provides the opportunity to develop many other activities and diversify pathways. It is a source of innovation in the field of energy, in collaboration with other renewable energy sources such as methane (CH_4) production through biological methanization from CO_2 derived from biogas and renewable hydrogen (H_2). This approach would enhance environmental yields and methane production. It is also possible to produce high-value molecules from inputs, volatile fatty acids, and biogas, notably CH_4. Besides, it is conceivable to produce synthetic fertilizers from biogas, particularly CH_4. Digestate can also produce fertilizers, amendments, or culture substrates tailored to the needs of each soil. Table 22.9 summarizes the advantages of the biomethanization process by transforming a product considered harmful, namely sewage sludge.

22.8 Summary of the Properties of Sewage Sludge

It is possible to summarize that wastewater treatment sludge is treated to reduce its volume to limit the quantities to be stored (especially for spreading), to stabilize its content to improve its chemical composition and consistency, and to reduce its biodegradation. Before being stabilized, the sludge is rich in water, and its high content of fresh organic matter makes it particularly fermentable, which leads to the production of unpleasant odors over extensive areas. Thus, stabilization allows the production of sludge with a high dry matter content, significantly reducing this nuisance for the environment. A specific treatment according to their final destination must be employed to these sludges, as wastewater treatment sludges have variable compositions depending on the wastewater treatment process adopted in the treatment plant: biological, physico-chemical, mixed, etc. Depending on their level of dry matter content, these sludges can either be treated and valorized for agricultural use or be disposed of by incineration or landfill. Thus, depending on the type of sludge from wastewater treatment plants and their ultimate use, they will not undergo the same treatment.

To determine the optimal treatment to reuse wastewater treatment sludge, it is necessary to consider several criteria: The origin of the sludge, do they come from urban or industrial wastewater? The physico-chemical composition of the sludge produced depending on the type of treatment adopted for wastewater, and finally the intended final destination for the sludge: recycling, valorization, disposal.

Thus, the destination of sewage sludge varies depending on the type of wastewater treatment adopted, and there are mainly three types of use for treated sewage sludge:

Utilization in agriculture: This use represents on average about 73% of the treated sludge on an international scale, mainly in Europe. These sludges can be transformed into fertilizers or compost to be spread on agricultural soils. This transformation into compost depends on many interdependent factors, such as the willingness of local authorities, their sensitivity to the issue, the characteristics of agricultural soils and the region, and the local climate. Some sludges can also produce biogas when dried and placed in thermal furnaces, in a biomethanization process. Other types of sludges are suitable for incineration, representing on average about 20% of the treated sludge. This practice is only common in developed countries, such as some European and American countries, and is widely used when other methods of use are impossible, such as agricultural use or biogas production.

Another aspect concerns the landfill process, also known as the technical term "Technical Landfill" or "Waste Storage Center". Approximately 8% of sludge is treated in this manner. This treatment is only present in developed countries with controlled landfills and strict monitoring, such as most European countries and some other developed countries. In general, there are three types of landfills: Class 1 landfills for hazardous or "special industrial" waste, Class 2 landfills for household and similar waste, and finally Class 3 landfills for inert waste.

As previously mentioned, and emphasized, the treatment of sewage sludge varies depending on the desired use: recycling, disposal, energy recovery, or landfilling.

Thickening is the first treatment step applied to sewage sludge. This process normally takes place before primary, secondary, or even tertiary sludge is mixed, which is produced successively at each stage of wastewater treatment. To help with the separation of dry matter and solid matter, flocculating agents can be added upstream. These agents can be synthetic organic substances or minerals such as aluminum salts, lime, or iron salts. Once prepared, the sludge thickens in two ways: by gravity, flowing into a silo above a draining net or fabric, and by flotation, injecting gas into the sludge to separate dry matter from liquid matter. This initial treatment increases the dry solids content of the sludge (between 6 and 8%) to reach quality sludge for optimizing subsequent treatments. The dehydration treatment increases the dryness of the sludge to have a pasty or solid consistency, with a dry solids content ranging from 15 to 40%. There are two dehydration techniques: mechanical dehydration by filtration or centrifugation, mainly used in large plants with several tens to several thousands of "Equivalent-Inhabitants" (EH), and dehydration by geo-synthetic membranes, a more recent technique adapted to small installations of 1000–2000 EH.

Drying is a process that allows sludge to solidify by completely drying it, which dehydrates it completely. This transformation into a solid state facilitates the storage and transportation of sludge by reducing its volume. There are two types of drying: thermal drying, which can be direct or indirect, using a heat transfer fluid. Direct drying is done by evaporation of sludge through convection, while indirect drying is done by heat conduction through a heated wall. Despite the drying method used, the recovered sludge can reach a dryness rate of up to 95%. It is important to emphasize that these techniques are very energy intensive, and solar drying, which is significantly more environmentally friendly, is preferred to reduce the environmental impact of sludge management. The drying bed, or solar drying, involves placing the sludge on a draining surface to dehydrate it. When these drying beds are placed under a greenhouse, the natural evaporation of sludge is accelerated by exposure to sunlight, which is called solar drying. Other variations exist not to depend on weather conditions, such as macrophyte drying beds.

The stabilization of sludge is an important step aimed at reducing the content of fresh organic matter to limit their reactivity and fermentability and thus reduce potential bad odors. This process only applies to fresh mixed sludge or secondary sludge. There are generally three techniques for stabilizing organic matter:

- Biological stabilization:
- Aerobic stabilization (in the presence of oxygen) mainly aims to transform sludge into compost for agricultural use.
- Anaerobic stabilization (without oxygen), known as "methanization", allows the conversion of sludge into biogas and the valorization of this biogas as heat, electricity, fuel, or fuel.
- Chemical stabilization: It consists of neutralizing the biological activity of sludge to reduce the content of organic matter. To do this, the sludge is mixed with lime (quicklime or hydrated lime) in calculated quantities based on the sludge dryness rate. This stabilization method can only be carried out after sludge dewatering, which is then called sludge hygienization.

Finally, sewage sludge is the result of wastewater treatment. They were once considered harmful waste, but in reality, these residues are exploitable resources, sources of energy and wealth. Composting sludge allows industries and communities to be self-sufficient in terms of energy, and even sell the energy produced to a local network. The use of sewage sludge also helps to combat the scarcity of raw materials, such as fertilizers for agricultural soils, by using the biogas produced through biomethanization. This biological process transforms sludge from urban wastewater treatment plants into green energy, biogas.

All the processes mentioned and studied in this chapter regarding sludge generated during wastewater treatment are governed by specific legislation. Once properly treated, they can be used in the agricultural sector as fertilizers. Thus, the quality of sludge varies depending on the composition of the original wastewater. The classification of sludge generated during wastewater treatment is based on the origin of the effluent to be treated, namely urban sludge and industrial sludge, as well as the stage of wastewater treatment: primary sludge, secondary sludge (biological), mixed sludge, and tertiary sludge (chemical or physico-chemical). Finally, the classification is also based on the type of treatment in the sludge line: thickening to obtain thickened sludge, stabilization to obtain stabilized (or digested) sludge, and dehydration to obtain dehydrated sludge (Figs. 22.5 and 22.6).

An important step in the bio-physical–chemical characterization of the produced sewage sludge involves conducting various analyses, such as chromatography, X-ray fluorescence spectroscopy, as well as classical bacteriological and chemical analyses like dryness, pH, nutrient contents, heavy metals, etc. These analyses will help establish the values of different parameters needed to determine the most suitable sludge treatment processes based on their final destination, notably.

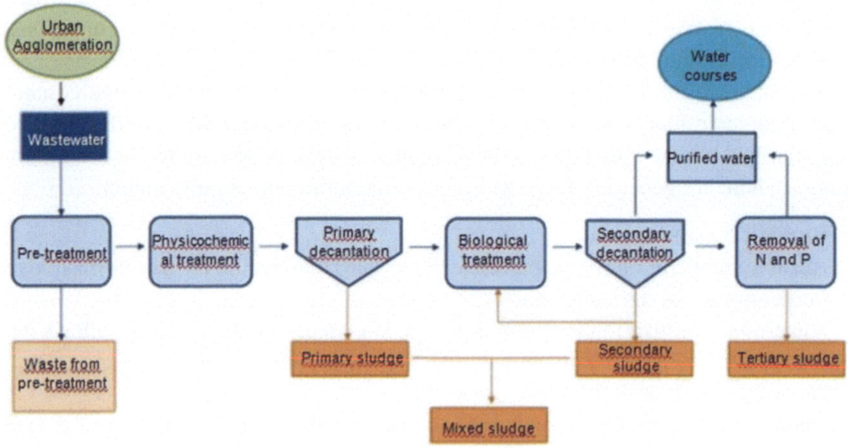

Fig. 22.5 Representation of a municipal sewage treatment plant illustrating the different processes of treating wastewater and the resulting sewage sludge

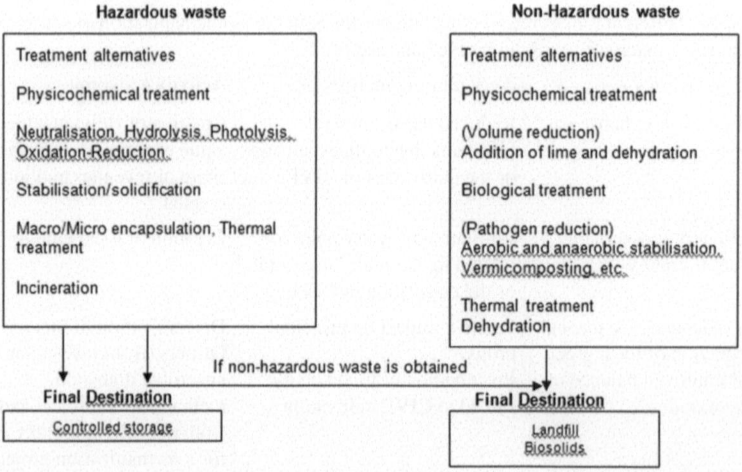

Fig. 22.6 Different treatment procedures used and recommended for the sludge produced in a municipal wastewater treatment plant

Thus, it is important to analyze the parameters that mainly influence the compliance of sludge for agricultural use. Before and after treatment, it is necessary to examine:

- Heavy metals: Cd, Cr, Ni, Hg, Pb, Zn, and Cu: Heavy metals are key elements for sludge characterization. Since the 1970s, a significant decrease in heavy metal content in sludge from wastewater treatment plants has been observed. This drastic reduction is mainly due to the various legislations implemented by countries to regulate and limit these elements due to their harmful environmental effects.
- Pathogenic microorganisms: Salmonella spp, *Escherichia coli*, helminth eggs, etc.
- Agronomic parameters: pH, electrical conductivity, organic matter, total nitrogen, organic nitrogen, phosphorus, calcium, magnesium, potassium, and iron.
- Organic pollutants: secondary organic aerosol (SOA), sandy clay loam (SCL), phthalates, nonylphenols, polycyclic aromatic hydrocarbons, polychlorinated biphenyls, dioxins and furans, brominated diphenyl ethers.

Through analysis of the sludge treatment process, it is possible to identify three primary treatment stages, each associated with distinct processes: thickening, biological stabilization, and chemical stabilization. These final sludge treatment processes represent a stage aimed at reducing volume and conditioning sludge for further use in terms of management, handling, storage, and valorization, with the main goal of optimizing the efficiency and economy of the various subsequent use processes.

Lastly, Table 22.10 summarizes the various advantages and disadvantages of the different treatment and reuse processes for municipal wastewater sludge produced in a wastewater treatment plant, and Table 22.11 presents the performance comparison between conventional and thermal digestion after pretreatment as reported by Valo et al. (2004).

Table 22.10 Different biogas recovery techniques through the biomethanization of sewage sludge and the positive externalities of the biomethane sector

Energy and wastes	Agricultural practices	Activity economic
Reduction of greenhouse gas emissions	Reduced use of mineral fertilizers due to digestate and/or the cultivation of CIVEs	Creation of Belgian added value experiment in Belgian share of revenues and jobs created
Production of non-variable and storable energy	Limitation of water pollution related to using digestate and/or the cultivation of CIVEs	Creation of local employment
Energy independence through the production of local green gas and improved balance of balance of trade	Positive impact on principal crops Preservation of biodiversity linked to CIVE cultivation	Diversification of income for farmers (from investment in an anaerobic digestion methanization project and/or from the cultivation of CIVEs for a methanization project)
Enhancement of existing gas network through the development of biomethanization	Reduction of odors caused by fielding of manure and slurry	
Participation in an energy transition and circular economy and circular economy	Development of an energy and energy and ecological transition within the farming community	
Biowaste treatment and recovery of biowaste food industry and local and local authorities		

An inter-cropped energy crop (CIVE) is a crop planted and harvested between two major crops in a crop rotation. Also known as "plant cover", intercrops are crops planted between two principal crops. They do not compete with food crops, whether for human or animal consumption. By being inserted between two priority crops, their vegetation cycle is usually too short for them to reach maturity

Table 22.11 Performance comparison between conventional digestion and digestion after thermal pretreatment

	Raw sludge	Digested sludge	Sludge pretreated by thermal hydrolysis
Total C.O.D. (mg O2/dm^3)	17,400	9761	5082
C.O.D. loss (%)		43.9	70.7
Increase (%)			61
TS (g/dm^3)	17.2	12.5	7.0
TS loss (%)		27.3	59.3
Increase (%)			117
VS (g/dm^3)	12	7.8	4.0
VS loss (%)		34.8	66.9
Increase (%)			92
Biogas production (dm^{3}CH$_4$/day/m^3)		78	118
Increase (%)			54
Biogas yield (dm^3 CH$_4$/kg VS introduced)		128	228
Increase (%)			78

Adapted from Valo et al. (2004)

22.9 Conclusion

Two types of sludge from wastewater treatment plants are generally distinguished, which are ideal substrates for anaerobic digestion, spreading on agricultural soils, and co-composting with other waste such as household garbage. Primary sludge comes from the direct settling of the raw effluent, rid of large solid debris, and fats after a process of sand and grease removal. They are generally easier to degrade through anaerobic digestion as they have a higher amount of organic matter than secondary sludge derived from clarification. The latter, also known as "biological sludge", comes from the clarification stage that follows the biological basin. Often, primary and secondary sludges are mixed to optimize biogas production, thus forming "mixed sludge".

The residual sludge, abundant in microorganisms, particularly bacteria, acts as a catalyst for the biological revitalization of agricultural soils when spread. This action stimulates the bio-physico-chemical processes of mineralization and restructuring, resulting in an overall improvement in soil fertility and crop yield. However, it is important to acknowledge that the mud may also contain pathogenic microorganisms such as salmonella, *E. coli*, and Klebsiella which can be harmful to both humans and animals.

The contamination of crops by pathogens has been clearly shown through various results obtained during significant research. The presence of fecal flora such as streptococci and fecal coliforms, including enteric pathogens, as well as typical pathogens such as Salmonella and Pseudomonas, has been observed on the aerial system of plants and crop roots. This pathogenic bacterial flora, present on crops such as forage, can pose a risk to the health of farmers and animals when consumed. However, this health risk usually decreases due to the generally adopted conditioning practices for these crops, such as forage, after harvest. These practices are materialized by a protocol of storage and short drying aimed at limiting losses due to forage respiration, which could change its nutritional value.

This promising method of spreading sewage sludge on agricultural soils can be considered a sustainable and beneficial solution for the environment and agriculture. The use and dispersal of these sludges have a positive impact on crop growth and yields, thus encouraging the development of this type of application. It is essential to consider different plant species and monitor the accumulation and transformation of other pollutants such as heavy metals to better understand the improvement of crops and overall soil health.

Furthermore, the energy recovery of sewage sludge through methanization allows the production of biomethane, thus contributing to the energy efficiency of sanitation systems. This technique can be applied to treatment plants with a minimum capacity of 15,000 Population Equivalents (PE).

Methanization is a solution that addresses the issues of waste treatment at the local level by allowing the valorization of sludge from urban and industrial wastewater treatment plants to produce energy. Studies have shown that these sludges have a maximum theoretical methane potential of 2.13 TWh/year. Therefore, sludge methanization reduces the quantities of sludge to be treated in the wastewater treatment plant by an average of 40%, eliminates bad odors, produces a "stabilized" digestate that is largely free of pathogenic germs, and has fertilizing properties for agricultural soils. Additionally, it helps reduce levels of generally harmful volatile organic compounds, produces a material (digestate) that is well accepted in agriculture, and helps with the planning of spreading or co-composting with other waste such as household waste. Lastly, it facilitates the recovery of essential fertilizing elements such as nitrogen and phosphorus contained in the sludge and allows for the energetic valorization of biogas through the use of a boiler, cogeneration (heat and electricity production), or injection.

References

ADEME/SYPREA/FP2E/INERIS (2007) Health risk assessment of wastewater treatment plant sludge spreading systems. Conventions 03 75 C 0093 et 06 75 C 0071 ADEME/SYPREA/FP2E/INERIS. Scientific basis for assessing health risks related to pathogens

AFNOR (2003) Soil quality—determination of copper, manganese, and zinc contents—extraction with ammonium acetate in the presence of EDTA—French standard NF X 31-120

Allen A (2001) Containment landfills: the myth of sustainability. Eng Geol 60(1–4):3–19

Annabi M, Houot S, Francou C, Poitrenaud M, Le Bissonnais Y (2007) Soil aggregate stability improvement with urban composts of different maturities. Soil Sci Soc Am J 71:413–423

Azarmanesh R, Qaretapeh MZ, Zonoozi MH, Ghiasinejad H, Zhang Y (2023) Anaerobic co-digestion of sewage sludge with other organic wastes: a comprehensive review focusing on selection criteria, operational conditions, and microbiology. Chem Eng J Adv 14:100453. https://doi.org/10.1016/j.ceja.2023.100453

European Commission DG Environment (2001) Evaluation of sludge treatments for pathogen reduction – final report

Fersi M, Smaoui Y, Mechichi T, Hachicha R (2018) Limits de tolérance de l'orge, des pois et de la laitue envers les composts riches en composés phénoliques et en substances lipidiques. Commun Sci Sol Anal Végétale 49:1418–1428. https://doi.org/10.1080/00103624.2018.1464181

Hachicha S, Chtourou M, Medhioub K, Ammar E (2006) Compost of poultry manure and olive mill wastes as an alternative fertilizer. Agron Sustain Dev 26(2):135–142

Hachicha S, Cegarra J, Sellami F, Hachicha R, Drira N, Medhioub K, Ammar E (2009) Elimination of polyphenols toxicity from olive mill wastewater sludge by its co-composting with sesame bark. J Hazard Mater 161(2):1131–1139

Hechmi S, Hamdi H, Mokni-Tlili S, Ghorbel M, Khelil MN, Zoghlami IR, Benzarti S, Jellali S, Hassen A, Jedidi N (2020a) Impact of urban sewage sludge on soil physico-chemical properties and phytotoxicity as influenced by soil texture and reuse conditions. J Environ Qual 49:973–986. https://doi.org/10.1002/jeq2.20093

Hechmi S, Hamdi H, Mokni-Tlili S, Zoghlami RI, Khelil MN, Benzarti S et al (2020b) Minéralisation du carbone, indicateurs biologiques et phytotoxicité pour évaluer l'impact des boues d'épuration urbaines sur deux sols à texture légère dans un microcosme. J Environ Qual 49:460–471.https://doi.org/10.1002/jeq2.20011

Hamdi H, Benzarti S, Manusadžianas L, Aoyama I, Jedidi N (2007) Essais biologiques en phase solide et activités microbiennes du sol pour évaluer l'écotoxicité du sol enrichie en HAP après un processus de biorestauration à long terme simulant l'agriculture terrestre. Chemosphère 70:135–143. https://doi.org/10.1016/j.chemosphere.2007.06.043

Hamdi H, Benzarti S, Aoyama I, Jedidi N (2012) Réhabilitation de sols dégradés contenant des HAP âgés basée sur la phytoremédiation avec de la luzerne (Medicago *sativa* L.). Biodétérioration Biodégradation Internationales 67:40–47

Hamdi H, Hechmi S, Khelil MN, Zoghlami IR, Benzarti S, Mokni-Tlili S, Hassen A, Jedidi N (2019) Repetitive land application of urban sewage sludge: effect of amendment rates and soil texture on fertility and degradation parameters. CATENA 172:11–20. https://doi.org/10.1016/j.catena.2018.08.015

Hassen A, Jedidi N, Cherif M, M'Hiri A, Boudabous A, Cleemput OV (1998) Mineralization of nitrogen in a clayey loamy-soil amended with organic wastes enriched with Zn, Cu, and Cd. Biores Technol 64:39–45. https://doi.org/10.1016/S0960-8524(97)00153-3

Hassen A, Mahrouk M, Ouzari H, Cherif M, Boudabous A, Damelincourt JJ (2000) UV disinfection of treated wastewater in a large-scale pilot plant and inactivation of selected bacteria in a laboratory UV device. Bioresour Technol 74(2):141–150

Hassen A, Belguith K, Jedidi N, Cherif A, Cherif M, Boudabous A (2001) Microbial characterization during composting of municipal solid waste. Biores Technol 80(3):217–225

Hassen W, Hassen B, Werheni R, Hidri Y, Jedidi N, Hassen A (2023) Processes of valorization and management of olive by-products: the pomace and olive mill wastewater. In: Wastewater from olive oil production, environmental impacts, treatment and valorization. pp 1–25. https://doi.org/10.1007/978-3-031-23449-1_1

Huang W, Li Y, Wang F, Feng L, Wang D, Ma Y, Yang W, Jingyang L (2023) Disinfectant sodium dichloro-isocyanurate synergistically strengthened sludge acidogenic process and pathogens inactivation: targeted upregulation of functional microorganisms and metabolic traits via self-adaptation. Water Res 247(3):120787. https://doi.org/10.1016/j.watres.2023.120787

Jedidi N, Hassen A, Van Cleemput O, M'Hiri A (1995) Quantification of nitrogen mineralization and organization processes in the soil in the presence of organic amendments. Can J Soil Sci 75:85–91

Miranda ARM, Antunes JEL, Araujo FF, Melo VMM, Bezerra WM, van der Brink PJ, Araujo ASF (2018) Les groupes bactériens moins abondants sont plus touchés que les groupes les plus abondants dans les sols traités avec des boues de tannerie compostées. Rap Sci 8:11755. https://doi.org/10.1038/s41598-018-30292-1

Mokni-Tlili S, Hechmi S, Ouzari Hadda-I, Mechergui N, Ghorbel M, Jedidi N, Hassen A, Hamdi H (2023) Co-occurrence of antibiotic and metal resistance in long-term sewage sludge-amended soils: influence of application rates and pedo-climatic conditions. Environ Sci Pollut Res Int 30(10):26596–26612. PMCID: PMC9652132. PMID: 36369449. https://doi.org/10.1007/s11356-022-23802-2

Oleszczuk P, Malara A, Jośko I, Lesiuk A (2012) Les changements de phytotoxicité des sols amendés par des boues d'épuration. Pollut L'eau L'air Sol 223:4937–4948

Picariello E, Pucci L, Carotenuto M, Libralato G, Lofrano G, Baldantoni D (2021) Compost and sewage sludge for the improvement of soil chemical and biological quality of mediterranean agroecosystems. Sustainability 13:26. https://doi.org/10.3390/su13010026

Reverdy AL, Baudez JC, Dieudé-Fauvel E (2011) La méthanisation des boues issues du traitement des eaux usées: comparaison des performances de 3 technologies différentes. IRSTEA, pp 40. ffhal-02598022ff

Smith S (1996) Agricultural recycling of sewage sludge and the environment. CAB International

Traore D, Nikiema M, Somda MK, Sawadogo1 JB, Dayeri D, Traore AS (2016) Contribution à la biométhanisation de la biomasse végétale : cas des résidus de légumes au Burkina Faso. Int J Biol Chem Sci 10(1):35–47, February, ISSN 1997-342X (Online), ISSN 1991-8631. http://ajol.info/index.php/ijbcs. http://indexmedicus.afro.who.int

Valo A, Carrère H, Delgenès JP (2004) Thermal, chemical and thermo-chemical pre-treatment of waste activated sludge for anaerobic digestion. J Chem Technol Biotechnol 79(11):1197–1203

Wray C (1975) Survival and spread of pathogenic bacteria of veterinary importance with environment. Vet Bull 45(8):543–550

Chapter 23
Gasification for Sustainable Valorization of Sewage Sludge

Simona Di Fraia, Nicola Massarotti, M. Rakib Uddin, and Laura Vanoli

Abstract Gasification is an attractive treatment for valorization of sewage sludge since it allows converting a waste, whose disposal is environmentally and economically challenging, into a resource. Indeed, during the last decades, sewage sludge disposal has become of the most important problems for wastewater treatment plants together with the high energy demand associated with the processes. In this context, gasification of sewage sludge aimed at energy recovery could be significantly advantageous. For this reason, this chapter aims at analyzing the current state of the art related to sewage sludge gasification focusing on its impact on sustainable valorization of the sludge. The work will deeply assess the main parameters that affect gasification efficiency, such as the type of reactor, gasifying agent, operating parameters of temperature and equivalence ratio, and the eventual impacts of catalysts application on process performances. Also the effect of these parameters on the process outcomes will be considered. Moreover, possible applications for the process outputs, syngas and biochar, will be examined, highlighting the environmental and economic benefits of their reuse and/or recovery. Then, to practically demonstrate the feasibility of sludge gasification as well as its advantages and drawbacks, a case study will be investigated. Finally, considering the main results of the analysis carried out, the challenges for widespread diffusion of sludge gasification will be illustrated together with the opportunities of implementing this process.

Keywords Wastewater sludge · Syngas · Biochar · Energy recovery · CHP · Thermochemical process

S. Di Fraia (✉) · N. Massarotti · M. Rakib Uddin · L. Vanoli
Department of Engineering, Università degli Studi di Napoli "Parthenope", Napoli, Italy
e-mail: simona.difraia@uniparthenope.it

M. Rakib Uddin
Department of Chemical Engineering and Polymer Science, Shahjalal University of Science and Technology, Sylhet, Bangladesh

S. Souabi et al. (eds.), *Wastewater Treatment Plants*, Water Science and Technology Library 130, https://doi.org/10.1007/978-3-031-87461-1_23

23.1 Introduction

Wastewater treatment is a crucial sector for preserving the quality of water sources and mitigating environmental impact. Due to continuous population growth and increasing environmental awareness, the need to address the challenges of wastewater management is becoming ever more pressing. Indeed, it not only ensures the safety of water resources but also focuses on the management of by-products, including sewage sludge (SS). SS is the solid by-product of wastewater treatment. Its production has significantly increased over the last decades due to the increase of people served by wastewater treatment plants and to the use of more intensive processes aimed at fulfilling the current stringent legislation (Neri et al. 2024). In the past, this by-product was often seen as unwanted to be eliminated as efficiently as possible. However, due to the evolution of environmental awareness instead of being viewed as problems to overcome, SS is emerging as a resource with potential, particularly relevant for energy and resources recovery (Singh and Agrawal 2008). The effectiveness of an energy recovery method is closely dependent on the specific physicochemical characterization of the sludge, especially in terms of moisture, nitrogen, and heavy metal content (Capodaglio and Olsson 2019). While incineration and anaerobic digestion represent established approaches for extracting energy potential of the sludge, thermochemical processes are emerging for better exploitation of the energy content of the sludge, due to their low overall environmental impact. Moreover, the thermochemical treatment of SS is becoming more and more attractive due to its rapid processing time, reduced footprint, and clear benefits in resource recovery (Li et al. 2023b). Thermal processes are also highly efficient in decreasing sludge volume and pollutants with a lower global warming potential than other treatment methods when they are coupled to energy recovery strategies (Teoh and Li 2020). Indeed, in wastewater treatment plants, recovering energy from sludge serves a dual purpose: it minimizes the amount of waste requiring disposal and lowers energy expenses, which typically represent 25–50% of the operational costs in WWTPs (Fraia et al. 2019). Consequently, this approach helps reduce greenhouse gas emissions associated with road transportation and energy generation, which are often reliant on fossil fuels (Fraia et al. 2020). Moreover, thermal processes are particularly indicated in major urban centers where other treatment and disposal methods may be challenging to apply due to space constraints and concerns over odor issues (Chang et al. 2022).

23.2 Sludge Characteristics

The properties of SS significantly depend on:

- the characteristics of the treated wastewater that can be municipal or industrial;
- the processes and technologies employed in the WWTPs.

Compared to industrial wastewater, the municipal one has a higher organic matter content, making it a more suitable option for resource recovery via thermochemical conversion (Djandja et al. 2021). In general, SS is a mixture of water, organic matter, nutrients, and several potentially hazardous substances, including inorganic and organic contaminants, and pathogens.

Depending on the processes of the WWTP, the main types of sludge that can be produced are (Manara and Zabaniotou 2012):

- primary sludge, which is generated during primary treatments of wastewater, where the heavy solids are separated through sedimentation;
- biological or secondary sludge that is generated in secondary biological treatments, where the organic matter is removed exploiting the activity of indigenous and/or water-borne microorganisms;
- mixed sludge that is a mix of primary and secondary sludge;
- tertiary sludge, which is formed in the tertiary or advanced wastewater treatment, aimed at removing nutrients, mainly nitrogen and phosphorus.

The sludge is then treated in the so called sludge line that aims at reducing the water content and the pathogen load of the sludge. The moisture content of SS has to be significantly reduced in order to facilitate the subsequent treatments and to decrease the transportation and disposal costs. Moreover, reducing the water content through drying before energy recovery is essential to guarantee an adequate heating value (Syed-Hassan et al. 2017). Pathogens contained into the sludge are destroyed through anaerobic and/or aerobic digestion as well as thermo-chemical processes. Mentioning anaerobic digestion, which is probably the most common process adopted for energy recovery from SS, it has to be underlined that it advised against coupling it to thermo-chemical processes since it decreases the volatile matter content of the substrate reducing the recoverable energy potential (Fraia et al. 2016).

When thermo-chemical processes, such as gasification, are employed for energy recovery of SS, its assessment through proximate and ultimate analyses is fundamental. Moreover, the significant content of macro- and micronutrients that can extracted from sludge for fertilizer production has to be considered (Kominko et al. 2024).

Beyond these valuable resources, SS contains hazardous substances, such as heavy metals that can be accumulated in soils and then absorbed by plants, reducing their growth, accessing the food chain and consequently compromising human health (Chang et al. 2024). Moreover, such an accumulation can be facilitated by the presence of micro-plastics which can be commonly found in wastewater and SS (Karami et al. 2016) (Table 23.1).

Table 23.1 Characteristics of SS expressed as wt.% dry matter (Djandja et al. 2021; Nahar et al. 2024; Hu et al. 2022; Thipkhunthod et al. 2005)

Proximate analysis			Ultimate analysis					Nutrients		
Volatile matter	Ash	Fixed carbon	C	H	O	N	S	Total nitrogen	Total phosphorus	Potassium
21.71–82.30	10–76.48	0.7–21.80	9–55	2–9.72	12.41–49.44	1.5–8.8	0.4–9.5	1.5–7	0.5–3.9	0.5–1

23.3 Gasification Process

Gasification converts the organic matter in a high-temperature and controlled oxygen atmosphere (Caputo et al. 2005). The main product is syngas, a gas primarily composed of hydrogen (H_2), carbon dioxide (CO_2), carbon monoxide (CO), and methane (CH_4), along with other substances such as sulfuric acid, hydrochloric acid, and ammonia. The percentage of each component varies significantly based on several factors, including the design of reactor, the moisture content, the type of agent employed, and the operating conditions (Brachi et al. 2014). Syngas can be used to produce energy through combustion (Sanlisoy and Carpinlioglu 2017) or raw materials for chemicals and liquid fuels (Yap and Nixon 2015). The raw syngas needs further cleaning and conditioning to ensure a product that can be used as a fuel (Brachi et al. 2014).

Gasification is a complex phenomenon that involves several interconnected stages, which, although generally divided into drying for water content evaporation, pyrolysis for devolatilization, oxidation for energy supply, and reduction, occur simultaneously (Wang et al. 2024).

The drying phase is the first stage of the gasification process and it prepares the feedstock for the following steps. This phase aims at removing the moisture from the organic material to be treated. During this phase, a thermal input is essential to heat and vaporize the water contained in the biomass (Singh Siwal et al. 2020). Thermal energy required depends on the humidity level of the incoming biomass, which is typically around 25%. During the drying process, heat is applied to the moist material, triggering the evaporation of moisture, and this heating persists until a temperature of 200 °C is reached (Ferreiro et al. 2024). The process requires energy, which can come from various sources, including residual heat from other stages of the gasification process, the reaction itself, or external sources. The drying process is essential as water can hinder the subsequent stages of the process, negatively affecting pyrolysis. The removal of water also helps to decrease the demand of thermal energy necessary to heat the material during the other stages of the process.

Pyrolysis takes place in an atmosphere without access to oxygen at high temperatures (350–600 °C) (Gao et al. 2020). High temperature is necessary as it enhances the rate of the chemical reactions involved in the decomposition of biomass, that allows converting the biomass into syngas and char, a concentrated form of black carbon (Ayub et al. 2024).

The oxidation stage is exothermic, resulting in a temperature rise of up to 1100 °C, providing heat to the other phases of the process that are endothermic. The oxidation of the condensable organic fraction generates low-molecular-weight products, which in turn decreases the amount of tar in the output (Manara and Zabaniotou 2012).

Finally, during the reduction step, the gas mixture and char generated from the earlier stages of pyrolysis and oxidation lead to the production of the final syngas (Molino et al. 2016).

As indicated in the literature (Basu 2010), gasification efficiency is influenced by the type of substrate, its size and moisture content, and the process temperature.

The high water content in the biomass (over 40% by weight) decreases the thermal efficiency of gasification process (Shakorfow 2016), requiring a preliminary thermal drying step.

The efficiency of gasification may be evaluated by calculating:

- carbon conversion efficiency (CCE) that represents the proportion of carbon in the substrate that is converted into carbon-containing gas (Shahbaz et al. 2020);
- cold gasification efficiency (CGE), which is the ratio between the chemical energy of the syngas and that of the substrate treated into the plant (Arena 2012).

23.3.1 Gasifier Type

The main focus of the gasification process is to ensure the fuel-rich condition by supplying a less quantity of oxidizing agent compared to the stoichiometric amount. This allows for maximizing the formation of CO and H_2 instead of CO_2 and H_2O, respectively, and less quantity of CH_4, tar, and lighter hydrocarbons (Klinghoffer and Castaldi 2013). To achieve this aim good mixing between the biomass and the gasifying agent is mandatory as well as the optimization of the operating parameters. Gasifier design is crucial to ensure the good mixing of biomass and gasifying agent to restrict the creation of an oxygen-rich zone to prevent combustion and improve the syngas quality. Based on the biomass and gasifying agent feeding system and directions of movement, gasifiers are classified depending on the gasifier design. The main configurations are described in the following subsections.

23.3.1.1 Fixed-Bed Updraft Gasifier

The fixed-bed updraft gasifier is the simplest form of reactor for energy recovery from SS with the lowest operating cost (Erakhrumen 2012). Biomass enters the reactor from the top whereas the gasifying agent is fed through a grate that is installed in the bottom section of the gasifier and acts as a bed to support the biomass. A simplified diagram of a fixed-bed updraft gasifier is presented in Fig. 23.1 (Warnecke 2000).

The gasifying agent is diffused in an upward direction whereas biomass moves in the opposite and for this reason, such kind of gasifier is also known as a fixed-bed countercurrent gasifier. The generated char during the gasification of biomass is combusted above the grate and forms H_2O, CO_2, and thermal energy. The hot gases with a temperature of around 1000 °C move in the upward direction and reduce in the reduction zone to form CO and H_2 through water gas reaction and the temperature drops to around 750 °C (Erakhrumen 2012; Sreejith et al. 2014). Ash generated through char combustion exits from the bottom of the gasifier. The gaseous stream generated through oxidation and reduction of biomass moves upward and exchanges heat with fed materials and reaches a temperature of around 400 °C which is sufficient to complete the pyrolysis (Fraia et al. 2021). Above the pyrolysis zone, the gas stream is in contact with the biomass and increases its temperature to around 200 °C reducing

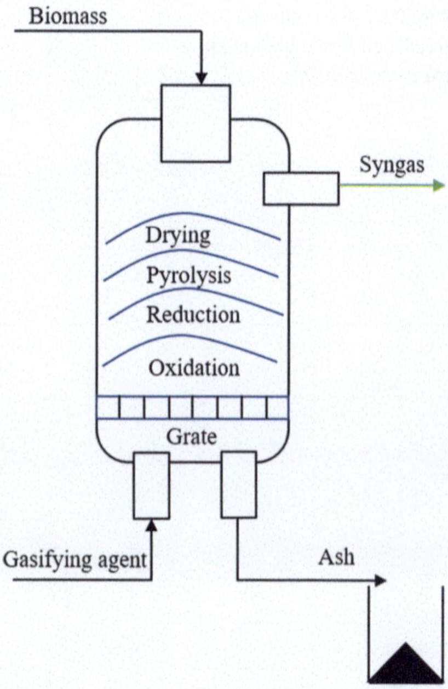

Fig. 23.1 Simplified diagram of a fixed-bed updraft gasifier (Warnecke 2000)

the moisture content and creating a drying zone in the reactor. The syngas exiting from the upper section of the gasifier has a temperature that varied between 200 and 300 °C (Erakhrumen 2012). Syngas generated in a fixed-bed updraft gasifier contains the highest quantity of CO_2 and tar content compared to other types of gasifier and consequently higher CCE (Brar et al. 2012; Jeong et al. 2019). The lower heating value (LHV) of syngas obtained from biomass through a fixed-bed updraft gasifier is low as it contains high CO_2 content which is the reason for a low CGE and in the range of 20–60% (Jeong et al. 2019; Zheng et al. 2019; Shi et al. 2016).

23.3.1.2 Fixed-Bed Downdraft Gasifier

The fixed-bed downdraft reactor is another simple form of gasifier whose operating principle is similar to updraft gasifier except for the substrate and the gasifying agent that enter from the top of the reactor and move downward. The generated syngas is extracted from the bottom of the gasifier. This kind of reactor is also known as a co-current gasifier as both biomass and GA diffuse in the same direction (Erakhrumen 2012). A schematic layout of a fixed-bed downdraft gasifier is illustrated in Fig. 23.2 (Warnecke 2000).

The energy required for drying and pyrolysis of biomass comes from hot syngas that heats the reactor wall. The tar compounds formed during the gasification treatment of biomass crack to 99.9% but the char content is not fully decomposed and

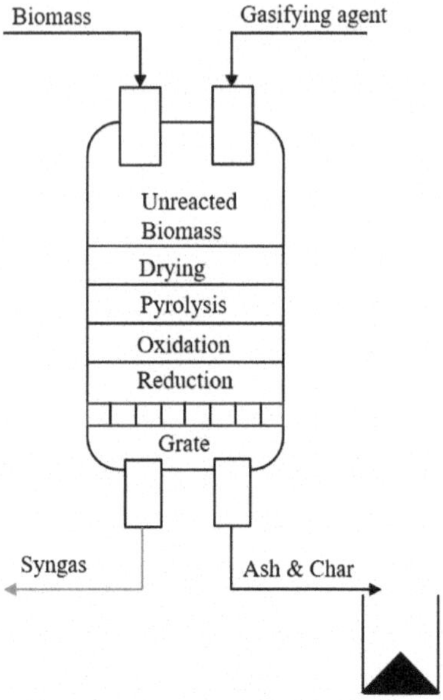

Fig. 23.2 Layout of a fixed-bed downdraft gasifier (Warnecke 2000)

exits with ash (Erakhrumen 2012). The syngas generated from biomass in a fixed-bed downdraft gasifier has a slightly lower quantity of CO_2 compared to the updraft process. Consequently, LHV of syngas, as well as CGE, is slightly higher in the downdraft process compared to the updraft process, varying between 30 and 80.91% whereas CCE shows a opposite trend due to the cracking of tar compounds (Brar et al. 2012; Jeong et al. 2019; Zheng et al. 2019; Shi et al. 2016).

23.3.1.3 Bubbling Fluidized-Bed Gasifier

Biomass is fed above the grate in a bubbling fluidized-bed gasifier and mixed with bed materials (usually sand or alumina) and the gasifying agent is fed from the bottom. Usually, 90% of the reactor volume is filled by bed materials (Erakhrumen 2012). Biomass particles with bed materials are suspended in upward-flowing gas, whose velocity increases inside the reactor generating bubbling conditions. At a certain value of gas velocity, the frictional force between the particles (biomass and bed materials) and the fluid counteracts the weight of particles, and the fluidization stage continues. Particles are mixed in all directions ensuring a uniform temperature profile inside the reactor. Generated syngas passes through the void space of bed materials with some ash particles to the freeboard zone and moves to a cyclone to separate the ash particles and the syngas that is collected as a top product. The layout of a bubbling fluidized-bed gasifier is presented in Fig. 23.3 (Loha et al. 2014).

The temperature is maintained at less than 950 °C to avoid the slagging of ash particles. The performance in a bubbling fluidized bed gasifier in terms of CGE varies

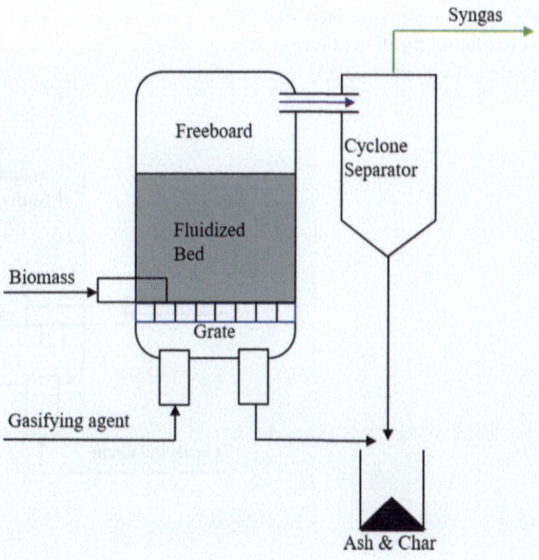

Fig. 23.3 Layout of a bubbling fluidized-bed gasifier (Loha et al. 2014)

in the range of 39.1–56.9% (Erakhrumen 2012; Jeong et al. 2019; Zheng et al. 2019; Shi et al. 2016).

23.3.1.4 Circulating Fluidized-Bed Gasifier

The operation and feeding system of both biomass and gasifying agent in a circulating fluidized-bed gasifier is the same as the bubbling fluidized-bed gasifier. The only exception is the higher fluidization velocity that ensures the circulation of the particles inside the full reactor volume. A schematic diagram of the circulating fluidized-bed gasifier is presented in Fig. 23.4 (Loha et al. 2014).

Due to the higher fluidization velocity, syngas is eluted out from the gasifier with particles of char, ash, biomass, and bed materials that are separated in a cyclone. The particles of ash, char, biomass, and bed materials are sent back to the gasifier above the grate whereas syngas elutes out as the top product from the cyclone. Syngas contains a higher concentration of H_2 and CO compared to bubbling fluidized-bed due to the higher mixing rate of biomass and gasifying agent. Thus, the generated syngas has higher LHV, CCE, and CGE compared to syngas generated in bubbling fluidized bed reactor (Erakhrumen 2012; Loha et al. 2014). CGE of the gasification process during syngas generation from biomass varies between 50 and 70% (Shi et al. 2016).

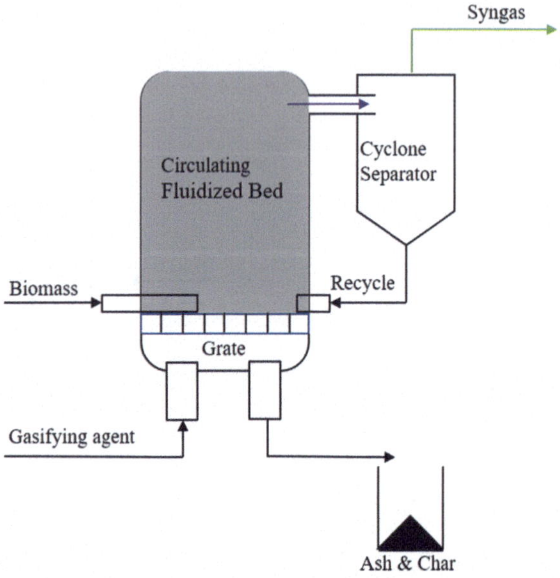

Fig. 23.4 Schematic view of a circulating fluidized-bed gasifier (Loha et al. 2014)

23.3.1.5 Entrained Flow Gasifier

The entrained flow gasifier is a special type of gasifier used for energy recovery as syngas from petroleum residues or coke. The temperature in the entrained flow gasifier is higher with respect to other gasifiers (higher than 1250 °C). The residence time of the biomass in the reactor is very short, in the range of a few seconds to minutes. Since the residence time is very short, particles should be in very small size or slurry form to finalize gasification. Due to the higher temperature, ash is transformed into slag stage, and the tar is broken down into H_2, CO, and CO_2 (Erakhrumen 2012; Watanabe and Otaka 2006). A representation the entrained flow gasifier is shown in Fig. 23.5 (Watanabe and Otaka 2006). The gasification process performance of entrained flow gasifiers in terms of CGE is the highest among all the gasifiers and varies between 58 and 88% (Chen et al. 2013).

23.3.2 Gasifying Agent

In order to generate inflammable gas from biomass including SS a gasifying agent is required (Pode et al. 2015). The nature of the gasifying agent determines the reaction path involved in the energy recovery from biomass and ultimately the syngas composition and its energy content (Bermudez and Fidalgo 2016). Air, steam, oxygen, carbon dioxide, and the combination of air and steam are frequently used as gasifying agent for the gasification of biomass (Gao et al. 2020; Fraia et al. 2021; Tezer et al. 2023; Milčák et al. 2024; Hosseingholilou et al. 2024).

Fig. 23.5 Schematic view of entrained flow gasifier (Watanabe and Otaka 2006)

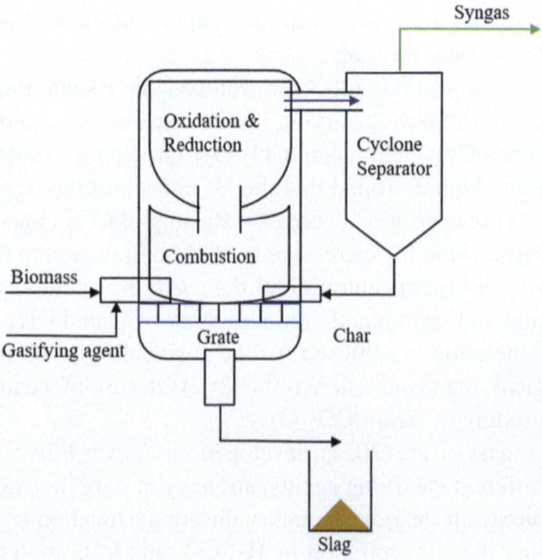

Li et al. (2023a) generated syngas from coal through gasification treatment in a two-stage entrained flow gasifier by applying three different gasifying agents, air and mixtures of O_2/CO_2 and O_2/H_2O. Researchers have investigated the tar and soot formation within the changes in gasifying agent composition. Tar formation was found to be highest for the gasification by using air and lowest for the case of O_2/H_2O mixtures due to the reduction in benzene formation with the increase of O_2 concentrations in the reactor. On the other hand, soot formation is highest for the gasification of coal by using air due to the domination of surface growth for hydrogen abstraction and acetylene addition.

Tezer et al. (2023) generated syngas from SS through gasification in a fixed bed reactor by using dry air and pure O_2 as gasifying agent. With the increase of O_2 concentration in the reactor, H_2 and CO_2 concentration in the syngas raised whereas the remaining components of CO and CH_4 decreased. Consequently, the syngas energy content in terms of LHV decreased with the increases in the flow rate of gasifying agent (either dry air or pure O_2). For a particular operating point, in which a similar O_2 content was supplied in the reactor by using air generated a higher H_2 containing syngas compared to pure O_2. However, the LHV was lower due to the dilution effect because of N_2 content of the air.

Milčák et al. (2024) generated syngas from the mixture of digestate and woodchips gasification in a fluidized bed reactor with three gasifying agents, air, air–steam, and O_2—steam to investigate the syngas quality and energy content. The energy content of the generated syngas in terms of LHV decreased continuously with gasifying agent concentration since the gasification zone moves to the combustion zone with O_2 content that increases in the reactor. Syngas had high H_2, CO, and CH_4 content in case of the steam gasification. However, the highest concentration of H_2, CO, and CH_4 components in the syngas and the highest LHV were used by employing O_2—steam mixture.

Xiang et al. (2020) developed an Aspen Plus model to show the effect of using different gasifying agents (steam, CO_2, and mixtures of steam/O_2, CO_2/O_2, steam/CO_2, and steam/CO_2/O_2) on syngas composition during biomass gasification. Authors found that the H_2 concentration in syngas rises with the steam while CO concentration decreases. Raising of CO_2 concentration in the gasifying agent is responsible for increasing the CO composition in the syngas. On the other hand, the rise of O_2 concentration in the gasifying agent slightly increases the H_2 concentration and significantly reduces that of CO and CH_4 in the syngas. The highest power generation is achieved for the gasification of biomass by applying CO_2 and O_2 as gasifying agent whereas the lowest in case of steam and O_2 and intermediate for the mixture of steam/CO_2/O_2.

Cao et al. (2024) developed an Aspen Plus model to investigate the effect of different gasifying agents, air, oxygen-enriched air, air/steam, and oxygen-enriched air/steam on syngas quality during gasification of sawdust. The authors concluded that the concentration of H_2, CO, and CH_4 in syngas as well as the LHV rises at a higher O_2 concentration in the gasifying agent. Char conversion as well as tar cracking increase with the rise of O_2 concentration in the gasifying agent. On the other hand, H_2 concentration in syngas as well as LHV of syngas and tar content decrease with the rise of steam concentration in syngas due to the promotion of water gas shift reaction.

Hosseingholilou et al. (2024) developed an Aspen Plus model to explore the variation of syngas quality with different gasifying agents, CO_2, steam, and air during the gasification of olive pomace. Researchers have determined that the H_2 content in the syngas increases as the flow rate of steam, O_2, and CO_2 concentration in the gasifying agent increases. The exergy efficiency increases with the decreasing of O_2 and CO_2 concentration but increases with that of steam in the gasifying agent.

23.3.3 Operating Parameters

The distribution of the components of syngas generated during the gasification of biomass and its energy content as well as the efficiency of the conversion process depends on the operating parameters, such as temperature, equivalence ratio (ER), steam-to-biomass ratio (SBR), and catalyst concentration (Bermudez and Fidalgo 2016; Cao et al. 2024; Yesilova et al. 2024; Carotenuto et al. 2023).

23.3.3.1 Temperature

Temperature significantly affects the decomposition of biomass into syngas and to control the syngas quality. When the temperature increases, endothermic reactions involved in the gasification of biomass move in the forward direction whereas exothermic reactions present a reverse trend. Syngas LHV, CCE, and CGE of gasification process increase with gasification temperature because of the increase of CO and H_2 concentration and decrease of CO_2 in syngas. Tar concentration also decreases with the increase in the gasification temperature (Sreejith et al. 2014; Pala et al. 2017; Vivanpatarakij et al. 2014). However, more energy is needed to increase the gasification temperature and at the same time, high temperature is responsible for the ash sintering, deposition, and corrosion of the gasifier (Brar et al. 2012). To avoid these problems, 700 to less than 1000 °C is identified as optimum range of temperature for biomass gasification (Brar et al. 2012; Abdelrahim et al. 2020; Andrés et al. 2019).

23.3.3.2 Equivalence Ratio

When air is employed as a gasifying agent, ER considerably influences the process. It is worth noticing that ER is the weight ratio between the air to biomass weight fed to the reactor and the air to biomass weight required to complete the combustion (Andrés et al. 2011). At high values of the ER, the combustion reaction is driven in the forward direction which determines a raise in the concentration of CO_2 in the syngas and a decrease of that of CO and H_2. Increasing ER increases the CCE but decreases the heating value of syngas and consequently the CGE. Optimum ER is identified in the range of 0.2–0.4 (Di Fraia et al. 2021; Abdelrahim et al. 2020; de Andrés et al. 2019; Mun et al. 2013; Mun et al. 2014; Fraia and Uddin 2022).

23.3.3.3 Steam to Biomass Ratio

Instead of ER, SBR is employed with steam employed as gasifying agent. It is a crucial parameter that influences the syngas composition as well as its energy content and conversion efficiency of the process. SBR is the ratio between the weight of steam to biomass fed to the reactor. An increase in the SBR increases the concentration of H_2 and CO due to the advancement of water gas shift and steam methane reforming reactions (Lasa et al. 2011). Consequently, LHV of syngas and CGE, as well as CCE increase (Zheng et al. 2019). The optimum SBR for the gasification of biomass (SS, wood, almond shell) is identified in the range of 0.5–1.5 (Bermudez and Fidalgo 2016; Andrés et al. 2019; Chen et al. 2021; Acar and Böke 2019).

23.3.3.4 Catalyst

Application of a catalyst in biomass gasification can substantially increase the reaction rate, reducing the processing time and temperature and increases the rate of tar components destruction which ultimately reduces the cost of syngas cleaning and increases the H_2 concentration in the product (Lv et al. 2007; Asadullah et al. 2002; Mun and Kim 2013). The application of dolomite and nickel-based catalysts significantly reduces the process temperature from 850 to 750 °C and increases the molar ratio of H_2/CO from 1.87 to 4.45 during syngas generation from pine sawdust by air–steam gasification in a fluidized bed reactor (Lv et al. 2007). Application of natural zeolite, olivine, and dolomite as bed materials and activated carbon as a catalyst during gasification of SS in a fluidized bed reactor can increase the H_2 content in syngas from 11.24 to 34.0 vol.% and reduce the tar content up to 98% compared to the gasification without catalyst (Mun et al. 2013; Mun et al. 2014; Mun and Kim 2013). The application of Aluminum–Nickel alloy catalyst increases 15 vol.% of H_2 content in the syngas compared to the red mud during the gasification of biomass mixtures containing green waste, SS, and olive pomace (Yesilova et al. 2024).

23.3.4 Output of the Gasification Process and Application

The principal output of gasification is syngas, but also other by-products, such as tar and ash containing heavy metals, are inevitably generated and their quantity is affected by the characteristics of the substrates entering the reactor or the conditions of the process (Śpiewak 2024).

Syngas is a versatile energy source that can be directly fed as a fuel in internal combustion engines and gas turbines to produce power and heat or it can be converted into various chemicals and biofuels (Rey et al. 2024). However, syngas produced from biomass often does not meet fuel quality standards and is not produced efficiently enough, requiring further treatments before its use (Suparmin et al. 2024). Alternatively, the gasification process itself may be modified to enhance syngas production. As an example, supercritical water gasification with the use of alkaline additives has been demonstrated to significantly reduce the sulfur concentration in syngas (Feng et al. 2021). Another strategy that was implemented to reduce the issues associated with sulfur emissions is represented by the co-gasification of coal and SS, through which the sulfur elements in solid products were found to exist mostly as nickel sulfide (Ma et al. 2024), especially at higher temperatures and reaction time. The combination of SS and bituminous coal was also investigated by employing chemical looping co-gasification in a fluidized bed reactor with natural hematite as the oxygen carrier (Han et al. 2023). The results of the experimental campaign highlighted that process efficiency and fuel conversion were higher compared to gasifying SS and bituminous coal separately, suggesting that chemical looping gasification is a promising method for producing syngas with a high hydrogen concentration. Chemical looping gasification was also employed to treat wet municipal sludge by using iron-aluminum composite oxygen carriers prepared from aluminum oxide particles and iron nitrate (An et al. 2024). Results demonstrated that the presence of water vapor enhanced hydrogen production during sludge gasification and prevented excessive reduction of the oxygen carriers. The released water vapor also improved the thermal conversion efficiency of the sludge. Higher temperatures initially favored the gasification reaction, but at 900 °C, slight sintering of the oxygen carriers and changes in the reaction atmosphere led to a decrease in hydrogen yield. Finally, controlling the mixing ratio of oxygen carriers to sludge was crucial for producing high-quality syngas.

Co-gasification with another biomass is more environmentally friendly compared to coal. As an example, municipal SS was co-gasified with cotton stalk in a fixed bed reactor by adding metal-rich texture dyeing sludge pyrolysis residues to enhance syngas production (Yan et al. 2023). The presence of active species, particularly sodium and iron, promoted tar cracking and char gasification, leading to higher syngas yields and lower solid and liquid byproducts. SS was also proposed to be co-gasified with spruce and birch residues, after torrefaction at different temperatures, to produce hydrogen-rich syngas (Huang et al. 2018). In this study, the additional advantage was the use of wet SS, eliminating the need for its separate drying and steam generation. Through numerical modeling, this study examined the impact of mixing ratios and torrefaction severity on carbon conversion, syngas composition, and hydrogen concentration, highlighting that biomass torrefied at high severity favored the generation of H_2 and CH_4 but suppressed that of CO_2 and CO.

As mentioned above, compared to conventional gasification, more innovative processes can be implemented to enhance the process efficiency in terms of syngas quality and quantity. Microwave-assisted gasification, which employs microwave energy to accelerate the process, was investigated using steam and carbon dioxide as gasification agents (Hu et al. 2024). The study focused on the impact of these agents on syngas production and hydrogen generation pathways. Results indicated that a steam-CO_2 mixture enhanced syngas yield and quality. The combination of steam and CO_2 demonstrated a synergistic effect in promoting H_2 production during and led to higher syngas yields compared to using pure steam or CO_2. CO_2 gasification was also investigated in combination with low temperature hydrothermal pretreatment to enhance syngas yield and quality (Huang et al. 2023). Hydrothermal carbonization with calcium oxide additives facilitated the conversion of pyridine-N to pyrrole-N and quaternary-N in the hydrochar, improving nitrogen stability. Additionally, CO_2 gasification of the hydrochar reduced tar formation, decreased oxygen and nitrogen content in tar, and increased syngas yield compared to direct SS gasification.

The main application for syngas is fueling of internal combustion engines due to their flexibility to be used in moving and stationary machineries and the advantages, compared to other technologies, of lower capital costs, reliability, strong part-load performance, high operating efficiency, modular design, and high safety record (Hagos et al. 2014). Despite these advantages, this field still lacks significant research (Dasappa et al. 2004). The major technical hurdle in developing commercially viable biomass gasification technologies for energy production is the high tar content in the produced raw syngas (Dudyński et al. 2015). Its formation reduces syngas yield and calorific value, and its condensation, occurring at low temperatures, can lead to clogs or blockages in pipes, filters, catalyst units, or engines (Heidenreich and Foscolo 2015). For these reasons, numerous studies on tar removal have resulted in the development of measures to minimize tar formation. Different strategies can be adopted, such as optimization of the operating conditions, modifications of the process itself, cleaning systems.

Regarding the operating conditions, it has been observed that increasing the ER reduces tar mass (Phuphuakrat et al. 2010). Also increasing the temperature significantly reduces gravimetric tar formation and at the same time it promotes combustible gas production, whereas increasing the throughput at a fixed temperature has the opposite effect (Andrés et al. 2016). The same study highlighted that higher temperatures lead to increased concentrations of polycyclic aromatic hydrocarbons in the tar. The gasifying agent as well as the addition of a catalyst may affect tar production. Using air and steam, particularly with dolomite as catalyst, significantly reduced gravimetric tar production across all tested throughputs with higher production of heavy polyaromatic hydrocarbons at higher throughputs (Roche et al. 2014).

It is worth noticing that these strategies positively affect syngas quality, also in terms of hydrogen content.

Another approach consists in modifications of gasification process. The staged gasification processes, which involve separate pyrolysis and gasification steps, can minimize tar content, beyond achieving efficient biomass conversion and producing high-quality syngas to be employed in small- and medium-scale systems for energy generation and as a replacement for natural gas in industrial applications (Wang et al. 2024). Moreover, using activated carbon in two stage gasification generally enhance hydrogen production and reduce ammonia, with a lower tar cracking and hydrogen production activity in case of acid-treated activated carbon (Choi et al. 2015). Two stage gasification has been further modified to include steam gasification into the first stage in order to promote tar decomposition (Yang et al. 2021). At the same time, also in this case, the hydrogen concentration was enhanced obtaining a H_2/CO ratio of the produced syngas between 0.9 and 4.7 by adjusting the calcium oxide (CaO) and steam content.

Regarding pretreatments, hydrothermal process has been found to reduce tar content but it did not favor the reduction of polycyclic aromatic hydrocarbons content in tar (Feng et al. 2018).

Finally, as concerns the cleaning systems, syngas can be purified by employing:

- wet gas cleaning technologies, like spray towers, cyclonic scrubbers, and venturi scrubbers, which use water to remove contaminants from syngas. However, these methods can be energy-intensive due to the need to cool the hot syngas (700–900 °C) before cleaning (Rey et al. 2024). Furthermore, there are additional costs associated with managing the contaminants captured in the cleaning process (Abdoulmoumine et al. 2015);
- dry gas cleaning that consist in mechanical and physical processes for particle removal, thermal processes for tar removal, and catalytic processes for removing tar, sulfur, nitrogen, alkali, and chlorine. This method has gained popularity due to its energy efficiency and environmental advantages, despite the low availability factor, sorbent degradation, and multi-stage separation processes (Sharma et al. 2010).

For syngas derived from SS, venturi scrubbers and sawdust adsorbers were evaluated for tar removal, observing that the system removed approximately 44% of light aromatic hydrocarbon tar, and no light polycyclic aromatic hydrocarbon, with tar removal efficiency dependent on gasification conditions, tar composition, and tar concentration (Phuphuakrat et al. 2010).

The solid by-product of SS gasification is biochar that can enhance soil quality, reduce crop uptake of heavy metals, and provide various agricultural benefits, such as improved soil nutrient content, increased cation exchange capacity, and reduced nutrient leaching (Singh et al. 2020). While due to its high organic matter content and carbon sequestration potential biochar represents a promising soil additive, its high ash content may limit its appeal as a solid biofuel (Trabelsi et al. 2021).

Moreover biochar from SS represents a versatile material with potential applications as an adsorbent, catalyst, and support material (Gopinath et al. 2021; Pelagalli et al. 2024; Sharmila et al. 2023, 2024).

23.4 Case Study

One of the most feasible applications of SS gasification is the use of syngas to produce energy to partially self-supply the demand of the wastewater treatment plant in which the sludge is generated. For this reason, the potentiality of combined heat and power (CHP) generation coupled to SS gasification (Fraia et al. 2021) is presented as an example of sustainable valorization of SS. The CHP system was modeled in Aspen Plus by using gasification coupled with an internal combustion engine (ICE). The gasification model was developed by applying Gibbs free energy minimization technique and the ICE system was simulated by using real engine operating conditions. The model was used to estimate CHP generation potentiality of SS in Italy as well as in 8 European countries.

23.4.1 Process Description

Energy recovery from SS was predicted by modeling gasification integrated with an ICE in Aspen Plus. There is no unique operation block in Aspen Plus to represent either gasification or ICE. Thus several built-in components eventually integrated with Fortran subroutine were connected to simulate the sub-processes of drying, pyrolysis, oxidation, and reductions of SS.

SS was considered as a nonconventional component due to its absence in the Aspen library, implementing its proximate and ultimate analysis. Air was chosen as a gasifying agent due to its availability in the nature to complete the gasification process. Specific gasification temperature was ensured by heating the incoming air through a heat exchanger that recovered thermal energy from syngas cooling. For raw syngas, removal of ash particles and cooling down to the room temperature of 30 °C were considered to ensure the maximum efficiency of ICE. All the details on the model development can be found in (Fraia et al. 2021; Abdelrahim et al. 2020; Brachi et al. 2022).

23.4.2 Gasification Model Development

The simulation model is based on a temperature-restricted equilibrium approach calibrated and validated using experimental data related to the gasification of SS generated in an Italian wastewater treatment plant in a bench-scale rotary kiln reactor. During model calibration, the average deviation of syngas composition predicted through the developed model compared to experimental data was 2.96% with a deviation of the single components lower than 5%. Regarding model validation, the average deviation was found to be 13.7% whereas that of the individual components was lower than 20%. Then the proposed integrated system was simulated by varying the main working parameters to define the values that maximize the performance.

23.4.3 Effect of Operating Parameters

ER and temperature were chosen as operating parameters to examine the variation of syngas composition and system performance to identify the optimum working values.

ER was varied from 0.15 to 0.40 at a fixed temperature of 850 °C to assess the effect on syngas composition, LHV, power obtained from syngas, and thermal power to be provided to complete the gasification process by heating the incoming gasifying agent. At increasing ER.

- the concentration of CO_2 was found to increase continuously whereas that of H_2 to raise gradually;
- the concentration of the other components (CO, CH_4, C_2H_6, and C_3H_8) decreased constantly due to the advancement of oxidation (combustion) reaction and reduction of the rate of water gas reaction. Consequently, syngas LHV and primary power obtained from the process decreased;
- thermal power required to sustain the process decreased due to the forward movement of the combustion reactions with O_2 concentration that are exothermic and generate heat.

Despite the highest net power was obtained at an ER of 0.15, the value of 0.20 was considered as the optimum since it generates a lower tar content thus it allows a higher performance and lifetime of the ICE when syngas is used as a fuel.

The effect of changing the temperature in the gasifier from 700 to 1000 °C was investigated in correspondence of an ER value of 0.20. Reaction rates are altered with the temperature e.g., endothermic reactions move to the forward directions whereas exothermic reactions experienced a reverse trend. Consequently, gasification product concentrations changed. As the temperature was increased:

- the content of CO and H_2 in the syngas increases continuously because of the endothermic nature of the water gas reaction whereas the remaining components of CO_2, CH_4, C_2H_6, and C_3H_8 experienced a reverse trend caused by the exothermic

nature of the other reactions involved in the gasification process, e.g., methanation, water gas shift, carbon, hydrogen, ethane, and propane combustion reactions;
- syngas LHV, power obtained from syngas, and thermal energy necessary as input to the process increased continuously.

900 °C was identified as optimal temperature for the gasification of SS, even though syngas LHV and power obtained from gasification product was higher at 950 °C than 900 °C. However, thermal power required to sustain the process was higher compared to the additional power obtained from the product at that temperature interval.

23.4.4 CHP Generation Potentiality of SS

The potential electrical and thermal energy production from SS was predicted to be 1.05 kWh_{el}/kg SS as dry solid (DS) and 1.68 kWh_{th}/kg SS as DS, respectively. In order to assess the benefit of SS gasification as a sustainable and energy efficient practice, especially if the produced energy is self-consumed within the plant that generates the sludge, CHP generation potentiality of SS produced in different European countries was predicted by using the model previously described. For SS production, data available for 2015 were used, whereas for the selection of the countries EU states in which landfilling and agricultural reuse are the main routes to manage the produced SS were considered. By using these assumptions, the following electrical energy potentials were derived: 3146 GWh/year in Spain, 1550 GWh/year in Poland, 1124 GWh/year in Italy, 575 GWh/year in Romania, 327 GWh/year in Greece, 160 GWh/year in Ireland, 157 GWh/year in Bulgaria, 154 GWh/year in Slovakia, and 25 GWh/year in Luxembourg.

In the EU countries, electrical energy consumption for wastewater treatment is predicted at around 1% of the national demand. Under this assumption, in the countries analyzed, the derived potentials could cover between 23.0 and 29.0% of the electricity employed by the wastewater treatment sector. It is worth noticing that wastewater treatment plants present also a high thermal energy demand in case of thermal drying, a process that is required when thermo-chemical processes are implemented. Indeed, initially water content of SS is 98.0–98.5 wt.% and through mechanical dewatering it can be reduced to around 70.0 wt.% (Carotenuto et al. 2022). For this reason, mechanically dewatered SS is usually subjected to thermal drying to decrease the water content to less than 10.0 wt.% as needed for gasification treatment. As mentioned above, thermal drying is an energy-consuming step, with a specific thermal demand of around 0.85 kWh_{th}/kg SS as DS. Recovering thermal energy produced by the proposed CHP system could cover 50.0–57.0% of the demand of thermal drying.

To conclude, CHP from SS through the proposed system of gasification coupled with an ICE offers the simultaneous advantages of increasing renewable energy generation which ultimately reduces the stress on fossil fuels as well as greenhouse gas emission caused by the current management techniques of landfilling and agricultural reuse.

23.5 Challenges, Opportunities, and Future Perspective

SS gasification offers a promising avenue for sustainable waste management and energy production. However, it also presents several challenges.

First of all, gasification, and in general thermo-chemical processes, requires a substrate with a low moisture content to be energy efficient. For this reason, in case of SS drying, which is a costly and very high energy demanding process, is needed (Bennamoun et al. 2013). In this context, employing renewables (Fraia et al. 2020; Fraia et al. 2018) or recovering energy from the same SS (Carotenuto et al. 2022) could increase the feasibility of the entire process.

Another critical issue is the low LHV of the produced syngas. Thus the research is mainly focusing on hydrogen-rich syngas characterized by a higher energy content and lower emissions, with the result that steam gasification has gained widespread popularity (Quan et al. 2022). At the same time, research efforts aim at optimizing gasifiers and catalysts to enhance tar cracking. Indeed, tar poses a significant health risk, due to the content of harmful teratogenic and carcinogenic compounds, and represents a substantial energy loss that could be transferred to the product gas (Hu et al. 2021). As mentioned in the previous sections, if tar condenses when syngas temperatures fall below the tar dew point, it leads to issues like clogging, fouling, corrosion, and reduced efficiency in downstream processes. The acceptable tar concentration in syngas varies based on its intended use, as illustrated in Table 23.2.

Tar removal can be achieved through two methods (Raheem et al. 2018):

- in-situ treatment that involves modifying gasifier parameters such as design, temperature, gasifying agent, and catalysts to minimize tar formation within the gasifier itself;

Table 23.2 Tar limits depending on the application (Syed-Hassan et al. 2017)

Application	Acceptable tar concentration ()
Fischer–Tropsch synthesis	< 0.1–1 mg/Nm^3
Methanol production	< 1 mg/Nm^3
Fuel cells	< 1 mg/Nm^3
Gas turbines	0.05–5 mg/Nm^3
Internal combustion engines	< 100 mg/Nm^3
Compressors	< 500 mg/Nm^3
Direct combustion	No tar limit

- ex-situ treatment, which involves installing specialized equipment downstream of the gasifier to capture and remove tar from the gaseous products.

The in-situ tar removal method is generally more cost-effective than the ex-situ approach, as it eliminates the need for expensive additional equipment like gas scrubbers (Campoy et al. 2010).

Beyond tar, also nitrogen and sulfur compounds represent a problem. Indeed, compared to traditional biomass, SS contains significantly higher levels of nitrogen and sulfur, that during thermal conversion are released as volatile compounds and then react to form toxic substances such as ammonia, hydrogen cyanide, and hydrogen sulfide (Syed-Hassan et al. 2017). Moreover, these compounds are oxidized when the produced syngas is combusted, forming nitrogen and sulfur oxides that contribute to environmental problems like smog, acid rain, and ozone depletion.

Finally, the high ash content in SS can significantly hinder gasifier operation, due to sintering, agglomeration, and clinker formation, which lead to frequent shutdowns and maintenance (Oladejo et al. 2019). Additionally, the low ash fusion temperature can block fuel flow, reducing heat transfer and gas quality (Seggiani et al. 2012).

Moving to the solid by-products of gasification, ash and char, the process can cause certain inorganic compounds, like phosphorus or zinc, to migrate from the SS into them, thus future research could explore the potential of valorizing solid gasification products by recovering valuable compounds or using them for the removal of toxic and hazardous substances from sewage (Werle and Dudziak 2014).

To maximize the benefits of SS gasification, a sustainable and efficient approach toward an integrated biorefinery is needed. Biorefineries focus on resource recovery and value creation rather than simple disposal, optimization of energy generation and material extraction, to significantly reduce environmental impact. The design of such systems should be tailored to local conditions, considering factors like sludge composition, economic context, geography, and climate (Campoy et al. 2010).

23.6 Conclusions

Global sewage sludge production is dramatically increasing and traditional disposal methods are becoming economically and environmentally unsustainable. Indeed, due to its high levels of organic matter, toxic substances, and heavy metals, sewage sludge disposal presents significant challenges. Therefore, also considering the current great concern for circular economy, effective sewage sludge management is crucial. Due to its energy potential, sewage sludge may represent a promising renewable energy source that can help address global energy supply and demand imbalances. As energy recovery option, gasification is a promising solution for sludge valorization, producing valuable energy products, especially for engine and thermal applications. Moreover, it effectively reduces sludge volume and eliminates harmful bacteria.

This chapter illustrates:

- the fundamentals of sewage sludge gasification, analyzing the effect of gasifier design, gasifying agent, and operating parameters;
- the outcomes of sewage sludge gasification and their main applications;
- a case study to show the principal benefits that the implementation of gasification can have on sludge treatment;
- the major challenges and opportunities of applying gasification for sewage sludge treatment.

The analysis of sewage sludge gasification as a promising alternative for waste treatment with energy recovery highlights that the value and usability of the products is crucial to determine the feasibility of the process. Thus the gasifier design and the operating conditions have to be carefully set. Particular attention has to be dedicated to emission control and operational challenges, such as those arising from the high content of inorganic compounds, heavy metals or nitrogen and sulfur. These issues can be addressed through careful reactor design, modern pollution control equipment, or sludge pre-treatment, but these measures will increase both capital and operating costs. A major obstacle to broader adoption of commercial advanced gasification technologies is related to syngas cleaning processes that can reliably meet stringent specifications and achieve higher electricity conversion efficiencies. To overcome these limitations and develop a more economically viable and environmentally sustainable technology, ongoing research should focus on process scale-up, tar reduction, and gas cleaning.

References

Abdelrahim A, Brachi P, Ruoppolo G, Di Fraia SDS, Vanoli L (2020) Experimental and numerical investigation of biosolid gasification: equilibrium-based modeling with emphasis on the effects of different pretreatment methods. 59:299–307.https://doi.org/10.1021/acs.iecr.9b03902

Abdoulmoumine N, Adhikari S, Kulkarni A, Chattanathan S (2015) A review on biomass gasification syngas cleanup. Appl Energy 155:294–307. https://doi.org/10.1016/J.APENERGY.2015.05.095

Acar MC, Böke YE (2019) Simulation of biomass gasification in a BFBG using chemical equilibrium model and restricted chemical equilibrium method. Biomass Bioenerg 125:131–138. https://doi.org/10.1016/j.biombioe.2019.04.012

An Z, Yuan Q, Liang H, Wang K, Yuan X, Wang C, Guo Q (2024) Investigation and optimization of syngas generation during chemical looping gasification of municipal sludge using Fe/Al oxygen carrier. J Energy Inst 117:101844. https://doi.org/10.1016/J.JOEI.2024.101844

Arena U (2012) Process and technological aspects of municipal solid waste gasification. A review. Waste Manag 32:625–639. https://doi.org/10.1016/J.WASMAN.2011.09.025

Asadullah M, Ito Si, Kunimori K, Tomishige K (2002) Role of catalyst and its fluidization in the catalytic gasification of biomass to syngas at low temperature. Ind Eng Chem Res 41:4567–4575. https://doi.org/10.1021/ie020112u

Ayub Y, Ren J, Shi T (2024) Exploring gasification process and technology for biomass-waste utilization: a comprehensive review on the path to sustainable energy. Process Saf Environ Prot 188:1489–1501. https://doi.org/10.1016/J.PSEP.2024.06.056

Basu P (2010) Biomass gasification and pyrolysis: practical design and theory. Academic Press. ISBN 0080961622

Bennamoun L, Arlabosse P, Léonard A (2013) Review on fundamental aspect of application of drying process to wastewater sludge. Renew Sustain Energy Rev 28:29–43. https://doi.org/10.1016/J.RSER.2013.07.043

Bermudez JM, Fidalgo B (2016) Production of bio-syngas and bio-hydrogen via gasification. ISBN 9780081004562

Brachi P, Chirone R, Miccio F, Miccio M, Picarelli A, Ruoppolo G (2014) Fluidized bed co-gasification of biomass and polymeric wastes for a flexible end-use of the syngas: focus on bio-methanol. Fuel 128:88–98

Brachi P, Di Fraia S, Massarotti N, Vanoli L (2022) Combined heat and power production based on sewage sludge gasification: an energy-efficient solution for wastewater treatment plants. Energy Convers Manag X 13:100171. https://doi.org/10.1016/j.ecmx.2021.100171

Brar JS, Singh K, Wang J, Kumar S (2012) Cogasification of coal and biomass: a review. Int J Res 2012:1–10. https://doi.org/10.1155/2012/363058

Campoy M, Gómez-Barea A, Fuentes-Cano D, Ollero P (2010) Tar reduction by primary measures in an autothermal air-blown fluidized bed biomass gasifier. Ind Eng Chem Res 49:11294–11301

Cao Y, Bai Y, Du J (2024) Parametric study on biomass gasification by using air, oxygen-enriched air, air/steam and oxygen-enriched air/steam agents: an ASPEN plus modeling. Int J Hydrogen Energy 73:265–273. https://doi.org/10.1016/j.ijhydene.2024.06.037

Capodaglio AG, Olsson G (2019) Energy Issues in sustainable urban wastewater management: use, demand reduction and recovery in the urban water cycle. Sustainability 12:266. https://doi.org/10.3390/su12010266

Caputo AC, Palumbo M, Pelagagge PM, Scacchia F (2005) Economics of biomass energy utilization in combustion and gasification plants: effects of logistic variables. Biomass Bioenerg 28:35–51

Carotenuto A, Di Fraia S, Massarotti N, Uddin MR, Vanoli L (2022) Combined heat and power generation from mechanically dewatered sewage sludge: numerical modelling. Chem Eng Trans 92:283–288. https://doi.org/10.3303/CET2292048

Carotenuto A, Di Fraia S, Massarotti N, Sobek S, Uddin MR, Vanoli L, Werle S (2023) Predictive modeling for energy recovery from sewage sludge gasification. Energy 263:125838. https://doi.org/10.1016/J.ENERGY.2022.125838

Chang H, Zhao Y, Zhao S, Damgaard A, Christensen TH (2022) Review of inventory data for the thermal treatment of sewage sludge. Waste Manag 146:106–118. https://doi.org/10.1016/J.WASMAN.2022.05.002

Chang X, Wu P, Chu Y, Zhou Y, Tang Y (2024) Pyrolysis-induced migration and transformation of heavy metals in sewage sludge containing microplastics. Waste Manag 189:401–409. https://doi.org/10.1016/J.WASMAN.2024.08.039

Chen WH, Chen CJ, Hung CI, Shen CH, Hsu HW (2013) A comparison of gasification phenomena among raw biomass, torrefied biomass and coal in an entrained-flow reactor. Appl Energy 112:421–430. https://doi.org/10.1016/j.apenergy.2013.01.034

Chen YH, Lan Thao Ngo TN, Chiang KY (2021) Enhanced hydrogen production in co-gasification of sewage sludge and industrial wastewater sludge by a pilot-scale fluidized bed gasifier. Int J Hydrogen Energy 46:14083–14095. https://doi.org/10.1016/j.ijhydene.2020.10.081

Choi YK, Cho MH, Kim JS (2015) Steam/oxygen gasification of dried sewage sludge in a two-stage gasifier: effects of the steam to fuel ratio and ash of the activated carbon on the production of hydrogen and tar removal. Energy 91:160–167. https://doi.org/10.1016/J.ENERGY.2015.08.027

Dasappa S, Paul PJ, Mukunda HS, Rajan NKS, Sridhar G, Sridhar HV (2004) Biomass gasification technology—a route to meet energy needs. Curr Sci 908–916

De Andrés JM, Narros A, Rodríguez ME (2011) Air-steam gasification of sewage sludge in a bubbling bed reactor: effect of alumina as a primary catalyst. Fuel Process Technol 92:433–440. https://doi.org/10.1016/j.fuproc.2010.10.006

De Andrés JM, Roche E, Narros A, Rodríguez ME (2016) Characterisation of tar from sewage sludge gasification. Influence of gasifying conditions: temperature, throughput, steam and use of primary catalysts. Fuel 180:116–126. https://doi.org/10.1016/J.FUEL.2016.04.012

de Andrés JM, Vedrenne M, Brambilla M, Rodríguez E (2019) Modeling and model performance evaluation of sewage sludge gasification in fluidized-bed gasifiers using Aspen Plus. J Air Waste Manag Assoc. https://doi.org/10.1080/10962247.2018.1500404

De Lasa H, Salaices E, Mazumder J, Lucky R (2011) Catalytic steam gasification of biomass: catalysts, thermodynamics and kinetics. Chem Rev 111:5404–5433. https://doi.org/10.1021/cr200024w

Di Fraia S, Uddin MR (2022) Energy recovery from waste paper and deinking sludge to support the demand of the paper industry: a numerical analysis. Sustainability 14:4669.https://doi.org/10.3390/su14084669

Di Fraia S, Massarotti N, Vanoli L, Costa M (2016) Thermo-economic analysis of a novel cogeneration system for sewage sludge treatment. Energy 115:1560–1571

Di Fraia S, Figaj RD, Massarotti N, Vanoli L (2018) Corrigendum to "An integrated system for sewage sludge drying through solar energy and a combined heat and power unit fuelled by biogas". Energy Convers Manage 171:587–603. (S0196890418306289) (https://doi.org/10.1016/j.enconman.2018.06.018)). Energy Convers Manag 185. https://doi.org/10.1016/j.enconman.2018.12.017

Di Fraia S, Macaluso A, Massarotti N, Vanoli L (2019) Energy, exergy and economic analysis of a novel geothermal energy system for wastewater and sludge treatment. Energy Convers Manag 195. https://doi.org/10.1016/j.enconman.2019.05.035

Di Fraia S, Macaluso A, Massarotti N, Vanoli L (2020) Geothermal energy for wastewater and sludge treatment: an exergoeconomic analysis. Energy Convers Manag 224:113180. https://doi.org/10.1016/J.ENCONMAN.2020.113180

Di Fraia S, Massarotti N, Uddin MR, Vanoli L (2021) Conversion of Sewage Sludge to combined heat and power: modeling and optimization. Smart Energy 5:100061. https://doi.org/10.1016/j.segy.2021.100061

Djandja OS, Yin LX, Wang ZC, Duan PG (2021) From wastewater treatment to resources recovery through hydrothermal treatments of municipal sewage sludge: a critical review. Process Saf Environ Prot 151:101–127. https://doi.org/10.1016/J.PSEP.2021.05.006

Dudyński M, Van Dyk JC, Kwiatkowski K, Sosnowska M (2015) Biomass gasification: influence of torrefaction on syngas production and tar formation. Fuel Process Technol 131:203–212. https://doi.org/10.1016/J.FUPROC.2014.11.018

Erakhrumen AA (2012) Biomass gasification: documented information for adoption/adaptation and further improvements toward sustainable utilisation of renewable natural resources. ISRN Renew Energy 2012:1–8. https://doi.org/10.5402/2012/536417

Feng Y, Yu T, Ma K, Xu G, Hu Y, Chen D (2018) Effect of hydrothermal temperature on the steam gasification performance of sewage sludge: syngas quality and tar formation. Energy Fuels 32:6834–6838. https://doi.org/10.1021/acs.energyfuels.8b00696

Feng H, Zhou Z, Hantoko D, Zhong L, Ar Rahim D, Fang W, Yan M (2021) Effect of alkali additives on desulfurization of syngas during supercritical water gasification of sewage sludge. Waste Manag 131:394–402. https://doi.org/10.1016/J.WASMAN.2021.06.029

Ferreiro AI, Ferreira AF, Fernandes EC, Coelho P (2024) Influence of process parameters on biomass gasification: a review of experimental studies in entrained flow reactors and droptube furnaces. Biomass Bioenerg 185:107217. https://doi.org/10.1016/J.BIOMBIOE.2024.107217

Gao N, Kamran K, Quan C, Williams PT (2020) Thermochemical conversion of sewage sludge: a critical review. Prog Energy Combust Sci 79:100843. https://doi.org/10.1016/J.PECS.2020.100843

Gopinath A, Divyapriya G, Srivastava V, Laiju AR, Nidheesh PV, Kumar MS (2021) Conversion of sewage sludge into biochar: a potential resource in water and wastewater treatment. Environ Res 194:110656. https://doi.org/10.1016/J.ENVRES.2020.110656

Hagos FY, Aziz ARA, Sulaiman SA (2014) Trends of Syngas as a fuel in internal combustion engines. Adv Mech Eng 6:401587. https://doi.org/10.1155/2014/401587

Han L, Wang Z, Wu P, Xu G, Li L, Zhang J, Shen Z, Mao S (2023) Fluidized bed chemical looping gasification of sewage sludge with bituminous coal to produce H_2 rich syngas: an examination on fuel synergy and reaction conditions. J Energy Inst 111:101442. https://doi.org/10.1016/J.JOEI.2023.101442

Heidenreich S, Foscolo PU (2015) New concepts in biomass gasification. Prog Energy Combust Sci 46:72–95. https://doi.org/10.1016/J.PECS.2014.06.002

Hosseingholilou B, Tavakoli N, Saidi M (2024) Cost-effectiveness and economic growth potential evaluation of olive pomace gasification process to sustainable fuel: comparison study of different gasifying agent. Process Saf Environ Prot 187:533–548. https://doi.org/10.1016/j.psep.2024.04.086

Hu M, Ye Z, Zhang H, Chen B, Pan Z, Wang J (2021) Thermochemical conversion of sewage sludge for energy and resource recovery: technical challenges and prospects. Environ Pollut Bioavailab 33:145–163

Hu M, Hu H, Ye Z, Tan S, Yin K, Chen Z, Guo D, Rong H, Wang J, Pan Z et al (2022) A review on turning sewage sludge to value-added energy and materials via thermochemical conversion towards carbon neutrality. J Clean Prod 379:134657. https://doi.org/10.1016/J.JCLEPRO.2022.134657

Hu M, Deng W, Su Y, Wang L, Chen G (2024) Production of hydrogen-rich syngas through microwave-assisted gasification of sewage sludge in steam-CO_2 atmosphere. Fuel 357:129855. https://doi.org/10.1016/J.FUEL.2023.129855

Huang YW, Chen MQ, Li QH, Xing W (2018) Hydrogen-rich syngas produced from co-gasification of wet sewage sludge and torrefied biomass in self-generated steam agent. Energy 161:202–213. https://doi.org/10.1016/J.ENERGY.2018.07.097

Huang W, Zhang R, Giannis A, Li C, He C (2023) Sequential hydrothermal carbonization and CO_2 gasification of sewage sludge for improved syngas production with mitigated emissions of NO_x precursors. Chem Eng J 454:140239. https://doi.org/10.1016/J.CEJ.2022.140239

Jeong YS, Choi YK, Park KB, Kim JS (2019) Air co-gasification of coal and dried sewage sludge in a two-stage gasifier: effect of blending ratio on the producer gas composition and tar removal. Energy 185:708–716. https://doi.org/10.1016/j.energy.2019.07.093

Karami A, Romano N, Galloway T, Hamzah H (2016) Virgin microplastics cause toxicity and modulate the impacts of phenanthrene on biomarker responses in African catfish (Clarias gariepinus). Environ Res 151:58–70. https://doi.org/10.1016/J.ENVRES.2016.07.024

Klinghoffer NB, Castaldi MJ (2013) Gasifi cation and pyrolysis of municipal solid waste (MSW). ISBN 9780857096364

Kominko H, Gorazda K, Wzorek Z (2024) Sewage sludge: a review of its risks and circular raw material potential. J. Water Process Eng 63:105522. https://doi.org/10.1016/J.JWPE.2024.105522

Li T, Choi C, Fukomoto K, Machida H, Norinaga K (2023a) Polycyclic aromatic hydrocarbons (PAHs) and soot formations under different gasifying agents: detailed chemical kinetic analysis of a two-stage entrained flow coal gasifier. Fuel 343:127876. https://doi.org/10.1016/j.fuel.2023.127876

Li H, Li M, Wang H, Tan M, Zhang G, Huang Z, Yuan X (2023b) A review on migration and transformation of nitrogen during sewage sludge thermochemical treatment: focusing on pyrolysis, gasification and combustion. Fuel Process Technol 240:107562. https://doi.org/10.1016/J.FUPROC.2022.107562

Loha C, Gu S, De Wilde J, Mahanta P, Chatterjee PK (2014) Advances in mathematical modeling of fluidized bed gasification. Renew Sustain Energy Rev 40:688–715. https://doi.org/10.1016/j.rser.2014.07.199

Lv P, Yuan Z, Wu C, Ma L, Chen Y, Tsubaki N (2007) Bio-syngas production from biomass catalytic gasification. Energy Convers Manag 48:1132–1139. https://doi.org/10.1016/j.enconman.2006.10.014

Ma M, Zhang S, Chen Y, Chen B, Guo L (2024) Optimization of hydrogen-rich syngas from coal and sewage sludge co-gasification in supercritical water. Chem Eng J 497:154792. https://doi.org/10.1016/J.CEJ.2024.154792

Manara P, Zabaniotou A (2012) Towards sewage sludge based biofuels via thermochemical conversion—a review. Renew Sustain Energy Rev 16:2566–2582. https://doi.org/10.1016/J.RSER.2012.01.074

Milčák P, Baláš M, Lisý M, Lisá H, Kracík P, Lachman J (2024) Digestate and woodchips gasification: a comparison of different gasifying agents. Fuel Process Technol 258. https://doi.org/10.1016/j.fuproc.2024.108091

Molino A, Chianese S, Musmarra D (2016) Biomass gasification technology: the state of the art overview. J Energy Chem 25:10–25. https://doi.org/10.1016/J.JECHEM.2015.11.005

Mun TY, Kim JS (2013) Air gasification of dried sewage sludge in a two-stage gasifier. Part 2: calcined dolomite as a bed material and effect of moisture content of dried sewage sludge for the hydrogen production and tar removal. Int J Hydrogen Energy 38:5235–5242. https://doi.org/10.1016/J.IJHYDENE.2013.02.073

Mun TY, Kim JWJS, Kim JWJS (2013) Air gasification of dried sewage sludge in a two-stage gasifier: part 1. The effects and reusability of additives on the removal of tar and hydrogen production. Int J Hydrogen Energy 38:5226–5234. https://doi.org/10.1016/J.IJHYDENE.2012.10.120

Mun TY, Cho MH, Kim JS (2014) Air gasification of dried sewage sludge in a two-stage gasifier. Part 3: application of olivine as a bed material and nickel coated distributor for the production of a clean hydrogen-rich producer gas. Int J Hydrogen Energy 39:5634–5643. https://doi.org/10.1016/j.ijhydene.2014.01.173

Nahar K, Thulasiraman AV, Vuppaladadiyam AK, Hakeem IG, Shah K (2024) Current understanding on the fate of contaminants during hydrothermal treatment of sewage sludge. Curr Opin Green Sustain Chem 49:100960. https://doi.org/10.1016/J.COGSC.2024.100960

Neri A, Rizzuni A, Garrone P, Cagno E (2024) Influence of policymakers and civil society stakeholders on sewage sludge management strategies: empirical results from European utilities. J Environ Manage 364:121396. https://doi.org/10.1016/J.JENVMAN.2024.121396

Oladejo J, Shi K, Luo X, Yang G, Wu T (2019) A review of sludge-to-energy recovery methods. Energies 12

Pala LPR, Wang Q, Kolb G, Hessel V (2017) Steam gasification of biomass with subsequent syngas adjustment using shift reaction for syngas production: an Aspen Plus model. Renew Energy 101:484–492

Pelagalli V, Matassa S, Race M, Langone M, Papirio S, Lens PNL, Lazzazzara M, Frugis A, Petta L, Esposito G (2024) Syngas-driven sewage sludge conversion to microbial protein through H_2S- and CO-tolerant hydrogen-oxidizing bacteria. Water Res 248:120698. https://doi.org/10.1016/J.WATRES.2023.120698

Phuphuakrat T, Nipattummakul N, Namioka T, Kerdsuwan S, Yoshikawa K (2010) Characterization of tar content in the syngas produced in a downdraft type fixed bed gasification system from dried sewage sludge. Fuel 89:2278–2284. https://doi.org/10.1016/J.FUEL.2010.01.015

Pode R, Diouf B, Pode G (2015) Sustainable rural electrification using rice husk biomass energy: a case study of Cambodia. Renew Sustain Energy Rev 44:530–542. https://doi.org/10.1016/j.rser.2015.01.018

Quan LM, Kamyab H, Yuzir A, Ashokkumar V, Hosseini SE, Balasubramanian B, Kirpichnikova I (2022) Review of the application of gasification and combustion technology and waste-to-energy technologies in sewage sludge treatment. Fuel 316:123199. https://doi.org/10.1016/J.FUEL.2022.123199

Raheem A, Sikarwar VS, He J, Dastyar W, Dionysiou DD, Wang W, Zhao M (2018) Opportunities and challenges in sustainable treatment and resource reuse of sewage sludge: a review. Chem Eng J 337:616–641. https://doi.org/10.1016/J.CEJ.2017.12.149

Rey JRC, Longo A, Rijo B, Pedrero CM, Tarelho LAC, Brito PSD, Nobre C (2024) A review of cleaning technologies for biomass-derived syngas. Fuel 377:132776. https://doi.org/10.1016/J.FUEL.2024.132776

Roche E, De Andrés JM, Narros A, Rodríguez ME (2014) Air and air-steam gasification of sewage sludge. The influence of dolomite and throughput in tar production and composition. Fuel 115:54–61. https://doi.org/10.1016/J.FUEL.2013.07.003

Sanlisoy A, Carpinlioglu MO (2017) A review on plasma gasification for solid waste disposal. Int J Hydrogen Energy 42:1361–1365

Seggiani M, Vitolo S, Puccini M, Bellini A (2012) Cogasification of sewage sludge in an updraft gasifier. Fuel. https://doi.org/10.1016/j.fuel.2011.08.054

Shahbaz M, Al-Ansari T, Inayat M, Sulaiman SA, Parthasarathy P, McKay G (2020) A critical review on the influence of process parameters in catalytic co-gasification: current performance and challenges for a future prospectus. Renew Sustain Energy Rev 134:110382. https://doi.org/10.1016/J.RSER.2020.110382

Shakorfow AM (2016) Biomass incineration, pyrolysis, combustion and gasification. Int J Sci Res 5:2319–7064

Sharma SD, Dolan M, Ilyushechkin AY, McLennan KG, Nguyen T, Chase D (2010) Recent developments in dry hot syngas cleaning processes. Fuel 89:817–826. https://doi.org/10.1016/J.FUEL.2009.05.026

Sharmila VG, Tyagi VK, Varjani S, Banu JR (2023) A review on the lignocellulosic derived biochar-based catalyst in wastewater remediation: advanced treatment technologies and machine learning tools. Bioresour Technol 387:129587

Sharmila VG, Shanmugavel SP, Banu JR (2024) A review on emerging technologies and machine learning approaches for sustainable production of biofuel from biomass waste. Biomass Bioenergy 180:106997. https://doi.org/10.1016/j.biombioe.2023.106997

Shi H, Si W, Li X (2016) The concept, design and performance of a novel rotary kiln type air-staged biomass gasifier. Energies 9. https://doi.org/10.3390/en9020067

Singh RP, Agrawal M (2008) Potential benefits and risks of land application of sewage sludge. Waste Manag 28:347–358. https://doi.org/10.1016/J.WASMAN.2006.12.010

Singh S, Kumar V, Dhanjal DS, Datta S, Bhatia D, Dhiman J, Samuel J, Prasad R, Singh J (2020) A sustainable paradigm of sewage sludge biochar: valorization, opportunities, challenges and future prospects. J Clean Prod 269:122259. https://doi.org/10.1016/J.JCLEPRO.2020.122259

Singh Siwal S, Zhang Q, Sun C, Thakur S, Kumar Gupta V, Kumar Thakur V (2020) Energy production from steam gasification processes and parameters that contemplate in biomass gasifier—a review. Bioresour Technol 297:122481. https://doi.org/10.1016/J.BIORTECH.2019.122481

Śpiewak K (2024) Gasification of Sewage sludge—a review. Energies 17:4476

Sreejith CC, Muraleedharan C, Arun P (2014) Performance prediction of steam gasification of wood using an ASPEN PLUS thermodynamic equilibrium model. Int J Sustain Energy 33:416–434. https://doi.org/10.1080/14786451.2012.755977

Suparmin P, Purwanti N, Oscar Nelwan L, Halomoan Tambunan A (2024) Syngas production by biomass gasification: a meta-analysis. Renew Sustain Energy Rev 206:114824. https://doi.org/10.1016/J.RSER.2024.114824

Syed-Hassan SSA, Wang Y, Hu S, Su S, Xiang J (2017) Thermochemical processing of sewage sludge to energy and fuel: fundamentals, challenges and considerations. Renew Sustain Energy Rev 80:888–913. https://doi.org/10.1016/J.RSER.2017.05.262

Teoh SK, Li LY (2020) Feasibility of alternative sewage sludge treatment methods from a lifecycle assessment (LCA) perspective. J Clean Prod 247:119495. https://doi.org/10.1016/J.JCLEPRO.2019.119495

Tezer Ö, Karabağ N, Öngen A, Ayol A (2023) Syngas production from municipal sewage sludge by gasification process: effects of fixed bed reactor types and gasification agents on syngas quality. Sustain Energy Technol Assessments 56. https://doi.org/10.1016/j.seta.2023.103042

Thipkhunthod P, Meeyoo V, Rangsunvigit P, Kitiyanan B, Siemanond K, Rirksomboon T (2005) Predicting the heating value of sewage sludges in Thailand from proximate and ultimate analyses. Fuel 84:849–857. https://doi.org/10.1016/J.FUEL.2005.01.003

Trabelsi ABH, Zaafouri K, Friaa A, Abidi S, Naoui S, Jamaaoui F (2021) Municipal sewage sludge energetic conversion as a tool for environmental sustainability: production of innovative biofuels and biochar. Environ Sci Pollut Res 28:9777–9791. https://doi.org/10.1007/s11356-020-11400-z

Vivanpatarakij S, Rulerk D, Assabumrungrat S (2014) Removal of tar from biomass gasification process by steam reforming over nickel catalysts. Chem Eng Trans 37:205–210. https://doi.org/10.3303/CET1437035

Wang F, Peng W, Zeng X, Sun D, Cui G, Han Z, Wang C, Xu G (2024) Insight into staged gasification of biomass waste: essential fundamentals and applications. Sci Total Environ 953:175954. https://doi.org/10.1016/J.SCITOTENV.2024.175954

Warnecke R (2000) Gasification of biomass: comparison of fixed bed and fluidized bed gasifier. Biomass Bioenerg 18:489–497. https://doi.org/10.1016/S0961-9534(00)00009-X

Watanabe H, Otaka M (2006) Numerical simulation of coal gasification in entrained flow coal gasifier. Fuel 85:1935–1943. https://doi.org/10.1016/j.fuel.2006.02.002

Werle S, Dudziak M (2014) Analysis of organic and inorganic contaminants in dried sewage sludge and by-products of dried sewage sludge gasification. Energies 7:462–476

Xiang Y, Cai L, Guan Y, Liu W, Cheng Z, Liu Z (2020) Study on the effect of gasification agents on the integrated system of biomass gasification combined cycle and oxy-fuel combustion. Energy 206:118131. https://doi.org/10.1016/j.energy.2020.118131

Yan J, Yan Y, Lai J, Jia D, Jiao Y, Zhao X, Yang L (2023) Co-gasification of municipal sewage sludge and cotton stalk enhanced by metal-enriched texture dyeing sludge additives for syngas production. Fuel 341:127669. https://doi.org/10.1016/J.FUEL.2023.127669

Yang X, Kan T, Kheradmand A, Xu H, Strezov V, Yu A, Jiang Y (2021) Tunable syngas production from two-stage sorption-enhanced steam gasification of sewage sludge. Chem Eng J 404:126069. https://doi.org/10.1016/J.CEJ.2020.126069

Yap HY, Nixon JD (2015) A multi-criteria analysis of options for energy recovery from municipal solid waste in India and the UK. Waste Manag 46:265–277

Yesilova N, Tezer O, Ongen A, Ayol A (2024) Enhancing biomass gasification: a comparative study of catalyst applications in updraft and modifiable-downdraft fixed bed reactors. Int J Hydrogen Energy 76:290–303. https://doi.org/10.1016/j.ijhydene.2024.05.075

Zheng X, Chen W, Ying Z, Huang J, Ji S, Wang B (2019) Thermodynamic investigation on gasification performance of sewage sludge-derived hydrochar: effect of hydrothermal carbonization. Int J Hydrogen Energy 44:10374–10383. https://doi.org/10.1016/J.IJHYDENE.2019.02.200

Chapter 24
Machine Learning and Deep Learning Applied to Wastewater Treatment

Nicolas Spogis, Juliana Neves, Sérgio Leonardo Butski Soares Santos, Natan Padoin, and Cíntia Soares

Abstract This book chapter delves into the application of machine learning techniques in wastewater treatment processes, examining the multifaceted challenges that often hinder seamless implementation. With an emphasis on comprehensiveness, the exploration encompasses a detailed analysis of technical, operational, and contextual hurdles. By dissecting these obstacles, the chapter aims to provide a nuanced understanding of the impediments faced by practitioners in the field. Additionally, strategic solutions and frameworks are proposed to overcome these challenges, fostering a more effective integration of machine learning methodologies in wastewater treatment systems. Through this comprehensive exploration, we contribute to the ongoing discourse on advancing sustainable water management practices, showcasing the potential of machine learning as a transformative tool in optimizing wastewater treatment processes.

Keywords Machine learning · CRISP-DM · Technical barriers · Contextual challenges · Operational hurdles

Industry 4.0, characterized by digitalization and automation, drives the integration of advanced technologies, such as big data and machine learning, across various sectors, including wastewater treatment. Based on this, the chapter presents the CRISP-DM (Cross-Industry Standard Process for Data Mining) methodology as a comprehensive framework for developing machine learning solutions for wastewater treatment, from business understanding to deployment. The aim is to outline the primary purposes of these technologies in the field, highlighting the practical advantages and challenges in implementing AI (Artificial intelligence) solutions in the sector.

J. Neves · S. L. B. S. Santos · N. Padoin · C. Soares (✉)
Laboratory of Materials and Scientific Computing (LabMAC), Chemical and Food Engineering Department, Federal University of Santa Catarina (UFSC), Florianópolis, Brazil
e-mail: cintia.soares@ufsc.br

N. Spogis
Faculty of Chemical Engineering, State University of Campinas, Campinas, Brazil

S. Souabi et al. (eds.), *Wastewater Treatment Plants*, Water Science and Technology Library 130, https://doi.org/10.1007/978-3-031-87461-1_24

As part of this revolution, machine learning enables the analysis of large volumes of data to extract valuable insights and optimize operations. In wastewater treatment, this translates into better water resource management, increased efficiency of treatment processes, and a quicker, more accurate response to variable conditions. The application of machine learning in wastewater treatment offers significant tangible benefits. First, this technology can considerably enhance process efficiency, enabling real-time operational adjustments based on accurate predictions. This results in a reduction in operational costs, such as the optimized use of chemicals and energy.

Additionally, machine learning can improve treated water quality, ensuring that effluents meet stringent environmental regulations. Another significant benefit is the contribution of this technology to more sustainable water resource management. By predicting and optimizing treatment process parameters, machine learning helps minimize environmental impacts, promoting greener and more efficient treatment practices. Implementing machine learning also paves the way for continuous innovation, encouraging the development of new methods and strategies to address emerging challenges in the wastewater treatment sector.

The steps of the CRISP-DM methodology applied to wastewater treatment will be detailed in the next section. It will present how each phase of this methodology contributes to developing effective predictive models, from business understanding to deploying solutions. In addition, typical data sources in wastewater treatment plants (WWTPs), data preparation and cleaning techniques, and modeling methods used to predict and optimize treatment processes will be described. The various practical applications of machine learning and deep learning in wastewater treatment, including water quality prediction, nutrient removal optimization, and analysis of patterns and anomalies, will also be explored. Additionally, the use of reinforcement learning for real-time operational optimization and the application of deep learning for complex process modeling and advanced automation will be discussed. Finally, the challenges and limitations of implementing these technologies in wastewater treatment will be discussed. This will include technical challenges, such as the need for large volumes of high-quality data, the complexity of modeling processes, and organizational challenges, such as resistance to adopting new technologies and specialized technical skills. Ethical and regulatory considerations associated with implementing AI-based solutions in wastewater treatment will also be addressed.

This chapter provides a comprehensive overview of the opportunities and challenges that machine learning and deep learning present in the context of wastewater treatment, contributing to a deeper and more advanced understanding of the topic.

24.1 Phases of the CRISP-DM Methodology in the Wastewater Treatment Context

The CRISP-DM (Cross-Industry Standard Process for Data Mining) methodology is widely recognized and adopted in data mining and machine learning projects. Developed in the 1990s by a consortium of companies led by Daimler-Benz, SPSS, and NCR, CRISP-DM has emerged as the de facto standard for executing data mining projects (Shearer et al. 2000; Schröer et al. 2021).

Structured into six distinct phases—business understanding, data understanding, data preparation, modeling, evaluation, and deployment—CRISP-DM provides an essential framework for ensuring the robustness and effectiveness of machine learning projects. This framework offers a clear pathway from problem definition to solution implementation, ensuring that all stages of the process are systematically addressed. Figure 24.1 illustrates the implementation diagram of the CRISP-DM methodology.

In the context of wastewater treatment, the application of machine learning has proven to be a powerful tool for optimizing processes, reducing costs, and improving treatment efficiency. The use of machine learning techniques allows for the analysis and interpretation of large volumes of operational data, helping to identify patterns and anomalies that are difficult to detect using traditional methods. This, in turn, enables the continuous optimization of treatment processes, ensuring that the plant operates efficiently and in accordance with environmental requirements.

Mowbray et al. (2022) discussed the importance of integrating process engineering principles with advanced machine learning techniques to solve complex

Fig. 24.1 CRISP-DM process diagram

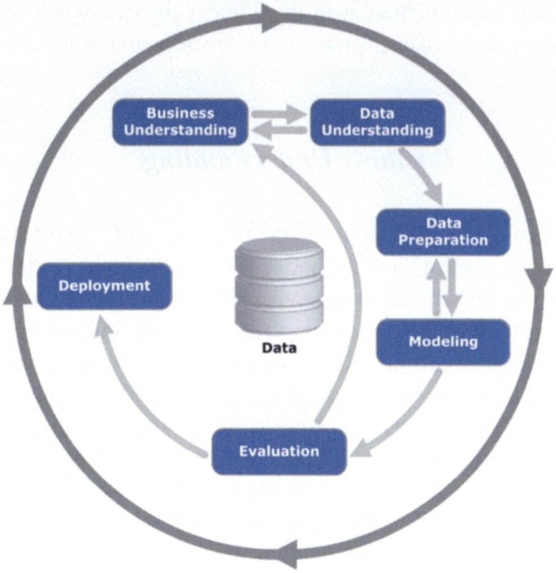

problems encountered in the chemical industry. He emphasized that the combination of specialized process knowledge with machine learning algorithms can lead to innovative and more effective solutions. In wastewater treatment, this integration is particularly valuable because it allows for the creation of predictive models that not only improve treatment efficiency but also ensure compliance with stringent environmental regulations and promote long-term sustainability.

The CRISP-DM methodology continues to be relevant and widely used due to its flexibility and comprehensiveness. It is applicable to a wide range of industries and types of projects, making it a valuable guide for data mining and machine learning professionals. The clear structure and detailed guidelines provided by CRISP-DM help ensure that projects are well planned and systematically executed, resulting in robust and efficient solutions.

In wastewater treatment, the application of the CRISP-DM methodology is crucial to ensure that machine learning projects are effective, well structured, and aligned with the plant's operational and environmental objectives. The CRISP-DM framework, with its six distinct phases, offers a systematic approach to executing data mining projects, from the initial problem definition to the implementation of solutions. This methodology not only facilitates the organization and execution of projects but also ensures that each stage is carefully considered, resulting in robust and efficient predictive models.

To understand how CRISP-DM can be applied to wastewater treatment, it is essential to explore each of its stages in the specific context of this sector. Each phase of CRISP-DM uniquely contributes to building solutions that not only improve treatment efficiency but also ensure compliance with environmental regulations and promote sustainability.

The stages of the CRISP-DM methodology will be detailed below, focusing on their implementation in the context of wastewater treatment and emphasizing the importance of each phase in the development of successful machine learning projects.

24.1.1 Business Understanding

Business understanding is the first stage of CRISP-DM and involves clearly defining the project's objectives and requirements. In the context of wastewater treatment, this initial phase is essential to ensure that subsequent machine learning efforts are aligned with the plant's specific operational and regulatory goals. This includes identifying precise targets and KPIs, such as the reduction of critical contaminants, which may include heavy metals such as mercury and cadmium; nutrients such as nitrates and phosphates; and other organic and inorganic pollutants.

The optimization of chemical usage is another crucial goal. Chemicals such as coagulants and flocculants are frequently used in the wastewater treatment process to help remove suspended solids and other contaminants. Precise dosing of these chemicals is vital not only for treatment efficiency but also for reducing operational costs and minimizing negative environmental impacts. Excessive chemical usage can result in undesirable chemical residues in the treated effluent, while insufficient dosing can compromise treatment quality.

Reducing energy consumption is another important consideration. Wastewater treatment is an energy-intensive process, especially in stages such as aeration, which is used to promote the biological degradation of organic pollutants. Machine learning techniques can be applied to optimize energy usage by adjusting operational parameters in real time, ensuring that energy consumption is minimized without compromising treatment efficiency.

Compliance with stringent environmental regulations is a constant priority. Wastewater treatment plants must adhere to specific standards set by environmental authorities, which determine the maximum allowable limits for different contaminants in treated effluent. Understanding these regulations and integrating these requirements into the project's objectives are crucial to ensure that the developed models help the plant stay within legal limits and avoid fines and penalties.

Process engineers, plant operators, and environmental specialists play essential roles in this phase. They collaborate to define success metrics and performance (KPIs) requirements, ensuring that all stakeholders are aligned with the project's objectives. Success metrics may include indicators such as contaminant removal efficiency, consistency in the quality of treated effluent, and the energy efficiency of the process.

This phase of business understanding is crucial for aligning machine learning efforts with the plant's operational and regulatory needs, ensuring that the results are relevant and applicable. For instance, a predictive model developed to optimize chemical dosing should not only improve treatment efficiency but also ensure that contaminant levels in the treated effluent consistently remain within regulatory limits.

Mowbray et al. (2022) emphasized the importance of understanding the operational context and specific challenges faced by treatment plants to effectively guide the application of machine learning techniques. A detailed understanding of the plant's processes and specific requirements allows for the creation of more accurate and useful models that can be integrated into existing control and monitoring systems to improve the overall operation of the plant.

24.1.2 Data Understanding

The data understanding phase is a fundamental step in CRISP-DM, where the collection and analysis of historical and operational data are performed to understand operational patterns and identify critical variables. This phase is particularly crucial in the context of wastewater treatment, given the complexity and variability of the processes involved.

During this phase, data are collected from various sources, including sensors and monitoring devices distributed throughout the treatment plant, as well as from physicochemical and laboratory quality analyses. These sensors capture a wide range of operational parameters, such as pH, turbidity, contaminant concentration, inflow and outflow rates, and temperature. The systematic collection of these data allows for a comprehensive analysis of plant performance over time.

Exploratory data analysis (EDA) is a key component of this phase. EDA techniques are used to identify correlations, seasonal patterns, and potential anomalies. For example, scatter plots, histograms, and heatmaps can reveal relationships between variables that are not immediately evident. Additionally, EDA can help detect outliers or abnormal values that may indicate operational issues or equipment failures (Jebb et al. 2017).

Sylvia and Murphy (2023) emphasized that data quality is essential for the success of predictive models. Incomplete, inaccurate, or inconsistent data can compromise the effectiveness of these models, leading to erroneous predictions and inadequate operational decisions. Therefore, ensuring data integrity and accuracy is a fundamental step in this phase. This may involve cleaning the data to remove or correct incorrect values, normalizing the data to ensure that it is on the same scale, and aggregating temporal data to capture long-term trends.

Moreover, it is important to document and understand the origin of the data, the context in which it was collected, and any transformations applied. This includes maintaining detailed metadata, which describes the characteristics of the data and any manipulations performed during the preparation process. Having a clear and detailed understanding of the available data allows for more accurate modeling and the construction of more robust predictive models (Andrienko and Gennady 2006).

24.1.3 Data Preparation

Data preparation is a critical step in the CRISP-DM process, where the collected raw data are cleaned, transformed, and structured for use in modeling. This step is essential to ensure that machine learning models can accurately capture the complex interactions between operational variables. During this phase, a series of procedures are performed to ensure the quality and consistency of the data.

First, the removal of outliers is a common practice. The outliers are extreme values that can distort the analysis and compromise the accuracy of the models. These values can arise due to measurement errors, sensor failures, or atypical operational conditions. Careful identification and removal of these outliers help create a more representative and reliable dataset.

However, it is important to exercise caution when doing this. Removing outliers indiscriminately can lead to the loss of valuable information and potentially bias the analysis. Therefore, this process should be carried out by professionals who thoroughly understand the underlying processes and the context of the data. They can distinguish between true anomalies and meaningful variations, ensuring that the dataset remains both accurate and representative of the real-world phenomena being studied.

Several techniques are commonly used for outlier detection and removal. The Z-score method is one approach, which measures the number of standard deviations a data point is from the mean. If the Z-score is greater than a certain threshold, usually set at 3 or −3, the data point is considered an outlier. This method is particularly suitable for data that is normally distributed. Another widely used technique is the interquartile range (IQR) method. The IQR is the range between the first quartile (Q1) and the third quartile (Q3) of the data. Outliers are typically defined as points that lie below Q1 minus 1.5 times the IQR or above Q3 plus 1.5 times the IQR. This method is effective for handling skewed data and non-normal distributions.

Additionally, the boxplot method is a visual technique that represents the distribution of data through quartiles and highlights outliers as points outside the "whiskers" of the plot. This method provides an intuitive way to identify outliers. It is crucial to remember that these techniques should be applied thoughtfully and with a deep understanding of the data and its context. Outlier detection and removal are not merely mechanical processes but require professional judgment to ensure the integrity and reliability of the dataset.

Handling missing values is another essential task in data preparation. Missing data can occur for various reasons, such as data collection failures or technical issues. Techniques such as data imputation, where missing values are replaced by estimates based on other available data, or the exclusion of incomplete records are applied to address this problem. The choice of method depends on the specific context and the amount of missing data.

Normalization and standardization of variables are essential steps to ensure that all variables are on the same scale, which is crucial in machine learning projects. The scale of variables can significantly influence algorithm performance, making these techniques indispensable for model effectiveness.

Normalization involves transforming the data to a specific range, such as 0 to 1. This process is particularly useful when variables have different units of measure or amplitudes, ensuring that all the variables contribute equally to the model. For example, temperatures evaluated in Celsius and pressures in Pascal have significantly different orders of magnitude.

Standardization adjusts the data so that it has a mean of zero and a standard deviation of one. This method is especially beneficial when the data follow a normal distribution, helping to stabilize variances and improve the performance of machine learning algorithms.

Both normalization and standardization techniques are fundamental for enhancing the effectiveness of predictive models, ensuring that no variable dominates the modeling process due to its original scale. Applying these techniques allows machine learning algorithms to process the data more efficiently and accurately, leading to better results and more robust insights.

In WWTPs, data preparation often involves the aggregation and analysis of time series, an essential process for capturing long-term trends and understanding operational patterns over time. Aggregating temporal data allows the identification of seasonal variations, annual trends, and other cyclical patterns that can significantly impact the efficiency of wastewater treatment.

For example, by analyzing aggregated data, it is possible to detect seasonal increases in pollutant loads due to climatic changes or specific events, enabling proactive adjustments to operational processes to maintain treatment effectiveness (Hewamalage et al. 2021).

The analysis of aggregated time series can also reveal important insights into equipment degradation and the need for preventive maintenance, contributing to more efficient plant management. Additionally, data aggregation facilitates the identification of underlying operational patterns that may not be evident in short-term analyses, allowing for a deeper and more comprehensive understanding of plant performance over time.

Feature engineering, the creation of new features derived from existing data, is an essential practice that can significantly increase the effectiveness of predictive models. This process involves transforming and combining raw data into new, more informative variables that help machine learning algorithms capture complex relationships and hidden patterns in the data (Zheng and Casari 2018).

In the context of wastewater treatment, feature engineering can include generating water quality indices that combine various measurements into a single representative metric. For example, by combining data on pH, turbidity, and contaminant concentration, a quality index can be created that provides a comprehensive view of water conditions at different points in the treatment process. This approach allows for a more holistic and accurate assessment of plant performance.

Another common technique is to transform variables to better capture nonlinear relationships. This can involve applying logarithms, exponential, or other mathematical transformations that help linearize complex relationships between variables. These transformations can reveal patterns that are not apparent in a direct analysis of the raw data.

Zheng and Casari (2018) emphasized that careful data preparation is fundamental to ensure that models can accurately capture the complex interactions between operational variables. Well-executed feature engineering allows predictive models not only to better fit historical data but also to generalize more effectively to new data. This results in more accurate predictions and actionable insights that can be used to optimize wastewater treatment processes and ensure compliance with environmental regulations.

Dimensionless variables are particularly useful in engineering, as they allow for the comparison and analysis of different physical quantities on a common scale. These dimensionless numbers often reveal underlying principles and scaling laws that govern the behavior of physical systems, facilitating more accurate modeling and prediction (Buckingham 1914). Incorporating dimensionless variables into feature engineering can further enhance the performance of predictive models by providing a more standardized and insightful representation of the data.

Moreover, creating new features can help reduce noise in the data and highlight the most relevant variables for modeling. By focusing on the most informative characteristics, machine learning models can operate more efficiently and robustly, providing a solid foundation for operational decision-making.

24.1.4 Modeling

In the modeling step, machine learning algorithms are applied to develop predictive models that are essential for optimizing and controlling wastewater treatment processes. This phase is crucial because the correct model selection and its appropriate application can significantly improve treatment efficiency and ensure compliance with environmental standards (Iratni and Chang 2019).

In the context of wastewater treatment, various types of models can be used depending on the nature and complexity of the problem to be solved. For example, regression models are widely employed to predict contaminant concentrations at the plant's output based on operational variables such as pH, temperature, and inflow rate.

These models are particularly useful not only when the relationship between variables is linear or can be approximated as such but also in nonlinear cases where advanced techniques can capture the complex interactions between variables (Rios Fuck et al. 2024).

For more complex issues where relationships between variables are nonlinear, more sophisticated algorithms such as decision trees and random forests (RF) are used. Decision trees split the data into subsets based on simple rules, helping to identify the most critical factors affecting treatment efficiency.

RF, extremely randomized trees, extreme gradient boosting (XGBoost), and light gradient boosting machine (LightGBM) improve prediction accuracy by reducing the risk of overfitting, a common problem where the model fits the training data too closely and loses the ability to generalize to new data (Chen and Guestrin 2016; Ke et al. 2017).

Neural networks, in turn, are capable of capturing complex nonlinear relationships between variables. These are particularly useful in systems where patterns are not immediately evident or are highly interdependent. Artificial neural networks can model these complex interactions and provide accurate predictions even in scenarios with high operational variability (Cechinel et al. 2024).

In addition to purely data-driven models, combining machine learning with first-principles models that incorporate physical and chemical laws is a recommended approach to improve both the accuracy and interpretability of predictive models. First-principles models are based on fundamental knowledge of the chemical and physical processes occurring in a WWTP, providing a solid foundation on which machine learning models can be built.

For example, when modeling the removal of a specific contaminant, known chemical reaction equations can be used in conjunction with historical data to train a machine learning model that predicts process efficiency under various operational conditions.

Sansana et al. (2021) demonstrated that integrating different types of models can lead to better results. This hybrid approach offers a more comprehensive and detailed view of the process, combining the robustness of first-principles models with the flexibility of machine learning algorithms.

Physics-informed neural networks (PINNs) can also be used in this context. PINNs integrate machine learning with first-principles models by embedding physical laws directly into the neural network architecture. This approach enhances model accuracy and robustness, especially when data are scarce or noisy, by ensuring that predictions adhere to known physical behaviors.

PINNs are valuable for optimizing complex processes in wastewater treatment, providing reliable and interpretable results by incorporating chemical and physical constraints.

For example, in a case study on wastewater treatment process optimization, using a combination of models not only allowed for highly accurate predictions of contaminant concentrations but also provided a better understanding of the interactions between different operational parameters and their influences on the final outcome (Shao et al. 2023).

When working with time series in machine learning, various specific techniques can be applied to extract valuable insights and improve the efficiency of wastewater treatment processes. Autoregressive integrated moving average (ARIMA) models are widely used to model and forecast univariate time series. These models are effective in capturing autocorrelation patterns in the data and making predictions based on past values, helping to anticipate changes in operational conditions (Hewamalage et al. 2021).

Recurrent neural networks (RNNs), particularly long short-term memory (LSTM) networks, are powerful for capturing long-term dependencies in time series. These networks can model sequential data with high temporal variability, making them especially useful for predicting fluctuations and complex behaviors over time (Gers et al. 2000; Sutskever et al. 2014; Sherstinsky 2020).

Exponential smoothing state space (ETS) models are also quite useful, as they allow time series to be decomposed into trend, seasonal, and error components. This provides a clear view of the underlying patterns and helps identify factors that may influence the efficiency of wastewater treatment during different periods (Hewamalage et al. 2021).

In addition to these techniques, the Prophet tool is another robust option for time series forecasting. Prophet is designed to handle large amounts of data and is particularly effective in identifying seasonal patterns and long-term trends, facilitating the prediction of future changes and the adaptation of operational processes as needed (Amdahl 1967).

Another advanced technique that has gained prominence is DeepAR, developed by Amazon Web Services (AWS) for time series forecasting. Unlike traditional methods, which treat each time series independently, DeepAR trains a single model across multiple related time series using additional and contextual data to enhance forecast accuracy (Salinas et al. 2020).

The application of DeepAR in wastewater treatment enables the prediction of critical parameters such as pollutant concentration, pH, chemical oxygen demand (COD), and biochemical oxygen demand (BOD). Capturing these complex and nonlinear dependencies between variables over time is essential in WWTPs, where operational conditions can vary significantly.

Additionally, DeepAR can incorporate contextual information, such as weather data and industrial consumption patterns, improving the accuracy of predictions and allowing real-time operational adjustments to ensure compliance with environmental regulations.

An additional advantage of DeepAR is its ability to handle sparse and intermittent time series, which are common in industrial environments with irregular data collection. It also generates probabilistic forecasts, providing not only predicted values but also an estimate of the associated uncertainty, which is crucial for informed decision-making.

Thus, these applications in wastewater treatment enable more proactive and efficient management, contributing to more sustainable and robust operations. By using these machine learning techniques to analyze time series, a more detailed and accurate understanding of the operational performance of a WWTP can be obtained. This not only improves treatment efficiency but also helps ensure compliance with environmental regulations and promotes long-term sustainability.

Machine learning modeling is not limited to predicting outcomes; it can also be used for process optimization and control. Predictive models can provide valuable insights into how to adjust operational parameters in real time to maximize treatment efficiency and minimize costs. Additionally, these models can be integrated into automated control systems, which continuously adjust plant operations based on generated predictions, ensuring that the treated effluent consistently meets regulatory standards.

24.1.5 Evaluation

The evaluation phase in CRISP-DM is crucial to ensure that the developed machine learning models are effective and meet the initially defined objectives. This stage involves rigorous validation of the models to ensure that they can generalize well to new data, meaning data that were not used during the training phase.

In WWTPs, the evaluation phase is especially important due to the variability of operational and environmental conditions. The models need to accurately predict the quality of treated water across a wide range of circumstances, which can vary significantly over time. These factors include seasonal variations, changes in the influent composition, temperature fluctuations, and other environmental factors.

To evaluate the models' generalization capability, techniques such as cross-validation are applied. In cross-validation, the dataset is divided into several parts, or "folds." The model is trained on some of these parts and tested on the remaining parts. This process is repeated multiple times with different divisions of the dataset to ensure that the model is robustly evaluated (Arlot and Celisse 2010).

This method helps identify whether the model is overfitting the training data, a problem where the model learns specific patterns from the training set that are not applicable to new data.

Evaluations of WWTPs can include testing under different operational conditions. For example, a model can be assessed for its ability to predict contaminant concentrations during periods of high hydraulic load versus low load periods or across different seasons when temperature variations may affect the efficiency of biological treatment processes.

In addition to cross-validation, it is important to use independent validation datasets to test the models. This involves reserving a portion of the data that is not used during initial training and is only used to evaluate the model's final performance. This approach provides an additional measure of the model's ability to generalize to unseen data.

The evaluation should also consider the models' robustness under extreme operational conditions. In emergency situations or equipment failures, it is crucial that models can still provide useful predictions and assist in decision-making to mitigate impacts.

Mowbray et al. (2022) emphasized that rigorous evaluation is essential for identifying model limitations and making necessary adjustments before deployment. This may include refining algorithms, adjusting model hyperparameters, or incorporating new data to improve prediction accuracy and robustness.

Continuous evaluation of the models after deployment is also fundamental. As operational and environmental conditions change, it is necessary to monitor the models' performance and recalibrate them as needed to maintain their effectiveness. This ensures that the models remain accurate and useful over time, providing ongoing support for the optimization and control of wastewater treatment processes.

Continuous evaluation of models after deployment is fundamental for maintaining their accuracy and effectiveness as operational and environmental conditions change. This involves monitoring the models' performance and recalibrating them as needed. Typically, datasets are split into training, validation, and test sets to build and evaluate models. A common approach is to allocate 60–70% of the data for training, 15–20% for validation, and 15–20% for testing. Regularly updating the models with new data helps ensure they adapt to changing conditions and continue to provide reliable support for optimizing and controlling wastewater treatment processes.

To achieve this, an initial model is developed using a split of the available data, where 60–70% is used for training the model, 15–20% for validation to fine-tune parameters, and 15–20% for testing the model's performance. Once the model is deployed, continuous monitoring is implemented to track its performance through key metrics such as accuracy, precision, recall, and F1 score. Additionally, techniques for detecting data drift are employed to identify changes in the data distribution that could affect model performance.

When performance degradation is detected, the model undergoes recalibration. This process involves retraining the model with new data to capture the evolving patterns and potentially adjusting its parameters to enhance performance. By incorporating regular updates and recalibrations, the models can maintain their effectiveness and continue to provide valuable support for the optimization and control of wastewater treatment processes over time. This proactive approach ensures that the models remain robust and adaptive to new conditions, safeguarding their utility and reliability in dynamic operational environments.

24.1.6 Deployment

In the final phase of CRISP-DM deployment, the developed predictive models are integrated into the control and monitoring systems of a WWTP. This stage is crucial, as it is here that the practical benefits of machine learning models are realized, translating into concrete improvements in plant operation.

Deployment involves automating control processes based on model predictions. For instance, predictive models (MPC) can provide real-time recommendations or control to adjust critical operational parameters, such as chemical dosing or aeration rate. These MPC controllers help to optimize the quality of treated water, ensuring that contaminant levels are maintained within regulatory limits. Precise chemical dosing, for example, is essential for the efficient removal of contaminants without wasting resources or causing adverse environmental impacts.

In addition to automation, deployment may also include implementing early warning systems. These systems are designed to detect anomalies and abnormal operational conditions that could indicate impending issues in the treatment process. By identifying these problems early, preventive measures can be taken to avoid treatment failures and minimize disruptions and maintenance costs. For example, if a predictive model detects an unexpected drop in the removal efficiency of a specific contaminant, it can trigger an alert for operators to investigate and correct the issue before it worsens (Wu et al. 2019).

The integration of predictive models into plant control systems can be performed in various ways. A common approach is the implementation of an advanced control system that uses model predictions to automatically adjust operational parameters. This system can operate continuously, monitor plant conditions and make real-time adjustments to optimize performance. Another approach is the use of monitoring dashboards, where plant operators can view model predictions and make informed decisions based on this information (Piga et al. 2019; Wu et al. 2019).

Iratni and Chang (2019) discussed how the successful deployment of machine learning models can transform the operation of WWTPs, providing significant benefits in terms of efficiency and sustainability. The application of machine learning enables more precise and efficient resource management, reducing energy and chemical consumption and improving the quality of treated effluent. This not only helps meet environmental regulations but also contributes to long-term sustainability by reducing the environmental impact of plant operations.

In addition to operational benefits, deploying predictive models can also improve the plant's responsiveness to changes in influent conditions. For example, during periods of heavy rainfall or periods of high industrial demand, the volume and composition of influent can vary significantly. Predictive models can help quickly adjust operational parameters to handle these variations, ensuring that plants maintain efficient performance even under challenging conditions (Cechinel et al. 2024; Rios Fuck et al. 2024).

The deployment phase also involves training plant operators to ensure that they can effectively use the new systems and technologies. This includes understanding how to interpret model predictions, respond to anomaly alerts, and make manual adjustments when necessary. Adequate training is essential to ensure that the benefits of predictive models are fully realized and that operators can rely on new tools to improve plant operation.

24.2 Applications of Machine Learning in Wastewater Treatment

Artificial intelligence techniques such as machine learning and deep learning have shown great potential in optimization analysis and parameter prediction in complex systems such as WWTPs (Zhao et al. 2020; Wang et al. 2023). Whether in industry or in a wastewater treatment process, these tools can be applied for a variety of purposes and are widely used for automation and decision support (Sundui et al. 2021).

In addition to these applications, it is also important to mention the possibility of improving the energy efficiency of processes (Torregrossa et al. 2018; Bagherzadeh et al. 2021; Alali et al. 2023; El Alaoui El Fels et al. 2023; Zhou et al. 2023), applying real-time operation control (Bernardelli et al. 2020; Chen et al. 2021), and predicting effluent quality (Mjalli et al. 2007; Guo et al. 2015; Zhou et al. 2019; El-Rawy et al. 2021; Ly et al. 2022; Shyu et al. 2023; Zhang et al. 2024), among other possibilities.

Machine learning has proven to be a tool of significant importance for wastewater treatment processes, mainly due to its ability to represent and predict the behavior of highly nonlinear systems, such as those found in WWTPs. Modeling a process can already benefit its operation, considering the various possibilities mentioned, which are closely related to economic cost issues. However, when machine learning models are compared to mechanistic models, the former stand out for their better predictive capacity, greater flexibility in their modeling structure, and continuous improvement of their capabilities through continuous learning from new data (Malviya and Jaspal 2021; Duarte et al. 2024).

For each objective that can be achieved by applying artificial intelligence tools, there are specific techniques that can be applied. For example, among machine learning techniques, models can be based on supervised learning, unsupervised learning or reinforcement learning. Additionally, as the volume of data and the complexity of processes increase over time, it is also possible to apply deep learning techniques, which are subfields of machine learning. For wastewater treatment, each of these techniques can be applied in different ways, and it is now possible to list the models usually applied in each area (Table 24.1) according to Duarte et al. (2024). In this section, each of these techniques is described in more detail in terms of its

Table 24.1 Categories of machine learning models and the main algorithms used in WWTPs

Category	Algorithms
Supervised learning	ANN, SVM, DTs, KNNs, GBMs
Unsupervised learning	K-means, PCA, GMM
Reinforcement learning	Q-learning, DDPG, policy gradient methods
Deep learning	RNN, DNN, CNN, LSTM, GAN

fundamentals, with descriptions of the main algorithms applied to each situation and discussions of real-world application cases.

24.2.1 Supervised Learning

When a machine learning algorithm is trained with data that have been previously labeled by a human, its learning is categorized as supervised (Géron 2019). This type of learning involves the accumulation of previous experiences from historical data of variables, which allows training with the ability to predict future behaviors or classify future records (Duarte et al. 2024). For this reason, a common application of this type of algorithm is the prediction of values for parameters of interest in a process (Géron 2019), such as the concentration of treated effluent or the setpoint of dissolved oxygen in a reactor.

In the literature, artificial neural network (ANN) models are among the most commonly used methods for predicting process performance in WWTPs (Duarte et al. 2024). Due to its high versatility, flexibility, and high prediction potential in complex systems, this method has advantages over other available machine learning algorithms and is considered one of the standard tools for parameter estimation in nonlinear systems (Mjalli et al. 2007; Guo et al. 2015; Géron 2019).

This algorithm can be based on either supervised or unsupervised learning, and it can also be categorized as deep learning. What differentiates its category is mainly whether the data are labeled or not and whether the number of hidden layers applied in its structure (as well as the amount of data provided for its training) is relatively large. In regard to deep learning, these numbers are quite significant, making models in this category more complex (Géron 2019).

To illustrate with examples from the literature, Guo et al. (2015) applied an ANN model based on supervised learning to assist in decision-making for the operations of a WWTP in South Korea, focusing on predicting the total nitrogen concentration in the treated effluent from other variables measured throughout the process. They compared their results with those of another methodology based on supervised learning, the support vector machine (SVM) model, which is also widely used in machine learning (Géron 2019) but is based on a risk minimization framework (Guo et al. 2015).

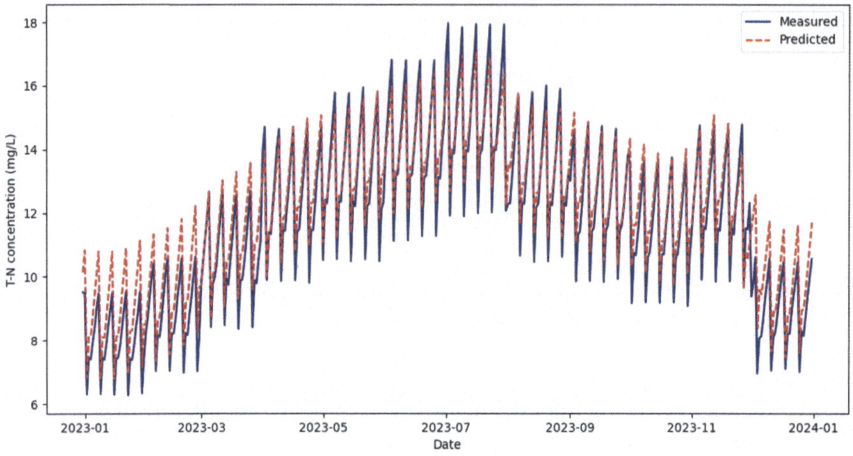

Fig. 24.2 Example of a comparison of the predicted and measured total nitrogen concentrations in the treated effluent

As a result of this study, they concluded that although the SVM model had better accuracy, the ANN model was better able to elucidate cause-and-effect relationships between total nitrogen concentration and other input parameters through sensitivity analysis, which proved to have a better physical meaning of the process with the results evaluated by this model. With this, the authors indicated that ANNs are the most recommendable for practical application in processes similar to those evaluated (Guo et al. 2015).

Figure 24.2 illustrates a comparative analysis of modeled and measured data using a supervised learning algorithm, and Fig. 24.3 shows a graphical result of a sensitivity analysis, demonstrating the prioritization of variables for predicting parameters of interest.

Among other supervised learning models applicable to wastewater treatment, Zhang et al. (2024) applied random forest (RF) and extreme gradient boosting (XGBoost) in addition to ANNs and SVMs. The aim of this study was to evaluate the relationship between the microbial community in activated sludge systems and process performance in terms of phosphorus and nitrogen removal. To achieve this goal, well-known RF models, which are based on the fundamental components of one of the most classic machine learning models, decision trees (DTs), have been proposed (Géron 2019; Alvi et al. 2023). In fact, RF is an ensemble of DTs, which makes it a powerful and robust algorithm with high predictive and classification capabilities (Zhou et al. 2019).

Similarly, the XGBoost model (applied in the study by Zhang et al. (2024)) is also based on DTs but applies the gradient boosting machines (GBMs) technique (Arif Ali et al. 2023). It is important to mention that although DT-based models may not capture as many interactions in highly complex systems as ANN models, they perform well in situations where a small to medium-sized dataset is available—unlike

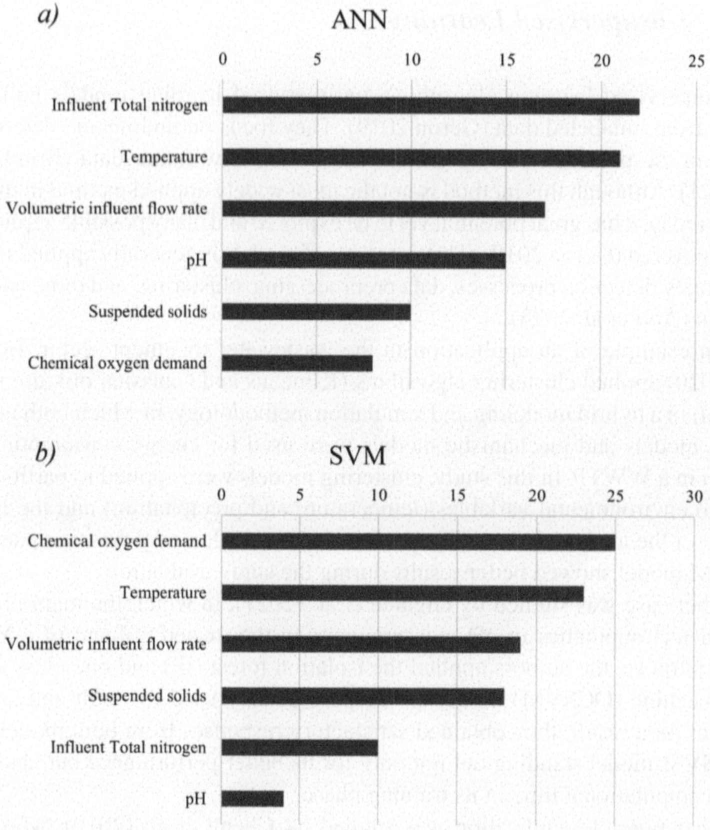

Fig. 24.3 Sensitivity analysis example for a case applying **a** ANN and **b** SVM

ANN models, which require a larger volume of data for their training (Arif Ali et al. 2023).

Other potential applications of supervised learning models in wastewater treatment found in the literature include energy consumption prediction based on different operational parameters (Torregrossa et al. 2018; Bagherzadeh et al. 2021; Alali et al. 2023), fouling prediction in membrane bioreactors to optimize maintenance strategies (Frontistis et al. 2023), soft sensor development for monitoring and predicting effluent quality parameters (Shyu et al. 2023), and framework development to indicate relationships between process variables to support process optimization (Wang et al. 2021), among other possible applications.

24.2.2 Unsupervised Learning

Unlike supervised learning algorithms, unsupervised learning models build their learning from unlabeled data (Géron 2019). They focus on finding and determining patterns in the provided data to group and classify new future data (Brunton and Kutz 2021). Although this method is not the most widely applied method in machine learning today, it has great potential yet to be explored and many possible applications to be discovered (Géron 2019). This category of models is generally applied for fault and anomaly detection processes, data preprocessing, clustering, and dimensionality reduction (Alvi et al. 2023).

As an example of an application in the wastewater treatment sector, Borzooei et al. (2020) applied clustering algorithms (K-means and Gaussian mixture models (GMMs)) in a hybrid modeling and simulation methodology, in which both machine learning models and mechanistic models were used for energy consumption optimization in a WWTP. In this study, clustering models were applied to partition data from two environmental variables (temperature and precipitation) and the influent flow rate of the analyzed plant. The performance of each model was compared, and the GMM model showed better results during the study evaluation.

Another case was studied by Gigante et al. (2021), in which the main objective was to detect anomalies in pH measurements in the treated effluent of a WWTP. For this purpose, the authors applied the isolation forest (IF) and one-class support vector machine (OCSVM) models, which are usually used for fault and anomaly detection. As a result, they obtained satisfactory responses from both models, with the OCSVM model standing out not only for its better performance but also for its shorter computational time in its training phase.

Another possible application was studied by Cheng et al. (2019), who aimed to develop an analysis methodology based on a system applying kernel principal component analysis (kernel PCA) with the OCSVM model to identify abnormal quality measurements of the influent in a WWTP. The authors justified this application by emphasizing the importance of good control and monitoring of the plant influent characteristics to prevent adverse occurrences in the process. After developing the models, the results showed good accuracy, corroborating the applicability of this methodology.

24.2.3 Reinforcement Learning

Machine learning models based on reinforcement learning (RL) are a category of techniques that use a training method quite different from those mentioned previously. Learning occurs based on a system of rewards and penalties. The agent (in this case, the learning system) makes decisions in response to situations in which it is placed, and based on its decision, it is then rewarded or penalized for it. As a result, the agent learns which decisions will have the highest rewards depending on the situations presented by the environment. This learning system is often associated with Markov decision processes (MDPs), which model decision-making in an environment and its subsequent actions based on a response action (reward or penalty) (Géron 2019; Alvi et al. 2023; Duarte et al. 2024).

In WWTPs, models based on this type of learning are usually applied to develop process control strategies (such as model predictive control (MPC)), improve pollutant removal efficiency, and optimize processes (Croll et al. 2023a). For the first case, Chen et al. (2021) and Croll et al. (2023b) applied different reinforcement learning algorithms with the aim of developing control systems in WWTPs.

Chen et al. (2021) focused on applying the multiagent deep deterministic policy gradient (MADDPG) model to control the concentration of dissolved oxygen in reactors and the chemical dosage. Croll et al. (2023b) applied deep Q-learning, proximal policy optimization (PPO), Advantage Actor Critic (A2C), and twin delayed deep deterministic policy gradient (TD3) models, all based on reinforcement learning, to apply a control system capable of minimizing aeration costs while maintaining the quality of treated effluent.

With respect to optimization objectives, Zhou et al. (2023) applied a different methodology of reinforcement learning-based modeling: the Reinforcement Self-Learning-based Multiobjective Particle Swarm Optimization (RSL-MOPSO) algorithm was proposed to optimize the energy consumption of WWTP operations. Initially, this isolated method (MOPSO) was not based on machine learning models; however, the authors developed a methodology applying machine learning techniques based on reinforcement learning to improve the results obtained by the algorithm, making it an interesting hybrid modeling strategy for solving problems in WWTPs.

24.2.4 Deep Learning

Finally, the last category focused on in this chapter is deep learning. Despite being a subfield of machine learning, the models categorized in this group can also be differentiated by their learning mode, whether supervised, unsupervised, or reinforcement learning. What distinguishes them as deep learning models and differentiates them from "shallow" models is their structure, the amount of data needed, and the complexity of their learning algorithms.

Deep learning is based on ANNs with structures containing several layers and is generally applied to highly complex and nonlinear systems. Due to this characteristic, models of this class are very useful for understanding the various relationships involved in wastewater treatment processes. They are currently used for performance prediction in WWTPs, for the development of soft sensors, for anomaly detection, and for process control, among other applications (Alvi et al. 2023).

However, despite having advantages over other models, methodologies using deep learning require large datasets to properly develop and train the models. In addition, there is a high computational cost to enable their execution, which can be a bottleneck in many cases (Alvi et al. 2023).

The main deep learning models currently applied to wastewater treatment systems, as indicated by the literature, include recurrent neural networks (RNNs), deep neural networks (DNNs), long short-term memory (LSTM), convolutional neural networks (CNNs), and generative adversarial networks (GANs). An example of an application is described by Wang et al. (2021) in a study aimed at developing a framework capable of indicating relationships between process variables to aid in the optimization of WWTP operations. The authors used different methodologies, including the RF model and the DNN. The DNN model was specifically applied as a benchmark to evaluate the performance of the RF model (that is not a deep learning model), which is described as a reference of good accuracy and reliability for the other tested models.

Another study that also applied DNNs was described by Khalil et al. (2023), where the objective was to model N_2O emissions in WWTPs using machine learning models. In addition to the DNN, the k-nearest neighbors (kNN) and adaptive boosting (AdaBoost) models were also applied to compare the achieved results. In this study, the authors chose these three models by evaluating them at different levels of complexity: kNN was evaluated as the least complex, and DNN was the most complex. Consequently, the results confirmed what has already been demonstrated by the literature: more complex models (such as those based on deep learning, in this case the DNN model) require greater computational capacity for processing, while simpler models are able to obtain results in less time and at a lower computational cost. However, at the same time, the DNN model achieved more accurate results than the kNN model. Therefore, a cost–benefit evaluation is necessary when applying more complex models.

A fictitious example of the application of model performance analysis based on metrics is shown in Table 24.2, where the kNN, AdaBoost, and DNN models were compared based on their obtained prediction results and processing times.

Another model also widely used in wastewater treatment systems is based on long short-term memory (LSTM)-type structures. Ly et al. (2022) and El-Rawy et al. (2021) applied this type of model to predict effluent quality parameters in different WWTPs. In both cases, the authors compared the performance of the method with that of other shallow and deep learning models. However, when evaluating the performance of the LSTM model, Ly et al. (2022) identified instability in the model structure, which ultimately impaired its predictive capacity. In contrast, El-Rawy et al. (2021) obtained satisfactory results, demonstrating that the success of deep learning applications depends on several factors.

Table 24.2 Fictitious examples of the accuracy metrics and processing times of the models

Model	Accuracy metrics			Processing time (s)	
	R^2	MAE	RMSE	Training	Prediction
kNN	0.88	19.2	40.3	0.2	0.1
AdaBoost	0.92	13.1	22.1	77	0.9
DNN	0.94	11.5	25.7	460	1.4

24.3 Challenges and Limitations in Implementing Machine Learning in Wastewater Treatment

The implementation of machine learning (ML) and deep learning (DL) techniques in wastewater treatment has consistently demonstrated exceptional performance in various applications, including system control and optimization, even in complex environments. In this context, ML is a field of artificial intelligence that focuses on developing algorithms and models that can learn from data, while DL is a subfield of ML that uses deep artificial neural networks to learn complex data representations.

The use of ML and DL in wastewater treatment is a research and application area of great interest. However, it faces technical and organizational challenges that must be overcome to ensure long-term success. These barriers not only hinder initial adoption but can also affect the effectiveness of these technologies in operational environments (Sundui et al. 2021; Abdar et al. 2021; Ho and Goethals 2022).

According to Imen et al. (2023), data are often stored on a government portal, and security limits data accessibility. This is one reason for the lack of data availability, but it can also be related to the cost of data collection and the technological limitations of sensors, among others. The main factors that can be considered more decisive in the implementation of ML and DL techniques are data quality, process complexity, scaling, scalability, model interpretability, adaptation to changes, cost, infrastructure, and acceptance (Doherty et al. 2017; Abdar et al. 2021; Imen et al. 2023). These key factors can be categorized into technical challenges and organizational (adaptive) challenges, as shown in Fig. 24.4.

Figure 24.4 shows that the implementation of machine learning in wastewater treatment faces significant technical and organizational challenges. From a technical perspective, data quality is critical, and wastewater data are usually noisy

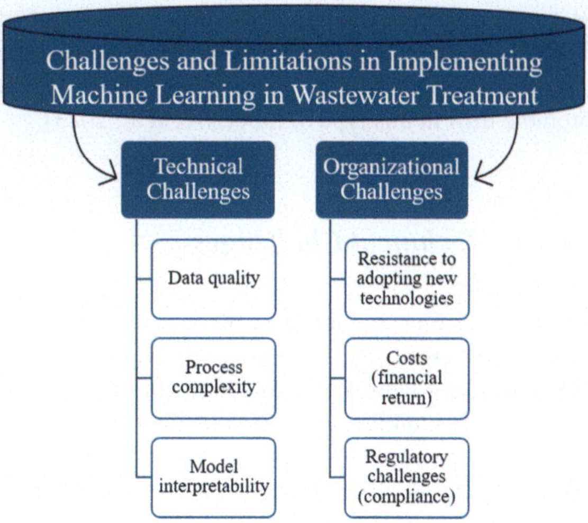

Fig. 24.4 Schematic representation of the challenges and limitations of ML models

and incomplete. In addition, the complexity of treatment processes makes accurate modeling difficult. The interpretability of models is also a challenge, especially in regard to ensuring the reliability of decisions. Organizational barriers include resistance to adopting new technologies, cost, and regulatory challenges. Collaboration between environmental engineers, data scientists, and regulators is essential to achieve effective and sustainable results.

Regarding the challenges faced, incomplete data ends up directly influencing the AI learning results. For each class of models, there are specific technical challenges faced that must be considered when applying the techniques. In the case of supervised learning, an important challenge to be faced would be the labeling of the data, which is often not organized and labeled in the correct way for models to read, making a careful pre-treatment stage of this information necessary (Géron 2019).

In the case of unsupervised learning, a possible difficulty may arise precisely because the data is not labeled, which may make it difficult to validate the results obtained, thus having a greater risk of obtaining inaccurate results. Furthermore, depending on the model, a large volume of data may be required for training, which may consequently require more computational time to train the algorithms (Géron 2019).

Regarding models based on reinforcement learning, the challenges are linked to issues such as obtaining adequate representation of real processes based on little data availability and the establishment of a reward function consistent with the objective of interest. For these cases, possible solutions would be increased data collection with adequate process sensing and a careful establishment of the reward function (Dulac-Arnold et al. 2021).

Deep learning-based models face similar problems regarding the availability of data of sufficient quantity and quality. Especially when evaluating the data available in WWTP, this is an important bottleneck to enable the application of models of this class. Furthermore, many deep learning models are considered "black box," making transparency and understanding of the applied models difficult, which may harm their reliability. However, to address this, developing methods for visualizing and interpreting model decision-making processes can be an effective strategy to promote trust in this type of tool (Alvi et al. 2023).

24.3.1 Technical Challenges

Technical challenges are typically considered easier to identify and solve with existing resources. They can be easily addressed by experts with prior knowledge who have dealt with similar situations involving technical issues. In the context of applying AI techniques to wastewater treatment, these challenges can be divided into two distinct points: data quality and process and modeling complexity (Wang et al. 2021; Lowe et al. 2022; Frontistis et al. 2023). Regarding data quality, the performance of ML or DL models is inextricably linked to the quality of the input data.

Poor data management leads to poor system representability. In wastewater treatment, it is essential to acquire accurate data representing the process, from chemical reactions to small system variations (Doherty et al. 2017; Sundui et al. 2021; Abdar et al. 2021).

Thus, the success of AI models depends heavily on the quality and quantity of available data. This is a problem because in many WWTPs, data collection mechanisms are outdated, or analyses are infrequent (not always collected daily). Therefore, a high quantity and quality of data are essential to ensure that this is not a limiting factor for the excellent applicability of ML and DL models, making them robust (Lowe et al. 2022; Frontistis et al. 2023).

In addition, data quality can be affected by the quality of the instruments used in the analysis and poor data maintenance, such as sensor noise, missing values, and biases in the way data collection is examined (Frontistis et al. 2023).

From the perspective of Lowe et al. (2022) and Doherty et al. (2017), many real-world systems face the problem of insufficient data and lack the necessary conditions for using AI models. Although they can benefit from the use of ML techniques, these models lack basic data management and storage resources (Doherty et al. 2017; Lowe et al. 2022).

Another important challenge is that even if the quality and quantity of the data are good, the data may not represent all the operating conditions of the plant. Wastewater treatment involves multiple stages/processes, such as settling, sedimentation, aeration, filtration, coagulation, flocculation, and disinfection. ML and DL models must account for these interrelated processes, increasing model complexity and requiring careful planning.

Thus, it has become one of the most indispensable stages of the process, along with good data quality. By relating data to processes, even small deviations in natural or treatment processes can be analyzed, and new information can be processed (Wang et al. 2021; Imen et al. 2023; Frontistis et al. 2023; Duarte et al. 2024). According to Alvi et al. (2023), most public datasets have low data frequencies (daily) and need important metadata, such as climatic information.

Furthermore, another problem can be related to overfitting (when a statistical model fits the observed dataset very well but is ineffective at predicting future outcomes). According to Xia et al. (2024), to reduce the risk of overfitting in wastewater treatment, cross-validation, feature selection, and data augmentation can be used when inaccurate predictions are available for new samples (Xia et al. 2024).

Thus, process complexity and modeling directly affect the performance of AI models. In wastewater treatment processes, the operating environment is complex, with seasonal changes, unanticipated events, significant disturbances, and load variations that do not exhibit linear behavior and involve various multidimensional interactions among variables.

These changing characteristics over time increase the complexity of the system and require a deeper study to make ML and DL models robust enough to adapt to these changes (Doherty et al. 2017; Abdar et al. 2021; Ho and Goethals 2022; Frontistis et al. 2023). According to Alvi et al. (2023), in real industrial plants, numerous disturbances impact the collected data. For a model to be robust, it must remain stable in the face of these perturbations.

Well-structured ML and DL models can still face problems due to the lack of detailed historical data, which leads to difficulties in accurate modeling (Doherty et al. 2017; Wang et al. 2021; Duarte et al. 2024). When issues of data quality and system complexity are addressed, robust models are achieved.

Model interpretability is also a very important factor. Correlating model-generated responses with the physical system is essential for understanding and explaining how an ML or DL model makes its decisions (Frontistis et al. 2023; Alvi et al. 2023; Duarte et al. 2024). According to Frontistis et al. (2023), one possible way to bridge the gap between predictions and human observers is to apply self-explanatory ML techniques.

24.3.2 Organizational (Adaptive) Challenges

Organizational challenges, also known as adaptive challenges, are often complex and require changes in habits, learning, and solutions. These challenges are usually difficult to recognize and typically involve people who tend to resist the necessary changes because they directly affect the workplace, whether due to values or responsibilities within the organization (Doherty et al. 2017; Frontistis et al. 2023; Alvi et al. 2023; Xia et al. 2024).

According to Maurya et al. (2024), an effective ML system largely depends on the combination of large amounts of data with human involvement. In the current scenario of applying ML and DL techniques in industry, adaptability in the workplace becomes an inescapable challenge, whether due to organizational or cultural restructuring changes.

Therefore, organizations need to drive internal transformation and ensure that their employees are ready to support new ways of working. In this context, the organizational challenges of applying AI techniques in wastewater treatment can be divided into three main points: resistance to adopting new technologies, cost (financial return), and regulatory challenges (compliance) (Maurya et al. 2024).

Resistance to change is also a common obstacle in organizations and businesses, especially regarding implementing new technologies and ensuring regulatory compliance. One of the best ways to prevent the consequences of change is to anticipate its impact and communicate with affected parties before implementing changes, explaining the benefits and reasons (Kumari et al. 2024). This helps to reduce shock and resistance when changes occur. According to Kumari et al. (2024), methodological barriers and the complexity of integrating stakeholder perspectives are issues that often arise in AI implementation.

Many organizations need more support for the adoption of new technologies in wastewater treatment. There is often a natural reluctance to adopt new technologies; for example, applying ML and DL in wastewater treatment involves integrating complex algorithms into existing systems. A lack of understanding of these technologies can lead to resistance from the professionals involved, and the need for continuous training and updating is also a factor.

Resistance may arise from a lack of trust in ML and DL models or a belief that traditional methods are more reliable (Maurya et al. 2024; Xia et al. 2024). According to Mallikarjuna (2024), AI can provide various benefits. Still, employees must acquire new skills and competencies to use these technologies effectively and consider ethical and social implications, such as intellectual freedom and bias.

From the perspective of Sahu et al. (2023), the lack of talent is a challenge in hiring qualified people because implementation will be limited without a well-defined AI strategy. Another major challenge is the need for more technical skills. Many industries need help finding professionals with the essential technical skills to handle data complexity and apply ML or DL techniques, requiring investment in training and development or hiring new specialized talent (Sahu et al. 2023).

Establishing an open dialog and informing everyone about the change process helps build trust and reduce resistance. Effective communication channels such as meetings, emails, intranets, or bulletin boards convey transparency in the process. In AI adoption, communication and transparency with employees are fundamental to human resource management (Mallikarjuna 2024).

One of the challenges that significantly influences the implementation of ML and DL techniques is investment (cost) since implementing these techniques requires resources in infrastructure (computers), specific training, and/or hiring specialized people. Therefore, there is a need for a high initial investment with no guarantee of immediate return, as there are risks such as potential model failures. In addition, these models require continuous maintenance and adjustments to adapt to changes in treatment processes or data patterns, which may result in additional operational costs (Iratni and Chang 2019; Sahu et al. 2023).

Regarding compliance challenges, data collection, storage, and processing in ML and DL may involve sensitive information. For this reason, there is often no data sharing, resulting in the need to create ML and DL models from scratch rather than adapting based on other stations that use good data.

The use of ML or DL strategies in wastewater treatment must adopt strict environmental and water quality standards to meet the minimum requirements of environmental regulators. In a general context, regulatory compliance in ML and DL is essential to protect users' rights and environmental regulations, prevent undesirable consequences, and promote confidence in the adoption of these innovative technologies (Iratni and Chang 2019; Ho and Goethals 2022).

24.3.3 Overcoming Barriers

To overcome these challenges, organizations can start by demonstrating partial implementations, pilot projects, success stories, and tangible results to reduce resistance. Improving over time can convince stakeholders of the effectiveness of using AI.

Therefore, it is essential to educate professionals about the benefits and limitations of ML and DL in wastewater treatment (Tufail et al. 2023; Xia et al. 2024).

Investments in training, capacity building, and partnerships with academic institutions and technology companies can also facilitate knowledge transfer and help overcome these challenges. Creating an environment fostering continuous learning and skills development can help mitigate cultural and technical skills barriers.

Therefore, the acceptance of ML and DL requires a holistic approach that includes education, organizational culture, and the demonstration of concrete results. After all, well-analyzed data are the foundation for valuable insights and intelligent decisions (Tripathi et al. 2021; Ho and Goethals 2022; Imen et al. 2023).

24.4 Conclusion

Integrating machine learning and deep learning into wastewater treatment represents a significant milestone in the ongoing digital transformation of Industry 4.0. By leveraging advanced technologies such as big data and machine learning, the wastewater treatment sector can achieve unprecedented efficiency, sustainability, and compliance with stringent environmental regulations. This chapter has presented the CRISP-DM methodology as a robust framework for developing machine learning solutions, guiding the process from business understanding to solution deployment.

Machine learning enables the analysis of large volumes of data, providing valuable insights that can optimize operations and enhance decision-making processes. In wastewater treatment, this translates into better water resource management, increased treatment process efficiency, and a more responsive approach to variable conditions. The practical benefits of implementing these technologies include reduced operational costs, optimized use of chemicals and energy, and improved treated water quality.

Furthermore, applying advanced algorithms, such as physics-informed neural networks (PINNs), has shown significant potential in combining data-driven models with first-principles models based on physical and chemical laws. This integration enhances the accuracy and robustness of predictive models, particularly in scenarios where data are scarce or noisy. PINNs and other hybrid approaches are valuable tools for optimizing complex processes in wastewater treatment, ensuring that predictions adhere to known physical behaviors and providing more reliable and interpretable results.

Despite these clear advantages, implementing machine learning and deep learning in wastewater treatment is challenging. Technical challenges must be addressed, such as the need for large volumes of high-quality data and the complexity of modeling processes. Additionally, organizational challenges, including resistance to adopting new technologies and the need for specialized technical skills, can hinder the successful deployment of these solutions. Ethical and regulatory considerations also play a crucial role, requiring careful attention to ensure the responsible use of AI-based technologies.

A common challenge is the need for more quality and effluent data in treatment plants. Regulatory bodies determine sampling points and the frequency of collection, which are influenced by various factors and lead to a need for more standardization in monitoring different treatment plants. This results in inconsistent collection intervals and insufficient data. Additionally, due to a lack of familiarity with daily operations, many professionals still need to understand the use of machine learning tools better. This creates significant barriers to their proper application and achieving satisfactory results. The need for qualified professionals in the field increases trust toward individuals with little or no practical experience. As a result, organizations in the sanitation sector often need to choose more modern alternatives, leading to substantial operational costs and resistance to change from both the company and employees.

Overcoming these challenges requires a strategic and collaborative approach. Investing in data quality, fostering continuous learning and skill development, and creating partnerships between environmental engineers, data scientists, and regulatory bodies are essential steps. Demonstrating tangible benefits through pilot projects and success stories can help build trust and reduce resistance to new technologies. The future of implementing machine learning in wastewater treatment is promising, with technological advancements poised to revolutionize the sector. Areas such as optimization of treatment processes, rapid contamination diagnostics, and predictive maintenance of equipment are fertile fields for research and development. The application of artificial intelligence already allows fast analysis of water samples, identification of bacteria, and identification of problems in less than 10 min.

Moreover, the integration of machine learning with Industry 4.0 suggests increased efficiency and accuracy of results, with control systems that continuously learn and adapt, improving decision-making and reducing human intervention. These advancements indicate a path of continuous innovation, where the precision and sustainability of wastewater treatment can reach new heights.

The successful integration of machine learning and deep learning into wastewater treatment holds promise for significant improvements in process efficiency, sustainability, and regulatory compliance. By adopting a systematic approach, such as the CRISP-DM methodology, and addressing the inherent challenges, the sector can fully harness the potential of these advanced technologies. This paves the way for future innovations and contributes to more effective and sustainable wastewater management practices, ultimately benefiting the environment and society.

References

Abdar M, Pourpanah F, Hussain S, Rezazadegan D, Liu L, Ghavamzadeh M, Fieguth P, Cao X, Khosravi A, Acharya UR, Makarenkov V, Nahavandi S (2021) A review of uncertainty quantification in deep learning: techniques, applications and challenges. Information Fusion 76:243–297

Alali Y, Harrou F, Sun Y (2023) Unlocking the potential of wastewater treatment: machine learning based energy consumption prediction. Water (Switzerland) 15. https://doi.org/10.3390/w15132349

Alvi M, Batstone D, Mbamba CK, Keymer P, French T, Ward A, Dwyer J, Cardell-Oliver R (2023) Deep learning in wastewater treatment: a critical review. Water Res 245

Amdahl GM (1967) Validity of the single processor approach to achieving large scale computing capabilities. In: AFIPS conference proceedings—1967 spring joint computer conference, AFIPS 1967

Andrienko N, Gennady A (2006) Exploratory analysis of spatial and temporal data: a systematic approach

Arif Ali ZH, Abduljabbar ZA, Tahir H, Bibo Sallow A, Almufti SM (2023) eXtreme gradient boosting algorithm with machine learning: a review. Academic J Nawroz University 12:320–334. https://doi.org/10.25007/ajnu.v12n2a1612

Arlot S, Celisse A (2010) A survey of cross-validation procedures for model selection. Stat Surv 4. https://doi.org/10.1214/09-SS054

Bagherzadeh F, Nouri AS, Mehrani MJ, Thennadil S (2021) Prediction of energy consumption and evaluation of affecting factors in a full-scale WWTP using a machine learning approach. Process Saf Environ Prot 154:458–466. https://doi.org/10.1016/j.psep.2021.08.040

Bernardelli A, Marsili-Libelli S, Manzini A, Stancari S, Tardini G, Montanari D, Anceschi G, Gelli P, Venier S (2020) Real-time model predictive control of a wastewater treatment plant based on machine learning. Water Sci Technol 81:2391–2400. https://doi.org/10.2166/wst.2020.298

Borzooei S, Miranda GHB, Abolfathi S, Scibilia G, Meucci L, Zanetti MC (2020) Application of unsupervised learning and process simulation for energy optimization of a WWTP under various weather conditions. Water Sci Technol 81:1541–1551. https://doi.org/10.2166/wst.2020.220

Brunton SL, Kutz JN (2021) Data-driven science and engineering: machine learning, dynamical systems, and control. Cambridge University Press

Buckingham E (1914) On physically similar systems. illustrations of the use of dimensional equations. Phys Rev 4:345–376

Cechinel MAP, Neves J, Fuck JVR, de Andrade RC, Spogis N, Riella HG, Padoin N, Soares C (2024) Enhancing wastewater treatment efficiency through machine learning-driven effluent quality prediction: a plant-level analysis. J Water Process Eng 58. https://doi.org/10.1016/j.jwpe.2023.104758

Chen T, Guestrin C (2016) XGBoost: a scalable tree boosting system. In: Proceedings of the ACM SIGKDD international conference on knowledge discovery and data mining

Chen K, Wang H, Valverde-Pérez B, Zhai S, Vezzaro L, Wang A (2021) Optimal control towards sustainable wastewater treatment plants based on multi-agent reinforcement learning. Chemosphere 279. https://doi.org/10.1016/j.chemosphere.2021.130498

Cheng T, Dairi A, Harrou F, Sun Y, Leiknes T (2019) Monitoring influent conditions of wastewater treatment plants by nonlinear data-based techniques. IEEE Access 7:108827–108837. https://doi.org/10.1109/ACCESS.2019.2933616

Croll HC, Ikuma K, Ong SK, Sarkar S (2023) Reinforcement learning applied to wastewater treatment process control optimization: approaches, challenges, and path forward. Crit Rev Environ Sci Technol 53:1775–1794

Croll HC, Ikuma K, Ong SK, Sarkar S (2023) Systematic performance evaluation of reinforcement learning algorithms applied to wastewater treatment control optimization. Environ Sci Technol 57:18382–18390. https://doi.org/10.1021/acs.est.3c00353

Doherty E, McNamara G, Fitzsimons L, Clifford E (2017) Design and implementation of a performance assessment methodology cognisant of data accuracy for Irish wastewater treatment plants. J Clean Prod 165:1529–1541. https://doi.org/10.1016/j.jclepro.2017.07.083

Duarte MS, Martins G, Oliveira P, Fernandes B, Ferreira EC, Alves MM, Lopes F, Pereira MA, Novais P (2024) A review of computational modeling in wastewater treatment processes. ACS ES T Water 4:784–804

Dulac-Arnold G, Levine N, Mankowitz DJ, Li J, Paduraru C, Gowal S, Hester T (2021) Challenges of real-world reinforcement learning: definitions, benchmarks and analysis. Mach Learn 110:2419–2468. https://doi.org/10.1007/s10994-021-05961-4

El Alaoui El Fels A, Mandi L, Kammoun A, Ouazzani N, Monga O, Hbid ML (2023) Artificial intelligence and wastewater treatment: a global scientific perspective through text mining. Water (Switzerland) 15. https://doi.org/10.3390/w15193487

El-Rawy M, Abd-Ellah MK, Fathi H, Ahmed AKA (2021) Forecasting effluent and performance of wastewater treatment plant using different machine learning techniques. J Water Process Eng 44. https://doi.org/10.1016/j.jwpe.2021.102380

Frontistis Z, Lykogiannis G, Sarmpanis A (2023) Machine learning implementation in membrane bioreactor systems: progress, challenges, and future perspectives: a review. Environments—MDPI 10

Fuck JVR, Cechinel MAP, Neves J, de Andrade RC, Tristão R, Spogis N, Riella HG, Soares C, Padoin N (2024) Predicting effluent quality parameters for wastewater treatment plant: a machine learning-based methodology. Chemosphere 141472. https://doi.org/10.1016/J.CHEMOSPHERE.2024.141472

Géron A (2019) Hands-on machine learning with scikit-learn, Keras & TensorFlow, 2nd ed. O'Reilly Media Inc.

Gers FA, Schmidhuber J, Cummins F (2000) Learning to forget: continual prediction with LSTM. Neural Comput 12(10):2451–2471. https://doi.org/10.1162/089976600300015015

Gigante D, Oliveira P, Fernandes B, Lopes F, Novais P (2021) Unsupervised learning approach for pH anomaly detection in wastewater treatment plants. In: Lecture notes in computer science (including subseries lecture notes in artificial intelligence and lecture notes in bioinformatics). Springer Science and Business Media Deutschland GmbH, pp 588–599

Guo H, Jeong K, Lim J, Jo J, Kim YM, Park J pyo, Kim JH, Cho KH (2015) Prediction of effluent concentration in a wastewater treatment plant using machine learning models. J Environ Sci (China) 32:90–101. https://doi.org/10.1016/j.jes.2015.01.007

Hewamalage H, Bergmeir C, Bandara K (2021) Recurrent neural networks for time series forecasting: current status and future directions. Int J Forecast 37. https://doi.org/10.1016/j.ijforecast.2020.06.008

Ho L, Goethals P (2022) Machine learning applications in river research: trends, opportunities and challenges. Methods Ecol Evol 13:2603–2621. https://doi.org/10.1111/2041-210X.13992

Imen S, Croll HC, McLellan NL, Bartlett M, Lehman G, Jacangelo JG (2023) Application of machine learning at wastewater treatment facilities: a review of the science, challenges and barriers by level of implementation. Environ Technol Rev 12:493–516

Iratni A, Chang NB (2019) Advances in control technologies for wastewater treatment processes: status, challenges, and perspectives. IEEE/CAA J Autom Sin 6(2):337–363. https://doi.org/10.1109/JAS.2019.1911372

Jebb AT, Parrigon S, Woo SE (2017) Exploratory data analysis as a foundation of inductive research. Hum Resour Manage Rev 27. https://doi.org/10.1016/j.hrmr.2016.08.003

Ke G, Meng Q, Finley T, Wang T, Chen W, Ma W, Ye Q, Liu TY (2017) LightGBM: a highly efficient gradient boosting decision tree. In: Advances in neural information processing systems

Khalil M, AlSayed A, Liu Y, Vanrolleghem PA (2023) Machine learning for modeling N_2O emissions from wastewater treatment plants: aligning model performance, complexity, and interpretability. Water Res 245. https://doi.org/10.1016/j.watres.2023.120667

Kumari S, Chowdhry J, Chandra Garg M (2024) AI-enhanced adsorption modeling: challenges, applications, and bibliographic analysis. J Environ Manage 351. https://doi.org/10.1016/j.jenvman.2023.119968

Lowe M, Qin R, Mao X (2022) A review on machine learning, artificial intelligence, and smart technology in water treatment and monitoring. Water (Switzerland) 14. https://doi.org/10.3390/w14091384

Ly QV, Truong VH, Ji B, Nguyen XC, Cho KH, Ngo HH, Zhang Z (2022) Exploring potential machine learning application based on big data for prediction of wastewater quality from different full-scale wastewater treatment plants. Sci Total Environ 832:154930. https://doi.org/10.1016/j.scitotenv.2022.154930

Mallikarjuna C (2024) An analysis of integrating artificial intelligence in academic libraries. DESIDOC J Library Inf Technol 44:124–129. https://doi.org/10.14429/djlit.44.02.18958

Malviya A, Jaspal D (2021) Artificial intelligence as an upcoming technology in wastewater treatment: a comprehensive review. Environ Technol Rev 10:177–187

Maurya BM, Yadav N, Iyer M, Yadav MK, Vellingiri B (2024) Artificial intelligence and machine learning algorithms in the detection of heavy metals in water and wastewater: methodological and ethical challenges. Chemosphere 353. https://doi.org/10.1016/j.chemosphere.2024.141474

Mjalli FS, Al-Asheh S, Alfadala HE (2007) Use of artificial neural network black-box modeling for the prediction of wastewater treatment plants performance. J Environ Manage 83:329–338. https://doi.org/10.1016/j.jenvman.2006.03.004

Mowbray M, Vallerio M, Perez-Galvan C, Zhang D, Chanona ADR, Navarro-Brull FJ (2022) Industrial data science—a review of machine learning applications for chemical and process industries. React Chem Eng 7

Piga D, Forgione M, Formentin S, Bemporad A (2019) Performance-oriented model learning for data-driven MPC design. IEEE Control Syst Lett 3. https://doi.org/10.1109/LCSYS.2019.2913347

Sahu S, Kaur A, Singh G, Kumar Arya S (2023) Harnessing the potential of microalgae-bacteria interaction for eco-friendly wastewater treatment: a review on new strategies involving machine learning and artificial intelligence. J Environ Manage 346. https://doi.org/10.1016/j.jenvman.2023.119004

Salinas D, Flunkert V, Gasthaus J, Januschowski T (2020) DeepAR: probabilistic forecasting with autoregressive recurrent networks. Int J Forecast 36. https://doi.org/10.1016/j.ijforecast.2019.07.001

Sansana J, Joswiak MN, Castillo I, Wang Z, Rendall R, Chiang LH, Reis MS (2021) Recent trends on hybrid modeling for Industry 4.0. Comput Chem Eng 151

Schröer C, Kruse F, Gómez JM (2021) A systematic literature review on applying CRISP-DM process model. In: Procedia computer science

Shao S, Fu D, Yang T, Mu H, Gao Q, Zhang Y (2023) Analysis of machine learning models for wastewater treatment plant sludge output prediction. Sustainability (Switzerland) 15. https://doi.org/10.3390/su151813380

Shearer C, Watson HJ, Grecich DG, Moss L, Adelman S, Hammer K, a Herdlein S (2000) The CRISP-DM model: the new blueprint for data mining. J Data Warehousing

Sherstinsky A (2020) Fundamentals of recurrent neural network (RNN) and long short-term memory (LSTM) network. Phys D: Nonlinear Phenom 404:132306. https://doi.org/10.1016/j.physd.2019.132306

Shyu HY, Castro CJ, Bair RA, Lu Q, Yeh DH (2023) Development of a soft sensor using machine learning algorithms for predicting the water quality of an onsite wastewater treatment system. ACS Environmental Au 3:308–318. https://doi.org/10.1021/acsenvironau.2c00072

Sundui B, Ramirez Calderon OA, Abdeldayem OM, Lázaro-Gil J, Rene ER, Sambuu U (2021) Applications of machine learning algorithms for biological wastewater treatment: updates and perspectives. Clean Technol Environ Policy 23:127–143. https://doi.org/10.1007/s10098-020-01993-x

Sutskever I, Vinyals O, Le QV (2014) Sequence to sequence learning with neural networks. In: Proceedings of the 28th international conference on neural information processing systems (NIPS'14), vol 2. MIT Press, Cambridge, MA, USA, pp 3104–3112. https://doi.org/10.48550/arXiv.1409.3215

Sylvia ML, Murphy S (2023) Exploratory data analysis. In: Clinical analytics and data management for the DNP, Third Edition

Torregrossa D, Leopold U, Hernández-Sancho F, Hansen J (2018) Machine learning for energy cost modelling in wastewater treatment plants. J Environ Manage 223:1061–1067. https://doi.org/10.1016/j.jenvman.2018.06.092

Tripathi S, Muhr D, Brunner M, Jodlbauer H, Dehmer M, Emmert-Streib F (2021) Ensuring the robustness and reliability of data-driven knowledge discovery models in production and manufacturing. Front Artif Intell 4

Tufail S, Riggs H, Tariq M, Sarwat AI (2023) Advancements and challenges in machine learning: a comprehensive review of models, libraries, applications, and algorithms. Electronics (Switzerland) 12

Wang D, Thunéll S, Lindberg U, Jiang L, Trygg J, Tysklind M, Souihi N (2021) A machine learning framework to improve effluent quality control in wastewater treatment plants. Sci Total Environ 784. https://doi.org/10.1016/j.scitotenv.2021.147138

Wang Y, Cheng Y, Liu H, Guo Q, Dai C, Zhao M, Liu D (2023) A review on applications of artificial intelligence in wastewater treatment. Sustainability (Switzerland) 15

Wu Z, Tran A, Rincon D, Christofides PD (2019) Machine learning-based predictive control of nonlinear processes. Part I: Theory. AIChE J 65. https://doi.org/10.1002/aic.16729

Xia L, Hao X, Zhou Y (2024) Machine learning-based global trends and the development prospects of wastewater treatment: a bibliometric analysis. J Environ Chem Eng 12. https://doi.org/10.1016/j.jece.2024.112732

Zhang Y, Wu H, Xu R, Wang Y, Chen L, Wei C (2024) Machine learning modeling for the prediction of phosphorus and nitrogen removal efficiency and screening of crucial microorganisms in wastewater treatment plants. Sci Total Environ 907. https://doi.org/10.1016/j.scitotenv.2023.167730

Zhao L, Dai T, Qiao Z, Sun P, Hao J, Yang Y (2020) Application of artificial intelligence to wastewater treatment: a bibliometric analysis and systematic review of technology, economy, management, and wastewater reuse. Process Saf Environ Prot 133:169–182

Zheng A, Casari A (2018) Feature engineering for machine learning: principles and techniques for data scientists. O'Reilly

Zhou P, Li Z, Snowling S, Baetz BW, Na D, Boyd G (2019) A random forest model for inflow prediction at wastewater treatment plants. Stoch Env Res Risk Assess 33:1781–1792. https://doi.org/10.1007/s00477-019-01732-9

Zhou P, Wang X, Chai T (2023) Multiobjective operation optimization of wastewater treatment process based on reinforcement self-learning and knowledge guidance. IEEE Trans Cybern 53:6896–6909. https://doi.org/10.1109/TCYB.2022.3164476

Chapter 25
Current Trends in Anaerobic Digestion and Renewable Gases Production: The Influence of Biomethane Under Technical and Economic Point of View

Águeda Bellver-Domingo, Francesc Hernández-Sancho, Alberto Torrecilla, Israel Díaz, and María Fdz-Polanco

Abstract Wastewater treatment is responsible for removing organic pollutants from wastewater producing sewage sludge as by-product that cannot be directly disposed into ecosystems due to its high presence of organic pollutants. To solve this problem, anaerobic digestion (AD) is commonly used to treat the sewage sludge, generating biogas that is converted into energy used on site by the wastewater treatment plants (WWTPs). However, the energy efficiency of biogas is not optimal due to the presence of CO_2. By removing CO_2, the methane content is higher, and its energy efficiency has been increased. This technological improvement is called biogas upgrading (BU), where biomethane is produced and used as renewable gas. Biomethane uses sewage sludge as raw material helping both the WWTPs to reduce wastewater treatments costs and reducing energy consumption of production processes. Specifically, biomethane has higher calorific value than biogas due to its percentage of methane (90%). This value makes biomethane as an energy source comparable to conventional natural gas, opening a wide range of possibilities to implement it in different areas. This chapter assesses the benefits of producing biomethane under different points of view, according to the economic sectors, the conservation degree of ecosystems and the human well-being. Throughout the identification of this benefits, the advantages of BU in WWTPs can be highlighted. Results obtained highlight the importance of biomethane for the decarbonization of societies through promoting high-quality renewable energies and ensuring the short-, medium-, and long-term environmental conservation thanks to the circular economy approach.

Á. Bellver-Domingo (✉) · F. Hernández-Sancho
Inter-University Institute for Local Development (IIDL-WATER), Water Economics Group, University of Valencia, Valencia, Spain
e-mail: Agueda.Bellver@uv.es

A. Torrecilla · I. Díaz · M. Fdz-Polanco
Department of Chemical Engineering and Environmental Technology, School of Industrial Engineering, University of Valladolid, Valladolid, Spain

Institute of Sustainable Processes, Valladolid, Spain

S. Souabi et al. (eds.), *Wastewater Treatment Plants*, Water Science and Technology Library 130, https://doi.org/10.1007/978-3-031-87461-1_25

Keywords Biomethane · Biogas · Circular economy · Sludge · WWTP

25.1 Introduction

The current climate situation resulting from ecosystem degradation associated with the economic growth of different territories has become a critical factor when defining future socioeconomic and environmental scenarios. The increase in raw materials consumption, energy requirements, and water demand highlight the need to implement protocols and technological improvements that allow maintaining the level of production without affecting the continuity of ecosystems. Energy consumption is a key aspect in this situation, since practically all of today's appliances and comforts require energy to function. At this point, energy is not only an input in the development of the daily activities of the territories, but also it has become a strategic good that define the relationships between countries, the development of their economy and the world politics. Until recently, fossil fuels and other energy sources such as natural gas, have generated clear dependencies between producing and consuming countries, dictating the world economy and causing a large part of the environmental problems that are currently being attempted to mitigate (Rosha et al. 2021).

Focusing the attention on water sector, there is an evident relationship between the urban hydrological cycle (capture, supply, collection, purification, and discharge) and the energy consumption. This sector is energy-dependent since water is a resource that requires pumping and piping to be transported to treatment facilities and end users. It is estimated that water collection requires around 50% of the energy consumed (considering both surface and underground extraction), followed by almost 40% of the energy consumed in wastewater treatment processes (IEA 2017; Modiri et al. 2023). These values demonstrate the importance of the hydrological cycle management regarding the emissions of greenhouse gases (GHG) and the energy dependency of territories. Specifically, wastewater treatment processes require high levels of energy resulting from the oxidation needs to eliminate organic matter, sludge management and pumping to move water masses. According to Alvarado et al. (2019), the energy consumption can overcome the 3% of total energy consumption of the industrial sector, causing GHG emissions—levels close to 200,000 kg CO_2/day (Zhuang et al. 2020)—by purifying water processes.

In response to this situation, different technological advances have been implemented, as well as policies and action plans in all productive sectors to reduce environmental impacts while seeking to maintain (and even improve) energy-use efficiency. The European Green Deal proposes an EU Energy System Integration Strategy. This strategy has the aim to reform the European energy system through increasing the efficient use of energy. As a result, a climate neutral EU by 2050 can be achieved, where the energy sources and facilities are linked. By then, the energy sector must be fully decarbonized, which means that there is no place for raw fossil fuels. In this context, non-electric renewable gases (RG) are essential since the resource and the energy used in the production process are renewable and focused on circular economy and sustainability. Biogas, biomethane, green H_2, and synthetic natural gas are all forms of RG (EBA 2024). Hence, to achieve decarbonization guaranteeing energy supply, this RG need to be increased to solve the problem of intermittency in wind and solar power, which are difficult to predict and forecast (Liebetrau et al. 2020).

The RG are essential energy vectors to solve this situation, contributing to both the circular economy development and the energy transition process. Furthermore, RG positively affects to carbon–neutral economy since can be injected into natural gas network (NGN), maximizing their use and guaranteeing security of supply. Considering this, a multidisciplinary analysis of RG is necessary to identify its impact on both the decarbonization of productive sectors and the improvement energy-use efficiency. It is important to highlight the need of multidisciplinary approach since RG has horizontal impact in economy, social well-being, and ecosystems dynamics. Hence, this chapter aims to assess the situation of RG (specifically biomethane) under technological point of view (technical options to satisfy the energy demand) and under an economic point of view (importance of the monetary benefit valuation) as a tool to decarbonize the water and energy sectors and ensure the response to climate change consequences.

25.2 Biogas Versus Biomethane: The Advantages of the BU Process

Biogas is produced by the AD process, using wide range of biological by-products (such as livestock manure, slurry, sewage sludge, municipal solid waste, and industrial waste). Specifically, AD produce two types of by-products: (i) biogas—composed by CH_4 (methane), CO_2 (carbon dioxide), and other contaminants (ammonia, water vapour, hydrogen sulphide (H_2S), nitrogen, and oxygen); (ii) digestate—biosolid used as fertilizer with positive impact on soil structure (EPA 2024). Focusing the attention on wastewater treatment plants (WWTPs), around the 60% of the organic matter removed from wastewater is collected as sewage sludge. WWTPs managers usually use AD to stabilize the organic matter of sludge to be finally disposed in landfills or used in agriculture. Traditionally, the biogas produced is burned in the same WWTP to obtain energy and reduce the energy consumption of wastewater treatment process (cogeneration).

Usually, raw biogas contains CO_2 and H_2S which negatively affect to the final calorific value, reducing the efficiency of energy production. When AD is used to obtain biogas, the formation of these compounds is inevitable. To solve this problem, both compounds can be removed to achieve better quality specifications (high calorific value and less risk of corrosion) and highest number of final users. This technical process is called BU (biogas upgrading), obtaining biomethane as final product, whose characteristics are identical to conventional natural gas since the quantity of CH_4 is close to 95%. This is the main advantage of biomethane over biogas, directly

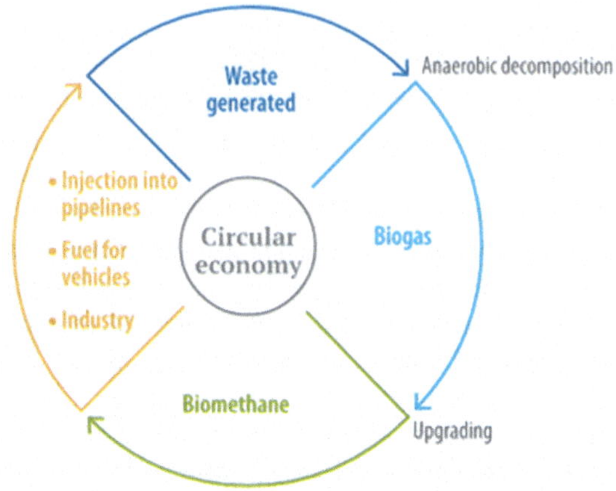

Fig. 25.1 Influence of biomethane in circular economy. *Source* Own elaboration

helping to both decarbonization strategy and circular economy (Fig. 25.1). Specifically, the main pathways for biomethane utilization are heat and steam production, electricity production, injection in NGN, and fossil fuels replacement (Zafar 2024).

As mentioned above, the characteristics of biomethane are identical to conventional natural gas, allowing to be injected into the NGN. As a result, biomethane can reach more users and promoting technological innovations and energy efficiency. However, the quality specifications (based on the methane content) for the biomethane obtained from an unconventional source are different between countries according to their legislative framework, for instance, 90% in Spain and 95% in Czech Republic or Germany (MITECO 2018 and Marcogaz 2018). Some other countries do not specify a CH_4 content but rather a range of higher calorific value that natural gas and biomethane must meet. These values indicate that it is necessary to upgrade biogas (composed mainly of 40–70% CH_4 and 30–60% CO_2) to reach appropriate values for the NGN injection (Ammenberg et al. 2021).

The BU technologies have been developed in natural gas industry, where removing impurities are essential to meet the legal and technical quality requirements. Recent and less developed technologies are based on biological upgrading through the application of hydrogenotrophic microorganisms to biomethanation process, i.e. conversion of CO_2 and H_2S into biomethane. Special mention should be made to electrolysis needed to remove H_2S. This process is based on separate one molecule of water to produce H_2. Electrolysis is a high-energy consumer, and current biomethanation techniques are focused on using variable renewable energy to supply this process, allowing to reduce the energy requirements and costs of biomethanation (Götz et al. 2016). Table 25.1 shows the main BU technologies distinguishing between biogas cleaning (removing impurities avoiding corrosion risk) and BU (specific elimination or capture of CO_2).

After analysing the available BU technologies, biological techniques have numerous advantages over physical/chemical technologies. The reason is that physical and chemical techniques produce CO_2 flow that could be treated or released or the atmosphere contributing to global warming. However, biological BU does not

Table 25.1 BU technologies available considering the difference between biogas cleaning and BU

Biogas cleaning	*H_2S elimination* H_2S is a natural compound found in biogas in small quantities produced from the microbial activity of sulphur reducing bacteria due to the presence of proteins and other sulphur compounds. There are three known ways of eliminating H_2S: biological processes, absorption through specific solvents, and adsorption with iron salts (Vuppaladadiyam et al. 2024; Nguyen and Khanal 2018)
	Water elimination Biogas leaves the anaerobic digester saturated with water that usually makes up about a 5% of biogas volume for mesophilic conditions (Angelidaki et al. 2019). Removing water from biogas can be achieved by decreasing temperature to force its condensation, using chemical absorption with zeolites, alumina or activated carbon, and, finally using diethylene glycol (Singhal et al. 2017)
	Siloxanes Siloxanes are organic compounds that contain silicon in their molecular structure, which can be linear or cyclical (Atelge et al. 2021). Their apparition in biogas is caused by their use in cosmetics, shampoos, detergents, and industrial paintings to serve as lubricating agents. Their removal from biogas prior to its combustion is necessary as they decompose at high temperatures in sand, which is abrasive and harmful for turbines and engines. Siloxanes can be eliminated via adsorption with silica gel or activated carbon after cooling down biogas to −30 °C. Chemical absorption with H_2SO_4 is the common way of eliminating these compounds from biogas
	Ammonia, O_2 and N_2 Ammonia is formed during the hydrolysis of proteins in the anaerobic process. O_2 and N_2 should be absent on the stream of biogas or at least in trace concentrations; otherwise, it indicates a possible introduction of air. Ammonia can be easily eliminated during BU if a water scrubber is used or during biogas drying. O_2 and N_2 are simultaneously eliminated through adsorption with activated carbon (Sun et al. 2015)
BU	*Water scrubbing* This process uses the difference solubility between CO_2 and CH_4 in water (Singhal et al. 2017). This technology is widely implemented in industry due to its ability to treat high volumetric flow rates of biogas, since 2000 Nm^3/h (Ahmed et al. 2021), and its use of water, a non-hazardous, and generally available solvent. The operation of a water scrubber typically involves introducing compressed biogas into scrubber (6–10 bar) at the lower section, while water is pumped into the upper section. As a result, a countercurrent flow is created, and CO_2 can be removed (Baena-Moreno et al. 2019)
	Organic physical scrubbing This technology uses an organic solvent instead of water. This solvent does not react with any component from biogas that is absorbed but they rather dissolve in them, belonging this technology to the physical category. The main reason to use an organic solvent to replace water is due to the higher selectivity of some of them to retain CO_2 and H_2S while keeping CH_4 undissolved. Most widely used solvent are ethers of polyethylenglycol, which is non-toxic and has a low vapour pressure, mixtures of methanol, and dimethyl ether as well as N-methyl-pyrrolidone. The operation in this case requires lower quantities of solvent compared to water, as they can solubilize more CO_2 resulting in a reduction in size of necessary equipment as well as pumping necessities (Khan et al. 2021)

(continued)

Table 25.1 (continued)

	Chemical scrubbing Using chemical solvents to absorb CO_2 from biogas has addition benefits since yielding chemical products that decompose with heat, which permits solvent regeneration. This technology is especially efficient when CO_2 concentrations are low (Ahmed et al. 2021). Most typically used solvents are amines, among which monoethanolamine, diethanolamine, triethanolamine, and methyl diethanolamine (MDEA) are the most common. The operational fundamentals of chemical scrubber are the same as described previously. A chemical scrubbing unit is composed of an absorber where biogas and the amine solution are introduced to form a counter current flow (Barbera et al. 2019)
	Pressure swing adsorption (PSA) Using the ability of some molecules of adsorbing onto the surface of porous materials through Van de Waal forces under high pressure. The chemical structure of the surface adsorbent and its molecular sieve play a crucial role to selectively capture impurities. Most typical materials employed as adsorbents possess high superficial area to maximize contact with the gas phase, namely activated carbon, zeolites, and metal organic framework (MOF). For biogas, multiple undesired impurities can be removed by the adsorbent: CO_2, O_2, N_2, and H_2S, the last can irreversibly bound to the adsorbent so it is essential to remove H2S in early stages of biogas treatment. CH_4 is not adsorbed as its size prevents its introduction into the porous material, which ultimately generates a 98% pure biomethane (Abd et al. 2021; Canevesi and Grande 2023)
	Cryogenic upgrading Biogas is composed of a range of different gases each of them possessing a determined boiling temperature at a given pressure, i.e. CO_2 sublimates at a temperature of -78°C, and CH_4 condenses at -161°C both at atmospheric pressure. To avoid freezing problems, it is common that other impurities beside CO_2 are eliminated previously to cryogenic upgrading. Generally, biogas is firstly cleansed from all impurities and compressed in a multistage compression unit, pre-cooled, and later distillate, obtaining methane with purities going up to 98% (Naquash et al. 2022; Yousef et al. 2019)
	Membranes Membranes have gained an increasing interest over the last years due to their low operational costs, high selectivity, and new materials which overall enhance their performance and their simple operation. Nevertheless, high investment costs and the need of multiple stages to obtain pure CH_4 are the main disadvantages of this technology. The underlying principle of separation consists in the diffusion of the CO_2 through the membrane, which is governed by Fick's Law, to obtain a permeate with high presence of CO_2 and a retentate of mostly CH_4. According to the purity of CH_4 desired, membranes can be implemented in a single stage or multistage system. The difference between them lies on the use of just one membrane module or multiple modules installed in series. The most common configuration consists in the installation of two sequential modules the second of which treats the retentate produced by the first to further increase final biomethane purity—close to 99.5% (Tomczak et al. 2024)

Source Own elaboration

separate CO_2 from biogas but rather transform it via biological reduction into CH_4 by using H_2 flow. This process—called electrolysis—solves the problem of generating pure streams of CO_2 and provides larger flows of biomethane (Lemmer et al. 2015; Valverde et al. 2016). Electrolysis is a high-energy consumer and can act as a barrier to promote biomethane production. However, thanks to the P2G process, the use of the power excess from intermittent renewable sources can be used to supply the electrolysis and make the biomethane process possible with less energy consumption. P2G is one of the main advantages of biomethane production, connecting the electricity system and NGN (David et al. 2019; Kalinci et al. 2015).

25.3 Identifying the Importance of Biomethane to Economy, Environment, and Well-Being

Current situation of water resources, the high energy consumption of the urban water cycle facilities, and the wastewater by-products management are some of the great challenges for administrations and operators. Any technological change aimed at improving water quality, purification efficiency, or obtaining usable by-products (such as biomethane) will require a high initial economic investment. However, each case needs to be analysed in depth through the identification of economic, environmental, and social aspects that can justify the technological investments to re-evaluate the sludge from the WWTPs, for instance.

Biomethane production is framed in two large issues, all of which are important within the current climate context: (i) the management of sewage sludge and circularity in WWTPs, and (ii) production of green energy and the transition towards sustainable energy sources (Idiano D'Adamo and Sassanelli 2022). Usually, the main destination of the sludge was landfill or direct use as fertilizer in agriculture. Both destinations have a strong environmental impact derived from highly polluting leachates to surface and groundwater bodies. AD makes it possible to stabilize its organic matter content while generating biogas, which also allows to produce energy. Therefore, upgrading biogas to biomethane allows this process to continue, but with greater efficiency. This situation reinforces the circularity of the WWTPs by-products and mitigates their environmental impacts.

Biomethane is a green energy that is more efficient than biogas, creating business opportunities within the energy system. The main advantage of biomethane is its similarity to natural gas, making it possible to introduce it into NGN and reaching more end users. This energy model reduces energy dependence, encourages local/national production of biomethane through a circularity scheme that revalue the sludge from the WWTPs, and improves the energy and environmental efficiency of the electrical network (Nguyen et al. 2021). For all these reasons, projects focused on BU need analysis tools for the associated economic, environmental, and social benefits to reinforce their implementation and identify the barriers according to the current sociopolitical context.

Therefore, **what are the aspects that must be considered regarding the influence of biomethane into economy, environment, and population well-being?** The identification of all these implications of biomethane is an essential point to achieve its positioning in the current energy market. At first it may seem that the economic effort necessary to produce biomethane is great; however, special importance must be given to the benefits that will be obtained over the time, such as energy independence and the sale of by-products obtained. These implications must be accompanied by a stable regulatory framework that provides security in the technological change, allowing to establish the bases for new energy production. For that purpose, this section presents the essential aspects of each implication with the aim to establish the foundations for the monetary valuation assessment.

25.3.1 Economic Implications

The economic dimension of biomethane production is a decisive factor to promote its use, since production costs must be lower than the production costs of the different alternatives, such as fossil fuels. The literature highlights that the general cost of biomethane production—for a plant able to generates between 200 and 300 m^3/h—is ranged 0.74–0.97 €/m^3 (Cucchiella et al. 2018), where 0.54 €/m^3 is the cost of injecting biomethane into NGN (Rotunno et al. 2017). These costs are influenced by different factors detailed below:

(a) Distance between raw material and AD plant since processing biowastes is not the same as processing sludge from the WWTP. Considering sewage sludge, these costs are reduced since the WWTPs capable of producing biomethane are WWTPs that have AD facilities to generated biogas.
(b) CAPEX and OPEX of the technologies to remove CO_2 and H_2S from biogas, since the different technical configurations have a different impact on the costs. Despite this, the literature establishes that, in general, the costs of BU represent between 10 and 20% of the global costs of AD (Iglesias et al. 2021).
(c) Technical implications of delivering biomethane to the end user through NGN, as well as its storage for later use. Although its storage can impact on costs, it must be considering that this fact is an advantage, since being able to store the biomethane generated in WWTPs is a technological advance (Angelidaki et al. 2018). Focusing attention on the injection into the NGN and the costs associated with it, the distance between the AD and the injection point must be considered as a determining factor. This is due to the costs of the technical equipment (pipes, sensors, among others) and the costs of the connection itself (IEA 2020). Generally, the injection costs are around 15 and 30 €/MWh depending on the BU technology (MITECO 2021). For this reason, detailed studies about the identification of barriers and strengths of biomethane injection into NGN are necessary. Allowing to reach many more users thanks to an existing infrastructure.

(d) Volume of sewage sludge available for AD and the influence of scale economies on production costs (Hengeveld et al. 2014). At this point, the biomethane production capacity of the largest WWTPs, as well as the possibility of centralizing (or not) the biomethane production on a smaller scale need to be analysed. This analysis must consider the sludge transport costs and the potential to deliver biomethane to remote territories (González et al. 2023; Hosseini et al. 2024; Schröer et al. 2024). Hence, the starting point for this analysis is the minimum biomethane production considered technically and economically feasible: 100 m^3/h (Cucchiella et al. 2018). Currently, there is no widespread biomethane production in WWTPs, being the largest WWTPs the responsible of biomethane. According to scale economies, these facilities can produce RG at a lower cost. However, once biomethane production is generalized throughout the territory, lower production cost can be achieved. Specifically, around a 25% reduction is achieved by 2040 (IEA 2020).

The study of these direct economic implications of biomethane production needs to be complemented with the identification of the global economic aspects related to biomethane as an energy source. Among these aspects, the energy independence that allows the generation of biomethane at the national level and the reduction in the production costs of using sewage sludge WWTPs are highlighted. This is because it is not necessary to build a biomethane generation plant, but rather the pertinent technological changes should be made in the WWTPs, and the AD should be adapted to the new conditions.

Considering that the European Union imports almost 60% of its fuels from other countries (Alberici et al. 2023), the possibility of generating biomethane at the national level would provide energy security during times of political changes. At the same time, it allows European countries to be resilient to the consequences of climate change on international energy sources. Another economic aspect to highlight regarding biomethane as an energy source is the sale of by-products generated in BU. The main by-product with a great economic impact is the digestate. This stream of stable organic matter can be used as fertilizer on agricultural farms, reducing production costs by avoiding the use of chemical fertilizers. Using digestate as fertilizer not only has a positive impact on the farmer, but also reduces the carbon footprint generated by the production and transportation of the chemical fertilizers traditionally used.

25.3.2 Environmental Implications

Biomethane has a clear positive environmental impact because it involves the revaluation of sewage sludge, which is usually stabilized and taken into a landfill. For this reason, the identification of environmental benefits of biomethane production are essential, acting as a tool to achieve the decarbonization of wastewater treatment sector. The future of energy sector depends on the production of RG. In this case, the possibility of using the sludge from the WWTP as raw material changes the energy scheme, allowing circular economy schemes reinforcing water-energy nexus:

(a) Reduction in CO_2 emissions associated with two main areas: all those activities that replace fossil fuels with biomethane, as well as the use of digestate as natural fertilizer in agriculture, can reduce the environmental impacts related to manufacture and use of chemical fertilizers. Specifically, the International Energy Agency report (2020) quantifies that by 2040 there will be around 1 billion tons of CO_2 avoided by producing and using biomethane at a global scale. Considering the current price per ton of CO_2 in Europe in July 2024 (67.86 €), the associated environmental benefit can be estimated around 68 billion €.
(b) Releasing crops since biomethane does not require processing of vegetables, as occurs with vegetal biodiesel. This reduction in demand resulting from the possibility of producing biomethane at the national level allows these lands to be used for food production (Hajjari et al. 2017; Zhang and Bai 2021). At the same time, environmental impacts related to the biodiesel transportation from producing countries (USA and Brazil) to consumer countries (EU) are avoided (Ardolino et al. 2018), which reinforces national energy independence thanks to biomethane production.
(c) Soil quality improvement using digestate as an organic matter source, recovering the soil structure and its water and nutrient retention capacity.
(d) Reduction in the use of non-renewable materials. Biomethane is generated in an anaerobic reactor and is subsequently used, stored, or injected into the NGN. However, other renewable technologies, such as solar or wind, require non-renewable materials for their construction (metals and plastics). At the end of their useful lives, these materials must be correctly managed, generating wastes despite are using wind and sunlight to generate energy (IEA 2020).
(e) Use of energy peaks from renewable energies (P2G). Renewable energy peaks cannot be introduced into the electrical grid and/or cannot be properly stored (Shapovalov et al. 2022). Hence, using these energy peaks to biomethane production—specifically for electrolysis—allows reducing production costs, ensuring technical and economic feasibility of BU and strengthening circular energetic production scheme (Cuff et al. 2020; Sun et al. 2022). This opportunity, as well as biomethane can be stored in the NGN for months, improves the efficiency of energy use and encourages the use of biomethane (Angelidaki et al. 2018). Studies show that intermittency in energy peaks does not negatively affect the biomethanation process, since the final composition of the biomethane and its characteristics are constant (Iglesias et al. 2021). For this reason, different technical combinations for electrolysis have been verified in the literature (Barbaresi et al. 2022; Wulf et al. 2020).

Quantifying these environmental benefits in monetary units is a fundamental part of the feasibility study of biomethane, since it allows to complement the technical and economic aspects related to BU. Environmental benefits provide robustness to feasibility studies as it addresses the quantification of the positive environmental impact that has been generated. The result is a complete analysis of the opportunities that biomethane has in order to address current environmental challenges. At the same time, the monetary quantification of environmental benefits allows to analyse two scenarios. On the one hand, there is the situation in which biomethane is successfully produced, and the environmental benefits mentioned above are achieved. While, on the other hand, there is the scenario where biomethane is not produced, consequently, the environmental impacts of the initial situation would continue. Those environmental benefits that are not achieved are part of the non-action costs related to biomethane: CO_2 emissions, loss of energy peaks from renewable energies, energy dependence and waste of the energy potential of sewage sludge and digestates, among others.

25.3.3 Social Implications

Current socioeconomic context forces to look for new solutions whose greater efficiency and lower environmental impact allow the development of society's productive activities while reducing the environmental consequences of growth. Considering the social implications of the production and use of biomethane are an important point when evaluating the feasibility of projects, since the significance that this RG has on society and its activities is analysed. Accessing to a fuel with less environmental impact and whose raw material is sewage sludge allows the establishment of circular economy in the territories, improving the well-being of the population and revaluing a by-product of purification in a more efficient way. Starting from the basis that all the implications discussed here are direct and indirectly related, the social aspects of biomethane are relevant when the importance of promoting sustainable energy as a social vector of territories is explored. Specifically, the following social implications are identified as follows:

(a) Biomethane is a sustainable energy whose use requires technological innovation to obtain the best results. This allows to attract knowledge, research, and financing opportunities to adapt technologies to the requirements of territories, exporting know-how and improving energy systems globally.
(b) Using biomethane as new renewable energy source, new jobs opportunities can be created. These new jobs are related by monitoring both the NGN injection and capability for the rural areas implementation. Jobs directly related to the biomethane production are based on the operation and maintenance of the WWTPs, as well as the design, construction, operation, and maintenance of biomethane production plants that use other biowastes, such as slurry and food remains. Indirect jobs created will arise from the execution of the previous ones, such as the manufacture of pipes and pumps, process controllers, the transport of biomethane if it is bottled, and the maintenance of productive activities that need biomethane, among others (EBA 2023).

(c) Biomethane is an energy source with good social acceptance, as it allows progress in energy transition processes thanks to easy access through NGN. Specifically, the study by Fubara et al. (2018) establishes that, with the current technological level, a substitution of natural gas with biomethane of between 50 and70% could be achieved for domestic energy supply. This substitution means a reduction in CO_2 emissions of between 65 and80%. This evidence makes biomethane a reliable and sustainable energy that allows to continue with the current productive activities at different levels, ensuring energy independence and associated socioeconomic improvements (Axon and Darton 2024).
(d) Benefits for human health by reducing CO_2 emissions, as well as the reduction in noise pollution since vehicles that use biomethane have a lower noise index. As a result, living conditions in large urban centres have been improved. Current technological level allows biomethane vehicles to be competitive when a constant flow of biomethane for refuelling is ensured. For this reason, changing the urban vehicle fleet is the first step to achieve the energy transition in urban areas (Patterson et al. 2011; Prussi et al. 2021). At the same time, reducing production costs of biomethane production by sewage sludge in WWTPs would provide the necessary security to begin using biomethane as an urban fuel.

25.4 Conclusions

Using biogas as new energy source for decarbonization of the economy, despite being a highly useful RG for the ecological transition, has a lower calorific value due to the presence of CO_2 and H_2S. Biomethane, which has approximately 90% CH_4 and does not contain CO_2 or H_2S, is a solution to this problem. Biomethane is a RG with a composition similar to natural gas that can be used by a wide variety of users, reducing the use of fossil fuels and generating environmental benefits not only at a national level, but also at a cross-border level. At the same time, positive social and economic synergies are generated, ensuring the production of biomethane at a lower cost and the utilization all the renewable energy produced by P2G.

This chapter has presented the technical foundations of biomethane, as well as the different technologies currently available for its production. Considering that the aim of biomethane production is to decarbonize both the energy sector and the productive sector, choosing a technology for BU must consider its influence on the ecological footprint of WWTPs. This chapter highlights the importance of biological technologies since they remove CO_2 by the addition of H_2 in the digester. This H_2 is obtained from the electrolysis of water, a highly energy-consuming process that uses the energy peaks of renewable energies as an energy source (P2G). This aspect gives more importance to biomethane as a RG, under a circular economy point of view, since the energy surpluses of conventional renewables energies can be used as energy source to biomethane production.

These advantages are closely related to the chapter aim, which is the identification of the economic, social, and environmental implications of biomethane. Technological change (including the adaptation for NGN injection) has an initial investment cost that must be considered in the research projects and pilot studies. Considering this, the monetary quantification of the importance of biomethane to economy, environment and social well-being can justify the initial economic barriers to implement it in the territory. The benefits identified in this chapter are the starting point for a detailed quantification of the monetary impact that the implementation of biomethane will have as a substitute of fossil fuels, as well as for promoting the revaluation of sewage sludge according to both the sustainability and the circular management scheme of the water cycle.

Acknowledgements This work is funded by the Ministry of Science and Innovation (Government of Spain) and the State Agency of Investigation (PID2020-112871RB-C21 and PID2020-112871RB-C22). Furthermore, this work was supported by the Regional Government of Castilla y León and the EU-FEDER (CL-EI- 2021-07, UIC 334 and 338).

References

Abd AA, Othman MR, Naji SZ, Hashim AS (2021) Methane enrichment in biogas mixture using pressure swing adsorption: process fundamental and design parameters. Mater Today Sustain 11–12:100063. https://doi.org/10.1016/j.mtsust.2021.100063

Ahmed SF, Mofijur M, Tarannum K, Chowdhury AT, Rafa N, Nuzhat S, Kumar PS, Vo DN, Lichtfouse E, Mahlia TMI (2021) Biogas upgrading, economy and utilization: a review. Environ Chem Lett 19:4137–4164. https://doi.org/10.1007/s10311-021-01292-x

Alberici S, Gräf D, Aurand M (2023) Beyond energy: monetising biomethane's whole-system benefits (European Biogas Association)

Alvarado VI, Hsu S, Wu Z, Lam C, Leng L, Zhuang H, Lee P (2019) A Standardized stoichiometric life-cycle inventory for enhanced specificity in environmental assessment of sewage treatment. Environ Sci Technol 53:5111–5123. https://doi.org/10.1021/acs.est.9b01409

Ammenberg J, Gustafsson M, O'Shea R, Gray N, Lyng K, Eklund M, Murphy JD (2021) Perspectives on biomethane as a transport fuel within a circular economy, energy, and environmental system. IEA Bioenergy Task 37

Angelidaki I, Treu L, Tsapekos P, Luo G, Campanaro S, Wenzel H, Kougias PG (2018) Biogas upgrading and utilization: current status and perspectives. Biotechnol Adv 36:452–466. https://doi.org/10.1016/j.biotechadv.2018.01.011

Angelidaki I, Xie L, Luo G, Zhang Y, Oechsner H, Lemmer A, Munoz R, Kougias PG (2019) Chapter 33—biogas upgrading: current and emerging technologies. In: Pandey A, Larroche C, Dussap C, Gnansounou E, Khanal SK, Ricke S (eds) Biofuels: alternative feedstocks and conversion processes for the production of liquid and gaseous biofuels (Second Edition). Academic Press, pp 817–843

Ardolino F, Parrillo F, Arena U (2018) Biowaste-to-biomethane or biowaste-to-energy? An LCA study on anaerobic digestion of organic waste. J Clean Prod 174:462–476. https://doi.org/10.1016/j.jclepro.2017.10.320

Atelge MR, Senol H, Djaafri M, Hansu TA, Krisa D, Atabani A, Eskicioglu C, Muratçobanoğlu H, Unalan S, Kalloum S, Azbar N, Kıvrak HD (2021) A critical overview of the state-of-the-art methods for biogas purification and utilization processes. Sustainability 13. https://doi.org/10.3390/su132011515

Axon CJ, Darton RC (2024) A systematic evaluation of risk in bioenergy supply chains. Sustainable Production and Consumption 47:128–144. https://doi.org/10.1016/j.spc.2024.03.028

Baena-Moreno FM, Rodríguez-Galán M, Vega F, Vilches LF, Navarrete B (2019) Review: recent advances in biogas purifying technologies. Int J Green Energy 16:401–412. https://doi.org/10.1080/15435075.2019.1572610

Barbaresi A, Morini M, Gambarotta A (2022) Review on the status of the research on power-to-gas experimental activities. Energies 15. https://doi.org/10.3390/en15165942

Barbera E, Menegon S, Banzato D, D'Alpaos C, Bertucco A (2019) From biogas to biomethane: a process simulation-based techno-economic comparison of different upgrading technologies in the Italian context. Renewable Energy 135:663–673. https://doi.org/10.1016/j.renene.2018.12.052

Canevesi R, Grande CA (2023) Biogas upgrading by pressure swing adsorption using zeolite 4A. Effect of Purge on Process Performance, Separation and Purification Technology 309:123015. https://doi.org/10.1016/j.seppur.2022.123015

Cucchiella F, D'Adamo I, Gastaldi M, Miliacca M (2018) A profitability analysis of small-scale plants for biomethane injection into the gas grid. J Clean Prod 184:179–187. https://doi.org/10.1016/j.jclepro.2018.02.243

Cuff G, Nelting K, Trautmann N, Mohammad-pajooh E (2020) Production and upgrading of biogas through controlled hydrogen injection for renewable energy storage. Bioresource Technol Rep 9:100373. https://doi.org/10.1016/j.biteb.2019.100373

David M, Ocampo-Martínez C, Sánchez-Peña R (2019) Advances in alkaline water electrolyzers: a review. J Energy Storage 23:392–403. https://doi.org/10.1016/j.est.2019.03.001

EBA (2023) European Biogas Association—Statistical Report 2023

EBA (2024) Biogas: Simply the best, 24

EPA (2024) How Does Anaerobic Digestion Work? 2024

Fubara T, Cecelja F, Yang A (2018) Techno-economic assessment of natural gas displacement potential of biomethane: a case study on domestic energy supply in the UK. Chem Eng Res Design 131:193–213. https://doi.org/10.1016/j.cherd.2017.12.022

González R, García-Cascallana J, Gómez X (2023) Energetic valorization of biogas, a comparison between centralized and decentralized approach. Renewable Energy 215:119013. https://doi.org/10.1016/j.renene.2023.119013

Götz M, Lefebvre J, Mörs F, McDaniel Koch A, Graf F, Bajohr S, Reimert R, Kolb T (2016) Renewable power-to-gas: a technological and economic review. Renewable Energy 85:1371–1390. https://doi.org/10.1016/j.renene.2015.07.066

Hajjari M, Tabatabaei M, Aghbashlo M, Ghanavati H (2017) A review on the prospects of sustainable biodiesel production: a global scenario with an emphasis on waste-oil biodiesel utilization. Renew Sustain Energy Rev 72:445–464. https://doi.org/10.1016/j.rser.2017.01.034

Hengeveld EJ, van Gemert WJT, Bekkering J, Broekhuis AA (2014) When does decentralized production of biogas and centralized upgrading and injection into the natural gas grid make sense? Biomass Bioenergy 67:363–371. https://doi.org/10.1016/j.biombioe.2014.05.017

Hosseini T, Culley SA, Zecchin A, Maier HR, Ashman PJ (2024) Impact of co-digestion and degree of centralization on the yield and viability of biomethane production: a case study in regional Australia. Energy Conversion Management: X 22:100585. https://doi.org/10.1016/j.ecmx.2024.100585

D'Adamo I, Sassanelli C (2022) A mini-review of biomethane valorization: managerial and policy implications for a circular resource. Waste Manage Res 40:1745–1756. https://doi.org/10.1177/0734242X221102249

IEA (2017) Water-Energy Nexus. World Energy Outlook Special Report

IEA (2020) Outlook for biogas and biomethane: Prospects for organic growth

Iglesias R, Muñoz R, Polanco M, Díaz I, Susmozas A, Moreno AD, Guirado M, Carreras N, Ballesteros M, (2021) Biogas from anaerobic digestion as an energy vector: current upgrading development. Energies 14. https://doi.org/10.3390/en14102742

Kalinci Y, Hepbasli A, Dincer I (2015) Techno-economic analysis of a stand-alone hybrid renewable energy system with hydrogen production and storage options. Int J Hydrogen Energy 40:7652–7664. https://doi.org/10.1016/j.ijhydene.2014.10.147

Khan MU, Lee JTE, Bashir MA, Dissanayake PD, Ok YS, Tong YW, Shariati MA, Wu S, Ahring BK (2021) Current status of biogas upgrading for direct biomethane use: a review. Renew Sustain Energy Rev 149:111343. https://doi.org/10.1016/j.rser.2021.111343

Lemmer A, Chen Y, Lindner J, Wonneberger AM, Zielonka S, Oechsner H, Jungbluth T (2015) Influence of different substrates on the performance of a two-stage high pressure anaerobic digestion system. Bioresour Technol 178:313–318. https://doi.org/10.1016/j.biortech.2014.09.118

Liebetrau J, Kornatz P, Baier U, Wall D, Murphy JD (2020) Integration of biogas systems into the energy system: technical aspects of flexible plant operation. IEA Bioenergy Task 37

Marcogaz (2018) Quality of biomethane required in European countries for injecting into natural gas grid

MITECO, Spanish Ministry for the Ecological Transition (2018) Resolution of October 8, 2018, of the general directorate of energy policy and mines, which modifies the technical management standards of the NGTS-06, NGTS-07 system and the detailed protocols PD-01 and PD-02. [In Spanish]

MITECO, Ministry of Ecological Transition and Demographic Challenge (2021) Plan Nacional de Depuración, Saneamiento, Eficiencia, Ahorro y Reutilización (DSEAR) [In Spanish]

Modiri M, Hasan AH, koloukhi HZ, Rostami F, Tafazzoli SM, Avami A (2023) Assessment of water-energy-emissions nexus in wastewater treatment plants using emergy analysis. Environ Dev Sustain 25:11905–11929. https://doi.org/10.1007/s10668-022-02559-1

Naquash A, Qyyum MA, Haider J, Bokhari A, Lim H, Lee M (2022) State-of-the-art assessment of cryogenic technologies for biogas upgrading: energy, economic, and environmental perspectives. Renew Sustain Energy Rev 154:111826. https://doi.org/10.1016/j.rser.2021.111826

Nguyen D, Khanal SK (2018) A little breath of fresh air into an anaerobic system: How microaeration facilitates anaerobic digestion process. Biotechnol Adv 36:1971–1983. https://doi.org/10.1016/j.biotechadv.2018.08.007

Nguyen LN, Kumar J, Vu MT, Mohammed JAH, Pathak N, Commault AS, Sutherland D, Zdarta J, Tyagi VK, Nghiem LD (2021) Biomethane production from anaerobic co-digestion at wastewater treatment plants: a critical review on development and innovations in biogas upgrading techniques. Sci Total Environ 765:142753. https://doi.org/10.1016/j.scitotenv.2020.142753

Patterson T, Esteves S, Dinsdale R, Guwy A (2011) An evaluation of the policy and techno-economic factors affecting the potential for biogas upgrading for transport fuel use in the UK. Energy Policy 39:1806–1816. https://doi.org/10.1016/j.enpol.2011.01.017

Prussi M, Julea A, Lonza L, Thiel C (2021) Biomethane as alternative fuel for the EU road sector: analysis of existing and planned infrastructure. Energ Strat Rev 33:100612. https://doi.org/10.1016/j.esr.2020.100612

Rosha P, Rosha AK, Ibrahim H, Kumar S (2021) Recent advances in biogas upgrading to value added products: a review. Int J Hydrogen Energy 46:21318–21337. https://doi.org/10.1016/j.ijhydene.2021.03.246

Rotunno P, Lanzini A, Leone P (2017) Energy and economic analysis of a water scrubbing based biogas upgrading process for biomethane injection into the gas grid or use as transportation fuel. Renewable Energy 102:417–432. https://doi.org/10.1016/j.renene.2016.10.062

Schröer D, Herlicka L, Heinold A, Latacz-Lohmann U, Meisel F (2024) A network design problem for upgrading decentrally produced biogas into biomethane. J Clean Prod 452:142049. https://doi.org/10.1016/j.jclepro.2024.142049

Shapovalov YB, Usenko SA, Salyuk AI, Tarasenko RA, Shapovalov VB (2022) Sustainability of biogas production: using of Shelford's law. IOP Conf Series: Earth Environ Sci 1049:012023. https://doi.org/10.1088/1755-1315/1049/1/012023

Singhal S, Agarwal S, Arora S, Sharma P, Singhal N (2017) Upgrading techniques for transformation of biogas to bio-CNG: a review. Int J Energy Res 41:1657–1669. https://doi.org/10.1002/er.3719

Sun Q, Li H, Yan J, Liu L, Yu Z, Yu X (2015) Selection of appropriate biogas upgrading technology-a review of biogas cleaning, upgrading and utilisation. Renew Sustain Energy Rev 51:521–532. https://doi.org/10.1016/j.rser.2015.06.029

Sun Z, Zhao L, Wu K, Wang Z, Wu J, Chen C, Yang S, Wang A, Ren N (2022) Overview of recent progress in exogenous hydrogen supply biogas upgrading and future perspective. Sci Total Environ 848:157824. https://doi.org/10.1016/j.scitotenv.2022.157824

Tomczak W, Gryta M, Daniluk M, Żak S (2024) Biogas upgrading using a single-membrane system: a review. Membranes 14. https://doi.org/10.3390/membranes14040080

Valverde L, Pino FJ, Guerra J, Rosa F (2016) Definition, analysis and experimental investigation of operation modes in hydrogen-renewable-based power plants incorporating hybrid energy storage. Energy Convers Manage 113:290–311. https://doi.org/10.1016/j.enconman.2016.01.036

Vuppaladadiyam AK, Jena MK, Hakeem IG, Patel S, Veluswamy G, Thulasiraman AV, Surapaneni A, Shah K (2024) A critical review of biochar versus hydrochar and their application for H2S removal from biogas. Revi Environ Sci Biotechnol 23:699–737. https://doi.org/10.1007/s11157-024-09700-8

Wulf C, Zapp P, Schreiber A (2020) Review of power-to-X demonstration projects in Europe. Front Energy Res 8

Yousef AM, El-Maghlany WM, Eldrainy YA, Attia A (2019) Upgrading biogas to biomethane and liquid CO_2: a novel cryogenic process. Fuel 251:611–628. https://doi.org/10.1016/j.fuel.2019.03.127

Zafar S (2024) An introduction to biomethane

Zhang L, Bai W (2021) Sustainability of crop–based biodiesel for transportation in China: barrier analysis and life cycle ecological footprint calculations. Technol Forecast Soc Chang 164:120526. https://doi.org/10.1016/j.techfore.2020.120526

Zhuang H, Guan J, Leu S, Wang Y, Wang H (2020) Carbon footprint analysis of chemical enhanced primary treatment and sludge incineration for sewage treatment in Hong Kong. J Clean Prod 272:122630. https://doi.org/10.1016/j.jclepro.2020.122630

Chapter 26
Recovery Methods for Treated Water

Wafa Hassen, Bilel Hassen, and Abdennaceur Hassen

Abstract Climate change characterized by recurring periods of drought, combined with the evolution of our water practices and population growth, strongly urges us to rethink our water management models to ensure sustainable and rational access to water for all. This is why the recovery of treated wastewater (RTWW) and its reuse represents a circular economy process that provides an excellent alternative to meet these new challenges sustainably. The practice of treated wastewater reuse (TWWR), also known as water reuse, offers many benefits at various levels. Three main benefits of treated wastewater can be distinguished: (i) Environmental benefits are reflected in the presence of beneficial substances in this treated wastewater, which helps limit the use of pesticides and other toxic products in nature, particularly in agricultural crops. This also contributes to maintaining a satisfactory level of groundwater, thus avoiding the deterioration of underground water sources. Finally, these waters can be discharged into coastal areas without risk to the coastal marine environment. (ii) The societal benefits are linked to the application of circular economy principles through the reuse of treated wastewater. This helps combat waste and creates new resources without relying on the extraction of drinking water. Additionally, it contributes to reducing the global water stress index, which is estimated to affect over 40% of the world's population. (iii) Economic benefits are ensured by the reuse of treated wastewater, allowing industries and agriculture, among others, to reduce their water consumption by 40–90%. This is significant savings for businesses and farmers. Nations disadvantaged by water scarcity could find a sustainable solution for water management in the reuse of treated wastewater. The recovery and recycling of treated

W. Hassen · A. Hassen (✉)
Laboratory of Treatment and Wastewater Valorization, Center of Research and Water Technologies (CERTE), Soliman, Tunisia
e-mail: abdohas@gmail.com

W. Hassen
e-mail: hassen.wafa@gmail.com

B. Hassen
Laboratories of Microbiology and Pathology of Aquatic Organisms, National Institute of Navy Sciences and Technologies (INSTM), Salammbô, Tunisia
e-mail: bilel.hassen@fst.utm.tn

S. Souabi et al. (eds.), *Wastewater Treatment Plants*, Water Science and Technology Library 130, https://doi.org/10.1007/978-3-031-87461-1_26

wastewater align with circular economy principles by avoiding unnecessary waste and enabling the valorization of resources for other purposes.

Keywords Wastewater · Recovery · Sustainable management · Water reuse · Environmental · Societal · Economic benefits

26.1 Introduction

Climate change is often marked by periods of recurring drought and/or episodes of flooding on a global scale, and both drought periods and flooding episodes impact all regions of the globe.

As for flooding phenomenon, the rise in temperatures leads to an increase in the risk of precipitation: the warmer the air, the more water vapor it contains which could potentially turn into intense showers. We are already seeing in this area, especially tropical, an intensification of rainfall during certain periods of the year, particularly in the autumn. This phenomenon exacerbates an already high-risk of flooding across all territories of the globe. Flooding encompasses several scenarios such as the overflowing of a river, either slowly or rapidly, the runoff of rainwater unable to infiltrate the soil anymore, bringing significant volumes of water over short durations and pouring into usually dry areas, the rise of groundwater to the surface of the soil, and especially the phenomenon of marine submersion after rapid and short floods lasting from a few hours to a few days along the coasts by the sea during unfavorable weather and oceanic conditions.

It is now well established that in a scenario of climate warming of + 1.5 °C, decadal precipitation, which occurs once every ten years, will occur 1.5 times more often. In a scenario of + 4 °C, the probability of this precipitation is 2.7 times higher. With high confidence, it is estimated that the intensity of such extreme precipitation events would increase by 7% for each degree of temperature rise.

Due to the strong link between extreme precipitation and urban runoff flooding, it can be concluded that urban runoff flooding will increase in frequency and intensity. This result is considered with a lot of confidence in mainland France, for example. It is important to act as floods can cause significant human and material damage, including loss of human life, population displacement, negative impacts on physical and mental health, degradation of property and electrical, telecommunication networks, disruption of public transportation services, waste collection and management, hospitals, etc. It is necessary to expect and prevent this risk.

As for drought phenomenon in parallel, drought is a lasting episode of water shortage leading to various impacts, particularly on the soil, vegetation, and aquifers as underground water reservoirs (NOAA 2022). There are three main types of drought: Meteorological drought, caused by a lack of rain when the amount of water is significantly lower than seasonal norms over an extended period. Soil drought also known as agricultural drought, caused by a lack of water in the soil, which hinders vegetation development; contrary to its name, it affects not only agricultural soil but all soils and plants (Abegunrin et al. 2016). Hydrological drought, occurring when water reserves in aquifers, rivers, and lakes fall below average. Drought can result from a lack of rain, but not only that: It can also result from high evaporation due to high temperatures. Drought can also be exacerbated by the intensive or inappropriate use of available water and poor soil management.

26.2 Major Effects of Climate Change

Climate change leads to higher temperatures and a decrease in summer precipitation, resulting in increased evaporation and water deficit in soils, bodies of water, and waterways, ultimately leading to three types of drought. The main impacts of climate change worldwide are multiple and notably affect water resources, with the decrease or increase in precipitation frequency, because of rising temperatures, intensity, and frequency of dry periods, and increased evapotranspiration having a direct and certain impact on water resources (Guo et al. 2015). This leads to soil drying out or becoming waterlogged, and surface and groundwater stocks decreasing or increasing, causing a disruption in overall water resources. Regarding agriculture and ecosystems, at the local level, the drier the soil, the faster it warms up and increases the temperature during heatwaves. This higher temperature in turn worsens evaporation and soil dryness. These negative feedback loops are increasingly observed in most areas around the globe. But inverse effects can occur during floods, leading to soil congestion and damage to crops. Higher temperatures, more frequent droughts and floods, and extreme weather events disrupt the life cycles of insects, making them vulnerable and disrupting their reproductive behaviors. The decline of insects has serious consequences for ecosystems as they play an important role in plant pollination, organic matter decomposition, and are an essential food source for many animal species, including birds, reptiles, and amphibians. As a result, the decline of insects sets off a chain reaction that affects the entire biodiversity.

Given this alarming situation, it is imperative to rethink our agricultural practices and our relationship with the environment by putting essential measures into practice such as reducing the use of pesticides, restoring natural habitats, and promoting biodiversity. On a global scale, over the long-term, another physical phenomenon is seen: Repeated droughts affect the vegetation's role as a "carbon sink" and make it less capable of absorbing CO_2, resulting in greenhouse gas production collecting in the atmosphere and indirectly exacerbating climate change and global climate change phenomena. Among the main consequences of drought on soils, after a period of drought, dried soils can no longer absorb precipitation properly, which can lead to floods, inundations, and landslides. Soils can suffer harmful consequences after a period of drought. Once dried out, they lose their ability to effectively absorb precipitation, which can result in floods, inundations, and landslides (Lindwall 2022). The decrease in water resources jeopardizes the yields of irrigated and rain fed crops.

Forest fires are also at risk of increasing due to the longer periods of drought, leading to a degradation and increased loss of biodiversity. Additionally, the decrease in natural forage resources could compromise the sustainability of pastoralism. Finally, wetlands are threatened by the increasing phenomena of salinization and eutrophication. With climate change, droughts are likely to intensify, which will have a significant impact on soil moisture and the survival of vegetation during the summer months.

During the summer, we are exposed to a double phenomenon, characterized by a decrease in precipitation and an increase in plant transpiration due to the heat. Consequently, drought is reinforced during the summer season. Despite climate change, rainfall could become more frequent in winter, but they may not be enough to compensate for the effects of summer. The repercussions on fauna and flora can manifest in the following way: For example, dried trees can die, and very dry vegetation can promote wildfires, emitting harmful gases that pollute the atmosphere and strengthen the greenhouse effect. Trees become much more sensitive to cold up to ten years after a period of drought. The consequences on fauna result in a decrease in the water levels of rivers, which affects the survival of fish and animals that drink from these water sources. They must then migrate to find water, leading to an imbalance in the ecosystem.

Coastal areas are facing a threat from rising sea levels, endangering over 3000 hectares of urban and industrial zones. Fishing practices are also at risk. The rise in sea levels, along with increasing temperatures and salinity levels, would further degrade ecological habitats along bathing coasts. The coastline, being a fundamental tourist resource, is at a major risk due to rising sea levels. Tourism and tourist activities are significantly affected by climate change, particularly rising temperatures, which would lead to a shift in seasons and regions suitable for tourism. In case of drought, the consequences for humans and vulnerable populations (babies, children, and the elderly) include a risk of dehydration, which could be fatal as they may not have the reflex to drink to compensate for the lack of water in their bodies.

In countries with limited access to water, drought exacerbates the risk of famine and epidemics due to malnutrition. Drought increases climate migrations of affected populations and can lead to conflicts between peoples. Another example of the impact of drought on human health is often overlooked and considered insignificant; changes in rainfall patterns and temperatures could result in the geographical redistribution of insect vectors of various viral diseases, such as mosquitoes or sandflies. Likewise, the decrease in water resources would lead to an increase in waterborne diseases: Limited access to clean water, sanitation systems, and hygiene would resurge pathogenic microorganisms. Respiratory diseases could be exacerbated by the rise in heatwaves and the deterioration of air quality. Imbalances in water resources, ecosystems, and agroecosystems linked to the increase in extreme weather events could have significant repercussions on food security in the country, leading to a rise in the number of people suffering from malnutrition.

For all these reasons, the lack of water is likely to become an important issue in the years and decades to come on a global scale. Agricultural practices will also need to be adapted. Unlike industry or households that use water but also discharge it, agriculture involves net consumption. Agriculture must evolve to gradually limit the production of the most water-intensive crops and replace them with others. This does not mean we will stop producing potatoes, beets, or corn. The impacts on agriculture, in the event of water stress materialized by an annual water availability per capita of around 1700 m^3 on average, are too low to properly irrigate crops, leading to a decrease in agricultural production and even crop losses. Therefore, it is high time to consider the role of agriculture in the overall water consumption balance. The impacts on drinking water supply in case of drought, with the water levels of rivers, lakes, streams, and aquifers being low, result in water restrictions for crop irrigation, domestic water uses (e.g., watering gardens), or industrial uses. In rural areas, water may be rationed or even cut off.

The impacts on water quality must also be considered. In the event of water scarcity, pollutants are less diluted, which directly increases the risk of contamination of water reserves and leads to poor quality water that can harm the health of populations. Drought episodes can affect various areas around the world, and drought can occur in any country that experiences these climatic conditions: decreased precipitation for a certain period accompanied by high temperatures. Some areas are more vulnerable than others, and only polar and subpolar areas are spared from drought because they are entirely covered in ice (Kirhensteine et al. 2016).

Among the most affected areas by periods of drought are tropical and subtropical zones at the tropics. These are the areas with the hottest and driest climates. The most significant droughts in recent years have affected countries in the Sahel, the Horn of Africa, Mexico, northeastern Brazil, certain areas of China, India, and Russia, as well as southeastern Europe. In arid areas, the most vulnerable countries are found, such as Africa and Western Asia.

The consequences of drought can extend over many years and are often followed by large-scale floods that make the already affected populations even more vulnerable. So it is important to expect these climatic phenomena through prevention measures and better water management. Among the solutions to consider to address the lack of water on a global scale are the restoration of wetlands, the modernization of water networks, and especially recycling in industries and the reuse of wastewater in agriculture. Reusing wastewater also means not losing the water that we collect. Therefore, it is also necessary to modernize the network that transports household wastewater to the treatment plant, so the reuse of wastewater will help clean our cities but they can also be used for certain forms of agriculture in certain areas.

As a conclusion, drought alongside floods has impacts on water quality, eutrophication, water temperature rise and evaporation, release of odor nuisances, etc., decrease and change in biodiversity at the scale of fauna and flora, and threatens our water consumption and food supply in many important areas of industries, agriculture, commerce, etc. Therefore, drought has significant repercussions on agriculture through crop losses, consequences on livestock, as well as on forests, hydropower and nuclear electricity production to operate nuclear power plants by water intake from rivers, streams, or the sea and the price of raw materials. Likewise, it contributes to increasing the risk of triggering fires. To combat and prevent the two major environmental threats,floods and drought,it is essential to act with vigilance. This includes encouraging individuals, professionals, and the general public to reduce environmental pollution, the main driver of climate change. Key actions include saving water, banning activities that pollute aquatic environments and the atmosphere, and prioritizing water use for agriculture and critical industrial sectors such as energy production. These efforts will also help control and limit excessive domestic and public water consumption.

26.3 Changing Water Practices and Population Growth

With the demographic growth and the socioeconomic development of the global populations known in recent years, the pressure of demand on water resources will become increasingly important, requiring better management of demand after the completion of the conventional resource mobilization phase. Water demand management is a key focus in future water policy to control consumption in various sectors, especially agriculture, which is the largest consumer of water resources. The tools for this management are primarily based on water conservation, its valorization, and rationalization of use, especially in the agricultural sector. To meet these goals, Tunisia must improve water legislation and help with its implementation, and establish institutions that allow for better management of water demand in various socioeconomic sectors. This policy will help control overall water demand for each country and ultimately satisfy their exploitable resources.

The global population could face pressures from intensified droughts, pollution issues, varying degrees of resource overexploitation, and even limited abundance risks in space and time, further complicating resource management. So each country must equip itself with a better understanding of these phenomena and resort to unconventional water sources to meet the demand for clean water in certain water-deficient regions and secure more resources for economic sectors development, especially in agriculture.

Most countries aware of these water scarcity issues have committed to appropriate and optimal water resource management policies, constituting the least costly and most sustainable solution. Significant efforts have been made to mobilize all identified resources, conserve water, improve network efficiency, and involve users in water management. This policy has enabled these countries to gain significant experience in water resource management and successfully meet the water demand during wet years and the minimum necessary during periods of scarcity and drought.

To address future water-related challenges, the global population must continue and strengthen this policy, especially focusing on water conservation and preserving resources in quantity and quality. This policy will inevitably let these countries delay the need for unconventional resources if possible and prevent unsustainable exploitation of these resources, particularly groundwater.

Beyond 2040, conventional water resources will tend toward their limits, and the demand will be increasingly important. Population growth and especially its urban part, the extension of comfort and hygiene enabled by income improvement, and finally the growth of agricultural, industrial, and tourist activities can lead to an imbalance between water supply and demand. Therefore, resorting to non-conventional resources such as desalination, treated wastewater, will be an important alternative to fill the deficit in traditional resources. To address the challenges of water and preserve the rights of future generations in terms of resources, the water sector must always be studied within a strategic framework based on carefully elaborated studies and in-depth research. All themes related to the sector must be considered, including the search for new mobilization techniques, using appropriate technologies for the production of non-conventional waters, the use of new economic approaches for optimal resource management, the control of water pollution and its impact on the environment, and finally the strengthening of the legislative and institutional framework. This document chapter has a thorough reflection on balancing water resources with demands and the prospects of the water use and recycling sector in future scenarios within a circular economy framework, aiming to identify the steps and approaches to be taken so that each country has the necessary means to meet these water needs, promote economic growth, and preserve the well-being of populations. This approach is primarily technical, economic, legislative, and institutional, coupled with the demographic growth and socioeconomic development of global populations experienced in recent years.

26.4 Global Water Resource Management

As mentioned several times before, the world has been subjected to recurrent climate changes in recent years, resulting in a climate characterized by its irregularity, manifested by the alternation of sequences of relatively rainy years leading to catastrophic floods and generally dry years. This characteristic has led to a deficit or imbalance in mobilizable resources, which can, in extreme cases, result in water overload and/or scarcity, or an imbalance between supply and demand. With conventional resources reaching their limits and an increase in demand in most countries, technical, economic, and legislative measures are necessary to better manage water resources and address this significant issue in the event of temporary scarcity.

Technical measures mainly encompass the interconnection of water mobilization structures such as the construction and operation of dams, especially drilling and water production in desalination plants for seawater or brackish water, wastewater treatment plants, to ensure the minimum necessary water supply during periods of regional or national scarcity. The next-generation dams can serve as a fundamental element in managing shortages by mobilizing excess water from exceptionally wet years. The economic justification for such structures should consider not only the usual economic parameters but also a coefficient specific to the scarcity of resources. Last, artificial recharge of groundwater aquifers from surface water surplus or treated wastewater is an important action within the framework of integrated resource management, based on intensive exploitation of these aquifers during dry years and their recharge during surplus years. The technical measures mentioned above cannot adequately meet the demand for water during periods of scarcity without appropriate legislation that considers the constraints of resource availability and priority sectors. This legislation must identify and define priorities by sector, assign allocations, and put measures into practice to promote water conservation, especially in the most water-intensive sectors.

26.5 Wastewater Reclamation and Recycling

The potential of wastewater in each country is generally proportional to its consumption and is estimated at millions or even billions of cubic meters per year depending on the size and population. This potential can reach high values, often exceeding 15% of the potential of conventional mobilizable resources. This volume would allow for the irrigation of thousands and thousands of hectares of agricultural land or the conservation of water in highly water-consuming industrial processes. This unconventional resource represents an opportunity for agriculture and industry, especially in a period where conventional resources are reaching their limits and can no longer meet all demands.

However, the use of this resource requires a lot of precautions and initiatives to preserve public health and the environment. Adequate treatments must be identified to safeguard the environment, and additional treatments such as tertiary treatment for disinfection, dephosphatation, etc., could be considered for its use in the agricultural or industrial sector. These treatments will incur additional costs that must be offset by economically profitable production.

The potential evolution of the exploitation of this resource must come with the strengthening of legislation related to this field through the development of new texts that consider the future use of this resource, the precautions to be taken for environmental protection, and the preservation of public health, as well as all the technical measures to be undertaken in the future to produce water suitable for discharge into the natural environment or for agricultural and industrial use.

A national action plan must be developed to enhance the value of this resource, while ensuring the protection of public health and the environment through appropriate measures and specific and evolving legislation. The reuse of wastewater is a political and socioeconomic issue for the future development of drinking water and sanitation services on a global scale. It has the major advantage of providing an alternative resource to limit water deficits, better preserve natural resources, and address water shortages caused by climate change. Already, many countries around the world, states, and major cities (Australia, California, Cyprus, Spain, Florida, Jordan, Malta, Singapore…) have ambitious goals to meet 10–30%, or even up to 60% of their water demand through the reuse of treated wastewater.

Agriculture, as previously emphasized many times, is a major consumer of water resources on a global scale, and agricultural irrigation has been, is, and will remain the largest consumer of recycled water with many well-recognized benefits and advantages, notably contributing to food security. The reuse and internal recycling of industrial wastewater have become common practices for many industries, with new trends such as the ambitious goal of zero liquid discharge and inter-sector water reuse, such as using urban wastewater for industrial purposes. Urban water reuse is characterized by rapid development due to its crucial role in the sustainable city of the future. In addition to urban irrigation of green spaces, other applications are gaining importance, such as industrial uses (cleaning, firefighting, cooling towers, etc.), building recycling, and environmental uses for the maintenance and restoration of water bodies, rivers, and wetlands.

Groundwater recharge, reservoir replenishment for indirect drinking water production, and even direct reuse of ultra-pure water to increase water supply have been implemented as sustainable solutions to the challenges of growing water deficits that some countries will face in the next 30 years.

The irrigation of agricultural soils and green spaces in certain urban centers represents the largest consumer of water, accounting for about 70% of global demand. In some arid and semi-arid countries, recycled water provides the majority of irrigation water. In addition to the conventional factor of chronic water scarcity, the need for alternative resources has been accelerated in recent years by increasingly severe and repetitive droughts. These droughts occur not only in traditionally arid areas such as certain parts of the United States, the Mediterranean region, the Middle East, and South Asia, but also in several temperate regions in Europe and North America (Kirhensteine et al. 2016).

The reuse of wastewater in agriculture is a millennia-old practice developed by ancient civilizations and used until the twentieth century as a wastewater treatment system in spreading fields. The practice of spreading wastewater, also known as percolation infiltration, is an underground spreading system for wastewater treatment including perforated and rigid spreading pipes installed in trenches or in a shallow bed filled with siliceous sand and gravel of various sizes. The existing soil serves as both a filter thanks to the bacteria present in the soil and to drain treated wastewater. The soil acts as an effective filter with up to one or two tons of "purifying" microorganisms per hectare.

26.6 Wastewater Reuse in Agriculture

The main interest in reusing wastewater in agriculture is to address water deficits and increase agricultural production yields through adequate irrigation water supply. Wastewater not only provides water for crops, but also contributes to improving yields by providing nutrients. From the perspective of agricultural and water policy, wastewater—as a water resource—offers several advantages: saving clean water, expanding cultivated areas, increasing agricultural productivity, reducing pressure on resources especially in arid and semi-arid regions, and mitigating polluting discharges through filtration processes in the soil.

To maintain food production, recent technical and research efforts have focused more on the efficiency of irrigation through the development of localized irrigation systems (e.g., drip irrigation) and/or networks equipped with sensors and computer-controlled systems. However, irrigation efficiency alone is not enough to solve water-deficit issues, and the provision of an alternative resource through the reuse of wastewater has become a development priority for many countries and regions (Al-Hazmi et al. 2023).

The fundamental principle of agricultural water reuse is the requirement for proper treatment of municipal wastewater to a specific quality level suitable for the intended use. Aside from the well-known benefits, the use of recycled water for irrigation can harm public health and the environment, depending on the treatment level, local conditions, and irrigation practices (Lazarrova and Brissaud 2007). In all cases, existing scientific knowledge, feedback from experience, and best practices help to reduce risks through effective planning, appropriate technological choice, and rigorous management of irrigation practices. The main risks associated with the reuse of treated wastewater for irrigation can be classified into three categories: health risks, agronomic and environmental risks, and operational risks related to the degradation of recycled water quality in distribution systems and clogging of irrigation equipment. Health hazards are mainly linked to two types of pollutants, namely pathogenic microorganisms and organic micropollutants. Pathogenic microorganisms are highly diverse, and water at the end of the conventional treatment chain requires disinfection treatment by chlorination, UV-254 irradiation, etc. Organic micropollutants, of various natures and compositions, are often considered more or less toxic substances, with a long list linked to the heterogeneity of the raw wastewater. Most of these micropollutants are poorly soluble, and wastewater treatment by sedimentation is effective in removing them. That means they are found more in sludge than in treated water. The minimal concentrations in urban effluents and their limited transfer to plants make the health risk low in the case of agricultural reuse (Cauchi et al. 1996), even if some accumulate in roots like PCB, or in soybean seeds for certain pesticides. In the worst case, they will only be an overload to the detoxification system of humans and animals (Faby and Brissaud 1997).

26.7 Controlling the Health Risks Associated with Wastewater Reuse

The control of health risks through additional tertiary treatment is often essential for high-risk health uses, such as the irrigation of raw-consumed vegetable crops and green spaces. Additional filtration is also necessary to prevent deposits in the distribution system and/or reduce the risk of clogging irrigation devices, especially sprinkler nozzles and drip systems. In parallel, the development and implementation of new wastewater treatment processes for irrigation and industrial cooling use have also contributed to improving the aesthetic quality of recycled water and eliminating odor and discoloration issues in recycled water, which has hindered several projects due to negative user perception.

As mentioned above, the major risk to public health associated with irrigation, and to a much lesser extent with recycling and industrial use of recycled water, is contaminated by pathogenic microorganisms, including viruses, bacteria, helminths, and protozoa. As highlighted by the World Health Organization (WHO) in its 2006 guidelines for the use of wastewater in agriculture, there is a risk of infection transmission when these conditions are met:

- The population, agricultural workers, or animals are exposed to pathogens in recycled water through direct or indirect contact in irrigated areas or by consuming contaminated crops (Priyashantha and Sugirtharan 2021);
- The number of ingested pathogens exceeds the infectious dose;
- The host becomes infected, resulting in the onset of disease and/or transmission of pathogens to other individuals or animals.

Therefore, the main objective of public health protection measures in water reuse projects is to prevent the first two conditions from occurring. This means that appropriate practices must be implemented to reduce the number of pathogens in recycled water and to implement multiple barriers and other measures aimed at reducing the likelihood of contact with potentially infectious microorganisms. The choice of these good practices for the protection of human and animal health depends on local conditions and economic concerns.

The main measures for controlling health risks in irrigation with recycled water include these four groups of good practices:

- Wastewater treatment and quality control during distribution and storage.
- Control of wastewater application through the choice of appropriate irrigation methods and agricultural practices.
- Restriction of the irrigated crops and restriction of public access, especially for green space irrigation.
- Other measures, including control of human exposure (e.g., protection measures for agricultural workers such as gloves and masks), harvest measures, education of the public concerned, and promotion of good hygiene practices. Finally, from a regulatory and legal point of view, the wastewater treatment process is universally recognized as the most effective measure to reduce health risks associated with the

use of recycled water for irrigation. For this reason, all national or international regulations set or recommend a quality level that wastewater treatment must ensure depending on the intended use and the degree of risk. Risk depends on several factors, including local endemic health status, farming methods, weather conditions, risk of direct contact, economic feasibility, etc. In principle, most newer standards require at least biological treatment of wastewater intended for water reuse in irrigation. In some cases, such as in Mexico, for example, priority may be given to preserving the fertilizing value of wastewater through advanced primary treatment (e.g., coagulation, flocculation, and lamellar settling), followed by filtration and/or disinfection to focus on the removal of microbiological pollution while preserving carbon and nutrients for irrigated crops.

The treatment of wastewater is widely recognized as the most effective measure to reduce health risks associated with the use of recycled water for irrigation and much less in industry. For this reason, all national or international regulations set or recommend a quality level that wastewater treatment must ensure depending on the intended use and the degree of risk. The degree of risk depends on several factors, including the local endemic health status, farming methods, climatic conditions, risk of direct contact, economic feasibility, etc.

In recent years, several countries and federal states have included water recycling in their national water resources management policy with rules, obligations, and financial incentives. In addition to a quality level in relation to microbiological and physicochemical indicators, some standards also advocate treatment processes and more barriers to reduce health risks.

In 2006, WHO completely revised its 1989 recommendations for agricultural irrigation with treated wastewater, drawing inspiration from the Australian approach to controlled health risk standards. The guideline values were maintained at the same level, thus limiting health risks to an acceptable level for public health protection, but with new ideas of more barriers ensuring the economic viability of reuse projects.

The United States is considered the global leader in wastewater reuse, with the first regulation adopted in 1918 in California, followed by several revisions, the latest of which dates back to December 2000. At the federal level, the United States Environmental Protection Agency (USEPA) revised its 2004 recommendations in 2012, mainly keeping two levels of recycled water quality guidelines and recommendations for better monitoring of treatment processes.

In most standards and regulations, the precautionary principle is applied for high-risk uses, such as the irrigation of crops consumed raw and unrestricted irrigation of green spaces in urban areas.

26.8 Advantages and Benefits of Reusing Wastewater on an Industrial Scale

The reuse of wastewater on an industrial or agricultural scale offers many advantages at different levels. Industrial reuse of wastewater and internal recycling are now a technical and economic reality. For some countries and types of industries, recycled water provides 85% of the overall needs. The largest water consumers are thermal and nuclear power plants, as well as paper industries, for cooling water. The quality of reused water is regulated and depends on the type of industrial application or production. Urban wastewater accounts for no more than 15% of the volume of reused water in industry. In the United States, for example, the volume of reused wastewater in industry is around 790,000 m^3/day, with 68% used for cooling. On the other hand, environmental benefits include beneficial substances in treated wastewater, which helps limit the use of pesticides, fertilizers, and other toxic products in nature, especially in agricultural crops. This reuse often results in multiple gains at different levels.

Thus, on an industrial scale, the reuse of wastewater as an essential valorization operation for industrial wastewater aims primarily to recover valuable resources present in liquid effluents, while reducing environmental impact and costs associated with waste management (Mustapha 2013). There are numerous ways to valorize an industrial effluent, including: (1) reintegrating water into the production process or other uses; (2) recovering chemical and biological products for reuse in the industry itself or in other processes; (3) recovering valuable salts, such as metal salts, nutrient salts, chemical salts, or saltwater salts, reducing the need to extract and treat salts from natural sources; (4) separating oils and fats that can be reused in the industry for the same purposes or transformed into by-products such as biodiesel or chemicals; (5) producing energy, mainly through anaerobic digestion of wastewater, to obtain usable biogas for heat and electricity production; (6) producing high-quality organic fertilizers, provided that the wastewater contains a high concentration of nutrients such as nitrogen and phosphorus; (7) recovering all types of metals, including precious metals such as gold, silver, or copper; (8) recovering acids and other materials from different types of spent baths, such as degreasing and dephosphating baths, anodic oxidation baths, pickling baths containing hydrochloric acid, zinc, and caustic soda, or baths containing other acidic compounds. These baths can be treated to regenerate the bath itself or for material valorization; (9) recycling organic solvents, including white spirit, acetone, ethanol, and ethyl acetate. Although their energy recovery is possible due to their high calorific value, recycling is a more economical and environmentally friendly option, ultimately by using the residual heat generated in certain industrial processes to preheat water before entering the treatment processes. Given this array of wastewater valorization technologies in industry mentioned, there are other different effective technologies for wastewater valorization, and the choice depends on the composition and volume of wastewater, available energy sources, valorization objectives, and specific industry conditions. It is common for wastewater treatment plants to include various treatment stages with different technologies, with membrane technologies and evaporation typically at the end of the process, enabling efficient recovery of valuable products. In addition to evaporation and crystallization, there are other more efficient technologies to treat wastewater, such as membrane systems including technologies like ultrafiltration, nanofiltration, and reverse osmosis. These technologies allow for the separation of solids, organic matter, and contaminants present in wastewater, which can help with water reuse and the recovery of valuable products. Anaerobic digestion involves a biological process of breaking down organic matter in wastewater in the absence of oxygen, generating biogas mainly composed of methane and carbon dioxide. Biogas can be used as an energy source. Similarly, advanced wastewater treatments include technologies that combine biological, chemical, and physical processes to remove specific contaminants from wastewater, such as nutrients and chemicals. Examples include denitrification and dephosphatation systems. Granular media filtration uses filtration beds to remove particles and organic matter from wastewater. These filtration media can be replaced periodically, and the removed material can be valorized or disposed of properly. Likewise, electrocoagulation and electro-oxidation technologies use electrochemical reactions to precipitate and remove substances from wastewater, contributing to disinfection and the elimination of organic compounds.

The reuse of wastewater on an agricultural scale offers many advantages at different levels. First, it improves crop yields by providing financial benefits while ensuring public health protection, which is essential due to the implementation of strict standards and regulations tailored to the specificity of different crops. In some countries where existing standards are very stringent, such as Australia, the United States, and some Middle Eastern countries, secondary treatment is required, and sometimes, in addition, tertiary disinfection treatment is required. In some European countries, the abundance of water resources does not promote the development of this wastewater reuse, limiting its application. The only significant application expanding in these European countries is the irrigation of golf courses. The experience of Mexico City appears as the most important wastewater reuse project globally. Almost 100% of the raw wastewater from the Mexican capital (from 45 to 300 m^3/s during rainy periods) is reused for the irrigation of over 85,000 ha of various agricultural crops. The United States, considered a model country, has widespread agricultural reuse with 34 states having regulations or recommendations, often very strict. These legislative measures, along with over thirty years of experience, make the United States a global leader in wastewater reuse. In Florida and California, respectively, 34% (340,000 m^3/day) and 63% (570,000 m^3/day) of the total volume of reused wastewater is used for agriculture.

The reuse of wastewater allows for a significant saving of water resources and the preservation of this valuable resource. It contributes to the protection of the natural environment by reducing effluent discharge into receiving bodies of water, creating new job opportunities, and promoting economic growth. Additionally, it helps limit the use of fertilizers in agriculture and provides water to water-stressed areas, indirectly reducing population migration. Studies conducted in developing countries such as India have shown that long-term irrigation with wastewater improves soil stability, retention capacity, hydraulic conductivity, and total porosity. The organic carbon content of soils irrigated with wastewater increases, regardless of soil type or agro-climatic conditions. However, this practice may have drawbacks, such as health and environmental risks, especially when the quality of this unconventional resource is not adequately monitored. Raw wastewater in peri-urban areas is not only of domestic origin; it may also be mixed with industrial wastewater, resulting in a wide range of quality levels. Consequently, untreated wastewater can pose health risks and contaminate various environments, including water bodies, soils, and plants.

Thus, significant health issues to consider are related to the presence and contamination by pathogens, whether through direct contact (skin or ingestion) with water and aerosols from spraying, or indirectly through the consumption of agricultural products. Therefore, raw and treated wastewater is a major source of health contamination. They can have various viruses, parasites, bacteria, chemical pollutants such as organic micropollutants, heavy metals, etc. (Priyashantha and Mahendranathan 2019). These wastewater carry many pathogens that can survive for long periods in the soil or on the surface of plants and enter the food chain. Wastewater reuse practices are well-known as potential sources of risks for public health, the environment, agronomic systems, soils, and crops. Reuse practices have been and still are one of the main causes of epidemics of diarrhea, cholera, typhoid, and shigellosis in Africa and Asia, with many people hospitalized and many deaths (WHO 2006). They are also a likely contributor to parasitic and skin infections. Farmers and their families, communities living near wastewater-irrigated areas, and especially consumers are all affected. Table 26.1 summarizes the global direct and undirected impacts of climate change on the globe.

Climate change and global warming have widespread effects across the globe, including higher average temperatures, shifts in rainfall patterns, elevated sea levels, and biodiversity loss. The rise in extreme weather events and alterations in ecosystems and land usage pose greater health hazards and lead to significant economic repercussions.

Table 26.1 Overview of the direct and indirect global effects of climate change

Direct consequences of climate change	Indirect consequences of climate change
1. Escalating maximum temperatures	1. A rise in food insecurity and water scarcity, particularly in developing nations
2. Escalating minimum temperatures	2. Growing threat to people's means of living from floods and wildfires
3. Escalating sea levels	3. Health hazards stemming from the rise in frequency and severity of extreme heat events
4. Elevated ocean temperatures	4. Economic consequences of addressing secondary damage linked to climate change
5. Surge in heavy precipitation (intense rain and hail)	5. Escalating spread of pests and diseases
6. Surge in the frequency of violent tropical cyclones	6. Decline in biodiversity due to the limited adaptability and speed of adaptation of plants and animals
7. Surge in aridity and drought	7. Ocean acidification caused by elevated HCO_3 levels in water due to increased CO_2 levels
8. Reduction in Arctic sea ice and snow cover	8. The necessity for adaptation across all sectors (e.g., agriculture, forestry, energy, infrastructure, tourism, etc.)
9. Glacier retreat and shrinkage	1. A rise in food insecurity and water scarcity, particularly in developing nations
10. Melting permafrost	2. Growing threat to people's means of living from floods and wildfires

26.9 Circular Economy Principles in the Reuse of Treated Wastewater

As mentioned many times before in this chapter, water resources on almost a global scale, apart from polar regions, are facing an increasing water level due to the consequences of climate change and greenhouse effects caused by growing pollution. They are experiencing a growing stress, with a worrying imbalance between the demand and availability of resources, both in temporal and geographical scales. Thus, water stress affects half of the global territory throughout the year, and the frequency and intensity of droughts have increased over the past thirty years. The overexploitation of water, especially for irrigation, but also for industrial use and urban development, is to blame. This is not only a problem for arid regions with low precipitation and high population density likely to increase water stress; temperate zones with intense agricultural, tourist, and industrial activities also suffer from frequent water shortages and/or costly supply solutions. Furthermore, climate change exacerbates these issues: water scarcity is expected to affect around half of Europe's river basins by 2030, for example. The increasing competition for water resources between different sectors is already emerging, and water prices do not yet adequately incorporate the externalities of use, which would allow for a much more holistic and sustainable resource management. So it is necessary to protect and reserve high-quality resources for human drinking water supply (Romano et al. 2020).

To cope with recurrent periods of drought and limit water withdrawals in a fragile natural environment, policymakers must mobilize their efforts, legislation, practices, etc., regarding the development of treated wastewater reuse (TWR). In this context of increasing recycling and reuse of urban and rural wastewater, particularly in water-intensive industrial processes and agricultural irrigation, it is crucial to evaluate the potential health and environmental risks. However, non-conventional solutions are increasingly being studied, and laws are beginning to regulate these new practices. The current transition toward a circular economy in most countries around the world, sensitive and aware of this water scarcity issue, could therefore become a stepping stone for TWR to be fully integrated into water management strategy plans. The recovery and recycling of treated wastewater align with the principles of the circular economy by avoiding unnecessary waste and enabling the valorization of resources for other purposes. This principle of the circular economy involves producing goods such as unconventional water resources and services such as sustainable water supply by limiting the consumption and waste of these resources and the production of residuals and waste (Ghangrekar et al. 2024). This model is based on creating positive value loops with each use or reuse of the material or product before final disposal. It particularly emphasizes new modes of design, production, and consumption, extending the product's lifespan, using rather than owning goods, and reusing and recycling parts.

Water is primarily used for agriculture, public water supply, electricity production, and industry. Agriculture is the most consuming category, representing 40% of the total annual water use, followed by public water supply at around 30%. Climate change, evolving practices, including changes in diet and hygiene, along with the exponential increase in population pose new challenges for water management. Drought and scarcity, which have been increasing in recent years due to stressful climate changes, are also closely linked to infrastructure systems, economic capacity to exploit and access resources, as well as the nature of demand. Therefore, the reuse of treated wastewater should be required in a circular economy process, which can become one of the key solutions to address consumption and use of water resources sustainably.

Like most practices related to the circular economy, it is necessary to change the perception of waste (from waste to resource) and explore new solutions to meet the needs of territories, to fundamentally transform the production system. This involves rethinking the entire framework of activity, from regulations to financing, through final uses and water production models. The main effects of climate change on human living conditions are shown in Fig. 26.1, and Table 26.2 provides an overview on the main benefits, risks, drawbacks, and advantages of reusing treated wastewater in industrial processes and different agricultural areas.

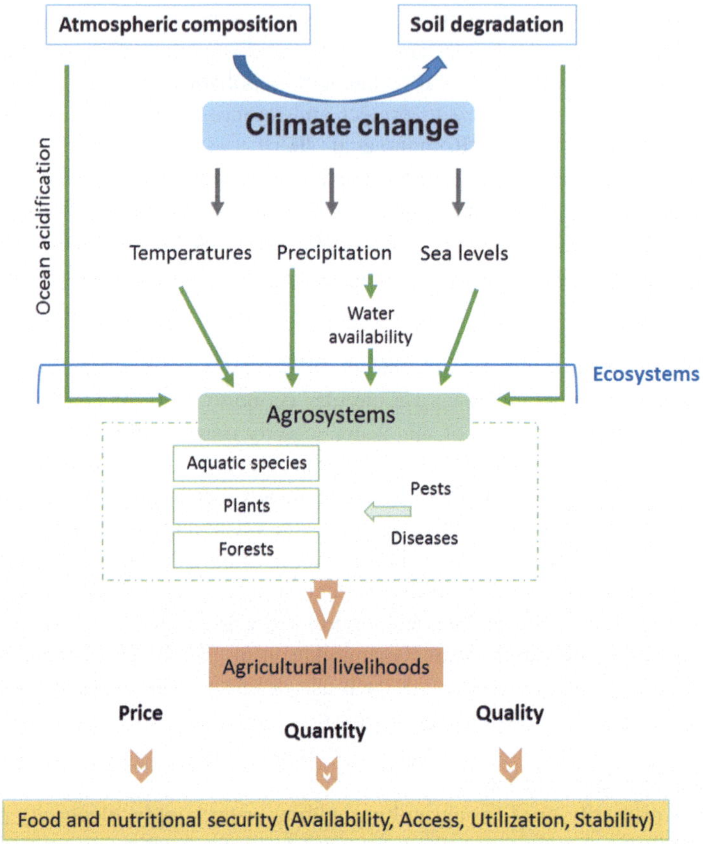

Fig. 26.1 Primary consequences of climate change on the natural environment and human well-being

26.10 Conclusion

The recovery, recycling, and reuse of residual water resources are all similar ideas aimed primarily at saving water resources, which have become increasingly expensive and costly for human life and the environment over the past twenty years due to disastrous community and climate changes. These increasingly expensive water resources require a significant change in our food habits and activities, as well as vigilance in our practical operational, regulatory, and perceptual ideas. This perception requires us to think and consider all waste or residues from any activity, especially those generated after using water resources, as a resource, meaning that all liquid or solid waste or residues are resources for human exploitation. This applies particularly to water resources, which are considered fundamental and too precious for human life on earth. The recycling and reuse of treated water in industry and agriculture offer a range of benefits for human health and the overall natural environment. However, water reuse often proves to be more expensive than using conventional resources, mainly due to the tertiary treatments required to follow stringent regulations and the distribution networks that need to be built and maintained.

The implementation of reuse of treated wastewater seems to be more complex than conventional resource supply, especially regarding water storage and distribution, which require significant infrastructure and represent a major financial challenge. Therefore, the geographical proximity of the wastewater treatment plant (WWTP) to potential uses remains the main factor driving the emergence of reuse of treated wastewater (RTWW) projects. Some perceive the (RTWW) as more risky than beneficial. There are environmental issues at stake, as the long-term pollution risk may seem concerning. However, apart from the agronomic impacts related to salinity, sodicity, and industrial fouling, scaling, and corrosion of cooling system circuits, there are no major environmental impacts to report. Finally, like any circular economy solution, RTWW must be able to adapt to the local context and needs. Currently, regulations

Table 26.2 Summary data on the advantages, benefits, disadvantages, and risks of reusing treated wastewater

	Benefits	Risks
Environment and health	• Quantitative targets for surface water bodies and environmental restoration (rebirth of aquatic ecosystems); • Increase natural and artificial flows in rivers and ponds, thus helping to achieve • Offers a less polluting alternative to desalination; • Recharge aquifers, which can help achieve good quantitative status and avoid groundwater deterioration (upstream aquifer characterization); • Valuation of TMEs is particularly important for WWTPs in coastal areas, which often discharge their treated effluent into the sea, making them more resilient to changes in demand triggered by demographic and climate change; • Reduces the need for artificial fertilizers by providing nutrients for irrigation. → Central condition for good management of environmental and health risks: set up a planned RTWW • Water savings in industrial processes, especially of the heavy type (cooling circuits in power and nuclear plants, etc.)	• Indirect risks of accumulation of pollutants, pathogens, and especially salts present in soils (salinization) and in process equipment; • Potential presence of pollutants, micropollutants, and pathogens, implying a potential direct risk to health and the environment. Certain substances such as polycyclic aromatic hydrocarbons (PAHs) and heavy metals are already regulated by most legislation, so it is important that the raw water used for RTWW complies with this legislation through regulatory directives.; • Lack of epidemiological studies on actual risks (health and environmental) and the diversity of pollutants present in TMEs (particularly trace elements)

(continued)

Table 26.2 (continued)

	Benefits	Risks
Economical	• Potential positive impact on land values through development in water-stressed areas (can encourage more productive agriculture); • Avoiding the loss of a resource can generate economic benefits: avoid economic losses in the event of restrictions on the use of drinking water during droughts • Promote innovative and dynamic industry through increased competitiveness; • Indirect benefit: encourage more appropriate pricing of water, which could lead to incentives to reduce water demand;	• Infrastructure costs may require significant public funding, and the economic viability of this project will depend on each specific situation • Costly option with a low return on investment (for the time being), particularly compared with water abstraction from water bodies, due in particular to infrastructure costs. Implies a need for direct/indirect subsidies to support the supply and demand of RTWWs. Note that the cost of conventional water resources is often subsidized
Social	• Some stakeholders will reduce their environmental footprint, making RTWW an attractive solution • The increased economic activities made possible would lead to social benefits such as employment, particularly for countries with important agricultural production threatened by droughts or tourism industries to protect (promoting sustainable tourism, sustainable water management during tourism peaks, ensuring good quality bathing water…) • Support food security by providing an alternative source for irrigation and support rural communities and businesses; • Encourage a more integrated approach to water management, considering both conventional water and wastewater in development plans, leading to greater social justice as both services are linked and assured. This is particularly true for rural areas, which could benefit from more flexible and less costly solutions (e.g., smaller pipes in a decentralized system and lower costs passed on to citizens)	• Although there are no studies to back this up, in some countries, public perception of water reuse seems negative, especially when it comes to potabilization. There may be mistrust of water reuse practices, which implies a need for education and support for both the general public and local authorities

only cover irrigation uses in a broad sense. Regulations focused on multi-use and multi-resource would help break down barriers between needs over time, providing outlets for each season and water quality suitable for various uses.

References

Abegunrin TP, Awe GO, Idowu DO, Adejumobi MA (2016) Impact of wastewater irrigation on soil physico-chemical properties, growth and water use pattern of two indigenous vegetables in Southwest Nigeria. Catena 139:167–178. Retrieved from https://doi.org/10.1016/j.catena.2015.12.014

Al-Hazmi HE, Mohammadi A, Hejna A, Majtacz J, Esmaeili A, Habibzadeh S, Saeb MR, Badawi M, Lima EC, Makinia, J (2023) Wastewater reuse in agriculture: prospects and challenges. Environ Res 236:116711. ISSN 0013-9351. https://doi.org/10.1016/j.envres.2023.116711

Cauchi H, Nakache S, Zagury B, Carre C, Denis D, Larbaigt D, Martigne S (1996) Feature: reusing wastewater after treatment. Tech Sci Methods 2:81–118

Faby JA, Brissaud F (1997) The use of purified wastewater in irrigation. International Office for Water, 76p

Ghangrekar Makarand M, Yadav S, Yadava Ram N (2024) Biological and hybrid wastewater treatment technology. Recent Developments in India, 638p. eBook ISBN 978-3-031-63046-0, Due: 10 September 2024, Series ISSN : 2730-6674

Guo LL, Perrie W, Long ZX, Toulany B, Sheng JY (2015) The impacts of climate change on the north Atlantic wave climate. Atmos Ocean 53:491–509. https://doi.org/10.1080/07055900.2015.1103697

Kirhensteine I, Cherrier V, Jarritt N, Farmer A, de Paoli G, Delacamara G, Psomas A (2016) EU-level instruments on water Reuse Final report to support the Commission's Impact Assessment, Luxembourg: Publications Office of the EU2 EEA (2009) Water resources across Europe—confronting water scarcity and drought (No2/2009)

Lazarrova V, Brissaud F (2007) Interest, benefits and constraints of wastewater reuse in France, Revue IEN l'Eau, l'Industrie, les Nuisances. N° 299

Lindwall C (2022) What are the effects of climate change? Reports and resources guide. NRDC. https://www.nrdc.org/stories/what-are-effects-climate-change#weather

Mustapha HB (2013) Management of industrial effluents: a review of the experiences in Kano, Northern Nigeria. Int J Adv Res 1(4):213–216

NOAA's Mission: science, service and stewardship (2022) Climate changes impacts. National Oceanic and Atmospheric Administration. Washington, DC 20230. https://www.noaa.gov/education/resource-collections/climate/climate-change-impacts

Priyashantha AKH, Mahendranathan C (2019) Heavy metal contamination and accumulation in groundwater and food crops in Sri Lanka: a review. Vingnanam J Sci 14(2):7–14

Priyashantha AKH, Sugirtharan M (2021) Heavy metals and pathogenic contamination in vegetable crops through wastewater irrigation and risk reduction in developing countries. Sci Eng Health Stud 15:21010008. ISSN (Online): 2630-0087 Review Article https://doi.org/10.14456/sehs.2021.31

Romano O et al (2020) Water reuse within a circular economy context 2. Educational, Scientific and Cultural Organization, Global Water Security Issues Series United Nations. Attribution-Share Alike 3.0 IGO (CC-BY-SA 3.0 IGO) license (http://creativecommons.org/licenses/by-sa/3.0/igo/) from the UNESCO Open Access Repository (http://www.unesco.org/new/en/unesdoc-open-access) and i-WSSM web page (http://unesco-iwssm.org)

WHO (2006) Guidelines for the safe use of wastewater, excreta and greywater. WHO Guidelines, Geneva, World Health Organization. http://www.who.int/water_sanitation_health/wastewater/gsuww/fr/

The manufacturer's authorised representative in the EU is Springer Nature Customer Service Centre GmbH, Europaplatz 3, 69115 Heidelberg, Germany. If you have any concerns regarding our products, please contact ProductSafety@springernature.com

Printed and bound by CPI Group (UK) Ltd, Croydon, CR0 4YY
01/10/2026
02225966-0018